34th UIT (Italian Union of Thermo-Fluid-Dynamics) Heat Transfer Conference 2016

AA006257

Journal of Physics: Conference Series Volume 796

Ferrara, Italy
4 – 6 July 2016

ISBN: 978-1-5108-3754-6
ISSN: 1742-6588

Printed from e-media with permission by:

Curran Associates, Inc.
57 Morehouse Lane
Red Hook, NY 12571

Some format issues inherent in the e-media version may also appear in this print version.

Printed by Curran Associates, Inc. (2017)

For permission requests, please contact the Institute of Physics
at the address below.

Institute of Physics
Dirac House, Temple Back
Bristol BS1 6BE UK

Phone: 44 1 17 929 7481
Fax: 44 1 17 920 0979

techtracking@iop.org

Additional copies of this publication are available from:

Curran Associates, Inc.
57 Morehouse Lane
Red Hook, NY 12571 USA
Phone: 845-758-0400
Fax: 845-758-2633
Email: curran@proceedings.com
Web: www.proceedings.com

Table of contents

Volume 796

34th UIT Heat Transfer Conference 2016
4–6 July 2016, Ferrara, Italy

Accepted papers received: 20 December 2016
Published online: 9 March 2017

Heat transfer in fire engineering

Multi-phase fluid dynamics, heat transfer and interface phenomena

Heat transfer in nuclear plants

IOP Conf. Series: Journal of Physics: Conf. Series **796** (2017) 011001 doi:10.1088/1742-6596/796/1/011001

34tht UIT Heat Transfer Conference
Ferrara, Italy, 4–6 July, 2016

The annual UIT Heat Transfer Conference of the "Unione Italiana di Termofluidodinamica" aims at promoting cooperation in the field of heat transfer and thermal sciences, by bringing together scientists and engineers working in related areas.
Several issues of interest are addressed, namely natural, forced and mixed convection, conduction, radiation, multi-phase fluid dynamics and interface phenomena, computational fluid dynamics, micro- and nano-scales, efficiency in energy systems, environmental technologies and buildings, heat transfer in fire engineering.
The 34th UIT Conference was held in Ferrara (FE), Italy, 4–6 July, 2015 in the spaces of the Scientific and Technological Center of The University of Ferrara.
The response has been enthusiastic: 61 abstracts, 36 oral and 18 poster presentations, 48 papers published on the Proceedings
To encourage the debate, the Conference Program has scheduled ample poster sessions and invited lectures from the best experts in the field along with a few of the most talented researchers.
Keynote Lectures were given by Professor Giovanni S. Barozzi (University of Modena), Professor Paolo Di Marco (University of Pisa) and Professor Nicola Bianco (University of Napoli Federico II).
This special volume collects a selection of the scientific contributions discussed during this conference; these works give a good overview of the state-of-the art Italian research in the field of Heat Transfer related topics.
I would like to thank sincerely the authors for presenting their works at the conference and in this special issue. I would also like to extend my thanks to the Scientific Committee and the authors for their accurate review process of each paper for this special issue.
Special thanks go to the organizing committee.

Professor Stefano Piva (president of The Organizing Committee)

IOP Conf. Series: Journal of Physics: Conf. Series **796** (2017) 011001 doi:10.1088/1742-6596/796/1/011001

About UIT (Unione Italiana Termofluidodinamica)

The Italian Union of Thermal-Fluid Dynamics (UIT) was founded in Bologna on December 19, 1984, following a proposal by Professor Enrico Lorenzini.

Mission

The Italian Union of Thermal-Fluid Dynamics (UIT) has the fundamental aim to contribute to the progress of the Italian research and application of thermal-fluid dynamics, by promoting and enhancing the scientific, professional and social values among its members, as well as the formation of new generations of young scientists towards the research work in the field of thermal sciences, heat transfer and their applications.
In order to achieve these objectives UIT undertakes:

↑←the organization of a heat transfer conference, on yearly basis

↑←the organization of international Conferences, Seminars, Courses, etc.

↑←the organization of a summer school, on yearly basis, for PhD students and young researchers

↑←the institution of a prize, on yearly basis, for the best Italian Masters thesis dealing with thermal-fluid dynamics

↑←the institution of a prize, on a two-year basis, for the best Italian PhD thesis dealing with thermal-fluid dynamics.

Governance

The governing bodies of the Italian Union of Thermal-Fluid Dynamics (UIT) are:

↑←the assembly (Assemblea)

↑←the steering committee (Comitato Direttivo)

↑←the president (Presidente)

↑←the auditors (Revisori dei conti)

↑←the arbiters (Probiviri)

↑←the secretary (Segretario).

The assembly of the UIT is made up by all ordinary members (natural persons, institutions and associations), honorary members, distinguished members and meets at least once a year. The assembly establishes the activity of the association, votes its budget and balance, appoints the governing bodies. At present more than 200 members from universities, national laboratories and industries take part in the UIT.
The scientific management of the UIT, subject to the limitations of the constitution (Statuto), is handled by a steering committee chosen from its membership. It is made up of seven officers (president, vicepresident, secretary, and four members). The steering committee is renewed every three years. President, vice-President and secretary cannot be appointed for the same chair more than twice on end.

IOP Conf. Series: Journal of Physics: Conf. Series **796** (2017) 011001 doi:10.1088/1742-6596/796/1/011001

UIT 2016

Steering committee Auditors
Vincenzo Naso, President
Giuseppe Vella, Vice-President
Sara Rainieri, Secretary
Gian Piero Celata
Davide Del Col
Paolo Di Marco
Alfonso Niro

Mario De Salve
Marco Spiga
Luca Tagliafico

Past-Presidents
Enrico Lorenzini
(First President and Founder)
Sergio Faggiani †
Claudio Pisoni †
Gian Piero Celata

Past-Secretaries
Sandro Salvigni
Piero Vigni †
Gian Piero Celata
Paolo Di Marco

Distinguished members
Gaetano Alfano †
Vittorio Betta
Maurizio Cumo
Giovanni Del Tin
Giovanni De Comelli †
Sergio Faggiani †
Enrico Lorenzini
Elio Oliveri
Claudio Pisoni †
Sandro Salvigni
Giorgio Sotgia

Honorary members
Arthur E Bergles †
Michael W Collins †
Marino Di Marzo
Frank P Incropera
Antonio C M Sousa

IOP Conf. Series: Journal of Physics: Conf. Series **796** (2017) 011001 doi:10.1088/1742-6596/796/1/011001

Scientific committee of the 34th UIT heat transfer conference

Stefano Piva – *Chair* - Università degli Studi di Ferrara, Italy
Giovanni Sebastiano Barozzi – Università degli Studi di Modena e Reggio Emilia, Italy
Gian Paolo Beretta – Università degli Studi di Brescia, Italy
Gian Piero Celata – Centro Ricerche ENEA Casaccia, Roma, Italy
Michele Ciofalo – Università degli Studi di Palermo, Italy
Mario De Salve – Politecnico di Torino, Italy
Paolo Di Marco – Università di Pisa, Italy
Marco Filippi – Politecnico di Torino, Italy
Marco Fossa - Università di Genova, Italy
Giovanni Latini – Università Politecnica delle Marche, Italy
Enrico Lorenzini – Università degli Studi di Bologna, Italy
Rita Mastrullo – Università degli Studi di Napoli Federico II, Italy
Oronzio Manca – Seconda Università degli Studi di Napoli, Italy
Vincenzo Naso – Università degli Studi di Napoli Federico II, Italy
Alfonso Niro – Politecnico di Milano, Italy
Enrico Nobile – Università degli Studi di Trieste, Italy
Carlo Nonino – Università degli Studi di Udine, Italy
Giorgio Pagliarini – Università degli Studi di Parma, Italy
Bruno Panella – Politecnico di Torino, Italy
Domenica Paoletti – Università degli Studi dell'Aquila, Italy
Sara Rainieri – Università degli Studi di Parma, Italy
Luisa Rossetto – Università degli Studi di Padova, Italy
Giovanni Tanda – Università degli Studi di Genova, Italy
Giuseppe Vella, – Università degli Studi di Palermo, Itlay

Keynote lecturers of 31st UIT heat transfer conference

Giovanni Sebastiano Barozzi – Università degli Studi di Modena e Reggio Emilia, Italy
Paolo Di Marco – Università di Pisa, Italy
Nicola Bianco – Università degli Studi di Napoli Federico II, Italy

IOP Conf. Series: Journal of Physics: Conf. Series **796** (2017) 011001 doi:10.1088/1742-6596/796/1/011001

Organizing committee of 34th UIT heat transfer conference

Stefano Piva (Chair)
Giovanni Casano (Co-chair)
Giacomo Bizzarri
Michele Bottarelli
Patrizio Fausti
Sante Mazzacane
Francesco Pompoli
Nicola Prodi

Administrative Secretariat

CONSORZIO FUTURO IN RICERCA
Conference and Training Activities Office

UIT conferences

1 1983 Trieste
2 1984 Bologna
3 1985 Palermo
4 1986 Genova
5 1987 Torino
6 1988 Bari
7 1989 Firenze
8 1990 Ancona
9 1991 Pisa
10 1992 Genova
11 1993 Milano
12 1994 L'Aquila
13 1995 Bologna
14 1996 Roma
15 1997 Torino
16 1998 Siena
17 1999 Ferrara
18 2000 Cernobbio (CO)
19 2001 Modena
20 2002 Maratea (PZ)
21 2003 Udine
22 2004 Genova
23 2005 Parma
24 2006 Napoli
25 2007 Trieste
26 2008 Palermo
27 2009 Reggio Emilia
28 2010 Brescia
29 2011 Torino
30 2012 Bologna
31 2013 Moltrasio (CO)
32 2014 Pisa
33 2015 L'Aquila

34 2016 FERRARA (FE)

Peer review statement

All papers published in this volume of *Journal of Physics: Conference Series* have been peer reviewed through processes administered by the proceedings Editors. Reviews were conducted by expert referees to the professional and scientific standards expected of a proceedings journal published by IOP Publishing.

Notes and experiments on the statics of capillary columns

G S Barozzi[1], M A Corticelli[1], D Angeli[2] and E Salerno[1]

[1] DIEF - Dipartimento di Ingegneria "Enzo Ferrari", Università di Modena e Reggio Emilia, Via P. Vivarelli 10 - 41125 Modena (Italy)
[2] DISMI - Dipartimento di Scienze e Metodi dell'Ingegneria, Università di Modena e Reggio Emilia, Via Amendola 2 - 42122 Reggio Emilia (Italy)

Email: giovanni.barozzi@unimore.it

Abstract. Experimental results are summarized, concerned with the statics of liquid columns in capillary tubes under non-standard conditions. Three configurations are considered: 1. inclined capillary tubes; 2. capillary effect in the horizontal branch of an L-shaped tube; 3. capillary columns in a vertical tube. The effect of inclination on capillary rise (1.) had already been explored in [1] using water in glass-tubes, and it was found that the vertical rise progressively reduces for increasing the tube inclination. This behaviour is now confirmed for n-Hexadecane ($C_{16}H_{34}$). For L-shaped capillaries (2.), the length of the horizontal branch of the tube occupied by the liquid is detected, as a function of the elevation of the branch itself over the feeding pool. The statics of suspended liquid columns (3.) is investigated for two configurations, namely: i. freely suspended columns, and, ii. edge-ending columns. In the latter case the evaporation transient is also tracked. Even if the experimental basis is limited, the results are sufficient to highlight some of the peculiar features of the statics of capillary columns under the above conditions. In particular, it is shown that the contact angle hysteresis plays a fundamental role in all the cases considered.

1. Introduction

Capillary effects play a crucial role in a number of technological applications, such as oil recovery, drug delivery, preparation of paints, pesticide production, food processing, micro-fluidics, wicking fabrics, dyeing of textile fabrics, and ink printing, just to mention a few. On the other hand, the rise of sap in plants, the locomotion of insects on the surface of water, or the wetting of the cornea are just examples of the ubiquitous importance of capillary forces in our daylife. This motivates the extremely huge research effort devoted to the analysis of the capillary forces and the interaction processes between liquids and solid surfaces throughout the world.

Capillary rise phenomena, in particular, have been widely investigated along more than three centuries. However, they encounter a widespread scientific interest, mainly dedicated in the recent years to the dynamic aspects of capillary rise, and, more specifically, to the time-rate evolution of the liquid column [2–15].

The theoretical bases of capillary statics in tubes are quite well grounded, and, for this reason, this topic is receiving much less attention in the present time. The recent extension of the capillary rise theory to arbitrarily shaped capillaries by Wang et al. [16] constitutes an exception in this context.

The rise (or fall) in a thin vertical tube immersed in a liquid pool at one end is in fact extremely well predicted by the Jurin's law, formally established in 1728 [17]. For circular tubes it reads:

$$H = \frac{2\,\gamma\,\cos\theta_e}{R\,\rho\,g} \tag{1}$$

Here, γ and ρ are the surface tension of the liquid, and the liquid-to-air density difference, respectively; H is the elevation of the meniscus apex over the free surface of the bath; R is the capillary inner radius; g is the gravitational acceleration, and θ_e is the equilibrium contact angle specific of the liquid-solid surface coupling. Equation (1) holds for both hydrophilic ($\theta_e < 90°$) and hydrophobic ($\theta_e > 90°$) couplings (in the latter case a negative rise of the liquid in the capillary occurs), subject to the condition $R << H$ (thin tubes). Equation (1) has been validated by hundreds of experiments, and can be theoretically derived following different and independent approaches. It is worth noting that the equation predicts an infinite H-value under zero gravity conditions. This result is consistent with the idea, underlying any of the derivation procedures of equation (1), that the hydrostatic equilibrium of the liquid column within a capillary tube derives by the balance of two actions, one of which is linked to the presence of the gravitational field, the other being provided by the capillary force.

This is, in fact, a statement of general validity to explain the static behaviour of a capillary column whatever it is the system geometry and orientation. In practical terms, however, a number of very basic questions remain unanswered as soon as we move from the basic case to which Jurin's Law refers. For example, no criterion is available to predict the capillary rise in an inclined tube, or the horizontal spread of the liquid in an L-shaped capillary with its short vertical leg immersed in a liquid pool. In the former case, equation (1) predicts no difference from the vertical orientation, thus omitting the effect of the gravitational field on the meniscus shape. In the second case, under the same assumption, the liquid will move indefinitely up to the open extremity of the horizontal branch of the tube. That this is not the case in practice is self-evident, but no criterion is available to predict the static asset of the system. In view of the frequent occurrence of similar situations in real world, it is curious that no reference to them can be retrieved from the open literature.

In this work we summarize the results of a series of simple experiments, relevant to three non-standard static conditions for liquid columns in cylindrical capillary tubes. These are:

1. Inclined capillary tubes
2. L-shaped capillaries
3. Suspended liquid columns in vertical capillary tubes

It is anticipated that for none of cases considered the experimental data to be presented can be considered as definitive, both in view of the elementary experimental tools available, and the limited number of cases considered. In particular, all the experiments were carried out with borosilicate-glass tubes, with inner diameters ranging from 1 to 2 mm, and distilled water was used in most of the cases.

2. The effect of inclination on capillary rise

A first mention to the effect of inclination on capillary rise can be found in [18], where, however, the final equilibrium state of the system was assumed to be coherent with equation (1).

Experimental data on this case were provided first by Barozzi and Angeli [1]. The vertical elevation of the meniscus on the feeding pool free-surface, H_α, was measured for water in a series of tubes for values of the declination angle, α, ranging from 0° (vertical orientation) to 88°. Tubes of inner diameter $\Phi = 1.92$ mm were used. The equilibrium contact angle, as estimated by equation (1) for $\alpha = 0°$ was $\theta_e = 47° \pm 2°$.

The experiment demonstrated that for this specific system H_α does not deviate from Jurin's law prediction, equation (1), up to α-values around 50°. Thereafter, however, the capillary rise was found to decrease monotonically for increasing α, the deviation from the reference level ($H_0 = 10.3$ mm) becoming quite well evident and definitely out of the uncertainty range of the data. The axial extension of the capillary column

$$L_\alpha = \frac{H_\alpha}{\cos\alpha} \tag{2}$$

reduced accordingly, and could be fitted by the equation

$$L_\alpha = \frac{H_0}{(\cos\alpha + k\,\sin\alpha)} \tag{3}$$

where $k = 0.021$. Results from equation (3) should be compared with predictions from the standard formula $L_\alpha = H_0/\cos\alpha$.

A successive series of experiments has been carried out with two tubes of the same series as above, using n-Hexadecane ($C_{16}H_{34}$). Relevant data for the operating fluid were directly measured by a Wilhelmy Plate Tensiometer (at 25 °C we had: $\rho = 766.25$ kg/m^3, and $\gamma = 22.657$ mN/m). The measuring technique was similar to the one reported in [1]. The capillary rise for the vertical orientation resulted to be $H_0 = 6.20$ mm ($\theta_e = 10°$), and $H_0 = 6.26$ mm ($\theta_e = 8°$) for capillary C1 and C2, respectively, with an estimated uncertainty of ± 0.06 mm. Results for the vertical capillary rise H_α as a function of the declination angle α, are reported in figure 1. The mean values of the capillary rise ratio H_α/H_0 for the four runs are plotted in figure 2, where results for water [1] and hexadecane are compared.

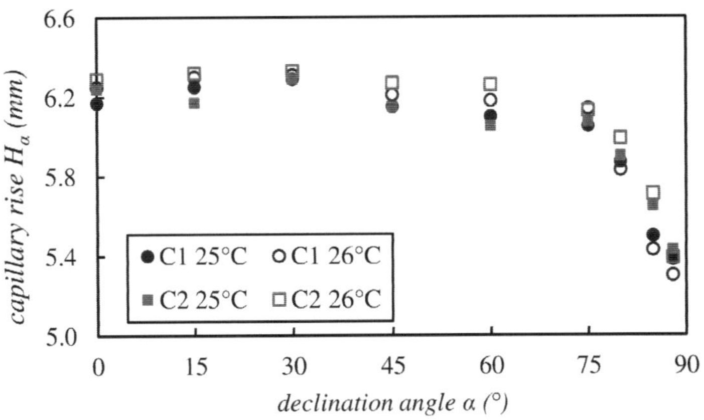

Figure 1. Capillary rise H_α vs. α for n-Hexadecane ($C_{16}H_{34}$).

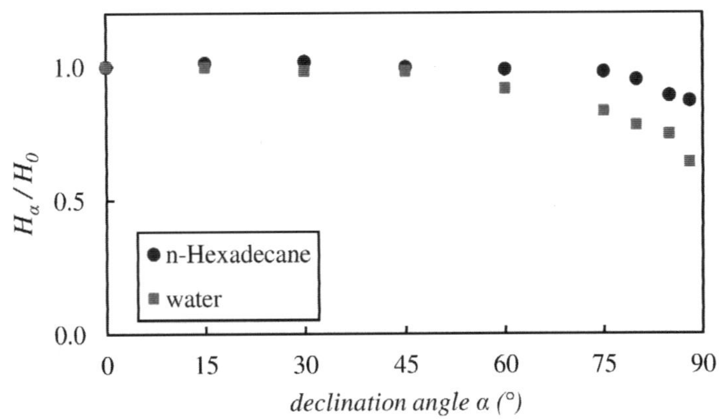

Figure 2. Capillary rise ratio H_α/H_0 vs. α for n-Hexadecane ($C_{16}H_{34}$) and water [1].

It can be noted that the effect of inclination is quite well evident in both the cases, even if it becomes detectable only for relatively high values of the inclination ($\alpha > 75°$), for which the capillary rise reduces of more than 10% towards the vertical reference.

This effect seems to be directly linked to the deformation of the meniscus produced by the component of the gravitational force normal to tube axis. This is made evident by the reproduction of the meniscus acquired with a Dataphysics contact angle goniometer (OCA) and given in figure 3, where the contact angle at the lower edge of the meniscus appears to be much lower than the one at the upper side. The loss of axial symmetry of the meniscus was already evident for $\alpha = 60°$, where, however, no measurable effect on H_α could be observed.

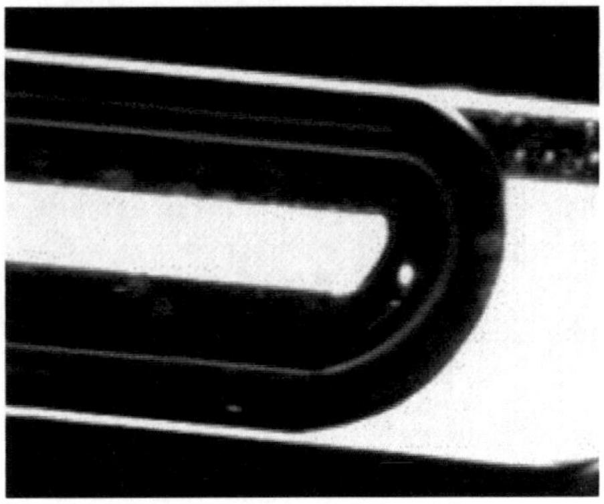

Figure 3. OCA image of the meniscus for n-Hexadecane, $\alpha = 85°$.

3. Statics of L-shaped capillary tubes

When immerging vertically the short-leg end of a L-shaped capillary tube in a liquid pool, the liquid rises along it, up to the H-value corresponding to Jurin's Law prediction, H_0, provided that $H_0 < H_B$, being H_B the elevation of the axis of the horizontal branch of the tube over the pool free surface.

When, on the contrary, the condition $H_B < H_0$ is fulfilled, the liquid will overcome the L-elbow and invade the horizontal tube-branch. It is then possible that the horizontal capillary elongation, L, be so high that the liquid reaches the open extremity of the tube. This would in fact be always the case if an ideal surface (rigid, perfectly smooth, non-adsorbing, and inert [19]) is considered, and the effects of gravity on the meniscus shape can be overlooked. The capillary force, in fact, could not be counteracted by any action.

It is intuitive that this does not happen in real world, provided that the horizontal branch of the tube is long enough.

An experiment was set up to have evidence of the above behaviour, using L-bended borosilicate-glass tubes directly provided by Hilgengerg Gmbh (Germany). Data provided by the producer were: $L_1 = 30$ mm (short leg), $L_2 = 300$ mm (long leg), $\Phi = 1.05 \pm 0.05$ mm (inner diameter), and S = 0.22 mm (wall thickness). The level of tolerance on the inner diameter was confirmed by SEM-measurements on samples taken by different capillaries in the series. Distilled water was used in the experiments. The vertical rise was preliminarily determined as $H_0 = 28.7 \pm 0.2$ mm, to which corresponds a mean value of the equilibrium contact angle $\theta_e \approx 15°$.

Experiments were carried out on an anti-vibration table. The experimental set-up is shown in figure 4. The free-surface level of water in the feeding pool was varied progressively, thus varying H_B. The metal scale shown was used to convert lengths into pixels of a digital photo-camera, and the horizontal lengths were thus retrieved by inverse conversion from digital images. Water was found to overcome the L-elbow for H_B-values just below the reference value H_0, and a short part the horizontal branch appeared to be invaded by the liquid already for $H_B = 28$ mm.

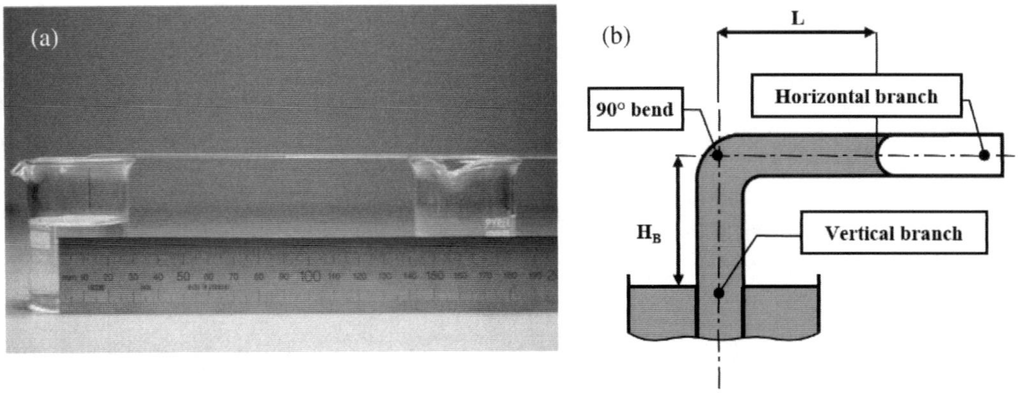

Figure 4. L-shaped capillary: snapshot (a) and schematic (b) of the experimental set-up.

The results of the experiment reported in figure 5 are exemplificative of the general trend of the results. The capillary elongation is found to reduce progressively for increasing H_B. However, this holds true only for $H_B > 22$ mm, and $H_B < 14$ mm, while in the intermediate range ($14 < H_B < 22$ mm) a plateau region appears, where L remains constant for decreasing H_B. In general terms, such a trend corroborates the idea that, when reducing the elbow height over the free surface of the bath, an increasing fraction of the capillary force potential is made available to push the liquid in the horizontal part of tube.

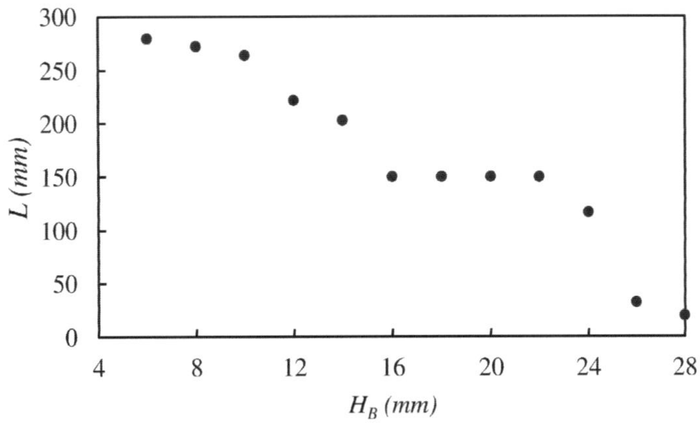

Figure 5. L-shaped capillary: L vs. H_B.

Analysing the statics of the system, if, as a first choice, we exclude pinning effects at the triple line, the vertical equilibrium of the liquid column can only be ensured by the creation of a vertical resultant of the capillary force at the triple line, F_{cv}, equivalent to the weight of the vertical liquid column, according to the equation:

$$F_{cv} = \rho\, g\, H_B\, \pi R^2 \tag{4}$$

Such a vertical force must necessarily correspond to a deformation of the meniscus, which loses its axial symmetry and produces a re-distribution of the contact angle values along the triple line. Higher values of the local contact angle, θ_ψ, are expected to occur in the upper part of the tube, so as to provide a vertical upward resultant.

The above mentioned meniscus deformation is not easy to track with the available visual tools. The sample OCA-images reported in figure 6 are in fact insufficiently resolved to allow for the quantification of the contact angles at the upper and lower edges of the section. However, they are sufficient to demonstrate that a meniscus deformation is actually present in figure 6.a (H_B = 26 mm, L = 32 mm), while it disappears in the plateau zone (H_B = 22 mm, L = 150 mm), in figure 6.b. It is evident that in the latter case, the statics of the systems is not led by the contact angle hysteresis, but rather by some pinning effect, since, in fact, the meniscus appears to remain axial-symmetric. This observation is of interest in itself, since it demonstrates that pinning can produce reactive forces in any direction.

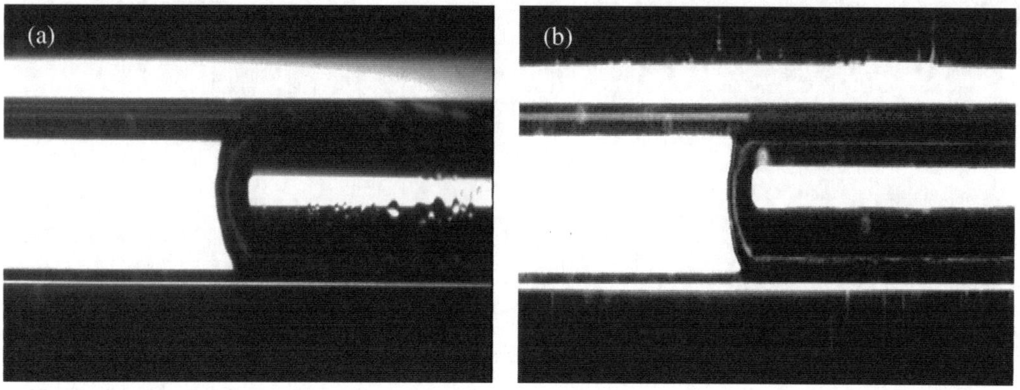

Figure 6. OCA images of the meniscus in the horizontal part of the capillary tube (water) in the proximity of the bend (a) and in the plateau zone (b).

4. Statics of a liquid column in a vertical capillary tube

A liquid column can remain suspended within a vertical capillary tube in virtue of the contact angle hysteresis. As schematized in figure 7.a, at the upper triple line the contact angle reduces, while it increases at the lower meniscus. The vertical equilibrium of the liquid column under gravity is ruled by the equation [20]

$$\frac{2\gamma}{R}(\cos\theta_1 - \cos\theta_2) = \rho\, g\, H_C \tag{5}$$

Here, R is the capillary radius, θ_1 and θ_2 are the non-equilibrium contact angles at the upper and lower triple line, respectively, and H_C is the height of the suspended column. In other words, an upward directed vertical force, $2\pi R\gamma\cos\theta_1$ balances the capillary downward force $2\pi R\gamma\cos\theta_2$ at the lower triple line, and the column weight, $\rho g\pi R^2 L$. The limit for stability is dictated by the conditions $\theta_1 \geq \theta_R$ and $\theta_2 \leq \theta_A$, where θ_R and θ_A are the recession and advancing contact angles, respectively.

The limiting condition to be necessarily satisfied is

$$\frac{2\gamma}{R}(\cos\theta_R - \cos\theta_A) \geq \rho\, g\, H_C \tag{6}$$

Equation (6) states that equilibrium is impossible for ideal surfaces, for which contact angle hysteresis is excluded. Rigid, high energy, non-adsorbing, and very smooth surfaces usually approximate the ideal behaviour (quasi-ideal surfaces), and glasses normally match these conditions. The hysteresis range should be very restricted for surfaces of this kind. This, however, ceases to be true for dirty or rough surfaces, in which cases the difference (θ_A - θ_R) can become as high as 50° [20]. The above considerations are convincing, but seem not to have been corroborated by experimental evidence in the open literature. In addition, equation (6) does not include any consideration of the effect of the discontinuity created by the ending edge of the tube, whose role is evident in everyday experience, but has never been quantified. The case is represented in figure 7.b.

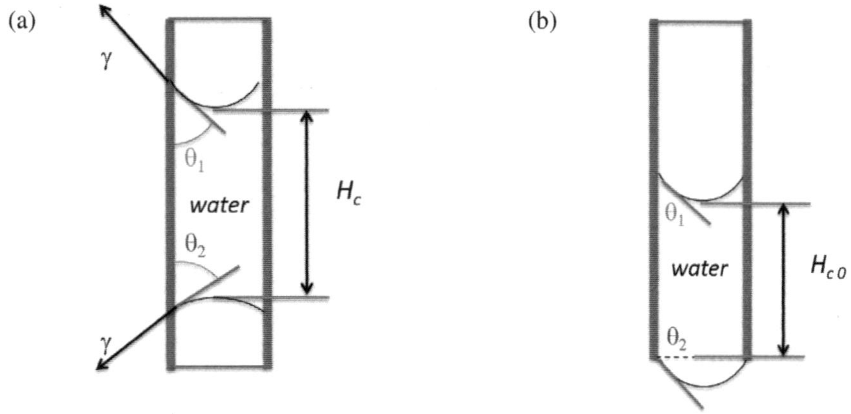

Figure 7. Statics of suspended capillary columns: freely suspended column (a), and edge-ending column (initial condition) (b).

Results of experiments for water in a borosilicate-glass capillary of inner diameter Φ = 1.04 mm are reported here. Two sets of experiments were carried out: i. freely suspended columns (figure 7.a), and, ii. edge-ending columns (figure 7.b). In the first case, a bubble of water was initially sucked by immersion in a water pool, and then reduced in volume by extracting a portion of mass with a syringe. The capillary tube containing the suspended liquid column was positioned vertically on the OCA optical bench, and the upper and lower contact angles were observed and directly measured. In the second series of experiments the liquid column was left to adhere to the lower capillary-end border. In this case, the lower meniscus results to be convex, and emerges from the capillary end. This specific arrangement was found to be prone to evaporation. The evolution of the upper and lower contact angles was followed over time periods ranging from 75 to 150 minutes, and the effect of the end-discontinuity on the equilibrium of the liquid column was observed in the course of the evaporation transients. Similar to the previous cases, the static contact angle was preliminarily measured for the capillary tube, by either direct observation at the OCA optical bench, or, indirectly, by inversion of equation (1). The estimated value of the static contact angle was θ_e = 62° ± 2°.

4.1. Freely suspended columns

Overall, seven series of data were collected for H_C ranging between 2 and 20 mm. The lowest H_C-values resulted to be very uncertain, since the relative error on H_C obviously increases for reducing the column height. Stable columns were found to be increasingly difficult to obtain for H_C over 14 mm. This was thus assumed to be the stability limit of the suspended column. Only results for H_C included within the range 4-14 mm were therefore considered to be acceptable, and these are reported in figure 8, in terms of the mean values of θ_1 and θ_2 vs. the measured H_C-value. It is worth to point out that the right and left values of the contact angles of the same meniscus, as estimated from the OCA images, could widely differ, probably due to the difficulties encountered in the manipulation and positioning of the tube. It was therefore decided to accept only the tests where the difference between the left and right θ-readings did not exceed 5°.

Figure 8 indicates that the contact angle at the lower meniscus increases slightly for increasing H_C from 4 to 14 mm, θ_2 passing from around 63° to around 68°. Over the same range, θ_1 decreases more rapidly, from 55° to 45°. The hysteresis range of θ is therefore not less than 23°. It is worth of note that the data-fittings of θ_1 and θ_2 converge to 60°, quite close to the measured equilibrium value, θ_e.

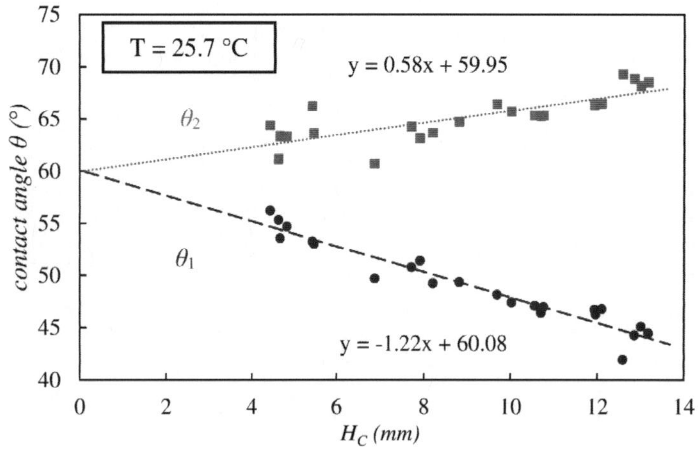

Figure 8. Statics of suspended capillary columns: upper (θ_1) and lower (θ_2) contact angles vs. H_C. The dashed lines indicate the linear fitting of the two datasets.

For the acceptable assays, the vertical force balance between the total capillary force, F_c, and the column weight, F_w, was finally checked using equation (7)

$$\Delta F = F_c - F_w \tag{7}$$

where

$$F_c = 2\pi R \gamma (\cos\theta_1 - \cos\theta_2) \tag{8}$$

$$F_w = \rho g H_C \pi R^2 + W_{1,2} \tag{9}$$

Here, ΔF is the closure error, and $W_{1,2}$ is a correction term accounting for the mass of liquid enclosed between the horizontal plane at the meniscus apex, and the meniscus surface. The results demonstrate that ΔF is relatively large, with a maximum deviation of order 40%, and is systematically negative. This suggests that significant pinning effects are present at the triple lines.

4.2. Edge-ending columns

A series of experiments was finally scheduled to observe the effect of the lower-end border of the capillary tube on the statics of the liquid column. As above, distilled water was used, leaving the suspended liquid column to emerge from the lower capillary end, as shown in figure 7.b.

As anticipated, during the experiments it was observed that a relatively fast evaporation process occurred, and it was decided to monitor it. Results are presented for a set of four experiments, carried out for values of the environmental temperature and the relative humidity within the range 22.4-22.6 °C, and 55-60%, respectively. The column height, and the upper and lower contact angles were measured every ten minutes, starting from different initial conditions (i.e. different values of H_{C0}).

The evolution of the water-column height is reported in figure 9 for all the runs. It can be noted that the trends are quasi-linear; the evaporation rate (50-60 μg/min) is therefore almost insensitive to the H_C-value, as well as the menisci's shape.

During the transient, the upper contact line moves downwards, while the lower contact line remains hooked to the tube edge. The time-evolution of the lower and upper contact angles is shown in figure 10.

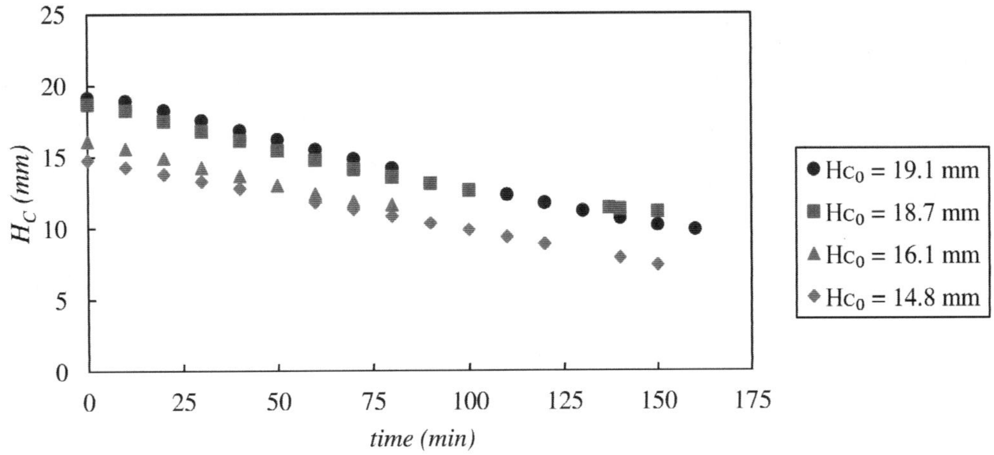

Figure 9. Edge-ending columns: water column height, H_C, vs. time.

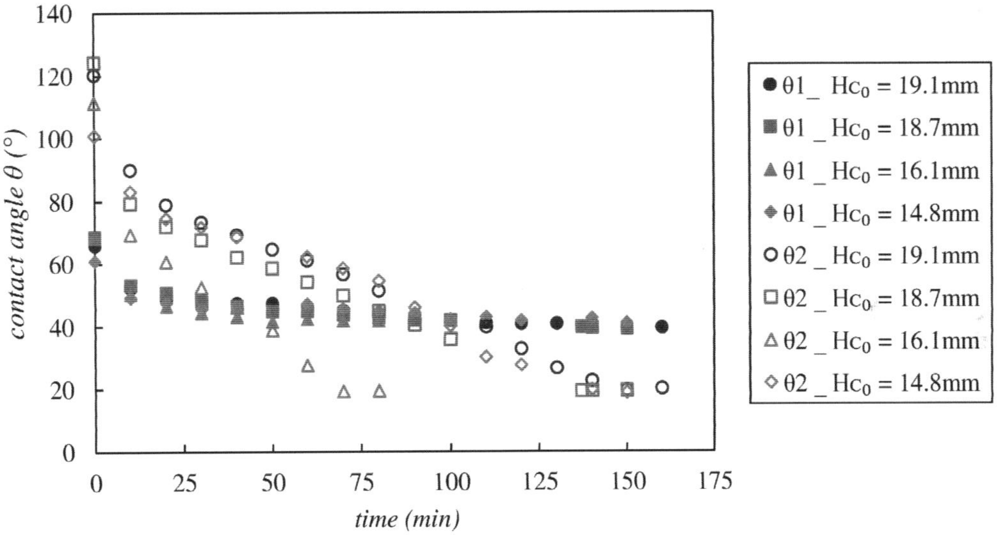

Figure 10. Edge-ending columns: upper (θ_1) and lower (θ_2) contact angles vs. time.

The data plots in figure 10 are very interesting; in fact, they highlight the different evolution of the two menisci during the evaporation transient: starting from an initial value between 61° and 68°, θ_1 slowly decreases, levelling down to a common value of 40°; θ_2, on the contrary, starts from values of order 100-120°. It decays very rapidly, reducing to less than 90° in the first ten minutes, then assuming a quasi-linear trend, eventually reaching a common threshold value of 20°.

This is well illustrated by the three last snapshots in figure 11, where a typical sequence of the lower meniscus evolution is reported.

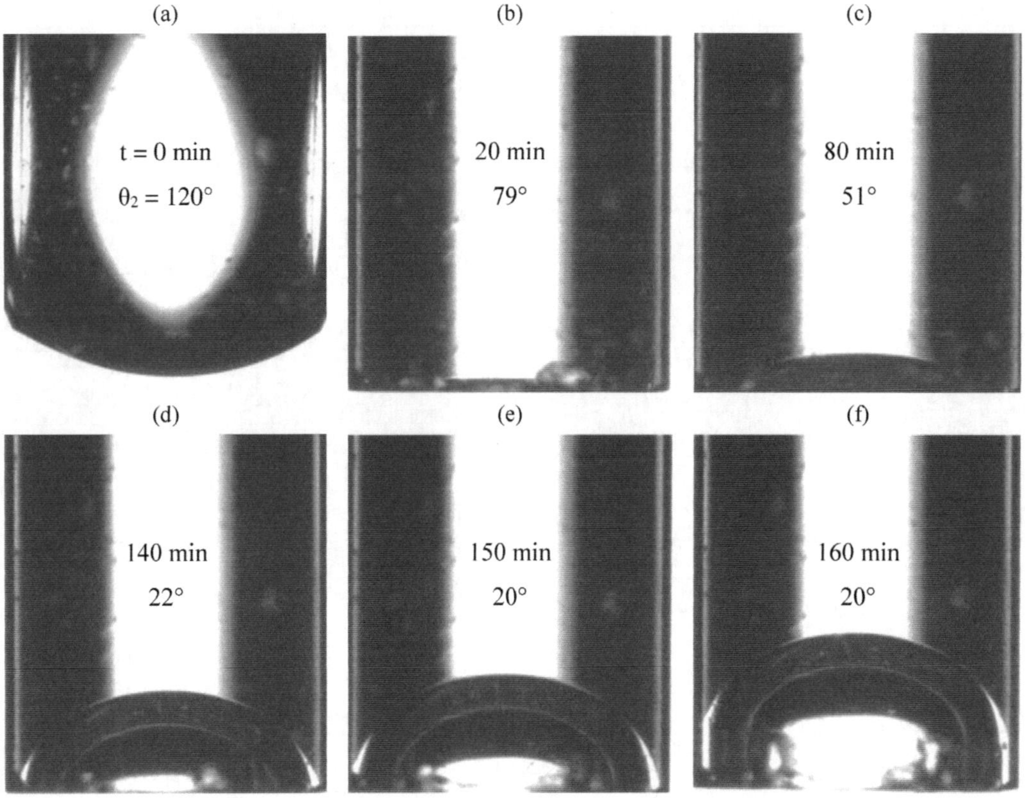

Figure 11. Edge-ending columns: time-evolution of the lower meniscus.

The last two images (for t = 150 min, and t = 160 min, respectively) also show that in the final stages of the observation, the meniscus rises along the capillary, θ_2 remaining unchanged at its end-value, 20°. Such an unexpected behaviour was observed systematically in all the experiments.

For the scope of checking the vertical force balance, equation (7) was enforced as in the previous case. Figure 12 reports $\Delta F/F_w$ vs. time. The data show that the unbalance between the capillary and gravitational components is quite high and positive at the beginning of the transient, the capillary force thus prevailing largely in this phase. Later on, however, due to the different evolution of the two contact angles (as from figure 10), not only the capillary force decays more rapidly than the column weight, but, at one point, it reverses in sign, giving rise to a very strong downward resultant.

All of this demonstrates that a third force must necessarily be present, aside the capillary and gravitational terms, to preserve the statics of the liquid column.

Such a balancing force is likely to be provided by the adhesion of the lower contact line to the capillary tube edge, and, by evidence, has a reactive nature, in that it can either be directed upward or downward, according to the direction of the capillarity-gravitational resultant to be counteracted. The most curious effect in this context is however represented by the fact that the border-adhesion-force continues to act even when the lower meniscus seems to detach from the tube edge. It is argued here that this detachment is only apparent, and that a thin liquid layer continues to connect the lower meniscus to the tube edge.

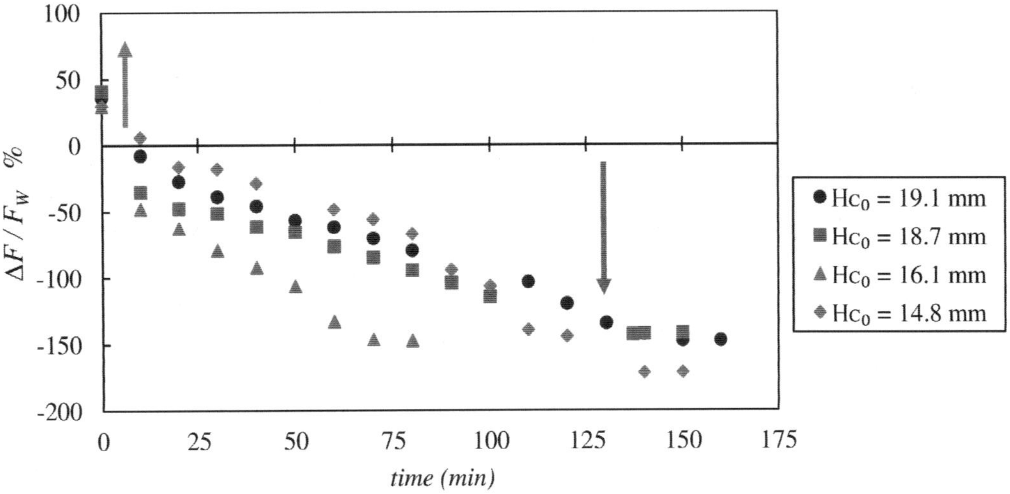

Figure 12. Edge-ending columns: unbalance between the capillary and gravitational forces (equation (7)) vs. time. Arrows indicate the resultant direction: upward (red) and downward (blue).

The experiment indicates that the triple line pins to the capillary edge, and such a pinning effect definitely controls the stability of the column by introducing a very strong stabilizing force.

Along the evaporation process both the contact angles undergo a progressive reduction, eventually approaching an asymptotic value. In the second half of the transient, in particular, the capillary resultant reverses in sign, and sums-up to the column weight. The above stabilizing force may therefore become as high as 150% of F_w or more. In the final stages of the observations the lower meniscus seems to detach from the edge, but the pinning force remains active anyway. All of this has little effect on the evaporation rate.

5. Final remarks

Only provisional conclusions can be drawn from the above experiments, due to the limits of the data collated, both in quantitative and qualitative terms. The experimental basis certainly needs to be extended for firm conclusions to be gained.

Keeping this premise in mind, all the results highlight anyway the effect of gravity on capillarity in tubes under non-conventional conditions. The capillary length scale [$l_C = (\gamma/\rho g)^{0.5}$] is normally used to decide whether gravitational effects on the meniscus shape can be overlooked. For water and n-Hexadecane $l_C \approx 2.7$ mm, and ≈ 1.7 mm respectively; these are therefore comparable to the tubes' diameter size (1-2 mm). Extension to lower-sized capillaries would then be useful, since it is expected that meniscus deformations observed in the inclined and the L-shaped tubes reduce for reducing the tube diameter.

The experiments presented here, in fact, demonstrate that gravity modifies the meniscus shape and affects the contact angle value or its distribution along the contact line, so as to provide the mechanical stability of the system. The contact angle variation should be included within the advancing-receding range typical of the solid-liquid couple. In the case of the edge-ending columns the upper contact angle was found to change of 20° or more, and a similar variability range was found for the freely suspended column. This would indicate that that glass surface was pretty far from being ideal. Note, however, that contact angle hysteresis has been demonstrated to be possible even for perfectly ideal surfaces, under partial wetting conditions [21, 22].

The thermodynamic stability of the above states remains questionable, since, in fact, only the equilibrium value, θ_e, is considered to correspond to a state of thermodynamic equilibrium. It is then possible that the observed states are meta-stable or non-unique.

The presence of pinning effects can obviously offset any other reasoning, and these have been observed (or can anyway be present) in the experiments. Pinning of the triple-line exercises reactive forces. When applied to the capillary edge, pinning has evident and even surprising effects.

As for the effect of the gravitational field on the capillary rise in inclined tubes, it can now to be stated with reasonable certainty that the vertical capillary rise reduces progressively for increasing the tube inclination from the vertical. The experiments for the L-shaped capillary show, in turn, that the horizontal elongation of the liquid column is related to the difference in height between the horizontal tube and the free surface of the feeding-pool.

A comment is now in order, on the possibility of modelling numerically cases of this kind, where the contact angle hysteresis plays a fundamental role in the mechanical stability of the hydrostatic system. An attempt was in fact carried out [23] to predict the statics of the inclined capillary used in [1], using a commercial CFD code, including the VOF algorithm for the tracking of the meniscus shape. Even if a reduction in the vertical rise of the column was actually obtained, it was also noted that the deviations of the contact angles from the static value were very limited, and fell within the numerical errors. This was later confirmed by analysing with the same method the case of an inclined capillary-slit. A slight meniscus deformation was found at the end of transient, but no deviation of the lower and upper contact angles from the equilibrium value. As a consequence, no reduction in the capillary rise could be observed. It can then be concluded that numerical tools not including a specific model for contact-angle hysteresis can not give any valid prediction in this field.

References

[1] Barozzi G S and Angeli D 2014 *Energy Procedia (Special Issue)* **45** 548–57
[2] Hadjiconstantinou N G 1999 *J. Comp. Phys.* **154** 245–65
[3] Zhmud B V, Tiberg F and Hallstensson K 2000 *J. Colloid Interface Sci.* **228** 263–69
[4] Erickson D, Li D and Park C B 2002 *J. Colloid Interface Sci.* **250** 422–30
[5] Xiao Y, Yang F and Pitchumani R 2006 *J. Colloid Interface Sci.* **298** 880–88
[6] Xue H T, Fang Z N, Yang Y, Huang J P and Zhou L W 2006 *Chem. Phys. Letters* **432** 326–30
[7] Lee S-L and Lee H-D 2007 *J. Fluid Eng.* **129** 957–65
[8] Fries N and Dreyer M 2008 *J. Colloid Interface Sci.* **327** 125–28
[9] Fries N and Dreyer M 2008 *J. Colloid Interface Sci.* **320** 259–63
[10] Barry D A, Wissermeir L, Parlange J-Y, Sander G C and Lockington D A 2009 *J. Colloid Interface Sci.* **338** 293–95
[11] Hilpert M 2009 *J. Colloid Interface Sci.* **337** 131–37
[12] Katoh K, Wakimoto T and Nitta S 2010 *JSEM* **16** 62–66
[13] Wolf F G, dos Santos L O E and Philippi P C 2010 *J. Colloid Interface Sci.* **344** 171–79
[14] Yamamoto Y, Ito T, Wakimoto T and Katoh K 2014 *Int. J. Multiphase Flow* **51** 22–32
[15] Yamamoto Y, Tokieda K, Wakimoto T, Ito T and Katoh K 2014 *Int. J. Multiphase Flow* **59** 106–12
[16] Wang Z, Chang C-C, Hong S-J, Sheng Y-J and Tsao H-K 2012 *Langmuir* **28** 16917–26
[17] Jurin J 1728 *Commentarii Academiae Scientiarum Imperialis Petropolitanae* **3** 281–92
[18] Fries N and Dreyer M 2008 *J. Colloid Interface Sci.* **327** 125–28
[19] Kwok D Y and Neumann A W 2003 Contact Angle Measurements and Criteria for Surface Energetic Interpretation *Contact Angle, Wettability and Adhesion* vol 3, ed K L Mittal (Utrecht: VSP) pp 117–59
[20] de Gennes P-G, Brochard-Wyart F and Quéré D 2004 *Capillarity and Wetting Phenomena: Drops, Bubbles, Pearls, Waves* (N.Y.: Springer Science)
[21] Starov V M and Velarde M G 2009 *J. Phys.: Condens. Matter* **21** 464121
[22] Kuchin I V, Matar O K, Craster R V and Starov V M 2014 *Colloid Interf. Sc. Comm.* **1** 18–22
[23] Barozzi G S, Corticelli M A and Agnani E 2015 *Proc. 33rd UIT Heat Transfer Conference (L'Aquila, Italy, 22-24 June 2015)*, paper n 116 pp 1–10

IOP Conf. Series: Journal of Physics: Conf. Series **796** (2017) 012002 doi:10.1088/1742-6596/796/1/012002

Numerical study of metal foam heat sinks under uniform impinging flow

A Andreozzi[1], N Bianco, M Iasiello and V Naso
Dipartimento di Ingegneria Industriale, Università degli studi di Napoli Federico II, P.le Tecchio 80, 80125, Napoli, Italy

[1] E-mail: assunta.andreozzi@unina.it

Abstract. The ever-increasing demand for performance improvement and miniaturization of electronics has led to a significant generation of waste heat that must be dissipated to ensure a reliable device operation. The miniaturization of the components complicates this task. In fact, reducing the heat transfer area, at the same required heat rate, it is necessary to increase the heat flux, so that the materials operate in a temperature range suitable to its proper functioning. Traditional heat sinks are no longer capable of dissipating the generated heat and innovative approaches are needed to address the emerging thermal management challenges. Recently, heat transfer in open-cell metal foams under an impinging jet has received attention due to the considerable heat transfer potential of combining two cooling technologies: impinging jet and porous medium. This paper presents a numerical study on Finned Metal Foam (FMF) and Metal Foam (MF) heat sinks under impinging air jet cooling. The analysis is carried out by means of the commercial software COMSOL Multiphysics®. The purpose is to analyze the thermal performance of the metal foam heat sink, finned or not, varying its geometric parameters. Results are presented in terms of predicted dissipated heat rate, convective heat transfer coefficient and pressure losses.

Nomenclature

A, b	Coefficients
C_p	Heat capacity at constant pressure (J/kg·K)
d_c	Cell size (m)
d_s	Strut thickness (m)
D	Inlet section diameter (m)
f	Inertial factor
h_c	Heat transfer coefficient (W/m^2·K)
h_v	Volumetric heat transfer coefficient (W/m^3·K)
H	Heat sink height (m)
k	Thermal conductivity (W/m·K)
k_p	Fin thermal conductivity (W/m·K)
K	Permeability (m^2)
$\mathbf{n}$	Normal vector
p	Pressure (Pa)
PPI	Pores per inch (1/in)
Pr	Prandtl number
$\dot{Q}$	Heat rate (W)
Re	Reynolds number
Re_c	Cell Reynolds number
s	Fin pitch (m)
t	Fin thickness (m)
T	Temperature (K)
$\mathbf{u}$	Velocity vector (m/s)
W	Heated plate side (m)

Greek symbols

ε	Porosity
μ	Dynamic viscosity (kg/m^2·s)
ρ	Density (kg/m^3)
τ	Shear stress (N/m^2)

Subscript

0	Ambient
eff	Effective
f	Fluid

IOP Conf. Series: Journal of Physics: Conf. Series **796** (2017) 012002 doi:10.1088/1742-6596/796/1/012(

s	Solid
Superscripts	
f	Fluid
s	Solid
Other symbols	
<>	Volume average

1. Introduction

With the advent of modern electronic computers, cooling electronic chips have become one of the principal subjects for heat transfer science and engineering. The technological progress in the field of microchips has long focused on miniaturization and on increasing the computational speed. The demand of high speed and miniaturization of electronic components results in increased power dissipation requirements for thermal management. Among various cooling techniques for electronic components, forced convective cooling with air features advantages of convenience and low cost. Limitation of the forced convective cooling with air, however, lies in the relatively low heat removal rates. In fact, traditional heat sinks are no longer capable of dissipating the generated heat and innovative approaches are needed to address the emerging thermal management challenges. The heat dissipation of a heat transfer medium is a complex function of many factors, e.g., available heat transfer area, heat conduction, heat transfer coefficient, and flow resistance characteristics. These factors are strongly dependent on the structure of the heat transfer medium; and they conflict with each other; for example, higher surface areas and stronger heat conduction capability usually lead to larger flow resistances.

Recently, heat transfer in open-cell metal foams under an impinging jet has received much attention due to the considerable heat transfer potential of combining two different cooling technologies: impinging jet and porous medium. Extensive studies have been carried out to enhance the forced convective cooling with air by utilizing heat sinks with different shapes, materials, flow patterns, etc... Aluminum metal-foam heat sinks have proven to be suitable in the thermal control of high-power electronic components, with excellent cooling performance under forced convective conditions [1]. Bhattacharya and Mahajan [2] proposed a heat sink made up by a finned metal foam heat exchanger. They inserted blocks of aluminium-foam between various parallel plate fins, and investigated the forced convection for foam samples with different porosities and PPI. The heat transfer coefficient turned out to be six times that in a conventional longitudinal finned heat sink. Kim et al. [3] investigated experimentally the heat transfer in a channel partially filled with an aluminium-foam heat sink, with an imposed uniform heat flux at its basis. Experiments were made for different PPI and Reynolds number, and results showed that the lowest thermal resistance was obtained with the lowest PPI foam. The comparison with a conventional parallel-plate heat sink showed the advantage to use an aluminium-foam based heat sink. Hsieh et al. [4] obtained Nusselt number correlations for different velocities, PPI and porosities, with an experimental apparatus similar to a processor cooling heat sink, with the heat source orthogonally invested by a stream of air. Shih et al. [5] showed that the performance of an aluminium-foam heat sink is affected by the height of a sink under impinging-jet flow conditions. The Nusselt number was no monotonically correlated with the height and an optimal height was found, as a function of PPI and porosity. Successively Shih et al. [6] demonstrated that the heat transfer performance of the same aluminium-foam heat sink improved by using an annular flow-restricting mask, that reduced the flow outlet section.

Recently, using 3-D printing and investment casting techniques, Krishnan et al. [7] fabricated and studied FMF heat sinks with regular foam structures, and concluded that on an equal pumping power basis, FMF heat sinks outperform conventional plate-fin heat sinks. Feng et al. [8] carried out a combined experimental and numerical study on finned metal foam (FMF) and metal foam (MF) heat sinks under impinging air jet cooling. Comparisons of experimental and numerical results reveal that using the laminar Darcy's extended model can predict fairly both the heat transfer and pressure drop of MF and FMF heat sinks under high Reynolds numbers. A numerical investigation was carried out to characterize the thermal performance of finned metal foam heat sinks subject to an impinging air flow by Feng et al [9]. The main objective of the study was to quantify the effects of all relevant configurational parameters (channel length, channel width, fin thickness, and fin height) of the heat sink upon the thermal performance. Various simulation cases for different combinations of channel parameters were carried out to obtain the Nusselt number correlation. Based on the inviscid impinging flow, a pressure drop correlation was derived for impinging flow in finned metal foam heat sinks. By using the above referred correlations, the thermal performance of finned metal foam heat sinks was compared with that of conventional plate-fin heat sinks. It was demonstrated that the finned metal foam heat sinks outperformed the plate-fin heat sinks on the basis of given weight or given pumping power.

This paper presents a numerical study on finned metal foam (FMF) and metal foam (MF) heat sinks under impinging air jet cooling. The numerical analysis is carried out by means of the commercial software COMSOL Multiphysics®. The purpose is to analyze the thermal performance of the metal foam heat sink, finned or not, varying its geometric parameters. Results are presented in terms of dissipated heat rate, convective heat transfer coefficient and pressure losses.

2. Mathematical model

The Finned Metal Foam (FMF) heat sink is represented in Fig.1. The heated plate is located at the bottom of the heat sink. The air flow comes from a circular section located on the top, takes heat from the plate and exits from the heat sink lateral sides. The Metal Foam (MF) heat sink has the same configuration but without fins. The heated plate is a 68 x 68 mm^2 square. The ratio of the impinging flow section diameter, D, to the heated plate side, W, is 0.25, 0.50, 0.75, 1.0. The height of the computational domain, H, is 10, 20, 30, 40 mm. The fins thickness, t, is 2 and 4 mm. The number of fins used in the computations is 1, 2, 4, 6, 8, and 10. The porosity and the PPI of the aluminum foam are 0.88, 0.91, 0.94, 0.97 and 5, 10, 20, 40, respectively.

2.1. Governing equations

The porous medium is treated as an equivalent homogeneous medium, and governing equations are written by using the Volume Averaging Technique (VAT). Due to the physics of the problem, a Local Thermal non-Equilibrium (LTNE) model is employed. The flow is assumed to be steady and laminar.
Effects of buoyancy, thermal dispersion and thermal tortuosity are neglected. Thermophysical properties are considered to be independent of the temperature. The governing equations for the porous domain are:

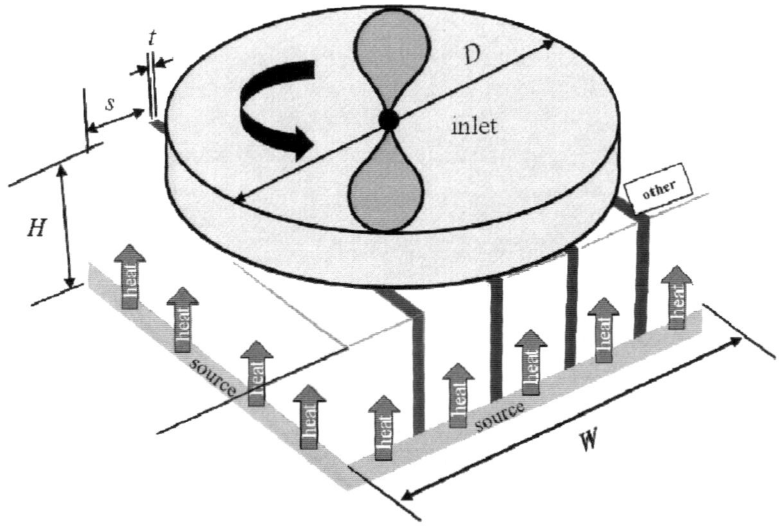

Figure 1. Finned Metal Foam (FMF) heat sink.

$$\nabla \cdot \langle \mathbf{u} \rangle = 0 \tag{1}$$

$$\frac{\rho}{\varepsilon^2}\langle \mathbf{u} \rangle \cdot \nabla \langle \mathbf{u} \rangle = -\nabla \langle p \rangle + \frac{\mu}{\varepsilon}\nabla^2 \langle \mathbf{u} \rangle - \frac{\mu}{K}\langle \mathbf{u} \rangle - \frac{\rho f}{\sqrt{K}}|\langle \mathbf{u} \rangle|\langle \mathbf{u} \rangle \tag{2}$$

$$\left(\rho C_p\right)_f \langle \mathbf{u} \rangle \cdot \nabla \langle T_f \rangle^f = \nabla \cdot \left(k_{eff,f} \nabla \langle T_f \rangle^f\right) + h_v\left(\langle T_s \rangle^s - \langle T_f \rangle^f\right) \tag{3}$$

$$0 = \nabla \cdot \left(k_{eff,s} \nabla \langle T_s \rangle^s\right) - h_v\left(\langle T_s \rangle^s - \langle T_f \rangle^f\right) \tag{4}$$

while the Laplace equation is used for the aluminium fins.

2.2. Closing coefficients

In order to solve porous medium governing equations, closing coefficients are required for permeability, K, inertial coefficient, f, thermal conductivities, $k_{eff,f}$ and $k_{eff,s}$, and volumetric heat transfer coefficient, h_v. Permeability and inertial coefficient are obtained from the following equations [10]:

$$K = \frac{d_c^2}{1039 - 1002\ \varepsilon} \tag{5}$$

$$f = \frac{0.5138\ \varepsilon^{-5.739}}{d_c}\sqrt{K} \tag{6}$$

where d_c is the cell size. For the energy equation, the effective thermal conductivities, $k_{eff,f}$, $k_{eff,s}$, and the volumetric heat transfer coefficient, h_v, are required. The effective thermal conductivity, k_{eff}, is [11]:

$$k_{eff} = k_{eff,s} + k_{eff,f} = \frac{k_s(1-\varepsilon)}{3} + k_f \varepsilon \tag{7}$$

The volumetric heat transfer coefficient, h_v, is obtained from the following correlation [12]:

$$h_v = \left(32.50\ \varepsilon^{0.38} - 109.94\ \varepsilon^{1.38} + 166.65\ \varepsilon^{2.38} - -86.98\ \varepsilon^{3.38}\right)\mathrm{Re}_c^{\ 0.438}\ \varepsilon^{0.438}\ \frac{k_f}{d_c^2} \tag{8}$$

where $\mathrm{Re}_c = |\mathbf{u}|\ \rho\, d_c/\mu$ is the cell Reynolds number.

2.3. Boundary conditions

Boundary conditions of the problem are resumed in Table 1. A rectangular coordinate system with the origin located at the bottom-left corner is employed for the three-dimensional model. For the heated plate, a 70 °C uniform temperature is imposed. The thermal conductivity of the aluminium fins is set equal to 170 W/m K. In the inlet flow section, a Robin boundary condition is used for both solid phase of the foam energy equation and aluminium fin equation. In particular, for the solid phase foam equation, a correlation for staggered cylinder in crossflow is used [13]:

$$h_c = A\,\mathrm{Re}^b\ \mathrm{Pr}^{0.37}\frac{k_f}{d_s} \tag{9}$$

where the Reynolds number is defined as Re = $\rho |\mathbf{u}| d_s / \mu$, with d_s the strut thickness, and the constants A and b depending on the Reynolds number [13].

Table 1. Boundary conditions of the problem.

	Momentum	Energy (fluid)	Energy (solid)	Fin
Inlet	$\mathbf{u} = -\lvert\mathbf{u}\rvert \cdot \mathbf{n}$	$T_0 = 20°C$	$-k_{eff,s}\dfrac{\partial \langle T_s \rangle^s}{\partial \mathbf{n}} = h_c\left(\langle T_s \rangle^s - T_0\right)$	$-k_p \dfrac{\partial T_s}{\partial \mathbf{n}} = h_c\left(T_s - T_0\right)$
Other	$\lvert\mathbf{u}\rvert = 0$		$\dfrac{\partial \langle T_f \rangle^f}{\partial \mathbf{n}} = \dfrac{\partial \langle T_s \rangle^s}{\partial \mathbf{n}} = \dfrac{\partial T_s}{\partial \mathbf{n}} = 0$	
Heat sink sides	$p = 101$ kPa $\tau_{i,j} = 0$		$\dfrac{\partial \langle T_f \rangle^f}{\partial \mathbf{n}} = \dfrac{\partial \langle T_s \rangle^s}{\partial \mathbf{n}} = \dfrac{\partial T_s}{\partial \mathbf{n}} = 0$	
Foam/fin Interface	$\lvert\mathbf{u}\rvert = 0$		$k_{eff,f}\dfrac{\partial \langle T_f \rangle^f}{\partial \mathbf{n}} + \dfrac{k_s(1-\varepsilon)}{3}\dfrac{\partial \langle T_s \rangle^s}{\partial \mathbf{n}} = k_p \dfrac{\partial T_s}{\partial \mathbf{n}}$	
Bottom	$\lvert\mathbf{u}\rvert = 0$		$\langle T_f \rangle^f = \langle T_s \rangle^s = T_s = 70°C$	

For the aluminium fin, the following equation is used [14]:

$$h_c = 0.6\,\mathrm{Re}^{0.5}\frac{k_f}{t} \tag{10}$$

where the Reynolds number is defined by using the fin thickness as the characteristic length.

3. Results

The solid-phase dimensionless temperature on the heated plate, for D/W = 1 and D/W = 0.25, are reported in Fig.2. It is shown that at higher D/W values the air reaches the whole heat sink, while, at lower D/W values not the whole heat sink is employed, since most of the heated plate remains at $\langle T_s \rangle^s / \langle T_s \rangle^s_{y=0}$ = 1. This means that, when the impinging section is larger, the efficiency of the heat sink is improved since all the heated plate contributes to heat transfer. However, it should be pointed out that reducing the impinging sections makes the central region of the heat sink colder than that of the larger impinging section case, since the air velocity locally increases.

Results are reported in terms of the heat transfer coefficient, evaluated with the following equation:

$$h_c = \frac{\dot{Q}}{W^2\left(\langle T_s \rangle^s - T_0\right)} \tag{11}$$

where $\dot{Q}$ is the heat rate and W^2 the area of the heated plate. The heat transfer coefficient is calculated on the heated plate.

Dissipated heat rate as a function of the velocity, for the FMF and MF heat sinks, together with data by Feng et al. [8], are reported in Fig. 3. The figure exhibits differences between predictions by the present model and Feng et al.'s model less than 8% for the FM model and less than 18% for the FMF.

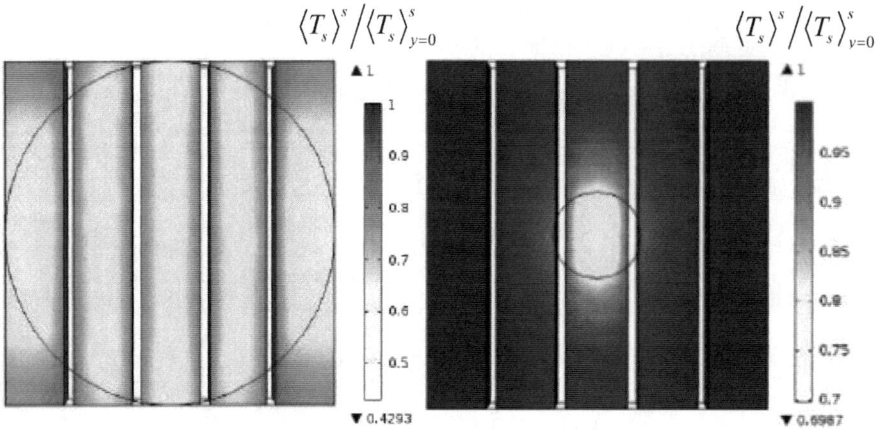

(a) (b)

Figure 2. Solid-phase dimensionless temperature on the heated plate: a) $D/W = 1$; b) $D/W = 0.25$.

Convection heat transfer coefficient and pressure drop are evaluated for $\varepsilon = 0.92$, PPI = 10, $|\mathbf{u}| = 0.5$ m/s, $t = 2$ mm, $H = 20$ mm and $D/W = 1$.

The heat transfer coefficient and the pressure drop as a function of the number of fins, for $t = 2$ mm and 4 mm, are presented in Fig. 4. The figure points out that, both for 2 and 4 fins, the heat transfer coefficient increases, attains a maximum and then decreases. This behaviour is due to a competition between two factors: the increase in the number of fins, which enhances the dissipated heat rate, and the reduction in the foam inserts volume. In the first part of the curves, the increase in the fins number improves the heat sink performance, but it reduces the foam volume, thus reducing the dissipated heat. As to the pressure drop, more fins means less foam volume, thus less pressure drop due to the reduction of foam inserts volume.

The heat transfer coefficient and pressure drop as a function of the sink height, for FMF and MF heat sinks with four fins, are reported in Fig.5. The heat transfer coefficient reduces with height because the impinging jet reaches the hot plate more difficultly, and also pressure drop reduces since a larger fraction of the fluid exits the heat sink from the side walls of the foam. The comparison of the MF heat sink with the FMF heat sink shows that the latter exhibits a higher heat transfer coefficient but also a higher pressure drop.

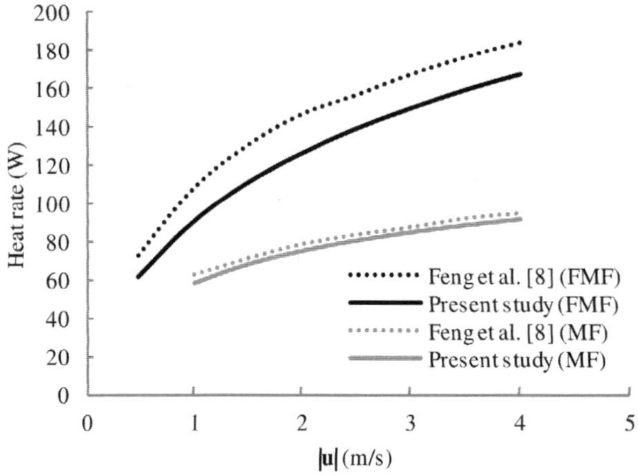

Figure 3. Dissipated heat rate *vs.* the velocity, for the FMF and MF heat sinks.

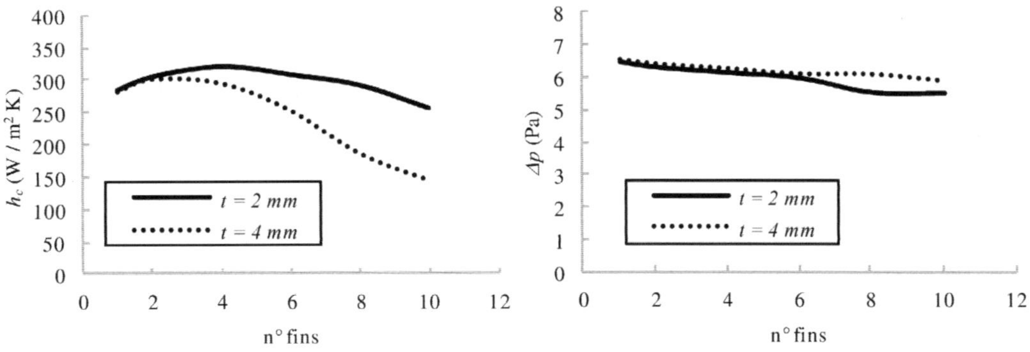

Figure 4. Heat transfer coefficient and pressure drop *vs.* the number of fins, for t = 2 mm and 4 mm.

The heat transfer coefficient and pressure drop as a function of the ratio of the impinging jet diameter to the side length of the heated plate, for FMF and MF sinks with four fins, are reported in Fig.6. We notice that the larger the section the larger the heat transfer coefficient, since increasing the plate size decreases the velocity of the fluid but allows the fluid to reach the plate in an easier way. The MF configuration performs better than the FMF one at low values of D/W because it is finned and has a smaller free area; thus the contribution of cold air is drastically reduced. Pressure drop in the MF configuration is unaffected by the size of the impinging section since the mass flow rate is unchanged, while pressure drop in FMF is affected by the size of impinging section because of the fins, that induce a concentrated pressure drop.

The heat transfer coefficient and pressure drop as a function of the porosity, for FMF and MF sinks with four fins, are reported in Fig.7. The figure shows that increasing the porosity reduces the heat transfer coefficient, because it is inversely proportional to the porosity, as well it reduces also the pressure drop, since the foam contains a decreasing fraction of solid.

The convection heat transfer coefficient and pressure drop as a function of PPI, for FMF and MF sinks with four fins, are reported in Fig.8. The figure exhibits a slight increase in heat transfer coefficient and a marked increase in pressure drop at increasing PPI. Heat transfer in MF configuration is almost independent of PPI, since at high values of the volumetric heat transfer coefficient a LTE condition is attained and enhancing PPI is useless.

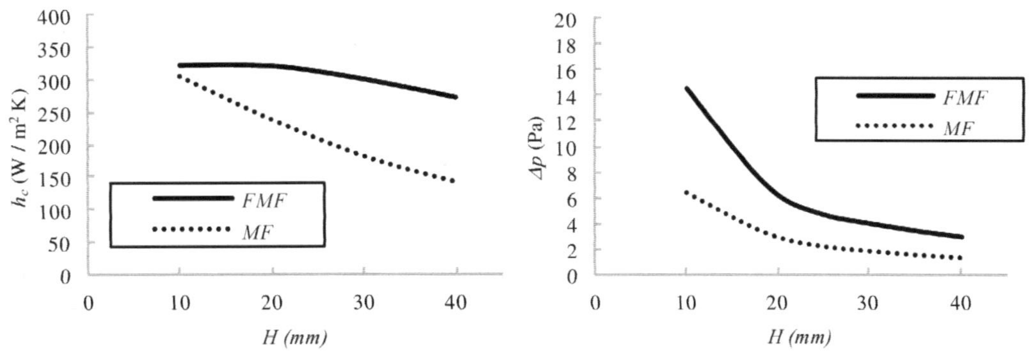

Figure 5. Heat transfer coefficient and pressure drop *vs.* the sink height, for FMF and MF heat sinks with four fins.

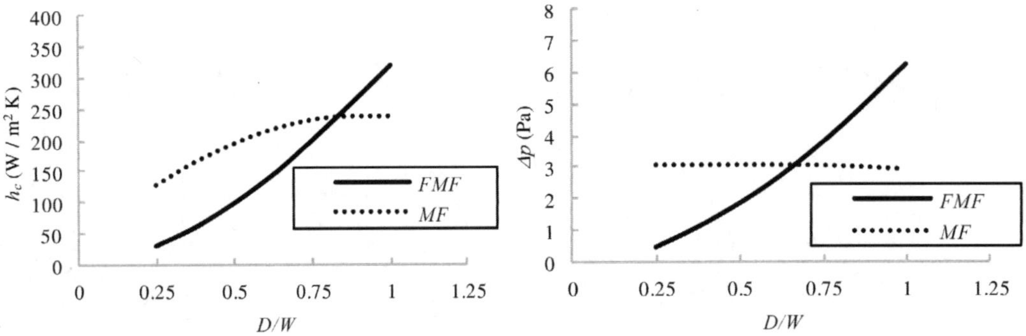

Figure 6. Heat transfer coefficient and pressure drop *vs*. the ratio of the impinging jet diameter to the side length of the heated plate, for FMF and MF sinks with four fins.

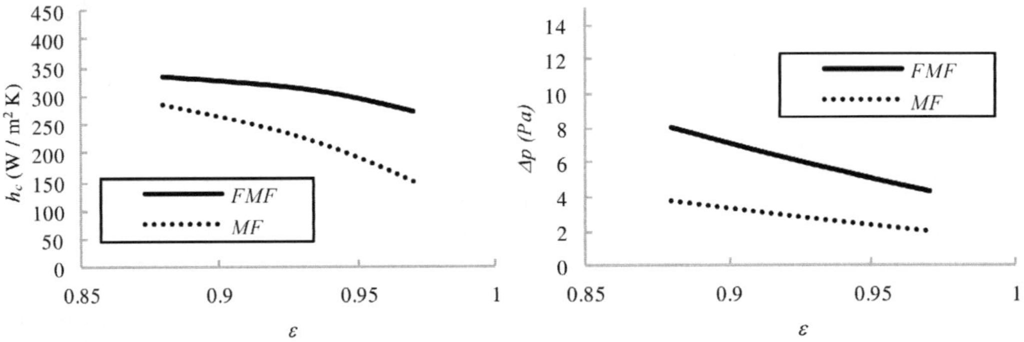

Figure 7. Heat transfer coefficient and pressure drop *vs*. the porosity, for FMF and MF sinks with four fins.

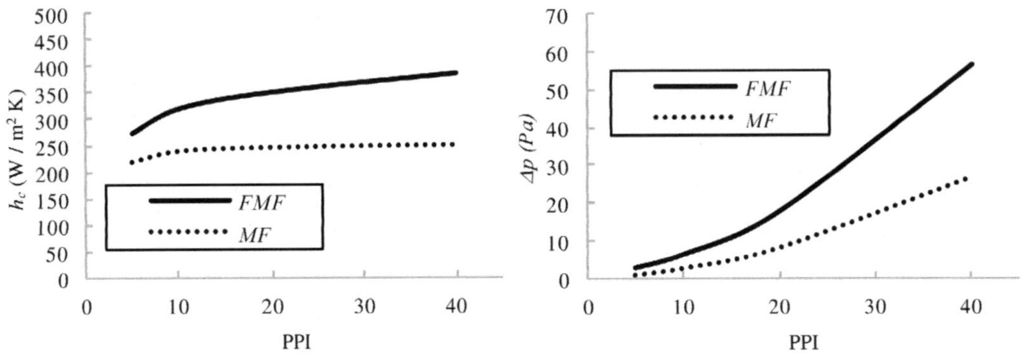

Figure 8. Heat transfer coefficient and pressure drop *vs*. PPI, for FMF and MF sinks with four fins.

4. Conclusions

In this paper a numerical analysis of heat transfer and pressure drop in a heat sink under impinging air jet cooling is carried out. Two different heat sink configurations have been investigated: a metal foam heat sink and a finned metal foam heat sink. The air-saturated foam is treated as an equivalent homogenous medium, by employing the Local Thermal non-Equilibrium model. The problem is numerically solved by means of the COMSOL Multiphysics® code. The effects of the ratio of the impinging jet diameter to the heated plate side, the height of the heat sink, the fins thickness, the number of fins, the foam porosity and PPI are analyzed. Results show that when the impinging jet diameter is equal to the heated plate side, both the heat transfer coefficient and pressure drop in the finned metal foam heat sink are larger than those in the metal foam heat sink. The metal foam heat sink performs better than the finned metal foam heat sink when the impinging jet diameter is smaller than the heated plate side.

5. References

[1] Sathe S and Sammakia B 1998 A review of recent development in some practical aspects of air cooled electronic packages *Trans. ASME, J. Heat Trans.* **120** 830–9

[2] Bhattacharya A and Mahajan R L 2002 Finned metal foam heat sinks for electronics cooling in forced convection *J. Electron. Pack.* **124** 155-63

[3] Kim S Y, Paek J W and Kang B H 2003 Thermal performance of aluminum-foam heat sinks by forced air cooling *IEEE T. Compon. Pack. T.* **26** 262-7

[4] Hsieh W H, Wu J Y, Shih W H and Chiu W C 2004 Experimental investigation of heat-transfer characteristics of aluminum-foam heat sinks *Int. J. Heat Mass Tran.* **47** 5149-57

[5] Shih W H, Chiu W C and Hsieh W H 2006 Height effect on heat-transfer characteristics of aluminum-foam heat sinks *J. Heat Trans-T.* ASME **128** 530-7

[6] Shih W H, Chou F C and Hsieh W H 2007 Experimental investigation of the heat transfer characteristics of aluminum-foam heat sinks with restricted flow outlet, *J. Heat Trans-T.* ASME **129** 1554-63

[7] Krishnan S, Hernon D, Hodes M, Mullins J and Lyons A M 2012 Design of complex structured monolithic heat sinks for enhanced air cooling *IEEE Trans. Compon. Packag. Manuf. Technol.* **2** (2) 266–77

[8] Feng S S, Kuang J J, Wen T, Lu T J and Ichimiya K 2014 An experimental and numerical study of finned metal foam heat sinks under impinging air jet cooling *Int. J. Heat Mass Trans.* 77 1063-74

[9] Feng S S, Kuang J J, Lu T J and Ichimiya K 2015 Heat transfer and pressure drop characteristics of finned metal foam heat sinks under uniform impinging flow *Journal of Electronic Packaging, Trans. ASME* **137** paper number 021014

[10] Wu Z, Caliot C, Bai F, Flamant G, Wang Z, Zhang J and Tian C 2010 Experimental and numerical studies of the pressure drop in ceramic foams for volumetric solar receiver applications *Appl. Energy* **87** 504-13

[11] Kamiuto K 2008 Modeling of Composite Heat Transfer in Open-Cellular Porous Materials at High Temperatures *Cellular and Porous Materials: Thermal Properties, Simulation and Prediction*, ed A Öchsner, G E Murch and M J S de Lemos (Weinheim: Wiley-VCH) chapter 6 pp. 165-98

[12] Wu Z, Caliot C, Flamant G and Wang Z 2011 Coupled radiation and flow modeling in ceramic foam volumetric solar air receivers *Sol. Energy* **85** 2374-85

[13] Zukauskas A. A. 1987 Convective heat transfer in cross-flow *Handbook of Single-Phase Convective Heat Transfer*, ed S Kakac, R K Shah and W Aung (New York, NY: Wiley)

[14] Eckert E R G and Drake R M 1972 *Analysis of Heat and Mass Transfer* (New York, NY: McGraw-Hill)

Unsteadiness and transition to turbulence in woven spacer filled channels for Membrane Distillation

M Ciofalo*, F Ponzio, A Tamburini, A Cipollina, G Micale
Dipartimento DICGIM, Università degli Studi di Palermo, Italy
*Corresponding author; email michele.ciofalo@unipa.it, tel. +39 091 232 225

Abstract. To characterize the performance of Membrane Distillation (MD) modules, channels filled with woven spacers were investigated by Computational Fluid Dynamics (including Direct Numerical Simulations and the use of the SST k-ω turbulence model) and by parallel experiments with Thermochromic Liquid Crystals. The cases considered here regard mutually orthogonal filaments with a spacer pitch to channel height ratio P/H=2, two spacer orientations θ with respect to the main flow (0° and 45°), and bulk Reynolds numbers Re from ~200 to ~2000, an interval of great interest in practical MD applications. For both values of θ, CFD predicted steady-state flow for Re up to ~300, and chaotic flow for Re larger than ~400. In the intermediate range Re≈300-400, periodic flow regimes were predicted for both orientations. These regimes were of particular interest and complexity, as they exhibited a slow global oscillation of the flow superimposed on high order harmonics corresponding to fast local oscillations. Experiments confirmed the appearance of unsteadiness for Re>~300. Heat transfer and friction were little affected by unsteadiness and exhibited a smooth behaviour with Re. The agreement with the experimental results was good using DNS, and acceptable using RANS.

1. Introduction and literature review

Spacers are widely used in a variety of membrane processes, among which Membrane Distillation (MD), both as structural elements and as mixing promoters, aimed at reducing polarization and enhancing heat or mass transport [1, 2, 3]. Unfortunately, they also cause pressure drop to increase [4]. Thus, much research effort has been devoted to the search for spacer configurations offering a compromise between heat / mass transfer and hydraulic losses [5, 6, 7, 8].

In recent years, Computational Fluid Dynamics (CFD) has increasingly been employed to this purpose. Once properly validated against experimental data, CFD provides valuable local information on both existing spacers and novel ones.

Early CFD efforts were limited to steady flow [9]. However, unlike other membrane processes such as Reverse Electrodialysis, characterized by very low Reynolds numbers and laminar or even creeping flow, MD usually involves Reynolds numbers up to a few thousand (see below for a precise definition of Re). Thus, an increasing number of CFD studies have been devoted to the simulation of unsteady flows. Limiting this brief survey to fully three-dimensional simulations (the only that can capture the complexity of the flow in a real spacer-filled channel), Koutsou *et al.* [10] performed Direct Numerical Simulations (DNS) with Fluent™, highlighting that unsteadiness may occur even at low Reynolds numbers, and Tamburini *et al.* [11] compared DNS results for early turbulent flow in channels filled with overlapped-type spacers, obtained by ANSYS-CFX™, with experimental measurements using Thermochromic Liquid Crystals. Mahdavifar *et al.* [12] performed DNS with OpenFOAM to investigate the effects of the spacer-to-wall clearance on the flow field; OpenFOAM and DNS were also used by Mojab *et al.* [13] to simulate unsteady flow in channels equipped with overlapped spacers. At high Reynolds number, DNS can become too computationally expensive, and the use of a turbulence model may be more suitable. Thus, Shakaib *et al.* [14] employed the Spalart-Allmaras model (based on a single transport equation for the turbulent viscosity) to predict temperature polarization in membrane distillation modules.

2. Configuration studied and main definitions

The spacer geometry considered in the present study is illustrated in Figure 1, and consists of mutually orthogonal cylindrical filaments of diameter δ, arranged in a woven pattern with pitch P. Figure 1(a) shows the physical spacer model used in the experimental tests, characterized by δ=5×10^{-3} m and P=2×10^{-2} m. Figure 1(b) shows the unit cell chosen as the computational domain in the numerical simulations; and Figure 1(c) shows a particular of the computational grid, consisting of both hexahedral and tetrahedral volumes. Since the pitch to filament diameter ratio is P/δ=4 and the channel height is H=2δ (neglecting the small deformation of the filaments at the contact points between one another and with the walls), the spacer pitch to channel height ratio is P/H=2.

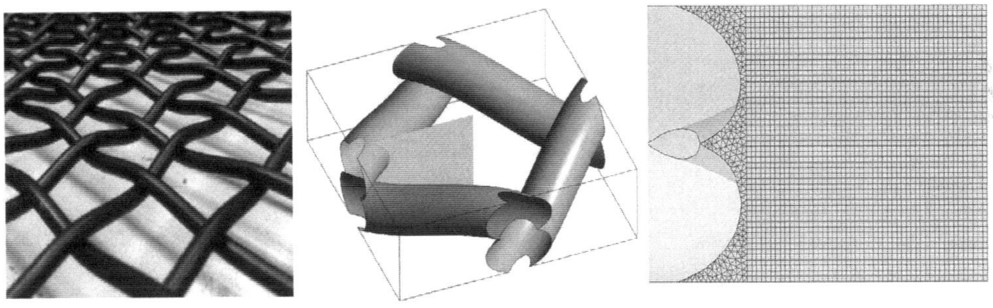

(a) (b) (c)

Figure 1. Woven spacer configuration examined in the present study. (a) Physical model used in the experiments; (b) unit cell for CFD; (c) computational grid in the region highlighted in (b).

As in our previous paper [11], dimensionless quantities are defined with reference to the void (spacerless) configuration, i.e. to a plane channel having height H and thus hydraulic diameter D_h=2H. Coherently, the bulk Reynolds number is

$$\mathrm{Re} = U\cdot 2H/\nu \tag{1}$$

in which $U = Q/(WH)$ is the mean "void channel" streamwise velocity, Q being the volume flow rate and W the channel's spanwise extent (in the experiments, H = 0.01 m and W = 0.22 m). We prefer this definition, adopted also by other authors [15], to that based on the actual hydraulic diameter and mean velocity, because it makes comparisons between different configurations more meaningful.

The friction velocity Reynolds number Re_τ is defined as

$$\mathrm{Re}_\tau = u_\tau (H/2)/\nu \tag{2}$$

in which $u_\tau=(\langle\tau_w\rangle/\rho)^{1/2}$ is the friction velocity. The quantity $\langle\tau_w\rangle$ is the wall-averaged wall shear stress which, in a void plane channel and under steady-state conditions, would balance the large-scale pressure gradient along the main flow direction ξ, defined as $p_\xi = -\Delta p/\Delta\xi$, so that $\langle\tau_w\rangle=(H/2)p_\xi$.

The Darcy friction coefficient (four times the Fanning friction factor) is defined as

$$f_D = \frac{4H}{\rho U^2} p_\xi \tag{3}$$

On the basis of the definitions given above for Re and Re_τ, one has also f_D=128(Re_τ/ Re)^2. The local Nusselt number is defined as

$$Nu = h\frac{2H}{\lambda} \tag{4}$$

in which $h=q''_w/(T_b-T_w)$ is the local heat transfer coefficient, T_b is the fluid bulk temperature in the unit cell and T_w is the local wall temperature. Wall-averaged values will be indicated as $\langle h\rangle$ and $\langle Nu\rangle$. The Prandtl number was assumed to be 4.63, representative of water at 39°C.

Woven spacers have been studied far less in depth than non-woven ones. They are often mentioned in connection with different membrane-based processes [6, 16, 17, 18], but no systematic analysis, either experimental or computational, of their performance regarding pressure drop and heat / mass transfer seems to have been performed. What can be said with some confidence is that woven spacers provide larger heat / mass transfer coefficients, but also higher pressure drops, than overlapped ones. Our group has recently performed experimental measurements and CFD simulations on woven spacers for Membrane Distillation with different values of the spacer pitch to channel height ratio, working at bulk Reynolds numbers up to 2000 [19]; this study follows a similar experimental and computational investigation on overlapped (non-woven) spacers [8, 11, 20, 21, 22, 23].

3. Experimental method

The experimental apparatus has been described in detail in previous papers [23]. Experiments were conducted on a scaled-up spacer-channel configuration, Figure 1(a), purpose manufactured from commercial black rubber wire 5 mm in diameter. The resulting distance between the plane channel walls was ~10 mm. The pitch of the wire net was ~20 mm, giving a pitch-to-height ratio *P*/*H* of ~2.

The use of Thermochromic Liquid Crystals (TLC) to obtain surface temperature distributions [24] or even three-dimensional temperature and flow fields [25] is amply documented in the scientific literature [24]. Images of the TLC-covered wall were split into HSV (Hue, Saturation, Value) components, of which only H was used to derive the wall temperature on the basis of an *in situ* calibration, performed by the isothermal method. For the TLC preparation used here (Hallcrest™ R30C5W) the nominal colour play (from red start to blue start) was 30-35°C while the actual resolved temperature range was ~31-46°C.

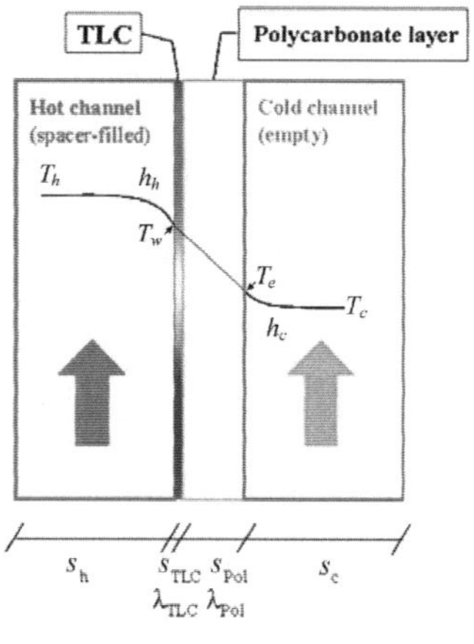

Figure 2. Sketch of the temperature profile across the different layers of the test section.

Figure 2 reports a schematic representation of the temperature profile across the different layers of the test section including a hot channel, a TLC sheet, a dividing wall made of 1 mm thick polycarbonate, and a cold channel. An elementary analysis of the heat transfer process occurring through the various layers in series, under the assumption of one-dimensional transport (demonstrated to be reliable in [23]), gives the local hot-side heat transfer coefficient *h* as a function of the locally measured temperature T_w of the TLC sheet once the bulk temperatures T_h, T_c and the cold-side heat transfer coefficient are known:

$$h = \frac{1}{r}\frac{T_w - T_c}{T_h - T_w} \tag{5}$$

Here, r is an overall external thermal resistance consisting of the conductive thermal resistances s/λ of the thermochromic liquid crystal film (*TLC*) and of the polycarbonate layer (*pol*), in series with the convective thermal resistance $1/h_c$ associated with the cold channel:

$$r = \frac{s_{TLC}}{\lambda_{TLC}} + \frac{s_{pol}}{\lambda_{pol}} + \frac{1}{h_c} \tag{6}$$

T_h and T_c were measured by Pt100 RTDs at the inlet and outlet of the hot and cold channels and a linear variation of T_h and T_c along the flow direction was assumed, as justified by the fact that estimated relaxation lengths are far larger than the test section length. The cold-side convection coefficient h_c was estimated by the Dittus-Bölter correlation (Nu=0.023 $Re^{0.8}$ $Pr^{0.4}$) written for turbulent flow in the cold channel; a sensitivity analysis showed that the value of h calculated by Eq. (5) is only a weak function of the value used for h_c. The value of $6.24\cdot 10^{-3}$ m^2K/W was computed for the overall thermal resistance r, and h was then expressed in dimensionless form as Nu=hd_{eq}/λ.

As an example of the information that can be extracted from the pictures by digital image processing, Figure 3 reports unit cell distributions obtained for P/H=2, θ=90° and a flow rate of ~60 l/h (yielding a bulk Reynolds number Re≈215). The flow is from left to right. Graph (a) is a picture of the active, TLC-covered wall, in which saturation was enhanced for clarity purposes; the arrow indicates the main flow direction. Graph (b) shows the distribution of the local Nusselt number Nu, derived by digitally processing the image in (a).

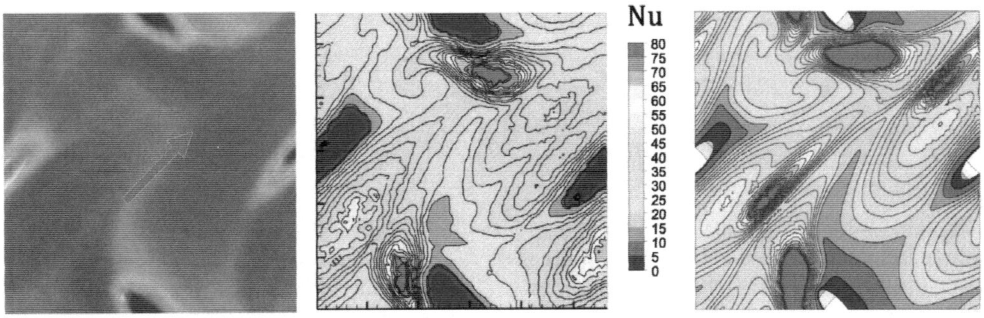

(a) (b) (c)

Figure 3. Unit cell of a woven spacer (P/H=2, θ=0-90°, Re≈215). (a) picture of the TLC-covered wall (the arrow indicates the flow direction); (b) local Nusselt number Nu; (c) DNS results for Re_τ=80, yielding Re≈240.

The analysis of short movies of the TLC-covered wall revealed that the colour (i.e., wall temperature) pattern was almost steady below a bulk Reynolds number of ~300 and clearly unsteady at larger values of Re. This is illustrated in Figure 4 by reporting three frames, separated by a few seconds, for each of the two Reynolds numbers 255 and 921.

Pressure drops in the hot fluid side were measured by using two longitudinally spaced tappings, located along the test section mid-line and 50 cm apart, and a Fuji Electric FCX-AII™ differential manometer with a response of ~7.4 μA/Pa. Pressure drops were generally very small (e.g. 25 Pa at Re=350) and thus could be measured only with a limited accuracy.

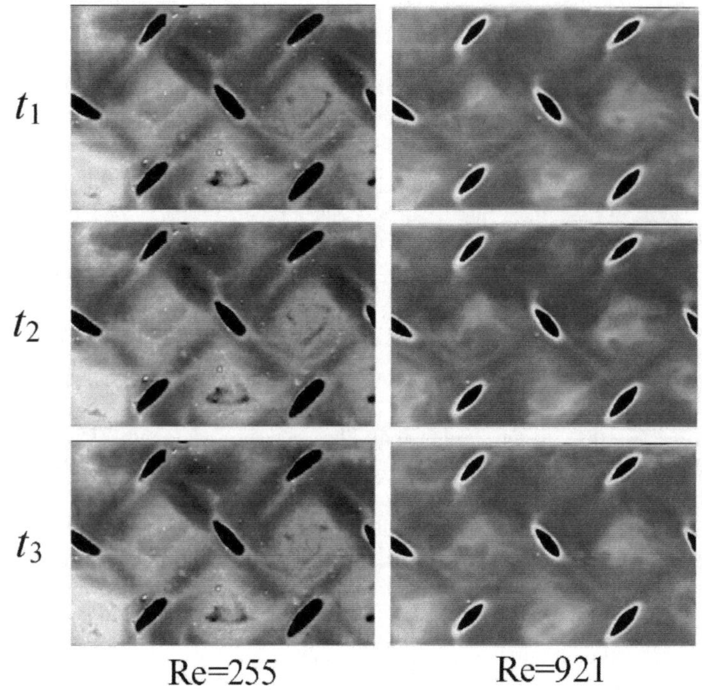

Figure 4. Photographs of the TLC-covered wall taken at three instants t_1-t_2-t_3 (separated by a few seconds) for θ=45° and two Reynolds numbers, showing substantial steadiness for Re=255 and large unsteadiness for Re=921. Images were processed so as to enhance the differences between frames.

4. Computational method

The computational domain was the unit, periodic cell shown in Figure 1(b). Mathematically, the problem was described by the continuity, Navier-Stokes and energy equations for a constant-property fluid. The "unit cell" treatment, first described by our group in ref. [26], allowed periodicity conditions to be adopted for all variables at the opposite faces of the computational domain. The friction Reynolds number Re_τ(i.e., the pressure drop per unit length) was imposed, while the bulk Reynolds number (i.e., the flow rate) was computed as part of the solution. All simulations were conducted by the finite volume ANSYS-CFX14™ code [27].

No slip conditions (u_i=0) were imposed at the walls and on the filaments' surface. In regards to the thermal boundary conditions, the bottom wall and the filaments were assumed to be adiabatic (q"=0), while at the top (thermally active) wall a third-type (mixed) condition was imposed:

$$T_w\text{-}T_c = r\, q"_w \qquad (7)$$

mimicking the experimental boundary condition. Values representative of those holding in the experiments were chosen for the outer temperature T_c and the thermal resistance r (19°C and $6.24 \cdot 10^{-3}$ m^2K/W, respectively). Also the spacer dimensions were the same as the experimental ones and the fluid was assumed to be water at 39°C with physical properties ρ=992.5 kg/m^3, c_p=4179 J/(kgK), λ=0.5985 W/(mK), μ=6.625×10^{-4} Pa·m, yielding Pr=4.63. Finally, as in the experiments, two orientations of the main flow with respect to the spacer filaments of the upper layer were considered (0° and 45°). In the present study, all simulations were run in time-dependent mode and attained either a steady-state or a time-dependent (time periodic or irregular, i.e. chaotic) regime.

Unlike the corresponding overlapped-spacer geometry [11], the present one could not be discretized with purely hexahedral volumes; a hybrid grid was thus built, in which tetrahedral volumes were used in the regions surrounding the filaments (~30% of the overall volume of the computational domain) and hexahedral volumes in the remaining regions. Figure 1(c) reports a detail of the mesh.

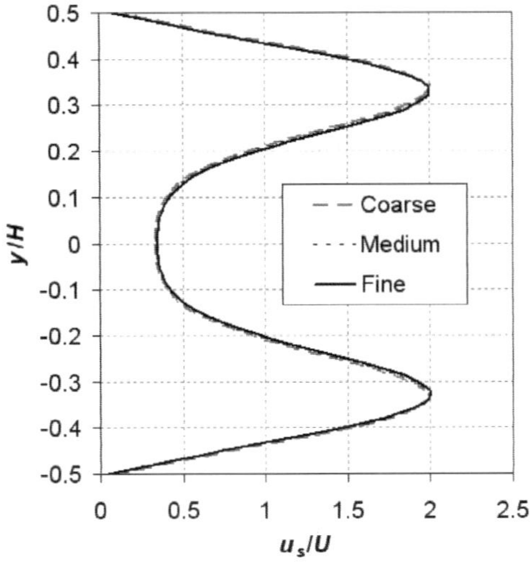

(a)

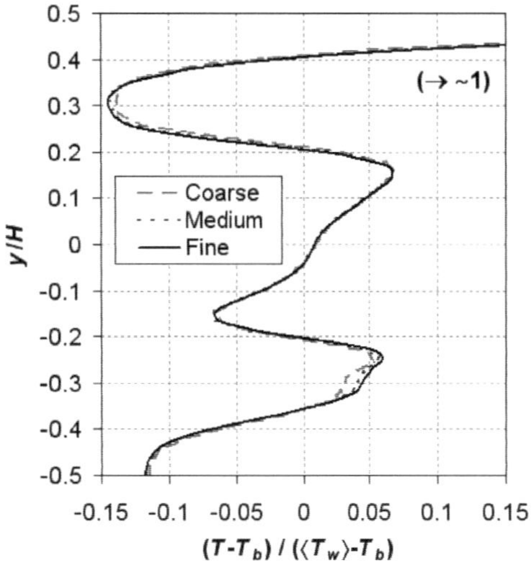

(b)

Figure 5. Grid sensitivity: profiles of normalized streamwise velocity u_s (a) and temperature T (b) along a line perpendicular to the walls and passing by the centroid of the unit cell, as computed for θ=45°, Re_τ=70 by using three different grids with ~0.84×10^6 (coarse), ~2.00×10^6 (medium) and ~3.49×10^6 (fine) volumes.

A grid sensitivity study was conducted for θ=45°, Re_τ=70 (steady-state flow at Re=273). Figure 5 reports the profiles of normalized streamwise velocity u_s (a) and temperature T (b) along a line perpendicular to the walls and passing by the centroid of the unit cell, as computed by using three grids with ~0.84×10^6, ~2.00×10^6 and ~3.49×10^6 volumes, respectively. As it can be observed, the results from the two finer grids are almost identical, and even the coarsest grid provides very similar results. For the final simulations, the intermediate grid of ~2×10^6 finite volumes was adopted.

5. Results

5.1 Case θ=0-90°

For this orientation (main flow parallel and orthogonal to the two filament layers), steady-state flow was predicted for Re_τ up to ~105 (Re up to ~340), periodic flow for $Re_\tau\approx110$ (Re≈350) and chaotic flow for $Re_\tau\approx120$ or larger (Re≥~390).

As an example of steady-state results, Figure 3(c) reports, along with experimental results obtained under similar conditions, the local Nusselt number distribution predicted for Re_τ=80, yielding Re≈240. A good general agreement can be observed. The main differences regard the values of Nu maxima, which are somewhat overestimated by the numerical simulations, and the narrow regions of high Nu predicted by CFD along the diagonal parallel to the main flow, not present in the experiments.

As an example of periodic behaviour, Figure 6(a) reports a time series of the streamwise velocity u_s at the centre of the unit cell for Re_τ=110, or Re≈352 (velocity is normalized by the mean streamwise velocity U, time by δ/U). The time interval showed includes three periods; the duration T_0 of each period is ~100 δ/U, corresponding to ~42 s at the scale of the experiments. Graph (b) reports an enhanced detail of the behaviour of u_s in the shorter interval highlighted in (a), showing a secondary "quasi-period" of ~ $T_0/100$ which corresponds to 1× δ/U, or ~0.42 s at the scale of the experiments.

The corresponding power spectrum is reported in Figure 7; here, frequency is normalized as $F=f/(U/\delta)$, i.e. as a Strouhal number based on mean velocity and filament diameter, while P_{uu} is in arbitrary units. Discrete peaks, corresponding to a base dimensionless frequency F_0 of ~0.01 and its harmonics, are visible; the very high order harmonic at 105 F_0 corresponds to the rapid oscillations in Figure 6(b).

IOP Conf. Series: Journal of Physics: Conf. Series **796** (2017) 012003 doi:10.1088/1742-6596/796/1/012003

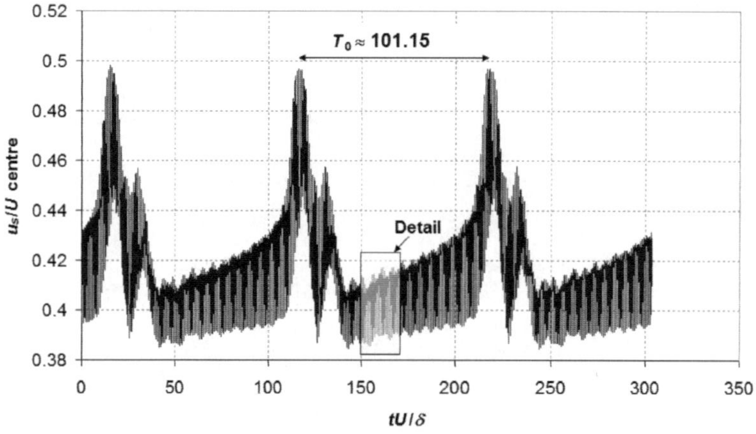

(a)

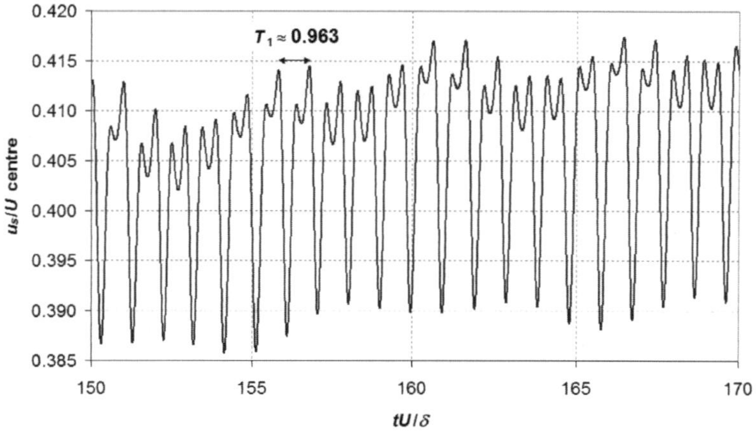

(b)

Figure 6. Time series of the streamwise velocity u_s at the centre of the unit cell for θ=0-90° and Re_τ=110 (Re=352, periodic flow). Velocity is normalized by the mean streamwise velocity U, time by δ/U. (a) Large scale behaviour over a time interval including three periods T_0. (b) Detail of the behavior of u_s in the interval highlighted in (a), showing a secondary "quasi-period" of ~ $T_0/100$.

Finally, as an example of fully chaotic behaviour, Figure 8 reports computational results for θ=0-90° and Re_τ=140 (Re=478): graph (a) shows a time series of the streamwise velocity u_s at the centre of the unit cell, while graph (b) is the corresponding time power spectrum. All quantities are normalized as in Figures 6 and 7. It can be observed that velocity fluctuations are irregular and that their spectrum includes a large number of independent frequencies, but a frequency interval exhibiting the characteristic -5/3 power law behaviour of full turbulence (inertial sub-range) is barely identifiable.

5.2 Case θ=45°

For this orientation (main flow bisecting the angle formed by the two filament layers), steady-state flow was predicted for Re_τ up to ~70 (Re up to ~225), periodic flow for Re_τ≈80-100 (Re≈270-365) and chaotic flow for Re_τ≈120 or larger (Re≥~465).

For the sake of brevity, representative steady-state results are not reported. As an example of periodic behaviour, Figure 9(a) reports a time series of the streamwise velocity u_s at the centre of the unit cell for Re_τ=100, or Re≈365. Figure 9(b) reports the corresponding power spectrum. Time, velocity and frequency are normalized as in the previous Figures 6-8.

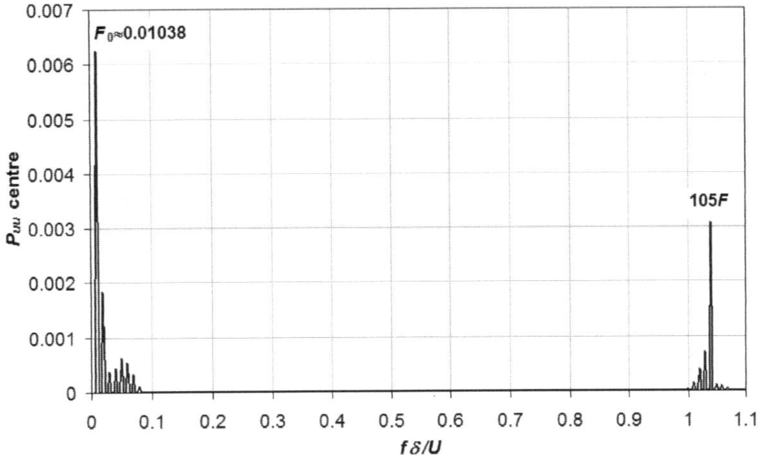

Figure 7. Power spectrum of the streamwise velocity u_s at the centre of the unit cell for θ=0-90° and Re_τ=110 (Re=352), reported as a function of time in Figure 6 (frequency is normalized by U/δ, i.e. as a Strouhal number, while P_{uu} is in arbitrary units).

The time interval considered in Figure 9 includes three periods, each of duration $T_0 \approx 6.5\delta/U$ (corresponding to ~2.65 s at the scale of the experiments). In the spectral density diagram, this periodicity corresponds to the base frequency $F_0 \approx 0.15$. However, most of the fluctuation intensity is associated with a secondary period $T_1 = T_0/5$ (i.e., ~$1.3\delta/U$, or ~0.53 s at the scale of the experiments), corresponding to the spectral peak at $F_1 = 5F_0$. Harmonics at 4, 6 and 9 × F_0 are also appreciable.

Finally, an example of fully chaotic behaviour is reported also for the 45° orientation in Figure 10, which is for Re_τ=120 (Re=465). The quantities reported are the same as in Figure 8, and time series and spectra are similar. In this latter case, the power spectrum P_{uu} in Figure 10(a) exhibits a marked peak at $f\delta/U \approx 2.3$ which does not have a clear counterpart in the case θ=0-90° of Figure 8.

5.3 Bifurcation diagram

On the basis of the simulations performed, the tentative bifurcation map in Figure 11 can be drawn. It reports the rms value of the fluctuating streamwise velocity, u_{rms}, at the centre of the computational domain (normalized, as usual, by U) as a function of the Reynolds number Re for both flow attack angles. Symbols correspond to actual computational results (S: steady-state flow; P: periodic flow; C: chaotic flow). Shaded circles indicate the presumable position of bifurcations from steady to periodic or from periodic to chaotic flow.

Of course, one has u_{rms}=0 for all steady-state solutions. For θ=45°, the periodic branch bifurcates from the steady-state one at Re≈225 and extends above Re=350; the amplitude of the fluctuations increases considerably in this range. A second bifurcation from time-periodic to chaotic flow must occur at slightly higher Re, and is accompanied by a reduction of the velocity fluctuation amplitude. For θ=0-90°, the periodic branch bifurcates from the steady-state one at some Re intermediate between ~330 and ~350 but is rather short, probably ending at Re=370-380; a further bifurcation leads to a chaotic flow branch exhibiting, at least in the range studied here, a monotonic increase of the velocity fluctuation amplitude. All bifurcations appear to be of the supercritical type, i.e. they are not associated to abrupt changes in the computed quantities.

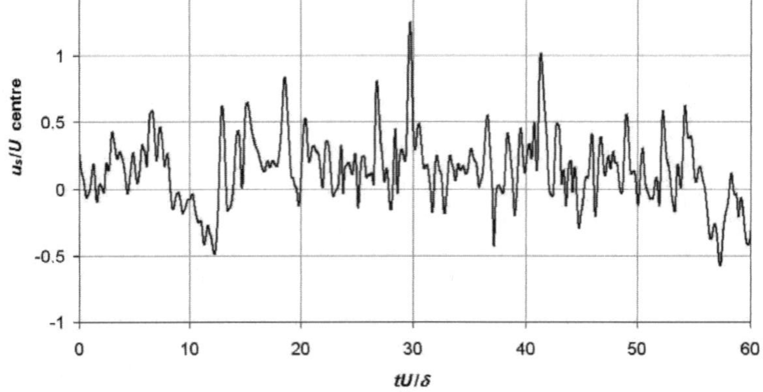

(a)

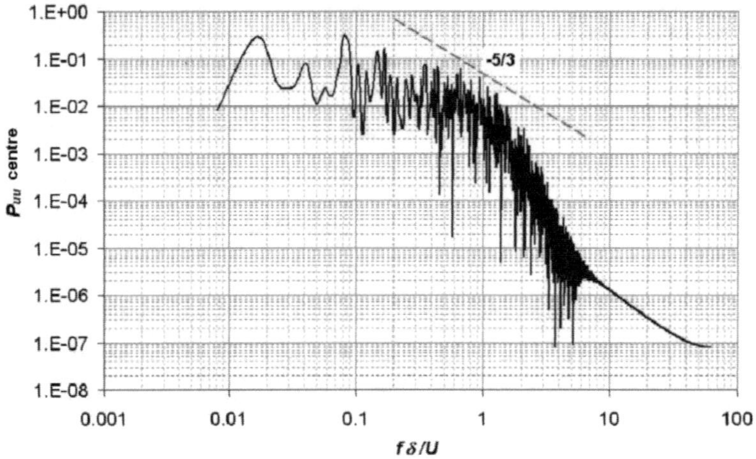

(b)

Figure 8. Computational results for θ=0-90° and Re_τ=140 (Re=478, chaotic flow). (a) Behavior of the streamwise velocity u_s at the centre of the unit cell (u_s is normalized by the mean streamwise velocity U, time by δ/U). (b) Corresponding power spectrum (frequency is normalized by U/δ, i.e. as a Strouhal number, while P_{uu} is in arbitrary units).

6. Comparison of experimental and computational results

6.1 Pressure drop

Figure 12 reports the Darcy friction coefficient f (normalized by that for parallel flow in a plane channel, 96/Re) both for θ=0-90° (a) and for θ=45° (b). Experimental results are shown along with computational ones, obtained either by direct numerical simulation or by a RANS turbulence model. In the case of time-dependent (periodic or chaotic) DNS solutions, f was computed as a time average. RANS simulations were run in steady-state mode; the SST model [28] was chosen not only because it is one of the best established turbulence models for general-purpose applications (at least among the two-equation, eddy diffusivity models), but also because, in a preliminary comparative study for the present spacer-filled channel configuration, it performed far better than alternative models including the two-equation RNG k-ε and the second-order SSG.

A first noteworthy feature of Figure 12 is that neither experimental results nor computational ones (DNS or RANS) exhibit any significant change of value or slope in correspondence with the transition from steady to unsteady flow (which, according to the DNS summary results reported in Figure 11, occurs at Re≈340 for θ=0-90° and at Re≈225 for θ=45°).

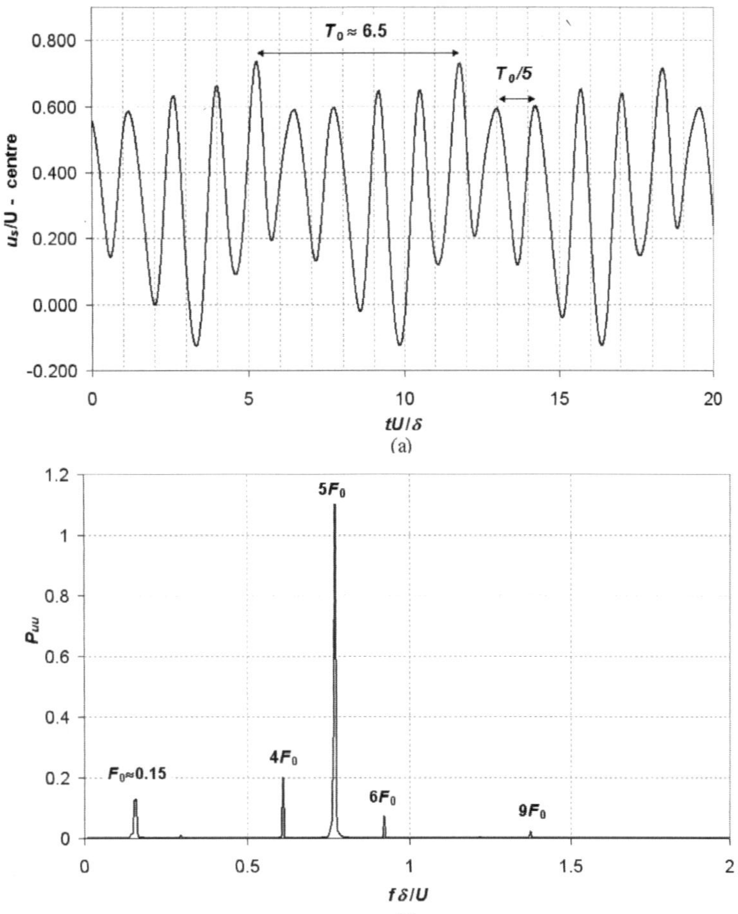

Figure 9. Computational results for θ=45° and Re_τ=100 (Re=365, periodic flow). (a) Behavior of the streamwise velocity u_s at the centre of the unit cell (u_s is normalized by the mean streamwise velocity U, time by δ/U). (b) Corresponding power spectrum (frequency is normalized by U/δ, i.e. as a Strouhal number, while P_{uu} is in arbitrary units).

Also the transition from periodic to chaotic flow (occurring at Re≈375 both for θ=0-90° and for θ=45° according to the map in Figure 11) is not reflected in any visible feature of the results. Note that the lack of large-scale effects (i.e., effects on the f *vs.* Re and ⟨Nu⟩ *vs.* Re curves) in correspondence with the loss of stability of stationary or periodic regimes is a feature common to many problems characterized (like the present one) by some geometrical complexity and by the existence of significant secondary flow even at low-Re. For example, it was observed both experimentally and by DNS in curved and helical pipes [29].

Both for θ=0-90° and for θ=45°, the agreement between experimental and DNS results is good in the limited ranges of overlap (Re up to ~500). For both values of θ, SST simulations underpredict the experimental results. For θ=0-90°, graph (a), the relative discrepancy remains limited (~10%) up to Re≈700, but becomes larger at the highest Re. For θ=45°, graph (b), the relative underprediction remains of the order of 20% over the whole range investigated but SST results exhibit a slight discontinuity at Re just above 1000, associated with the transition from a poor and slow convergence at lower Re to a better and faster convergence at higher Re.

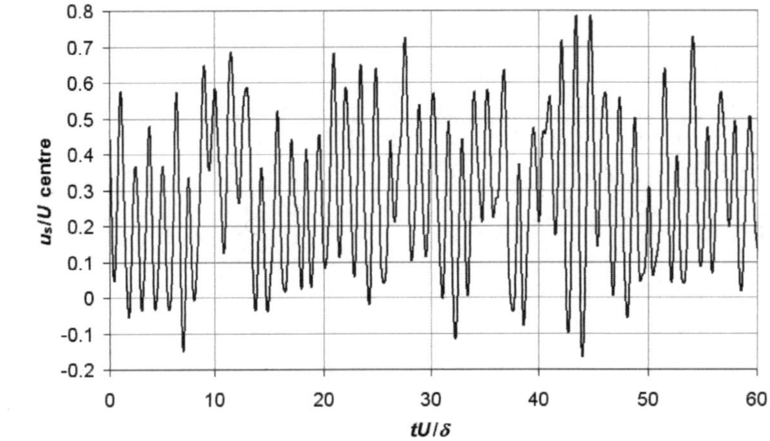

(a)

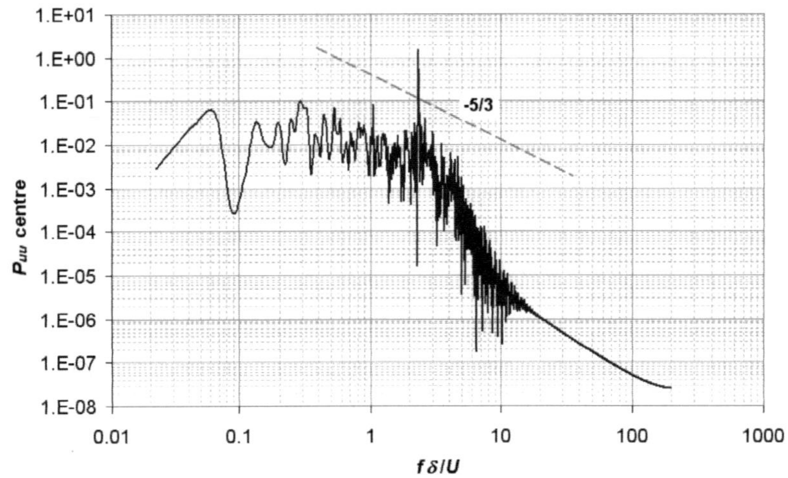

(b)

Figure 10. Computational results for θ=45° and Re_τ=120 (Re=465, chaotic flow). (a) Behavior of the streamwise velocity u_s at the centre of the unit cell (u_s is normalized by the mean streamwise velocity U, time by δ/U). (b) Corresponding power spectrum (frequency is normalized by U/δ, i.e. as a Strouhal number, while P_{uu} is in arbitrary units).

This discontinuity seems to have little to do with the transition from steady to unsteady or from periodic to chaotic flow, discussed a few lines above; it occurs at much higher Re and regards only RANS results which are all for steady-state conditions by definition. Moreover, no discontinuity was observed in the RANS results for θ=0-90°. Either computational results for creeping flow or an extrapolation of the experimental results to Re=0 indicate that, for very low Re, the friction coefficient is enhanced about 14.6 times by the presence of the spacer.

6.2 Heat transfer

Figure 13 reports the wall-averaged Nusselt number ⟨Nu⟩ both for θ=0-90° (a) and for θ=45° (b). As in Figure 12, experimental results are shown along with computational ones, obtained by direct numerical simulation or by steady-state SST simulations. As for f, in the case of time-dependent DNS results ⟨Nu⟩ was computed as a time average.

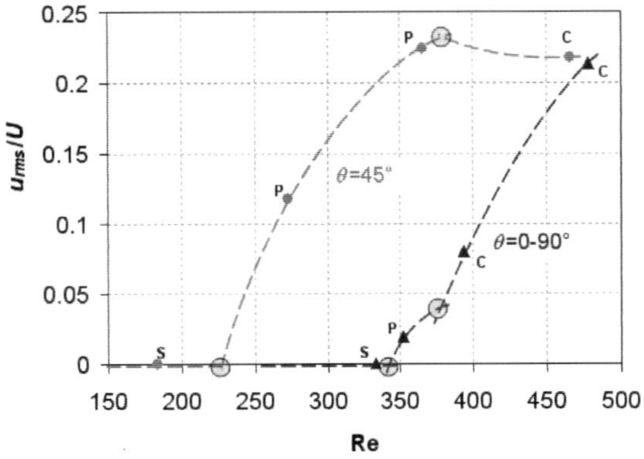

Figure 11. Tentative bifurcation diagram reporting the rms streamwise velocity fluctuation u_{rms} at a monitoring point (normalized by the mean streamwise velocity U) as a function of the Reynolds number Re. Symbols correspond to computational results (S: steady-state flow; P: periodic flow; C: chaotic flow). Shaded circles indicate bifurcations from steady to periodic or from periodic to chaotic flow.

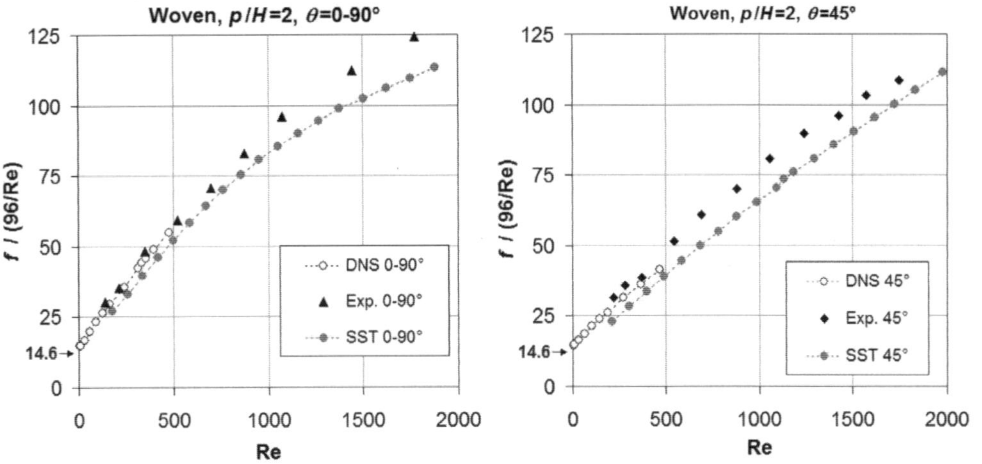

(a) (b)

Figure 12. Friction coefficient (normalized by that for parallel flow in a plane channel, 96/Re): experimental results along with computational results obtained by DNS and by the SST model. (a) θ=0-90°; (b) θ=45°.

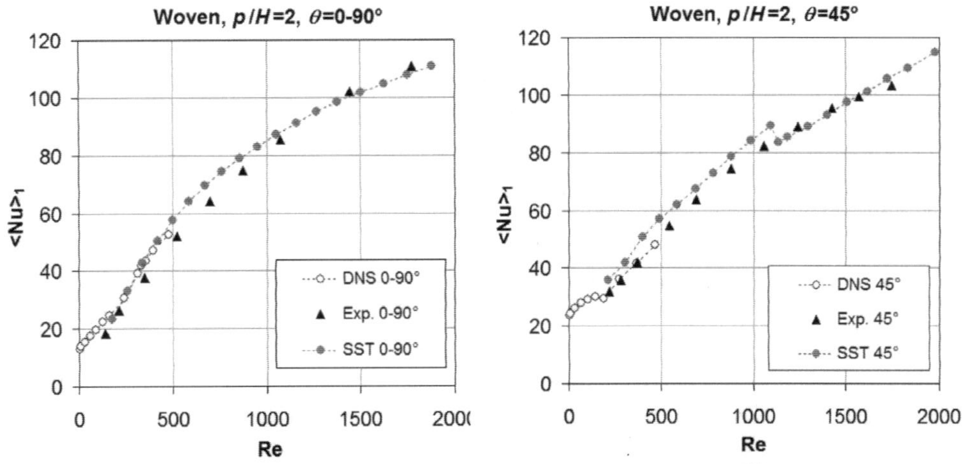

(a) (b)

Figure 13. Wall-averaged Nusselt number: experimental results along with computational results obtained by DNS and by the SST model. (a) θ=0-90°; (b) θ=45°.

The remarks on the lack of features associated with the transitions from steady to unsteady flow and from periodic to chaotic solutions, made for the friction coefficient (see Figure 12 and Section 6.1 above), apply also to the Nusselt number in Figure 13.

In the limited range of overlap (Re up to ~500), the agreement between experimental and DNS results is excellent for θ=45° (b) and good, with some slight overprediction, for θ=0-90° (a). For both angles, SST simulations yield some overprediction (max. 15%) of the experimental results up to Re=1000-1300, while a better agreement is obtained for larger Re. As in Figure 12, the SST results for θ=45° (but not those for θ=0-90°) exhibit a small discontinuity for Re just above 1000, in correspondence with the transition from a poor convergence at low Re to a better one at high Re.

7. Conclusions

The aim of the present work was to answer the following questions, which are all of some considerable importance for the optimum design of Membrane Distillation modules and their spacers:

I) In the complex geometry of a woven spacer filled channel, what are the possible flow regimes, the associated Reynolds number ranges and the modes of transition from a regime to another?

II) How do these transitions affect friction and heat transfer?

III) Is there a computational approach capable of reproducing experimental data through the whole Reynolds number range of interest for MD applications (say, 200 to 2000)?

To the first of these questions, the answer is provided by DNS (with the partial support of experimental work). This shows that *transition from steady to unsteady flow occurs via a first bifurcation from steady to time-periodic flow, followed by a second one from periodic to chaotic flow.* Both bifurcations are of the supercritical type, so that they do not imply either abrupt jumps in the flow quantities or hysteresis phenomena. The critical values of Re, and the relative extent of the different regime ranges, vary with the flow attack angle; in particular, periodic flow was found to exist in a small range of Re (340~375) for θ=0-90°, but in a much broader range (225~375) for θ=45°.

To the second question, the answer is that *unsteadiness and turbulence affect only marginally and indirectly both friction and heat transfer*, which are dominated by the features of the mean flow and, in particular, by the spacer-induced recirculation. The intensity of this latter increases with Re, but does not exhibit any abrupt jump in correspondence with flow regime transitions, as proved by the fact that both the f *vs.* Re and the $\langle Nu \rangle$ *vs.* Re curves (as derived either from experiments or from DNS) are smooth in the whole Reynolds number range investigated (0~2000).

Finally, the answer to the third question is that *the only computational approach providing an excellent prediction of experimental results is direct numerical simulation*, which, however, becomes highly computationally demanding in turbulent flow (Re>~400). Among the various RANS models tested (including RNG k-ε, SST k-ω and second-order Reynolds stress transport models), *SST provided the best results.* Even SST, however, exhibited shortcomings including a significant underprediction (10-20%) of the friction coefficient, a poor convergence for Re<~1000, and an abrupt, although limited, jump in f and $\langle Nu \rangle$ across Re≈1000 for θ=45°. Despite these drawbacks, SST allowed acceptably accurate results to be computed through the whole Re range of interest, and correctly predicted laminarization at low Re without the need for external "*ad hoc*" input.

References

[1] L. Martínez-Díez, M.I. Vázquez-González, Temperature and concentration polarization in membrane distillation of aqueous salt solutions, *J. Membr. Sci.*, 156 (1999) 265-273.

[2] J. Schwinge, P.R. Neal, D.E. Wiley, D.F. Fletcher, A.G. Fane, Spiral wound modules and spacers: Review and analysis, *J. Membr. Sci.*, 242 (2004) 129–153.

[3] M. Amokrane, D. Sadaoui, C.P. Koutsou, A.J. Karabelas, M. Dudeck, A study of flow field and concentration polarization evolution in membrane channels with two-dimensional spacers during water desalination, *J. Membr. Sci.*, 477 (2015) 139-150.

[4] G. Schock, A. Miquel, Mass transfer and pressure loss in spiral wound modules, *Desalination*, 64 (1987) 339-352.

[5] J. Phattaranawik, R. Jiraratananon, A.G. Fane, C. Halim, Mass flux enhancement using filled channels in Direct Contact Membrane Distillation, *J. Membr. Sci.*, (2001) 193-201.

[6] F. Li, W. Meindersma, A.B. de Haan, T. Reith, Optimization of commercial net spacers in spiral wound membrane modules, *J. Membr. Sci.*, 208 (2002) 289–302.

[7] Y. Yun, J. Wang, R. Ma, A.G. Fane, Effects of channel spacers on Direct Contact Membrane Distillation, *Desalination and Water Treatment*, 34 (2011) 63-69.

[8] A. Tamburini, A. Cipollina, S. Al-Sharif, M. Albeyrutty, L. Gurreri, G. Micale and M. Ciofalo, Assessment of temperature polarization in Membrane Distillation channels by Liquid Crystal Thermography, *Desalination and Water Treatment*, 55 (2015) 2747-2765.

[9] S.K. Karode, A. Kumar, Flow visualization through spacer filled channels by computational fluid dynamics I. Pressure drop and shear rate calculations for flat sheet geometry, *J. Membr. Sci.*, 193 (2001) 69–84.

[10] C.P. Koutsou, S.G. Yiantsios, A.J. Karabelas, Direct numerical simulation of flow in spacer-filled channels: effect of spacer geometrical characteristics, *J. Membr. Sci.*, 291 (2007) 53–69.

[11] A. Tamburini, M. Renda, A. Cipollina, G. Micale, M. Ciofalo, Investigation of heat transfer in spacer-filled channels by experiments and direct numerical simulations, *Int. J. Heat Mass Transfer*, 93 (2016) 1190-1205.

[12] A. Mahdavifar, A. Pollard, J.G. Pharoah, S.B. Beale, Wall proximity effects on flow over a simple membrane spacer, *Computers & Fluids*, 88 (2013) 180–188.

[13] S. M. Mojab, A. Pollard, J.G. Pharoah, S.B. Beale, E.S. Hanff, Unsteady Laminar to Turbulent Flow in a Spacer-Filled Channel, *Flow Turbulence Combust.*, 92 (2014) 563–577.

[14] M. Shakaib, S.M.F. Hasani, I. Ahmed, R.M. Yunus, A CFD study on the effect of spacer orientation on temperature polarization in Membrane Distillation modules, *Desalination*, 284 (2012) 332–340.

[15] M. Gimmelshtein, R., Semiat, Investigation of flow next to membrane walls, *J. Memb. Sci.*, 264 (2005) 137-150.

[16] A. R. Da Costa, A. G. Fane, D. E. Wiley, Spacer characterization and pressure drop modelling in spacer filled channel for ultrafiltration, *J. Membr. Sci.*, 87 (1994) 79-98.

[17] J. Liu, A. Iranshahi, Y. Lou, G. Lipscomb, Static mixing spacers for spiral wound modules, *J. Membr. Sci.*, 442 (2013) 140–148.

[18] T. Melin, New ED - Advanced Bipolar Membrane Processes for Remediation of Highly Saline Waste-water Streams, *Proc. Nano4water Workshop*, 26 October 2010.

[19] F. Ponzio, A. Tamburini, A. Cipollina, G. Micale, M. Ciofalo, Experimental and computational investigation of heat transfer in channels filled by woven spacers, Submitted to *Int. J. Heat Mass Transfer*, 2016.

[20] A. Tamburini, A. Parlapiano, A. Cipollina, M. Ciofalo, G. Micale, Temperature distribution analysis in spacer filled channels for Membrane Distillation, Procs. 7th Int. Symp. *Turbulence, Heat and Mass Transfer*, Palermo, Italy, 24-27 September 2012, pp. 299-302, K. Hanjalic, Y. Nagano, D. Borello and D. Jakirlic, eds, Begell House Inc., New York, 2012.

[21] S. Al-Sharif, M. Albeirutty, A. Cipollina, G. Micale, Modelling flow and heat transfer in spacer-filled Membrane Distillation channels using open source CFD code, *Desalination*, 311 (2013) 103-112.

[22] A. Tamburini, G. Micale, M. Ciofalo, A. Cipollina, Experimental analysis via Thermochromic Liquid Crystals of the temperature local distribution in Membrane Distillation modules, *Chem. Eng. Trans.* 32 (2013) 2041-2046.

[23] A. Tamburini, P. Pitò, A. Cipollina, G. Micale, M. Ciofalo, A Thermochromic Liquid Crystals image analysis technique to investigate temperature polarization in spacer-filled channels for membrane distillation, *J. Membr. Sci.*, 447 (2013) 260-273.

[24] M. Ciofalo, I. Di Piazza, J. A. Stasiek, Investigation of flow and heat transfer in corrugated-undulated plate heat exchangers, *Heat and Mass Transfer*, 36 (2000) 449-462.

[25] M. Ciofalo, M. Signorino, M. Simiano, Tomographic Particle Image Velocimetry and Thermography in Rayleigh-Bènard convection using suspended Thermochromic Liquid Crystals and Digital Image Processing, *Experiments in Fluids*, 34 (2003) 156-172.

[26] A. Tamburini, G. La Barbera, A. Cipollina, M. Ciofalo, G. Micale, CFD simulation of channels for direct and reverse Electro Dialysis, *Desalination and Water Treatment*., 48 (2012) 370-389.

[27] Ansys Inc., *Ansys-CFX Reference Guide, Release 14.5* (2012).

[28] F. R. Menter, Two-equation eddy-viscosity turbulence models for engineering applications, *AIAA Journal*, 32 (1994) 269–289.

[29] I. Di Piazza, M. Ciofalo, Numerical prediction of turbulent flow and heat transfer in helically coiled pipes, *Int. J. Thermal Sci.*, 49 (2010) 653-663.

IOP Conf. Series: Journal of Physics: Conf. Series **796** (2017) 012004 doi:10.1088/1742-6596/796/1/012004

Numerical study of laminar magneto-convection in a differentially heated square duct

A Tassone, F Giannetti and G Caruso

Department of Astronautical, Electrical and Energy Engineering – Nuclear Section, Sapienza, Università di Roma, Corso Vittorio Emanuele II 244, Roma, Italy, 00186

Corresponding author: alessandro.tassone@uniroma1.it

Abstract Magnetohydrodynamic pressure drops are one of the main issues for liquid metal blanket in fusion reactors. Minimize the fluid velocity at few millimeters per second is one strategy that can be employed to address the problem. For such low velocities, buoyant forces can effectively contribute to drive the flow and therefore must be considered in the blanket design. In order to do so, a CFD code able to represent magneto-convective phenomena is required. This work aims to gauge the capability of ANSYS© CFX-15 to solve such cases. The laminar flow in a differentially heated duct was selected as validation benchmark. A horizontal and uniform magnetic field was imposed over a square duct with a linear and constant temperature gradient perpendicular to the field. The fully developed flow was analyzed for $Gr = 10^5$ and Hartmann number (M) ranging from 10^2 to 10^3. Both insulating and conducting duct walls were considered. Strong dampening of the flow in the center of the duct was observed, whereas high velocity jets appeared close to the walls parallel to the magnetic field. The numerical results were validated against theoretical and numerical results founding an excellent agreement.

1. Introduction

Liquid metals are often considered as the working fluid for tokamak fusion reactors thanks to their excellent thermo-hydraulics and tritium breeding properties. The eutectic alloy of lithium and lead (LiPb) is the leading candidate and it is actually being studied in three out of the four blanket concepts considered for the PPPT EUROfusion Breeding Blanket project. Although attractive, LiPb has many issues still to be addressed. One of the most important is the interaction between the fluid and the containment magnetic field which sensibly modifies the features of the flow inside the blanket, i.e. electromagnetic drag, enhanced corrosion rate, turbulence suppression, etc. [1]. In order to reduce the MHD contribution to the pressure drops, the molten metal can be employed exclusively as a tritium breeder, whereas the role of coolant for the blanket and the first wall is assumed by a non-conductive fluid, usually water or helium. In this configuration, the liquid metal velocity is reduced to a few millimeters per second and therefore the MHD pressure drops are greatly limited. However, for such low velocities, the buoyancy forces arising from the temperature distribution inside the fluid can play an important role and thus the phenomenon of magneto-convection must be considered during the design of the blanket.

One of the most studied case, both theoretically [2][3][4] and experimentally [5][6], is the buoyant flow which arises in a differentially heated vertical duct when a strong transverse magnetic field is applied. Similar geometries are often encountered in helium-cooled blankets. To support the blanket design, a

CFD code able to tackle the complexity of MHD problems for the values of the fundamental parameters expected is required. Despite the many efforts of the fusion community spent in the development, a mature code specifically tailored to simulate MHD flows is still unavailable. Main-purpose codes are often employed to realize MHD analyses, using available built-in models (which usually require an additional license) or modifying the governing equations of the ordinary hydrodynamic flow [7]. In the past, a user-modified version of ANSYS CFX was studied and found to represent successfully the main features of MHD buoyant flows [8]. Since 2009, a MHD module has been available as an add-on of the main code and this paper aims to assess the capability of ANSYS© CFX-15 to reproduce the phenomena of magneto-convection.

2. Formulation

In this section the governing equations of the MHD flow will be outlined following the ϕ-formulation employed by the CFX solver [9]. A sketch of the problem geometry is provided in Figure 1a.

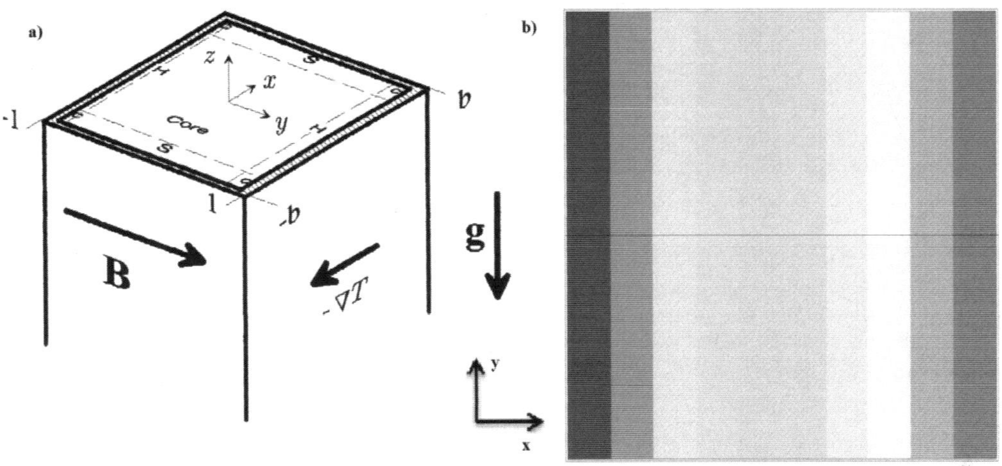

Figure 1 a) Sketch of the benchmark geometry [8] (left), **b)** Temperature contour (right)

The duct is filled with LiPb, initially at rest. Two opposite walls of the duct are kept at fixed temperature values, whereas the other duct walls are assumed as adiabatic. A constant temperature gradient $\nabla \boldsymbol{T} = \hat{x}$ is considered as the source of momentum for the fluid. A uniform and constant transverse magnetic field $\boldsymbol{B} = \hat{y}$, perpendicular to the temperature gradient, is applied. The pair of walls parallel to the magnetic field are called *side walls*, whereas the pair orthogonal to it are called *Hartmann walls*. The gravitational acceleration $\boldsymbol{g} = -\hat{z}$ is aligned with the duct axis. The physical proprieties of the fluid are assumed constant in the range of temperature considered and evaluated at the reference temperature T_0. The only exception to this assumption is the density in the gravitational body force term which, employing the Boussinesq approximation, it is a linear function of the temperature. Since the fluid considered is LiPb, which is characterized by a low value of the magnetic Reynolds number ($R_m \ll 1$), the induction-less approximation can be employed and then the induced magnetic field neglected [10]. Under these conditions, a steady flow would be governed by the non-dimensional MHD momentum equation

$$\frac{Gr}{M^4}(\boldsymbol{v}\cdot\boldsymbol{\nabla})\boldsymbol{v} = -\nabla p + \frac{1}{M^2}\nabla^2\boldsymbol{v} + \boldsymbol{j}\times\boldsymbol{B} + T\hat{\boldsymbol{x}} \tag{1}$$

and the continuity equation

$$\nabla\cdot\boldsymbol{v} = 0 \tag{2}$$

The vectors $\boldsymbol{v}$ and $\boldsymbol{j}$ are the fluid velocity and the electric current density. These quantities would be scaled respectively with $w_0 = \nu Gr/L\,M^2$ and $j_0 = \sigma w_0 B$ where ν is the fluid kinematic viscosity and σ the electrical conductivity. The symbols *Gr* and *M* identify the non-dimensional parameters *Grashof number* and *Hartmann number*. The former quantifies the intensity of the buoyancy forces compared with the viscous forces in the fluid, whereas the latter is correlated to the ratio between the electromagnetic forces and the viscous forces. They are expressed with the relations

$$Gr = g\beta\,\Delta T\,L^3/\nu^2 \quad (3)$$

$$M = BL\sqrt{\sigma/\rho\nu} \quad (4)$$

where β is the thermal expansion coefficient, ΔT is a characteristic temperature difference and L is the length scale of the flow, usually the half-width of the duct in the direction of the magnetic field. The ratio Gr/M^2 which appears in the velocity scale can be considered as an analogue of the Reynolds number, whereas the ratio Gr/M^4 is the buoyant MHD equivalent of the interaction parameter N. The temperature T is the difference between the local and the reference temperature T_0 divided by ΔT. The scaled pressure p is the difference between the local and the hydrostatic pressure evaluated at T_0 divided by $\sigma w_0 B^2 L$. The electric current density is obtained by the Ohm's law and the charge conservation

$$\boldsymbol{j} = -\nabla\phi + \boldsymbol{v} \times \boldsymbol{B} \quad (5)$$

$$\nabla \cdot \boldsymbol{j} = 0 \quad (6)$$

where ϕ is the electric potential, which is scaled by Lw_0B. Combining (5) and (6), a Poisson equation for the potential is obtained that, once resolved, gives the distribution of the electric potential and then the one of the current density

$$\nabla^2\phi = \nabla \cdot (\boldsymbol{v} \times \boldsymbol{B}) \quad (7)$$

The temperature distribution will be obtained by the non-dimensional energy equation

$$Pe(\boldsymbol{v} \cdot \nabla)\boldsymbol{T} = \nabla^2\boldsymbol{T} + Q \quad (8)$$

where $Pe = Gr\,Pr/M^2$ is the Peclet number and Q is the non-dimensional volumetric heat source term. Since the fluid is a liquid metal (Pr $\ll 1$), for a laminar flow the condition $Pe \ll 1$ is satisfied and then the distribution of temperature is independent by the velocity field (Figure 1b). In the following, the source term Q will be considered as negligible. Then, Eq. (8) will reduce to a Laplace equation, the solution of which will be a linear temperature distribution [3]. The flow features are affected by the duct walls electric conductivity, which is assumed to have an arbitrary value. To denote that, the parameter wall conductance ratio is introduced with the following expression

$$c = (\sigma_w t)/(\sigma L) \quad (9)$$

The symbol σ_w identifies the wall electrical conductivity and t the wall thickness. In the following it will arise the necessity to distinguish between the conductance ratio of the side walls (c_S) and of the Hartmann walls (c_H).

3. Numerical strategy

The dimensionless parameters considered for the study are available in Table 1. Two coupled computational domains (one for the LiPb, the other for the walls) were employed. The fluid properties were evaluated at $T_0 = 600\ K$ and according to the correlations available in [11]. A linear temperature profile was deduced from the simplified expression of (8) and imposed on both the domains through the setup of fixed temperature boundary conditions on the duct walls parallel to the x-axis. For the other walls the adiabatic boundary condition ($\partial T_n = 0$) was assumed. Therefore, a purely conductive heat transfer regime is established in the duct. Eq. (1), (2) and (7) are solved by CFX in the fluid domain, whereas just Eq. (1) and (7) are considered in the solid one. The code results were validated employing the analytical solutions developed in [3]. These were obtained with an asymptotic analysis for a fully developed flow that, therefore, must satisfy the conditions

$$Gr/M^4 \ll 1 \tag{10}$$

$$c_S \gg M^{-\frac{1}{2}} \tag{11}$$

The inertia-less condition (10) is satisfied for a sufficiently long duct. Since Eq. (11) requires that no currents close through the side wall boundary layers, thus ruling out the limiting case of perfectly insulating walls ($c = 0$), for these cases the numerical results presented in [8] and [12] were used for validation purpose.

Table 1 Dimensionless groups

Gr	Pr	Pe	M	c
10^5	$2.68 \cdot 10^{-2}$	$2.68 \cdot 10^{-1}$	10^2	0
				∞
		$1.67 \cdot 10^{-2}$	$4 \cdot 10^2$	0
				∞
		$2.68 \cdot 10^{-3}$	10^3	0
				∞

The quality of the numerical results was gauged with two indices: a local error, using as reference the flow peak velocity, and an integral error, evaluated with the root squared deviation from the analytical value of the integral of the velocity profile. For the latter, the velocity profile plotted at $y = 0$ was considered. For the integral error a maximum value of 2% was deemed acceptable, whereas for the peak this condition was relaxed to 5%. The fully developed flow profile was obtained by rendering only a tiny slice of the duct and imposing periodic boundary conditions for all the variables on the top and bottom surfaces, thus simulating an infinitely tall duct with a small computational effort. The no-slip boundary condition was applied at the walls for the velocity, together with the conservation of potential and current density. At the external wall surface, a zero-flux boundary condition for J was imposed to satisfy Eq. (6). For each *M*, test cases with perfectly conducting ($c = \infty$) and insulating walls were performed. Moreover, for $M = 10^2$, simulations with wall conductance ratio ranging from $c = 1$ to $c = 10^{-3}$ were realized in order to test the capability of the code in dealing with finite wall conductivity flows.

3.1. Computational grid

Due to the presence of the magnetic field, the boundary layers of a MHD flow have a different behavior depending on the relative orientation between the wall and the field. For a Hartmann wall, the layer thickness is calculated with the scaling law $\delta_H = 1/M$, whereas for a side wall the law $\delta_S = 1/\sqrt{M}$ is valid. This difference of scale has been addressed by the benchmark mesh employing custom inflation settings. Since it is characterized by a steep velocity gradient, a sufficient number of nodes in the Hartmann layer are necessary to properly render the profile and avoid the under-estimation of the core velocity. A minimum of 3 nodes is required for the perfectly conducting case, whereas more points would be needed for the finite conductivity and insulating cases in which the currents close through the layer. For this reason, all the benchmark simulations were performed with a mesh provided with 15 nodes in the Hartmann layer region. Outside of this tiny region, no significant change of the flow variables is expected along the magnetic field lines, therefore only 35 nodes are employed here. A different treatment must be applied in the direction perpendicular to the magnetic field where many more nodes are needed to follow the velocity and electric potential profile. In Figure 2, it can be seen the comparison of the results obtained by three grids of increasing resolution.

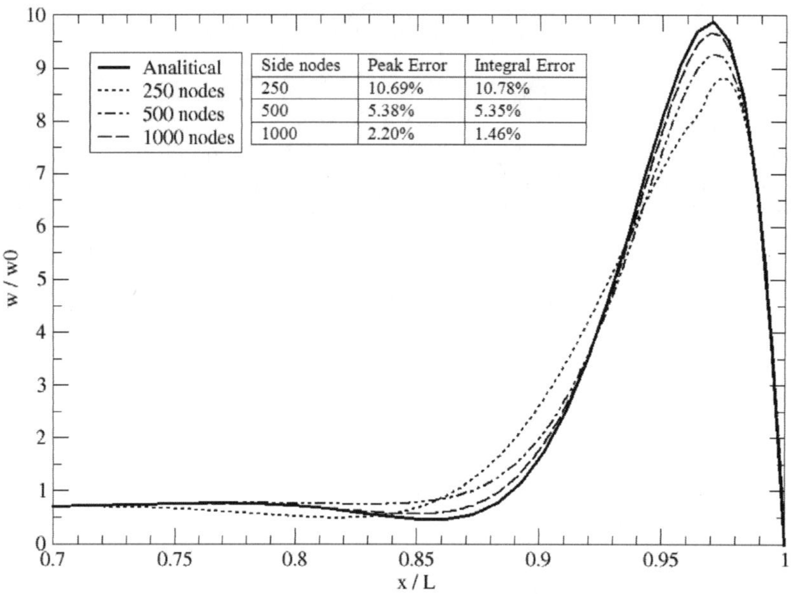

Figure 2 Grid sensitivity analysis for the side wall velocity profile at $\boldsymbol{y} = \mathbf{0}$ for the $\boldsymbol{c} = \infty$ and $\boldsymbol{M} = \mathbf{10^3}$ case

The number of nodes needed for a constant accuracy increases with the Hartmann number: for the $M = 10^2$ test case, one hundred nodes on the half-width of the duct were sufficient to meet the peak error criterion, whereas for the $M = 1000$ case this number jumped to one thousand.

The benchmark is a 2D MHD flow case, since the induced currents are confined in the cross-section plane. Unlike other CFD codes as FLUENT or OpenFOAM, it is not possible to run a 2D simulation in CFX therefore a certain number of "useless" nodes must be allocated in the axial direction. To avoid the arise of numerical instability a minimum of 4 nodes are usually enough but, especially for high values of M, more nodes can be useful to foster the convergence and prevent "phantom" axial currents to appear. In order to proper satisfy the conservation of charge, a conformal mesh between the solid and the liquid computational domain was realized. The only exception was represented by the perfectly insulating case mesh where, since no current closes through the walls, the solid domain was not rendered.

All the simulations were performed with the "high resolution" advection scheme which is a local-weighted version of the upwind discretization scheme [10]. During the performing of the benchmark simulations it became clear that not all the methods provided by CFX for the calculation of the time-step are equally suitable for MHD purposes. In particular, the Local Timescale option, although the most robust, has been incapable to obtain the expected electric potential distribution and was prone to produce unphysical results. This difficulty was often overcome by performing a first run employing the Auto Timescale option to produce a distribution to use as an initial condition for the Local Timescale run which was then able to reach a meaningful solution.

4. Results and discussion

No driving pressure gradient was considered; the motion of the flow was sustained completely by buoyancy forces arising from the temperature distribution inside the duct, i.e. a linear profile $\boldsymbol{T} = T(x)\hat{\boldsymbol{x}}$ with the limiting values $T(-b) = -1$ and $T(b) = +1$.

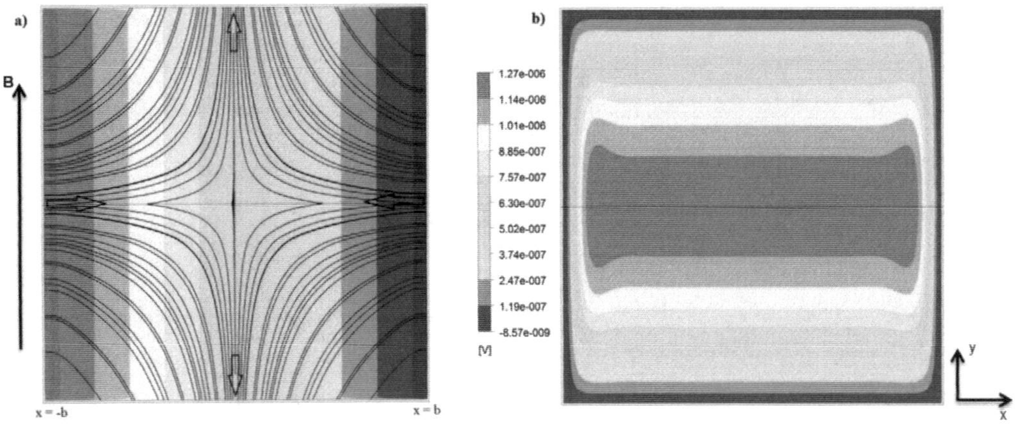

Figure 3 a) Electric current paths and Lorentz force contour (left), **b)** Electric potential contour for $\boldsymbol{c} = \infty$ (right)

4.1. Perfectly conducting walls

A single circulation cell appears with a rising jet near the hot wall and a descending one close to the cold wall. The flow is symmetric across the center line of the duct and the profile velocity is flattened in the core. An inflection point is present where the linear core profile meets the jet. Electric currents are induced in the core region of the duct. Since the conductivity of the side wall is much higher than the conductivity of the side walls boundary layer (11), the current pass through the latter and close in the former. In Figure 3a the current paths are highlighted: the reversed motion of the jets causes the induction of opposite currents in the two halves of the duct which, to satisfy the charge conservation, form 4 loops with a saddle point in the duct center.

Since viscous and inertia effects are negligible in the core region, the electromagnetic and buoyancy forces balance each other and it is found that the component $j_x = O(1)$ is a linear function $j_x = -x$. From (5) and (6), it is obtained that $j_y = y$ and $\phi = \phi_H(x) + (1/2)(1 - y^2)$, where the potential of the Hartmann wall can be expressed as $\phi_H = 1/2\ (1/M + c_H)^{-1}x^2$ [12].

For perfectly conducting walls, the Hartmann wall potential is zero and then the potential in the core region is constant in the direction perpendicular to the magnetic field. This result can be clearly observed in Figure 3b along with the parabolic profile characteristic of the field lines direction. From the Ohm's law (5) it can be found that the only non-zero component of the velocity in the core is the axial one, which it can be written as

$$w = -\partial_x\phi - j_x = [(1/M + c_H)^{-1} + 1]x \tag{12}$$

The velocity is therefore linear and independent by the Hartmann number. If the potential at the side wall is called ϕ_S, it is obvious that $\phi_S = \phi_H = 0$ and then a proper side layer solution has to be found in order to match with the core plateau. The result is a steep potential gradient which acts as an electromagnetic pump, driving the fluid from the strong dampening region of the core to the side layer, and forms a jet [3]. Velocity profiles for increasing value of M are provided in Figure 4 where it can be observed as the peak velocity scales with $M^{1/2}$.

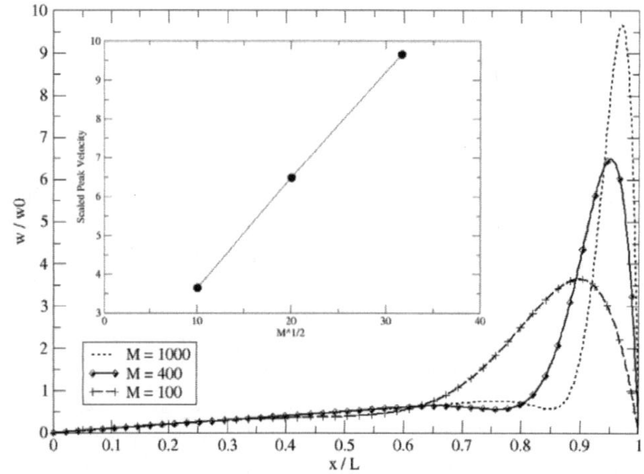

Figure 4 Side wall velocity profile for $\boldsymbol{c} = \infty$ and increasing $\boldsymbol{M}$. On the diagram, $\boldsymbol{w_{MAX}}$ is plotted against $\boldsymbol{M^{1/2}}$

4.2. Perfectly insulating walls

If the case of perfectly insulating duct ($c = 0$) is considered, the currents close through the boundary layer (Figure 5b). A gradient $\partial_x \phi_H = O(M)$ will appear in the core region alongside with a parabolic potential profile (Figure 5a). The Hartmann number term will dominate in (12) and the core velocity takes the form $w \cong Mx$.

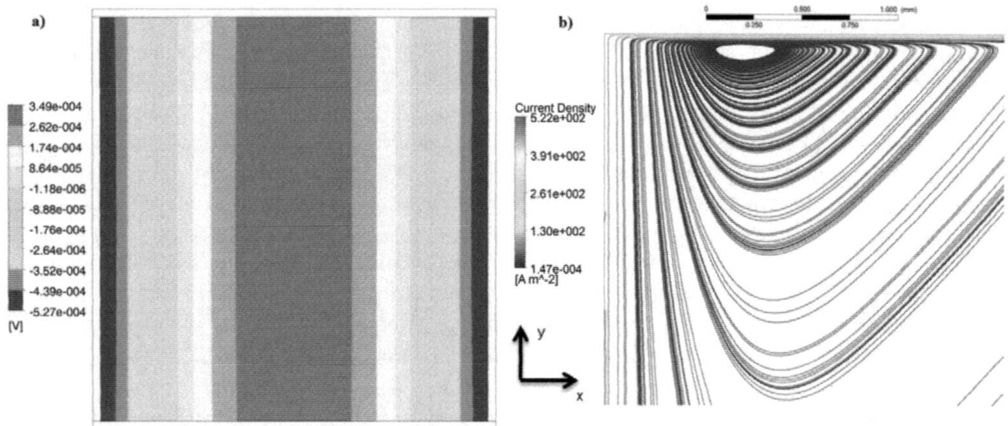

Figure 5 a) Electric potential contour for $c = 0$ (left), **b)** Current closure through the Hartmann boundary layer (right)

In Figure 6a the side wall profile for $c = 0$ and increasing M is shown. For a wall of finite electric conductivity, the behavior of the flow would be included inside the variable space defined by the two limiting cases discussed. Starting from $c = \infty$, a decrease of the wall conductance ratio will force the current to gradually shunt from the wall to the boundary layer. This will cause a net decrease of the damping effect exerted by the Lorentz force in the side layer, due to the more resistive path and the turning of the currents toward the magnetic field direction. The formation of stronger jets due to the intense electromagnetic drag still present in the core is promoted. When $c \rightarrow 0$, the flow rate tends to normalize with the fluid flowing back into the core from the side layers thanks to the decrease of the MHD damping (Figure 6b).

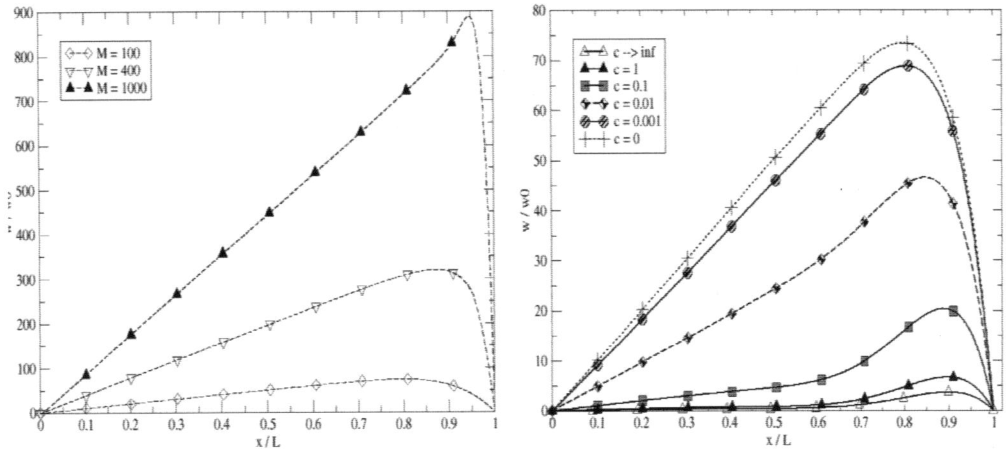

Figure 6 a) Side wall velocity profile for $\boldsymbol{c} = \mathbf{0}$ and increasing $\boldsymbol{M}$ (left), **b)** Side wall velocity profile for c ranging from 0 to ∞ and $\boldsymbol{M} = \mathbf{100}$ (right)

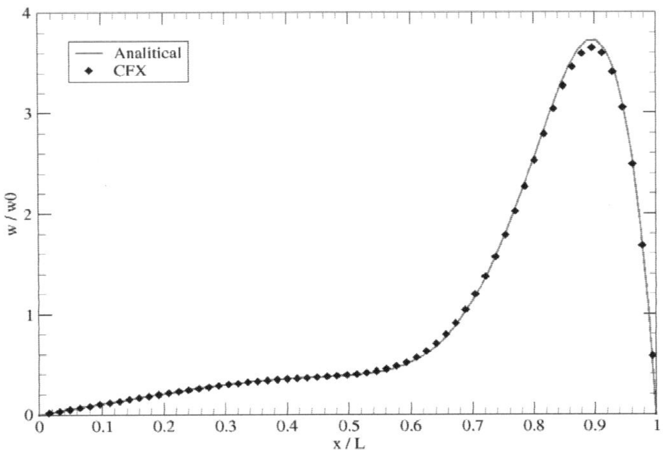

Figure 7 Validation of numerical results for the $\boldsymbol{c} = \infty$, $\boldsymbol{M} = \mathbf{100}$ test case

4.3. Validation results

ANSYS-CFX was able to represent correctly all the features described for every value of conductance ratio and Hartmann number. In the Figures 7, 8a and 8b; a comparison between the numerical results produced by the code and the analytical solution is provided. Although good quality results were produced in every test case, as it can be seen in Table 2, a slowing convergence rate was observed for high values of the Hartmann number due to the stronger coupling between the velocity and the magnetic field. Moreover, this issue was worsened when a low value of the conductance ratio was also considered, since the closure of the currents through the boundary layer requires the use of more resources than those needed for perfectly conducting walls.

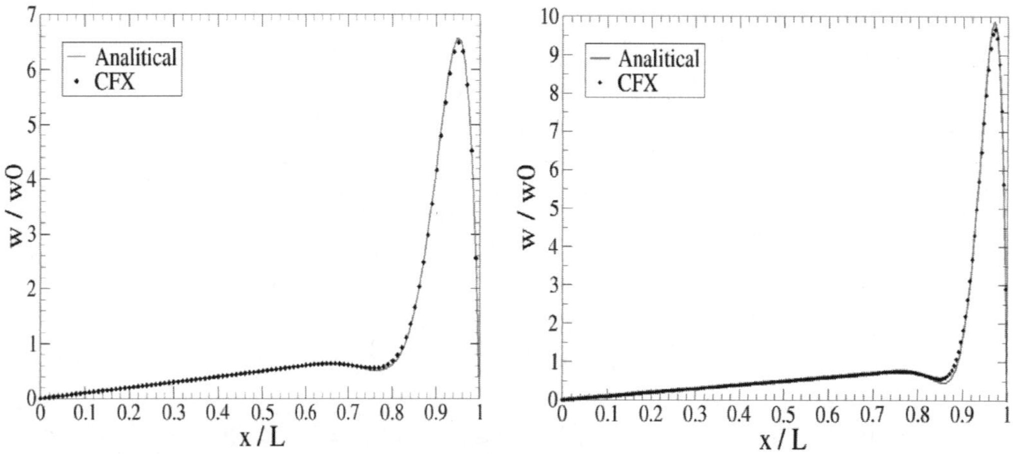

Figura 8 a) Validation of numerical results for the $\boldsymbol{c} = \infty$, $\boldsymbol{M} = \mathbf{400}$ test case (left), **b)** validation of numerical results for the $\boldsymbol{c} = \infty$, $\boldsymbol{M} = \mathbf{1000}$ test case (right)

5. Conclusions

In this paper a numerical study of the buoyant flow in a vertical square duct with differentially heated walls and a transversal magnetic field applied has been presented. The code ANSYS© CFX-15 was employed to perform the simulations, thanks to its MHD add-on. A description of the reference flow was provided for values of the conductance ratio ranging from 0 to ∞ and for M ranging from 10^2 to 10^3. The code was able to reproduce the features described in regard of electric potential, Lorentz force, current paths and velocity profile. The quality of the numerical results was assessed employing two indices, a local error and an integral error, and it was found in excellent agreement both with theoretical solutions in [2] and previous calculations presented in [8] and [12]. Although not included in this paper, analogous simulations for the benchmark of pressure-driven laminar flows have been performed which showed the same degree of accuracy [13].

Table 2 Overview of validation indices

Test Case M	c	*Peak Error [%]*	*Integral Error [%]*
100	0	0.68	-
	∞	2.01	0.79
400	0	4.08	-
	∞	1.26	1.49
1000	0	2.08	-
	∞	2.20	1.46

Slower convergence was observed for high value of M and low conductivity walls and for $M > 10^3$, cases not presented here: severe oscillations of the residuals also occurred in these cases. Particular attention should be paid in the realization of the computational grid and in the choosing of the solver options in order to produce good quality results.

Nevertheless, ANSYS© CFX-15 can be considered as a reliable tool to perform analyses for 2D MHD buoyant flows. Further validation studies are needed in order to assess the behavior of the code in more complex geometries and 3D MHD flows, a necessary condition to satisfy in order for a code to being deemed useful in aiding the design of tokamak fusion reactors blankets.

List of symbols

Physical quantities

b	Channel half-width in x-direction
B	Magnetic induction [T]
g	Gravitational acceleration [m s^{-2}]
j	Current density [A m^{-2}]
l	Channel half-width in y-direction
L	Characteristic length [m]
p	Pressure [Pa]
Q	Heat source
T	Temperature [K]
v	Velocity [m s^{-1}]

Greek symbols

β	Thermal expansion coefficient [K^{-1}]
δ	Boundary layer thickness
ν	Kinematic viscosity [m^2 s^{-1}]
ρ	Density [kg m^{-3}]
σ	Electrical conductivity [S m^{-1}]
ϕ	Electric potential [V]

Dimensionless parameters

Gr	Grashof number
M	Hartmann number
Pe	Péclet number
Pr	Prandtl number
Re	Reynolds number
R_m	Magnetic Reynolds number
c	Wall conductance ratio

Acronyms and abbreviations

MHD	Magneto-hydrodynamics
CFD	Computational fluid-dynamics
LiPb	Lithium-lead eutectic alloy
PPPT	Power Plant Physics & Technology

References

[1] Smolentsev S, Moreau R, Bühler L and Mistrangelo C, 2010 MHD thermofluid issues of liquid-metal blankets: phenomena and advances *Fusion Eng. Des.* **85** 1196-1205
[2] Bülher L, 1997 Laminar buoyant magnetohydrodynamic flow in vertical rectangular ducts *Phys. Fluids* **10** 223-236
[3] Tagawa T, Authié G and Moreau R, 2002 *Eur. J. Mech. B-fluids* **21** 383-398
[4] Aleksandrova S and Molokov S, 2004 *Fluid Dyn. Res.* **35** 37-66
[5] Authié G, Tagawa T and Moreau R, 2003 *Eur. J. Mech. B-fluids* **22** 203-220
[6] Burr U, Barleon L, Jochmann P and Tsinober A, 2003 Magnetohydrodynamic convection in a vertical slot with horizontal magnetic field *J. Fluid Mech.* **475** 21-40
[7] Smolentsev S, Badia S, Bhattacharyay R and Bühler L, 2015 An approach to verification and validation of MHD codes for fusion applications *Fusion Eng. Des.* **100** 65-72
[8] Di Piazza I and Bühler L, 2000 *Fusion Technol.* **38** 180-189
[9] ANSYS© CFX., 15.0, CFX-Solver Theory Guide, 2013, ANSYS Inc.
[10] Davidson P A, 2001 *An Introduction to Magnetohydrodynamics* (Cambridge: Cambridge University Press)
[11] De les Valls E M, 2008 *J. Nucl. Mater.* **376(3)** 353-357
[12] Mistrangelo C and Bühler L, 2012 *IEEE T. Plasm Sci.* **40(3)** 584-589
[13] Tassone A, 2015 CFD simulation of the MHD flow in the WCLL breeding blanket module MSc final thesis (Università di Roma "La Sapienza", Rome, Italy).

IOP Conf. Series: Journal of Physics: Conf. Series **796** (2017) 012005 doi:10.1088/1742-6596/796/1/012005

Application of Recurrence Analysis to the period doubling cascade of a confined buoyant flow

D Angeli[1], M A Corticelli[2], A Fichera[3] and A Pagano [3]

[1] Dipartimento di Scienze e Metodi dell'Ingegneria, Università degli Studi di Modena e Reggio Emilia, Via Amendola 2, 42122 Reggio Emilia (Italy)

[2] Dipartimento di Ingegneria Enzo Ferrari, Università degli Studi di Modena e Reggio Emilia, Via Vivarelli 10, 41125 Modena (Italy)

[3] Dipartimento di Ingegneria Elettrica, Elettronica e Informatica, Università degli Studi di Catania, Viale Andrea Doria 6, 95125 Catania (Italy)

E-mail: diego.angeli@unimore.it

Abstract. Recurrence Analysis (RA) is a promising and flexible tool to identify the behaviour of nonlinear dynamical systems. The potentialities of such a technique are explored in the present work, for the study of transitions to chaos of buoyant flow in enclosures. The case of a hot cylindrical source centred in a square enclosure, is considered here, for which an extensive database of results has been collected in recent years. For a specific value of the system aspect ratio, a sequence of period doublings has been identified, leading to the onset of chaos. RA is applied here to analyse the different flow regimes along the route to chaos. The qualitative visual identification of patterns and the statistics given by the quantitative analysis suggest that this kind of tool is well suited to the study of transitional flows in thermo-fluid dynamics.

1. Introduction

Transition to chaos of thermal convection flows in enclosures is one of the most intriguing topics in the study of the onset of fluid turbulence [1]. Yet, the diversity of patterns and scenarios that could arise as a consequence of the interplay between the temperature and velocity fields is far from being completely unfolded. To this aim, different techniques can be employed for the analysis of nonlinear behaviours encountered throughout transition.

This study is concerned with the 2D numerical analysis of the dynamics of an air-filled isothermally cooled square cavity with a concentric horizontal cylindrical heat source, investigated with respect to the governing parameters of the non-dimensional description of the problem: the ratio between the cavity side length L and the minimum enclosure to cylinder gap width H, $A = L/H$, and the Rayleigh number Ra, for a fixed value of the Prandtl number $Pr = 0.7$. Recent results [2] showed that at least two distinct types of dynamics can be observed: (i) a thermal plume arising from the cylindrical source for low A-values ($A \leq 3.3$), and (ii) Rayleigh-Bénard-like cellular structures above the heat source for higher A-values ($A \geq 5$).

For the case $A = 2.5$ a thorough description was reported in [3–5] of the period-doubling route to chaos that the thermal plume undergoes for increasing Ra. Moreover, this bifurcation path was shown to be determined by the nonlinear interplay of two fundamental modes of the thermal fluid-dynamic system.

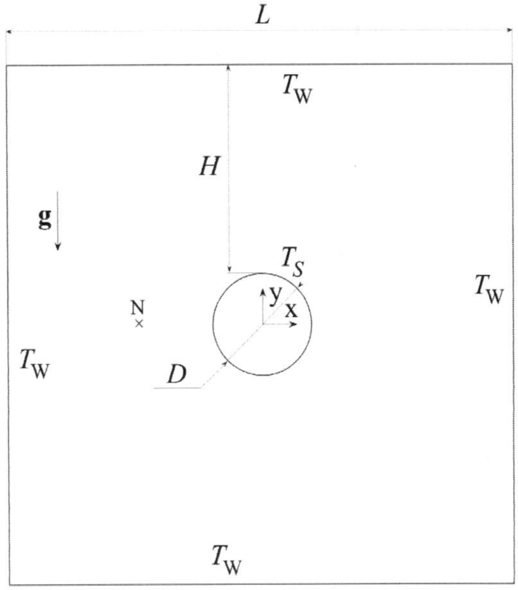

Figure 1. Schematic view of the system under consideration.

Substantially different is the mechanism involved for the case of the Rayleigh-Bénard cellular patterns that characterise the flow for $A = 5$. In fact, in this case the growth of Ra causes the progressive winding of the trajectory on an attractor resembling the features of a French horn, typical of the blue-sky catastrophe driving the transition from periodic dynamics to chaos in the Shilnikov's scenario [6]. Due to the complexity of this scenario and of the observed attractor, more sophisticated tools are required to identify unequivocally the system behaviour. With this respect, promising seems the adoption of the Recurrence Analysis (RA) [7].

With this purpose, the present study is devoted to test the tools made available by RA through their preliminary application to the description of the recurrent patterns observed in the dynamics of the swaying thermal plume for the case of sufficiently wide gap, at $A = 2.5$, so as to take advantage of the high level of refinement already reached in previous studies. In particular, the low order dynamics arising from the period-doubling cascade undergone by the basic periodic limit cycle will be followed by means of the Recurrence Plot of the intermediate regime solutions encountered on the way to chaos.

2. Problem statement

The problem is stated in terms of the incompressible Navier-Stokes formulation. The Oberbeck-Boussinesq approximation is enforced, all the fluid properties being consistently assumed as constant, apart from density in the buoyancy term.

The governing equations are tackled in their non-dimensional form. The gap between the top of the cylinder and the upper cavity wall is indicated in [8] as the most suitable scale length, $H_{ref} = H$. In fact, it reduces the dependence of the solution ranges on the aspect ratio, in particular, for what concerns the heat transfer rate and the first transitions between different regimes. Moreover, the region above the cylinder is subject to the maximum inverse thermal gradient, *i.e.* to the highest buoyancy force acting on the fluid system, and this is again related to the reference length H.

Temperature is non-dimensionalized according to a reference temperature T_{ref}, *i.e.* the temperature difference between the cylinder and the cavity walls ($T_{ref} = T_S - T_W$), and the following velocity scale is chosen:

$$U_{ref} = \sqrt{g\beta T_{ref} H_{ref}} \tag{1}$$

where g denotes the gravitational acceleration and β is the thermal expansion coefficient of the fluid. Coherently, time is made dimensionless by choosing a reference time scale $t_{ref} = H/U_{ref}$. The continuity, momentum, and energy equations are then given the following form:

$$\nabla \cdot \mathbf{u} = \mathbf{0} \tag{2}$$

$$\frac{\partial \mathbf{u}}{\partial t} + \mathbf{u} \cdot \nabla \mathbf{u} = -\nabla p + \frac{Pr^{1/2}}{Ra^{1/2}} \nabla^2 \mathbf{u} + T\hat{\mathbf{g}} \tag{3}$$

$$\frac{\partial T}{\partial t} + \mathbf{u} \cdot \nabla T = \frac{1}{(RaPr)^{1/2}} \nabla^2 T \tag{4}$$

where t, $\mathbf{u}$, p and T represent the dimensionless time, velocity vector, pressure and temperature, respectively, and $\hat{\mathbf{g}}$ is the gravity unit vector. The Rayleigh and Prandtl numbers are defined as:

$$Ra = \frac{g\beta T_{ref} H_{ref}^3}{\nu\alpha} \tag{5}$$

$$Pr = \nu/\alpha \tag{6}$$

where ν and α represent the momentum and thermal diffusivity, respectively. A constant value $Pr = 0.7$ is assumed for air. With reference to figure 1, the following non-dimensional boundary conditions are imposed:

$$T = 0, \quad \mathbf{u} = \mathbf{0} \tag{7}$$

at the enclosure walls, and:

$$T = 1, \quad \mathbf{u} = \mathbf{0} \tag{8}$$

on the cylinder surface.

3. Numerical methods

The numerical technique adopted is based on a Finite Volume implementation of a second order Projection Method, following [9]. Time-discretizations of the conservation equations are performed according to a three-level scheme, which is fully implicit for the diffusive terms, and explicit Adams-Bashforth for the advective terms. Such a practice is second order accurate in time.

Spatial derivatives are approximated with second order central differences on staggered, non-uniform Cartesian grids. A direct resolution of the discrete momentum and energy equations at each time-step is made possible by means of Approximate Factorization, while the Poisson problem associated with the pressure-velocity coupling [9] is solved through a fast Poisson solver, based on Matrix Decomposition.

The 2D modelling of arbitrarily irregular boundaries on Cartesian grids is achieved thanks to the original scheme developed in [10]. The technique involves a local modification of the 5-point computational stencil where boundary segments intersect the stencil arms. The variables on the modified stencil are mapped on the global grid, by means of a linear operator determined by geometrical features and boundary conditions. The overall accuracy of the method is virtually preserved, as well as the computational efficiency of the Cartesian approach.

Details on the grid size, time step choice, initial conditions, are outlined in detail in [3], while a grid independency test for a periodic flow is reported in [8].

For the purposes of recurrence analyses, the time histories of dimensionless temperature T sampled at location N (as reported in figure 1) were considered. The simulated time series, originally collected with a fixed time step size $\Delta t = 8.25 \times 10^{-3}$, were resampled at a larger interval $\Delta t = 3.3 \times 10^{-2}$ in order to decrease the computational effort required by recurrence quantification.

4. Recurrence Analysis

Previous analyses of the simulated dynamics have been devoted to the topological description of the attractors in the state space spanned by the system variables, namely the dimensionless temperature and the horizontal and vertical components of the dimensionless velocity, sampled at specified points of the domain. The present study instead proposes a different approach aiming at characterising the observed recurrent behaviour of the state variables, which is a fundamental and typical property of nonlinear systems. In particular, the sequence of bifurcating attracting solutions arising for increasing Rayleigh number will be distinguished on the basis of the differences between the patterns in their recurrence plots.

The construction of the recurrence plot (RP) allows for the observation of the existence of recurrent states $\mathbf{x}_i$ of the attractor representation in an original m-dimensional phase space. The RP is a 2D representation space, with time spanning both axes, defined as [7]:

$$R_{i,j} = \Theta\left(\varepsilon - \|\mathbf{x}_i - \mathbf{x}_j\|\right), \; \mathbf{x}_i, \mathbf{x}_j \in \mathbb{R}_m, \; i, j = 1, \ldots, N \tag{9}$$

where N is the number of considered states $\mathbf{x}_i$, ε is a threshold distance defining the transvers dimension of the cylinder centred on the trajectory, $\|\cdot\|$ is the norm operator and $\Theta(\cdot)$ is the Heaviside function. This corresponds to representing the recurrence of states at two different time i and j with a black dot in the recurrence plot. The RP is symmetric with respect to the main diagonal, which corresponds to the line of identity (LOI). The application of statistical analyses of the distribution of points in the RP offers a variety of tools for the characterisation of the dynamics of the system under consideration. These tools are collectively addressed as recurrence quantification analysis (RQA) and will be synthetically described in the following.

The choice of the embedding dimension does not usually play a fundamental role and, hence, especially if the RP is used as the basis for RQA, it is often preferable simply to set it at 1, i.e. no embedding is required [11]. Similarly, also the time delay for the embedding is not determinant and, therefore, it can also be set at 1 [12]. It is worth observing that both the embedding dimension and the time delay are critical parameters in the application of traditional nonlinear tools based on attractor characterisation on phase space and, hence the analysis of the time series by means of RPs and RQA appears much simpler in this regard. Accordingly, in this study, both the embedding dimension m and the time delay τ have been set to 1.

Far more relevant is the choice of the threshold radius ε, both for the determination of the RP and for the application of the RQA tools; in fact, if the threshold is too small, only a limited number of recurrent state can be detected, whereas if it is too large false recurrence and even consecutive points of a trajectory will populate the RP. Various criteria have been proposed for an appropriate choice [7, 12, 13]. In particular, Webber and Zbilut [12] suggest to choose the radius threshold looking for a scaling region in the recurrence point density (RR).

Different topologies arise depending on the distribution of the black points in the RPs and, in fact, RPs allow for the identification of transitions and interrelations within the system dynamics. In fact, the topology of the RPs has been related [7] to the system dynamics. In particular, for the aims of the present study it is of interest the identification either of periodic behaviours, through the observation of diagonal lines in the RP, or of deterministic chaotic behaviours, which manifest themselves with isolated points close to periodic diagonal lines, as a consequence of the existence of unstable periodic orbits. Also of interest is the observation of horizontal or vertical lines in the RP, as these are due to laminar states, i.e. states for which no appreciable change in the trajectory occurs within a certain time window. Beyond the interest of the present study, the study of Marwan et al. [7] also reports a detailed description of the RP topology of other classes of dynamical states, such as nonstationary, intermittent or strongly fluctuating phenomena.

As previously mentioned, the recurrence quantification analysis, RQA, represents a set of statistical tools for the synthetic description of a RP or for the comparison of RPs characteristic of different conditions [7, 12]. The following section reports a brief synthesis of these tools and their possible relation with the dynamics invariants typically used for the characterisation of nonlinear dynamics in phase space. The analytical definition of the reported parameters is omitted for brevity and can be found in the reported literature.

The parameter known as recurrence rate, RR, is a measure of the tendency of a system to return nearby to a previous state. RR is calculated as the percentage of recurrence points in the recurrence plot and, therefore, it depends on the value of the threshold ε. Though nonlinear, the dependence of RR on ε usually presents a linearly scaling region, within which it is appropriate to search for an optimal value of the threshold.

The determinism DET is the statistical quantification of the predictability of a system and, in fact, is defined as the ratio of the recurrence points aligned along diagonals parallel to the main diagonal, LOI, to the entire set of recurrent points.

In a similar fashion, the parameter laminarity LAM evaluate the number of points lying on vertical (or horizontal) lines, which are due to the permanence of the system in a given dynamical state for a time corresponding to the number of aligned points in the RP.

The mean and values of the lengths of diagonal lines and of vertical lines are respectively expressed by L, predictability time of the system, and TT, trapping time, i.e. the mean time scale (in terms of sampling intervals) for which small changes are observed in time series. Another relevant parameter is the divergence DIV, i.e. the inverse of the maximum length of diagonal lines in the plot.

The mean period of the dynamics of the system has been related in [7, 14] with the parameter defined as recurrence of the first kind, T_1, calculated as the average of empty space between the points of the RP. Taking this value as a reference, it is possible to calculate the probability distributions of periods t separating the diagonal lines and, also, the recurrence period density entropy, $ENTR_t$. Both the latter parameters, T_1 and $ENTR_t$, might be of particular interest when the system may express a variety of periodic behaviours, such as when bifurcations as period doubling and intermittency are expected.

5. Results and discussion

This section reports the results of the analyses performed on the time series of the temperature at point N in the scheme in figure 1, for a set of simulations that have been chosen as representative of the solutions found for increasing values of the Rayleigh number in the period doubling route to chaos. The cases considered are summarized in Table 1. Parenthetically, for the sake of comparisons, such solutions are the same discussed in [3–5].

In accordance with the methodological guidelines described in the previous section, the embedding dimension for the reconstruction of the phase space has been set to 1 (i.e. no embedding has been considered). The choice of the radius threshold, ε, has been performed on the basis of the sensitivity analysis on parameters RR and DET for the solution P_{32} at $Ra = 1.94975 \times 10^5$. figure 2 (a) and (b) reports the values of the two parameters for increasing ε; both show the existence of a scaling region and allow to justify the choice of an optimal threshold corresponding to $\varepsilon = 0.03$.

Figure 3 (a)-(d) and figure 4 report the temperature time series and the RPs for the cases P_1, QP, P_2, P_4 and P_8. The RPs for the higher order periodic solutions, up to P_{128}, have been

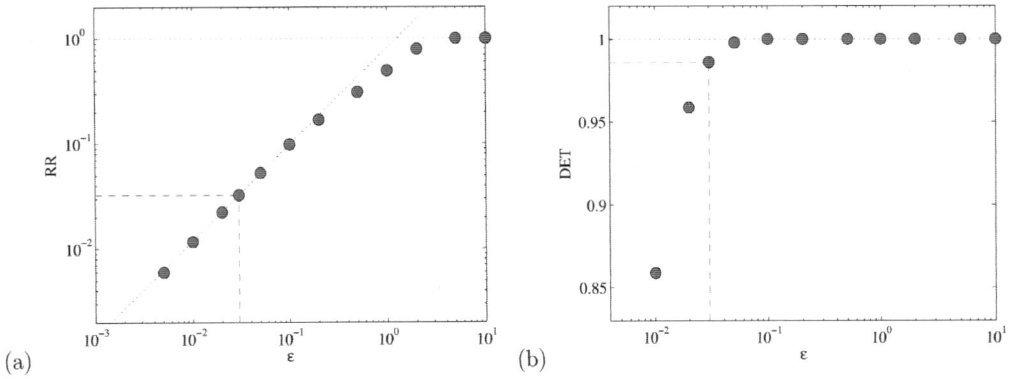

Figure 2. Plots of recurrence rate RR and determinism DET as a function of threshold ε for $Ra = 1.94975 \times 10^5$ (P_{32} regime). Statistics are computed on a sequence of 50000 samples ($\Delta t = 3.3 \times 10^{-2}$) of the normalized temperature series, with embedding dimension and time delay set to $m = 1$ and $\tau = 1$, respectively. The chosen threshold value $\varepsilon = 0.03$ is highlighted in the plots.

calculated but are not reported for the sake of brevity. For each of the periodic solutions in figure 3 (a), (c), (d) and figure 4, the pattern that corresponds to a complete limit cycle in phase space has been encircled and, in fact, it is possible to observe its identical repetition along the main diagonal of the RP.

Such a behaviour cannot be observed in figure 3 (b), as a consequence of the quasiperiodic dynamics of the time series, giving rise to a trajectory in phase space that continuously whirls on a T_2 torus but never closes on it. Moreover, it is possible to observe that the patterns in the RP of the quasiperiodic case can be considered as originated by a sort of weak modulation of the basic sub-patterns that correspond to the two modes of the P_2 solution; in fact, such modulation ends as soon as the latter two modes, incommensurate in the quasiperiodic regime, lock together in the P_2 limit cycle. More details corroborating this observation are reported in [3–5].

Table 1. *Ra*-values and corresponding flow regimes analysed in the present work.

flow regime		$Ra \times 10^{-5}$
periodic	P_1	1.71
quasiperiodic	QP	1.76875
periodic	P_2	1.8
periodic	P_4	1.93645
periodic	P_8	1.945
periodic	P_{16}	1.94925
periodic	P_{32}	1.94975
periodic	P_{64}	1.95
periodic	P_{128}	1.95008
incipient chaos	CH_1	1.955
fully-developed chaos	CH_2	2

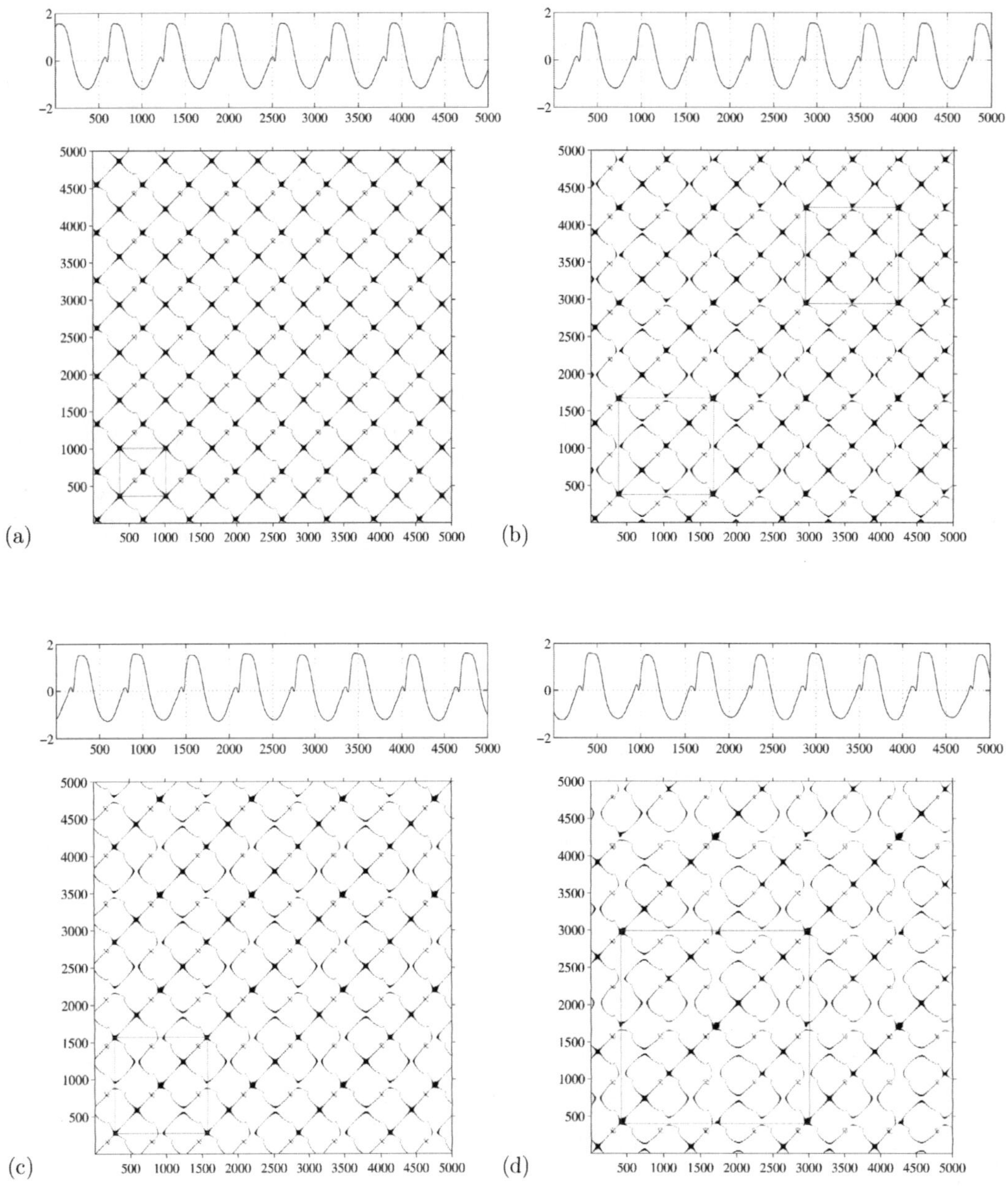

Figure 3. Normalized temperature time series and corresponding recurrence plot for regimes (a) P_1, (b) QP, (c) P_2, (d) P_4. Plots have been obtained on a sequence of 5000 samples ($\Delta t = 3.3 \times 10^{-2}$) with embedding dimension $m = 1$, time delay $\tau = 1$ and threshold $\varepsilon = 0.03$. Recurrent patterns for each regime are framed in red, except for plot (b), where two analogous but slightly different patterns are highlighted with red and blue frames.

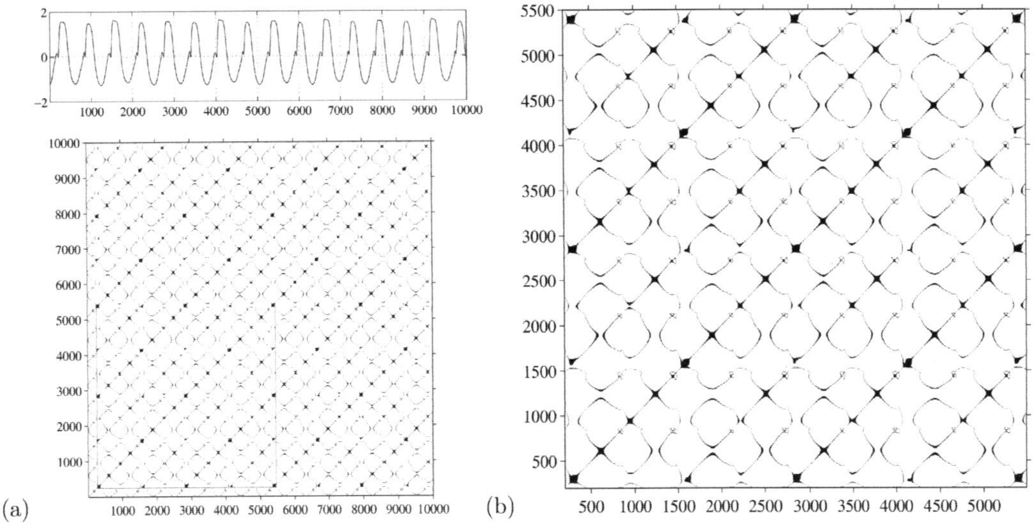

Figure 4. Normalized temperature time series and corresponding recurrence plot (a) for regime P_8, alongside with a zoomed view (b) of the recurrent pattern. The plot has been obtained on a sequence of 5000 samples ($\Delta t = 3.3 \times 10^{-2}$) with embedding dimension $m = 1$, time delay $\tau = 1$ and threshold $\varepsilon = 0.03$.

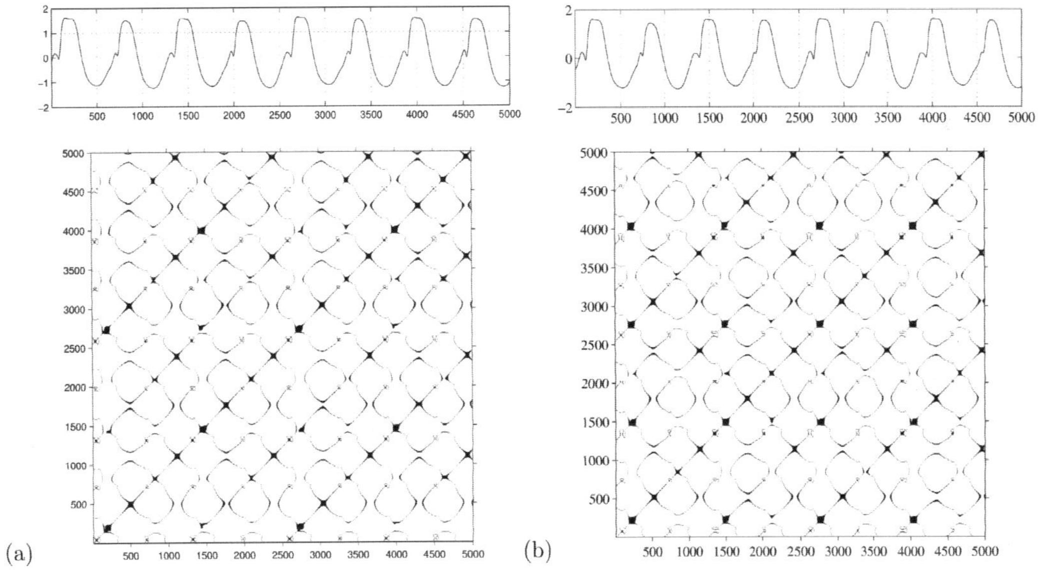

Figure 5. Normalized temperature time series and corresponding recurrence plot for regimes (a) CH_1 and (b) CH_2. The plots have been obtained on a sequence of 5000 samples ($\Delta t = 3.3 \times 10^{-2}$) with embedding dimension $m = 1$, time delay $\tau = 1$ and threshold $\varepsilon = 0.03$.

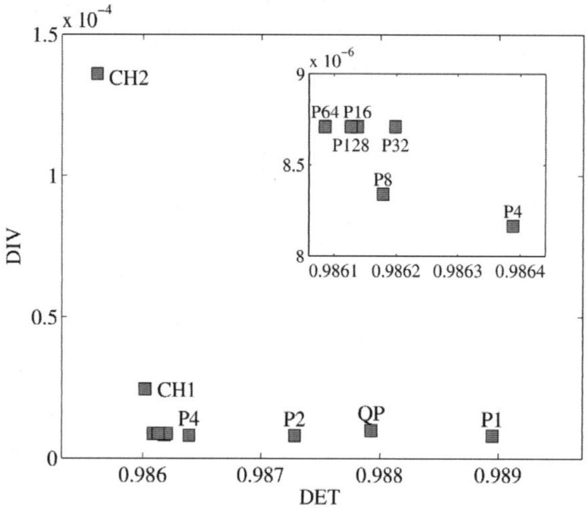

Figure 6. Plots of divergence DIV against determinism DET for all the regimes considered here. Statistics are computed on sequences of 50000 samples ($\Delta t = 3.3 \times 10^{-2}$) of the normalized temperature series, with embedding dimension and time delay set to $m = 1$ and $\tau = 1$, and with a threshold value $\varepsilon = 0.03$.

It is worth observing that, as a consequence of the period doubling from P_1 to P_2, the first single pattern in figure 3 (a) is doubled in a couple of different sub-patterns in figure 3 (c); the same observation can be drawn for the following period doublings, from P_2 to P_4 and from P_4 to P_8, each corresponding to a duplication in the high order dynamics of the basic sub-patterns observable in the low order one, with progressively smaller differences among them. Such a behaviour has been identically observed also for the higher order period doublings, not reported here. This progression clearly determines a rapid growth of the complexity of the RP global structure and, as it is typical in a period doubling cascade, ends with the appearance of chaotic dynamics.

In fact, figure 5 (a) and (b), report the RPs for two chaotic solutions. In particular, they are representative of the case of incipient chaos at $Ra = 1.955 \times 10^5$ and of developed chaos at $Ra = 2 \times 10^5$, respectively. From the observation of these plots, it is easy to identify the growing influence of chaotic mixing on the complexity of the RP structure, that can be recognised in the higher variety and in the lack of repeatability of the basic patterns.

Reported discussion on the topology of the RPs show that recurrence analysis represents a powerful tool for the characterisation of bifurcation path such as the period doubling cascade discussed in the present study. Therefore, the interest of the study has been extended to the extraction of the RQA statistical parameters discussed in the previous section.

All of the parameters previously mentioned have been evaluated for the entire set of simulations along the period doubling cascade. From their preliminary analysis, it has emerged that not all of the parameters are useful for the dynamics under consideration in the present study. For this reason, and for the sake of conciseness, a different approach has been chosen here. In particular, each of the analysed simulations has been described by means of the couple of parameters (DET, DIV), i.e. those showing the highest sensitivity, and are reported in the parameter space in figure 6.

The separation in the chosen parameter space between periodic and chaotic solutions is evident; this is true both for the case CH_1, corresponding to incipient chaos, and, markedly more, for the case CH_2. These results demonstrate the capabilities and flexibility of RPs and of RQA in the characterisation of transitional phenomena in nonlinear dynamics and will be used in the extension of the present study to the analysis of the more complex dynamics observed in [2] for a cavity aspect ratio $A = 5$.

6. Concluding remarks

The suitability and applicability of Recurrence Analysis to the identification of successive flow bifurcations of a natural convection flow from an enclosed horizontal cylinder was investigated here. In the case under consideration, the scenario leading to chaos consists in a sequence of period-doubling bifurcations, preceeded by a window of quasiperiodic flow. The analyses benefitted of the existence of an extensive database of numerical simulations already described and discussed in recent works.

First, an optimal value of the threshold ε for the identification of recurrence points was sought for and identified on the basis of a single case. Successively, recurrence plots were carried out for a set of simulated time series, corresponding to the different regimes encountered throughout the route to chaos. A clear distinction between the regimes could be easily assessed by visual inspection of the plots. The features of period-doublings were clearly displayed, as well as the mixing characteristics of deterministic chaos.

Secondly, recurrence quantification analysis tools revealed that some of the most relevant statistics computed on recurrence plots allowed for a rather simple discrimination between periodic and chaotic solutions.

It is concluded that recurrence analysis is indeed a powerful tool for the characterisation of transitional dynamics in buoyancy-induced flows. Further application of the technique on more complex scenarios occurring on the same case (but for different values of the system main parameters) is thus envisaged.

Acknowledgments

The authors wish to acknowledge the use of the software "CRP toolbox" and to thank the Authors who provided it at `http://tocsy.pik-potsdam.de/CRPtoolbox`.

References

[1] Yang K T 1988 *J. Heat. Transf.* **110** 1191–1204
[2] Angeli D, Pagano A, Corticelli M A, Fichera A and Barozzi G S 2011 *Front. Heat Mass Transf.* **2** 023003
[3] Angeli D and Pagano A 2013 *Int. J. Therm. Sci.* **68** 20–31
[4] Pagano A and Angeli D 2016 *Int. J. Therm. Sci.* **99** 195–203
[5] Angeli D and Pagano A 2016 *Int. J. Therm. Sci.* **109** 231–241
[6] Kuznetsov Y 1998 *Elements of Applied Bifurcation Theory* 2nd ed (*Applied Mathematical Sciences* vol 112) (New York: Springer-Verlag)
[7] Marwan N, Romano M, Thiel M and Kurths J 2007 *Physics Reports* **438** 237–329
[8] Angeli D, Levoni P and Barozzi G S 2008 *Int. J. Heat Mass Transf.* **51** 553–565
[9] Gresho P M 1990 *Int. J. Num. Meth. Fluids* **11** 587–620
[10] Barozzi G S, Bussi C and Corticelli M A 2004 *Num. Heat Transf. B* **46** 56–77
[11] March T, Chapman S and Dendy R 2005 *Physica D* **200** 171–184
[12] Webber C and Zbilut J 2005 *Tutorials in Contemporary Nonlinear Methods for the Behavioral Sciences* ed Riley M and Van Orden G pp 26–94
[13] Marwan N 2011 *Int. J. Bifurcat. Chaos* **21** 1003–1017
[14] Thiel M, Romano M and Kurths J 2004 *Phys. Lett. A* **330** 343–349

IOP Conf. Series: Journal of Physics: Conf. Series **796** (2017) 012006 doi:10.1088/1742-6596/796/1/012006

Numerical analysis of heat conduction problems on irregular domains by means of a collocation meshless method

R Zamolo and E Nobile

Dipartimento di Ingegneria e Architettura, Università degli Studi di Trieste, via A. Valerio 10, 34127 Trieste, Italy

E-mail: nobile@units.it

Abstract. A Least Squares Collocation Meshless Method based on Radial Basis Function (RBF) interpolation is used to solve steady state heat conduction problems on 2D polygonal domains using MATLAB® environment. The point distribution process required by the numerical method can be fully automated, taking account of boundary conditions and geometry of the problem to get higher point distribution density where needed. Several convergence tests have been carried out comparing the numerical results to the corresponding analytical solutions to outline the properties of this numerical approach, considering various combinations of parameters. These tests showed favorable convergence properties in the simple cases considered: along with the geometry flexibility, these features confirm that this peculiar numerical approach can be an effective tool in the numerical simulation of heat conduction problems.

1. Introduction

The accuracy of standard numerical methods used in Computational Fluid Dynamics (CFD), such as Finite Element, Finite Volume and Spectral Element Methods among others, rely on a high quality discretization of the spatial domain. This discretization process, known as meshing, heavily affects both the overall time consumption and the accuracy of the simulation, with higher costs when dealing with problems that require continuous remeshing (moving/deformable domains); finally, meshing can't always be fully automated.

To circumvent the need of a mesh, several *meshless* approaches have been proposed [1–3]: the general idea is to employ only one set of points, distributed over the domain, to approximate the unknown field and its derivatives, without any geometrical discretization of the domain.

A wide class of meshless methods which has been the object of recent developments is the one based on Radial Basis Function (RBF) interpolation [4–7]: this technique accepts only the relative (Euclidean) distance between the points as interpolation parameters. An important distinction exists between *globally supported* RBF and *locally supported* RBF, as the former produces a fully populated interpolation matrix while the latter produces a sparse matrix. We'll focus our attention on locally supported RBF because, in general, the solution of a sparse linear system can be much more efficient than the solution of a full matrix linear system; the gain in computational efficiency is balanced by the loss of spectral convegence that is a property of some globally supported RBF [8].

IOP Conf. Series: Journal of Physics: Conf. Series **796** (2017) 012006 doi:10.1088/1742-6596/796/1/012006

The final set of equations is then obtained through a *collocation* method using a number of collocation points larger than the number of the unknowns, i.e., the resulting coefficient matrix is rectangular (and sparse); therefore the least squares solution can be efficiently computed through a QR decomposition of the matrix using the standard MATLAB® built-in routines.

Several computations are carried out considering various combinations of parameters and different types of polygonal 2D domains for a simple *Poisson* equation to outline the properties of this numerical approach, comparing the computed solution to the corresponding analytical solution. These tests showed that this method can be an effective and robust tool in the numerical simulation of heat conduction problems even in complex shaped domains.

2. Governing equation and boundary conditions

Let us consider the following 2D Poisson equation in the unknown temperature field ϕ:

$$\nabla^2\phi = q \tag{1}$$

defined on the domain Ω; equation (1) is representative of steady state heat conduction problems with internal heat generation q, in the case of a constant $k = 1$ thermal conductivity.

For simplicity let us consider only *Dirichlet* boundary conditions (fixed temperature) along the domain boundary $\Gamma = \partial\Omega$:

$$\phi = \bar{\phi} \tag{2}$$

Since the computed solution will be compared to the corresponding analytical solution ϕ_{an}, we'll set $\bar{\phi} = \phi_{an}$; the domain Ω and the terms q and ϕ_{an} will be defined case by case.

3. Domain definition

Let us consider for simplicity a 2D simply-connected polygonal domain Ω defined by a set of vertices $\mathbf{v}_i$, $i = 1,\ldots,V$. The sides Γ_i , $i = 1,\ldots,V$ connecting a vertex to the following one (except for the last side which connects the last vertex to the first) compose the domain boundary $\Gamma = \cup\Gamma_i$.

4. Point distribution

The point distribution process is here described; since two sets of points will be needed (*field* and *collocation* points, see §5.1), in this section we'll refer to a generic distribution of points $\mathbf{x}_i^{pt}$, $i = 1,\ldots,N_{pt}$, where pt stands for *point type*: for field points $pt = f$ while for collocation points $pt = c$.

4.1. Uniform distribution

First of all we create a uniform cartesian distribution of $N_{U,pt}$ points covering the whole Ω with spacing h for each of the two dimensions; any spatial shift or rotation of this distribution has negligible influence on the results. Points outside Ω or lying on Γ are discarded (inside/outside Ω test is performed using a simple but efficient horizontal right hand boundary intersection).

4.2. Thickened distributions around vertices

For each vertex $\mathbf{v}_i$ where a thickened point distribution is required, two parameters are defined: a thickening parameter $\tau_i \geq 1$ and a de-thickening factor $f_i > 0$; τ_i defines the point thickening around $\mathbf{v}_i$, while f_i defines the thickening decreasing factor from $\mathbf{v}_i$.

More precisely, if $\tau_i > 1$ we define a sequence of radii r_j:

$$r_{j+1} = r_j + \frac{h}{1 + (\tau_i - 1)\exp(-r_j^2 f_i^2)} \tag{3}$$

for $j = 0, \ldots, J$ with $r_0 = 0$ till a final radius r_J such that $(r_{J+1} - r_J)/h > 0.75$.

For each radius r_j starting from r_1 we create a uniform tangential distribution of points centered in $\mathbf{v}_i$ with angular spacing $\Delta\vartheta = r_{j+1}/r_j - 1$ such that the tangential spacing is approximately equal to the radial spacing. Again, points outside Ω or lying on Γ are discarded; the total number of points for this distribution is $N_{V,pt}$.

4.3. Boundary distribution

For each side Γ_i with length L_i and vertices $\mathbf{v}_p, \mathbf{v}_q$, we define a sequence of coordinates s_j:

$$s_{j+1} = s_j + \frac{h}{1 + (\tau_p - 1)\exp(-s_j^2 f_p^2) + (\tau_q - 1)\exp(-s_j^2 f_q^2)} \tag{4}$$

for $j = 0, \ldots, J$ with $s_0 = 0$ till a final coordinate $s_{J+1} > L_i$; normalizing s_j with respect to s_{J+1}, the boundary distribution on side Γ_i is:

$$\mathbf{x}_j^{pt} = (1 - s_j)\mathbf{v}_p + s_j \mathbf{v}_q \tag{5}$$

The total number of points for this distribution is $N_{B,pt}$ and we have $N_{pt} = N_{U,pt} + N_{V,pt} + N_{B,pt}$.

4.4. Refinement process

At this point the final distribution $\mathbf{x}_i^{pt}$ may exhibit point clustering because of the superposition of the uniform distribution and the thickened distributions around vertices; this phenomenon has to be avoided because points close to each other will produce bad conditioned interpolation matrices.

To overcome this problem we employed a simple point refinement process based on the reciprocal repulsion of points: each point is subjected to the radial repulsion force of the closest 9 points (the neighbor search is done using a simple cartesian partitioning scheme).

The repulsion force magnitude is chosen to be the following:

$$F(d) = \alpha h_c \left[4\left(\frac{d}{h_c}\right)^2 + 1 \right]^{-2} \tag{6}$$

where d is the distance between the points, $h_c = 1/\sqrt{\rho_i}$ is the local point spacing around $\mathbf{x}_i^{pt}$ ($\rho_i \approx \#points/\Delta area$ is the local point density) and $\alpha = 0.1 \div 0.6$ is an adjustable parameter.

This refinement process is iterated typically $4 \div 10$ times over the whole point set except for the boundary points which are fixed.

5. Numerical method

5.1. Point sets

Since a least squares collocation method is employed [1, 9], two point distributions are defined: the field points $\mathbf{x}_i^f$, $i = 1, \ldots, N_f$, where the N_f temperature field unknowns ϕ_i are defined, and the collocation points $\mathbf{x}_i^c$, $i = 1, \ldots, N_c$, where the discretized version of equation (1) will be written in a collocation fashion with $N_c \geq N_f$.

5.2. RBF interpolation

RBF interpolation approximates the field ϕ around $\mathbf{x}$ through the following expansion [10]:

$$\phi(\mathbf{x}) = \sum_{i=1}^{n} a_i \varphi(\|\mathbf{x} - \mathbf{x}_i^f\|) + \mathbf{b} \cdot \mathbf{x} + c \tag{7}$$

IOP Conf. Series: Journal of Physics: Conf. Series **796** (2017) 012006 doi:10.1088/1742-6596/796/1/012006

i.e., a linear combination of RBFs φ centered at the n field points $\mathbf{x}_i^f$ plus a linear polynomial in $\mathbf{x}$. If a locally supported interpolation is chosen, only the n closest points to $\mathbf{x}$ are considered.

There are many possible choices for the RBFs φ [11, 12]; the generalized Multiquadrics (MQ) have been chosen since they seem to offer the better results if appropriate parameters are employed [8, 13–15]:

$$\varphi(r) = \left[(\varepsilon r)^2 + 1\right]^{p/2} \tag{8}$$

where ε is the *shape factor* and p is an odd integer.

The coefficients a_i, $\mathbf{b}$ and c are computed writing equation (7) for the n neighbor points $\mathbf{x}_i^f$:

$$\phi(\mathbf{x}_i^f) = \phi_i \tag{9}$$

The following additional relations are needed to get a symmetric and square local interpolation coefficient matrix $\mathbf{G}$ [10]:

$$\sum_{i=1}^{n} x_i^f a_i = 0 \quad , \quad \sum_{i=1}^{n} y_i^f a_i = 0 \quad , \quad \sum_{i=1}^{n} a_i = 0 \tag{10}$$

Collecting the n coefficients a_i and the n unknown values ϕ_i in column vectors $\mathbf{a}$ and $\boldsymbol{\phi}$, respectively, the interpolation system, in compact form, is the following:

$$\mathbf{G}\begin{Bmatrix}\mathbf{a}\\ \mathbf{b}\\ c\end{Bmatrix} = \begin{Bmatrix}\boldsymbol{\phi}\\ \mathbf{0}\end{Bmatrix} \tag{11}$$

and finally, since $\mathbf{G}$ is symmetric, we have:

$$\mathbf{a}^T = \boldsymbol{\phi}^T \mathbf{G}_a^{-1} \tag{12}$$

where $\mathbf{G}_a^{-1}$ is the upper $n \times n$ submatrix of $\mathbf{G}^{-1}$.

5.3. Least squares collocation

Defining $r_i(\mathbf{x}) := \|\mathbf{x} - \mathbf{x}_i^f\|$ and putting equation (7) into equation (1) we have:

$$\sum_{i=1}^{n} a_i \nabla^2 \varphi\big(r_i(\mathbf{x})\big) = q(\mathbf{x}) \tag{13}$$

The Laplacian L of the MQ is therefore:

$$L(r) := \nabla^2\varphi(r) = 2p\varepsilon^2\left[(\varepsilon r)^2 + 1\right]^{p/2-1}\left[\frac{p/2-1}{(\varepsilon r)^2+1}(\varepsilon r)^2 + 1\right] \tag{14}$$

Writing equation (13) in a generic collocation point $\mathbf{x} = \mathbf{x}_j^c$ will give:

$$\sum_{i=1}^{n} a_i L\big(r_i(\mathbf{x}_j^c)\big) = q(\mathbf{x}_j^c) \tag{15}$$

Collecting the n distances r_i in the column vector $\mathbf{r}$, equation (15) can be written as:

$$\mathbf{a}^T L\big(\mathbf{r}(\mathbf{x}_j^c)\big) = q(\mathbf{x}_j^c) \tag{16}$$

Finally, recalling $\mathbf{a}^T$ from equation (12) we have:

$$\boldsymbol{\phi}^T \mathbf{G}_a^{-1} L\big(\mathbf{r}(\mathbf{x}_j^c)\big) = q(\mathbf{x}_j^c) \tag{17}$$

Equation (17) is the final scalar equation for the generic collocation point $\mathbf{x}_j^c$; the column vector $\mathbf{G}_a^{-1} L\big(\mathbf{r}(\mathbf{x}_j^c)\big)$ contains the n coefficients of this equation.

The final system is obtained writing equation (17) for all N_c collocation points:

$$\mathbf{A}\boldsymbol{\phi} = \mathbf{q} \tag{18}$$

where $\mathbf{A}$ is the $N_c \times N_f$ coefficient matrix, $\mathbf{q}$ is the column vector of generation term q evaluated at all N_c collocation points and $\boldsymbol{\phi}$ refers to all N_f field unknowns ϕ_i.

The imposition of Dirichlet boundary conditions, equation (2), is straightforward; partitioning $\mathbf{A} = [\mathbf{A}_I \mid \mathbf{A}_B]$ and $\boldsymbol{\phi} = \{\boldsymbol{\phi}_I; \boldsymbol{\phi}_B\}$ through Internal ($N_{U,f} + N_{V,f}$) and Boundary ($N_{B,f}$) field point indexes, equation (18) becomes:

$$\mathbf{A}_I \boldsymbol{\phi}_I = \mathbf{q} - \mathbf{A}_B \boldsymbol{\phi}_B \tag{19}$$

Since $N_c > N_{U,f} + N_{V,f}$, $\mathbf{A}_I$ is a tall rectangular matrix and the solution $\boldsymbol{\phi}_I$ must be computed using a least squares solver; we employed the standard MATLAB® backslash (\) operator which, in this case, performs a QR decomposition of the matrix.

5.4. Remarks on RBF interpolation

It is known that the interpolation matrix $\mathbf{G}$, equation (11), tends to be severely ill-conditioned when small shape parameters ε are employed, or when the interpolation points are close to each other, or both [16].

To alleviate this problem we employed an improved LDL solver which takes advantage of the form of the upper $n \times n$ entries of $\mathbf{G}$ ($g_{ij} = \varphi(r_{ij}) = 1 + m(r_{ij})$, where $m(\cdot)$ is the non-costant part of MacLaurin expansion of the RBF φ) to gain some numerical precision with no additional costs.

5.5. Error norm

The comparison between the computed solution ϕ_{comp} and the corresponding analytical solution ϕ_{an} will be done computing the following RMS error norm e:

$$e = \sqrt{\frac{1}{A(\Omega)} \int_\Omega (\phi_{comp} - \phi_{an})^2 \, d\Omega} \tag{20}$$

where $A(\Omega)$ is the area of Ω. The 2D integral in equation (20) is computed by numerical quadrature via a Delaunay triangulation of the field point set.

6. Numerical results and discussion

6.1. Test case A: uniform distribution, harmonic q

The problem is defined on a square domain $\Omega = [0,1]^2$ and solved using only uniform distributions of points ($N_{V,f} = N_{V,c} = 0$) and $n = 9$ as number of local interpolation points.

The chosen reference solution is harmonic:

$$\phi_{an} = \sin(2\pi x)\sin(2\pi y) \tag{21}$$

and its Laplacian is therefore also harmonic:

$$q = \nabla^2 \phi_{an} = -8\pi^2 \sin(2\pi x)\sin(2\pi y) \tag{22}$$

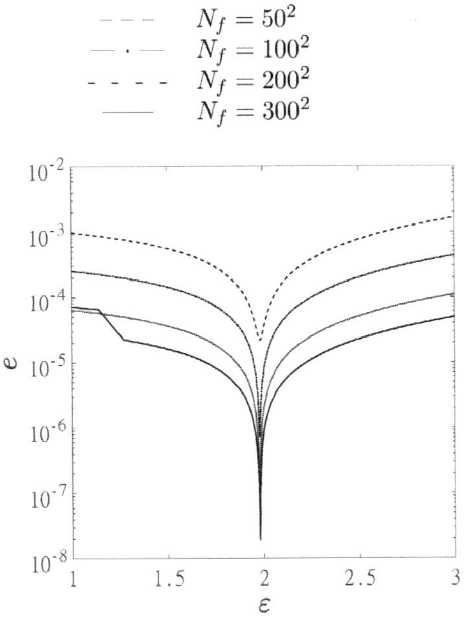

Figure 1. Test case A. Total error e versus shape parameter ε; $p = 1$ and $N_c/N_f = 2$.

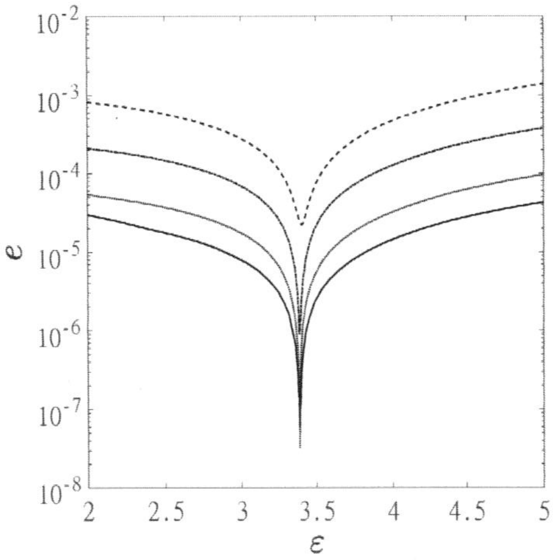

Figure 2. Test case A. Total error e versus shape parameter ε; $p = 3$ and $N_c/N_f = 2$.

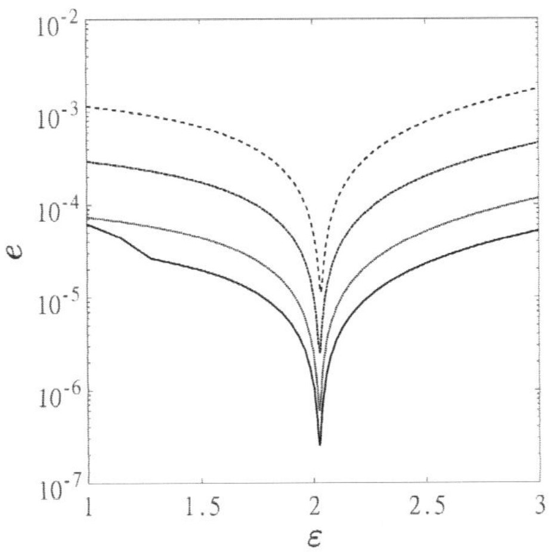

Figure 3. Test case A. Total error e versus shape parameter ε; $p = 1$ and $N_c/N_f = 4$.

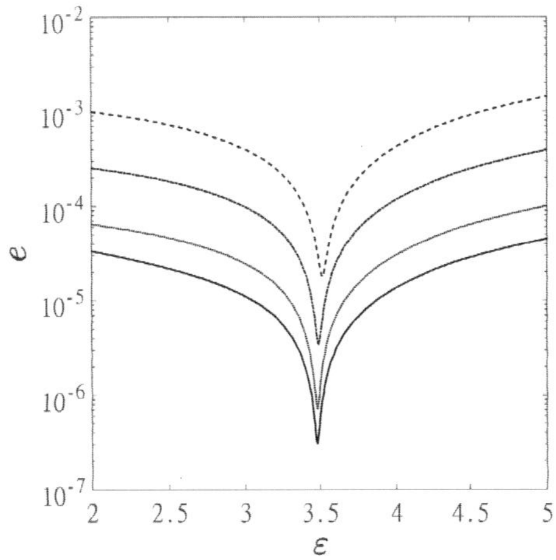

Figure 4. Test case A. Total error e versus shape parameter ε; $p = 3$ and $N_c/N_f = 4$.

First af all we investigated the influence of MQ shape parameter ε on the solution error e considering two MQ exponent factors $p = 1, 3$, two ratios $N_c/N_f = 2, 4$ and various numbers of field points N_f (Figures 1-4).

These figures show that for each MQ exponent factor p there exists a particular shape parameter ε^* which minimizes the error e; for $p = 1$, ε^* lies around 2 and for $p = 3$, ε^* lies around 3.5. However, ε^* depends upon the specific solution ϕ, so the optimal choice for ε has to be made case by case.

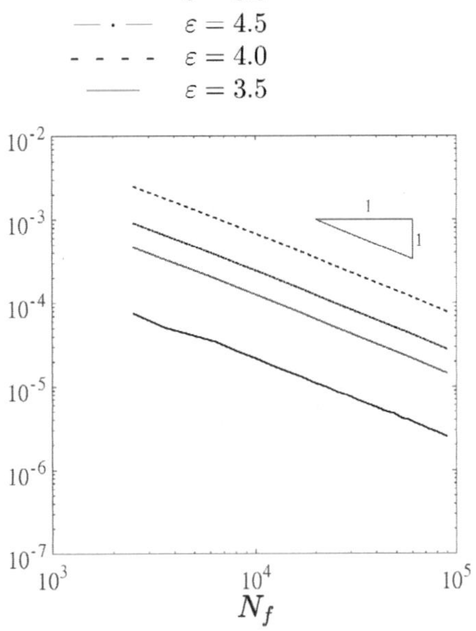

Figure 5. Test case A. Covergence curves for $p = 3$ and $N_c/N_f = 2$.

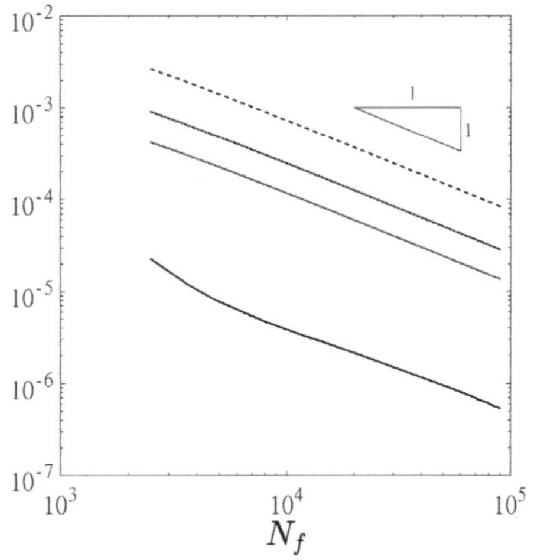

Figure 6. Test case A. Covergence curves for $p = 3$ and $N_c/N_f = 4$.

On the other hand, from the same figures it is possible to see how the influence of MQ exponent factor p is negligible on the error curves; for this reason from now on we'll consider only $p = 3$.

Finally, the influence of N_c/N_f ratio for both $p = 1$ and $p = 3$ seems to be positive on the error e only for shape parameters around ε^* but, surprisingly, the smaller error $e(\varepsilon^*)$ is observed with $N_c/N_f = 2$ instead of $N_c/N_f = 4$, i.e., employing less collocation points; for ε far from ε^*, however, this behaviour disappears.

Convergence curves for different shape parameters ε and for two ratios $N_c/N_f = 2, 4$ are reported in Figures 5 and 6. We first observe that each curve has an asymptotic slope of -1, i.e., the order of the method is 2 (in 2D the average field point spacing is $h_s \propto N_f^{-1/2}$); again, these curves clearly show that a right choice of the shape parameter ε heavily affects the error: more than two orders of magnitude of difference are observed passing from $\varepsilon = 6.0$ to $\varepsilon = 3.5$ in Figure 6.

As stated before, a comparison of previous convergence curves confirm that the influence of N_c/N_f ratio is negligible for ε far from ε^*, i.e., the upper curves (Figures 5 and 6) have similar or identical magnitude; for this reason from now on we'll consider only $N_c/N_f = 2$, as it is the cheaper choice in a computational sense.

6.2. Test case B: thickened corner, q with singularity

This problem is defined again on a square domain $\Omega = [0, 1]^2$, but both field and collocation point distributions are thickened in the corner around the $x - y$ coordinates origin $O =: \mathbf{v}_1$ ($\tau_1 > 1$), as reported in Figure 7. We employed a thickened distribution around $\mathbf{v}_1$ because the following reference solution, defined in polar coordinates, is singular in that point:

$$\phi_{an} = \sqrt{r} \tag{23}$$

Figure 7. Field point distribution for test case B: $\tau_1 = 3.0$, $f_1 = 3.0$, $N_f = 17462$.

($r = \|\mathbf{x}\|$); its Laplacian is therefore :

$$q = \nabla^2 \phi_{an} = \frac{r^{-3/2}}{4} \tag{24}$$

The singularity of this solution is twofold: ϕ_{an} has infinite first derivatives in $\mathbf{v}_1$ and its Laplacian, which is the RHS of the problem, goes to infinity with order $3/2$ in the same point[1]; thus, this is a very severe problem. The number of local iterpolation points was again $n = 9$.

No ε^* test has been conducted in this case, thus convergence curves of Figures 8-9 have been computed only for two distinct shape factors $\varepsilon = 1.5, 3.5$, but various combinations of thickening parameters τ_1, f_1 have been considered.

In theses figure the convergence curves for a uniform point distribution are also shown; the comparison of the thickened distribution curves with the uniform distribution ones clearly reveals that in this case the employment of a thickened distribution around the singularity allows a much better resolution of the singular problem, maintaining a second order accuracy.

The influence of point distribution thickening parameters τ_1, f_1 around $\mathbf{v}_1$ is also visible in the same Figures (8-9); despite their little influence, it can be appreciated how an increase in τ_1

[1] The q term for the boundary collocation point $\mathbf{x}_k^c$ lying on the corner $\mathbf{v}_1$, where the Laplacian is infinite, should be corrected as $q_k = (h/2\tau_1)^{-3/2}$.

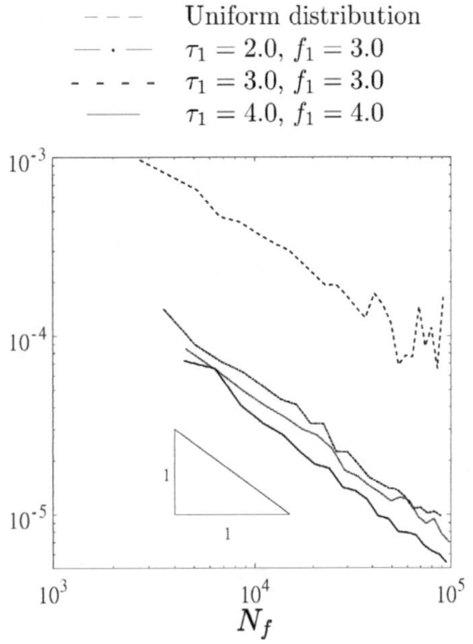

Figure 8. Test case B. Convergence curves for $\varepsilon = 1.5$.

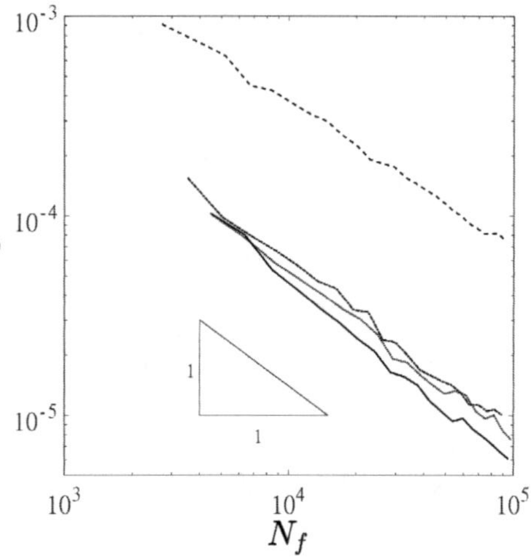

Figure 9. Test case B. Convergence curves for $\varepsilon = 3.5$.

and f_1, i.e., a thicker and localized distribution around the corner, helps in the resolution of the singularity.

Finally, the influence of shape factor ε (Figures 8-9) seems not to be of primary importance in this case, but other computations should be performed to confirm this fact; however, greater shape parameters ε reduce the ill-conditioning problems [16] when dealing with finest point distributions (i.e. high N_f / large thickening parameters τ_i).

6.3. Test case C: irregular domain, q with singularities

This problem is defined on the polygonal domain reported in Figure 11 with thickened point distributions around vertices $\mathbf{v}_2$ ($\tau_2 = 3.0, f_2 = 6.0$), $\mathbf{v}_8$ ($\tau_8 = 2.5, f_8 = 12.0$) and $\mathbf{v}_9$ ($\tau_9 = 2.5, f_9 = 12.0$). These vertices correspond to the points where the following reference solution is singular:

$$\phi_{an} = \|\mathbf{x} - \mathbf{v}_2\| + \frac{1}{4}\left(\|\mathbf{x} - \mathbf{v}_8\| + \|\mathbf{x} - \mathbf{v}_9\|\right) \tag{25}$$

Its Laplacian is therefore :

$$q = \nabla^2\phi_{an} = \frac{1}{\|\mathbf{x} - \mathbf{v}_2\|} + \frac{1}{4}\left(\frac{1}{\|\mathbf{x} - \mathbf{v}_8\|} + \frac{1}{\|\mathbf{x} - \mathbf{v}_9\|}\right) \tag{26}$$

This reference solution is less severe than the one of test case B because ϕ_{an} has finite first derivatives, but its Laplacian goes to infinity with order 1 in the three vertices $\mathbf{v}_2, \mathbf{v}_8$ and $\mathbf{v}_9$[2].

[2] The q term for the three boundary collocation points corresponding to these three vertices $\mathbf{v}_k$, where the Laplacian is infinite, should be corrected as $q = 2a_k(h/2\tau_k)^{-1}$, with a_k the respective coefficient as in (26).

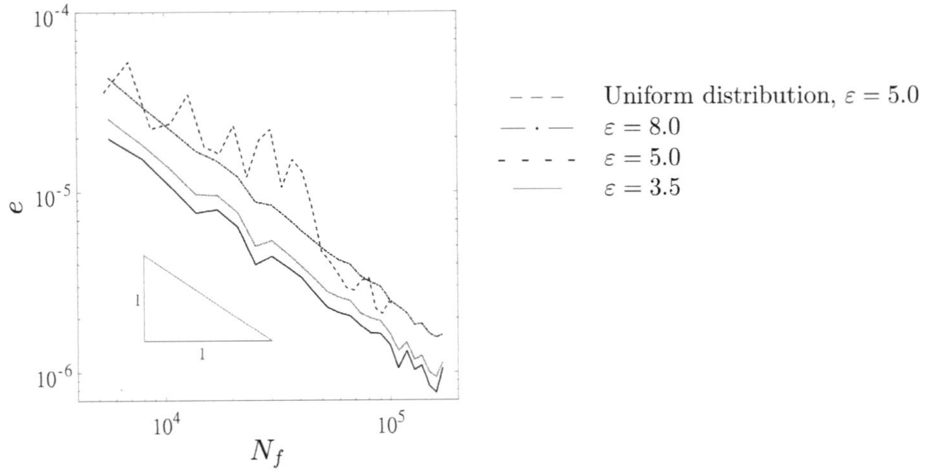

Figure 10. Test case C. Convergence curves for complex shaped domain of Figure 11.

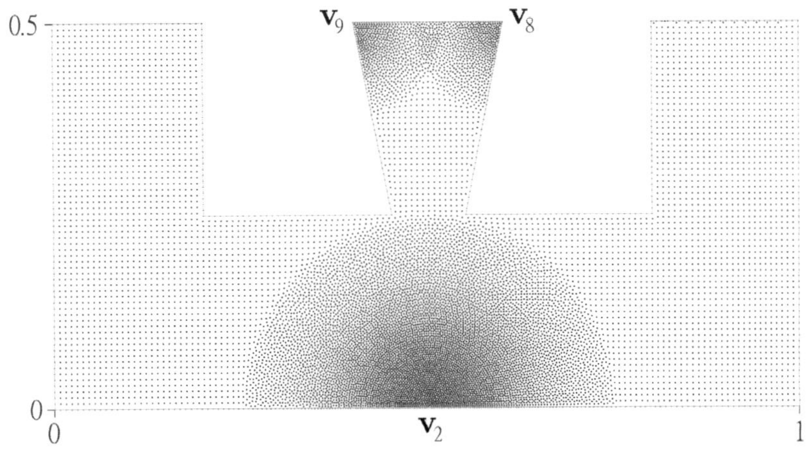

Figure 11. Field point distribution for test case C: $N_f = 13793$.

Again, $n = 9$ neighbor interpolation points are employed. Convergence curves are shown in Figure 10, where the convergence curve for a uniform distribution ($\varepsilon = 5.0$) is also reported for comparison; three shape factors $\varepsilon = 3.5, 5.0$ and 8.0 are considered.

From the previous Figure (10) it can be observed how, again, the lowering of shape factor ε reduces the total error e, maintaining a second order accuracy; this is observed at least till $N_f = 10^5$, beyond which the use of small ε (3.5,5.0) with fine point distribution causes ill-condition instabilities.

The advantage in the use of a thickened point distribution around singolarities is not so evident in this case, although some improvement exists; this little difference is due to the less severe singularities employed in this test case.

6.4. CPU times

Typical CPU time for a single complete computation, including point distribution processes, coefficients computation (requiring a space partitioning scheme with associated data structures) and least square solution (MATLAB® built-in solver), was in the order of seconds even for the cases where large number of points were employed ($N_f \approx 10^5$, $N_c \approx 4 \cdot 10^5$) on a modern laptop (quad-core Intel® i7 2.6GHz).

7. Conclusions and future work

In this work a Least Squares Collocation Meshless Method is used for the numerical simulation of heat transfer problems, in the specific case of 2D steady state conduction. The peculiarity of this approach is the flexibility against complex shaped domains because the traditional mesh/grid is no longer required.

Three particular test cases have been considered and for each test case several convergence curves have been computed to highlight the numerical properties of this approach. Each test showed good convergence properties (second order in space), even when dealing with singular data and irregular domains; in these cases thickened point distributions around singularities have been employed to improve the solution.

These numerical properties confirm that this method is an efficient and versatile tool for the numerical simulation of practical heat conduction problems in complex shaped domains.

The reported activity will be carried on in order to outline advantages and disadvantages of this specific approach for a wider range of cases. Further analyses will be conducted considering other domain geometries (not necessarily polygonal, 3D), various boundary conditions and transient (unsteady) problems; a significant step from a computational efficiency point of view will be a parallel Graphics Processing Unit (GPU) implementation of the procedure, currently under study.

References

[1] Li H and Mulay S 2013 *Meshless Methods and Their Numerical Properties* (CRC Press)
[2] Belytschko T, Krongauz Y, Organ D, Fleming M and Krysl P 1996 *Computer Methods in Applied Mechanics and Engineering* **139**(1–4) 3–47
[3] Katz A 2009 *Meshless methods for computational fluid dynamics* Ph.D. thesis Stanford University
[4] Fasshauer G 1997 *In: Surface Fitting and Multiresolution Methods* ed Mehaute A L, Rabut C and Schumaker L (Vanderbilt University Press) pp 131–138
[5] Sarler B and Vertnik R 2006 *Computers and Mathematics with Applications* **51**(8) 1269–1282
[6] Zerroukat M, Power H and Chen C 1998 *Numerical Methods in Engineering* **42**(7) 1263–1278
[7] Lai S, Wang B and Duan Y 2010 *Progress in Electromagnetics Research B* **24** 351–367
[8] Sarra S A and Kansa E 2009 *Multiquadric Radial Basis Function Approximation Methods for the Numerical Solution of Partial Differential Equations* (*Advances in Computational Mechanics* vol 2) ed Atluri S N (Tech Science Press)
[9] Zhang X, Liu X, Song K and Lu M 2001 *Numerical Methods in Engineering* **51**(9) 1089–1100
[10] Golberg M, Chen C and Bowman H 1999 *Engineering Analysis with Boundary Elements* **23**(4) 285–296
[11] Larsson E and Fornberg B 2003 *Computers and Mathematics with Applications* **46**(5–6) 891–906
[12] Hu H, Li Z and Cheng A 2005 *Computers and Mathematics with Applications* **50**(1-2) 289–320
[13] Franke R 1982 *Mathematics of Computation* **38**(157) 181–200
[14] Kansa E 1990 *Computers and Mathematics with Applications* **19**(8-9) 127–145
[15] Kansa E 1990 *Computers and Mathematics with Applications* **19**(8-9) 147–161
[16] Fornberg B and Wright G 2004 *Computers and Mathematics with Applications* **48**(5-6) 853–867

Optimization of a single-stage double-suction centrifugal pump

Aljaž Škerlavaj[1,2], Mitja Morgut[1,2], Dragica Jošt[1] and Enrico Nobile[2]
[1] Kolektor Turboinštitut d.d., Rovšnikova 7, 1210 Ljubljana (Slovenia)
[2] Dipartimento di Ingegneria e Architettura, Università degli Studi di Trieste, via A. Valerio 10, 34127 Trieste (Italy)
E-mail: `aljaz.skerlavaj@kolektor.com`

Abstract. In this study, the objective of the optimization of a double-suction pump is the maximization of its hydraulic efficiency. The optimization is performed, by means of the modeFRONTIER optimization platform, in steps. At first, by means of a DOE (Design of Experiments) strategy, the design space is explored, using a parameterized CAD representation of the pump. Suitable metamodels (surrogates or Response Surfaces), which represent an economical alternative to the more expensive 3D CFD model, are built and tested. Among different metamodels, the evolutionary design, radial basis function and the stepwise regression models seem to be the most promising ones. Finally, the stepwise regression model, trained on a set of 200 designs and constructed with only five the most influential input design parameters, was chosen as a potentially applicable metamodel.

1. Introduction

Centrifugal pumps are widely used in industrial applications. Double-suction (also called: double-entry) centrifugal pumps allow transportation of greater flow rates than single-entry pumps [1] because they are less prone to cavitation problems (smaller NPSHR). Another advantage is counter-balancing of axial hydraulic forces due to double-entry design [1]. The pump geometry considered in this study, together with a sketch of the impeller, is depicted in figure 1.

Usually, hydraulic turbomachines are designed by performing modifications of a nearly-suitable and experimentally validated baseline geometry. The modifications are either performed directly in a CAD software by the designer, or by modifying input variables in in-house codes using direct or inverse singularity methods. Nowadays, modified geometries are compared by means of Computational Fluid Dynamics (CFD) systems. Only the best design is then usually manufactured and experimentally tested. There are numerous benefits of using optimization techniques to perform modifications in automatized way instead of using traditional human-based 'trial-and-error' technique. First of all, no occasional hard-to-identify human-based errors are present, which guarantees the same design and CFD procedure for each design. Secondly, after the (large) initial effort is performed to include the design creation in an optimization loop, the human interaction is reduced and a lot of variants can be tested, which in turn allows a designer to understand the effect of the design variables on the objective (e.g., pump efficiency) much better. In turbomachines, usually multiple objectives have to be optimized. One of the first multi-objective optimization study was performed by Lipej and Poloni [2]. Due to importance of centrifugal pumps in industrial applications, as well as of pump-turbines, many recent studies

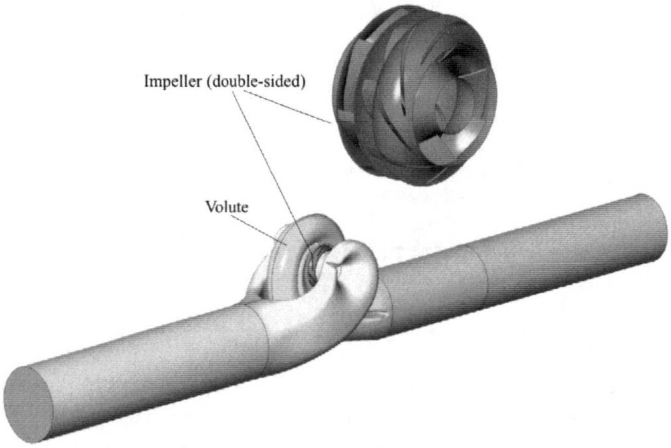

Figure 1. Pump geometry.

aim to optimize their geometry to improve their performance. Shingai et al.[3] performed a multi-objective optimization of pump-turbine, where the objectives were efficiency of turbine, pump efficiency, cavitation characteristics for both types of operation, torque to hold runner blades and total pump head. Xuhe et al. [4] optimized a pump turbine for efficiency in both modes of operation whereas Zhang et al. [5] performed optimization of a centrifugal pump for vibration, using fluid-structure interaction. The purpose of this work is to compare different RSM (Response Surface Methods) for the case of optimization of a centrifugal pump. We recall that RSM (also called metamodels or surrogates) represent an economical alternative to the more expensive 3D CAE models, CFD in this case. In particular, the following RSM algorithms were considered: polynomial singular value decomposition with the second order of polynomial degree, stepwise regression [6], K-nearest RSM [7], Kriging RSM [8], anisotropic Kriging [9], Gaussian processes [10], Hardy's multiquadric radial basis function [11], neural networks [12] and evolutionary design [13]. In this study, the specific speed $n_q = n\sqrt{Q/2}/H^{3/4}$ of the pump was equal to 62 (specified per impeller side). Only the impeller geometry was modified while the rest of the geometry was considered as fixed, according to the limitations given in a commercial project, where only the impeller could be modified.

2. Numerical procedure

In figure 2, the logic of the optimization strategy, where the CFD simulations were performed to evaluate the different designs, is sketched.

For the impeller, the shape of the meridional channel was based on past experience and was not modified during the optimization procedure. On the other hand, the blade geometry was generated from a certain set of input parameters (figure 3) provided to an in-house code that uses an inverse-singularity method. The data was exported to ANSYS DesignModeler, where a 3-D geometry of an impeller was automatically created. For each geometry, a tetrahedral-based, unstructured mesh of the impeller was automatically created in ANSYS ICEM, which also provided the generation of 15 prism layers near the walls. The CFD model consisted of a single-channel impeller (using a rotational periodicity boundary condition) and a symmetric half of a volute with an outlet pipe, as depicted in figure 4. During the optimization, only the single-channel impeller mesh was being modified. At the impeller inlet, the velocity components were prescribed as a function of radius around rotation axis. The components were obtained

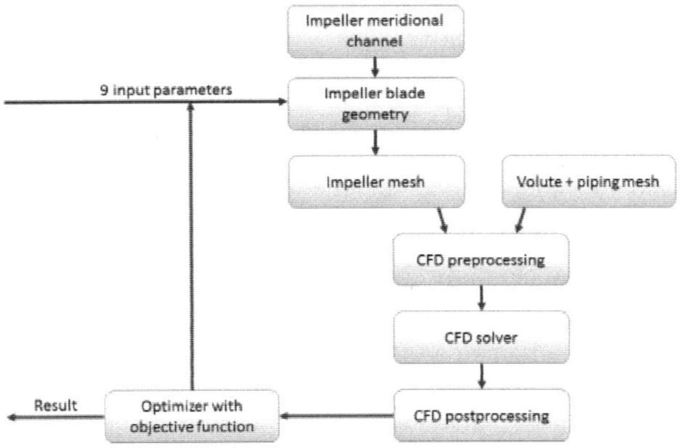

Figure 2. Logic of the numerical procedure.

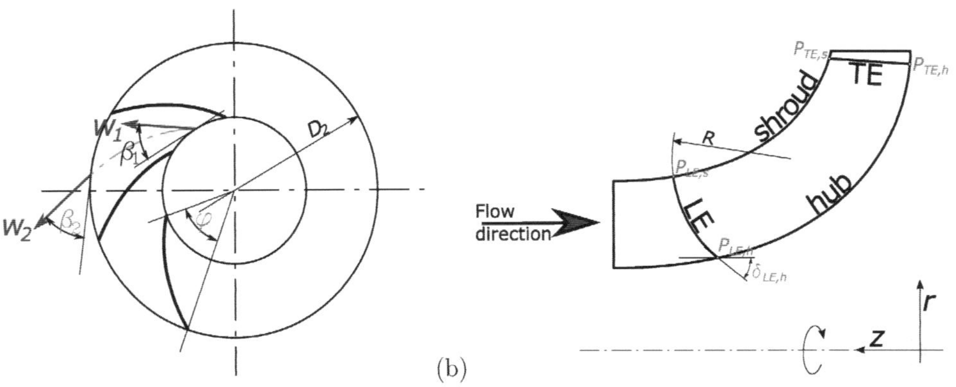

Figure 3. Optimized geometric parameters (in red): (a) front view; (b) side view.

from a previously performed numerical simulation in a 'traditional' way (without optimization, using a full geometry model). At the outlet, average pressure boundary condition was used. A stage condition [14] was used as a frame change / mixing model at the general grid interface between the impeller and the volute. A symmetry condition was used to take into account both sides of the impeller and another half of the volute.

The ANSYS CFX software was used for CFD simulations. These were performed only for the design point, as steady-state single-phase simulations with shear-stress-transport $k-\omega$ turbulence model [15] with a curvature correction [16] (SST-CC). The SST model uses blending between the $k-\omega$ turbulence model close to walls and the $k-\varepsilon$ turbulence model elsewhere, while limiting the eddy viscosity at the boundary layer and the turbulence production term. The *high-resolution* advection scheme was used, which is an upwind adaptive scheme, based on the Barth and Jespersen's limiter [17], and limits the numerical advection correction in order to suppress possible oscillations due to large gradients. The scheme tends to be locally as close as possible to the second order accuracy, and provides a compromise between robustness and accuracy. A local time-scale factor equal to 10 was used for 550 iterations. The result data (head, torque and efficiency) were averaged over the last 50 iterations to produce meaningful result. The

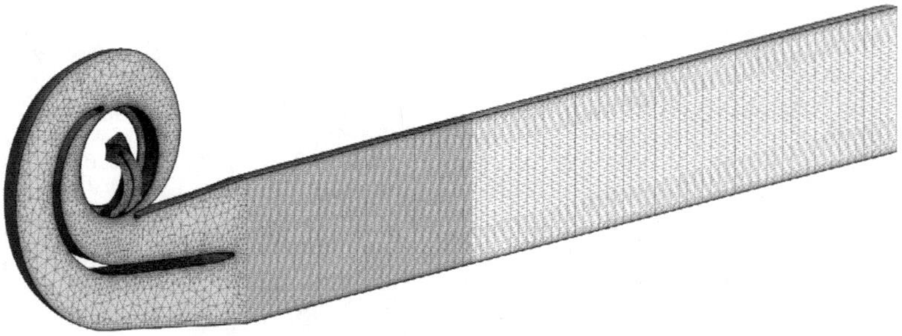

Figure 4. Example of numerical geometry with mesh.

optimization process was driven by modeFRONTIER 2014 [18] optimization platform. It is worth to remark that the objective of the optimization was the pump efficiency. Ten input variables were used to define the pump shape: length of the blade (φ in figure 3) in polar coordinate system for hub, middle and shroud chord; three parameters to define the position and the shape of the impeller blade leading edge in the meridional view ($P_{LE,s}$, $P_{LE,h}$ and $\delta_{LE,h}$ in figure 3); two parameters to define the position of the blade trailing edge in the meridional view ($P_{TE,s}$ and $P_{TE,h}$ in figure 3); and two parameters that influence the impeller trailing edge angle inclination (in the polar view), at hub and at shroud (β_2 in figure 3). The length of the blade in polar coordinate system for middle chord was defined as an average of the length of the hub and the shroud chord, so only nine effective variables were used. Design of experiments (DOE), which fills the design space with the initial set of design variables, was based on the Latin Hypercube [19] samplings. The latter guarantees uniform random distribution of points over the variable range. The initial generation represented 20 sets of variables. Afterward, the Multi-Objective Genetic Algorithm (MOGA-II) [20] based on generational evolution was used for 10 generations. Probability of directional cross-over, probability of selection and probability of mutation were set to 0.5, 0.05 and 0.1, respectively. After performing CFD simulations on 200 cases, each one taking half an hour to complete, the design space was approximated by training a response surface method (RSM). Several RSM algorithms were evaluated for this particular case: polynomial singular value decomposition with the second order of polynomial degree (SVD2), stepwise regression (STEP), K-nearest RSM (KN), Kriging RSM (KR), anisotropic Kriging (AKR), Gaussian processes (GP), Hardy's multiquadric radial basis function (RBF), neural networks (NN) and evolutionary design (ED). For the most promising RSM algorithms, a virtual optimization was performed. After the final verification, performed with CFD simulations, a suitable RSM was proposed.

3. Results and discussion

3.1. Optimization with MOGA-II algorithm

After the CFD simulations were performed on a set of initial design parameters (DOE), subsequent CFD simulations were performed during optimization process with MOGA-II algorithm for 10 generations. The objective of the optimization was overall (hydraulic) pump efficiency. The hydraulic efficiency of the pump is the ratio between the absorbed hydraulic power in fluid (velocity and pressure) and the provided power to the rotor, and is defined as $\eta_{pump} = Q\rho gH/(M\omega)$, where Q is flow rate (in m^3/s), ρ is fluid density (in kg/m^3), g is standard acceleration due to gravity (in m/s^2), H is pump head (in m), M it torque on the shaft (in Nm) and ω is angular velocity (in rad/s). As the purpose of this work was to identify eventual RSM procedure for our future research, no constraints and additional objective variables were used. It

Table 1. Error estimate for different RSM: evolutionary design (ED), radial basis function (RBF), stepwise regression (STEP), polynomial singular value decomposition (SVD2), K-nearest (KN), neural networks (NN), Kriging (KR), anisotropic Kriging (AKR) and Gaussian processes (GP).

RSM algorithm	ED	RBF	STEP	SVD2	KN	NN	KR	AKR	GP
Mean absolute error	0.097	0.135	0.139	0.157	0.199	0.450	0.571	0.573	0.819

should be stressed that for a real-case optimization the pump head should be constrained, as the efficiency was increasing in parallel with increasing head, which was noticed also by Kim et al. [21]. The history chart of optimization procedure for the efficiency is presented in figure 5. The relative efficiency is defined as the ratio of actual efficiency to the efficiency of the first design. The efficiency of the first (baseline) design was close to the values of efficiency usually obtained by the first traditional design. During the first 20 designs, the highest relative efficiency of 100.89 % was obtained. During the MOGA-II optimization, the highest relative efficiency was equal to 101.12 %. The trend in figure 5 shows that efficiency could still be increased by further MOGA-II optimization process.

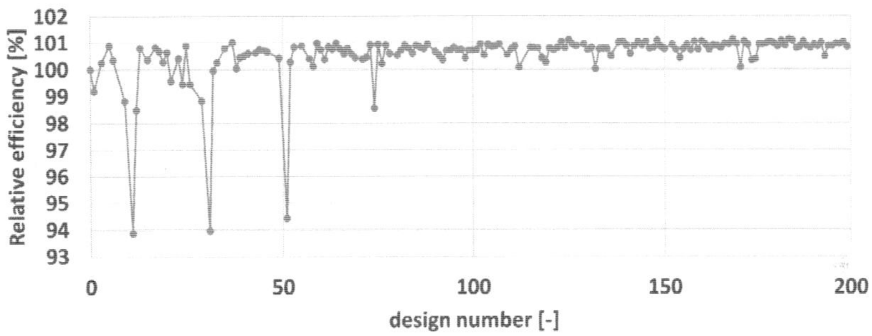

Figure 5. Improvement of efficiency with MOGA-II algorithm.

3.2. Training and validation of RSM algorithms

As described previously, numerous RSM algorithms, available within the modeFRONTIER optimization software, were trained on nine initial variables. For the training, 80 % of all previously computed designs were used (160 designs, out of which 23 were faulty). The trained singular value decomposition polynomial was of the second order (SVD2). For the validation of RSM algorithms, 20 % of all previously computed designs were used (40, out of which 5 were faulty). Results of validation are presented in Table 1.

In figure 6, relative efficiency predicted by four most promising algorithms is compared to the relative efficiency obtained by CFD simulations.

Although the most promising algorithm by means of absolute error was ED, it can be noticed in figure 6 that at large values of efficiency the curve seemed to be flat, meaning that it might not distinguish high values of efficiency from the lower ones, which might in turn make it useless for the virtual optimization process. On the contrary, in case of the RBF, STEP and SVD algorithms, the high efficiency values predicted by CFD resulted in high efficiency values predicted by these algorithms as well (higher than the rest of the curve). The ED algorithm was used for virtual

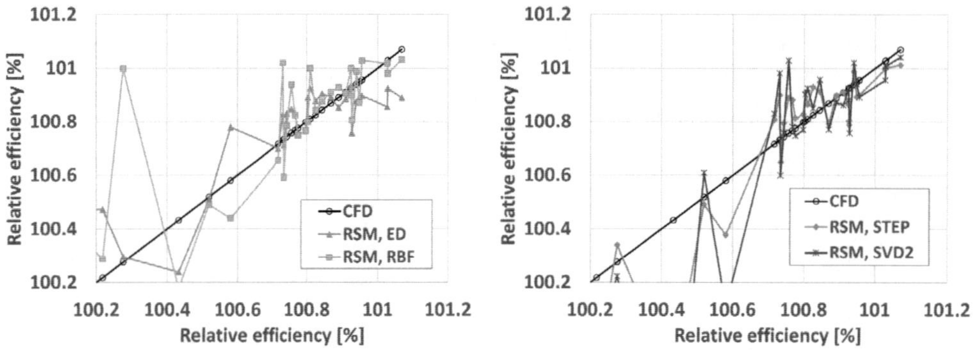

Figure 6. Validation of RSM algorithms with CFD results. Left: ED and RBF algorithms. Right: STEP and SVD2 algorithms.

optimization for 40 generations (virtual designs between 200 and 1000), as presented in figure 7. The highest relative efficiency obtained was equal to 101.064 %, which is smaller than that obtained in the first 200 designs with CFD.

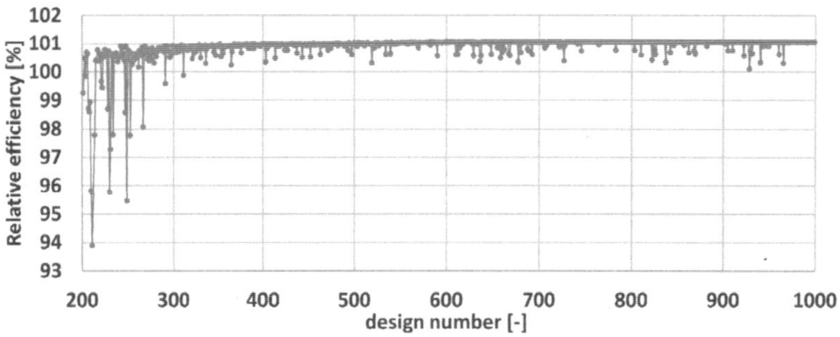

Figure 7. Virtual optimization of efficiency with ED algorithm.

Nine designs with values of efficiency greater than 100.95 % with distinctly different input variable combinations were used for validation purposes, and this is presented in figure 6. It can be noticed that some of the points were quite accurately predicted, especially at the right-hand side of the curve. However, the highest values of efficiency was predicted for a different setup of input variables than the CFD simulations. It can be concluded that although the ED algorithm appeared as the best choice, it has to be trained on more than 160 CFD simulation cases. The largest noticed difference in efficiency between CFD and ED model was approximately 1 %. It is interesting to note that for one of the cases in [4], where RSM algorithm was trained on 40 cases, such difference was approximately 2.6 % for a case on a Pareto front. The ED model was improved by training it on 200 designs (marked as ED200) in figure 8. However, the results were similar as with training on 160 designs, which means that the same conclusions can be drawn.

The behavior of the other three RSM algorithms (RBF, STEP and SVD2) looks quite similar to figure 6. Therefore, the second-best model in table 1, the RBF algorithm, was tested. As evident from figure 9, the relative efficiency after 40 generations reached 101.38 %, which is higher than that obtained by ED algorithm in figure 7. The reason might be due to the fact that in figure 6 the RBF algorithm was able to generate higher values of efficiency than the ED. It

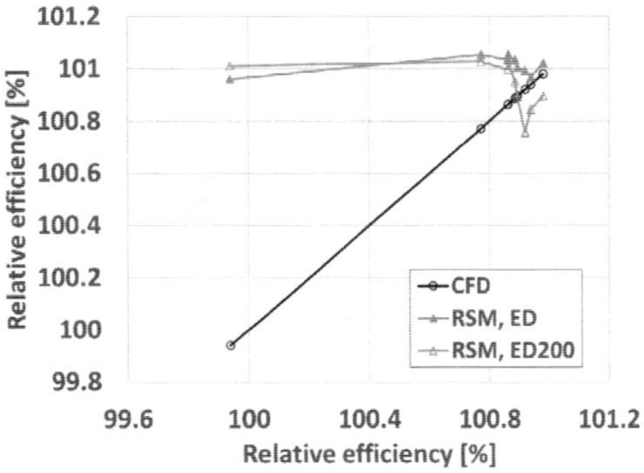

Figure 8. Validation of ED algorithm.

should be noted that the efficiency predicted by RBF is also higher than that obtained during the first 200 designs.

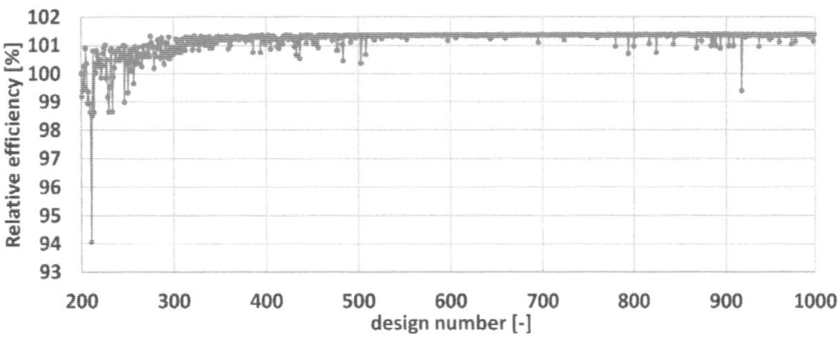

Figure 9. Virtual optimization of efficiency with RBF algorithm.

Nine designs with values of efficiency greater than 101.28 % with different input variable combinations were used for validation purposes. Results are presented in figure 10. In general, the results represent almost a flat curve over the observed range of efficiency obtained by CFD results. Secondly, the values (of curve RBF) are much higher than thos obtained by CFD, from 1.2 % to 1.84 %. The results improved when RBF was trained on 200 designs instead of on 160 CFD simulations (curve RBF200). In addition, when only the five most influential input variables were used for training on 200 designs (curve RBF200_5var), the results seemed to improve.

The last algorithm tested was STEP algorithm. After 40 generations the highest relative efficiency obtained was equal to 101.35 (figure 11), which is similar to that of the RBF.

From the validation with CFD results (figure 12), similar conclusions as for RBF function can be drawn: the curve STEP is flat and too high. Trained on 200 designs, the curve STEP200 is closer to CFD simulations. However, when only 5 design variables were used for training on 200 designs, the result was improved considerably (curve STEP200_5var).

Finally, in figure 13, the most promising STEP and RBF models were compared to all CFD simulations that were performed in figures 8, 10 and 12. It can be noticed that the prediction for both algorithms improved when only 5 variables were used for training. In addition, it seems that the STEP200_5var curve follows the CFD curve very well, which means that it has a potential

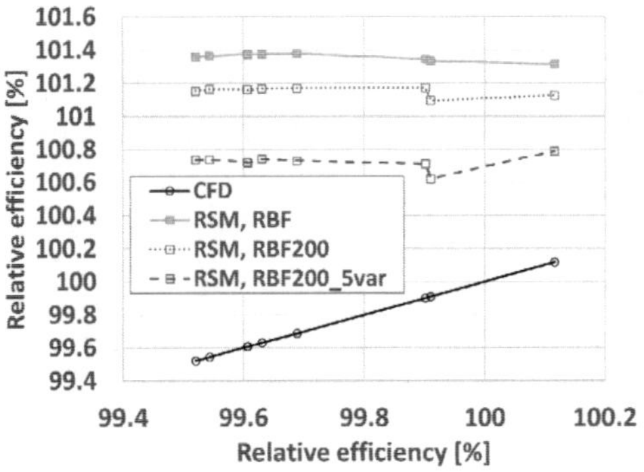

Figure 10. Validation of the RBF algorithm.

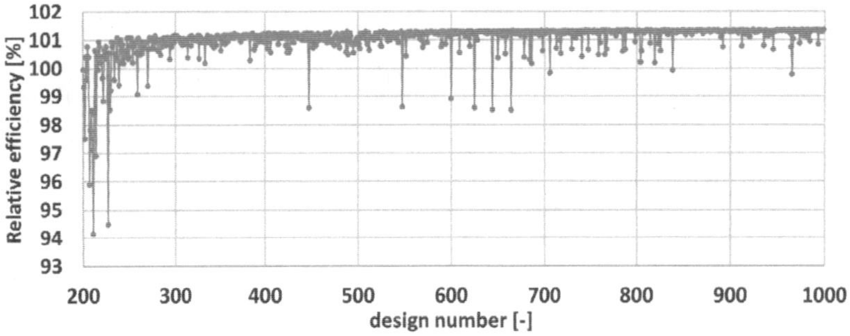

Figure 11. Virtual optimization of efficiency with the STEP algorithm.

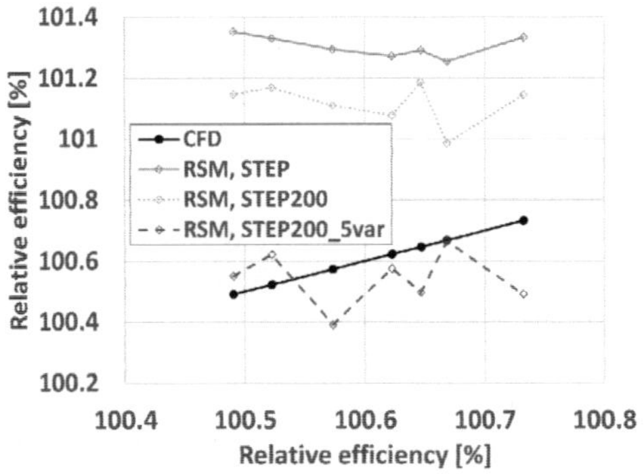

Figure 12. Validation of the STEP algorithm.

to be used for virtual optimization of the pump case with 200 initial design points simulated with CFD.

The efficiency prediction of the RSM STEP algorithm, trained on a set of 200 designs by using five variables, could not be improved by using just the high-efficiency values in training phase. Even when only three low-efficiency designs were discarded (when efficiency was smaller than 95 %), as presented by the curve STEP200_5var_95 in figure 12, large discrepancy occurred at

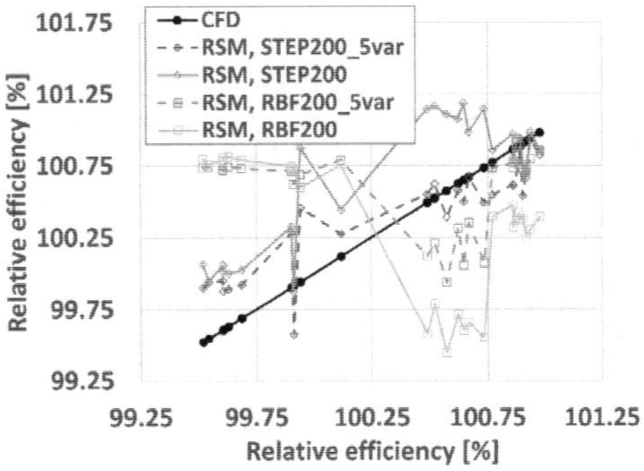

Figure 13. Comparison of STEP and RBF models with CFD results.

efficiency value 99.94 %. The predicted value was approximately 100.8 %, which is very close to the highest efficiency values at the right side of the curve. Such discrepancies might lead to improper virtually-optimized designs.

Finally, the initial (baseline) and optimized geometries of the impeller, characterized by 100 % and 100.9 % relative efficiencies respectively, are illustrated in figure 14.

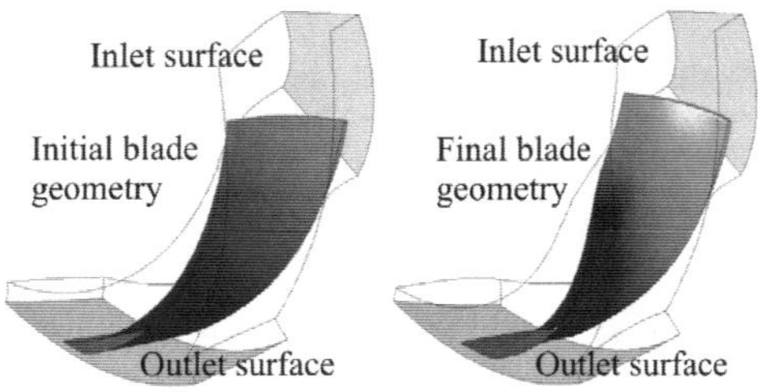

Figure 14. Impeller geometries. Left: initial (baseline); Right: optimized

The results obtained by the simplified steady CFD model were confirmed by running a full 3D unsteady simulation, using the SAS-SST (Shear Stress Transport Scale Adaptive Simulation) DES-like turbulence model of Egorov and Menter [22].

The experimental validation, by measuring the performances of the optimized impeller, is in progress at Kolektor Turboinštitut.

4. Conclusions

Several metamodels, which represent an economical alternative to the more expensive 3D CFD model, were built and tested. Based on comparison with CFD simulations, the following conclusions were drawn:

- The most promising metamodels were the ED, the RBF and the STEP;
- A set of 200 designs was not sufficient for the three RSM, when built on 9 design variables;
- For such set, the STEP metamodel, built on five most influential input design variables, seemed a potentially applicable metamodel for virtual optimization of a centrifugal pump;
- Prediction of metamodels was improved by increasing the number of initial (CFD) designs and by building/training the metamodels on only the most influential design variables;

- The five most influential design variables for this case, by order of importance, were: blade trailing edge inclination angle at hub, location of leading edge at hub in the meridional view, length of the blade at hub in the polar view, location of leading edge at shroud in the meridional view, length from the axis of rotation to the impeller blade trailing edge, at hub.
- Discarding low-efficiency designs in the training phase of the metamodels can lead to false high-efficiency design predictions.

Acknowledgments
This research has received funding from the People Programme (Marie Curie Actions) of the European Union's Seventh Framework Programme FP7/2007-2013/ under REA grant agreement n° 612279 and from the Slovenian Research Agency ARRS - Contract No. 1000-15-0263.

References

[1] Gülich JF 2008 *Centrifugal Pumps* (Berlin: Springer-Verlag) p 926
[2] Lipej A and Poloni C 2000 Design of Kaplan runner using multiobjective genetic algorithm optimization *J. Hydraul. Res.* **38(4)** pp 73-9
[3] Shingai K, Omura Y, Tani K, Tominaga H and Ito S 2008 Multi-objective optimization of diagonal flow type reversible pump-turbines *Proc. 24th Symp. on Hydraulic Machinery and Systems* 27-31 Foz do Iguassu (Madrid: IAHR)
[4] Xuhe W, Baoshan Z, Lei T, Jie Z and Shuliang C 2014 Development of a pump-turbine runner based on multiobjective optimization *J. Phys. Conf. Series: Earth Environ. Sci.* **22** 012028
[5] Zhang Y, Hu S, Zhang Y, Chen L 2014 Optimization and analysis of centrifugal pump considering fluid-structure interaction *Scientific World J.* 2014 131802
[6] Venables WN and Ripley BD 2002 Modern Applied Statistics with S (New York: Springer-Verlag) p 498
[7] Alfeld P 1989 Scattered data interpolation in three or more variables *Mathematical Methods in Computer Aided Geometric Design* ed T Lyche and LL Schumaker (New York: Academic Press) pp 1-34
[8] Deutsch CV and Journel AG 1992 *GSLIB: Geostatistical Software Library and User's Guide* ed AG Journel (New York: Oxford University Press) p 340
[9] Rasmussen CE and Williams CKI 2006 *Gaussian Processes for Machine Learning* ed T Dietterich, C Bishop et al (Cambridge: MIT Press) p 262
[10] Gibbs M and MacKay DJC 1997 *Efficient Implementation of Gaussian Processes* (Cambridge: Cavendish Laboratory) p 17
[11] Buhmann MD 2003 *Radial Basis Functions: Theory and Implementations* (Cambridge: Cambridge University Press)
[12] Hagan MT and Menhaj MB 1994 Training feedforward networks with the Marquardt algorithm *IEEE Trans. on Neural Networks* **5** 989-93
[13] Fillon C 2008 *New Strategies for Efficient and Practical Genetic Programming* (Trieste: Università degli Studi di Trieste) p 145
[14] ANSYS 2016 *ANSYS CFX-Solver Theory Guide* (ANSYS: Canonsburg) p 362
[15] Menter F R Kuntz M and Langtry R 2003 Ten years of industrial experience with the SST turbulence model *Turbulence, Heat and Mass Transfer* 4. ed Hanjalič K Nagano Y and Tummers M (Begell House, New York) pp 625-632.
[16] Smirnov P E and Menter F 2009 Sensitization of the SST turbulence model to rotation and curvature by applying the Spalart-Shur correction term. *J. Turbomachinery* **131**, no. 4, p 041010.
[17] Barth T J and Jespersen D C 1989 The design and application of upwind schemes on unstructured meshes *27th Aerospace Sciences Meeting* AIAA Paper 89-0366.
[18] ESTECO 1999-2016 *modeFRONTIER User Manual* (ESTECO: Trieste)p 504
[19] McKay MD, Conover WJ and Beckman RJ 1979 Latin hypercube sampling: a comparison of three methods for selecting values of input variables in the analysis of output from a computer code *Technometrics* **21** 239-45
[20] Poles S 2003 MOGA-II *An improved Multi-Objective Genetic Algorithm* (Trieste: ESTECO)
[21] Kim JH, Oh KT, Pyun KB, Kim CK, Choi YS and Yoon JY 2012 Design optimization of a centrifugal pump impeller and volute using computational fluid dynamics *J. Phys. Conf. Series: Earth Environ. Sci.* **15** 032025
[22] Egorov Y and Menter F 2008 Development and application of SST-SAS turbulence model in the DESIDER project. *Advances in Hybrid RANS-LES Modelling* ed Peng S -H and Haase W (Springer, Heidelberg), pp 261-270.

IOP Conf. Series: Journal of Physics: Conf. Series **796** (2017) 012008 doi:10.1088/1742-6596/796/1/012008

Adjoint optimal control problems for the RANS system

Andrea ATTAVINO*, Daniele CERRONI*, Roberto DA VIÀ*, Sandro MANSERVISI*, Filippo MENGHINI*

*University of Bologna, DIN, Lab. of Montecuccolino, Via dei Colli 16, Bologna

E-mail: `filippo.menghini3@unibo.it`

Abstract. Adjoint optimal control in computational fluid dynamics has become increasingly popular recently because of its use in several engineering and research studies. However the optimal control of turbulent flows without the use of Direct Numerical Simulation is still an open problem and various methods have been proposed based on different approaches. In this work we study optimal control problems for a turbulent flow modeled with a Reynolds-Averaged Navier-Stokes system. The adjoint system is obtained through the use of a Lagrangian multiplier method by setting as objective of the control a velocity matching profile or an increase or decrease in the turbulent kinetic energy. The optimality system is solved with an in-house finite element code and numerical results are reported in order to show the validity of this approach.

1. Introduction

The optimization of industrial devices in which the flow is turbulent is one of the main problems in current engineering research and development issues. Turbulent flows are encountered commonly in engineering studies and when performing Computational Fluid Dynamics simulations there are several ways to deal with this problem. To perform studies at a very deep and precise level Direct Numerical Simulations are needed in order to resolve all the time and space scales of the turbulent flow. However these methods are computationally prohibitive for industrial applications and therefore several models have been developed to overcome this problem, with the main employed in the last years being Reynolds Averaged Navier Stokes models. Some interesting studies are currently being performed to use Large Eddy Simulations or hybrid LES-RANS models but common practice is still to use RANS models in industrial applications [1, 2].

Continuous adjoint optimal control is a very interesting and promising optimization technique that can be used in several research fields [3, 4]. In this method one defines an objective functional, the control and the constraints of the problem, which are the state equations. Then through a Lagrangian multiplier technique the adjoint system is derived and it can be solved to obtain a local optimal solution starting from a reference state. Many objectives can be set when studying optimal control problems such as velocity matching profile, lift enhancement or drag reduction, decrease of the viscous dissipation and many others. We should remark that continuous adjoint methods are not widely used in practice while discrete or automatic differentiation tools are preferred in industrial design, since they are available in many commercial and open-source CFD packages. This is also due to the fact that these optimization methods are employed usually to improve the shape of devices, such as airplane wings or turbine blades. However the continuous adjoint method that we study in this work could improve the design of industrial devices when the optimization is performed on other parameters than shape, such as the control of a source term in the domain or the value of a boundary condition as an inflow velocity and direction from a specified boundary [4, 5]. Another big advantage of adjoint methods over automatic differentiation tools that are very commonly used is that in order to obtain the sensitivities with respect to all the parameters one needs to solve only one adjoint system at the computational cost of a CFD solution [6, 7]. Moreover in this work we take into account the turbulence modeling without considering a frozen turbulent viscosity as it is usually done in optimization methods, see [6, 7, 8, 9] for a discussion on these subjects.

To summarize, in this work we study the optimal control problem with distributed and boundary control for a turbulent flow modeled with a two-equation RANS model [10]. Interesting features of this approach include the presence of turbulence modeling included explicitly in the derivation of the adjoint equations, the possibility of setting as objective of the control both a function of the velocity or of the turbulence intensity, which can be useful when a laminarization of the flow is sought, and the extension to boundary control with respect to the work already done in [8, 9]. The optimality system is derived for both the distributed and boundary control and numerical results obtained with the distributed control are reported. These results show the effectiveness of our approach in the optimization of turbulent flows.

2. Optimality system

For an incompressible fluid the conservation of mass and momentum is defined by the Navier-Stokes system. The RANS equations are very similar to the Navier-Stokes ones since they differ in one unknown term, the Reynolds Stress tensor, that needs to be modeled. These equations together with Wilcox k-ω turbulence model are [10]

$$\nabla \cdot \mathbf{v} = 0\,, \tag{1}$$

$$\rho(\mathbf{v} \cdot \nabla)\mathbf{v} = \nabla \cdot (\mathbf{T} + \rho \mathbf{T_t}) + \mathbf{f}\,, \tag{2}$$

$$(\mathbf{v} \cdot \nabla)k = \nabla \cdot [(\nu + \sigma_k \nu_t)\,\nabla k] + P_k - \beta^* \, k\,\omega\,, \tag{3}$$

$$(\mathbf{v} \cdot \nabla)\omega = \nabla \cdot [(\nu + \sigma_\omega \nu_t)\,\nabla \omega] + \alpha \frac{\omega}{k} P_k - \beta \omega^2 + \frac{\sigma_d}{\omega} \nabla k \cdot \nabla \omega\,, \tag{4}$$

where ρ is the fluid density, $\mathbf{f}$ a force acting on the flow and the viscous stress tensor $\mathbf{T}$ can be defined for a Newtonian fluid as

$$T_{ij} := -p'\delta_{ij} + \mu S_{ij} \qquad S_{ij} := \frac{\partial v_i}{\partial x_j} + \frac{\partial v_j}{\partial x_i}\,, \tag{5}$$

where p' is the fluid pressure, μ the dynamic viscosity and $\mathbf{S}$ the deformation tensor. The production term of the turbulence kinetic energy in (3-4) is

$$P_k = \frac{\nu_t}{2} \|\mathbf{S}\|^2\,. \tag{6}$$

The unknown Reynolds stress tensor $\mathbf{T_t}$ can be computed by introducing the eddy viscosity ν_t,

$$T_{tki} = -\frac{2}{3} k \delta_{ki} + \nu_t S_{ki}\,. \tag{7}$$

The eddy viscosity ν_t is defined in the turbulence model in the following form [10]

$$\nu_t = min\left[\frac{k}{\omega}, \frac{k}{C_{lim}\sqrt{\frac{\mathbf{S}:\mathbf{S}}{2\beta^*}}}\right]. \tag{8}$$

Table 1. Coefficients employed in the turbulence model.

$\sigma_k = 0.6$	$\sigma_\omega = 0.5$
$\beta^* = 0.09$	$\alpha = \dfrac{13}{25}$
$C_{lim} = \dfrac{7}{8}$	$\beta_0 = 0.0708$
$\beta = \beta_0 \dfrac{1 + 85\,\chi}{1 + 100\,\chi}$	$\chi = \left\| \dfrac{\Omega_{ij}\,\Omega_{jk}\,S_{ki}}{(\beta^*\,\omega)^3} \right\|$

We also report the other coefficients needed by the model in Table 1. In this Table Ω_{ij} is the velocity rotation tensor. It should be noted that in a two-dimensional problem the function χ vanishes. As we study only two-dimensional problems in this work we can neglect this term in the following. For details see [10] and citations therein.

The state equations presented above are considered the constraints in the optimal control setting. To complete the definition of the problem we define the objective functional on the computational domain Ω and controlled surface Γ_c

$$\mathcal{J}(\mathbf{v}, k, \mathbf{f}, \mathbf{g}) = \frac{1}{2}\int_\Omega a\,\|\mathbf{v} - \mathbf{v}_d\|^2\, d\Omega + \frac{1}{2}\int_\Omega b\,|k - k_d|^2\, d\Omega + \frac{\beta}{2}\int_\Omega \|\mathbf{f}\|^2\, d\Omega + \frac{\beta_1}{2}\int_{\Gamma_c} \|\mathbf{g}\|^2\, d\Gamma + \frac{\lambda}{2}\int_{\Gamma_c} \|\nabla\mathbf{g}\|^2\, d\Gamma\,. \tag{9}$$

In (9) $\mathbf{f}$ is the force that acts on the fluid as reported in (2) while $\mathbf{g}$ is a Dirichlet condition for the velocity, $\mathbf{v} = \mathbf{g}$ on Γ_c. The weight functions a and b can be used to better specify the region on which we want to obtain the velocity profile or turbulence kinetic energy [5]. Moreover they can be used to study a velocity matching problem by setting $b = 0$ or a turbulence kinetic energy matching by setting $a = 0$. We note that the distributed or boundary control can be studied by considering $\beta_1 = 0$, $\lambda = 0$ or $\beta = 0$, respectively. The adjoint equations obtained are the same for the two types of control but the problem changes completely. The last terms of (9) are regularization terms needed to obtain a non-singular control $\mathbf{f} \in L^2(\Omega)$ and $\mathbf{g} \in H^1(\Gamma)$ [4]. The numerical values of β, β_1 and λ are important for the solution of the optimal control problem since high values can result in poor optimization results while low values may lead to convergence issues.

The total Lagrangian of this problem, neglecting for simplicity the limiting values on ν_t, can be written as

$$\mathcal{L} = \mathcal{J}(\mathbf{v}, k, \mathbf{f}, \mathbf{g}) + \int_\Omega p_a \nabla\cdot\mathbf{v}\, d\Omega + \int_\Omega \mathbf{v}_a \cdot \left[(\mathbf{v}\cdot\nabla)\mathbf{v} - \nabla\cdot\left(\frac{\mathbf{T}}{\rho} + \mathbf{T_t}\right) - \mathbf{f}\right] d\Omega + \int_{\Gamma_c} \boldsymbol{\tau}\cdot(\mathbf{v} - \mathbf{g})d\Gamma + \int_\Omega k_a\left[(\mathbf{v}\cdot\nabla)k - \nabla\cdot\left[(\nu + \sigma_k\nu_t)\,\nabla k\right] - P_k + \beta^*\,k\,\omega\right] d\Omega + \int_\Omega \omega_a \left[(\mathbf{v}\cdot\nabla)\omega - \nabla\cdot\left[(\nu + \sigma_\omega\nu_t)\,\nabla\omega\right] - \alpha\frac{\omega}{k}P_k + \beta\omega^2 - \frac{\sigma_d}{\omega}\nabla k\cdot\nabla\omega\right] d\Omega\,. \tag{10}$$

The total Lagrangian (10) is composed of the objective functional (9) and the RANS equations with the turbulence model (1-4), each of them multiplied by a proper Lagrange multiplier. The term with $\boldsymbol{\tau}$ is present only if we choose a boundary control so that the velocity can vary on the controlled boundary Γ_c during the optimization process. The first order necessary conditions needed to obtain the optimal solution can be set by considering the total variation of the Lagrangian null, $\delta\mathcal{L} = 0$. By taking the Fréchet derivatives of (10) with respect to all the variables involved and setting them to zero one obtains the optimality system [4, 8, 9]. When the derivatives are taken with respect to the Lagrangian multipliers we obtain back the state system (1-4) in weak form, while when we varies with δp and $\delta\mathbf{v}$ we get

$$\int_\Omega \nabla\cdot\mathbf{v}_a\,\delta p\,d\Omega = 0 \qquad \forall\delta p\in L^2(\Omega)\,, \tag{11}$$

$$\begin{aligned}\int_\Omega \Big[&-\nabla p_a\cdot\delta\mathbf{v} + \mathbf{v}_a\cdot(\delta\mathbf{v}\cdot\nabla)\mathbf{v} - [(\mathbf{v}\cdot\nabla)\mathbf{v}_a]\cdot\delta\mathbf{v} - \nabla\cdot[(\nu+\nu_t)\mathbf{S}(\mathbf{v_a})]\cdot\delta\mathbf{v} + \\ &k_a(\delta\mathbf{v}\cdot\nabla)k + \omega_a(\delta\mathbf{v}\cdot\nabla)\omega + a\,(\mathbf{v}-\mathbf{v}_d)\cdot\delta\mathbf{v} + \\ &2\,\nabla\cdot[(k_a\nu_t + \alpha\,\omega_a)\mathbf{S}(\mathbf{v})]\cdot\delta\mathbf{v}\Big]\,d\Omega = 0 \qquad \forall\delta\mathbf{v}\in H^1(\Omega)\,,\end{aligned} \tag{12}$$

and with δk

$$\begin{aligned}\int_\Omega \Big[&-(\mathbf{v}\cdot\nabla)k_a - \nabla\cdot[(\nu+\sigma_k\nu_t)\nabla k_a] + \beta^* k_a\omega + \nabla\cdot\Big[\sigma_d\frac{\omega_a}{\omega}\nabla\omega\Big] + \\ &\frac{\nu_a}{\omega} + b\,(k-k_d)\Big]\,\delta k\,d\Omega = 0 \qquad \forall\delta k\in H^1(\Omega)\,,\end{aligned} \tag{13}$$

and finally for $\delta\omega$

$$\begin{aligned}\int_\Omega \Big[&-(\mathbf{v}\cdot\nabla)\omega_a - \nabla\cdot[(\nu+\sigma_\omega\nu_t)\nabla\omega_a] + \beta^* k_a k\delta\omega + 2\beta\omega_a\omega + \\ &\sigma_d\,\nabla\cdot\Big[\frac{\omega_a}{\omega}\nabla k\Big] + \frac{\sigma_d\omega_a}{\omega^2}\nabla k\cdot\nabla\omega - \frac{\nu_a\,k}{\omega^2}\Big]\,\delta\omega\,d\Omega = 0 \qquad \forall\delta\omega\in H^1(\Omega)\,.\end{aligned} \tag{14}$$

These are the adjoint equations of the state system (1-4) written in weak form. The boundary conditions are already incorporated in this formulation. The interested reader can find a complete derivation and discussion on the boundary conditions to be set for this problem in [8, 9]. By considering the variations with respect to the turbulent viscosity ν_t one obtains an algebraic equation in ν_a that is solved by setting

$$\nu_a = k_a\,\mathbf{S}:\mathbf{S} - \nabla\mathbf{v_a}:\mathbf{S}(\mathbf{v}) - \sigma_k\nabla k_a\cdot\nabla k - \sigma_\omega\nabla\omega_a\cdot\nabla\omega\,. \tag{15}$$

Therefore this expression can be substituted in the equations (13-14).

The last terms that need to be considered are the variations with respect to the controls $\mathbf{f}$ and $\mathbf{g}$. If we are considering a distributed control the equation in the variation $\delta\mathbf{f}$ is

$$\int_\Omega (\beta\mathbf{f} - \mathbf{v_a})\cdot\delta\mathbf{f}\,d\Omega = 0 \qquad \forall\delta\mathbf{f}\in L^2(\Omega)\,, \tag{16}$$

so that the force $\mathbf{f}$ on the fluid is equal to the adjoint velocity $\mathbf{v_a}$ scaled by the parameter β. On the other hand, if a boundary control is chosen the variation in $\delta\mathbf{g}$ gives rise to a differential equation on the controlled boundary Γ_c

$$\int_{\Gamma_c} [\lambda\nabla\mathbf{g}:\nabla\delta\mathbf{g} + \beta_1\mathbf{g}\cdot\delta\mathbf{g} - [(\nu+\nu_t)\mathbf{S}(\mathbf{v}_a)\cdot\mathbf{n}]\cdot\delta\mathbf{g}]\,d\Gamma = 0 \qquad \forall\delta\mathbf{g}\in H^1(\Gamma)\,, \tag{17}$$

that needs to be solved in order to obtain the controlled velocity on the boundary Γ_c as $\mathbf{v} = \mathbf{g}$.

Some remarks can be done on this optimality system. First we notice that the source terms in the adjoint equations are the terms that we have defined in the objective functional as the differences $(\mathbf{v} - \mathbf{v}_d)$ and $(k - k_d)$. So the driving forces of the adjoint variables are big if the objective is far from being achieved while are null if the objective is reached, as it can be expected. Moreover the control, being it distributed or boundary, is a function of the adjoint velocity or of its derivatives. In the following we study two problems with distributed control, one to obtain a matching velocity and the other to control the value of the turbulent kinetic energy in some specific region of the domain. This can be achieved by setting properly the weight functions a and b.

3. Numerical Results

In this Section we report the numerical results obtained by implementing the optimality system as described in the previous Section in an in-house finite element code. This code is a parallel mpi-based code which implements a multigrid solver and which is being developed at the laboratory of Montecuccolino of the University of Bologna [11]. It is based on several open-source libraries for parallel computing and linear algebra, namely the openmpi, PETSc and MED libraries [12, 13, 14]. We use standard quadratic-linear finite elements for the velocity-pressure solution for both the state and adjoint systems in order to fulfill the inf-sup condition. The turbulent variables k-ω and their adjoints are approximated with quadratic elements. We use a steepest descent gradient based algorithm to solve the optimality system in a segregated way because a one-shot solver is not affordable for such a complex system [3, 8, 9].

We have studied a backward-facing step problem with Reynolds number $Re = 10000$ based on the inlet diameter. In the results we report the same color scale and iso-lines are used for all the variables among different cases in order to better compare the effect of the optimization with respect to the reference case. In Figure 1 the geometry of the test case is reported together with the flow field in the reference case. With reference to Figure 3 the segment AB is the inlet, EF the outlet and the other boundaries are solid walls. The total length of the channel is 6 and the inlet and the step CD are 1 in non dimensional length. The length of the entrance channel before the step BC is 2. From the flow field as depicted in Figure 1 one can see that the vortex that develops after the step has a reattachment point after around 2.5 in non-dimensional length. In Figure 2 the reference results for the turbulent kinetic energy are reported with iso-lines. One can see that the main source of turbulence is the step. The high values of turbulent kinetic energy produced here diffuse along the channel before exiting from the outlet.

We now report the results obtained by studying the distributed optimal control of this test case with three different objectives. The first is to control the velocity in a specific region of the domain near the outlet. In the second and the third we aim at increasing or decreasing

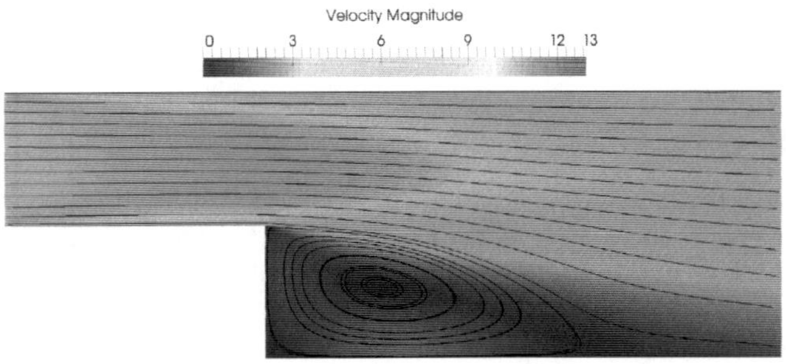

Figure 1. Backward facing step geometry and velocity streamlines, reference case.

the turbulence intensity in the region after the step where the vortex develops. We study these problems with $\beta = 0.01$.

In Figure 3 the weight function a is reported on the whole domain with the values $a = 0$ and $a = 1$ for the velocity matching case. The desired velocity is set to $\mathbf{v}_d = (8, 0)$ in the region near the outlet. In Figure 4 the adjoint velocity is reported with streamlines and magnitude. By considering that this is the force that is applied to the fluid in order to control the fluid velocity, one can see that the control is pushing downward the flow in order to obtain a constant

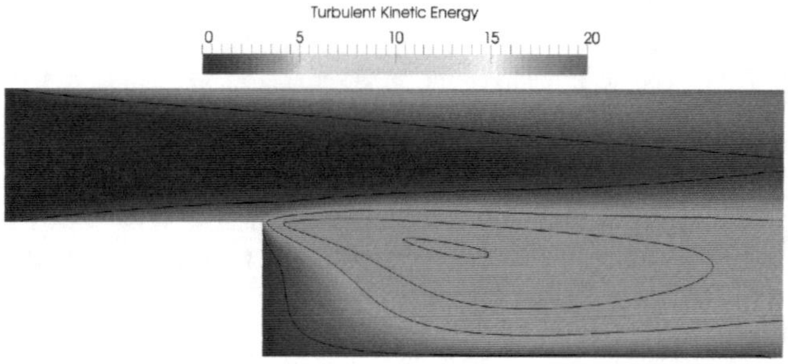

Figure 2. Iso-lines of the turbulent kinetic energy in the reference case.

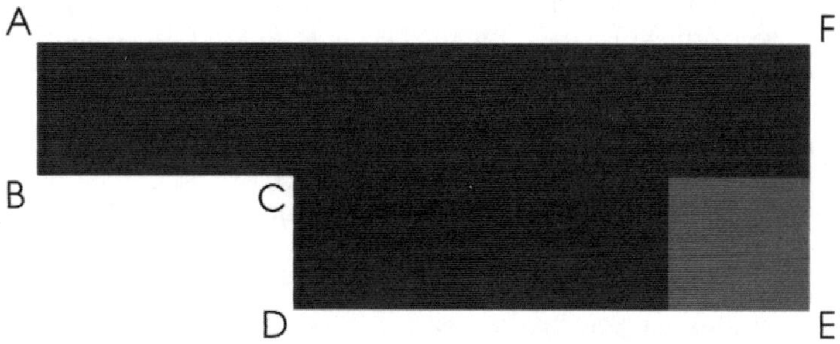

Figure 3. Velocity matching case, representation of the a weight function with values $a = 0$ (blue) and $a = 1$ (red). AB is the inlet, EF the outlet, the other boundaries are solid walls.

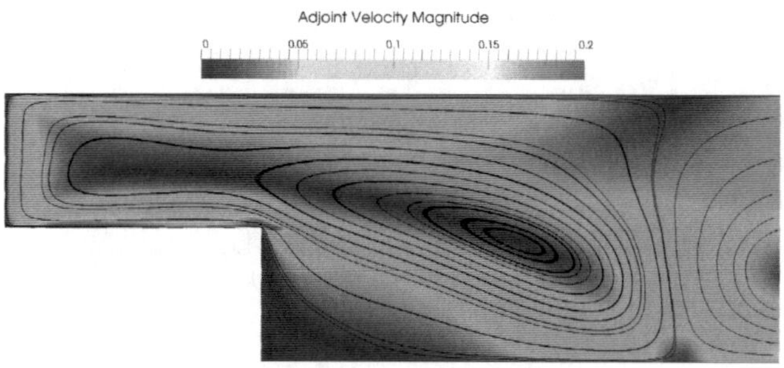

Figure 4. Velocity matching case, adjoint velocity streamlines and magnitude.

horizontal velocity in this region. In Figure 5 the controlled velocity is reported with streamlines and magnitude after the control is applied to the fluid. This result should be compared with the reference velocity reported in Figure 1. As one can see in the optimal solution the flow is horizontal and with a constant velocity of $(8, 0)$ in the region near the outlet where the objective functional was set.

We report in Figure 6 the weight function b as set for the two turbulence control problems. The region with $b = 1$ is set after the step. In the second test case of turbulence increase we set $k_d = 20$. In this case we show the adjoint turbulent kinetic energy which is the main adjoint variable in this setting. In Figure 7 this variable is reported for the optimized case with iso-lines. One can see that k_a is negative where the turbulence is too high while is positive near the wall where the turbulence cannot be increased too much. In Figure 8 the turbulent kinetic energy is reported in the optimized case with iso-lines. One can compare this result with the reference case in Figure 2 in order to better understand the optimized result obtained. This very strong increase in the turbulent kinetic energy is achieved with the formation of a vortex by the force $\mathbf{f}$ applied to the fluid, as it is shown in Figure 9 where the streamlines and the magnitude of the fluid velocity in this increase turbulence test case are reported.

We now report the results obtained for the decrease of turbulence test case that has been studied by setting $k_d = 0$. The weight function b is the same as reported for the previous test case in Figure 6. The contours and the magnitude of the turbulent kinetic energy are reported in Figure 10 for this test case. One can see that the turbulence intensity decreases in the region where $b = 1$ and that the plume of turbulence produced by the step is not diffusing in this

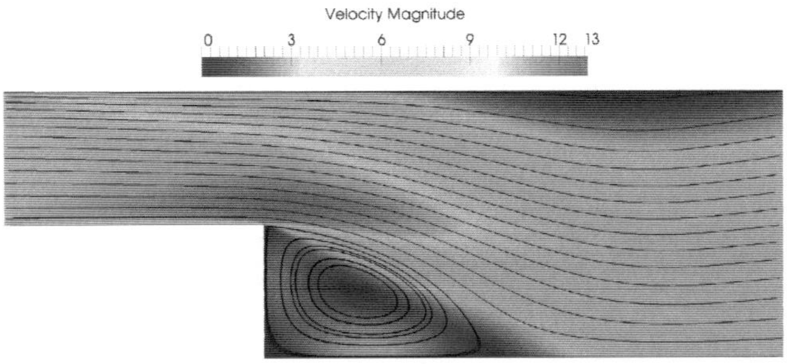

Figure 5. Velocity matching case, velocity streamlines and magnitude.

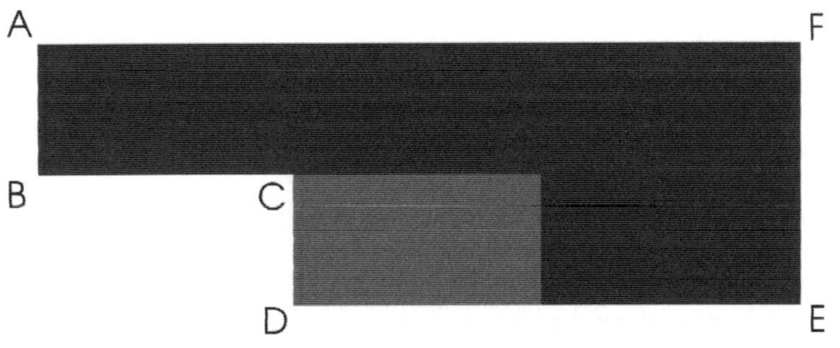

Figure 6. Turbulence control case, representation of the b weight function with values $b = 0$ (blue) and $b = 1$ (red).

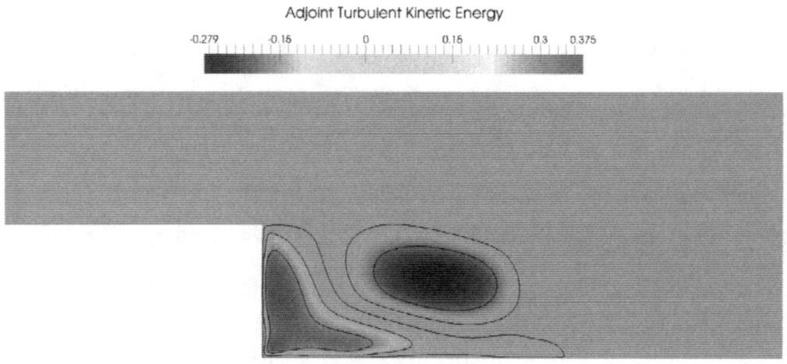

Figure 7. Turbulence increase case, adjoint turbulent kinetic energy iso-lines and magnitude.

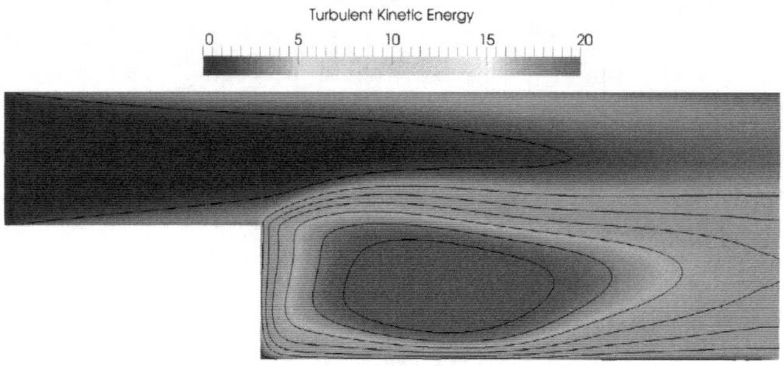

Figure 8. Turbulence increase case, turbulent kinetic energy iso-lines and magnitude.

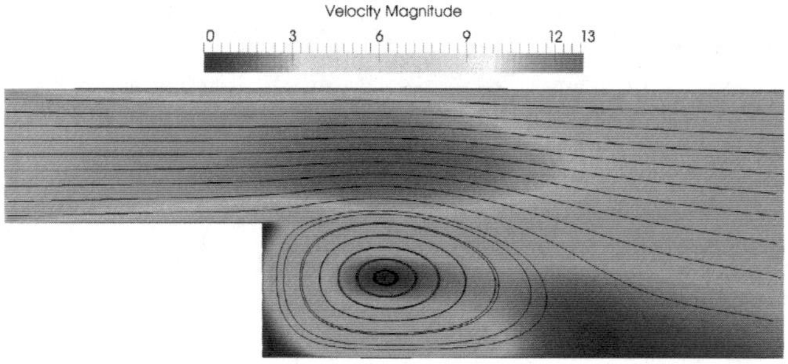

Figure 9. Turbulence increase case, velocity streamlines and magnitude.

region but it is advected further on the channel. This result is achieved by changing the velocity pattern as reported in Figure 11. The formation of the vortex after the step is now shifted in a further region with respect to the main flow, as the reattachment length becomes different.

Finally in Table 2 we report the values of the objective functionals as computed from the definition (9) for the three test cases considered. The reference case is always reported for a proper comparison of the obtained values. As one can see, by applying the optimal control a strong reduction of the objective functional is achieved.

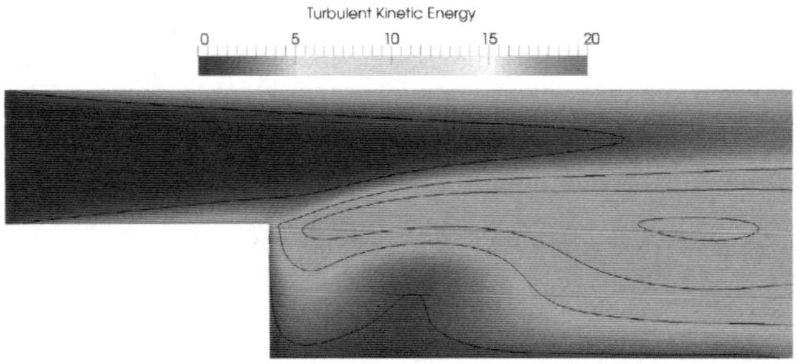

Figure 10. Turbulence decrease case, turbulent kinetic energy iso-lines and magnitude.

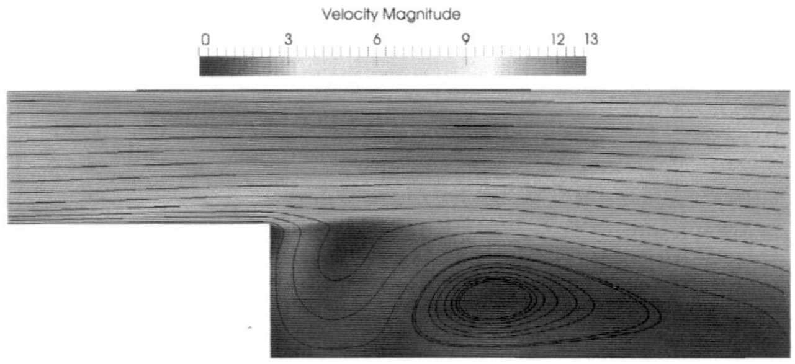

Figure 11. Turbulence decrease case, velocity streamlines and magnitude.

Table 2. Objective functional values computed for the three test cases of velocity matching, turbulence increase and decrease.

Test case	Velocity matching	Turbulence increase	Turbulence decrease
Reference	15.64	50.18	191.63
Optimal	2.42	21.06	35.16

4. Conclusion

In this work we studied the adjoint optimal problem for the RANS system closed with a two-equation turbulence model. The optimality system was derived for a distributed and a boundary control and two objectives were considered, namely a matching velocity and a matching turbulent kinetic energy. We studied the backward facing step problem with distributed control and three different objective definitions. The first is to control the value of the velocity field in a specific region and the other two to increase or decrease the turbulence intensity in the vortex region after the step. The results obtained have shown that the method is reliable and that the objective functional can be decreased even of one order of magnitude in this test case. For these reasons this approach could be of great interest for the study and optimization of industrial devices in which the flow is turbulent and CFD-RANS simulations can be performed on the system. More work needs to be done in order to test this method in practical applications where the flow pattern is more complex but these results on the backward facing step problem are very promising.

IOP Conf. Series: Journal of Physics: Conf. Series **796** (2017) 012008 doi:10.1088/1742-6596/796/1/012008

References

[1] Fröhlich J and Terzi D 2008 Hybrid les/rans methods for the simulation of turbulent flows *Prog. Aerosp. Sci.* **44** 349-377

[2] Sagaut P, Deck S and Terracol M 2006 *Multiscale and multiresolution approaches in turbulence* (Imperial College Press)

[3] Chinchuluun A, Pardalos P, Enkhbat R and Tseveendorj I 2010 *Optimization and optimal control: theory and applications* (Springer Optimization and Its Applications: Springer)

[4] Gunzburger M 2003 *Perspectives in flow control and optimization* (Advances in Design and Control: SIAM)

[5] Cerroni D, Manservisi S and Menghini F 2015 Numerical comparison of different solution methods for optimal boundary control problems in thermal fluid dynamics *Int. J. Mech.* **9** 1-10

[6] Zymaris A, Papadimitriou D, Giannakoglou K and Othmer C 2009 Continuous adjoint approach to the SpalartAllmaras turbulence model for incompressible flows *Comput. Fluids* **38** 1528-38

[7] Stück A and Rung T 2011 Adjoint rans with filtered shape derivatives for hydrodynamic optimisation *Comput. Fluids* **47** 22-32

[8] Manservisi S and Menghini F 2016 Optimal control problems for the navier-stokes system coupled with the k-ω turbulence model *Comput. Math. Appl.* **71** 2389-06

[9] Manservisi S and Menghini F 2016 Numerical simulations of optimal control problems for the reynolds averaged navier-stokes system closed with a two-equation turbulence model *Comput. Fluids* **125** 130-43

[10] Wilcox D 2008 Formulation of the k-omega turbulence model revisited *AIAA Journal* **46** 2823-38

[11] Aulisa E, Manservisi S and Seshaiyer P 2006 A computational multilevel approach for solving 2D Navier-Stokes equations over non-matching grids *Comput.Methods in Appl.Mech.Eng.* **195** 4604-4616

[12] Gabriel E et al. 2004 Open MPI: Goals, Concept, and Design of a Next Generation MPI Implementation *11th European PVM/MPI Users' Group Meeting* 97-104

[13] Balay S et al. 2015 PETSc Users Manual *Argonne National Laboratory* ANL-95/11 - 3.6

[14] SALOME platform developed by CEA-EDF, http://www.salome-platform.org/

IOP Conf. Series: Journal of Physics: Conf. Series **796** (2017) 012009 doi:10.1088/1742-6596/796/1/012009

CFD analysis of turbulent heat transfer and thermal striping phenomena in T-junctions with liquid sodium

P Ferrara[1] and P Di Marco[1]
[1]Dipartimento Dipartimento di Ingegneria dell'Energia, dei Sistemi, del Territorio e delle Costruzioni, Università di Pisa, Largo L. Lazzarino 1, 56100 Pisa, Italy.
E-mail:paolo.ferrara@ino.it

Abstract. This paper describes the methodology adopted to perform a CFD analysis with the aim to study thermal striping in T-junctions with liquid sodium; in particular, a portion of the Phenix reactor's secondary circuit has been taken into account, by implementing a detailed 3D model of both fluid and solid domain; RANS and LES turbulence models are used and their results are compared, then frequency analysis of temperature fluctuations has been performed. One of the main results of this analysis is that LES turbulence models are not yet completely reliable in evaluating.

1. Introduction

"Thermal mixing" is one of the causes of thermal fatigue failure in nuclear power plants. Thermal mixing characterizes the phenomenon where hot and cold flow streams join, mix and result in temperature fluctuations. Temperature fluctuations cause cyclic thermal stresses and, therefore, they may lead to fatigue cracking of the pipe wall. Thus prediction of thermal field in piping system is an important aspect from the nuclear reactor safety point of view. In order to assess the structural strength, stability and life of such T-junctions, it is essential to know some data with high precision: (i) magnitude of the temperature fluctuations, (ii) characteristic frequencies of temperature fluctuations, (iii) regions of pipe wall where the temperature fluctuations are higher, (iv) attenuation of the temperature fluctuations in the boundary layer near the pipe wall [1]. High thermal fatigue cycles are caused by "thermal striping", that is a complex thermal hydraulics phenomenon, which generates random fast temperature fluctuations, originating from the incomplete mixing of hot and cold jets of fluid sodium, in the vicinity of adjoining structural wall surface. Thermal striping occurs at a few locations in the hot and cold sodium pools in the reactor assembly, predominantly on the core cover plate of control plug and at mixing 'Tee' junctions in the secondary sodium pipelines. The range of frequency of oscillations under thermal striping is found to be 0.1÷10 Hz [2]. A sodium leak detected in a Tee junction of the fast breeder reactor Phenix was the starting point of an IAEA international benchmark on thermal fatigue [3]. Hot sodium flow is injected by a small pipe in a large cold pipe, the temperature fluctuations due to the mixing downstream of the Tee have initiated a crack. Nowadays CPU resources are much higher than 10-15 years ago and new turbulence models have been developed, so it is interesting to assess them in the given problem.

Therefore a portion of the Phenix reactor's secondary circuit has been taken into account, by implementing a detailed 3D model (figure 1) [3]. The purpose of this study is: (i): to build a computational grid that allows to reach a good compromise between reduced computational time and high quality results; (ii): to evaluate turbulent heat transfer from liquid sodium to solid wall; (iii): to study the time-averaged behaviour of the fluxes in the mixing region for this specific case, evaluating thermal striping phenomena; (iv): to perform very accurate calculations, in order to compare numerical results with experimental data coming from the real nuclear facility, these data are listed in IAEA TECDOC 1318, [3]. Simulation results must be compared with experimental data, that are detected in real Tee by using 15 thermo-couples (called TCs for the sake of simplicity); but, in present work, only 5 of them are considered, because they are the only ones that detect some temperature fluctuations with respect to an unperturbed condition (i.e.: $T_{ref} = 340°C$).

Our simulations will be performed by using both RANS and LES turbulence models: the goal of RANS calculations is mainly to find: (i) the best grid and (ii) the best turbulence model to perform the final step of the analysis, so, at this point, we just want to make a *relative* comparison between results and it is crucial to do it as fast as possible. In other words: at this point it not important if the results of the such calculations are or not close to experimental data. On the contrary, when we perform the final step of the analysis, (i.e. a LES calculation with the best grid and the most suitable LES model), the goal is to obtain temperature data comparable with those contained in IAEA TECDOC 1318.

2. Fluid model geometry

The first step of a CFD calculation is the creation of a geometry, that allows to handle accurately the case to study; in particular, the T-junction of [3] can be modelled as in figure 1. The reference system is oriented as described in [3]: X axis is vertical and coincident with main pipe axis in the mixing region, Z axis is horizontal. Different colours of surfaces in figure 1 underline that those surfaces can be handled separately during the definition of the CFD problem, so that different boundary conditions can be assigned in different zones of the fluid domain. In figure 1, X axis is vertical and Z axis is horizontal: they lie on the symmetry plane.

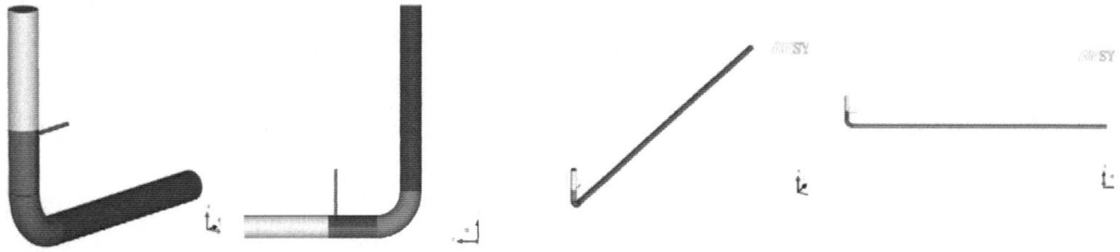

a b c d

Figure 1. Geometry for CFD calculation: a: isometric view; b: lateral view. Geometrical model with long horizontal pipe: c: isometric view; d: lateral view.

When performing RANS simulations, geometrical model of figure 1.a and 1.b suits our purposes, whereas, when performing very accurate LES calculations, [4], more realistic initial and boundary conditions are needed, in particular at the inlet, because, inside the mixing region, turbulent flux should be completely developed. If we use ANSYS CFX 12, it is not possible to simply add a random velocity distribution to the mean fluid velocity at the inlet (i.e.: to the laminar inlet conditions). Therefore, for the last calculation, the upstream region of the main pipe has been modified, by increasing the horizontal tube length to let the turbulent flow be fully developed. To estimate such length increment, we proceed in this way: it is known that the entrance length (here called $L_{entrance}$) needed to reach a fully developed turbulent flow (from an uniform flow at the pipe inlet) can be evaluated as follows [5]: $L_{entrance} \simeq 4.4 \cdot D \cdot (\mathrm{Re})^{1/6}$, where D is the hydraulic diameter of the pipe; thus, if $\mathrm{Re} \simeq 6.8 \cdot 10^6$, as in our case, then: $L_{entrance} \simeq 55D$, therefore the horizontal pipe should be at least 28m long, instead of 4m. Hence, to be more conservative, we decided to use a 32m long horizontal pipe (i.e. 8 times the original length), in order to let the turbulent flow fully develop, starting from laminar conditions at the "new" inlet, that is now far downstream. "Laminar conditions", in this case, means: uniform velocity profile and constant temperature along the inlet surface.

3. Computational grids of fluid and solid domain for RANS and LES calculations

The fluid domain of mixing Tee is meshed with Hexa-type elements, in order to better control the mesh quality. ANSYS ICEM-CFD allows to create Hexa meshes, using "blocks", namely splitting the entire domain into simpler parts, each of them can be seen as a *big*-Hexa-type element, so that this can be easily split into many *smaller* Hexa-type elements. In Phenix Tee case, the fluid volume is divided into three parts, as one can see in figure 2. The first goal of the present work is to create a global mesh, that may lead to a compromise between good quality calculation results and low computational cost, thus saving CPU time. To reach this goal many attempts were done, in particular to assess the parameter y^+ in the mixing region, by varying the parameters "*n*" (number of nodes along an edge) and "*r*" (ratio between consecutive nodes), for near-wall blocks.

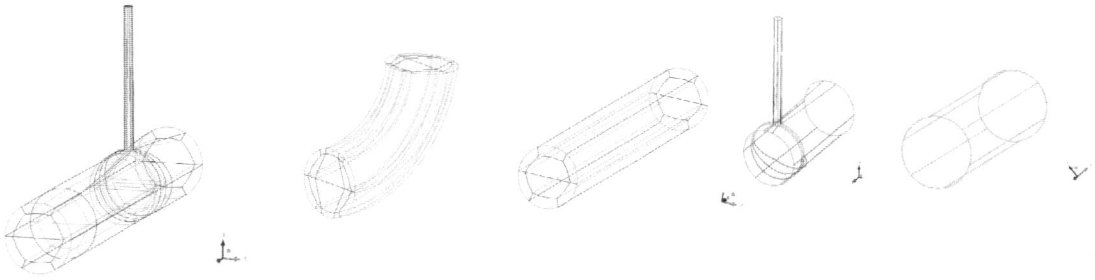

a b c d e

Figure 2. Mesh blocks: a: mixing region: vertical pipe and branch pipe. Upstream region: b: elbow; c: horizontal pipe. Solid wall blocks: d: injection; e: downstream zone.

In other words, mesh refinement in circumferential and axial directions of main pipe can be coarser than refinement in radial direction, but, at the same time, it is needed to satisfy requirements for computational grids to be used in LES calculations. In particular, y^+ should be close to 1 in near-wall region, al least in the zone of interest, i.e. the mixing region, near the OXZ plane, after the injection. In order to improve simulation results, it is necessary to reach a good *mesh quality* [6] in the mixing region. Conduction inside pipe solid wall is calculated, but only in the region downstream hot and cold fluids intersection, i.e. only where such analysis is really needed, so that a more accurate analysis of the T-junction problem can be performed, thanks to the Conjugate Heat Transfer (CHT) tool [7]. Main features of these grids, y^+ distribution and results of preliminary RANS calculations are summarized in [8,9]: looking at figure 7 of [8], that is referred to the finest grid for RANS and LES calculations, we can see that average value of y^+ is essentially around unity, in proximity of the junction. Besides, main features of final grids (called A1, A2 and A3) can be found also in this paper, in table 2. In RANS and LES run, thermal conduction inside pipe solid wall has been calculated (in the mixing region) using a 3D domain of the AISI 304 stainless steel wall. In order to guarantee a good mesh quality, the solid volume is divided into two sub-volumes: injection zone and downstream zone, as one can see in figure 2.d and figure 2.e; see also [8,9].

4. RANS Calculations

4.1 RANS Calculations planning

RANS calculations are divided in 3 groups: (i) the first one is aimed to compare the results of the calculations performed with 16 grids (described in [8]), in order to build the final grids for LES calculations (called A1, A2 and A3); (ii) the second one is performed by using the best grid of the previous group, in order to compare different turbulent models: Reynolds Stress Model (RSM), Shear Stress Transport (SST) and k–ε method with standard coefficients, (one low-resource consuming RANS calculation is performed for each of the final grids); (iii) the third one is constituted by a single calculation performed with the final grid to be used for LES runs, in order to describe the mean flow: this is useful also to find a good initial condition for LES calculations.

It is well known that an unsteady RSM approach should be useful to verify the natural steadiness of the phenomenon, but we also know that this capability is better verified by LES models, so we decided to limit the use of RANS simulations to steady-state cases, in order to evaluate only the average temperature profile. The computational domain of the first group is constituted only by the fluid mesh, in order to save CPU time, whereas the computational domain of the other 2 groups are constituted by both solid and fluid mesh, in order to provide better results. Only the last RANS calculation (group (iii)) has been performed by using the 32m upstream horizontal tube, see figure 1.c and 1.d.

4.2 Main features of RANS calculations

In first RANS calculations, non-dimensional wall distance (y^+) is evaluated for a steady-state case of a simplified model, where solid wall of the pipe is not included in the model, but the action it produces near wall in sodium temperature profile is taken into account, by assessing the wall heat transfer coefficient on the cylindrical surfaces of the fluid domain. Main features of the RANS calculations may be summarized as follows: (i) RANS turbulence models: steady-state RSM, SST and k–ε method with standard coefficients (see sec. 4.1); (ii) main inlet has a mass flow of 800kg/s and a temperature of 340°C; (iii) branch-pipe inlet has a mass flow of 7kg/s and a temperature of 430°C; (iv) outlet has a mass flow of 807kg/s; (v) cylindrical surface of fluid domain is considered as a solid wall, with heat transfer coefficient (h) of 16 W/m^2 K and outside temperature of 25°C.

Material properties for liquid metal and solid are considered constant and they are extrapolated from data of International Nuclear Society Council (INSC) at an intermediate value between hot and cold sodium temperature: for first calculations it is chosen to use the mean value between 340°C and 430°C, that is 385°C. Convergence criteria for each steady-state calculation is that the normalized Root Mean Square (RMS) values of residuals must be under 10^{-4}. Turbulent heat flux is modelled by using the so called "thermal energy" model implemented in ANSYS CFX 12.1, see [10]. Concerning numerical schemes features, ANSYS-CFX adopts an "unstaggered" grid scheme, RANS equations are discretized by using a first order backward Euler scheme and the convective term is modelled by using a first order "upwind" scheme. Sometimes first order schemes are not very accurate, but, in our case, results of RSM simulations are very close to experimental data (see figure 3.a), thus this approximation suits our purposes, allowing also to save CPU time. For further details, see [10].

5. LES Calculations

5.1 LES Calculations planning

First of all, a sensitivity analysis has been performed by comparing different computational grids and different subgrid models, only to make only a *relative* comparison between grids and numerical models, even if the calculated values could result still far from experimental data. On the contrary, the final step of LES calculation campaign consists of an extension of simulation time of the best case (that is: the best computational grid with the best subgrid model), in order to collect enough data for *statistics*, to make a good comparison between calculated and experimental data. Main features of computational grids for sensitivity analysis can be found in table 2 and in [9].

At the end of these simulations, the best subgrid model between Dynamic, WALE and Smagorinsky model can be chosen, so that its simulation time can be *extended* up to 60s: to be more precise, the final LES calculation must be *repeated* from the beginning, because the initial condition for the final run should be as accurate as possible. The long duration of simulation time is absolutely needed, because: (i) it allows the thermal wave to completely cross the solid wall of the pipe in radial direction, that is: about 12s, (because the TCs are placed outside such wall), therefore the total time extension of the simulation must be much greater than 12s; (ii) it is necessary to collect enough data for *statistics* to make a good comparison between calculated and experimental data; besides, it is necessary to *cut* the initial transient from the analysis, i.e. at least the first 12s, if looking at figure 9. So, in our case, the transient due to the approximate initial conditions seems to be much greater than *one* residence time (i.e. the amount of time to cut, as recommended in literature [11]).

Therefore, it is useful to define a simple criterion in order to distinguish between such *initial transient* and the *stable solution*. It was chosen to apply this criterion to *each* TC: (i) the time-averaged temperature value has been evaluated, starting from 57 different initial instants of time, thus obtaining 57 values (named: "value 1", "value 2", ... , "value 57"), calculated as follows: "value 1" is the time-averaged temperature between "simulation time = 1s" and "simulation time = 60s" (i.e. the last one), for a given TC; "value 2" is the time-averaged temperature between "simulation time = 2s" and "simulation time = 60s", for a given TC; and so on. (ii) Such mean values must be normalized by dividing them for an appropriate mean temperature value, related to a given TC (because this value is different for each TC).This is called *asymptotic* value and it can be approximated with the mean temperature value between the 57 values previously described ("value 1", "value 2", ... , "value 57"). (iii) The time interval of the *stable* solution could be defined as the time interval where fluctuations of normalized mean temperature are less than 0.5% with respect to the *asymptotic* value; (iv) To be conservative, the longest initial transient between those relative to the 5 TCs should be chosen.

At this point temperature fluctuations profile should be correctly evaluated, thermal striping phenomena occurring in T-junctions with liquid sodium can be accurately studied and also frequency analysis can be correctly developed.

5.2 Main features of LES calculations

Main features of LES calculations in present work may be summarized as follows: (i) LES subgrid model is implemented in ANSYS CFX, [10]: classic Smagorinsky, Dynamic or WALE: they are all used for sensitivity analysis; then, the best one is chosen for the final run; (ii) total simulation time: 5s (for sensitivity analysis) or 60s (for the final simulation), as it was explained in the previous section; (iii) boundary conditions are the same used in RANS calculations. Time-step size should be less than 0.00005s, to maintain Courant Friedrichs Lewy (CFL) number below 1, in order to capture small scale phenomena, but, if this value would be adopted for the entire transient, computational cost would be huge, so this value has been used only at the beginning of each transient, that is: between 0s and 0.1s, whereas time-step has been increased to 0.0001s between 0.1s and 0.3s, then it has been increased to 0.0002s until 12s and, finally, time-step has been increased up to 0.001s until the end of the simulation. Doing so, CPU time for the final calculation has been over 1 month. Initial condition for *each* LES calculation (*but the last one*) is the result of a steady-state RANS k–ε calculation performed with the *same grid* of the corresponding LES calculation, in order to avoid interpolation errors at the beginning of the LES transient; whereas, initial condition for the *last* LES calculation is given by the result of a calculation performed using the RANS turbulence model that better simulates the mean flux of the mixing Tee. Here the convective term is modelled by using a central difference scheme. The unsteady term in LES calculations is modelled by using a second order backward Euler scheme. Convergence criteria for each time-step is that the normalized RMS values of residuals must be under 10^{-4}. *Experimental* data of temperature are taken every 0.001s, therefore, *computed* temperature data must be extrapolated with the same frequency (i.e. 0.001s), in order to make a correct comparison.

In sec. 1, it has been already remarked that the comparison between simulation results and experimental data should be done in correspondence of 15 TCs, even if, in present work, only 5 of them are considered, because they are the only ones that detect some temperature fluctuations with respect to an unperturbed condition (i.e.: T_{ref} = 340°C) [3]; their position is reported in table 4 of [8]. Besides, an "*interface point*" (called "Interf TC") is defined in correspondence to each TC, namely a point that lies on the interface between fluid and solid, in such a way that it has the same X and Y coordinates of the corresponding TC, but Z = 0.247m at the interface, whereas: Z = 0.254m, for each TC, in fact TCs lie on the external surface of the pipe wall.

6. Results and discussion

In our case, the best RANS model to simulate the *mean flux* inside T-junction is a steady-state RSM, because it provides mean temperature values that are very close to the experimental data (figure 3.a), in fact the difference between experimental and calculated data is less than 1°C over about 400°C, for each TC but the first one. So RANS RSM turbulence model seems to be the best initial condition (IC) to start the final LES calculation, even if it has a very high computational cost, when compared to other RANS models. In RANS and LES calculations, material properties can be considered constant with temperature, in particular this approximation is valid for sodium density, because Richardson number is less than 10^{-3} inside the mixing region: it means that buoyancy effects are negligible, according to the criterion suggested by Jackson for liquid metals in ascending pipes [12]; so, in this case, forced convection is dominant. In fact, if we consider a steady-state k–ε case, where density is variable with temperature, we can notice that density variations are negligible, because their maximum variation (normalized with medium density) is less than 2.5% (see figure 4.a). Now main results of a steady-state calculation performed with RSM turbulence model are briefly summarized and, when necessary, they are compared with the corresponding results obtained with LES subgrid turbulence models.

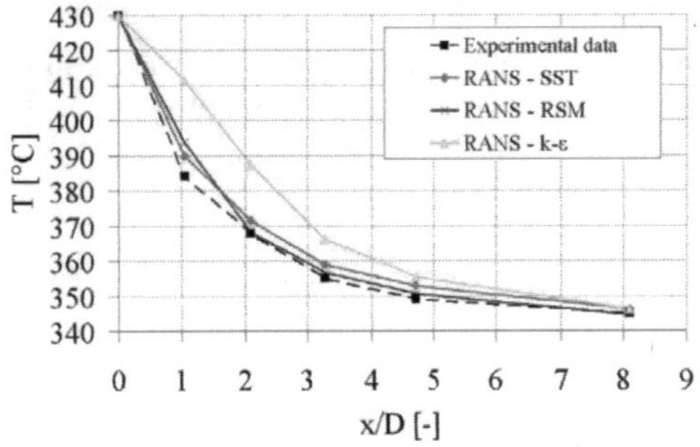

a

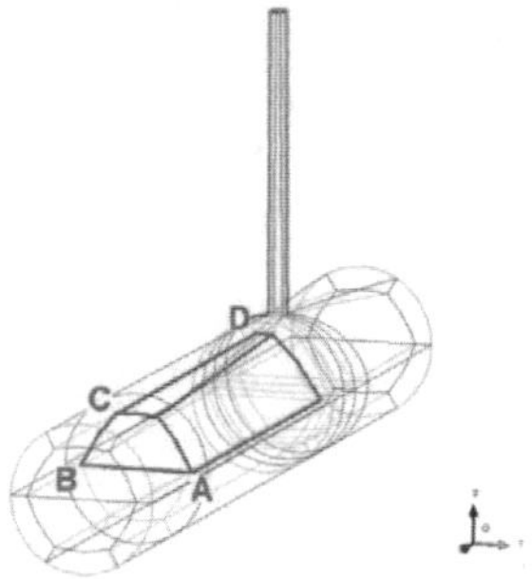

b

Figure 3. a: Comparison between 3 RANS turbulence models: RSM, SST and k–ε; b: Main inner block in mixing region, useful to summarize most important features of final grids, listed in Table 2.

Table 1. Classification of mixing flow types as a function of M_R.

Jet type	M_R value
Wall jet	$90 < M_R$
Re-attached jet	$20 < M_R < 90$
Turn jet	$2.5 < M_R < 20$
Impinging jet	$M_R < 2.5$

Table 2. Main features of the 3 final grids in the mixing region; **(*):** n and r are relative to the near-wall block above the segment CD.

Case name		A1	A2	A3
N° of elements (x 10^6)		2.25	3.14	2.60
n (*)		28	30	29
r (*)		1.25	1.25	1.25
N° of elements in main inner block, see figure 3.b	AB	24	24	24
	BC	24	40	34
	CD	147	168	147

Comparing figure 4.b and figure 4.c, it is possible to notice a non-negligible temperature difference between the solid-liquid interface and the outside wall (about 10°C for corresponding axial and azimutal positions), because this thin wall made of stainless steel (only 0.007 m thick) acts as a damper for internal temperature variations, being them either in time or in space. Such variations are damped by a quasi-exponential law through the wall thickness: it means that TCs, placed outside wall thickness, are capable of detecting only strongly attenuated temperature fluctuations with time. In fact, for the final LES simulation, the time evolution of temperature in 5 TCs is shown in figure 9.a, whereas the same variables in 5 corresponding "interface points" are plotted in figure 10, showing how big the damping can be. In fact, the peak-to-peak temperature difference can be greater than 100°C at the solid-fluid interface, whereas it rarely becomes greater than 5°C outside solid wall.

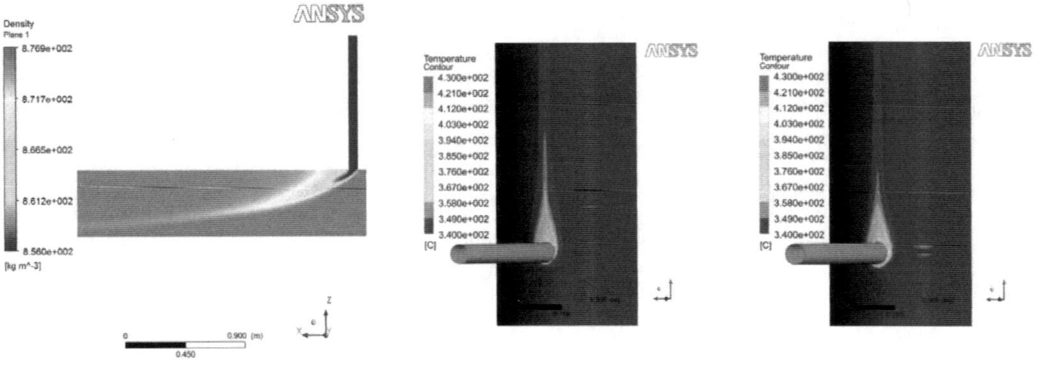

a b c

Figure 4.a: Sodium density in OXZ plane; b: Temperature on the interface between solid and fluid domain; c: Temperature outside solid wall.

According to [13], the pipe section in the mixing region can be divided in 4 parts (see figure5.a): region "A", called "*pure tap water*", where RMS temperature values are low; region "B", called "*mixing zone*", where there is a mixture with high RMS temperature values; region "C", called "*pure DI*", adjacent to region "B", where RMS temperature values are low, (similar to region "A"); region "D", called "*separation region*", placed in the region close to the branch pipe: it contains 2 counter-rotating vortexes which rule the recirculation region. Such vortexes can be clearly observed in Phenix Tee case, see figure 5.b.

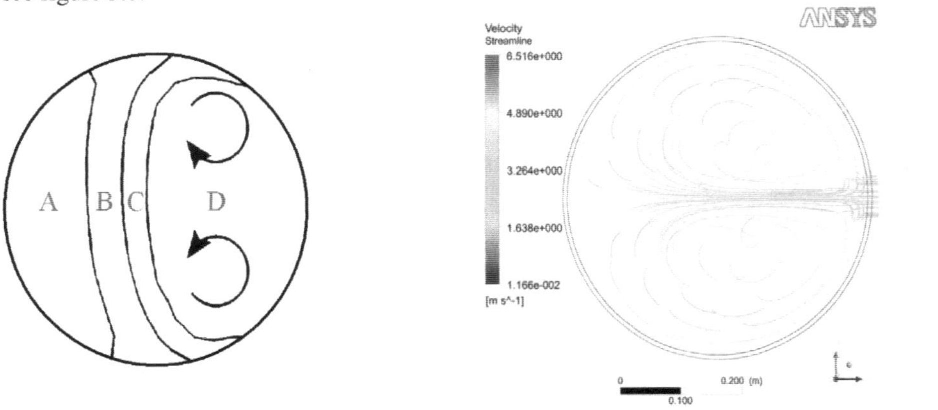

a b

Figure 5.a: the mixing region can be divided into 4 parts; b: streamlines projected on OYZ plane; they start from a geometrical OYZ plane located at X = 0.9 m.

According to [14], if we consider a T-junction with 90° bend, mixing flow depends on the value of a non-dimensional parameter, called "momentum ratio", defined as:

$$M_R = \frac{\rho_m U_m^2 \cdot (D_m \cdot D_b)}{\rho_b U_b^2 \pi \cdot (D_b / 2)^2}, \tag{1}$$

where ρ_m and ρ_b are, respectively, "main" and "branch" flow densities, U_m and U_b are the mean flow velocities of "main" and "branch" fluxes, whereas D_m and D_b are the hydraulic diameters of "main" and "branch" fluxes. Different values of M_R define different types of mixing flow, as it is shown in table 1; one can have either a "*wall jet*", if the branch flux remains attached to the wall of the main pipe, or a "*re-attached jet*", if the branch flux first separates from the wall of the main pipe and, then, it re-attaches to it, or a "*turn jet*", if the branch flux separates from the wall of the main pipe and it remains close to the axis of the main pipe, or an "*impinging jet*", if the branch flux hits the wall on the opposite site of the injection zone. Using equation (1), where: $\rho_m \approx 876.9$ kg/m^3, $\rho_b \approx 856$ kg/m^3, $U_m \approx 4.6$ m/s, $U_b \approx 2.1$ m/s, $D_m = 0.494$ m and $D_b = 0.068$ m, it is found that roc.≈ 43, so, according to table 1, flux type is in the range of "*re-attached jets*". Anyway, we have to remark that the threshold values used in table 1 are referred to the case of a T-junction without any elbow, according to [14], whereas the T-junction considered in IAEA benchmark shows an elbow just downstream the secondary flux injection; therefore such threshold values can only be considered as a first reference to distinguish between different types of mixing regions, in fact they could be shifted up or down, by the presence of an elbow. So it is difficult to predict exactly the type of mixing flow, even if it should be closer to a *turn jet*, because the presence of an elbow bends the streamlines through the extrados of the curved pipe, therefore such streamlines are deviated to the opposite side of the injection zone, so they are likely to push also secondary flow away from the wall that is close to the branch pipe. Besides, the calculated flow can have different features depending on different turbulence models adopted, because of their different capacity to evaluate mixing, in fact, mean flux calculated by using RANS turbulence models behaves as a *turn jet* (figure 6.a, figure 7.a, figure 7.c, figure 8.a), [14], because the hot plume arranges near the main pipe axis, downstream the secondary injection. Therefore the mixing region gets away from the main pipe wall and its shape is very massive and almost cylindrical: this is due to the low capacity of RANS turbulence models to evaluate mixing phenomena between hot and cold fluids (figure 7.a, figure 7.c). We remind that "mixing zone" can be defined as an iso-surface with a temperature equal to T_{min} plus 5% of maximum ΔT between hot and fluid, therefore, in this case: iso-surface with T ≈ 345°C. This can be useful to better understand figure 7.a and figure 7.b. On the contrary, instantaneous flow calculated by using LES subgrid models behaves as a *wall jet* (figure 6.b,figure 7.b, figure 7.d, figure 8.b, [14], because of the greater capacity of such models to evaluate mixing phenomena between hot and cold fluids, in fact the mixing region is located very close to the solid wall and its shape is very diffused and constituted by many small structures (figure 7.b, figure 7.d). So, it seems that neither RANS nor LES simulations can reproduce correctly the shape of the mixing region, because, as it was anticipated above, the real behaviour of the mixing region should be a way in the middle between the results found with different turbulent models, that is: it should be closer to a "*re-attached jet*".

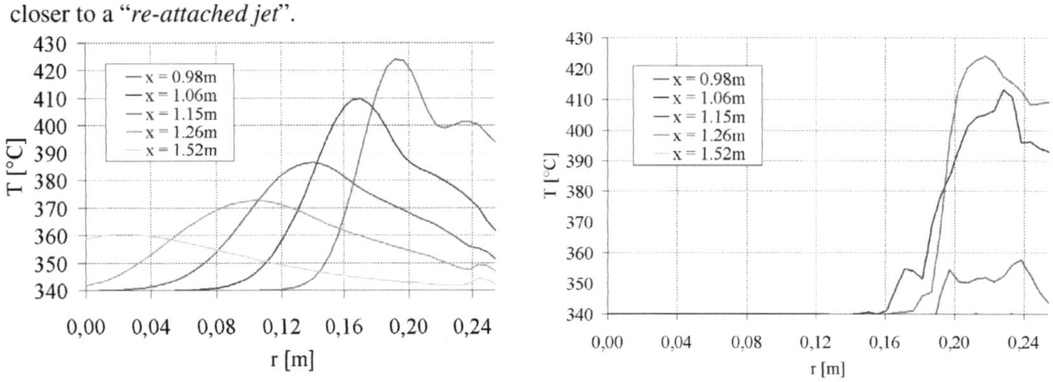

Figure 6. Temperature profile versus pipe radius for 5 TCs: a: RANS RSM turbulence model. Here mixing zone penetrates into the central part of the main pipe (see green curve); b: LES Dynamic subgrid model, medium values: here mixing zone is only in near-wall region.

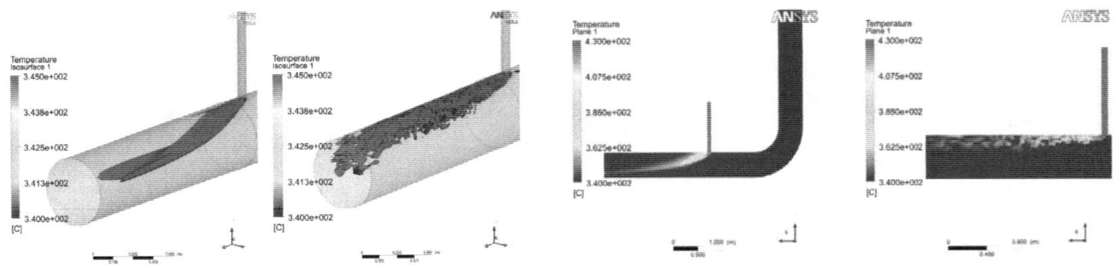

a b c d

Figure 7. Mixing zone: a: RANS RSM turbulence model; b: LES Dynamic subgrid model, after 12s. Temperature in OXZ plane: c: RANS RSM turbulence model; d: LES Dynamic model, after 12s.

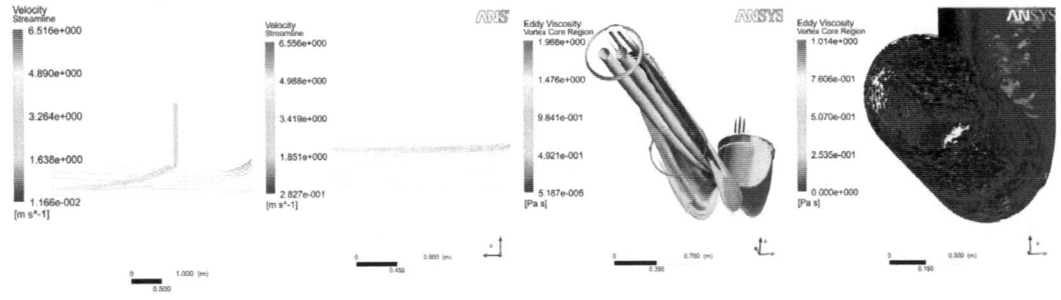

a b c d

Figure 8. Streamlines in OXZ plane: a: RANS RSM turbulence model; b: LES Dynamic model, after 12s.c:Vortices shape, RANS RSM turbulence model; Q-criterion [10]: Q = 22 s^{-2}, time-averaged values; d: Vortices shape, LES Dynamic subgrid model; Q-criterion: Q = 68 s^{-2}, instantaneous values after 12s.

In figure 6.a, referred to a RANS simulation, we can see that the bright green line, referred to TC 6 (i.e.: the TC placed farther from the secondary injection) shows a temperature increment of about 20°C close to the main pipe axis (that is: where the radius "r" is close to 0 m); it means that, in such region, a hotter plume of liquid sodium is flowing through the colder bulk of fluid, and it is another prove that the RANS models underestimate mixing phenomena. This does not happen in LES simulation, as it can be shown in figure 6.b, where the bright green line, referred to TC 6, lies in the

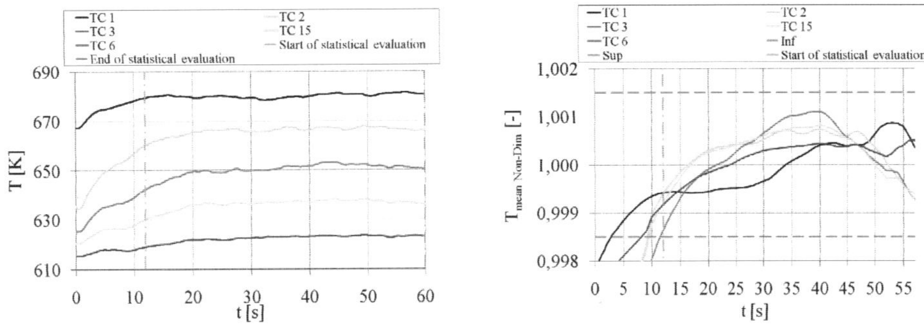

a b

Figure 9.a: Time evolution of temperature in 5 TCs: initial transient is at least 12s long; b: Visual representation of a simple criterion to distinguish between the initial transient and the stable solution: for further details, see sec. 5.1.

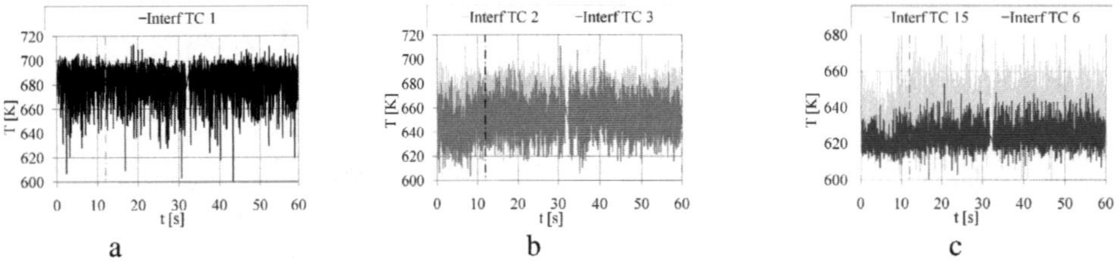

Figure 10. Time evolution of temperature in 5 interface points: a:Interf TC 1; b: Interf TC 2 and Interf TC 3; c: Interf TC 15 and Interf TC 6. Statistical evaluation: from t = 12s to the end.

horizontal axis, when the flow is close to the main pipe axis, therefore the mean temperature is very close to 340°C, so it means that mixing phenomena are confined in the near-wall region, as effectively happens in the real case. Also the vortices shape is very different between RANS and LES turbulence models, in fact, in the first case, time-averaged values are taken into account, so we can find 2 big counter-rotating vortices over the mixing region (figure 8.c), while, in the second case, instantaneous values are taken into account, so we can find many small eddy structures, that are very close to the solid wall (figure 8.d). The time-step extension is very important in the evaluation of the temperature fluctuations, because, if it does not satisfy the Courant-Friedrich-Levy condition, turbulence is damped, therefore temperature peaks can be attenuated, so time-step should be very small. On the other side, it is necessary to simulate a long transient in order to collect enough data for a correct evaluation of statistics, therefore, a good compromise must be found, if we want to fulfil 2 opposite requirements. Using the criterion described in sec. 5.1, the initial transient extension is about 12s (see figure 9.b), that is, the stable solution seems to start after about six residence times, in fact, after that time, mean normalized temperature values in correspondence to each TC are very close to 1; it means that the first 12s of this LES simulated transient shall not be taken into account to evaluate statistics. Besides, it is needed to wait that the thermal wave completely crosses the solid wall of the pipe, if we want to correctly evaluate the statistic values, because TCs are placed outside such wall.

Results of calculations performed with LES turbulence models *for sensitivity analysis* show some difference with experimental data, in fact RMS temperature fluctuations are under-estimated by some degrees, whatever is the grid and/or the LES subgrid model adopted. RMS temperature fluctuations are more sensitive to the extension of horizontal pipe, i.e.: 4m vs. 32m length (figure 12.a), than to grid refinement (figure 11.a) or to LES subgrid model (figure 11.b); in fact enough space shall be included in the upstream region of the mixing Tee to let the turbulent flow to be completely developed. In this problem, the finest grid (A3, see table 2) shows the best *qualitative* profile of RMS temperature fluctuations (figure 11.a), so it is chosen to perform the subsequent runs; besides, the

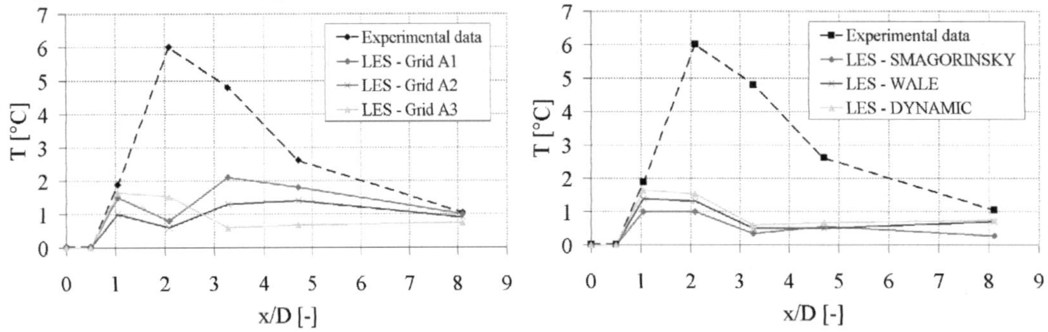

Figure 11. RMS values of temperature fluctuations: comparison between experimental data and calculated data: a: different grids; b:different LES subgrid turbulence models.

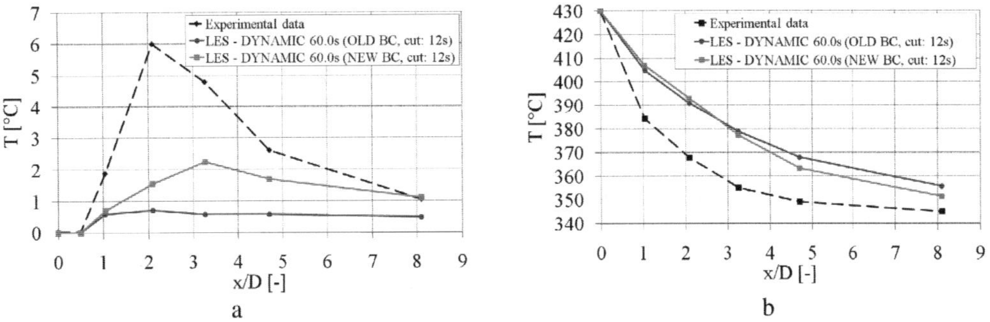

Figure 12. Different BCs: "OLD BC" means: simulation with 4m long horizontal pipe; "NEW BC" means: simulation with 32m long horizontal pipe. Comparison between experimental data and calculated data: a: RMS values of temperature fluctuations; b: mean temperature values.

Dynamic subgrid model appears to be the best one to correctly evaluate RMS temperature fluctuations, even if the results produced by the 3 subgrid models are very similar (figure 11.b). Looking at figure 12.a, it is useful to remember that RMS values of temperature fluctuations could be much higher, if we would consider also the initial transient, because, in that case, the solution would be *quickly* changing to evolve from ICs to quasi-stationary values, whereas we are only interested in to the stable solution. Unfortunately, the code provides data that are quite lower than experimental measurements; it means that RMS values of temperature fluctuations are *under-estimated* by the code, if compared with experimental data (see red line in figure 12.a); on the contrary, mean temperature values are *over-estimated* by the LES model of about 15°C, see figure 12.b. Anyway, we can observe that the adoption of a more correct boundary condition at the inlet, (called "NEW BC", i.e. simulation with a "32m horizontal pipe"), is useful to improve results of both RMS and mean values of temperature, with respect to the case with "4m horizontal pipe", (called "OLD BC"), i.e. the case with the non-correct inlet condition, (see, the blue line in figure 12.a and figure 12.b). Nevertheless, the differences with experimental data still remain appreciable, probably because of the following reasons: *first*: the extension of time-step is high in order to save CPU time, but it probably damps too much turbulent oscillations, *second*: it is possible that the LES subgrid model adopted is not capable of correctly modelling thermal striping phenomena, in particular: the fluctuations of turbulence structures; *third*: the grid resolution is very fine in radial direction, but it is still quite rough in axial and azimutal direction, (again, this choice has been taken to save CPU time).

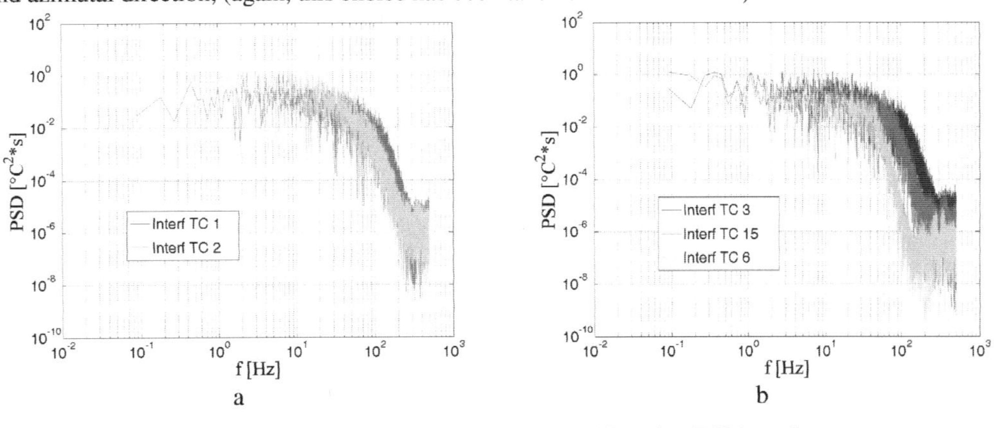

Figure 13. Frequency analysis: Power Spectral Density (PSD) vs. frequency in various positions at the interface between solid and fluid domains: a:Interf TC 1 and Interf TC 2; b: Interf TC 3, Interf TC 15 and Interf TC 6.

Frequency analysis of calculated data shows that the greatest energy content lies between 0.5Hz and 20Hz, which is a frequency interval where thermal striping may take place, being in good agreement with data detected in the real pipe; anyway frequency spectrum of calculated data does not show any characteristic frequency (see figure 13).

7. Conclusions

We have to remark that main results calculated in this work are quite close to those found during IAEA benchmark [3], demonstrating: (i): the reduction of computational cost, because we got results that are comparable with those obtained by participants to the IAEA benchmark performed in the 90', but only using a single powerful PC. (ii): the fact that some RANS turbulence models seem to be reliable in evaluating mean temperature values, even if the quality of such results is very sensitive to the chosen RANS model, (whereas each LES turbulence model provides more or less the same results, limiting the data dispersion). (iii): considering LES runs, the most important improvement in the temperature results has been provided by the adoption of a very long horizontal pipe in the upstream region of the mixing Tee, rather than using different grids or LES subgrid models. Nevertheless, the differences with experimental data are still appreciable, so we think that LES turbulence models should be used very carefully, when they are applied to complex industrial applications. Finally, further developments can be suggested to improve results quality: *first*: it is mandatory to enhance mesh quality (especially in azimutal and axial direction, in order to decrease elements aspect ratio) and *second*: it could be useful to develop a specific subgrid model for the given problem.

References

[1] Naik-Nimbalkar V S, Patwardhan A, W Banerjee I, Padmakumar G and Vaidyanathan G 2010 *Chemical Engineering Science* **65** 5901
[2] Chellapandi P, Chetal S C, and Baldev R 2009 *Nuclear Engineering and Design* **239** 2754
[3] IAEA-TECDOC-1318 2002 *Final report of a co-ordinated research project*
[4] Kuczaj A K, Komen E M, J and Loginov M S 2010 *Nuclear Engineering and Design* **240** 2116
[5] Munson B R, Young D F, Okiishi T H and Huebsch W W 2012 *Fundamental s of Fluid Mechanics* 6th ed John Wiley & Sons Inc;
[6] Menter F R 2002 *Report EVOL-ECORA-D01*
[7] Kuhn S, Braillard O, Niceno B, and Prasser H-M 2010 *Nuclear Engineering and Design* **240** 1548
[8] Ferrara P and Di Marco P 2011 *Proc. XXIX Congresso Nazionale UIT sulla Trasmissione del Calore* Torino
[9] Ferrara P and Di Marco P 2012 *Proc. Turbulence Heat and Mass Transfer 7* Palermo
[10] *ANSYS CFX-121* 2009 *User Manual* ANSYS Inc.
[11] Hu L-W and Kazimi M S 2006 *International Journal of Heat and Fluid Flow* **27** 54
[12] Jackson J D 1983 *International Journal of Heat and Fluid Flow* **4** 107
[13] Walker C, Simiano M, Zboray R, and Prasser H-M 2009 *Nuclear Engineering and Design* **239** (1) 116
[14] Hosseini S M, Yuki K and Hashizume H 2008 *International Journal of Heat and Mass Transfer* **51** 2444

Experimental investigation on the convective heat transfer enhancement in tubes with cross-helix profile wall corrugation

Luca Cattani[b], Fabio Bozzoli[a,b], Sara Rainieri[a,b], Giorgio Pagliarini[a,b]

[a] Department of Industrial Engineering, University of Parma,
Parco Area delle Scienze 181/A I-43124 Parma, Italy
[b] SITEIA.PARMA Interdepartmental Centre, University of Parma,
Parco Area delle Scienze 181/A, I-43124 Parma, Italy

E-mail address of the corresponding Author (Luca Cattani): luca.cattani@nemo.unipr.it

Abstract. Wall corrugation is a popular heat transfer enhancement technique since it acts as a disturbance source in the flow that significantly enhances the thermal performance of the tube section with a limited pressure drop augmentation if compared to other passive techniques, such as the ones based on insert devices. In the hereby presented study nine pipes characterised by a cross-helix type corrugation were tested: this kind of corrugation, that was obtained by rolling twice the same tube with two helical corrugations evolving along opposite directions, showed a performance that exceeded the single helix-type corrugation behaviour and moreover presented an earlier transition to unstable regime. In the analysis the effect of the cross-helix type corrugation profile on the forced convection heat transfer mechanism was experimentally investigated in the Reynolds and Prandtl number range 25÷1000 and 115÷150 respectively. In particular, the effect of both the corrugation depth and the corrugation pitch were analysed. The results were compared with other types of wall corrugation and with the predictions for the smooth tube in order to point out the achieved heat transfer enhancement.

Key words: Convective heat transfer enhancement, cross-helix corrugated tubes, passive heat transfer enhancement technique.

1. Introduction

Among passive heat transfer enhancement techniques, wall corrugated tubes certainly represent the most widely adopted solution in heat exchangers industrial applications. In fact, wall corrugation acts as a disturbance source in the flow that significantly enhances the thermal performance of the tube section by limiting the pressure drop augmentation if compared to other passive techniques, such as the ones based on insert devices [1]. The main augmentation mechanisms activated by wall corrugation are: the interruption of the thermal boundary layer, the flow mixing due to the onset of secondary swirl flow components, the increase of the heat transfer surface area and, above all, the early onset of a transitional turbulent flow regime [2,3]. Practically these phenomena are able to activate a significant heat transfer enhancement only if a critical condition can be reached [4-6]. Under this critical value, they normally don't present a significant heat transfer augmentation over the smooth wall behaviour. On the other hand, an important increase is achieved with the transition to the unsteady regime. This is a results of complex interactions between the core fluid and the boundary layer fluid through shear-layer self-sustained oscillations [7]. Guzman and Amon [8] investigated the transition from laminar to unsteady state for converging-diverging conducts by numerical simulations in Reynolds number range from 10 to 850. They stated that the transition occurred through a sequence of intermediate state of self-sustained periodic, quasi-periodic and finally aperiodic or chaotic regime.

Wang and Vanka [9] numerically investigated the heat transfer for flow through a periodic array of wavy passages: the Authors observed that above a critical value of Reynolds number of about 180 a transition appeared, and that it was characterised by a destabilization of thermal boundary layers replacing the fluid close to the wall with that of the core region thus producing a mechanism of heat and mass transfer enhancement. Niceno and Nobile [7] studying different types of wavy channels observed the appearance of the unsteady regime for a range of Reynolds number values between 80 and 200.

Garcia et al. [10] analysed the thermal behaviour of three types of enhancement technique based on artificial roughness: corrugated tubes, dimpled tubes and wire coil inserts. The Authors identified the onset of the transition to the unsteady regime in the range $100 < Re < 200$ for the wire coils inserts and $900 < Re < 1000$ for the other two types of roughness. The type of transition (e.g., smooth or sudden) and the Reynolds number value at which it occurs were found to be strongly depending on the roughness profile and geometrical characteristics.

Regarding wall corrugation, the helical profile is the most widely adopted in industrial applications, since these tubes can be quite easily manufactured by a continuous cold rolling process for both inward or outward corrugation. In this case, the main geometrical parameters that characterize the corrugation profile are: the tube diameter, the corrugation pitch, the corrugation depth, the number of helix starts and the corrugation profile [2]. For spirally inward corrugations, the corrugation depth was proved to be the most significant parameter that affects the critical Reynolds number: the critical Reynolds number decreased as the corrugation depth increased [4]. Moreover, for a given corrugation depth also the sharpness of the corrugation profile had a non-negligible effect [5]. This phenomenon was even amplified by a cross-helix type corrugation, for the first time investigated in literature in [11]. This kind of corrugation, that was obtained by rolling twice the same tube with two helical corrugations evolving along opposite directions, showed a performance that for a given corrugation depth exceeded the single helix-type corrugation behaviour and presented an earlier transition to unstable regime [11].

In the present investigation the effect of the cross-helix type corrugation profile on the forced convection heat transfer mechanism is experimentally further investigated in the Reynolds and Prandtl number range 25÷1000 and 115÷150 respectively. In particular, the effect of both the corrugation depth and of the corrugation pitch is analysed, by pointing out the effect of heat transfer enhancement and of promotion of the transition to unstable regime.

2. Tubes Geometry, Experimental Setup and Data Processing

The tubes under test show a wall corrugation with a cross-helix profile and they belong to the general category usually referred to as spirally enhanced tubes. In the present investigation nine tubes, having different corrugation pitch l and depth e (as specified in table 1) were tested.

Table 1. Corrugated tubes characteristics.

	l (mm)	e (mm)	l/D	e/D	Ω
Tube No. 1	13	0.6	0.928	4.29×10^{-2}	1.98×10^{-3}
Tube No. 2	13	0.8	0.928	5.71×10^{-2}	3.52×10^{-3}
Tube No. 3	13	1.0	0.928	7.14×10^{-2}	5.49×10^{-3}
Tube No. 4	18	0.6	1.29	4.29×10^{-2}	1.43×10^{-3}
Tube No. 5	18	0.8	1.29	5.71×10^{-2}	2.54×10^{-3}
Tube No. 6	18	1.0	1.29	7.14×10^{-2}	3.97×10^{-3}
Tube No. 7	29	0.6	2.07	4.29×10^{-2}	0.89×10^{-3}
Tube No. 8	29	0.8	2.07	5.71×10^{-2}	1.58×10^{-3}
Tube No. 9	29	1.0	2.07	7.14×10^{-2}	2.46×10^{-3}

They are obtained by embossing a smooth tube made of stainless steel having an internal diameter D of 14 mm and a wall thickness s of 1 mm. In Table 1 also the dimensionless pitch and corrugation depth are specified together with the corresponding severity index Ω, defined as $e^2/l \cdot D$ according to Withers and Habdas [12]. Two representative tubes are shown in figure 1.

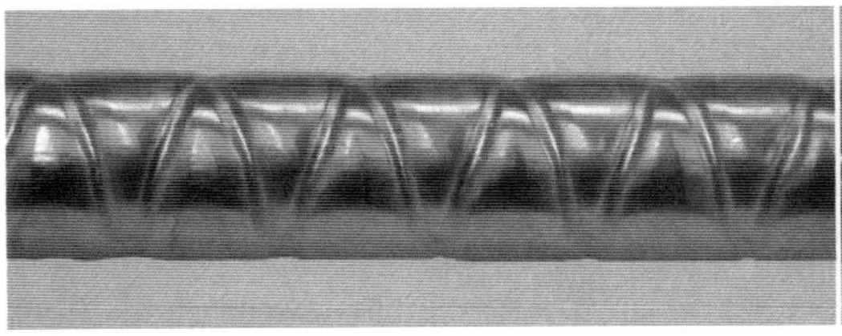

Figure 1a: Tube No.1.

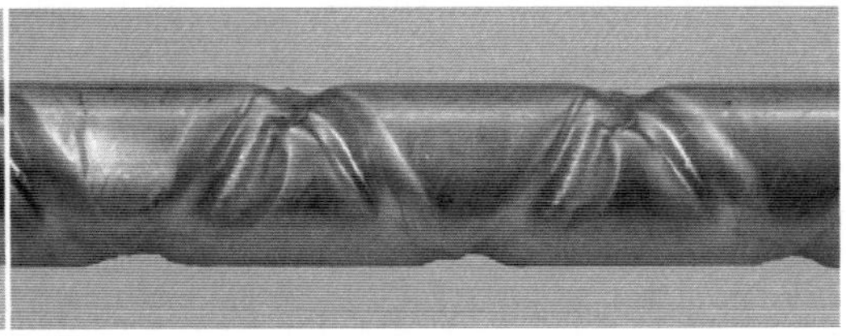

Figure 1b: Tube No.9.

The tubes were tested by using the experimental set-up and data acquisition system sketched in figure 2. The working fluid was conveyed by a volumetric pump to a holding tank, and it entered the test section equipped with stainless-steel fin electrodes, which were connected to a power supply, type HP 6671A. This setup allowed the investigation of the heat transfer performance of the tube under the prescribed condition of uniform heat flux generated by the Joule effect in the wall. The heat flux provided to the fluid was selected to make the buoyancy forces negligible compared to inertial ones for the fluid velocity values investigated here. The tested section was inserted horizontally in a loop completed by a secondary heat exchanger, fed with city water, to keep the working fluid temperature constant at the test section inlet. The heated section was preceded by an unheated development section of about 1 m. To minimise the heat exchange with the environment, the heated section was thermally insulated. The inlet and the outlet fluid bulk temperatures were measured with type-T thermocouples. The bulk temperature at any location in the heat transfer section was then calculated from the power supplied to the tube wall assumed distributed uniformly per unit length over the heat transfer surface area, decreased by the heat losses through the insulation. Volumetric flow rates were obtained by measuring the time needed to fill a volumetric flask placed at the outlet of the test section.

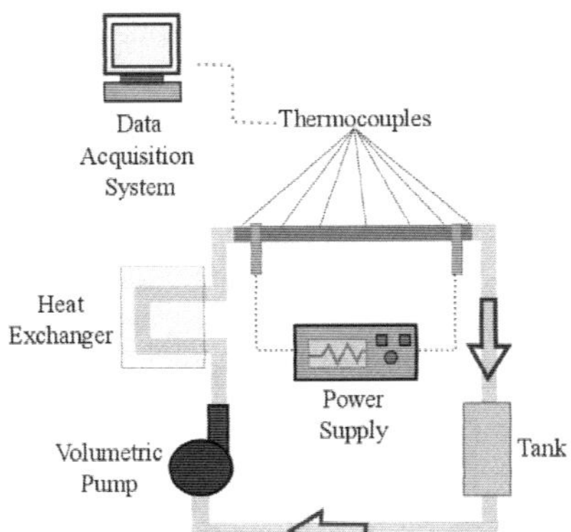

Figure 2. Sketch of the experimental setup.

The thermal entrance problem under the uniform wall heat flux boundary condition was then studied by using mixtures of water and ethylene glycol. In the temperature range characterizing the experimental conditions, the fluid Prandtl number range was 115÷150 while the effect within the boundary layer of the variation of the fluid properties with temperature was considered to be negligible.

The experimental set-up enabled to measure the local Nusselt number along the axial coordinate and its average values along the heated section as follows:

$$Nu_x = \frac{h_x \cdot D}{\lambda} \tag{1}$$

$$Nu = \frac{1}{L_H}\int_0^{L_H} Nu_x dx \tag{2}$$

where λ is the fluid thermal conductivity, x the axial coordinate, L_H the length of the heated section while the local convective heat transfer coefficient h_x is defined as follows:

$$h_x = \frac{q}{(\overline{T}_w - T_b)} \tag{3}$$

where $\overline{T}_w$ and T_b are the circumferentially averaged wall temperature and local bulk fluid temperature respectively. The circumferentially averaged wall temperature was obtained by averaging the values measured by the thermocouples placed on the external surface of the pipe wall (i.e., two sensors at each axial position), under the insulating material. In evaluating the heat exchanged per unit surface area, q, the heat transfer area was assumed equal to the envelope cylinder surface area. In the data reduction the local bulk temperature was used for evaluating fluid properties. The Nusselt number value has been evaluated by varying the Reynolds number, which is defined as follows:

$$Re = \frac{w \cdot D}{\nu} \tag{4}$$

being ν and w the fluid kinematic viscosity and mean fluid velocity over the tube's cross section, respectively. Regarding the relevant dimensionless parameters, the uncertainties have been calculated by using the usual propagation of errors procedure [13] and the maximum uncertainty for Nusselt number and the Reynolds number was estimated to be respectively ±10% and ±3%.

3. Results

The local Nusselt number along the dimensionless abscissa, defined in equation 5, is reported in figure 3 for tubes No. 1, 2, 3, 7, 8, 9 for two representative Reynolds number values. In the figure, it is reported also the analytical solution holding for the smooth wall pipe with uniform heat flux boundary condition [14] which was previously proved to be representative for the present experimental set-up [6,11,15].

$$x^* = \frac{x}{Re \cdot Pr \cdot D} \tag{5}$$

The effect of the corrugation is evident: for all tubes here reported, by increasing the Reynolds number the heat transfer characteristics of the enhanced tube depart from the smooth wall behavior by providing a significant heat transfer augmentation.

For the lowest investigated Reynolds number values, the thermal behavior of all the pipes is similar to each other and close to smooth wall behavior while for the highest Reynolds number values their performance changes depend on the geometrical parameters. For the two tubes with the lowest corrugation depth (tubes No.1 and 7, e = 0.6 mm) the enhancement is not as much significant. For the tubes with the corrugation depths e = 0.8 mm (tubes No.2 and 8) and e = 1 mm (tubes No.3 and 9), the enhancement over the smooth wall behavior becomes very important. This behavior could be interpreted as a consequence of a departure from the steady laminar flow induced by the cross-helix wall corrugation by pointing out that the corrugation depth is a fundamental parameter in promoting this transition.

IOP Conf. Series: Journal of Physics: Conf. Series **796** (2017) 012010 doi:10.1088/1742-6596/796/1/012010

In figure 3 it is also possible to observe that, in agreement with the findings of Rainieri et al. [11], in corrugated pipes in the fully developed conditions there is not a flat distribution of the local Nussel number but it presents peaks and descents due to the continuous interruption of the boundary layer promoted by the wall corrugation.

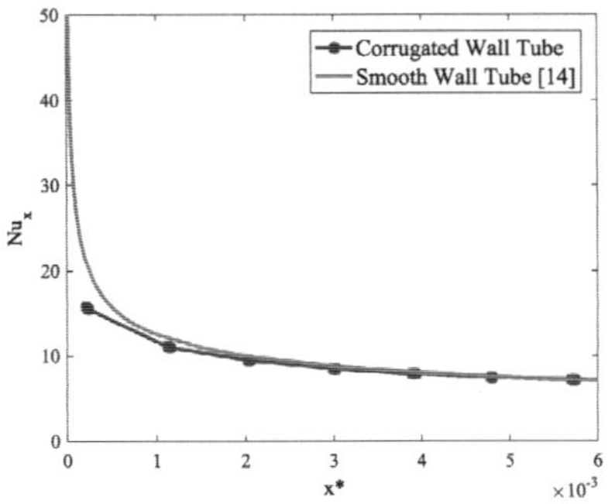

Figure 3a: Local Nusselt number versus the dimensionless abscissa for *Re* = 89 - Tube No.1.

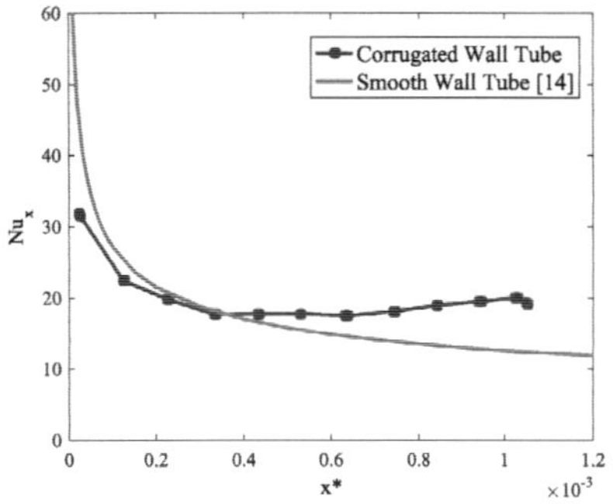

Figure 3b: Local Nusselt number versus the dimensionless abscissa for *Re* = 978 - Tube No.1.

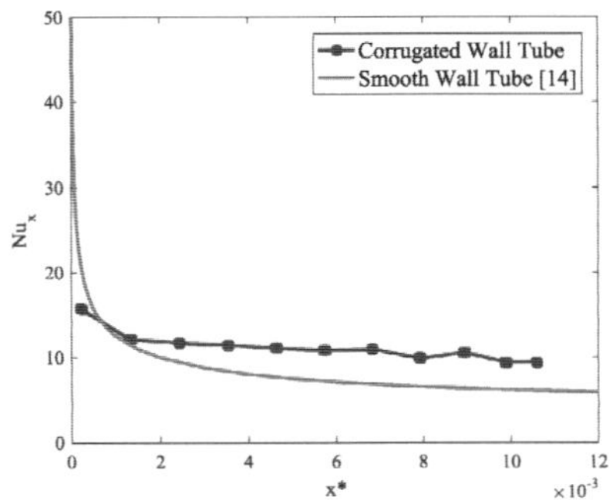

Figure 3c: Local Nusselt number versus the dimensionless abscissa for *Re* = 88 - Tube No.2.

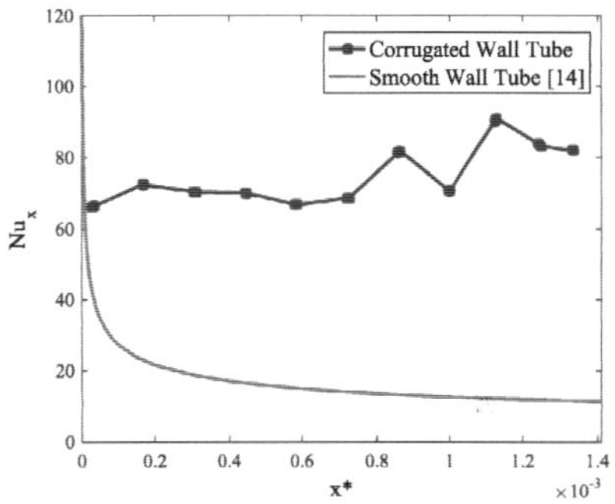

Figure 3d: Local Nusselt number versus the dimensionless abscissa for *Re* = 791 - Tube No.2.

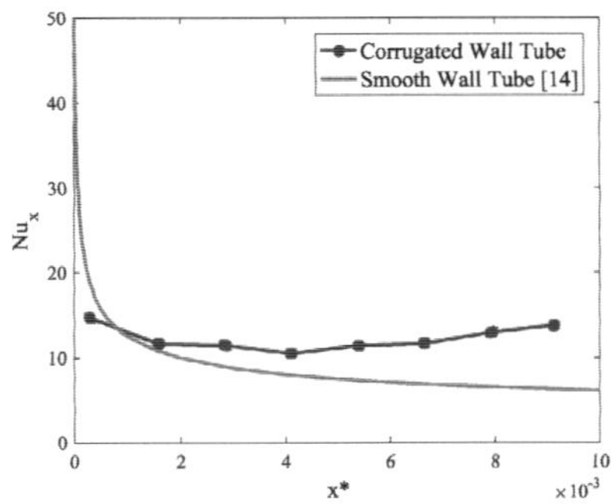

Figure 3e: Local Nusselt number versus the dimensionless abscissa for *Re* = 89 - Tube No.3.

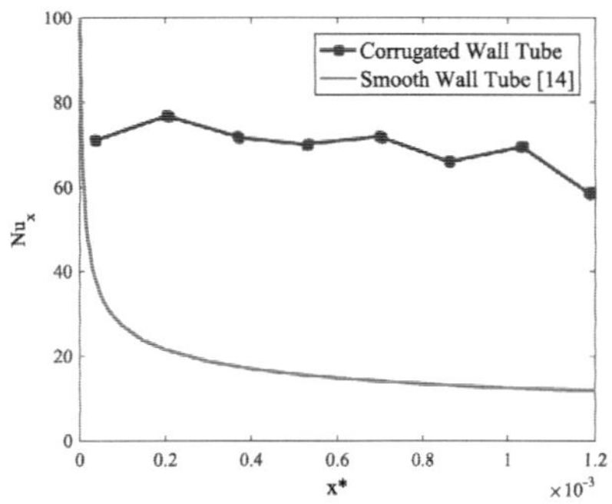

Figure 3f: Local Nusselt number versus the dimensionless abscissa for *Re* = 763 - Tube No.3.

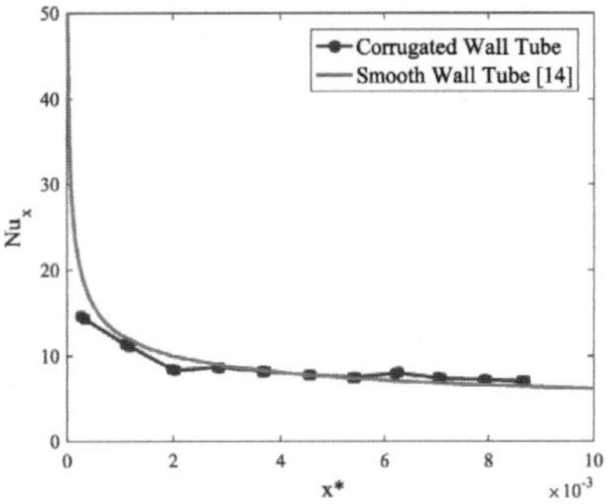

Figure 3g: Local Nusselt number versus the dimensionless abscissa for *Re* = 100 - Tube No.7.

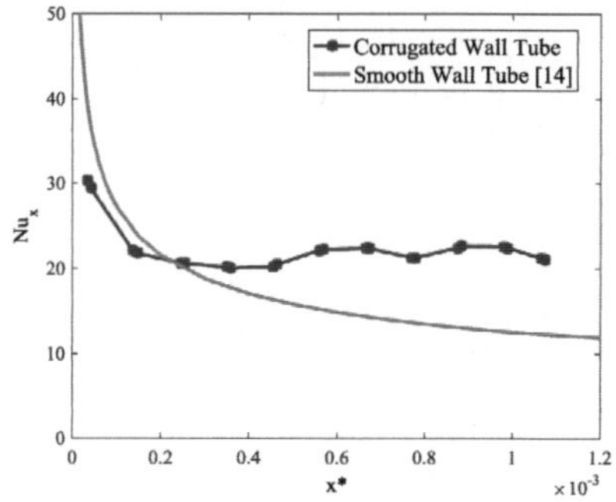

Figure 3h: Local Nusselt number versus the dimensionless abscissa for *Re* = 909 - Tube No.7.

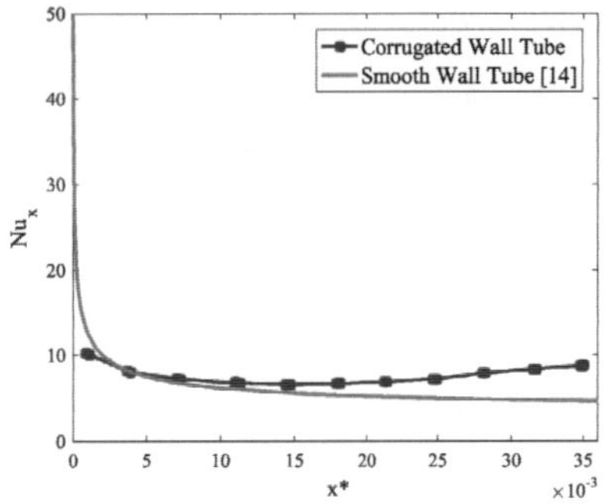

Figure 3i: Local Nusselt number versus the dimensionless abscissa for *Re* = 26 - Tube No.8.

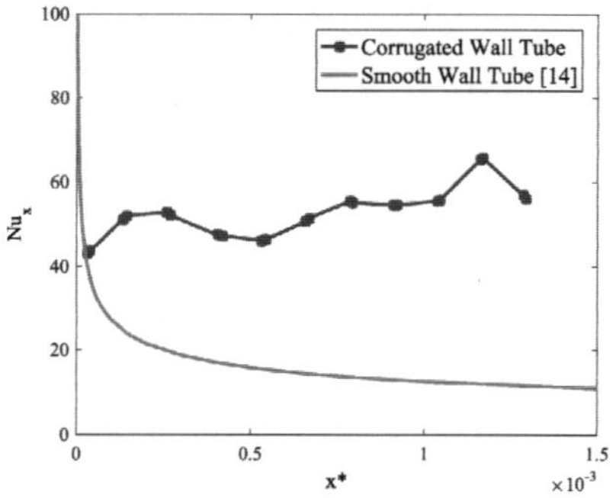

Figure 3l: Local Nusselt number versus the dimensionless abscissa for *Re* = 735 - Tube No.8.

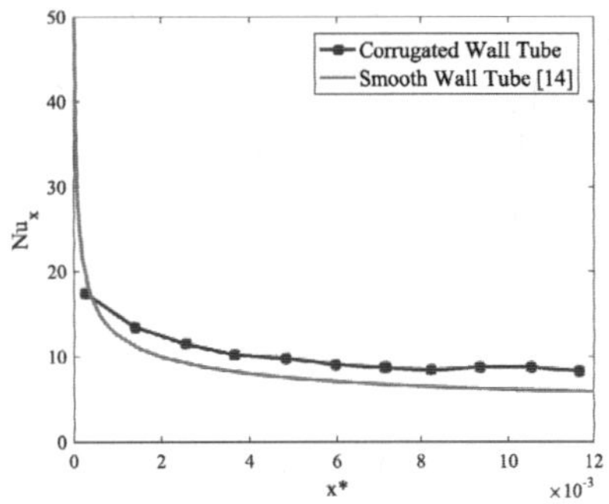

Figure 3m: Local Nusselt number versus the dimensionless abscissa for *Re* = 74 - Tube No.9.

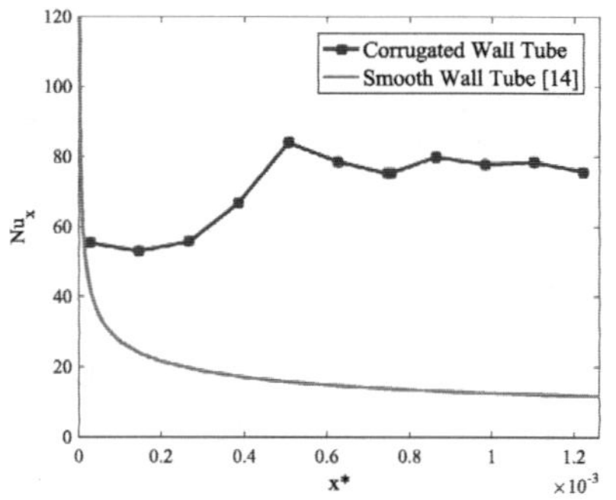

Figure 3n: Local Nusselt number versus the dimensionless abscissa for *Re* = 762 - Tube No.9.

In order to highlight the effect of the corrugation depth, it is meaningful to compare the heat transfer performance among pipes having the same corrugation pitch and different corrugation depths. In figure 4 the Nusselt number averaged over the heated length versus Reynolds number is reported for the tubes having the same corrugation pitch (l=13 mm (a), l=18 mm (b) and l=29 mm (c)) and different corrugation depths (e = 0.6, 0.8 and 1 mm). These data confirm that the corrugation depth is a crucial parameter in increasing the heat transfer performance for the pipes under test. The results also highlight that for a given corrugation pitch, the critical condition can be described by a precise value of the corrugation depth. Under a specific value of the corrugation depth *e*, the wall corrugation effects are almost negligible, while above that value the enhancement does not depend on the severity of the wall corrugation. Moreover, it is possible to notice that for the pipes with corrugation depth equal to 0.8 and 1 mm a much anticipated onset of transition from the laminar to an unstable flow regime occurs. In particular, a clearly departure from the smooth wall behavior due to the onset of an unstable regime occurs in the range $100 < Re < 250$; in addition, it is also possible to observe that the transition from laminar to unstable regime probably occurs through a sequence of intermediate state as found in [8]. In fact, a possible second transition it is present in the range $600 < Re < 800$.

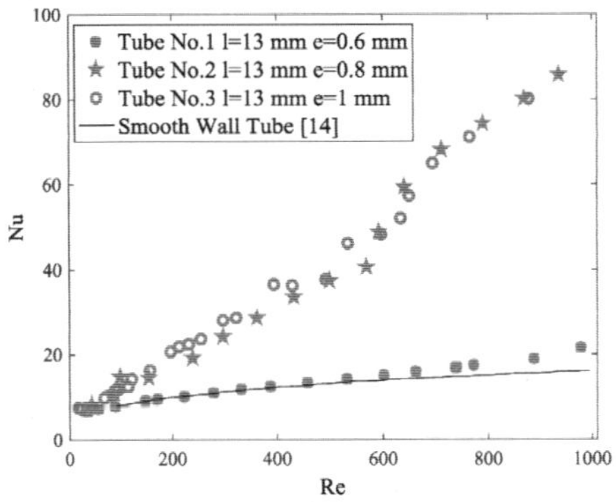

Figure 4a: Average Nusselt number for tubes No.1, 2 and 3 (*l*=13 mm) versus Reynolds number and comparison with the smooth wall tube behavior [14].

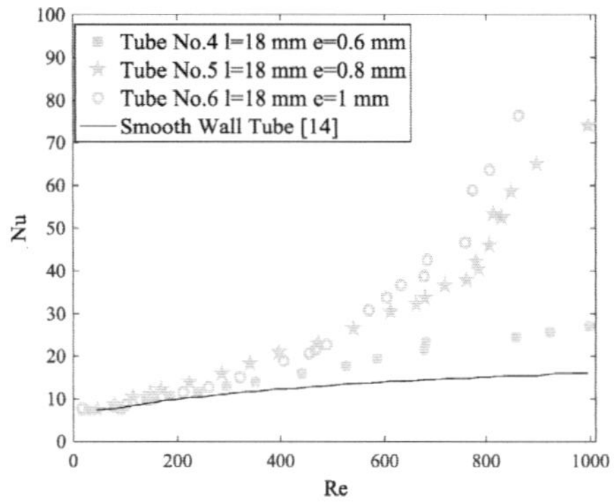

Figure 4b: Average Nusselt number for tubes No.4, 5 and 6 (*l*=18 mm) versus Reynolds number and comparison with the smooth wall tube behavior [14].

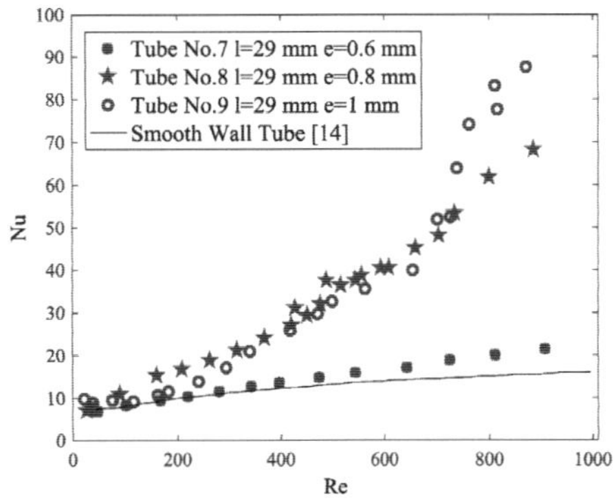

Figure 4c: Average Nusselt number for tubes No.7, 8 and 9 (*l*=29 mm) versus Reynolds number and comparison with the smooth wall tube behavior [14].

Regarding the effect of the corrugation pitch instead, in figure 5 the average Nusselt number versus the Reynolds number is reported for groups of tubes characterized by the same corrugation depth (*e*=0.6 mm (a), *e*=0.8 mm (b) and *e*=1 mm (c)) and by different corrugation pitches (*l*=13, 18 and 29 mm). The effect of the corrugation pitch *l* on the heat transfer performance of the pipes under test is less defined. For the lowest corrugation depth value hereby considered (*e*=0.6 mm) the best performance are associated to intermediate pitch value (*l*=18) while for the highest corrugation depth values (*e*= 0.8mm and *e*= 1mm) the pipe with the intermediate corrugation pitch presents the worst performance while the best performance is associated to the lowest pitch value (*l*=13 mm). Thus, from these data, it is not possible to establish a clear connection between the corrugation pitch effect and the heat transfer performance.

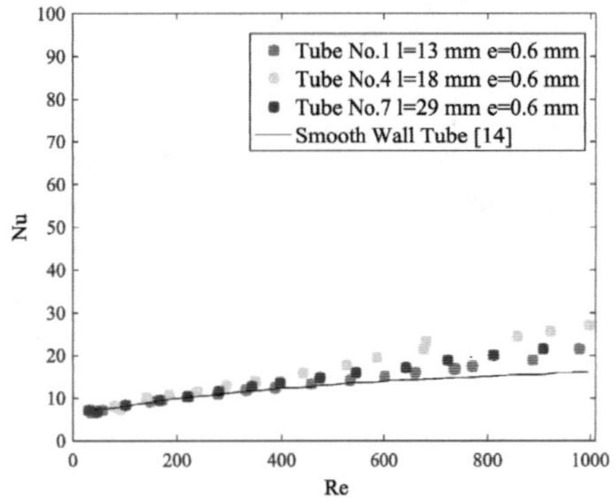

Figure 5a: Average Nusselt number for Tubes No.1, 4 and 7 (*e*=0.6 mm) versus Reynolds number and comparison with the smooth wall tube behaviour [14].

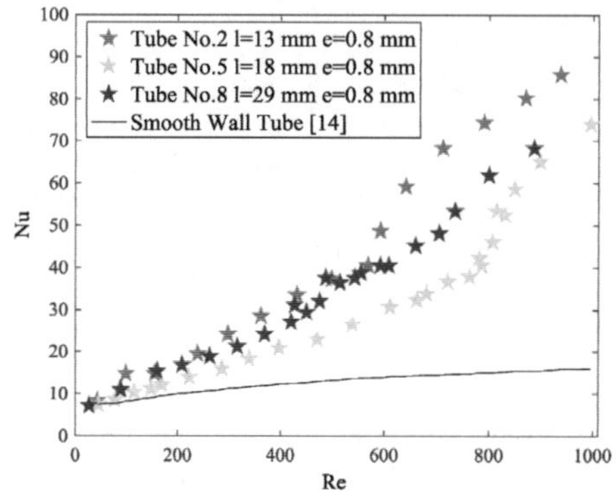

Figure 5b: Average Nusselt number for Tubes No.2, 5 and 8 (*e*=0.8 mm) versus Reynolds number and comparison with the smooth wall tube behaviour [14].

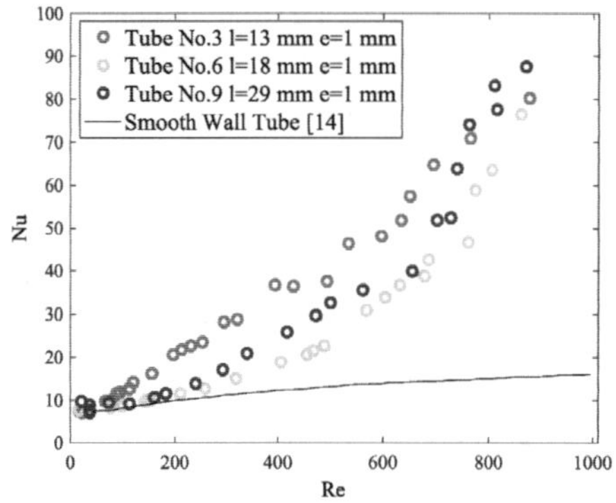

Figure 5c: Average Nusselt number for Tubes No.3, 6 and 9 (*e*=1 mm) versus Reynolds number and comparison with the smooth wall tube behaviour [14].

The limit of considering the average Nusselt number over the heated length is that the results obtained are valid only for the specific length under test: in order to make the results more general the asymptotic Nusselt number Nu_a was considered in the data processing. In fact, for the experimental configurations here studied, the thermally fully developed conditions were expected to be reached in the downstream region of the heated section. Regarding the measurements showing spatial oscillations in the local Nusselt number distribution, an average value in the fully developed region was considered.

The asymptotic thermal performances of all the nine tubes tested are compared in figure 6. The data confirm the predominance of the effect of the corrugation depth in promoting the heat transfer enhancement. Moreover, the data highlight the significant improvement of the convective heat transfer coefficient achievable with the cross-helix corrugation profile

The performance of the enhanced geometry was appropriately defined by the asymptotic heat transfer enhancement defined as follows:

$$\varepsilon_{h,\infty} = \frac{Nu_a}{Nu_{a,0}} \quad (6)$$

where the subscript 0 refers to reference problem, i.e. the analytical solution available for the thermally and hydrodynamically fully developed laminar flow problem [14] in straight smooth wall tubes under the uniform wall heat flux boundary condition.

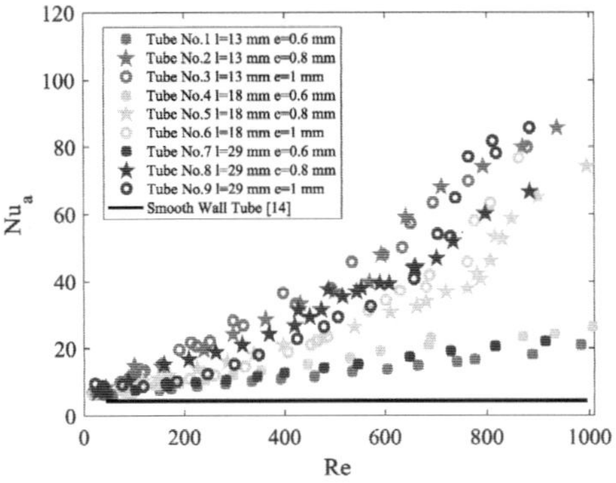

Figure 6: Asymptotic Nusselt number for the tubes under test and comparison with the smooth wall tube behaviour.

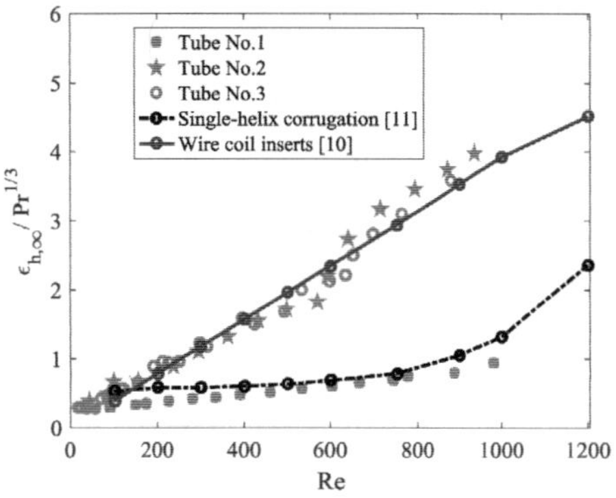

Figure 7: Heat transfer enhancement for the tubes No.1, 2 and 3 and comparison with the results obtained by Rainieri et al. [11] and by Garcia et al. [10].

In figure 7 the heat transfer enhancement performances are reported for the three tubes with corrugation pitch l equal to 13mm: in the same figure the results obtained by Rainieri et al. [11] for similar roughness (i.e. single-helix). It is interesting to compare these behaviors with the effects promoted by wire coil inserts since they have many points in common with severe wall corrugation (Garcia et al. [10]). Table 2 reports the geometric characteristics of these pipes. The results point out that the heat transfer enhancement obtained for the two cross-helix corrugated tubes with the two highest corrugation depth here tested exceeds the ones obtained for the single-helix corrugation by Rainieri et al. [11]. These data confirm the more effective capability of the cross-helix corrugation profile with respect to the single-helix one in anticipating the transition from the laminar flow regime. Moreover, there is a good matching between the results obtained for the cross-helix corrugation and the data achieved in [10] for the wire coils.

Table 2: pipes characteristics analysed in [11] and [10].

	l (mm)	e (mm)	l/D	e/D	Ω
Single-helix [11]	13	1.0	1.08	8.33×10^{-2}	6.41×10^{-3}
Wire coil inserts [10]	18	0.6	1.29	4.29×10^{-2}	1.43×10^{-3}

4. Conclusions

The forced convective heat transfer in cross-helix corrugated wall tubes was studied in the Reynolds and Prandtl number ranges 25÷1000 and 115÷150 respectively, by adopting ethylene glycol and water-ethylene glycol mixtures as working fluids. The aim of the research was to deepen the heat transfer enhancement achievable by using this passive technique. In particular, the effect of both the corrugation depth and of the corrugation pitch was analysed, by pointing out the correlated heat transfer enhancement effect: nine tubes, having different corrugation pitches and depths were tested. The results are expressed in terms of local, average and asymptotic Nusselt number and of heat transfer enhancement over the smooth wall tube behavior. The data show that the corrugation depth is a crucial parameter in increasing the heat transfer performance of the pipes under test. Under a specific value of the corrugation depth e the wall corrugation effects are almost negligible, while above that value the enhancement doesn't depend on the severity of the wall corrugation. For the pipes with corrugation depth equal to 0.8 and 1 mm a very early onset of transition from the laminar to an unstable flow regime occurs. In particular, a clearly departure from the smooth wall behavior occurs in the range $100 < Re < 250$ and a probable second transition is present in the range $600 < Re < 800$. These data confirm that the transition from laminar to unstable regime can occur through a sequence of intermediate state as observed by Guzman and Amon [8]. On the other hand, the effect on heat transfer of the corrugation pitch, resulted instead less clear for the nine pipes under test. The heat transfer enhancement performances were also compared to the results obtained by Rainieri et al. [11] and by Garcia et al. [10] for pipes with single-helix corrugation and wire coil inserts. The results confirm the more effective capability of the cross-helix corrugation profile with respect to the single-helix one to enhancing the transition from the laminar flow regime by significantly pushing the heat transfer tube's performance over the smooth wall tube behavior. In addition, they show a good matching with the data achieved in [10] with the wire coil inserts. In order to complete the analysis on this passive heat transfer enhancement technique, a further study that accounts for the pressure drop augmentation will be addressed.

Acknowledgments

This work was partially supported by the Emilia-Romagna Region (POR-FESR 2014-2020). MBS S.r.l. (Parma, Italy) is gratefully acknowledged for the set-up of the experimental apparatus.

References

[1] Webb R and Kim H H 2005 Principles of Enhanced Heat Transfer (New York: Taylor & Francis)
[2] Kareem Z S, Mohd Jaafar M N, Lazim T M, Abdullah S and Abdulwahid A F 2015 Passive heat transfer enhancement review in corrugation *Exp. Therm Fluid Sci.* **68** 22
[3] Ji W T, Jacobi A M, He Y L and Tao W Q 2015 Summary and evaluation on single-phase heat transfer enhancement techniques of liquid laminar and turbulent pipe flow *Int. J. Heat Mass Transfer* **88** 735
[4] Vicente P G, Garcıa A and Viedma 2002 A Experimental study of mixed convection and pressure drop in helically dimpled tubes for laminar and transition flow, *Int. J. Heat Mass Transfer* **45** 5091
[5] Rainieri S and Pagliarini G 2002 Convective heat transfer to temperature dependent property fluids in the entry region of corrugated tubes *Int. J. Heat Mass Transfer* **45** 4525
[6] Rainieri S and Pagliarini G 1997 Convective heat transfer to orange juice in smooth and corrugated tubes *Int. J. Heat Technol.* **15 (2)** 69
[7] Niceno N and Nobile E 2001 Numerical analysis of fluid flow and heat transfer in periodic wavy channels *Int. J. Heat Mass Transfer* **22** 156
[8] Guzman A M and Amon C H 1996 Dynamic flow characterization of transitional and chaotic regimes in converging-diverging channels *J. Fluid Mech.* **321** 25
[9] Wang G and Vanka S P 1995 Convective heat transfer in periodic wavy passages *Int J. Heat Mass Transfer* **38** 3219
[10] Garcia A, Solano J P, Vicente P G and Viedma A 2012 The influence of artificial roughness shape on heat transfer enhancement: corrugated tubes, dimpled tubes and wire coils *Appl. Therm. Eng.* **35** 196
[11] Rainieri S, Farina A and Pagliarini G 1996 Experimental investigation of heat transfer and pressure drop augmentation for laminar flow in spirally enhanced tubes *Proceedings of the 2nd European Thermal-Sciences and 14th UIT Heat Transfer Conference* **1** 203
[12] Withers J G and Habdas E P 1974 Heat transfer characteristics of helical-corrugated tubes for in-tube boiling of refrigerant R-12 *AIChE Symposium Series* **70** 98
[13] Kline S J and McClintock F A 1953 Describing uncertainties in single-sample experiments *Mech. Eng.* **75** 3
[14] Shah R K and London A L 1978 *Laminar Flow Forced Convection in Ducts* (New York: Academic Press)
[15] Rainieri S, Bozzoli F, Cattani L and Pagliarini G 2013 Compound convective heat transfer enhancement in helically coiled wall corrugated tubes *Int. J. Heat Mass Tran.* **59** 353

Sensitivity coefficients for thermal properties measurements using a boundary condition of the 4th kind

G D'Alessandro[1] **and F de Monte**[1]

[1] Department of Industrial and Information Engineering and Economics, University of L'Aquila, via G. Gronchi 18, 67100 L'Aquila, Italy

giampaolo.dalessandro@graduate.univaq.it

Abstract. In the parameter estimation the sensitivity analysis of the temperature to the unknown parameter (such as thermal conductivity, heat capacity) plays a fundamental role. Considering the experimental configuration for thermal properties measurements of solid materials, the heater in contact with the specimen can be modeled through an high conductivity thin layer at which an heat flux is applied. In such a case the one dimensional finite Cartesian body, representing the sample, is subject to a boundary condition of the 4th kind at the heated boundary. The transient temperature within the finite body, assuming the back side surface insulated, is derived by means of the Green's Function Solution Equation. Then the scaled sensitivity coefficients are computed in part analytically and in part numerically, for two different location: at the interface between the heater and the sample, and at the sample back side. The results show that the sensitivity coefficients with respect to the thermal conductivity and to the heat capacity of the sample are uncorrelated.

1. Introduction

Sensitivity coefficients have many applications such as parameter estimation, optimal experimental design and uncertainty or error analysis [1]. In parameter estimation, considerable insight is provided by carrying out a sensitivity analysis of the temperature to the unknown parameters, such as thermal conductivity, volumetric heat capacity or thermal diffusivity. In particular the knowledge of sensitivity coefficients not only allows a deep understanding of the phenomenon but it can represent an important design tool. In fact sensitivity coefficients provide useful design information concerning the effect of parameters on the model response [1].

As described in [2], the parameter estimation technique requires measured temperature values. It is essential that the measured response be sensitive to the parameters of interest: the more sensitive the temperature (or large the sensitivity coefficient) is, the more valuable the temperature measurements are [1]. Also, in order to gain much insight and information from the results as possible, it must be that the sensitivity coefficients with respect to the parameters of interest are uncorrelated and large in magnitudes [3]. For these reasons, the experimental apparatus have to be properly designed.

The sensitivity coefficients allow the experimental apparatus to be made closer to the optimal. For example, in an apparatus for which the sensitivity coefficients had not been investigated it is possible that additional materials in the experimental configuration (for instance the thin heater giving up heat to the sample) can have a large impact on the temperature than the material of interest (sample), with direct consequences on the quality of the results [1]. Definitely, on the one hand the sensitivity coefficients make possible a preliminary evaluation of the goodness of the experimental results (at least from a qualitative point of view), on the other hand are directly involved in the estimation of the parameters when minimizing the ordinary least square norm [4-6].

The focus of the current paper is to provide sensitivity coefficients in the event that the effect of the thin heater in contact with the specimen is simulated through an high-conductivity thin layer to which a jump in the heat flux is applied. The heat conduction problem thus addressed consists in a one dimensional finite Cartesian body, which represents the sample, subject to a boundary condition of the fourth kind [7, Chap. 2] at the heated boundary $x=0$, and insulated at the back side ($x=L$). The governing equations are solved by means of the Green's Function Solution Equation (GFSE) [7]. Once the thermal field is obtained, the so-called "scaled" sensitivity coefficients are computed by performing the partial derivative of the temperature with respect to the parameters of interest. Their computation is carried out in part analytically and in part numerically by using a two-point central difference scheme [8]. The coefficients are presented in a graphical form and they are calculated for two different location: at the interface between the heater and the sample, and at the sample back side. The results of the analysis show that the sensitivity coefficients with respect the thermal properties of the sample are uncorrelated, and both the thermal conductivity and the heat capacity can be estimated.

2. Analytical formulation

Figure 1a shows the schematic of an experimental apparatus for the thermal properties measurements of an orthotropic solid specimen: a thin layer heater located at the interface of two samples of the same material and thickness and gives heat up at surface of both samples. For sake of thermal symmetry, the three layer configuration (specimen-heater-specimen) reduces to the simplified configuration depicted in figure 1b. In particular the finite body, initially at uniform temperature T_{in}, is in perfect thermal contact with a high-conductivity layer at the boundary surface $x=0$. The thin layer, which a step change in heat flux is applied, it is at the same initial temperature T_{in} ; its thickness and heat capacity are denoted by L_f and C_f , respectively;. Also the finite body is insulated at the back side $x=L$, and its properties are considered temperature-independent.

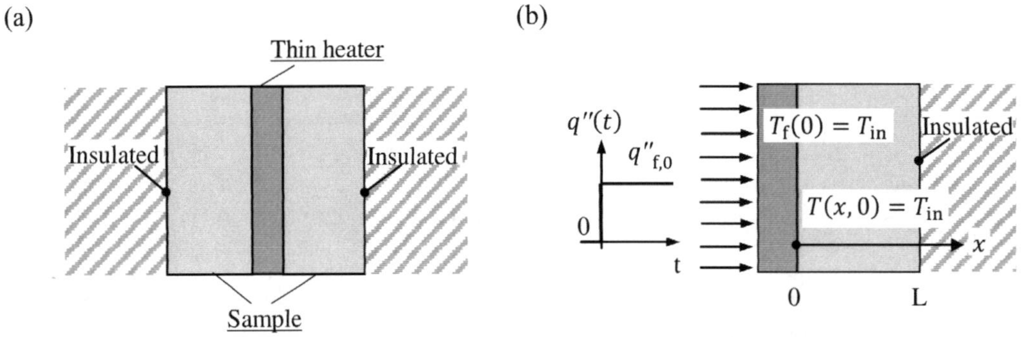

Figure 1. Schematic of the experimental apparatus for thermal properties measurements (a) and simplified schematic for the addressed problem (b).

The analytical formulation of this transient, linear, Cartesian heat conduction problem (denoted by X42B10T00 according to the numbering system devised in [9]) is defined in dimensionless form as follows.

$$\frac{\partial^2 \tilde{T}}{\partial \tilde{x}^2} = \frac{\partial \tilde{T}}{\partial \tilde{t}} \qquad (0 < \tilde{x} < 1; \tilde{t} > 0) \tag{1a}$$

$$-\left(\frac{\partial \tilde{T}}{\partial \tilde{x}}\right)_{\tilde{x}=0} + P\frac{\partial \tilde{T}_{\mathrm{f}}}{\partial \tilde{t}} = 1\,; \quad -\left(\frac{\partial \tilde{T}}{\partial \tilde{x}}\right)_{\tilde{x}=1} = 0 \qquad (\tilde{t} > 0) \tag{1b}$$

$$\tilde{T}_{\mathrm{f}}(\tilde{t}) = \tilde{T}(0,\tilde{t}) \qquad (\tilde{t} > 0); \quad \tilde{T}_{\mathrm{f}}(0) = 0 \tag{1c}$$

$$\tilde{T}(\tilde{x},0) = 0 \qquad (0 < \tilde{x} < 1) \tag{1d}$$

The dimensionless variables appearing in the above equations are defined as:

$$\tilde{T} = \frac{T - T_{\mathrm{in}}}{q''_{\mathrm{f,0}} L / k}, \quad \tilde{x} = \frac{x}{L}, \quad \tilde{t} = \frac{\alpha t}{L^2}, \quad \tilde{T}_{\mathrm{f}} = \frac{T_{\mathrm{f}} - T_{\mathrm{in}}}{q''_{\mathrm{f,0}} L / k}, \quad P = \frac{C_{\mathrm{f}} L_{\mathrm{f}}}{CL}$$

3. Temperature solution

The solution to the problem defined by equations (1) is not available in heat conduction literature. In particular the dimensionless governing equations (1a)-(1f) are solved by means of the GFSE method. As the series solution obtained by means of the standard GFSE requires a large number of terms to reach the numerical convergence, an alternative temperature solution is determined.

3.1. *Solution with GFSE*

The Green's function for the X42 problem is provided in literature [7, p 609, #X42]. In dimensionless form it results in:

$$\tilde{G}_{X42}(\tilde{x},\tilde{t} \,|\, \tilde{x}',\tilde{\tau}) = \frac{1}{P+1} + 2\sum_{m=1}^{\infty} \frac{X_m(\tilde{x}) X_m(\tilde{x}')}{\eta_m} \exp\left[-\beta_m^2(\tilde{t} - \tilde{\tau})\right] \tag{2}$$

Where β_m is the m-th eigenvalue whose computation is discussed in Subsection 3.3, and

$$\eta_m = (P\beta_m)^2 + P + 1 \tag{2a}$$

$$X_m(\tilde{x}) = \cos(\beta_m \tilde{x}) - P\beta_m \sin(\beta_m \tilde{x}) \tag{2b}$$

By applying the GFSE to the current problem, considering that the only non-homogeneous term is the boundary condition of the fourth kind, the dimensionless temperature solution $\tilde{T}(\tilde{x},\tilde{t})$ can be obtained performing the following integral:

$$\tilde{T}(\tilde{x},\tilde{t}) = \int_{\tilde{\tau}=0}^{\tilde{t}} \tilde{G}_{X42}(\tilde{x},\tilde{t} \,|\, 0,\tilde{\tau})\, d\tilde{\tau} \tag{3}$$

Then substituting equation (2) into the above integral yields

$$\tilde{T}(\tilde{x},\tilde{t}) = \frac{\tilde{t}}{P+1} + 2\sum_{m=1}^{\infty} \frac{\cos(\beta_m \tilde{x}) - P\beta_m \sin(\beta_m \tilde{x})}{\eta_m \beta_m^2} - 2\sum_{m=1}^{\infty} \frac{\cos(\beta_m \tilde{x}) - P\beta_m (\beta_m \tilde{x})}{\eta_m \beta_m^2} \exp\left(-\beta_m^2 \tilde{t}\right) \tag{4}$$

The first summation in equation (4) exhibits an algebraic convergence, that is it requires a large number of terms. For this reason, an alternative solution with better convergence properties is determined in the following subsection.

3.2. *Alternative solution*

An alternative solution to the problem defined by equations (1) may be derived by means of the Alternative GFSE (AGFSE) [7, Chap. 3]. In this case the temperature solution is considered as the sum of two terms: the former, denoted by $\tilde{T}^*(\tilde{x},\tilde{t})$, which satisfies only the boundary conditions, and the latter which represents the transient solution, denoted by $\tilde{T}'(\tilde{x},\tilde{t})$.

$$\tilde{T}(\tilde{x},\tilde{t})=\tilde{T}^*(\tilde{x},\tilde{t})+\tilde{T}'(\tilde{x},\tilde{t}) \tag{5}$$

Then, by substituting equation (5) into the governing equations (1), the original problem can be split up into a set of two simpler problems:

- a non-homogeneous problem defined for $\tilde{T}^*(\tilde{x},\tilde{t})$, as:

$$\frac{\partial^2\tilde{T}^*}{\partial\tilde{x}^2}=\frac{\partial\tilde{T}^*}{\partial\tilde{t}} \tag{6a}$$

$$-\left.\frac{\partial\tilde{T}^*}{\partial\tilde{x}}\right|_{\tilde{x}=0}+P\frac{\partial\tilde{T}^*(0,\tilde{t})}{\partial\tilde{t}}=1; \qquad \left.\frac{\partial\tilde{T}^*}{\partial\tilde{x}}\right|_{\tilde{x}=1}=0 \tag{6b}$$

- and a transient problem defined for $\tilde{T}'(\tilde{x},\tilde{t})$, with homogeneous boundary conditions:

$$\frac{\partial^2\tilde{T}'}{\partial\tilde{x}^2}=\frac{\partial\tilde{T}'}{\partial\tilde{t}} \tag{7a}$$

$$-\left.\frac{\partial\tilde{T}'}{\partial\tilde{x}}\right|_{\tilde{x}=0}+P\frac{\partial\tilde{T}'(0,\tilde{t})}{\partial\tilde{t}}=0; \qquad \left.\frac{\partial\tilde{T}'}{\partial\tilde{x}}\right|_{\tilde{x}=1}=0 \tag{7b}$$

$$\tilde{T}'(\tilde{x},0)=-\tilde{T}^*(\tilde{x},0) \tag{7c}$$

As the temperature solution, equation (4), contains a time dependent term which dominates temperature for large times, that is $\tilde{t}/(P+1)$, the $\tilde{T}^*(\tilde{x},\tilde{t})$ solution should be:

$$\tilde{T}^*(\tilde{x},\tilde{t})=f(\tilde{x})+\frac{\tilde{t}}{P+1} \tag{8}$$

Where the $f(\tilde{x})$ function can be obtained by substituting equation (8) in equations (6). It results in:

$$f(\tilde{x})=\frac{\tilde{x}^2}{2(P+1)}-\frac{\tilde{x}}{(P+1)} \tag{9}$$

Instead the problem defined by equations (7) is conveniently solved by means of the GFSE: in this case the only non-homogeneous term is the initial condition. The solution $\tilde{T}'(\tilde{x},\tilde{t})$ can be obtained by the following equation:

$$\tilde{T}'(\tilde{x},\tilde{t})=-\int_{\tilde{x}'=0}^{1}\tilde{G}_{X42}(\tilde{x},\tilde{t}\,|\,\tilde{x}',0)f(\tilde{x}')d\tilde{x}'-\underbrace{P\tilde{G}_{X42}(\tilde{x},\tilde{t}\,|\,0,0)f(0)}_{=0} \tag{10}$$

In the above equation we have considered that $\tilde{T}^*(\tilde{x},0)=f(\tilde{x})$, and the Green's function is defined by equation (2). Then by performing the integral in equation (10) one can obtain the $\tilde{T}'(\tilde{x},\tilde{t})$ solution:

$$\tilde{T}'(\tilde{x},\tilde{t})=\frac{1}{3(P+1)^2}-2\sum_{m=1}^{\infty}\frac{\cos(\beta_m\tilde{x})-P\beta_m\sin(\beta_m\tilde{x})}{\eta_m\beta_m^2}\exp(-\beta_m^2\tilde{t}) \tag{11}$$

Finally by substituting equations (8) and (11), bearing in mind equation (9), into equation (5), an alternative form for the temperature solution is obtained:

$$\tilde{T}(\tilde{x},\tilde{t})=\frac{\tilde{t}}{P+1}+\frac{\tilde{x}^2}{2(P+1)}-\frac{\tilde{x}}{P+1}+\frac{1}{3(P+1)^2}-2\sum_{m=1}^{\infty}\frac{\cos(\beta_m\tilde{x})-P\beta_m\sin(\beta_m\tilde{x})}{\eta_m\beta_m^2}\exp(-\beta_m^2\tilde{t}) \tag{12}$$

3.3. Computation of the eigenvalues

The eigenvalues appearing in the temperature solutions, equations (4) and (12), are computed as roots of the eigencondition (transcendental equation) defined as [7],

$$\beta_m \cot(\beta_m) = -P^{-1} \tag{13}$$

Its roots may be computed by using explicit approximate relations based on the second-order modified Newton method [10]. These relations provide an approximate value of the exact eigenvalue with high accuracy (8-decimal place after one iteration, and 15-decimal place after two iterations) for the P range $[0,\infty)$.

3.4. Computational solution

The infinite series appearing in the alternative temperature solution, equation (12), exhibits an exponential convergence at any time except near $\tilde{t}=0$. However as we cannot take into account infinite terms a convergence criterion for this series is needed. Following the procedure given in [11], the maximum number of required terms $m_{\max}$ to get a truncation error of 10^{-A} (A=2,3,..15) in the series solution may be taken as:

$$m_{\max} = \text{ceil}\left\{\frac{1}{2}+\frac{1}{\pi}\left[\frac{A\ln(10)}{\tilde{t}}\right]^{1/2}\right\} \tag{14}$$

In particular, the above equation has been obtained by using $(m-1/2)\pi$ as a conservative estimate for β_m. In addition, according to equation (14) a large number of terms are required for early times. However for times less than the so-called deviation time $\tilde{t}_{\mathrm{d}}$, the temperature solution can be replaced by a semi-infinite transient solution (X40B1T00) with an error less than 10^{-A} (A=2,3,..15) [9, 12]. The solution for the X40B1T00 case is a well-established exact analytical solution available in the heat conduction literature [13, p 306, #12] and discussed in [14]:

$$\tilde{T}(\tilde{x},\tilde{t})=\sqrt{\frac{4\tilde{t}}{\pi}}\exp\left(-\frac{\tilde{x}^2}{4\tilde{t}}\right)-\tilde{x}\,\mathrm{erfc}\left(\frac{\tilde{x}}{\sqrt{4\tilde{t}}}\right)-P\left[\mathrm{erfc}\left(\frac{\tilde{x}}{\sqrt{4\tilde{t}}}\right)-\mathrm{Am}(\tilde{x},\tilde{t},P)\right]\quad\left(0\le\tilde{t}\le\tilde{t}_{\mathrm{d}}=\frac{(2-\tilde{x})^2}{10A}\right) \tag{15}$$

where $\mathrm{Am}(\tilde{x},\tilde{t},P)$ is the so-called Amos function defined as

$$\mathrm{Am}(\tilde{x},\tilde{t},P)=\exp\left(\frac{\tilde{x}}{P}+\frac{\tilde{t}}{P^2}\right)\mathrm{erfc}\left(\frac{\tilde{x}}{\sqrt{4\tilde{t}}}+\frac{\sqrt{\tilde{t}}}{P}\right)=\exp\left(-\frac{\tilde{x}^2}{4\tilde{t}}\right)\mathrm{erfcx}\left(\frac{\tilde{x}}{\sqrt{4\tilde{t}}}+\frac{\sqrt{\tilde{t}}}{P}\right) \tag{15a}$$

A computer code in Matlab ambient for calculating the solution of the current problem is provided in [10]. Also, in this reference, a more detailed derivation of the solution is given, and a numerical intrinsic verification of the above code is carried out.

For sake of completeness, a plot of the dimensionless temperature as a function of time for different P values is given in figure 2.

4. Sensitivity coefficients

The scaled sensitivity coefficients are defined as partial derivatives of the temperature with respect to the model parameter γ of interest (e.g. $q''_{\mathrm{f}0}, k, C, C_{\mathrm{f}}$) multiplied by the parameter itself:

$$X_\gamma = \gamma\frac{\partial T}{\partial \gamma} \tag{16}$$

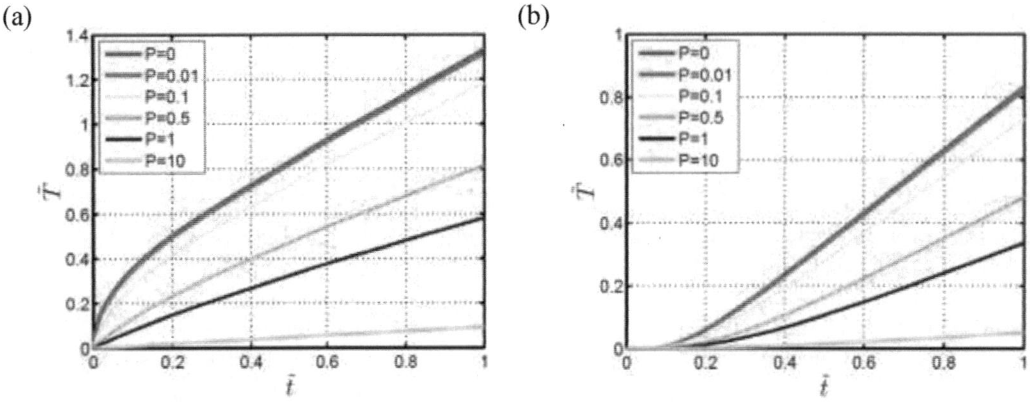

Figure 2. Dimensionless temperature as a function of the dimensionless time, for different P values, at $\tilde{x}=0$ (a) and at $\tilde{x}=1$ (b).

Thus the scaled sensitivity coefficients have units of °C and their calculation requires the use of the chain rule [15]. They can be expressed in a dimensionless form as follows:

$$\tilde{X}_\gamma = \frac{X_\gamma}{q''_{f0} L / k} = \gamma \frac{\partial \tilde{T}}{\partial \gamma} \tag{17}$$

4.1. Heat flux as a parameter

If the parameter of interest is the heat flux applied to the boundary surface $\tilde{x}=0$, the corresponding dimensional scaled sensitivity coefficient is:

$$X_{q''_{f0}} = q''_{f0}\frac{\partial T}{\partial q''_{f0}} = q''_{f0}\frac{\partial}{\partial q''_{f0}}\left(\tilde{T}\frac{q''_{f0}L}{k}\right) = q''_{f0}\left[\frac{q''_{f0}L}{k}\underbrace{\frac{\partial \tilde{T}}{\partial q''_{f0}}}_{=0} + \tilde{T}\frac{\partial}{\partial q''_{f0}}\left(\frac{q''_{f0}L}{k}\right)\right] = \frac{q''_{f0}L}{k}\tilde{T} \tag{18}$$

In dimensionless form it results in $\tilde{X}_{q''_{f0}} = \tilde{T}$, where $\tilde{T}$ is defined by equation (12). Thus figure 2 also shows the sensitivity coefficient to the applied heat flux.

4.2. Thermal conductivity and volumetric heat capacities as parameters

As $\tilde{T}$ depends on thermal conductivity of the specimen k, and also it depends on the volumetric heat capacities of both specimen and heater, C and C_f respectively, three coefficients should be considered.

When the parameter of interest is the thermal conductivity, the relative sensitivity coefficient can be expressed as follows

$$X_k = k\frac{\partial T}{\partial k} = k\frac{\partial}{\partial k}\left(\tilde{T}\frac{q''_{f0}L}{k}\right) = k\left[\frac{q''_{f0}L}{k}\frac{\partial \tilde{T}}{\partial k} + \tilde{T}\frac{\partial}{\partial k}\left(\frac{q''_{f0}L}{k}\right)\right] = k\left[\frac{q''_{f0}L}{k}\frac{\partial \tilde{T}}{\partial k} - \frac{q''_{f0}L}{k^2}\tilde{T}\right] \tag{19}$$

where $\tilde{T} = \tilde{T}[\tilde{t}, P, \beta_m(P)]$, with $\tilde{t}[\alpha(k, C)]$, $P(C, C_f)$. Thus, by using the chain rule:

$$\frac{\partial \tilde{T}}{\partial k} = \frac{\partial \tilde{T}}{\partial \tilde{t}}\frac{\partial \tilde{t}}{\partial \alpha}\frac{\partial \alpha}{\partial k} \tag{20}$$

By substituting equation (20) in equation (19), one can obtain the scaled sensitivity coefficient with respect to the thermal conductivity. In dimensionless form it results in:

$$\tilde{X}_k = \tilde{t}\frac{\partial \tilde{T}}{\partial \tilde{t}} - \tilde{T} \tag{21}$$

Following the same procedure described above, the scaled sensitivity coefficient with respect to the volumetric heat capacity C is obtained.

$$\tilde{X}_C = C\frac{\partial \tilde{T}}{\partial C} = -\tilde{t}\frac{\partial \tilde{T}}{\partial \tilde{t}} - P\frac{\partial \tilde{T}}{\partial P} - P\sum_{m=1}^{\infty}\frac{\partial \tilde{T}}{\partial \beta_m}\frac{d\beta_m}{dP} \tag{22}$$

In similar manner the scaled sensitivity coefficient with respect to the volumetric heat capacity of the heater C_f in dimensionless form, results in:

$$\tilde{X}_{C_\mathrm{f}} = C_\mathrm{f}\frac{\partial \tilde{T}}{\partial C_\mathrm{f}} = P\frac{\partial \tilde{T}}{\partial P} + P\sum_{m=1}^{\infty}\frac{\partial \tilde{T}}{\partial \beta_m}\frac{d\beta_m}{dP} \tag{23}$$

It is worth to note that all scaled sensitivity coefficients discussed above sum to zero for all values of time and position.

$$\tilde{X}_{q^{\bullet}_{\mathrm{f},0}} + \tilde{X}_k + \tilde{X}_C + \tilde{X}_{C_\mathrm{f}} = 0 \tag{24}$$

4.3. Computation of the scaled sensitivity coefficients $\tilde{X}_k, \tilde{X}_C$ and $\tilde{X}_{C_\mathrm{f}}$

The sensitivity coefficient $\tilde{X}_k$ may be computed analytically through equation (12) and its partial derivative with respect to time. As regards the coefficients $\tilde{X}_C$ and $\tilde{X}_{C_\mathrm{f}}$, they cannot be evaluated analytically as the derivative $d\beta_m/dP$, appearing in equations (22) and (23), cannot be given in an algebraic form (see equation (13)). For this reason, they will be computed numerically by using a two point central difference (CD) scheme [8, p 440, #12-6c] for the derivatives $\partial\tilde{T}/\partial C$ and $\partial\tilde{T}/\partial C_\mathrm{f}$, as suggested by the following equation.

$$\frac{\partial \tilde{T}}{\partial \gamma} = \frac{\tilde{T}[\tilde{x}, \tilde{t}(\gamma+\Delta\gamma), P(\gamma+\Delta\gamma), \beta_m(\gamma+\Delta\gamma)] - \tilde{T}[\tilde{x}, \tilde{t}(\gamma-\Delta\gamma), P(\gamma-\Delta\gamma), \beta_m(\gamma-\Delta\gamma)]}{2\Delta\gamma} + O(\Delta\gamma^2) \tag{25}$$

Bearing in mind that $\tilde{t}[\alpha(k,C)], P(C, C_\mathrm{f})$ and $\beta_m(P)$, the above equation can be evaluated for $\gamma = C$ and $\gamma = C_\mathrm{f}$.

Also for early times (less than the deviation time $\tilde{t}_\mathrm{d}$) the temperature field within the specimen may be evaluated accurately by using equation (15). Therefore the sensitivity coefficients can be computed analytically by performing the derivatives of equation (15) with respect to $\tilde{t}$ and to P. While the coefficient $\tilde{X}_{q^{\bullet}_{\mathrm{f},0}}$ is obtained immediately, the sensitivity coefficients $\tilde{X}_k$, $\tilde{X}_C$ and $\tilde{X}_{C_\mathrm{f}}$ can be evaluated through the following algebraic expressions:

$$\tilde{X}_k = \tilde{t}\frac{\partial \tilde{T}}{\partial \tilde{t}} - \tilde{T} = (P+\tilde{x})\mathrm{erfc}\left(\frac{\tilde{x}}{\sqrt{4\tilde{t}}}\right) + \left(\frac{\tilde{t}}{P} - P\right)\mathrm{Am}(\tilde{x}, \tilde{t}, P) - 2\sqrt{\frac{\tilde{t}}{\pi}}\exp\left(-\frac{\tilde{x}^2}{4\tilde{t}}\right) \tag{26}$$

$$\tilde{X}_C = -\tilde{t}\frac{\partial \tilde{T}}{\partial \tilde{t}} - P\frac{\partial \tilde{T}}{\partial P} = P\,\mathrm{erfc}\left(\frac{\tilde{x}}{\sqrt{4\tilde{t}}}\right) - 2\sqrt{\frac{\tilde{t}}{\pi}}\exp\left(-\frac{\tilde{x}^2}{4\tilde{t}}\right) + \left(\tilde{x} + \frac{\tilde{t}}{P} - P\right)\mathrm{Am}(\tilde{x}, \tilde{t}, P) \tag{27}$$

$$\tilde{X}_{C_\mathrm{f}} = P\frac{\partial \tilde{T}}{\partial P} = 2\sqrt{\frac{\tilde{t}}{\pi}}\exp\left(-\frac{\tilde{x}^2}{4\tilde{t}}\right) - P\,\mathrm{erfc}\left(\frac{\tilde{x}}{\sqrt{4\tilde{t}}}\right) + \left(P - \tilde{x} - \frac{2\tilde{t}}{P}\right)\mathrm{Am}(\tilde{x}, \tilde{t}, P) \tag{28}$$

5. Results

Figures 3-5 show the dimensionless sensitivity coefficients $\tilde{X}_k$, $\tilde{X}_C$ and $\tilde{X}_{C_r}$ as a function of time for different P values respectively. Two different locations ($\tilde{x}=0$ and $\tilde{x}=1$) are also considered. In these figures a curve for $P=0$ is plotted as a reference case; in fact, in such a case, the thin heater is completely neglected and the solution matches to that of the X22B10T0 problem. For sake of brevity, this solution is not given here but it is available in [7, 16].

Figure 3 shows the temperature sensitivity to the thermal conductivity k. In particular, for $\tilde{x}=0$, it reduces when the heat capacity for unit area ratio P increases; while for $\tilde{x}=1$ this behavior cannot be observed. It's interesting to note that the sensitivity to k at the backside ($\tilde{x}=1$) is less affected by the P parameter than the sensitivity at $\tilde{x}=0$, when large times are considered. This aspect suggests that for estimating the thermal conductivity, in the event that the heat capacities ratio (P) is unknown, the temperature sensor should be placed at $\tilde{x}=1$. In addition; assuming that the P parameter is known, according to the figure 3 if $P<0.5$ the sensitivity to the thermal conductivity is higher at $\tilde{x}=0$; on the contrary if $P>0.5$ the highest sensitivity occurs at $\tilde{x}=1$.

The sensitivity coefficient $\tilde{X}_C$ is depicted in figure 4. By observing the figure it is possible to note an appreciable difference between the sensitivity at $\tilde{x}=0$ and $\tilde{x}=1$ (not at early times) only for large P values: in this case the sensitivity is higher at the sample backside.

The temperature sensitivity to the heat capacity of the heater is shown in figure 5. The higher sensitivity occurs at the interface between the heater and the sample ($\tilde{x}=0$). By comparing figures 4 and 5 it is possible to state that the temperature of the sample is more sensitive to its heat capacity than to the heat capacity of the heater.

Finally, figure 6 compares the sensitivity coefficients $\tilde{X}_k$, $\tilde{X}_C$ and $\tilde{X}_{C_r}$ computed for $P=0.1$ and for the locations $\tilde{x}=0$ and $\tilde{x}=1$. For time less than the deviation time (defined by equation (15)) the coefficients $\tilde{X}_k$ and $\tilde{X}_C$ respond in the same way: in particular they are equal for $\tilde{x}=0$ (see equations (26) and (27)). For larger times the temperature sensitivity to the thermal conductivity k approaches to a constant value; while the sensitivity to the heat capacity of the sample C increases linearly with the time. Then the coefficients $\tilde{X}_k$ and $\tilde{X}_C$ are uncorrelated, and both the thermal properties, k and C, can be estimated.

(a) (b)

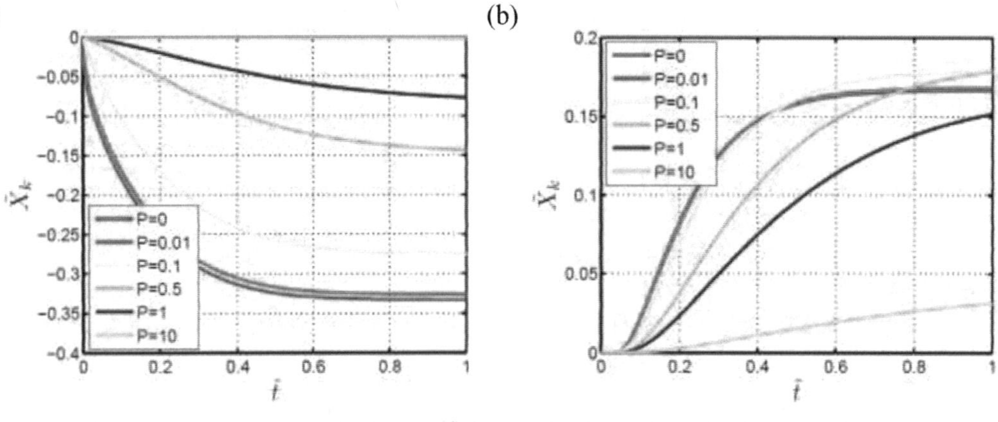

Figure 3. Dimensionless scaled sensitivity coefficient to the thermal conductivity as a function of the dimensionless time, for different P values, at $\tilde{x}=0$ (a) and at $\tilde{x}=1$ (b).

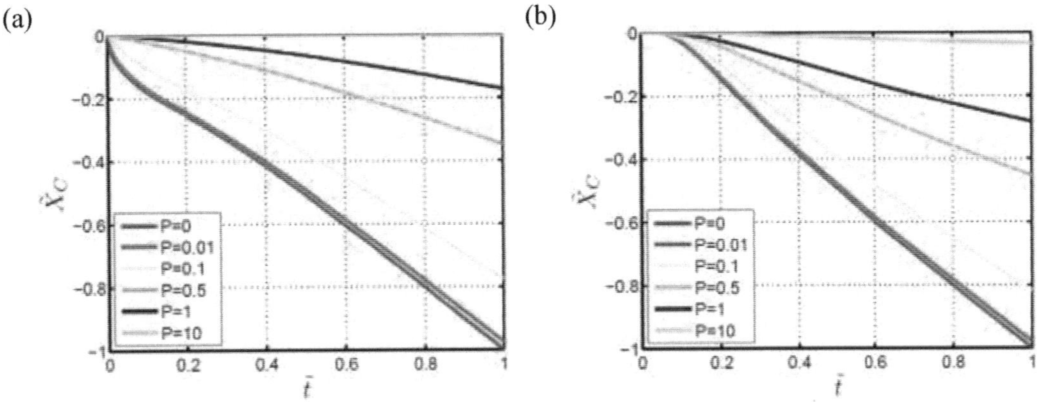

Figure 4. Dimensionless scaled sensitivity coefficient to the heat capacity of the sample as a function of the dimensionless time, for different P values, at $\tilde{x}=0$ (a) and at $\tilde{x}=1$ (b).

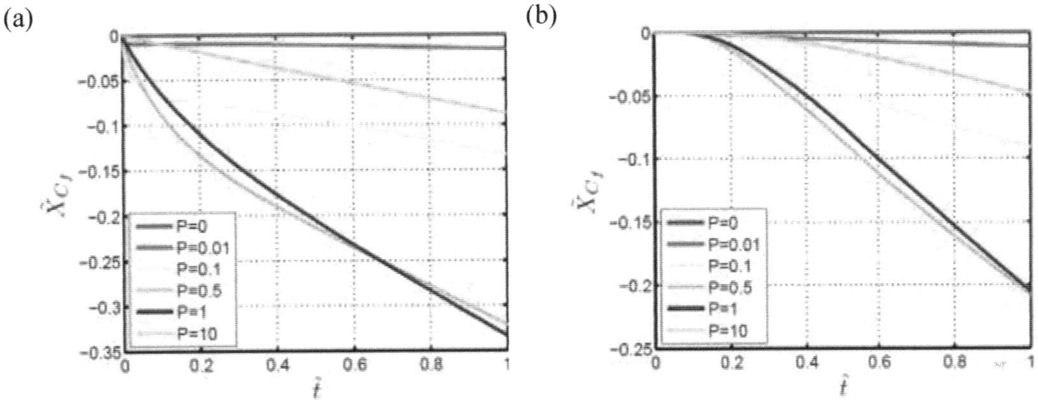

Figure 5. Dimensionless scaled sensitivity coefficient to the heat capacity of the heater as a function of the dimensionless time, for different P values, at $\tilde{x}=0$ (a) and at $\tilde{x}=1$ (b).

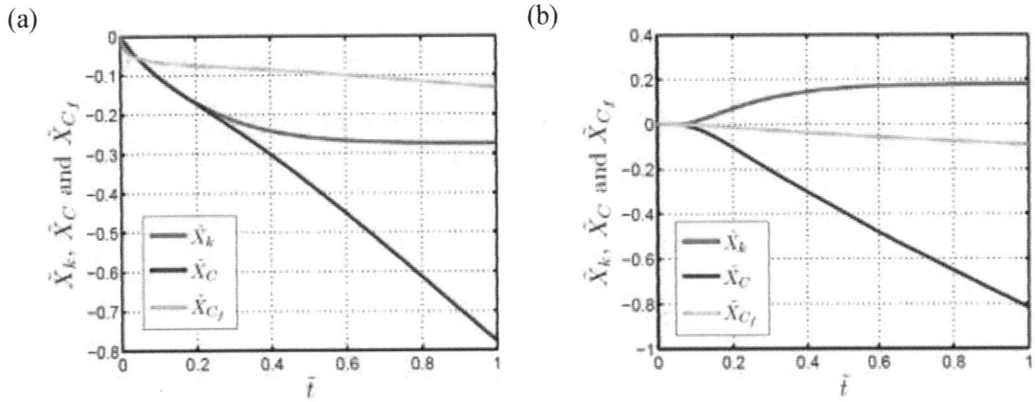

Figure 6. Comparison among dimensionless scaled sensitivity coefficients for $P=0.1$ at $\tilde{x}=0$ (a) and at $\tilde{x}=1$ (b).

6. Conclusions

The temperature solution for the X42B10T00 problem has been derived by means of the GFSE. Then the sensitivity coefficients with respect to the thermal properties (thermal conductivity and heat capacity of the specimen, and heat capacity of the heater) have been studied in order to establish the best location for the temperature sensor.

The analysis shows that the sensitivity coefficients to the thermal conductivity and to the heat capacity of the sample are uncorrelated and both these thermal properties can be estimated. In particular, when the parameter to estimate is the thermal conductivity the optimal location of the temperature sensor is determined by the heat capacities for unit area ratio (P). However when P is unknown the temperature sensor should be placed at the back side of the sample. This is also the best sensor location for estimating the heat capacity of the sample. Moreover, at the back side, the measured temperature is less affected by the heat capacity of the heater.

References

[1] Dowding K J, Blackwell B F and Cochran R J 1999 Application of sensitivity coefficients for heat conduction problems *Numerical Heat Transfer Part* B **36** 33-55

[2] Beck J V and Arnold K J 1977 *Parameter Estimation in Engineering and Science* (New York: John Wiley and Sons)

[3] Marzola A, de Monte F and Mishra D *An Efficient Technique for Computing Sensitivity Coefficients of Solid Materials for Thermal Property Measurements 8th Int. Conf. On Inverse Problems in Engineering (ICIPE),* Cracow, May 12-15 2014

[4] Garnier B, Delauney D and Beck J V 1992 Estimation of thermal properties of composite materials without instrumentation inside the sample *Int. J. Thermophys.* **13** 1097-111

[5] Sakly A, Jemni A, Lagonotte P and Petit D 2011 Estimation of the effective thermal properties in a matallic medium by an inverse technique using temperature measurements *J. Appl. Fluid Mechanics* **4** 23-9

[6] Aviles-Ramos C and Haji-Sheikh A 2001 Estimation of thermophysical properties of composites using multi-parameter estimation and zeroth-order regularization *Inverse Problems Engineering* **9** 507-36

[7] Cole K D, Beck J V, Haji-Sheikh A and Litkouhi B 2011 *Heat Conduction Using Green's Function 2nd Edition* (Boca Raton FL: CRC Press Taylor&Francis)

[8] Ozisik N M 1993 *Heat Conduction* (Hoboken NJ: John Wiley and Sons)

[9] Cole K D, Beck J V, Woodbury A and de Monte F 2014 Intrinsic verification and a heat conduction database *Int. J. Thermal Sciences* **78** 36-47

[10] D'Alessandro G and de Monte F 2016 X42B10T00 *Exact Analytical Conduction Toolbox*, exact.unl.edu

[11] de Monte F, Beck J V and Amos D E 2011 A heat-flux based building block approach for solving heat conduction problems *Int. J. Heat Mass Transfer* **54** 2789-800

[12] de Monte F, Beck J V and Amos D E 2012 Solving two-dimensional Cartesian unsteady heat conduction problems for small values of the time *Int. J. Thermal Sciences* **60** 106-13

[13] Carslaw H S and Jaeger J C 1959 *Conduction of Heat in Solids 2nd edition* (New York: Oxford University Press)

[14] de Monte F and Beck J V 2013 X40B10T00 *Exact Analytical Conduction Toolbox*, exact.unl.edu

[15] Blackwell B F, Cochran R J and Dowding K J 1999 Development and implementation of sensitivity coefficient equations for heat conduction problem *Numerical Heat Transfer Part* B **36** 15-32

[16] de Monte F and Beck J V 2013 X22B10T0 *Exact Analytical Conduction Toolbox*, exact.unl.edu

IOP Conf. Series: Journal of Physics: Conf. Series **796** (2017) 012012 doi:10.1088/1742-6596/796/1/012012

Numerical investigation on laminar round-jet impinging on a surface at uniform heat flux in a channel partially filled with a porous medium

B Buonomo[1], A Diana[2], O Manca[1,*], S Nardini[1]

[1]Dipartimento di Ingegneria Industriale e dell'Informazione, Seconda Università degli Studi di Napoli, Via Roma 29, 81031 Aversa, Italy
[2]Dipartimento di Ingegneria Meccanica, Energetica, Gestionale e dei Trasporti, Università degli Studi di Genova, Genova (GE), Italy

*Corresponding author: oronzio.manca@unina2.it

Abstract. Horizontal channel partially filled with porous media and a single round jet impinging on the porous medium are numerically investigated. The wall facing the round jet is partially heated at uniform heat flux. A two-dimensional axial symmetric flow in the channel is assumed to evaluate the thermal behavior within the channel. The analysis in the porous medium is accomplished in local thermal equilibrium conditions and under the Brinkman–Forchheimer-extended Darcy law assumption. The problem is solved employing the Ansys-Fluent code. Results are given in terms of stream function and temperature fields of fluid and solid matrix, wall temperature profiles, air velocity and temperature profiles along the transversal section of channel. The Peclet number ranges from 1 to 1000 and Rayleigh number values are 10, 50, 100 and 1000. Reynolds jet number, solid wall distance and wall heat flux effects on thermal and fluid dynamic behaviors are investigated. Results indicate that Nusselt number has the highest value for the channel with a porous medium of thickness equal to the channel gap, whereas it presents very small changes increasing the porous medium length on the heated wall. Correlations among average Nusselt, Peclet and Rayleigh numbers are proposed.

1. Introduction

Heat transfer enhancement technologies play an important role in several applications, in terms of technical advantages and cost savings. Jet impingement of a cold fluid is an efficient cooling method of a heated surface, especially if jets are employed with porous media, like aluminum foams. As indicated in literature, jets are becoming important in many engineering applications, as turbine blade internal cooling, electronic cooling systems, solar collectors, food and drying technologies, combustion and anti-icing systems, etc [1-6]. Several parameters, like porosity and thermal conductivity, have effects on the fluid and thermal behaviors and they are investigated to evaluate optimal configurations in order to enhance the heat transfer. First studies on laminar impinging jets on a porous medium are carried out numerically and experimentally, considering the channel totally or partially filled in [7-12]. In [13-16], an impinging slot jet is numerically studied considering mixed convection, Darcy and local thermal equilibrium (LTE) and local thermal non-equilibrium (LTNE) models. In [17-21], the effect of opposite mixed convection on the jet impingement cooling of a partially heated surface is investigated at assigned temperature in a confined porous channel. In [22], impinging laminar jet on an array of discrete protruding heat sources without and with a porous layer is studied thanks to generalized Darcy–Forchheimer–Brinkman and LTE models. In [23, 24] experimental studies of a round, impinging-jet flowing through metallic foam with restricted flow outlets are performed, considering different values of porosity, pore density, length of sample, air velocity and flow outlet height. In [25, 26] a confined, round, laminar impinging jet, fed upward against the gravity, is numerically and experimentally studied, under the influence of an impingement-surface heating.

In this work, a parallel-plate channel partially filled with a high permeability porous medium with a single round jet impinging on the porous medium is numerically investigated. The opposite wall to the air round jet is partially heated at uniform heat flux. The fluid flow in the channel is assumed as two dimensional and the porous medium is modeled using the Brinkman–Forchheimer-extended Darcy model, taking account of viscosity and inertia effects. The structure of the porous medium is homogenous and isotropic, the thermophysical properties of the air and the porous medium are considered temperature independent and the fluid flow is steady state, laminar and incompressible. The analysis in the porous medium is accomplished under LTE conditions and a two-dimensional numerical axial symmetric model is developed to evaluate the hydrodynamic characteristics of heat transfer within the channel. Moreover, the buoyancy effect is considered, whereas the radiation effect is neglected. The problem is solved employing the Ansys-Fluent code. Results are given in terms of stream function, fluid and solid matrix temperature fields, wall temperature profiles, air velocity and temperature as well as solid profiles along the transversal section of porous medium. *Pe* varies from 1 to 1000 and four *Ra* values are considered: 10, 50, 100 and 1000. The study evaluates the effects of *Re*, the effect of the solid wall distance and of wall heat flux on thermal and fluid dynamic behaviors. Further, Nusselt numbers and pressure drop are estimated.

2. Geometrical Description and Mathematical Formulation

A sketch of the investigated system is reported in Fig. 1.

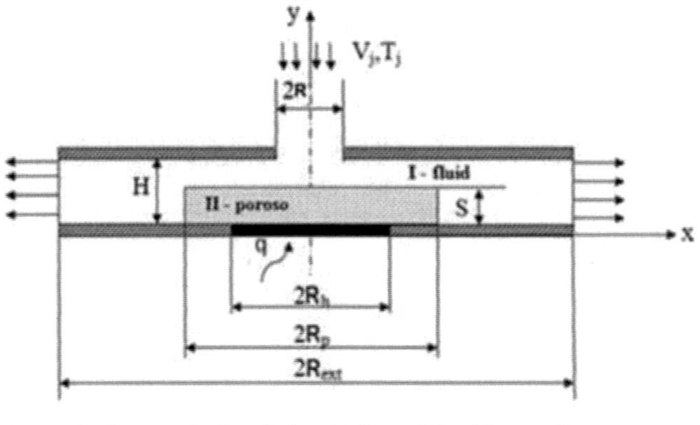

Figure 1. Geometrical and physical model with coordinate system.

The wall facing the air round jet is partially heated at uniform heat flux and the upper wall is considered adiabatic. The round jet section is equal to $2R_j$ and the channel height is H. The heat source radius is equal to R_h and the length of the lower plate is $2R_{ext}$. The unheated zone of the lower wall is assumed adiabatic.

V_j and T_j are, respectively, the velocity and the temperature of the inlet round jet. The impinging jet enters in the channel, which is partially filled by a porous medium with a thickness equal to *S*. In the outlet section, the flow is assumed hydro-dynamically and thermally fully developed. Air is the working fluid (*Pr*=0.71). Past works demonstrate that spatial variation of porosity near solid walls causes possible channeling effect in forced and in mixed convection. Actually, wall channeling increases viscous and inertial effects and the heat transfer rate, but it is particularly important for packed-sphere beds with high particle diameters. Assuming constant values of porosity and permeability, even close to a solid boundary, seems valid for metal foams and fibrous media, thanks to their relatively homogeneous structures [27-29]. The governing equations, in cylindrical coordinates, are [30]:

Continuity equation:

$$\frac{\partial u}{\partial x}+\frac{1}{r}\frac{\partial v}{\partial r}=0 \tag{1}$$

x-momentum equation:

$$\rho_f\left(\frac{1}{\varphi}\frac{\partial u}{\partial t}+\frac{u}{\varphi^2}\frac{\partial u}{\partial x}+\frac{v}{\varphi^2}\frac{\partial u}{\partial r}\right)=-\frac{\partial p}{\partial x}+\frac{\mu_f}{\varphi}\left[\frac{\partial^2 u}{\partial x^2}+\frac{1}{r}\frac{\partial}{\partial r}\left(\frac{r\partial u}{\partial r}\right)\right]-\frac{\mu u}{K}-\frac{C_F}{K^{1/2}}\rho_f\sqrt{u^2+v^2}\,u$$
$$+\rho_f g\beta_f\left(T_f-T_\infty\right) \tag{2}$$

r-momentum equation:

$$\rho_f\left(\frac{1}{\varphi}\frac{\partial v}{\partial t}+\frac{u}{\varphi^2}\frac{\partial v}{\partial x}+\frac{v}{\varphi^2}\frac{\partial v}{\partial r}\right)=-\frac{\partial p}{\partial x}+\frac{\mu_f}{\varphi}\left[\frac{\partial^2 v}{\partial x^2}+\frac{1}{r}\frac{\partial}{\partial r}\left(\frac{r\partial v}{\partial r}\right)-\frac{v}{r^2}\right]-\frac{\mu v}{K}-\frac{C_F}{K^{1/2}}\rho_f\sqrt{u^2+v^2}\,u \tag{3}$$

- Fluid phase:

Energy equation:

$$\varphi\rho_f c_{p,f}\frac{\partial T_f}{\partial t}+\rho_f c_{p,f}\left(u\frac{\partial T_f}{\partial x}+v\frac{\partial T_f}{\partial r}\right)=\varphi k_f\left[\frac{\partial^2 T_f}{\partial x^2}+\frac{1}{r}\frac{\partial}{\partial r}\left(\frac{r\partial T_f}{\partial r}\right)\right]+h_{sf}a_{sf}\left(T_s-T_f\right) \tag{4}$$

- Solid phase:

Energy equation:

$$(1-\varphi)\rho_s c\frac{\partial T_f}{\partial t}=(1-\varphi)k_f\left[\frac{\partial^2 T_s}{\partial x^2}+\frac{1}{r}\frac{\partial}{\partial r}\left(\frac{r\partial T_s}{\partial r}\right)\right]-h_{sf}a_{sf}\left(T_s-T_f\right) \tag{5}$$

where *u* and *v* are velocity components in cylindrical coordinates (*x*, *r*).

The boundary conditions are:

- Inlet section: V_j and T_j are velocity and temperature values respectively for the round jet.

- Outlet section: $\phi=u=v=T$; $\quad \partial\varphi/\partial n=0$ (6)

- Unheated adiabatic wall: $u=v=0$; $\partial T/\partial n=0$ (7)

The dimensionless variables are defined as:

$$\theta=\frac{T-T_\infty}{\frac{q_w H}{k_f}} \qquad E_c=\frac{p_r}{\rho_f V_r^2}\rightarrow E_c=1\rightarrow p_r=\rho_f V$$

$$Re=\frac{\rho_f V_r L_r}{\mu};\quad Da=\frac{K}{L_r^2};\quad Gr=\frac{g\beta\Delta T_r L_r^3}{v_f^2\alpha_f};\quad Ri=\frac{Gr}{\mathrm{Re}^2};\quad Pe=PrRe \tag{8}$$

The average Nusselt number along the heated surface is defined as:

$$Nu_{avg}=\frac{hD_j}{k_f} \tag{9}$$

where D_j is the inner diameter of the round jet and h is the heat transfer coefficient, calculated as:

$$h=\frac{Q_{net}}{A_b(T_s-T_e)} \tag{10}$$

where Q_{net} is the net heat input, A_b is the substrate area, T_s is the average temperature of the substrate and T_e is the jet exit temperature.

3. Results and discussion

Wall temperatures for solid and fluid phases are equal, as assumed in Eq. (7). The porous medium is saturated by air (Pr = 0.71, k_f= 0.025 W/m K). Simulations are performed for assigned dimensionless lengths of the channel and the heated surface lengths. They are R/H = 5 and R_h/H =1 respectively. Da is equal to 10^{-3} and κ is equal to 10^{-4}. The dimensionless half-width of the round jet varies in the range $0.1 \leq R/H \leq 1$. Results are discussed for $1 \leq Pe \leq 1000$ and for Ra equal to 10, 50, 100, 1000. Dimensionless parameters are calculated as:

$$S=\frac{s}{H}\quad R_p=\frac{r_p}{H}\quad R_h=\frac{r_h}{H}\quad R_j=\frac{r_j}{H}\quad R_{ext}=\frac{r_{ext}}{H} \tag{11}$$

In the following, several configurations are analyzed, considering different geometrical and thermal parameters.

3.1 Configuration 1: S equal to 0.2, R_p equal to 5.

In Figure 2(a), Nu_{avg} varies in function Pe, considering four values of Ra. When Pe and Ra arise, Nu_{avg} grows. When Pe>100, curves have the same trend: values are equal changing Ra. Curves demonstrate that when Ra>100, forced convection is predominant; instead, when Ra<100, buoyancy forces are predominant. Observing temperature trend along x-direction in Figure 2(b), it is notable that near the spot, values are constant thanks to the high thermal conductivity of the porous medium.

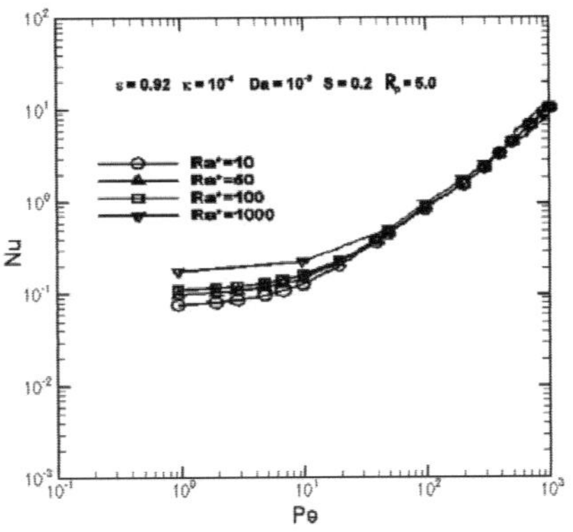

(a) Average Nusselt number as a function of Peclet number for different Rayleigh numbers

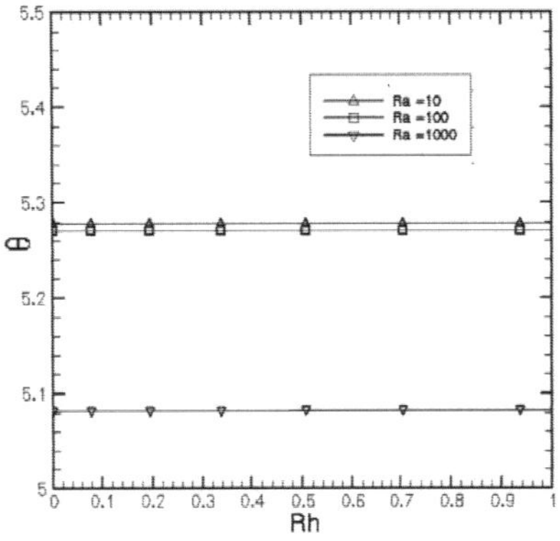

(b) Wall temperature profile as a function of R_h for different Rayleigh numbers.

Figure 2. Configuration 1: Thermal characteristics.

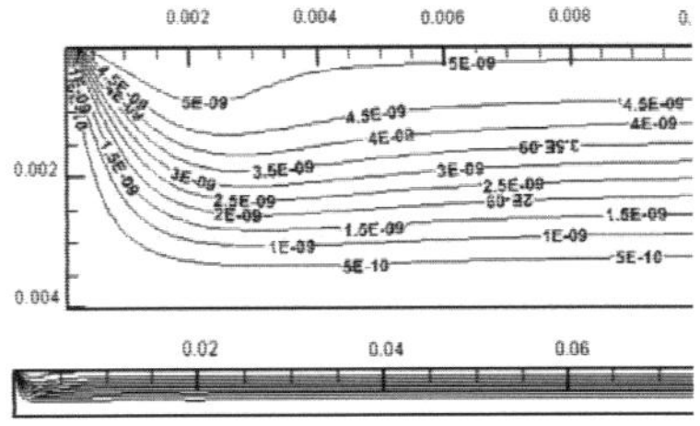

(a) Ra = 10.

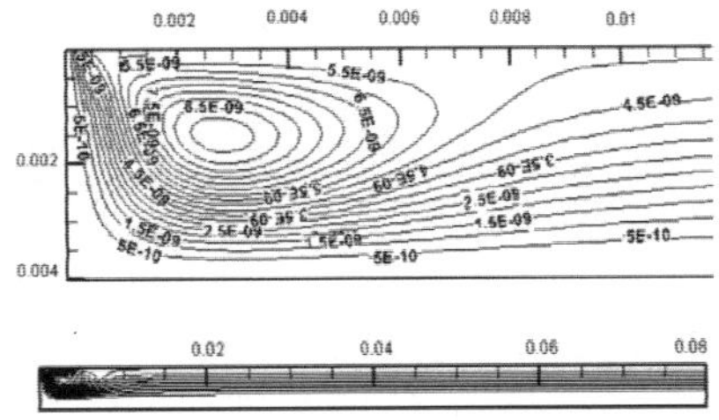

(b) Ra = 100

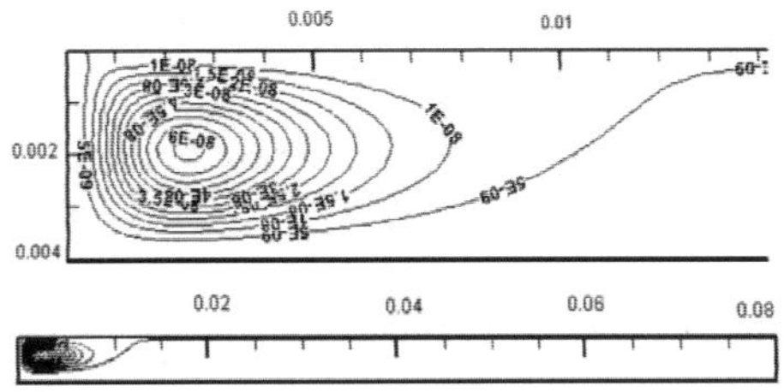

(c) Ra = 1000

Figure 3. Streamlines and isothermal patterns in proximity of the heated spot.

Figures 3(a), 3(b) and 3(c) show streamlines for the fixed value *Pe* =10, and for three different values of *Ra*. Greater is *Ra*, more variable is the flow, above all in proximity of the porous medium and greater is the swirling zone, due to buoyancy forces.

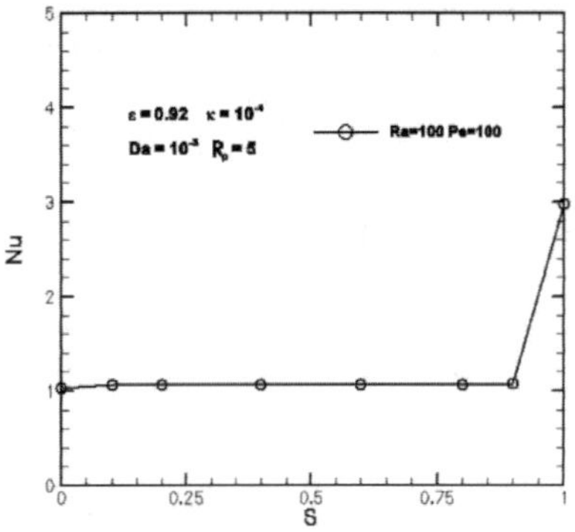

(a) Average Nusselt number as a function of *S* for assigned Rayleigh and Peclet numbers.

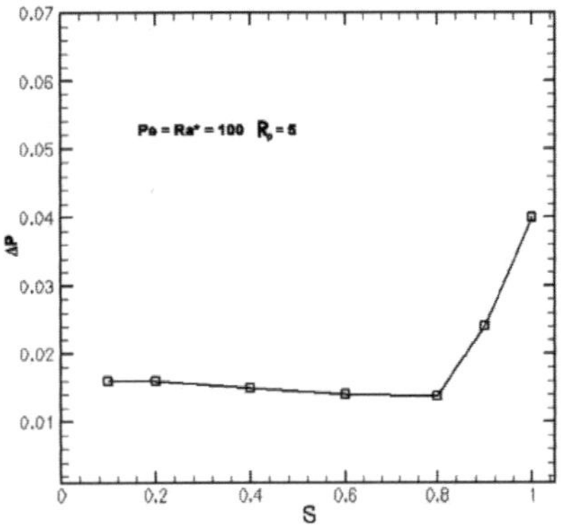

(b) Pressure drop profile as a function of *S*.

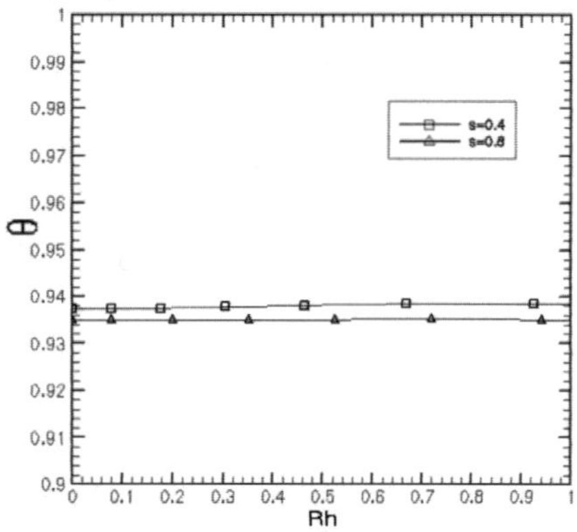

(c) Wall temperature profile as a function of R_h for different Rayleigh numbers.

Figure 4. Configuration 2: Thermal characteristics.

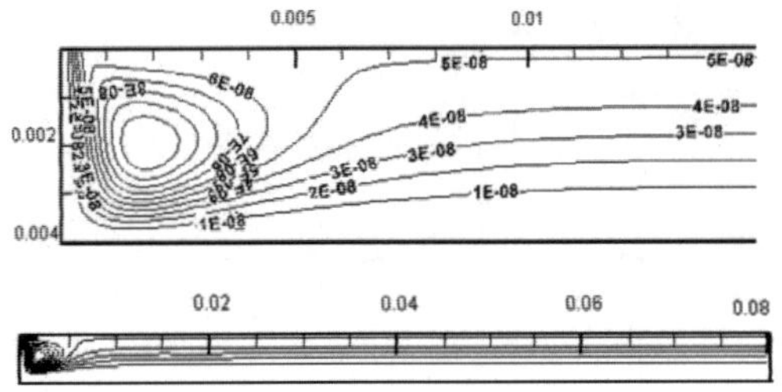

(a) $S = 0.4$.

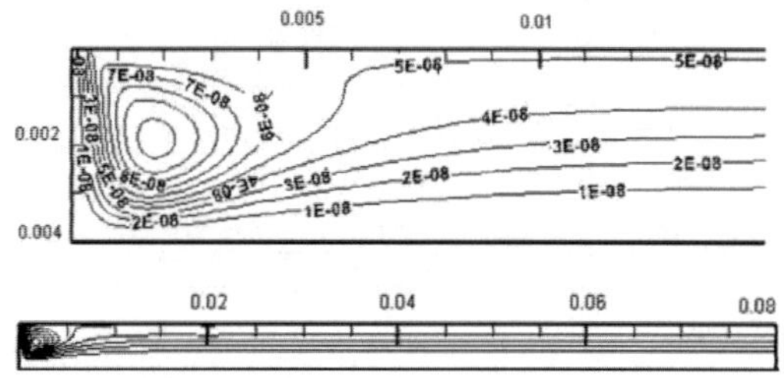

(b) $S = 0.8$.

Figure 5. Streamlines and isothermal patterns in proximity of the heated spot for (a) *S*=0.4 and (b) *S*=0.8.

IOP Conf. Series: Journal of Physics: Conf. Series **796** (2017) 012012 doi:10.1088/1742-6596/796/1/012012

3.2 Configuration 2: S variable, R_p equal to 5.

To optimize heat transfer, it is essential to know the *Nu* trend as a function of porous medium thickness. In Figure 4(a), *Ra* and *Pe* are constant and equal to 10; *S* is variable from 0.1(clean channel) to 1.0 (channel totally filled by porous medium). Nu_{avg} is almost constant when *S* varies from 0.1 to 0.9 and it is maximum when *S* = 1.0. The porous medium grants a higher convective heat transfer coefficient for any *S* value and this is confirmed by ΔP, as a function of *S*. In Figure 4(b), ΔP raises with *S* when *S*>0.8. These results indicate that the heat loss is the least when *S* varies from 0.4 to 0.8. Figure 4(c) shows an almost constant temperature when *S* varies from 0.4 to 0.8.

Figures 5(a) and 5(b) show streamlines and the vortex is the same for different *S* values.

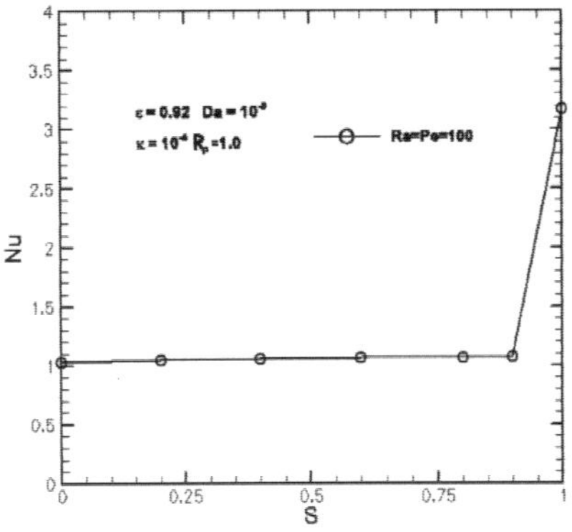

(a) Average Nusselt number as a function of *S* for assigned Rayleigh and Peclet numbers

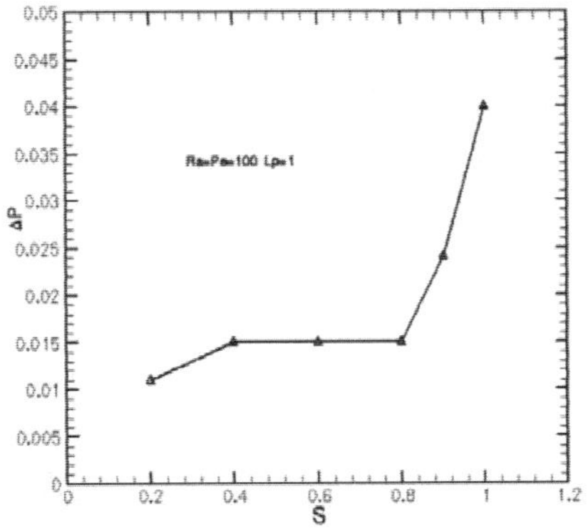

(b) Pressure drop profile as a function of *S*.

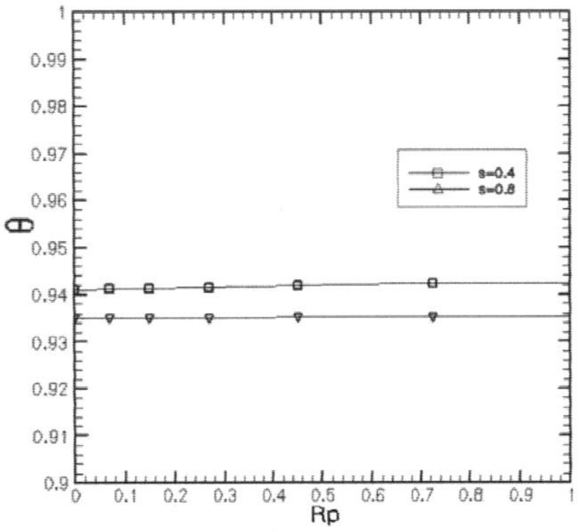

(c) Wall temperature profile as a function of R_p for different Rayleigh numbers.

Figure 6. Configuration 3: Thermal characteristics.

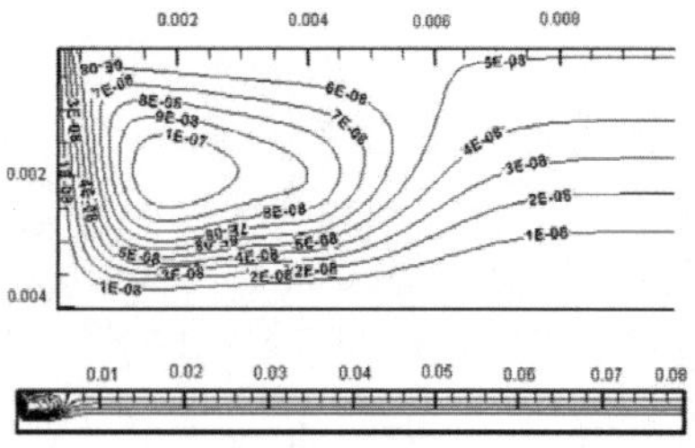

(a) $S = 0.4$.

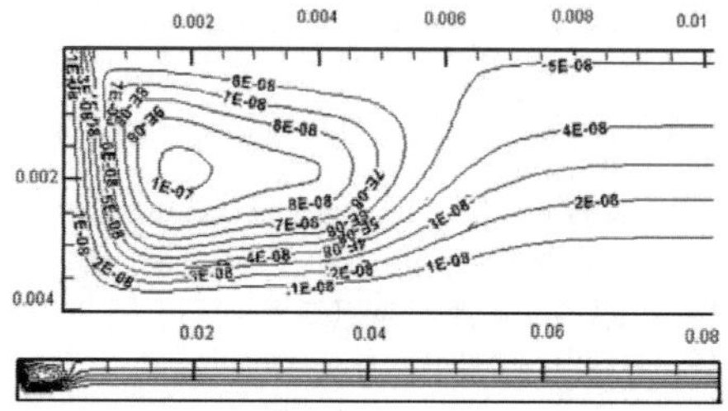

(b) $S = 0.8$

Figure 7. Streamlines and isothermal patterns in proximity of the heated spot for (a) S=0.4 and (b) S=0.8

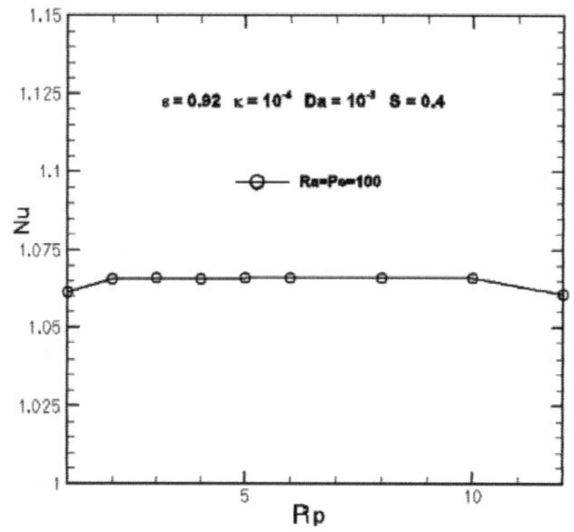

(a) Average Nusselt number as a function of R_p for assigned Rayleigh and Peclet numbers.

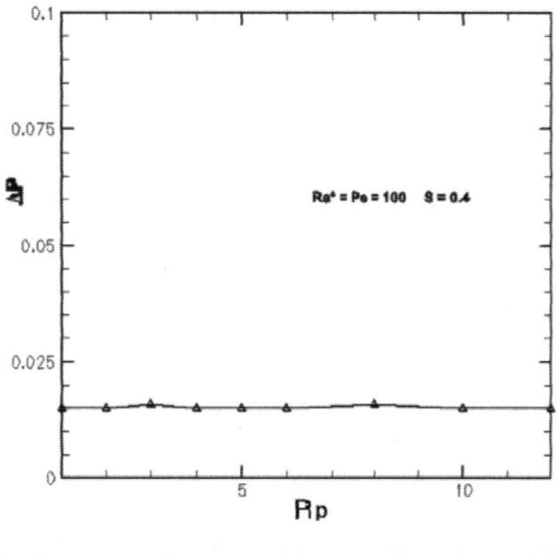

(b) Pressure drop profile as a function of R_p

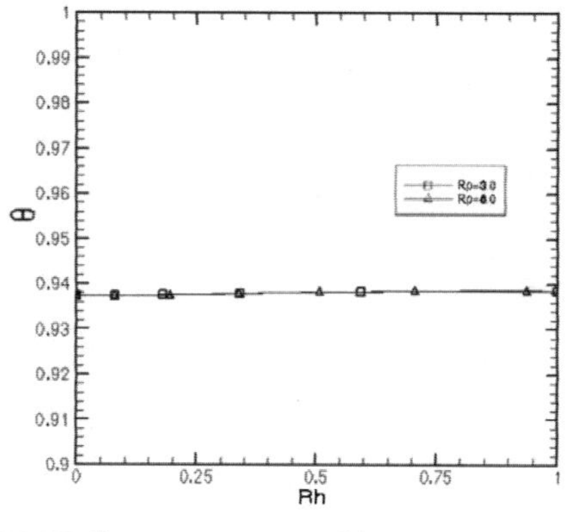

(c) Wall temperature profile as a function of R_h for different R_p

Figure 8. Configuration 4: Thermal characteristics.

3.3 Configuration 3: S variable, R_p equal to 1.

In this case, the porous medium length is equal to the heated wall. *Ra* and *Pe* are equal to 100. In Figure 6(a), Nu_{avg} is almost linear, when *S* varies from 0.1 to 0.8 and it is maximum when $S = 1.0$. ΔP, in Figure 6(b), has a considerable growth when $S > 0.8$, so it is appropriate to choose a smaller *S* value. Figure 6(c) shows an almost constant temperature when *S* is equal to 0.4 and to 0.8.

Figures 7(a) and 7(b) show streamlines and the vortex is the same for different *S* values.

3.4 Configuration 4: S equal to 0.4, R_p variable.

Now, the *S* value is fixed and R_p varies from 1 to 12, considering $Ra = Pe = 100$. In Figure 8(a), Nu_{avg} stabilizes when $R_p>8$, so the porous medium is not necessary. When $R_p = 12$, Nu_{avg} decreases, because vortexes make worse the heat change, so the most suitable value of R_p is smaller than 10. ΔP, in Figure 8(b), does not change considerably with R_p. These results prove that, for $S = 0.4$, the most suitable value of R_p is 2. Figure 8(c) shows an almost constant temperature when R_p varies from 3 to 8.

Figures 9(a) and 9(b) show streamlines and the vortex is the same for different *S* values.

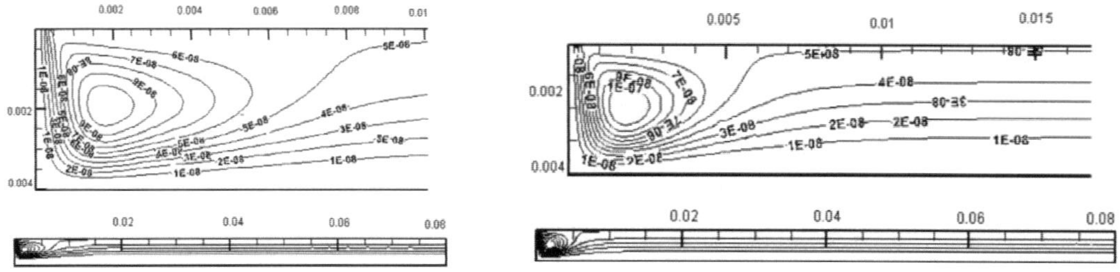

(a) $R_p = 3$. (b) $R_p = 8$.

Figure 9. Streamlines and isothermal patterns in proximity of the heated spot for (a) R_p=3 and (b) R_p=8.

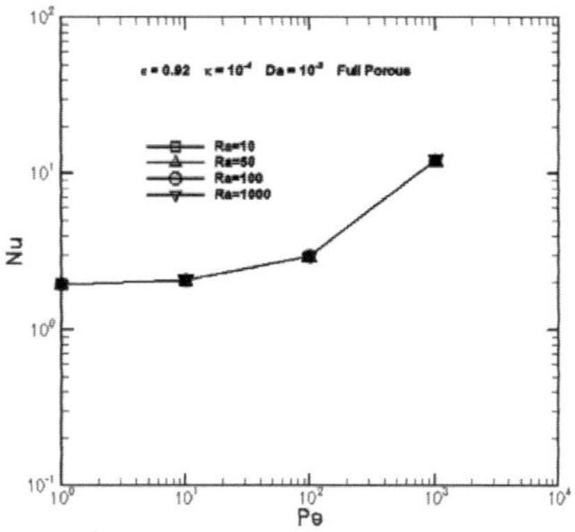

Figure 10. Average Nusselt number as a function of *Pe* for different Rayleigh numbers.

3.5 Configuration 5: S equal to 1 (channel totally filled).

In this chase, the channel is totally filled by the porous medium. Figure 10 shows that curves coincide for different values of *Ra*: Nu_{avg} is not influenced by *Ra*, because solid medium heat conductivity is greater then fluid medium ($k = 10^{-4}$), so solid temperature drops are negligible and buoyancy forces are weak. Figures 11(a) and 11(b) show that streamlines are the same for different values of *Ra*.

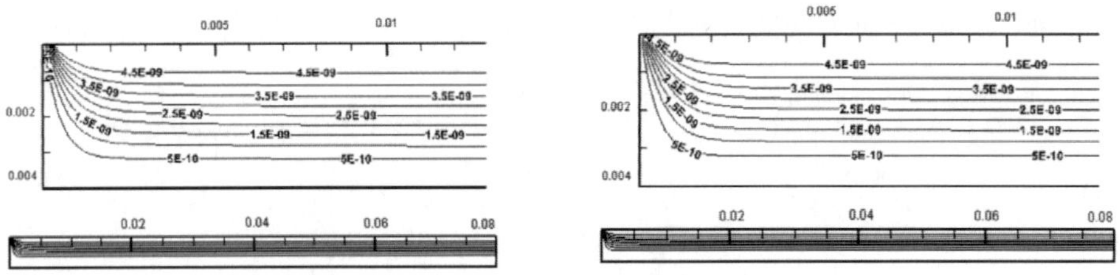

(a) $Ra = 10$. (b) $Ra = 100$.

Figure 11. Streamlines and isothermal patterns in proximity of the heated spot for (a) *Ra*=10 and (b) *Ra*=100.

Furthermore, the heat transfer coefficient ratio, h_m/h_c, is analyzed in the following, as a function of *S* and R_p, for assigned *Ra* and *Pe* ($Ra = Pe = 100$). As shown in Figure 12(a), if *S* varies from 0.2 to 0.8, h_m/h_c does not arise considerably with or without porous medium but it is noticeable when *S* is equal to the channel height. In Figure 12(b), if $R_p < 3$, h_m/h_c arise considerably with or without porous medium, so in this range heat losses are the smallest.

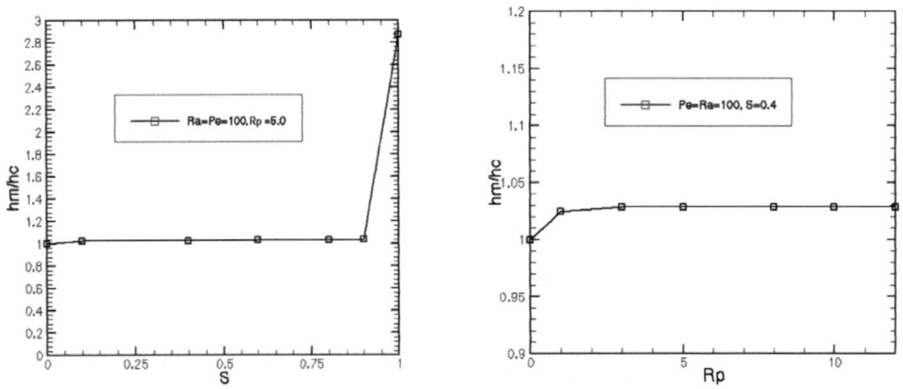

(a) In function of *S* for assigned *Ra, Pe*, R_p. (b) In function of R_p for assigned *Ra, Pe, S*.

Figure 12. Heat transfer coefficient ratio.

4. Conclusions

A parallel-plate channel filled partially with a high permeability porous medium and a single round jet impinging on the porous medium are numerically investigated. The opposite wall to the air round jet is partially heated at uniform heat flux. The fluid flow in the channel is assumed two dimensional and the porous medium is modeled using the Brinkman–Forchheimer-extended Darcy model. The porous medium structure is homogenous and isotropic, the thermophysical properties considered temperature independent. The analysis in the porous medium is accomplished under local thermal equilibrium conditions.

Analyzing five different configurations, studies reveal the correlation between Nusselt number and Rayleigh and Peclet numbers. For assigned *S* and R_p (Configuration 1), when *Pe* and *Ra* arise, Nu_{avg} grows, but if $Pe > 100$, Nu_{avg} values are equal changing *Ra*. Varying *S* (Configurations 2 and 3), results indicate that the heat loss is the least when *S* varies from 0.4 to 0.8 and the vortex zones are similar for each *S*. For assigned R_p and varying *S* (Configuration 4), the average Nusselt number, Nu_{avg}, stabilizes when $R_p > 8$, so the porous medium is not necessary. The study of the correlation between average Nusselt number and porous medium thickness (Configuration 5) optimizes heat transfer system. In particular, the presence of porous medium increases the heat transfer near the heated bottom wall, so when the channel is totally filled by the porous medium, Nu_{avg} is not influenced by *Ra*, cause solid medium heat conductivity is greater then fluid medium, solid temperature differences are negligible and buoyancy forces are weak.

5. Nomenclature

a_{sf}	specific surface area, m^2/m^3
Bi	Biot-like number
C_F	inertia coefficient
D	width of the jet, m
Da	Darcy number
g	gravitational acceleration, ms^{-2}
H	width of the channel, m
R	round slot radius, m
h_{sf}	interstitial heat transfer coeff., Wm^{-2}K^{-1}
k	thermal conductivity, Wm^{-1}K^{-1}
K	permeability, m^2
R_h	half of the heat source length, m
Ra	Rayleigh number
Nu_{avg}	average Nusselt number
Pr	Prandtl number
q	heat flux, Wm^{-2}
Pe	Peclet number
T	temperature, K
u,v	velocity components, ms^{-1}
U,V	dimensionless velocity components
x,y	Cartesian coordinates, m
X,Y	dimensionless Cartesian coordinates
ΔP	pressure drop

Greek Symbols

α	thermal diffusivity, m^{2}s^{-1}
β	volumetric expansion coefficient, K^{-1}
γ	modified conductivity ratio
ϑ	dimensionless temperature
κ	effective thermal conductivity ratio
ν	kinematic viscosity, m^{2}s^{-1}
φ	porosity
ρ	density, kg m^{-3}

Subscripts

eff	effective
f	fluid
j	jet
s	solid
w	wall
avg	average

6. References

[1] Han B and Goldstein RJ 2001 *Ann. New York Acad. Sciences* **934** 147.
[2] Wright WB 2004 *AIAA Paper* 4153.
[3] Agostini B, Fabbri M, Park JE, Wojtan L, Thome JR and Michel B 2007 *Heat Transfer Eng.* **28** 258.
[4] Ebadian MA and Lin CX 2011 *ASME J. Heat Transfer* **133** art. no. 110801.
[5] Gururatana S 2012 *Am. J. App. Sciences* **9** 436.
[6] Andrei L, Carcasci C, Da Soghe R, Facchini B, Maiuolo F, Tarchi L and Zecchi S 2013 *ASME J. Turbomach.* **135** art. no. 031020.
[7] Viskanta R 1993 *Exp Therm. Fluid Science* **6** 111.
[8] Polat S 1993 *Drying Tech* **11** 1147.

[9] Iacovides H and Launder BE 1995 *Int. J. Heat Fluid Flow* **16** 454.
[10] Baukal CE Jr. and Gebhart B 1996 *Int. J. Heat Fluid Flow* **17** 386.
[11] Fu WS and Huang HC 1997 *Int. J. Heat Mass Transfer* **40** 2261.
[12] Jeng TM and Tzeng SC 2005 *Int. J. Heat Mass Transfer* **48** 4685.
[13] Saeid NH and Mohamad AA 2006 *Int. J. Heat Mass Transfer* **49** 3906.
[14] Saeid NH 2007 *Int. J. Heat Mass Transfer* **50** 4265.
[15] Sivasamy A, Selladurai V and Kanna PR 2010 *Int. J. Heat Mass Transfer* **53** 5847.
[16] Sivasamy A, Selladurai V and Kanna PR 2010 *Int. J. Therm. Sciences* **49** 1238.
[17] Wong KC and Saeid NH 2009 *Int. Comm. Heat Mass Transfer* **36** 45.
[18] Wong KC and Saeid NH 2009 *Int. J. Therm. Sciences* **48** 96.
[19] Wong KC and Saeid NH 2009 *Int. Comm. Heat Mass Transfer* **36** 155.
[20] Sivasamy A, Kanna PR and Selladurai V 2011 *Int. J. Heat Mass Transfer* **54** 4127.
[21] Buonomo B, Manca O and Nardini S 2012 *4th Int. Conf. Porous Media and its Applications in Science and Engineering, ICPM4*, June 17-22, Potsdam, Germany.
[22] Lam PAK and Prakash KA 2015 *Energy Conv. Manag.* **89** 626.
[23] Shih WH, Chou FC and Hsieh WH 2007 *ASME J. Heat Transfer* **129** 1554.
[24] Yakkatelli R, Wu Q and Fleisher AS 2010 *Exp Thermal Fluid Science* **34** 1008.
[25] Feng SS, Kuang JJ, Wen T, Lu TJ and Ichimiy K 2014 *Int. J. Heat Mass Transfer* **77** 1063.
[26] Lee CH, Lim KB, Lee SH, Yoon YJ and Sung NW 2007 *Exp Thermal Fluid Science* **31** 559.
[27] Hunt ML and Tien CL 1988 *Int. J. Heat Mass Transfer* **31** 301.
[28] Jeng TM and Tzeng SC 2008 *Int. Comm. Heat Mass Transfer* **35** 30.
[29] Lee DY and Vafai K 1999 *Int. J. Heat Mass Transfer* **42** 423.
[30] Buonomo B, Manca O and Nardini S 2015, *Handbook of Porous Media, Third Edition, edited by Kambiz Vafai*, CRC Press 631.

Numerical investigation on forced convection in rectangular cross section micro-channels with nanofluids

B Buonomo*, L Cirillo, O Manca, S Nardini, S Tamburrino

Dipartimento di Ingegneria Industriale e dell'Informazione, Seconda Università degli Studi di Napoli (SUN), Via Roma 29, Aversa (CE), 81031, Italia

*Corresponding author: buonomo.bernardo@unina2.it

Abstract. In this paper a numerical investigation on laminar forced convection flow of a water–Al_2O_3 nanofluid in a rectangular microchannel is accomplished. A constant and uniform heat flux on the external surfaces has been applied and a single-phase model approach has been employed. The analysis has been performed in steady state regime for particle size in nanofluids equal to 38 nm. The CFD commercial code Fluent has been employed in order to solve the 3-D numerical model. The geometrical configuration under consideration consists in a duct with a rectangular shaped crossing area. A steady laminar flow and different nanoparticle volume fractions have been considered. The base fluid is water and nanoparticles are made up of alumina (Al_2O_3). The length the edge and height of the duct are 0.030 m, 1.7 x10^{-7} and 1.1 x10^{-7} m, respectively. Results are presented in terms of temperature and velocity distributions, surface shear stress and heat transfer convective coefficient, Nusselt number and required pumping power profiles. Comparison with results related to the fluid dynamic and thermal behaviors are carried out in order to evaluate the enhancement due to the presence of nanoparticles in terms of volumetric concentration.

1. Introduction

The heat transfer in fluids is very crucial for industrial applications, especially for heating or cooling operation. Convective heat transfer can be enhanced passively by changing flow geometry, by enhancing the thermal conductivity of the working fluid, or putting another material in the system with high value of the thermal conductivity. An innovative method of enhancement of thermal conductivity of base fluids is to insert nanoparticles. In fact, several investigations have been accomplished on the microchannel heat sink using nanofluids [1]. The nanofluid in a microchannel heat exchanger was studied by Lee and Choi [2] and they confirmed that the nanofluid enhances the cooling rate respect to the pure water and liquid nitrogen.

A numerical study on the laminar forced convection in microchannel with nanofluid is carried out in Koo and Kleinstreuer [3] considering two types of nanofluids, CuO/ethylene glycol and CuO/water. They found that nanoparticles with a high value of thermal conductivity enhance the heat transfer. An experimental investigation with nanofluid in microchannel is realized by Chied and Chuang [4]. The results showed that there is a slight increase of the pressure drop due to the presence of the nanoparticles but the heat transfer is improved. The turbulent forced convection in duct with nanofluid was studied in numerical way in Behzadmehr et al. [5]. The nanofluid is modelled employing the two phase mixture model and a comparison with the experimental values is made. The

IOP Conf. Series: Journal of Physics: Conf. Series **796** (2017) 012013 doi:10.1088/1742-6596/796/1/012013

results showed that the two phase mixture model is more accurate than the single phase model. Moreover the use of nanofluid improves the Nusselt number and it seems that the friction factor is not affected. A study on the rectangular microchannels is accomplished by Jung et al. [6], where they found an improvement of the convective heat transfer coefficient but a worsening of the friction factor, in fact the convective heat transfer coefficient of nanofluid is increased of 32% respect to the convectional water. A numerical study on Al_2O_3/water nanofluid in microchannel is carried out in Bhattacharya et al. [7]. The study showed that in the laminar flow the heat transfer coefficient for nanofluid increase with the Reynolds number. Chen and Ding [8] studied the thermal performance of a microchannel heat sink using a porous medium model with Forchheimer-Brinkman-Darcy equation for the fluid flow and the Local thermal non-equilibrium model for the heat transfer. The viscous dissipation effect is taken into account and the results showed that the heat transfer decreases as the particle's diameter increases.

A numerical 3D simulation for incompressible, steady fluid flow of a trapezoidal microchannel using CuO/water nanofluid is accomplished in Yang et al. [9]. They found that the thermal resistance of nanofluid is smaller than the water and the pressure drop increase slightly using the nanofluid respect to the pure water.

After this short review, it seems that there is a lack of work on forced convection in rectangular microchannels in nanofluids. In this paper, a numerical investigation on laminar forced convection in rectangular microchannel with nanofluids is accomplished, in order to assess thermo-dynamic behavior of the considered geometry by using Al_2O_3/water based nanofluids. The nanofluid is modelled with the single phase model and the system is heated by a constant and uniform heat flux on of the top walls of the duct. Results are presented for different nanoparticle volume concentrations and Reynolds number values.

2. Geometrical model and data reduction

The 3-D numerical model consists of a duct with a rectangular shaped crossing area, showed in Fig. 1a, with length, edge and height respectively equal to 0.030 m, 1.7×10^{-7} m and 1.1×10^{-7} m. The flow in the duct is three-dimensional, steady-state, laminar and incompressible. Thermo-physical properties of the nanofluid are considered constant with temperature except for the density where is assumed the Boussinesq approximation because of the presence of buoyancy force. A constant and uniform heat flux q on the top wall is applied; the others are adiabatic and for the inlet section is assumed a uniform velocity and temperature profile. The analysis has been performed for particle size in nanofluids equal to 38 nm. The geometrical configuration of the cross section is showed in Fig. 1b and the geometrical quantities are summarized in Table 1.

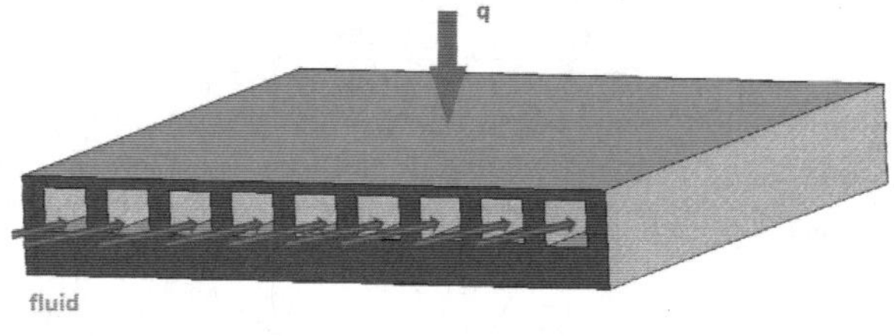

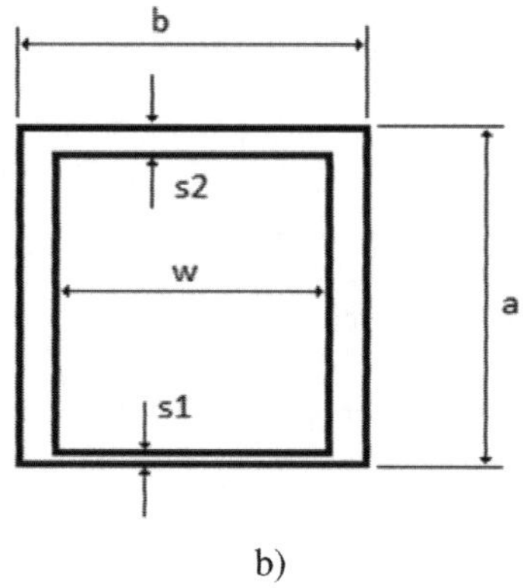

a) b)

Figure 1. a) Schematic presentation of the microchannel heat sink; b) cross-section of microchannel used in the computations.

Table 1. Dimensions of geometrical quantities

a (μm)	*b* (μm)	*w* (μm)	*s1* (μm)	*s2* (μm)
280	447	430	6	10

Continuity:

$$\frac{\partial u}{\partial x}+\frac{\partial v}{\partial y}+\frac{\partial w}{\partial z}=0 \tag{1}$$

Momentum:

$$\left[u\frac{\partial u}{\partial x}+v\frac{\partial u}{\partial y}+w\frac{\partial u}{\partial z}\right]=-\frac{1}{\rho_{nf}}\frac{\partial P}{\partial x}+\nu_{nf}\left[\frac{\partial^2 u}{\partial x^2}+\frac{\partial^2 u}{\partial y^2}+\frac{\partial^2 u}{\partial z^2}\right] \tag{2}$$

$$\left[u\frac{\partial v}{\partial x}+v\frac{\partial v}{\partial y}+w\frac{\partial v}{\partial z}\right]=-\frac{1}{\rho_{nf}}\frac{\partial P}{\partial y}+\nu_{nf}\left[\frac{\partial^2 v}{\partial x^2}+\frac{\partial^2 v}{\partial y^2}+\frac{\partial^2 v}{\partial z^2}\right] \tag{3}$$

$$\left[u\frac{\partial w}{\partial x}+v\frac{\partial w}{\partial y}+w\frac{\partial w}{\partial z}\right]=-\frac{1}{\rho_{nf}}\frac{\partial P}{\partial z}+\nu_{nf}\left[\frac{\partial^2 w}{\partial x^2}+\frac{\partial^2 w}{\partial y^2}+\frac{\partial^2 w}{\partial z^2}\right] \tag{4}$$

Energy for fluid phase:

$$\left(\rho c_p\right)_{nf}\left[u\frac{\partial T_{nf}}{\partial x}+v\frac{\partial T_{nf}}{\partial y}+w\frac{\partial T_{nf}}{\partial z}\right]=\lambda_{nf}\left[\frac{\partial^2 T_{nf}}{\partial x^2}+\frac{\partial^2 T_{nf}}{\partial y^2}+\frac{\partial^2 T_{nf}}{\partial z^2}\right] \tag{5}$$

Energy equation in the solid domain:

$$\frac{\partial^2 T_s}{\partial x^2}+\frac{\partial^2 T_s}{\partial y^2}+\frac{\partial^2 T_s}{\partial z^2}=0 \tag{6}$$

The following boundary conditions are assumed:

- inlet section: uniform velocity and temperature profile;
- outlet section: outflow condition with velocity components and temperature derivatives equal to zero;
- external duct walls: assigned uniform and constant heat flux on the top wall, the others are adiabatic;
- solid-fluid interface: velocity components equal to zero and continuity of the temperature and heat flux.

The dimensionless parameters Reynolds number, Nusselt number and friction factor are considered for the data reduction and they are expressed by the following relations:

$$Re=\frac{Vd_h}{\nu} \tag{7}$$

where V is the average inlet velocity, d_h is the hydraulic diameter.

$$Nu_{av} = \frac{qd_h}{(T_w - T_m)\lambda_{nf}} \tag{8}$$

where q is the heat flux, T_w and T_m represent the average wall and bulk fluid temperatures, respectively.

$$f = 2\Delta P \frac{d_h}{4L} \frac{1}{\rho_{nf} V^2} \tag{9}$$

$$Pr = \frac{\mu c_p}{\lambda} \tag{10}$$

3. Grid independence analysis

The governing equations (1)-(6) with the assumed boundary conditions were solved by means of Fluent 6.3 [10] code. A segregated method was chosen to solve the stationary equations and a second-order upwind scheme was employed for energy and momentum equations. Pressure and velocity were coupled by using SIMPLE coupling scheme. Simulations were considered converged by assuming the convergence criteria of 10^{-4} and 10^{-5} and 10^{-8} for the residuals of continuity, velocity components and energy, respectively. At the inlet section, flow was considered laminar, with a velocity corresponding to the assigned Reynolds number, at a temperature of 293K and ambient pressure conditions. The no-slip condition was applied on the channel walls, which were heated by a uniform and constant heat flux.

A grid dependence test is accomplished to realize the more convenient grid size by monitoring average Nusselt for rectangular microchannel system. Four different unstructured mesh distributions, were tested with water as fluid on the rectangular sectioned duct for *Ti*=293 K, *V*=0.6m/s (*Re*=50) with a uniform heat flux equal to about 75.6 kW/m^2 imposed on the upper surface of the channel in order to perform the grid-independence analysis. They have 70000, 144000, 280000, 480000 and 739000 nodes, respectively. To evaluate the computational grid the Richardson extrapolation was used in terms of Nusselt number. The third grid case (280000 nodes) was adopted because it ensured a good compromise between the computational time and the accuracy requirements. The validation has been then performed by comparing the numerical results, in terms of Nusselt number and friction factor, for *Re*=50 with data from Shah and London [11] for uniform and constant thermal heat flux boundary condition. The reference analytical value is equal to 3.11, while the computation Nusselt number value is 3.18. $Rexf_{avg}$ value for is equal to 13.20. For numerical simulation $Rexf_{avg}$ value is 13.01. The percentage differences between the two solution are acceptable, for Nusselt number and for friction factor too.

4. Thermo-physical properties of nanofluids

In this analysis both pure water and Al_2O_3/water based nanofluids with a particle diameter of 38 nm are considered as working fluid. The single-phase model approach is adopted in order to describe the nanofluid behavior because small temperature differences and small particle volume fractions are considered. In Table 2 the values of density, specific heat, dynamic viscosity and thermal conductivity, are reported for water and Al_2O_3 particles at the reference temperature of 293 K. In this paper different values of volume concentrations, equal to 0.1%, 0.5%, 1.0%, 2.0%, 4.0% and 4.0%, are investigated. Thermal properties are evaluated at a temperature equal to 293 K. Density was evaluated by using the classical formula valid for conventional solid-liquid mixtures while the specific heat and thermal expansion coefficient values are calculated by assuming the thermal equilibrium between nanoparticles and surrounding fluid. Thermal expansion coefficient values are evaluated by adopting the relation for different volume particle concentrations.

Density:

$$\rho_{nf} = (1-\varphi)\rho_{bf} + \varphi\rho_p \tag{11}$$

Specific heat:

$$\left(\rho c_p\right)_{nf} = (1-\varphi)\left(\rho c_p\right)_{bf} + \varphi\left(\rho c_p\right)_p \tag{12}$$

Thermal expansion coefficient:

$$\beta_{nf} / \beta_{bf} = \frac{1}{\left(\frac{1-\varphi}{\varphi}\frac{\rho_{bf}}{\rho_p}\right)}\frac{\beta_p}{\beta_{bf}} + \frac{1}{\left(\frac{\varphi}{1-\varphi}\frac{\rho_{bf}}{\rho_p}+1\right)} \tag{13}$$

Table 2. Material properties at the reference temperature of 293 K.

Material	$\rho\,(\mathrm{kgm^{-3}})$	$Cp\,(\mathrm{Jkg^{-1}K^{-1}})$	$\beta\,(\mathrm{K^{-1}})$	$\mu\,(\mathrm{Pas})$	$\lambda\,(\mathrm{Wm^{-1}K^{-1}})$
Al_2O_3	3880	773	//	//	36
Water	998.2	4128	2.100x10^{-4}	993x10^{-4}	0.597

Viscosity and thermal conductivity are evaluated by means following equations, which are adopted because they are expressed as a function of particle volume concentration and diameter.

Dynamic viscosity:

$$\frac{\mu_{nf}}{\mu_{bf}} = \frac{1}{1-34.87\left(d_p / d_f\right)^{-0.3}\varphi^{1.03}} \tag{14}$$

with $d_f \;\; 0.1\left(\frac{6M}{N\pi\rho_{f,0}}\right)$, in which M is the molecular weight of the base fluid, N is the Avogadro number and $\rho_{f,0}$ is the mass density of the base fluid calculated at $T = 293$ K.

Thermal conductivity:

$$\frac{\lambda_{nf}}{\lambda_{bf}} = 1 + 4.4\,\mathrm{Re}^{0.4}\,\mathrm{Pr}^{0.66}\left(\frac{T}{T_{fr}}\right)^{10}\left(\frac{\lambda_p}{\lambda_{bf}}\right)^{0.03}\varphi^{0.66} \tag{15}$$

with $\mathrm{Re} = \frac{2\rho_{bf}k_b T}{\pi\mu_{bf}^{\;2} d_p}$, in which k_b is the Boltzmann constant, equal to 1.36x10^{-26}, and T_{fr} is equal to 273.15 K.

Validation has been accomplished by comparing the results at Re = 100 and Gr = 0 with literature, thermal boundary condition referring to constant axial wall heat flux with constant peripheral wall temperature [11]. In particular, the comparison in terms of local and average Nusselt number has been accomplished by considering data obtained by Wilbulswas [12] and also reported in [11]. Data are presented as a function of the axial coordinate for the thermal entrance region, z^*, defined as $z^*=z/(dh\ Re)$. Furthermore, the validation in terms of friction factor has been performed by comparing data from Fleming and Sparrow [13], Miller and Han [14], and Aggarwala and Gangal [15], also reported in [11], with the obtained results for fully developed laminar flow in forced convection. Data reduction has been performed by considering the axial coordinate for the hydrodynamic entrance region, such as z^+ parameter, defined by $z^* = z/(dh\ Pe)$. Figure 2 shows these comparisons and it is clear that the present numerical results are in good agreement with the literature data. In particular, a maximum difference of 1.5% is observed for local and average Nusselt number at most, as shown in Figure 2a, while a maximum error of 2% is evaluated in terms of friction factor, as reported in Figure 2b. Thermal properties are evaluated at a temperature equal to 293 K and they are reported in Table 3

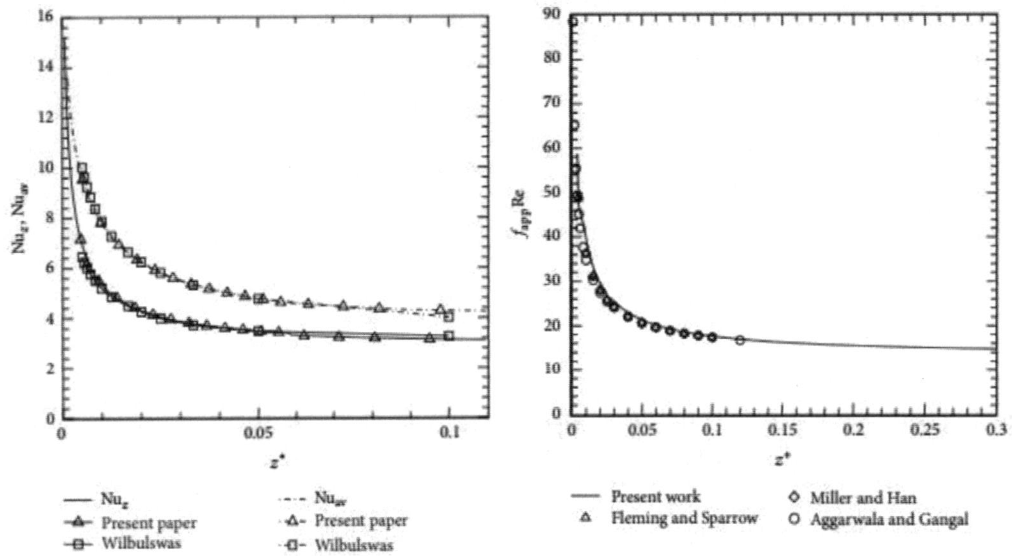

Figure 2. Validation pure water at Re=100 and Gr=0: a) average and local Nusselt number; b) f_{avg}xRe=f_{app}xRe

Table 3. Nanofluids properties at the reference temperature of 293 K.

Φ (%)	ρ (kgm^{-3})	Cp (Jkg^{-1}K^{-1})	λ (Wm^{-1} K^{-1})	μ (kgm^{-1} s^{-1})
0	998.2	4182	0.597	9.93x10^{-04}
0.1	1002	4179	0.599	1.00 x10^{-03}
0.5	1013	4117	0.611	1.03x10^{-03}
1.0	1027	4053	0.619	1.08x10^{-03}
2.0	1056	3931	0.632	1.18x10^{-03}
4.0	1113	3707	0.653	1.46x10^{-03}

5. Results

The results of the numerical simulations are carried out for q=7.56x10^{4}W/m^2, Reynolds number, Re, in the range 50-1000, Prandtl number in the range 6.96-8.29, and the nanoparticle volume fractions, ϕ, in the 0%-4.0% range. Different values of Grashof number, ranging from 0 to 5x10^6, were considered. Results are reported in terms of average convective heat transfer coefficient, average Nusselt number and required pumping power profiles.

In Fig. 3a average convective heat transfer coefficient profiles as a function of Reynolds number values for different nanoparticle volume fractions and thickness *S* are reported. For every Reynolds number nanofluids give an increment in the average convective coefficient respect to the one for the base fluid and this advantage is as greater as volume concentration of nanoparticle is. Besides, for each thickness *S* the convective coefficient increases approximately linearly as the Reynolds number increases. The effect of the value of the thickness *S* is very weak as expected because the microchannel wall is made of aluminum. Increment of average convective heat transfer coefficient due to the presence of nanoparticles is clear in Fig. 3b where the ratio between convective coefficients with and without nanoparticles inside the base fluid is reported as function of the Reynolds number. For each volume concentration the enhancement increases as the Reynolds number increases but this effect becomes weaker and weaker for high Reynolds numbers.

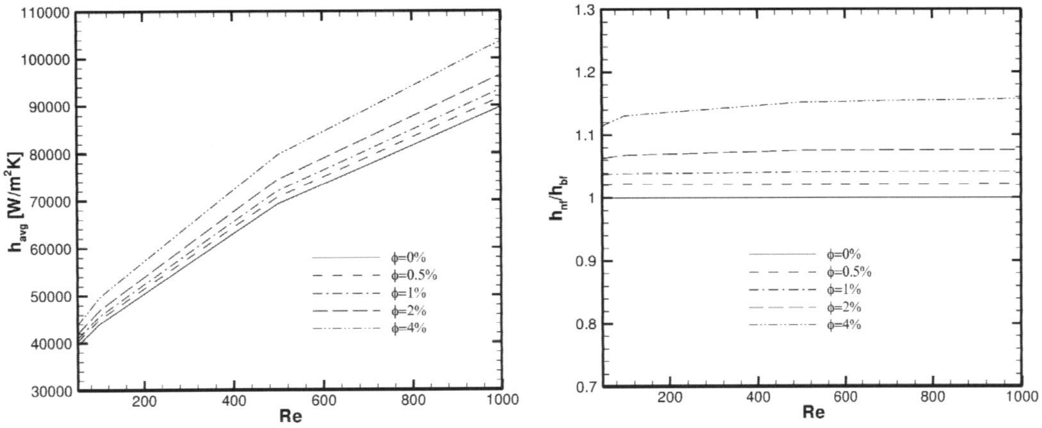

a) b)

Figure 3. Convective heat transfer coefficient for different volume fractions at varying of Reynolds number: a) absolute coefficient; b) relative coefficient

In Fig.4a average Nusselt number, which is defined in Eq.(8), as a function of the Reynolds number for S=6μm and different volume concentration is shown.

When the value of Reynolds number increases average Nusselt number increases too. For $\varphi \leq 0.5\%$

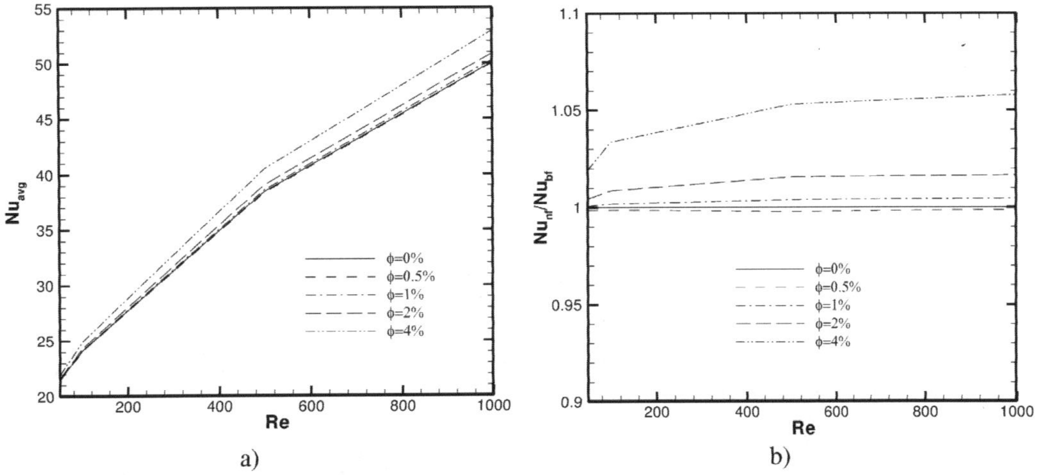

a) b)

Figure 4. Nusselt number for different volume fractions at varying of Reynolds number: a) absolute Nusselt number; b) relative Nusselt number.

Nusselt profiles are nearly overlapping. In addition, for $Re \leq 100$ average Nusselt numbers are very similar for all the investigated volume concentration values.

Indications of the effect of nanoparticles volume concentration on the augmentation of Nusselt number are given in Fig.4b. When volume concentration is lower 0.5%, average Nusselt number is nearly independent on the presence of nanoparticles and Reynolds number so it is very close to the value for pure water. For higher concentrations average Nusselt number increases as Reynolds number increases and this effect is stronger for $Re \leq 500$. Anyway the enhancement of the average Nusselt number is not marked. Maximum value of the Nu_{nf}/Nu_{bf} ratio is equal to 1.055 which is attained for φ=4.0% and Re=1000.

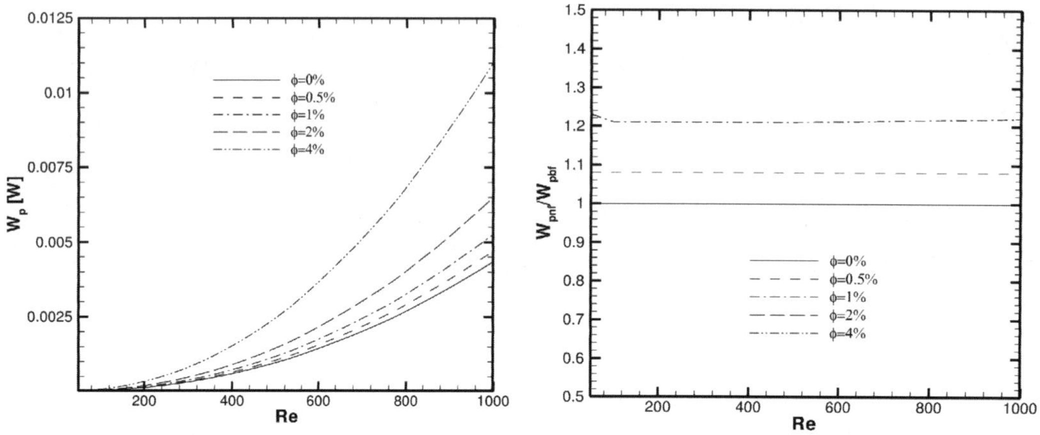

a) b)

Figure 5. a) Required pumping power as a function of Reynolds number; b) relative required pumping power as a function of Reynolds number

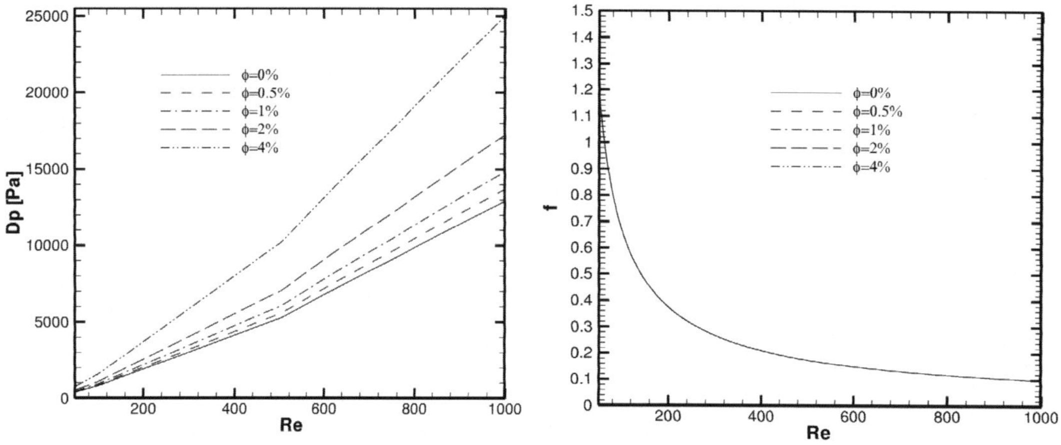

a) b)

Figure 6. a) Load losses as a function of the Reynolds number and for different volume concentrations of particles; b) friction factor as a function of Reynolds number

An interesting question is if the increase in the heat transfer performance due to the presence of nanoparticles does not require excessive pumping power. As an effect also pumping power strongly depends on particle volume concentration as shown in Fig.5a, particularly for $\varphi \leq 0.5\%$. Pumping power enhancement as function of Reynolds number is shown in Fig.5b for several volume concentrations. This ratio for an assigned concentration is not dependent on Reynolds number. For φ=4.0% pumping power is 1.2 times the pumping power with pure water.

In Fig.6a pressure drop in the channel is reported as a function of Reynolds number for S=6μm and for different volume concentrations of nanoparticles. It is worth noticing that at an assigned Reynolds number pressure drop increases very quickly with particle volume concentration.

As far as the friction factor defined by Eq.(9), for a duct with rectangular cross section and for laminar flow and hydro-dynamically and thermally fully developed flow:

$$f=13.3/Re \tag{16}$$

In Fig.6b the friction factor as function of Reynolds number is reported for different volume concentrations. It is worth noticing that the friction factor is nearly independent on volume concentration. In conclusion the nanofluids have a higher capability to remove heat than pure fluid but this costs in terms of pumping power.

6. Conclusions

Results about the numerical analysis on laminar forced convection with nanofluids flowing in rectangular cross-sectioned ducts are presented in this paper. One of the walls is heated at uniform and constant heat flux. The investigation was carried out for $50 \le Re \le 1000$ and Al_2O_3/water based nanofluids at volume concentrations equal to 0%, 0.1%, 1%, 3% and 4%. The single-phase model was adopted in order to describe the behaviour of nanofluids as working fluids and thermo-physical properties were evaluated by means of equations, available in literature. The results showed the increase of the convective heat transfer coefficients, in particular, for high concentration of nanoparticles and for increasing values of Reynolds number.

The highest enhancement is evaluated for φ = 4% with the average convective heat transfer coefficient values equal to 1.14, 1.145, 1.147 times greater than the ones obtained for pure water, at Re =50, 500 and 1000, respectively. However, the disadvantages are represented by the growth of the wall shear stress and the required pumping power, observed in particular, at high particle concentrations.

Nomenclature

C_f	specific heat	Eq.(8)
d	Diameter	m
f	friction factor	Eq.(10)
H	duct height	m
l	duct edge length	m
L	total duct length	m
Nu	Nusselt number	Eq.(7)
P	pressure	Pa
Pr	Prandtl number	Eq.(11)
Re	Reynolds number	Eq.(6)
V	average velocity	m/s
$\dot{V}$	volume flow rate	m^3/s
W_p	pumping power	W
x, y, z	spatial coordinates	m
Greek symbols		
α	thermal diffusivity	m^2/s
β	volumetric expansion	1/K

	coefficient	
φ	nanoparticle volumetric concentration	
λ	thermal conductivity	W/m K
ν	kinematic viscosity	m^2/s
μ	dynamic viscosity	Pa s
ρ	density	kg/m^3
τ	wall shear stress	kg/m
Subscripts		
avg	average	
b_f	base fluid	
f	fluid	
h	hydraulic	
I	inlet	
m	mass	
n_f	nanofluid	
p	solid particle	
w	wall	

References

[1] Mohammed H A, Bhaskaran G, Shuaib N H, and Saidur R 2011 *Renew. Sust. Energy Rev.* **15** 1502.
[2] Lee S and Choi S U S, 1996 *ASME , Material Division (Publication) MD* **72** 227.
[3] Koo J, and Kleinstreuer C 2005 *Int. J. Heat Mass Transfer* **48** 2652.
[4] Chein R Y and Chuang J 2007 *Int. J. Therm. Science* **46** 57.
[5] Behzadmehr A, Saffar-Avval M and Galanis N 2007 *Int. J. Heat Fluid Flow* **28** 211.
[6] Jung J Y, Oh H S and Kwak H Y *2009 Int. J. Heat Mass Transfer* **52** 466.
[7] Bhattacharya P, Samanta A N and Chakraborty S 2009 *Heat Mass Transfer* **45** 1323.
[8] Chen C H and Ding C Y 2012 *Int. J. Therm. Science* **50** 378.
[9] Yang Y T, Tsai K T, Wang Y H and Lin S H 2014 *Int. Comm. Heat Mass Transfer* **57** 27.
[10] Fluent Inc. 2006 *FLUENT Computational Fluid Dynamic Code Version 6.3 User Guide*, www.fluent.com
[11] Shah R K and London A L 1978 Suppl. 1 *Advances in Heat Transfer.*
[12] Wilbulswas P 1966 Laminar-flow heat transfer in non-circular ducts [Ph.D. thesis], London University,
[13] Fleming D P and Sparrow E M 1969 *J. Heat Transfer* **91** 345.
[14] Miller R W and Han L S 1971 *J. Appl. Mech* **38** 1083.
[15] Aggarwala B D and Gangal M K 1975 *Trans. Can. Soc. Mech. Eng* **3** 231.

Second principle approach to the analysis of unsteady flow and heat transfer in a tube with arc-shaped corrugation

G Pagliarini[a,b,c], P Vocale[a], A Mocerino[a], S Rainieri[a,b,c],
[a]Department of Industrial Engineering, University of Parma, Parco Area delle Scienze 181/A, 43124 Parma – Italy
[b]SITEIA.PARMA Interdepartmental Centre, University of Parma, Parco Area delle Scienze 181/A, 43124 Parma – Italy
[c]CIDEA Interdepartmental Centre, University of Parma, Parco Area delle Scienze 181/A, 43124 Parma – Italy

E-mail: giorgio.pagliarini@unipr.it

Abstract. Passive convective heat transfer enhancement techniques are well known and widespread tool for increasing the efficiency of heat transfer equipment. In spite of the ability of the first principle approach to forecast the macroscopic effects of the passive techniques for heat transfer enhancement, namely the increase of both the overall heat exchanged and the head losses, a first principle analysis based on energy, momentum and mass local conservation equations is hardly able to give a comprehensive explanation of how local modifications in the boundary layers contribute to the overall effect. A deeper insight on the heat transfer enhancement mechanisms can be instead obtained within a second principle approach, through the analysis of the local exergy dissipation phenomena which are related to heat transfer and fluid flow. To this aim, the analysis based on the second principle approach implemented through a careful consideration of the local entropy generation rate seems the most suitable, since it allows to identify more precisely the cause of the loss of efficiency in the heat transfer process, thus providing a useful guide in the choice of the most suitable heat transfer enhancement techniques.

1. Introduction

Convective heat transfer enhancement in ducts may be obtained by modifying the fluid-dynamics and then the thermal boundary layer. In many industrial sectors this goal is frequently achieved by suitably modifying the duct wall (mainly throughout periodic corrugation performed by means of cold working operations) or by inserting in the main flow devices which are able to modify the flow regime (turbulators, wire coil inserts, etc...) [1,2]. These techniques, known as passive intensification techniques, mainly rely on their ability to anticipate the transition to the unsteady flow regime and hence to the turbulent one. Passive techniques are usually applied based on the experience of the designer and their effect is verified *a posteriori*, through the determination of the heat actually exchanged.

Bejan [3] pointed out that heat transfer and fluid flow mechanisms are intimately related to thermodynamics through entropy generation due to the non-ideality of the transfer process. He also pointed out that the entropy generation due to the convective heat transfer in internal flow is inversely proportional to the convection coefficient. This result can be easily understood if one considers that, for the same amount of heat exchanged, to a higher convection coefficient a lower wall to fluid temperature difference is associated and, therefore, a lesser degree of irreversibility in the heat transfer process.

In spite of the extensive work done on the topic of entropy generation due to convection heat transfer, the knowledge of the overall entropy generation given by the entropy balance of the heat transfer section is of limited use in the engineering practice. Otherwise, the knowledge of the entropy generation on a local scale, could be a useful tool to identify the zones of the heat transfer device, possibly away from the wall-fluid interface, where the entropy generation may be high. Finally, the detailed map of the local entropy generation might suggest possible geometry or operational optimal changes intended to increase the overall system efficiency.

Sciacovelli et al. [4] have presented a critical review of the contributions to the local entropy generation analysis to systems involving convective heat transfer to which the early work by Pagliarini and Barozzi [5] can be added.

An example of the heat transfer efficiency optimization based on the local entropy generation rate analysis was provided by Giangasparo and Sciubba [6], who recently applied a heuristic approach to the optimization of a solar heat exchanger.

Calculating the local entropy generation rate requires the resolution of the mass, momentum and energy conservation equations and the subsequent post-processing of the velocity and temperature data as shown by De Groot and Mazur [7]. It is a challenging task, which requires significant computing resources, especially when the unsteady flow regime is considered.

The transition from the laminar to the unsteady flow regime in a two dimensional periodic arc-shaped profile channel has been studied by Ničeno and Nobile [8]. Rainieri et al. [9] have confirmed their critical Reynolds number value, and they have extended the analysis to the case of the axisymmetric duct, by simulating the heat and mass transfer phenomenon in three spatial dimensions.

In this paper the second principle approach, based on the analysis of the local entropy generation, is applied to the study of convective heat transfer in the transition from the laminar to the unsteady flow regime in a two dimensional channel with a periodic arc-shaped profile.

2. Problem statement

The two-dimensional channel shown in figure 1 has been hereby considered. A Newtonian fluid in steady flow condition exchanges heat with the duct wall, on which a uniform heat flux condition is applied.

Assuming constant fluid properties, the heat transfer problem is mathematically formulated as follows:

$$\nabla \cdot \mathbf{v} = 0 \tag{1}$$

$$\rho \frac{D\mathbf{v}}{Dt} = -\nabla p + \mu \nabla^2 \mathbf{v} \tag{2}$$

$$\rho c_p \frac{DT}{Dt} = \lambda \nabla^2 T + \mu \Phi \tag{3}$$

being $\mathbf{v}$ the velocity field, ρ the fluid density, t the time, p the pressure, μ the fluid dynamic viscosity, T the fluid temperature, α the thermal diffusivity and Φ the viscous dissipation function.

Because of the periodicity of the channel geometry, the integration domain of the previous equation set was limited to a single channel module (figure 2), on whose inlet and outlet sections a periodic boundary condition is set, for both velocity and temperature distribution, while the bulk fluid temperature is incremented from the inlet to the outlet section based on the overall energy balance of the heat transfer module.

Equations (1) to (3) were solved by adopting the finite element method, implemented within the COMSOL Multiphysics environment. For Reynolds number values higher than the critical value, for which the flow becomes unsteady, the simulation stop time was set to 22,000 seconds, in order to obtain meaningful mean value for all the local variables.

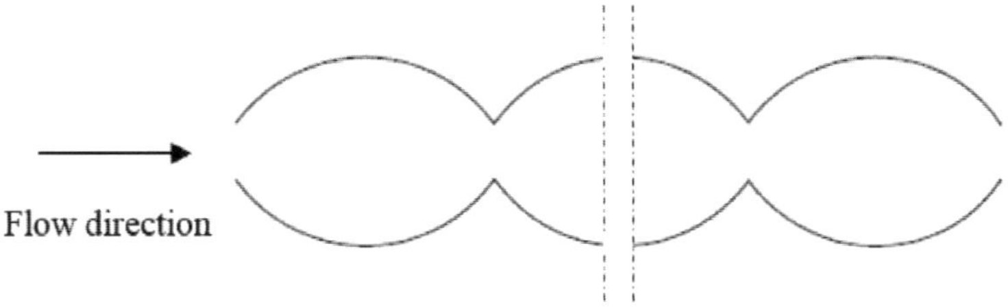

Figure 1. Sketch of the arc-shaped profile.

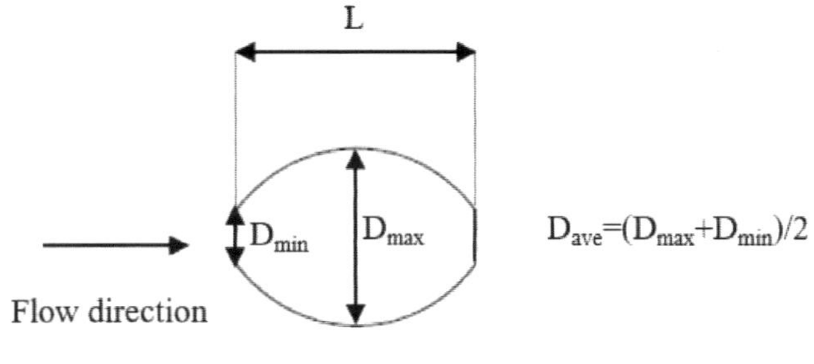

Figure 2. Single module of the corrugated channel under investigation.

At the end of each simulation, the local entropy generation, due to both heat transfer and viscous dissipation, was computed as follows:

$$s_T = -\frac{1}{T^2}\mathbf{q}\cdot\nabla T \tag{4}$$

$$s_u = \frac{\mu}{T}\Phi \tag{5}$$

Integrating s_T and s_u all over the volume of the considered channel module, gives the overall entropy generation, S_T and S_u.

The computed overall entropy generation rate was then compared with the value given by the entropy balance equation:

$$\Delta S_t = \int_2 \rho s(\mathbf{v}\cdot\mathbf{n})dA - \int_1 \rho s(\mathbf{v}\cdot\mathbf{n})dA - \int_{A_w} \frac{q}{T_w}dA_w \tag{6}$$

where s indicates the specific entropy, A the cross-sectional area, A_w the heat transfer surface area, q the wall heat flux and T_w the wall temperature. The subscript 1 and 2 refer to the inlet and outlet sections, respectively.

The entropy fluxes, through the inlet and outlet sections and through the solid wall as well, were computed by post-processing the results of the numerical simulations.

The comparison between the overall entropy generation rate given by eq. (6) and the value obtained by integrating the entropy generation distributions, s_T and s_u, over the channel module volume, showed deviations included within ± 1%.

The accuracy of the numerical model was checked by performing several simulations on different grid sizes; the grid independence analysis was carried out by monitoring both the averaged Nusselt number and the overall entropy generation rate.

It was observed that for a number of grid element higher than 6000, a mesh-independent solution can be achieved.

3. Results

The numerical simulations were performed for Reynolds number values ranges between 30 and 403 and for Prandtl number equal to 0.7, keeping T_I (i.e. the bulk temperature at the inlet section) fixed at 300 K. The channel under investigation is characterized by L/D_{max}= 1.4 and D_{min}/D_{max}=0.23.

It has to be pointed out that the Reynolds number was evaluated by considering the average cross-section diameter as characteristic length (D_{ave}), as shown in figure 2:

$$\mathrm{Re} = \frac{\rho W_{ave} D_{ave}}{\mu} \tag{7}$$

being W_{ave} the average value of the fluid velocity evaluated at the average cross-section.

The results confirmed that the flow regime becomes unsteady for Re higher than about 60, as highlighted by Ničeno and Nobile [8] and by Rainieri et al. [9].

For all the simulations the Bejan number, $(S_T + S_u)/S_T$, was higher than 0.999. Therefore, the analysis of the entropy generation rate was limited to the heat transfer effects, while the effects of viscous dissipation were disregarded.

Figure 3 shows the non-dimensional overall entropy generation rate, as a function of the Reynolds number, which can be defined as follows:

$$\Sigma_T = \frac{S_T}{Q/T_m} \tag{8}$$

where T_m is the thermodynamic mean temperature, which, neglecting pressure variations, is expressed as follows:

$$T_m = \frac{(h_2 - h_1)}{(s_2 - s_1)} = \frac{(T_2 - T_1)}{\ln\left(T_2/T_1\right)} \tag{9}$$

being h_1 and h_2 the specific averaged enthalpy evaluated on the inlet and on the outlet sections, respectively. T_1 and T_2 indicate the inlet and outlet bulk temperature, respectively.

As long as the flow regime is laminar ($Re < Re_c$) the overall entropy generation slightly increases; when the flow becomes unsteady, the entropy generation rate decreases monotonically with the Reynolds number.

In the same figure the Nusselt number averaged on all over the interface, is shown; it is obtained, by surface integration, from the local Nusselt number which is defined as follows:

$$Nu = \frac{q}{(T_w - T_b)} \frac{D_{ave}}{\lambda} \tag{10}$$

being T_b the arithmetic mean of the inlet and the outlet bulk temperature, λ the fluid thermal conductivity and D_{ave} is the characteristic length defined in figure 2.

As expected, it has an opposite trend to the entropy generation, since it increases with *Re* in the unsteady flow regime.

Indeed, from the entropy balance equation (eq. (6)), the relation between Σ_T and Nu is as follows:

$$\Sigma_T = 1 - \frac{1}{\mathrm{Re}\,\mathrm{Pr}} \frac{1}{\ln\left(T_2/T_1\right)} \int_{A_w*} Nu \frac{(T_w - T_b)}{T_w} dA_w * \qquad (11)$$

in which A_w* is the dimensionless heat transfer surface area.

The non-dimensional entropy generation ranges between 0 and 1; the lower limit can be reached for $T_w \rightarrow T_1$, while the upper one for $T_w \rightarrow \infty$.

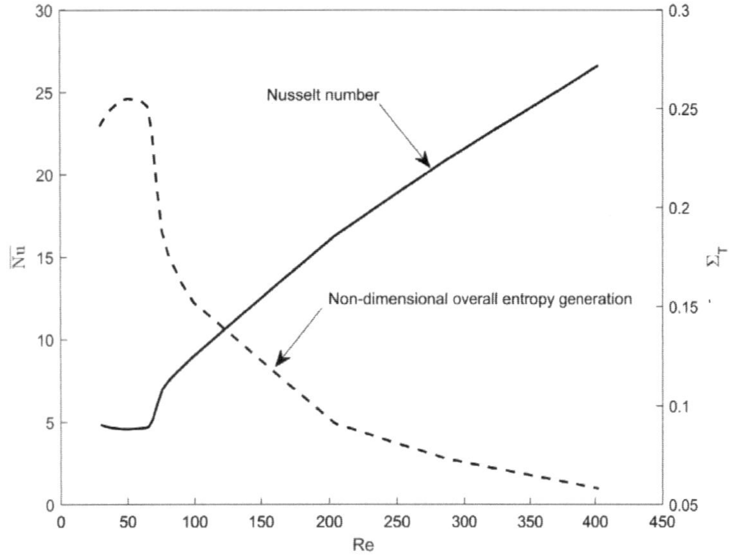

Figure 3. Overall entropy generation rate caused by heat transfer and averaged Nusselt number as a function of the Reynolds number.

3.1. Velocity and temperature distribution

The flow pattern and the temperature distribution are visualized by the streamlines and the temperature contours in figures 4 and 5, respectively. The maps show the solution obtained for the simulation stop time. As expected, owing to the symmetry conditions, the solution is symmetrical with respect to the channel symmetry axis in the laminar flow regime, while no longer being symmetrical in case of unsteady flow.

The streamline pattern points out that in the laminar flow regime two large vortices between the mean flow and the wall are formed. When the flow regime becomes unsteady the vortices increase and they produce a strong interaction between the mean flow and the wall. It results a very thick thermal boundary layer in the laminar flow regime, which occupies the entire flow recirculation zone.

On the contrary in the unsteady flow regime the thickness of the thermal boundary layer reduces as the size of the vortices adhering to the wall reduces.

Furthermore, since the value of the inlet temperature is the same for all values of *Re* here considered, the wall temperature considerably reduces as the Reynolds number increases.

IOP Conf. Series: Journal of Physics: Conf. Series **796** (2017) 012014 doi:10.1088/1742-6596/796/1/012014

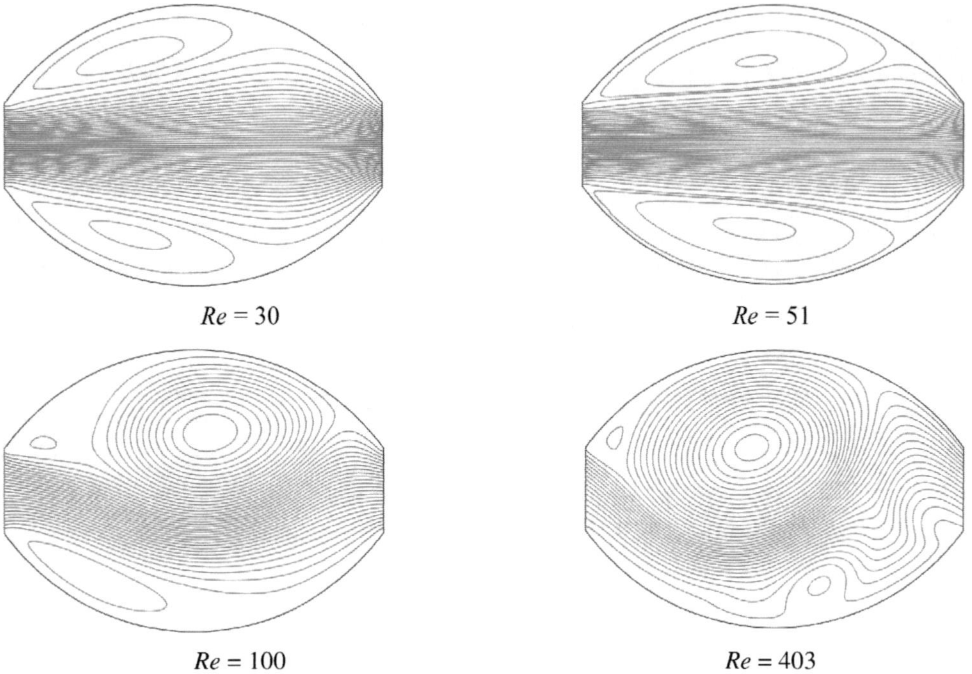

Figure 4. Streamlines for different values of *Re*.

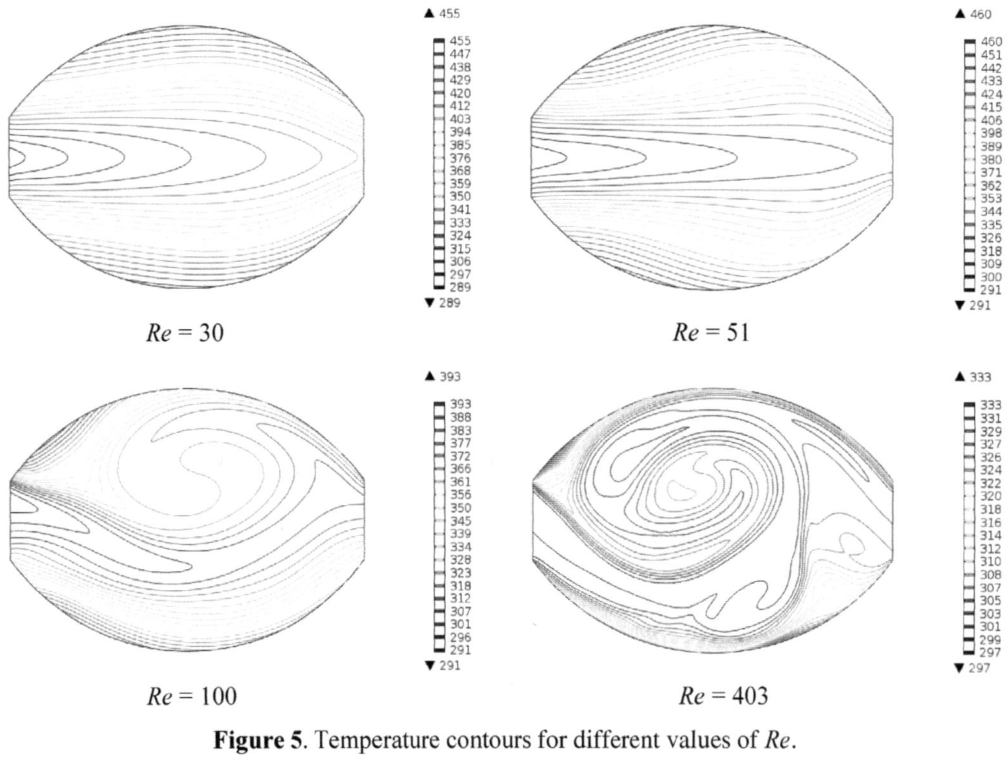

Figure 5. Temperature contours for different values of *Re*.

3.2. Local entropy generation

The local entropy generation due to heat transfer only is shown in figure 6, for the same values of the Reynolds number shown in the previous figures.

In the laminar flow regime ($Re < Re_c$) the entropy generation is particularly high away from the wall, where the temperature gradients are useless to the convective heat transfer. When, instead, the flow becomes unsteady ($Re > Re_c$), the entropy generation is concentrated near the wall.

At the same time the area near the channel axis, in which entropy generation rate almost vanishes, amplifies. Despite the maximum entropy generation value grows by one order of magnitude from the laminar to the unsteady flow regime, the total generation rate reduces, as shown by the graph in figure 3. Due to the relation between entropy generation and heat transfer efficiency, figure 6 highlights the critical zones, where a useless entropy generation rate affects the heat transfer efficiency.

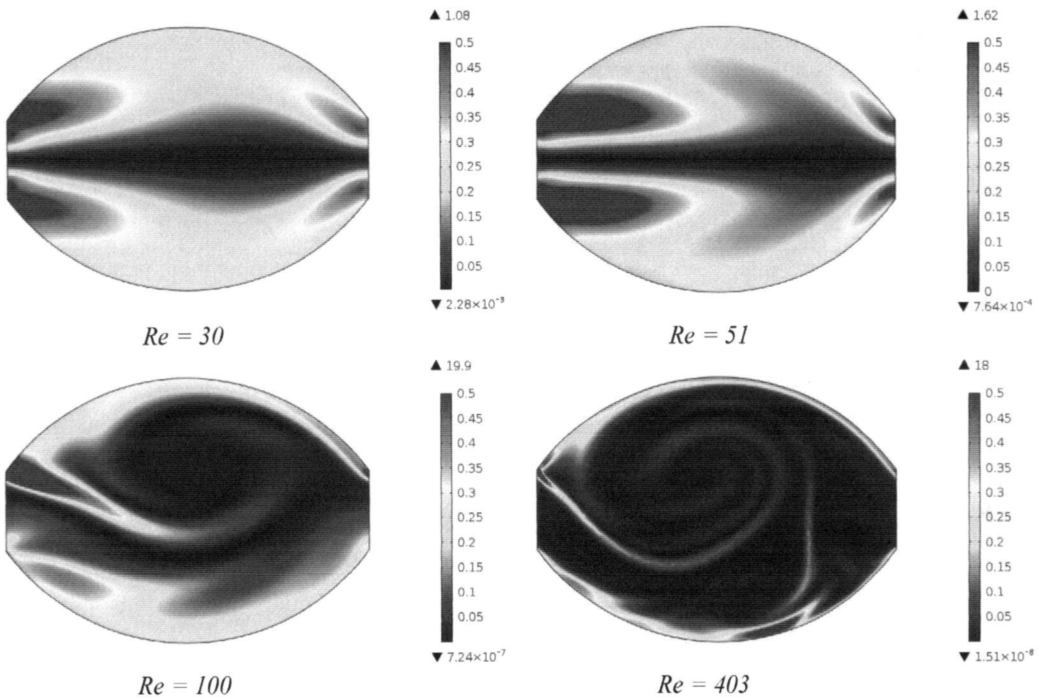

Figure 6. Entropy generation due to heat transfer for different values of *Re*.

4. Conclusions

The convective heat transfer problem for a $Pr = 0.7$ fluid in a two dimensional periodic arc-shaped profile channel has been considered in a Reynolds number range which includes both the laminar and the unsteady flow regime. The local entropy generation due to heat transfer has been calculated by post-processing the temperature data, to assess the heat transfer efficiency of the considered wall profile within a second principle approach.

On the basis of the relationship between entropy generation rate and heat transfer coefficient in internal flow, the maps of entropy generation in the flow have allowed to identify the zones which are the main responsible for the loss of heat transfer efficiency.

In conclusion, the second principle approach has demonstrated that optimizing the convective heat transfer at the wall-fluid interface is equivalent to optimize the local entropy generation in the system volume, by reducing the excess entropy generation rate where useless for the heat transfer process.

Acknowledgments
This work was partially supported by the Emilia-Romagna Region (POR-FESR 2014-2020).

References

[1] Webb R L and Kim N H 2005 *Principles of Enhanced Heat Transfer* (New York: Taylor & Francis)
[2] Rainieri S, Bozzoli F and Cattani L 2014 *Journal of Physics: Conference Series* **547** 012002
[3] Bejan A 1996 *Entropy Generation Minimization* (Boca Raton, FL: CRC Press)
[4] Sciacovelli A, Verda V and Sciubba E 2015 *Renewable Sustainable Energy Rev* **43** 1167
[5] Pagliarini G and Barozzi G 1991 *J. Heat Transfer* **113** 526
[6] Giangasparo G and Sciubba E 2013 *Energy* **58** 52
[7] De Groot S R and Mazur P 1962 Non-equilibrium Thermodynamics (Amsterdam: North-Holland Publishing Company)
[8] Ničeno B and Nobile E 2001 *Int. J. Heat and Fluid Flow* **22** 156
[9] Rainieri S, Bozzoli F and Pagliarini G *Proc. 4th Int. Conf. on Computational Heat and Mass Transfer* (*Paris-Cachan, 2005*)

New experimental results on local heat transfer inside a rectangular channel with rib-roughened surfaces

D Fustinoni, P Gramazio, L Vitali and A Niro
Politecnico di Milano, Department of Energy,
Campus Bovisa, Via Lambruschini 4, 20156 Milano, Italy

Corresponding author email: alfonso.niro@polimi.it

Abstract. In this paper we present new experimental results on local heat transfer characteristics of a forced air-flow through a 12-mm-height, rectangular channel of 1:10 aspect ratio, with square-cross-section ribs mounted onto the lower surface. Data are collected on a completely redesigned test section. Specifically, the electric heater is made of very thin copper tracks, in direct contact with the air flow and covering at 97.5% the channel lower surface to guarantee a very uniform heat flux. The copper tracks are laminated onto a 2-mm thick board of FR-4 glass epoxy to provide negligible heat conduction inside the plate and heat losses from its sides. Finally, the channel walls are in XPS and, into the upper one, a double glazing consisting of two 120 mm x 120 mm Germanium windows is mounted to allow optical access to the IR camera and to reduce local heat dispersions. Data here presented refer to convection over 4 mm x 2 mm ribs in transverse configuration for Reynolds numbers, based on the duct hydraulic diameter, ranging between 700 and 8000. Preliminary tests show how the new apparatus has significantly improved the quality, the ease and the quickness of the measurements.

1 Introduction

Forced convection inside rectangular channels with one or more rib-roughened surfaces, with a large variety of configurations, is a subject particularly considered in literature [1-4], as ribs are an efficient and cost-effective way to improve heat transfer in a broad spectrum of applications, ranging from gas-turbine blades cooling to plate compact heat exchangers enhancement. Moreover, channels with only one heated and ribbed surface have been employed as solar air heaters during the last decade [5-8], following the trend of increasing interest around renewable energy technologies.

In this context, local heat transfer measurements over surfaces with complex geometries are very useful from both a conceptual and a practical standpoint, as they provide insight into fluid-dynamic enhancement mechanisms and can be employed as a tool to guide geometry optimization. Moreover, infrared thermography allows a not-intrusive, full-field surveying of the surface temperature with a high spatial resolution, and it can be effectively used to determine the distribution of convective heat transfer coefficient once the local generated and dissipated heat fluxes are known. In fact, several other research group adopted IR thermography-based techniques [9-11], but the experimental procedures are far from being well established, as they strongly depend on the trial setup and on the modeling strategies to calculate the non-convective heat flux components.

Our research group is experienced in both measuring global and local heat transfer in internal geometry [12-13], but local heat transfer, enhanced by square ribs, for transitioning flow regimes, i.e. for Reynolds numbers ranging from 2500 to 25000, need more investigation to characterize the effects of turbulence onset and to identify the flow structures generated by the ribs. These considerations have led to fully redesign our former experimental apparatus and measurement procedures in order to obtain higher precision. In this paper we present the new experimental apparatus, along with the results for the reference flat channel and the configuration with transverse ribs. In both cases, the channel has a 1:10 aspect ratio rectangular channel and only one heated surface, for steady state conditions.

2 Experimental setup

A new experimental setup, which consists of an open loop wind tunnel, has been designed (figure 1) and built aimed at improving the precision of local convective heat transfer coefficient measurements. In particular, the new test section has been thought to reduce dispersive heat flux components from the heater surface, and to reduce the system's thermal inertia. Therefore, the resulting apparatus should allow a higher accuracy in determining the fraction of convective heat flux from the total input, i.e. the known electrical heating by joule effect, and should reach steady-state conditions faster than the previous bulky channel.

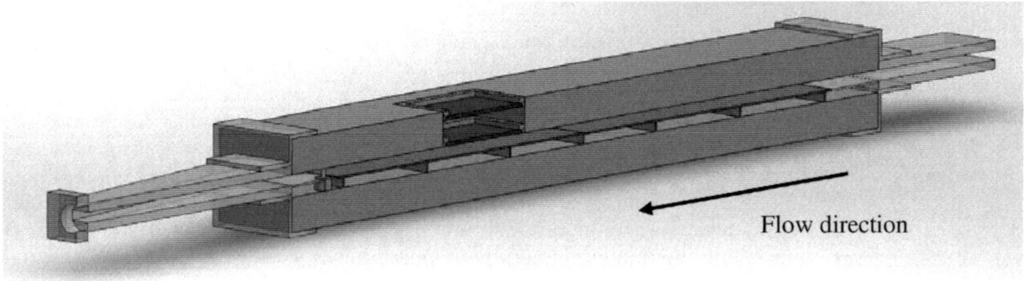

Figure 1. Test section design

In detail, the flow control section, which consists of a 7-stage, 30 kPa-head, 5.5 kW-power blower operating in suction mode and a valve system, has been improved since our previous works with a Hendress Hauser T-Mass 65F15 flow meter for automatic and precise mass-flow readings. Air flows from a convergent inlet and a non-heated entry section, 120 mm wide, 12 mm high, 40 hydraulic diameters long, into a 840 mm long duct with the same rectangular section size, whose external walls are made of 41,6 mm thick extruded polystyrene (XPS), which assures both thermal insulation due to its low thermal conductivity (0.033 W/m/K) and low mass, coupled with a 1 mm HPL laminate skin for damage protection. The channel inner section consists of an aluminum-tape coated upper wall, two Plexiglas side walls and two adjacent 420 mm x 120 mm black-painted heater breadboards on the lower wall, covering the whole test surface.

Each heater is made of a 17.5 μm thick copper circuit deposited on a FR-4 glass epoxy surface, providing a 1.1 Ω resistance, and it is specifically designed to increase planar conduction. The two breadboards are independently powered and controlled with a 5A Carroll & Meynell variac, whose tension is reduced by a 10x, 240 VAC, Legrand transformer. The current provided to each heater is measured by means of a shunt circuit connected to an Agilent 34970A data logger, which is also used to collect tension drops and temperature data. Between the heaters and the bottom external wall, an 11-mm-hight gap provides both thermal insulation and room for wiring.

The optical access for IR surface measurements takes place through a top opening where a 130 mm x 130 mm, 4 mm thick, Plexiglas framed, germanium window is mounted in order to ensure the top surface continuity. However, preliminary heat transfer FEM simulations showed that a single-window arrangement leads to a discontinuity of thermal properties of the channel wall due to the higher germanium thermal conductivity (about 60 W/m/K), which could affect local measurements with respect to an ideal non-windowed channel. To reduce this effect, a double germanium window arrangement has been designed, with a 20 mm air gap between each glazing. Furthermore, three interchangeable top sections with different opening locations have been made in orderto allow surface measurements in different channel positions.

The IR-camera is a FLIR T650sc, 640 x 480 pixel, microbolometric-type, with a declared NETD of 20 mK at 30 °C, and it is mounted on a Manfrotto tripod with an orientable head. The camera points the test surface at a 20° angle with respect to the normal direction, in order to avoid self-reflections.

The apparatus has also been equipped with several temperature sensors as listed in table 1. In particular, the air temperature is measured with PT100 RTDs, one placed at the channel inlet, and three at the outlet, just behind the test section, for bulk temperature measurements. Each of these is embedded in a vertical-mounted, 2-mm diameter, aluminum cylinders, which provide a physical integral action. The bulk temperature is a weighted average of these three measurements.

T-type thermocouples are employed to measure temperature of germanium window, channel side wall, and channel bottom wall beneath the heater in five spots at 200 mm, 360 mm, 520 mm, 690 mm, and 820 mm, respectively, from the first breadboard edge, as shown in figure 2.

Table 1: thermocouple and Pt-thermometre arrangement

T sensor	Type	Placement
T1	RTD	Inlet temperature
T2, T3, T4	RTD	Bulk temperature
T5	T-type thermocouple	Germanium window, channel-side
T6,T7,T8,T9,T10	T-type thermocouple	Channel wall, beneath the heater

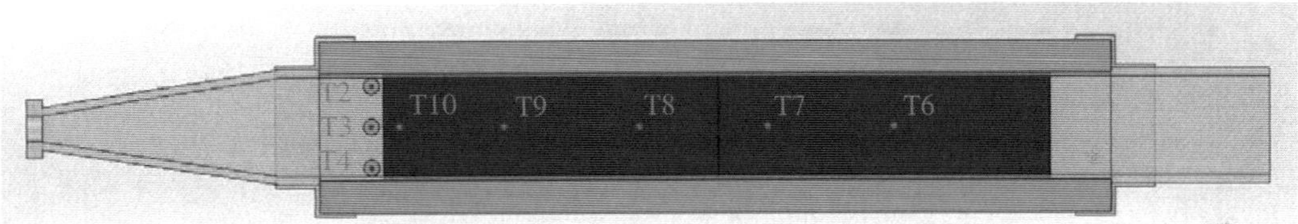

Figure 2. Temperature probes, test section.

Finally, the Plexiglas ribs are laser-carved from a Plexiglas plate and kept in position by Plexiglas rails, a solution that allows quick configuration changes, high placement accuracy, and prevent from gluing on the heater's delicate surface.

3 Data processing and uncertainty analysis

The local convective heat transfer can be estimated by an appropriate modeling of the dispersed non-convective heat flux components, which are based on the experimental measurements.

In detail, the local convective heat transfer coefficient is evaluated as

$$h(x,y) = \frac{q_c(x,y)}{T_w(x,y) - T_b(x)} \tag{1}$$

where $q_c(x,y)$ and $T_w(x,y)$ are the convective heat flux and the temperature, respectively, on the surface test at position x,y, according to the reference system shown in figure 3, and $T_b(x)$ the air bulk temperature. As the channel is operated at constant heat flux, temperature $T_b(x)$ is assumed to linearly vary between the measured values at the test section inlet and outlet. Temperature distribution $T_w(x,y)$

is obtained by IR-thermographic measurements, as will be discussed in detail. Finally, the convective heat flux $q_c(x,y)$ is evaluated as

$$q_c(x,y) = q_{in} - q_{kz}(x,y) + q_{kxy}(x,y) - q_r(x,y) \tag{2}$$

where $q_{in}=(Q_{el}/A)$ is the heat flux delivered by the electric heater assuming a uniform generation by Joule effect; $q_{kz}(x,y)$ the heat losses from the breadboard back side evaluated by the temperature drop measured across the underlying air gap; $q_{kxy}(x,y)$ the heat flux diffusing inside the copper tracks in *x-y* direction; $q_r(x,y)$ the heat losses by radiation towards the channel walls, and through the germanium windows. With regards to our experimental setup, such terms require further specifications about how they are evaluated, that are illustrated in the following subsections.

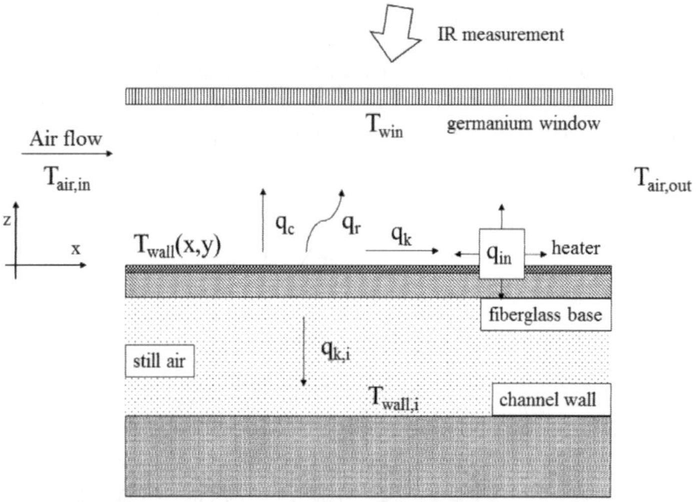

Figure 3. Schematic heat-transfer model of the test section, side view

3.1 Backside breadboard heat losses.

The conductive heat flux towards the lower wall $q_{k,i}$ is calculated by a simple 1D conduction model as showed in equation 3, under the hypothesis of negligible thermal resistance of the fiberglass heater breadboard with respect to the still air layer, and of uniform lower wall temperature. It should be pointed out that radiative heat transfer between the heater plate backside and the channel wall is considered negligible, since their walls have been coated to this end with low-emissivity aluminum foils.

$$q_{kz}(x,y) = \frac{k_{air}}{s_{air}}[T_w(x,y) - T_{si}] \tag{3}$$

3.2 PCB copper tracks conductive heat flux

The heat flux $q_{kxy}(x,y)$ diffusing inside the copper tracks in *x-y* direction is evaluated by means of equation 4, that is obtained by solving a steady-state energy balance inside the copper tracks and assuming negligible temperature variations in their z-direction The derivatives are calculated by interpolating the measured temperature distribution on the surface with a 2D polynomial surface, see paragraph 3.5. The black paint surface layer is supposed to have negligible effect on the surface temperature distribution. Finally, on the basis of FEM simulations heat diffusion inside the breadboard in x-y direction as well as temperature drop across its thickness are assumed negligible too.

$$q_{kxy}(x, y) = k_{Cu} s_{Cu} \left(\frac{\partial^2 T_w}{\partial x^2} + \frac{\partial^2 T_w}{\partial y^2} \right) \tag{4}$$

3.3 Radiative heat flux

The radiative heat flux is calculated for each value of the x coordinate with a Matlab script, which solves the radiative heat transfer problem in a 2D cavity with known temperatures, as schematically shown in figure 4. In detail, the germanium window temperature T_{win} is measured and considered uniform due to the high germanium conductivity, the heater temperature distribution $T_w(x,y)$ is known by means of IR measurements, and on the side walls a linear trend between the top and bottom known temperatures is imposed, as found out from FEM simulations in design stage. Heat dispersions due to the high germanium window transmittance in IR region are also taken into account.

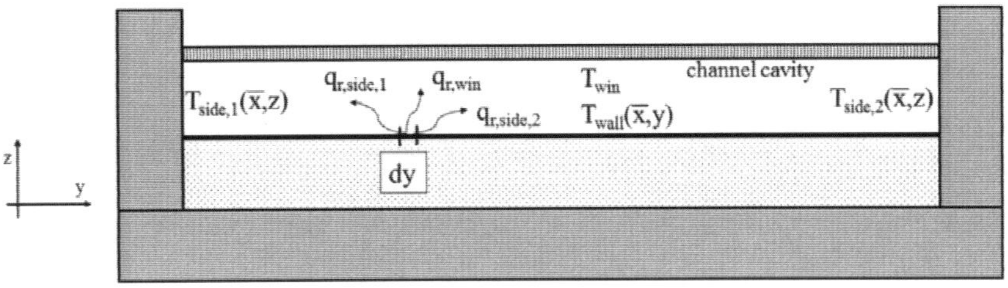

Figure 4. Schematic radiative model of the test section, front view

3.4 Calibrations

As described in the section 3, the sensing instruments adopted in this work are six type T thermocouples, four RTDs, and an IR-camera. Both kind of temperature sensors have been calibrated at once according to the procedure described in detail in [13], i.e., by means of a ThermoHaake heat bath as reference over four uniformly spaced points in the temperature range 30 – 60 °C. The resulting standard deviation is 0.02 K for the thermocouples, and of 0.01 K for the RTDs.

The IR-camera requires a peculiar in-situ calibration in order to evaluate the correct emissivity of the target surface and the eventual loss due to the double germanium window between the camera and the target itself. To this end, a custom test rig has been specifically built. A blackbody, which is a hollow aluminum rod, blackened with the same paint used for the heater surface, is connected to the ThermoHaake thermal bath and heated to fixed temperature. An embedded type T thermocouple measures the blackbody temperature, while two thermocouples record the temperature of the germanium windows. Once stationary conditions have been reached, the target emissivity is set to match the IR camera reading with the measured surface temperature. This procedure has been repeated for three temperatures settings, namely, 40 °C, 50 °C and 60 °C, obtaining a highly linear calibration curve, with $R^2 > 0.999$.

The mass flow meter has been calibrated by the "Istituto Nazionale di Ricerca Metrologica (INRiM)" in Turin, and a 2% accuracy is proven for the range of Reynolds numbers exploited in this research.

3.5 Thermogram processing

A thermogram processing procedure is required (i) to achieve white noise reduction, (ii) to correct the spatial distortion introduced by the angled framing and non-uniform focus over the target surface, and (iii) to allow the computation of the second derivatives, which are needed to estimate the surface conductive heat flux.

The white noise reduction in steady-state conditions is achieved by time-averaging the thermography sequence, obtaining one IR image from a sequence. In particular, the sequences have been acquired both at a 20 Hz frame rate for 1 s, and at a 1 Hz frame rate for 30 s. The output difference between the two acquisition procedures is around the IR-camera NETD, so the quickest has been chosen for this analysis.

The spatial distortion has been corrected by means of a projective transform [14], which restores a perpendicular top-view of the target surface from the angled image, as shown in figure 5. Due to the geometric symmetry of the framed scene, the focus issue has been solved by considering only the top half of the image, and by mirroring it over the centerline. This procedure also overcomes the viewing limitations given by the germanium windows frames. For ribbed surfaces, one surface portion between ribs at a time is taken into account.

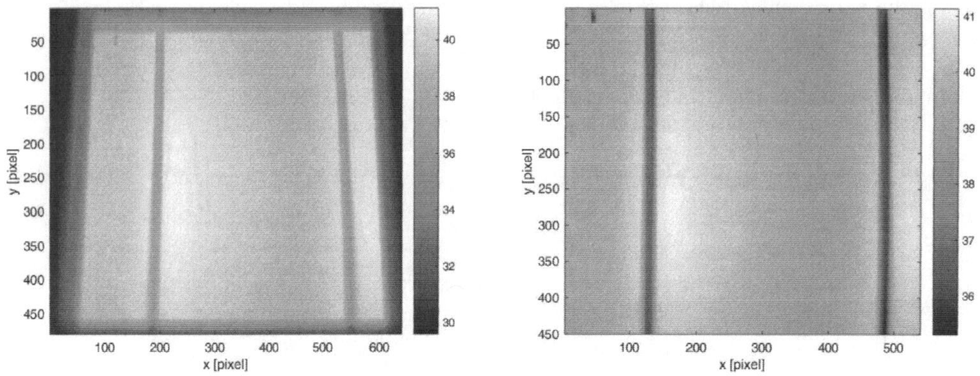

(a) (b)

Figure 5. Time-averaged thermogram, original (a) and after projective transform (b), for the 90° ribbed channel

The last procedure is a filtering of temperature data resulting from time-averaged, projective-transformed, centerline-mirrored and cut thermogram, which allows to calculate the second derivative. To this end, many tests were preliminary carried out with two kinds of techniques, namely, polynomial interpolations in the sense of least squares of degree from 4 to 12 in spanwise direction, and the Wiener filter for 5 different values of mask size applied twice, i.e., a first time to the temperature field and then to its first derivative.

Except for the Wiener filter with mask sized up to 9x9, all tested filtering procedures bring about spanwise distributions of convective coefficient that, in the central zone, i.e., between 15 mm and 105 mm, overlap with differences among them within 5% at most, as well as if compared with the value estimated by the Gnielinski correlation. However, near the sidewalls, the differences between the heat transfer coefficient and the mean value referred to the central zone increase with the order of the fitting polynomial, whereas the values of the heat transfer coefficient obtained by means of the Wiener filter dramatically oscillate with growing amplitude as the mask size increases.

For this reason, the Wiener filter was replaced with a polynomial fitting (4th-order spanwise, 2nd-order streamwise for the flat-surface; 4th-order spanwise, 3rd-order streamwise for the 90° ribbed configuration; R^2 value larger than 0.98 in both cases), resulting in a behaviour of the heat transfer coefficient without inflections, and with values near the sidewalls only slightly increasing above the mean central one.

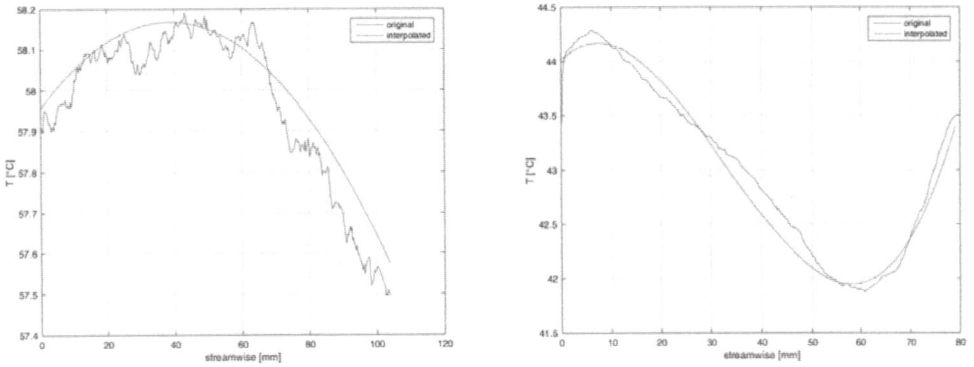

(a) (b)

Figure 6. Streamwise temperature interpolation, flat plate (a) and ribbed (b) configurations

4 Results

In order to assess the experimental apparatus, and to validate the data logging and post-processing procedures, heat transfer coefficients were preliminary measured in a reference configuration, i.e., a rectangular channel with a flat bottom wall. Heat transfer data here presented are for air flows entering the test section at a temperature around 25 °C, for eight Reynolds numbers, based on the duct hydraulic diameter, ranging from 650 to 7280.

Figure 7a displays the time-mean wall temperature field for Re=5400, whereas the corresponding field of heat transfer coefficient *h*, obtained with the procedures previously described, is shown in figure 7b. As can be seen, quite small *h* variations occur in streamwise direction, while spanwise they are within about 10% with respect to its mean value, and for all other Re they never exceed 15%.

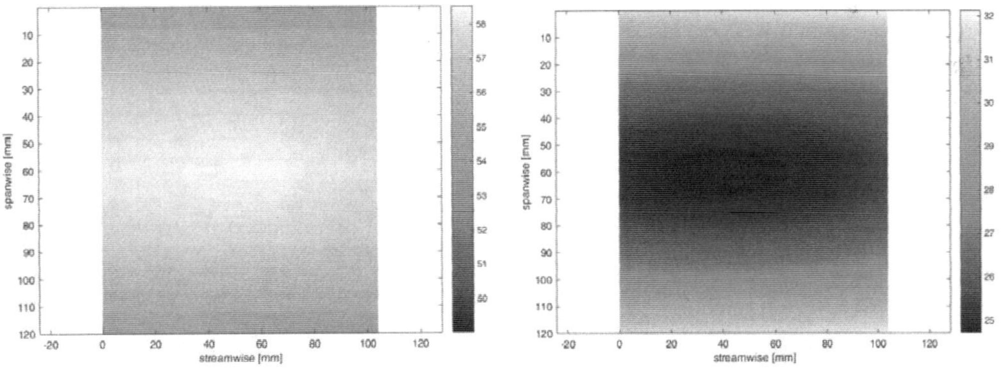

(a) (b)

Figure 7. Temperature (a) and *h* (b) distribution for Re = 5400, flat reference surface

The ribbed configuration has transverse ribs, only on the channel bottom wall, with a square section of 4 mm in side, and a pitch-to-rib height ratio p/e=20. Figures 8a and 8b display the time-mean wall temperature and heat transfer coefficient fields, respectively, for Re=5400. With reference to local effects, the temperature and convective heat transfer fields shown in figure 8 highlight the consequence of the ribs on fluid-dynamics and heat transfer inside the channel. As it can be seen, the temperature distribution clearly shows the rib effect on flow separation as well as on the reattachment zone. In particular, for the case reported here, the temperature shows a minimum at near one fourth of pitch downstream any ribs, which corresponds also to the region of maximum local h.

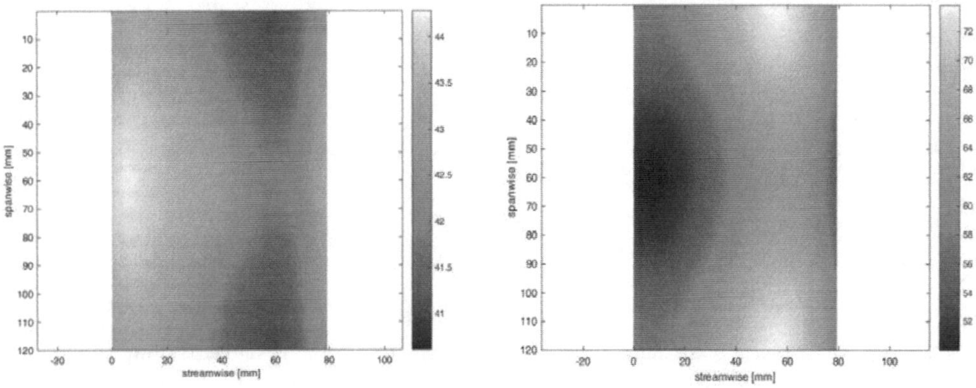

(a) (b)

Figure 8. Temperature (a) and h (b) distribution for Re = 5400, 90° rib configuration

Spanwise h variations can be still observed, but they continue to be within 15%. We conjecture this outcome is due to a residual conductive effect of side-walls that, however, does not significantly affect the values of the spanwise average heat transfer coefficient.

Indeed, Figure 9a displays the average Nusselt numbers on the considered frame for the flat channel and, as it can be seen, they are in very good agreement with values previously found by our research group in a similar rectangular channel [13] but operated at constant wall temperature. They also agree well with the values calculated by the correlations of Shah and London for Re<2300, and of Gnielinsky for higher values, reported in the same figure as solid lines. Figure 9b displays the average Nusselt numbers for the ribbed wall along with the results previously obtained in a configuration with ribs of the same shape, dimensions and pitch but mounted also on the channel upper wall, operated at constant wall temperature.

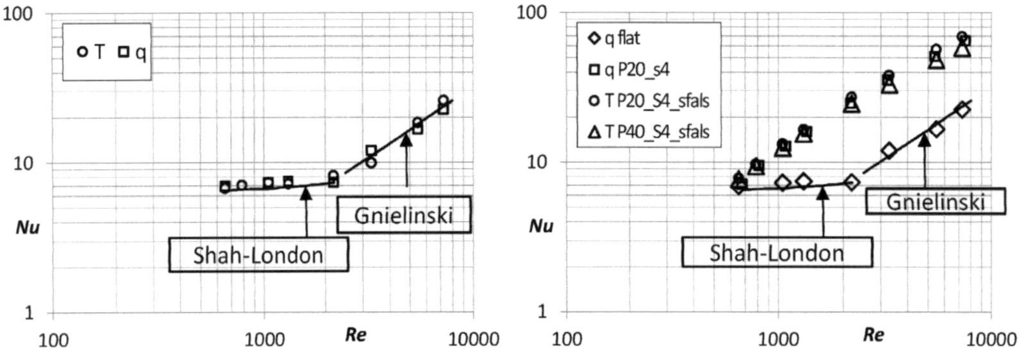

(a) (b)

Figure 9. Temperature (a) and h (b) distribution for Re = 5400, flat reference surface

Steady state conditions have been reached in about 1 hour since the test rig activation, which is the expected improvement compared to the previous test rig, which took around 6 hours.

5 Conclusions

In this paper, a new test section and post-processing procedure to measure the local convective heat transfer inside channels has been presented. Preliminary tests show how the new apparatus has significantly improved the quality, the ease and the quickness of the measurements, although the spanwise h variations are slightly larger than expected, and hence require further refinements and investigations in order to identify more surely local small scale phenomena. In particular, the insulation around the breadboard and its high planar conductivity give low conductive heat flux components in the balance. Thus, a good estimation of h requires a good local radiative heat transfer modeling, especially since the germanium window is spectrally-selective. A great care has been given in this aspect, by measuring the germanium spectral transmittance and by evaluating pixel-by-pixel local view factors as inputs for the estimation of local heat flux, but improvements like 3D effects simulations and dispersion reduction by covering the window with a ceiling having the same temperature as the breadboard are at study. The spanwise h variation should also be investigated, either by new FEM analysis of the conduction between the breadboard and the side walls, and by CFD modeling to better understand the spanwise vortex structure. Further work should also take into account more complex rib configurations, and, given the sensitivity and resolution of the IR-camera, smaller turbulators can be tested. Finally, the procedure should be repeated by varying the IR-camera viewing point, to prove the independency of the results from the framing angle and position.

References

[1] Won S Y and Ligrani P M 2004 *Int. J. of Heat and Mass Trans.* **47** 1573-1586
[2] Tanda G 2004 *Int. J. of Heat and Mass Trans.* **47** 229-243
[3] Wang L and Sunden B 2007 *Heat and Mass Trans.* **8** 759-766
[4] Gao X and Sunden B 2001 *Therm. Fluid Sci.* **24** 25-34
[5] Tanda G 2011 *Energy* **36** 6651-6660
[6] Bhushan B and Singh R 2010 *Energy* **35** 202-212
[7] Hans V S, Saini R P and Saini J S 2010 *Sol. En.* **84** 898-911
[8] Patil A K, Saini J S and Kumar K 2010 *Int. J. Ren. Energy Res.* **2** 1-15
[9] Stafford J, Walsh E and Egan V 2009 *Meas. Sci. and Tech* **20** 1-11
[10] Akafuah N K, Hall III C A and Wang T 2004 *Eng. Summer Conf. HT/FED 2004* vol. 1 pp. 367-375
[11] Belmiloudi A 2011 *Heat Transfer - Theoretical Analysis, Experimental Investigations and Industrial Systems* ed. Intech, chapter 21, available online
[12] Ciminà S, Wang C, Wang L, Niro A and Sunden B 2015 *J. of En. Heat Trans.* **22** 29-45
[13] Fustinoni D, Gramazio P, Colombo L P M and Niro A 2012 *J. Phys.: Conf. Ser.* **395** 012042
[14] Szeliski R 2011 *Computer Vision: Algorithms and Applications* ed. Springer, chapter 3, draft available online

Application of close-range aerial infrared thermography to detect landfill gas emissions: a case study

G Tanda[1,2,*], M Migliazzi[3], V Chiarabini[4] and P Cinquetti[5]

[1] DIME, University of Genoa, Genoa, Italy
[2] RIT4Waste Center – University of Genoa, Genoa, Italy
[3] GLOBI Hi-Tech Srl, Genoa, Italy
[4] KIM-RemoteSensing GmbH, Klagenfurt am Wörthersee, Austria
[5] AMIU, Genoa, Italy

*Corresponding Author. E-mail: giovanni.tanda@unige.it

Abstract. Monitoring waste disposal sites is important to check that the produced biogas, potentially explosive, is properly collected by the biogas extraction system of the landfill site and to evaluate the residual biogas flow escaping from upper surface of the landfill. As the biogas migrates to the surface, the soil through which it flows is expected to reach a higher temperature than the surrounding environment; thus, measuring the thermal footprint of the landfill soil surface could allow the detection of biogas leakages and spots suitable for the gas extraction. Close-range aerial infrared thermography is an innovative approach able to identify thermal anomalies with a good resolution over a large region of the landfill surface. A simple procedure to deduce the biogas flow rate emerging from the soil into the atmosphere, based on infrared thermography measurements, is presented. The approach has been applied to a case study concerning a large landfill located in Genoa (Italy). Aerial infrared photographs taken during different days and seasons showed the presence of thermal anomalies over regions along the peripheral boundary of the landfill still not interested in biogas extraction.

1. Introduction

Solid wastes collected in urban landfills contain a significant percentage of organic substances whose degradation produces a large amount of biogas that, due to its large methane content, is a precious energy source with high global warming potential. Few landfills are thought to recover more than 60% of the available gas [1]; the rest is allowed to escape to the atmosphere. Biogas leakages have a large impact on the environment even at a local scale, due to risks of explosions, odour, and adverse effects on human health and vegetation. The landfill biogas is a product of anaerobic digestion of organic compounds and this reaction is accompanied by release of heat. As the biogas migrates to the surface, the soil through which it flows is expected to reach a higher temperature than the surrounding environment. Measuring the thermal print of the landfill soil surface could allow the detection of biogas leakages and spots suitable for the gas extraction. In principle, the evaluation of the flow rate of the biogas emerging from the soil into the atmosphere can be deduced from a steady-state heat balance. The thermal power produced by digestion is dissipated to the environment by radiation, convection and conduction; thus, heat transfer dissipated to the environment is basically a function of heat transfer surface area and the soil surface temperature.

Infrared thermography (IRT) is an innovative diagnostic tool able to detect thermal anomalies on the landfill surface. It constitutes a non-invasive technique that registers the temperature distribution by means of a thermal camera that receives and processes the infrared radiation emitted from the target

IOP Conf. Series: Journal of Physics: Conf. Series **796** (2017) 012016 doi:10.1088/1742-6596/796/1/012016

surface. A first review of application of IRT to monitor waste disposal sites has been provided by Lewis et al. [1], where most of the mentioned previous studies concerned the use of ground based infrared (IR) cameras. The IRT inspections of landfills as viewed from nearby sites have been recently documented in refs. [2–4]. The main limitation of a ground based monitoring system is that the target surface is viewed largely away from the normal direction. As known, the directional emissivity of dielectric materials tends to be markedly dependent upon the viewing angle (i.e. the angle between direction of emission and the normal to the emitting surface) when it exceeds about 60°, while a quite constant and relatively high emissivity is typically registered for angles less than about 60° [5]. Therefore, a ground based IR imaging could lead to a markedly increased uncertainty in the surface temperature measurements due to the large uncertainties in the ground directional emissivity.

Aircraft-mounted cameras improve the potential of the technique, since a large area, viewed with the sensor perpendicular to the target, can be monitored in a relatively short time [6–9]. Aerial IRT has its root in military applications but it can be successfully applied to many fields such as inspection of environment (to monitor sea, river and coastal pollution, waterways and drainage systems, forest fires, landfills, etc.), buildings (to detect roof moisture and heat losses), industrial sites (to identify leaks from heating systems or faults in electric distribution and transmission systems) and agriculture (to investigate physical transport phenomena between vegetation and atmosphere). Aerial infrared applications can be divided into two types: those where a straight-down view and/or a large area view is needed and those where long distances must be covered in a limited amount of time [7]. Both helicopters and light airplanes can be used to perform aerial infrared surveys. In the case of landfill surveys aerial IRT might miss small scale details (i.e., small biogas leaks) due to large distances from the target.

The use of unmanned aerial vehicles (UAVs) represents the added value to the thermographic inspection of waste disposal sites. In fact, UAVs permit to gather IR images with a viewing angle close to zero and to vary the scale of the imaging by properly arranging the distance between the UAV and the target (typically from 10 to 70 m) in order to provide an adequate resolution of the surface thermal map. The IR camera, mounted on UAVs, is typically synchronized with the position log of the UAV navigation system and with a built-in visual camera in order to obtain the precise thermal mapping of the landfill.

Applicability and limitations of IRT to detect gas leakages from landfills are examined in ref. [1]: generally speaking, results may be markedly affected by weather and soil conditions. Windy and warm days are usually not indicated as well as the presence of surface wetness or different types of ground materials; all these factors could give rise to anomalies without connections to landfill gas leakage or dilute the associated thermal anomaly. Conversely, night-time surveys reduce light condition influences while cloudy conditions mitigate the effect of the sky radiation. It is apparent that the variability of environmental conditions (e.g., air temperature and solar/sky radiation) and of the observed targets (e.g., modifications in the soil coverage and vegetation) makes quantitative IRT a non-trivial issue, requiring specific expertise by the user as well as an analysis performed in comparative terms (i.e., by looking at differences between targets with known radiometric properties and targets under study). Combination of different techniques, in order to get more elements for the interpretation and validation of results by cross-checking the information, is also advisable [3].

2. Monitoring of waste disposal sites

The digestion process to produce biogas is accompanied by a release of heat. This thermal power is dissipated to the environment by radiation, convection and conduction. Assuming for the sake of simplicity, steady-state conditions and neglecting thermal conduction, the energy balance is given by the following equation:

$$Q = \sigma\,\varepsilon\,A\,(T_w^{\,4} - T_{surr}^{\,4}) + h\,A\,(T_w - T_{air}) \tag{1}$$

where T_w [K], T_{surr} [K] and T_{air} [K] are the temperatures of the soil surface, the surroundings, and the ambient air , respectively, σ (=5.67 $\times 10^{-8}$ W/m^2·K^4) is the Stefan-Boltzmann constant, ε is the total hemispherical emissivity of the soil, h [W/ m^2·K] is the convective heat transfer coefficient and A [m^2]

is the soil surface area. In equation (1), the first term on right-hand side represents the heat transfer rate from the soil surface to the surroundings by radiation (evaluated assuming an enclosure formed by a gray body at emissivity ε surrounded by a very large surface), while the second term is the convective heat transfer rate from the soil surface to the environmental air. In order to identify the heat transfer rate associated with the biogas production, the surroundings and the ambient air temperatures can be replaced by the undisturbed soil temperature $T_{w,ref}$ [K], as follows:

$$Q = \sigma \varepsilon A \left(T_w^4 - T_{w,ref}^4\right) + h A \left(T_w - T_{w,ref}\right) \quad (2)$$

Literature empirical equations, which take into account the effect of wind, can be used for the convective heat transfer coefficient h. For instance, the following equation, taken from ref.[10], can be assumed for h:

$$h = 5.7 + 3.8\, V \quad (3)$$

where V is the wind velocity [m/s].

As the soil emissivity ε is a known parameter, typically assumed to be 0.9-0.95, the heat produced by digestion and dissipated to the environment is basically a function of heat transfer surface area A and the soil surface temperatures T_w and $T_{w,ref}$. Dividing Q by the heat associated with the production of methane (about 7325 kJ / m^3, ref. [6]) leads to the estimation of the methane flow rate reaching the soil surface. The calculated methane flow rate from the above procedure is just a rough estimation due to the assumptions done in equation (1); weather conditions may induce not-steady surface temperature distributions, uneven material properties (surface emissivity and thermal conductivity) may affect the energy balance as well. Moreover, small gas leakages flowing through long pathways to reach the surface could not be recognized since insufficient to maintain the soil at a surface temperature relatively higher than that of the undisturbed soil.

2.1 Test case for validation

A simple experiment has been designed in order to validate the previously described procedure. An aluminum plate (150 mm × 110 mm, thickness 8 mm), connected to an electric plane heater on the bottom side, has been placed inside a large container (400 mm × 500 mm) made of 20mm-thick extruded polystyrene, as shown in figure 1. The container has then been filled with a 10mm-thick sand layer. Several fine-gauge thermocouples have been placed inside the heated plate and along the sand layer. Once electric power is delivered to the heater, the plate, at steady-state, attains a uniform temperature value, detected by thermocouples and controlled by the DC power supply. The surface temperature of the sand is then detected by an infrared camera (FLIR T335, 320 x 240 pixels) assuming a thermal emissivity ε for sand of 0.95. Experiments were conducted in a laboratory room with ambient and wall temperatures equal to 26.5 ± 0.3 °C, without air currents (V = 0 m/s in equation 3). The delivered electric power, measured by means of a voltmeter and an amperometer, was adjusted in order to maintain a prescribed temperature difference between the plate and the ambient air.

A preliminary experiment has been conducted to evaluate the heat conduction losses from the container polystyrene walls. The sand layer facing the ambient air was covered by a 20mm-thick polystyrene layer in order to inhibit any convection and radiation heat transfer from the sand surface. At the steady-state, a thermal power of 1.6 W was dissipated by the heater to maintain a uniform temperature of 53°C over the aluminum plate (ambient air temperature equal to 26.5°C). Due to the symmetrical arrangement of the insulation layers, it was assumed that an equal quantity of 0.8 W has been dissipated by thermal conduction across the bottom and the top insulation layers. Once the top insulation layer was removed (thus allowing convective and radiant heat exchanges between the sand surface and the surroundings), the dissipated thermal power required to maintain the same temperature (53°C) over the heated plate was 5.1 W. If the conduction losses (0.8 W) through the bottom insulation layer are subtracted, the convective and radiant heat transfer rate Q_{meas} from the sand surface to the surrounding was measured to be 4.3 W.

Figure 2 displays the thermal map of the sand surface under the above described conditions. Two different areas have been identified: the larger one (Ar1) encompasses the entire thermal anomaly registered by the IR camera, the smaller one (Ar2) includes the surface region where the surface temperature is significantly higher with respect to boundaries. The imaging process has been performed using a dedicated software for thermal images elaboration (Flir QuickReport©, FLIR Systems) and equation (1) for the estimation of heat transfer rate. Results for the two regions of interest identified were:

Ar1) Q_{calc} = 4.6 W (52% radiation, 48% convection)
Ar2) Q_{calc} = 3.5 W (53% radiation, 47% convection)

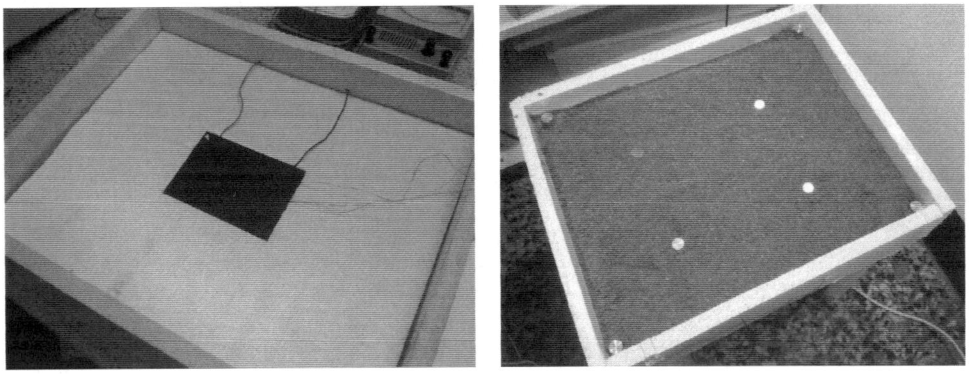

Figure 1. The container simulating a landfill with biogas production: the thermal source (left) and the covering sand layer (right).

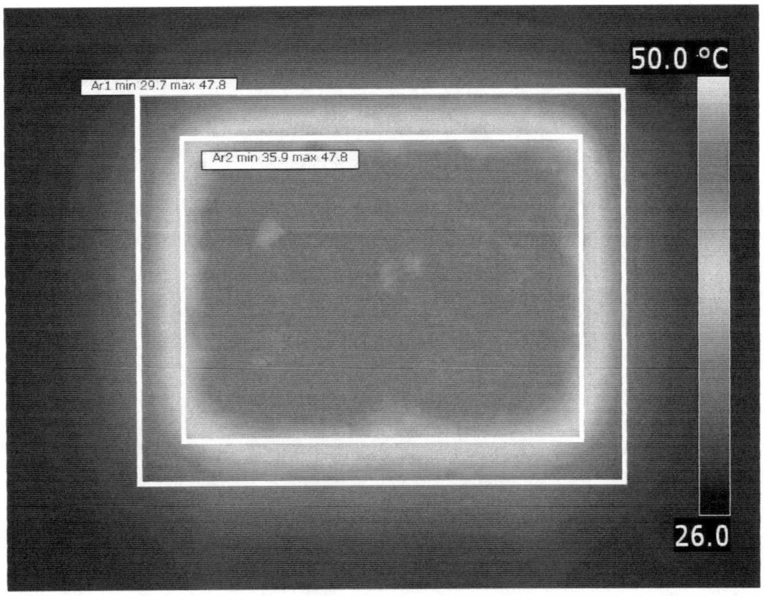

Figure 2. Temperature map of the sand surface, provided by the IR camera, for the simulated landfill experiment.

By comparing the calculated heat transfer rates (Q_{calc}) with the measured one (Q_{meas} = 4.3 W), it can be argued that the heat transfer rated from the sand surface is correctly estimated (within a 7% error) by averaging the wall temperature over the region of area Ar1 and it is slightly underestimated (by 19%) when a smaller area (Ar2) is used for the analysis. Further experiments performed by varying the heated plate-to-ambient temperature difference (from 17 to 35 K) and the thickness of the sand layer (from 1 to 2.5 cm) gave rise to similar qualitative results, i.e. an agreement within 20% between the heat transfer rate from the sand surface to the environment based on the measurement of dissipated electric power and that deduced by IR imaging over a significant surface area.

2.2 Close-range aerial IR inspection of a landfill site
The procedure has then been applied to the study of biogas emissions from the soil of a waste disposal site. To accomplish this task, several flights with UAVs equipped with infrared and visual cameras have been performed over the landfill of Mount Scarpino, located in Genoa, northern Italy and one of the widest European mountain landfill. This landfill, covering approximately an area of 58 hectares, has been used to dispose and manage the solid waste of the town of Genoa and its province for over forty years. A landfill gas power plant has been operating on site since 2007. The biogas (with methane concentration at 50% by volume) actually extracted from this site is about 50×10^6 m^3 per year, giving rise to an electrical energy production of about 65×10^6 kWh per year, saving more than 300.000 tons/year of CO_2.

Figure 3 shows an aerial image of the investigated landfill: the figure reports the distribution of wells and pipes for the biogas extraction. The investigation was conducted by performing several UAV flights over a period from February 2015 to March 2016, using different IR cameras and namely FLIR T460 (320 x 240 pixels), FLIR A65 (640 x 512 pixels), and FLUKE Ti 300 (240 x 180 pixels). Dedicated software programs (Flir Reporter Professional©, Flir Systems and SmartView©, Fluke Corporation, depending on the type of IR camera employed) and equation (2) were used to process the experimental data.

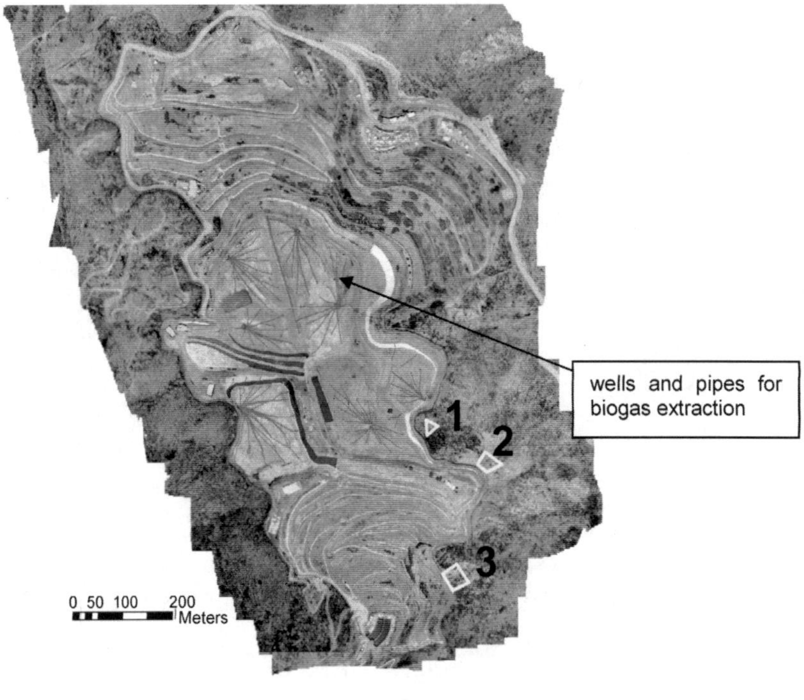

Figure 3. Topographic view of the landfill. Wells and pipes for the biogas extraction as well as the inspected sites (No.1, 2, and 3) by aerial infrared thermography are reported.

Flights were performed under different environmental conditions: in the mornings of one day of February, 2015 and one day of August 2015 (both with clear sky), and during the late evening (without solar radiation) of two different days in the month of March 2016.

Environmental conditions (ambient air temperature and relative humidity, wind velocity) were detected by a local weather station. For all cases, wind was moderate (typically 5 m/s or less) and the terrain was dry, except for one of the flights performed in March 2016, when the site was wet and muddy in some spots. A typical IR image of the landfill taken from the UAV is reported in figure 4, together with the visible image of the same site. The IR image clearly shows the pipes carrying hot biogas from the extraction well.

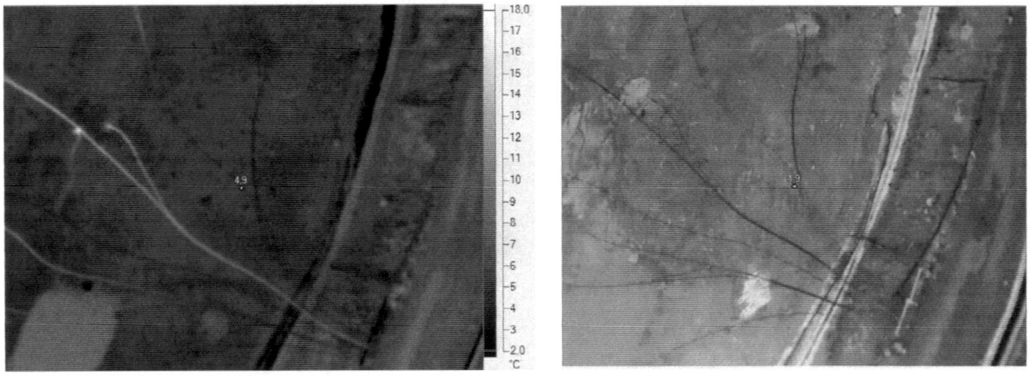

Figure 4. An example of application of aerial infrared thermography to the monitoring of the landfill: pipes carrying hot biogas extracted from the well are clearly visible in the left-hand IR image.

The main objective of the landfill survey was to investigate the boundaries of the landfill, not interested in biogas extraction, in order to identify residual biogas superficial emissions and to infer, through a thermal balance, a rough estimation of the amount of escaping biogas flow rate. Figure 5 shows the landfill peripheral surface mainly interested in aerial IR imaging, highlighted with color codes related to the detected surface temperatures.

Attention was turned to three specific sites deemed to be suitable for IR inspection: they were covered by bush vegetation, sufficiently uniform to minimize the risk of false identification of gas leaks, i.e., the occurrence of anomalies not connected to landfill gas leakage but associated with different types of ground materials and consequent different interactions with weather conditions. The investigated sites showed thermal anomalies occurring during all the flights. The three inspected sites are clearly identified in figure 3 (namely No.1, 2, and 3). Since the flights performed in March 2016 were featured by the most suitable conditions for aerial thermographic investigation (low ambient temperature, no direct solar radiation), only data acquired during these flights were processed, whereas those recorded during the morning flights (with direct solar radiation) were considered only for a qualitative comparison of the thermal footprints over the sites of interest.

Infrared and visible images taken during the flight were matched by a custom-made software to produce a mosaic image giving at the same time the topographic view and the thermal map of the inspected sites, as shown in figures 6–8. In each figure, only surface temperature data exceeding by more than 2°C the minimum surface temperature were displayed in coded colours.

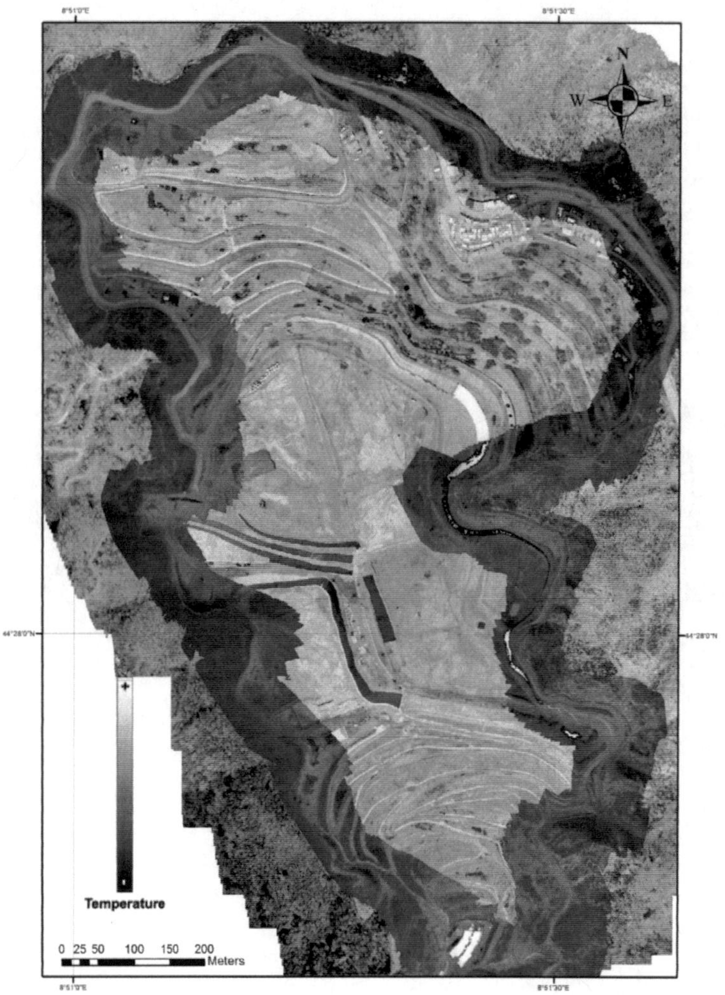

Figure 5. Mosaic image of the landfill containing the topographic view of the entire site and the IR imaging of the peripheral region.

Thermal anomalies were clearly detected over the three investigated sites. Namely, surface area framed within the boundaries of site No.1 registered a temperature ranging from 3.5 to 10.3°C (average 5.2°C), area of site No.2 showed a surface temperature between 4.2 and 14.1°C (average 8.4°C) and site No.3 was featured by a surface temperature from 4.2 and 13.3°C (average 7.4°C). Table 1 reports the estimated heat flux associated with the biogas digestion from each site and the relevant methane flow rate actually dispersed into the atmosphere. Repeated measurements over the same sites by using different images captured during the same flight or comparisons among measurements over the same site by processing data taken in different flights revealed differences in the estimated methane flow rate within ± 30%. This figure can be realistically assumed as the uncertainty of this method, whose order of magnitude is comparable to that reported in similar studies [11] or associated with other measuring techniques, such as the tracer gas method or the static chamber method [12]. Taking into account the relatively large estimated uncertainty in the measurement, the amount of methane escaping from the inspected sites was found to be comparable to that actually collected by the existing wells and pipes conveying the extracted biogas to the thermal power station for energy conversion.

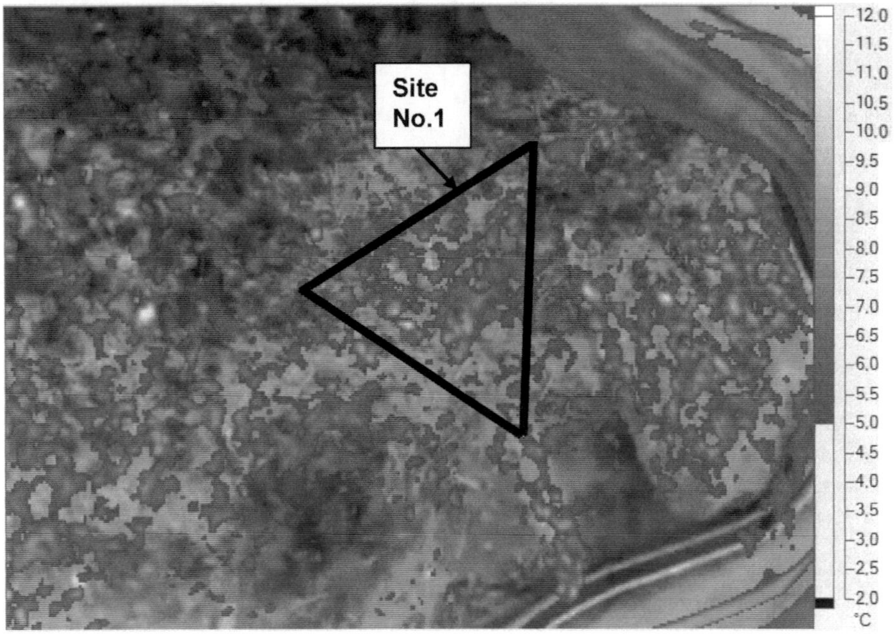

Figure 6. Thermographic data gathered for site No.1.

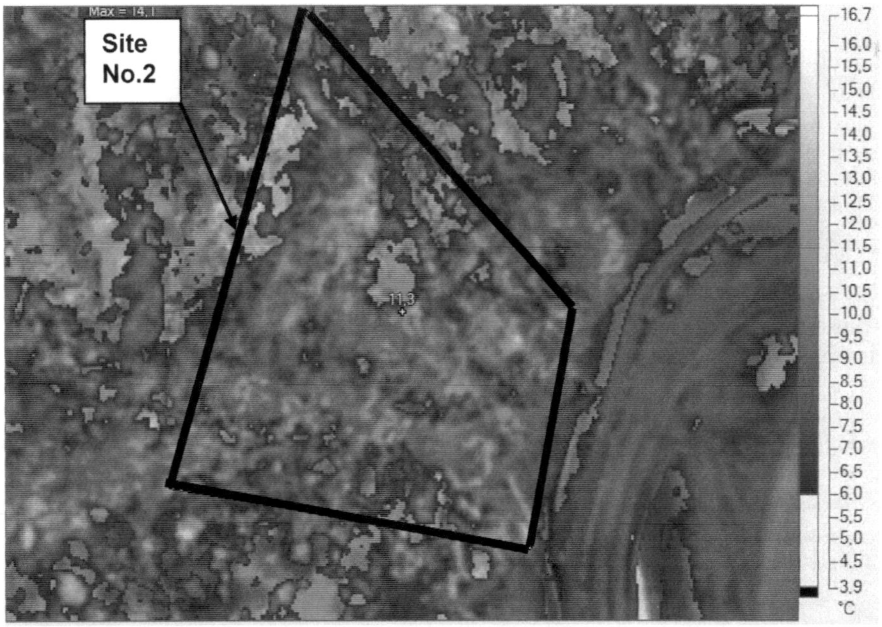

Figure 7. Thermographic data gathered for site No.2.

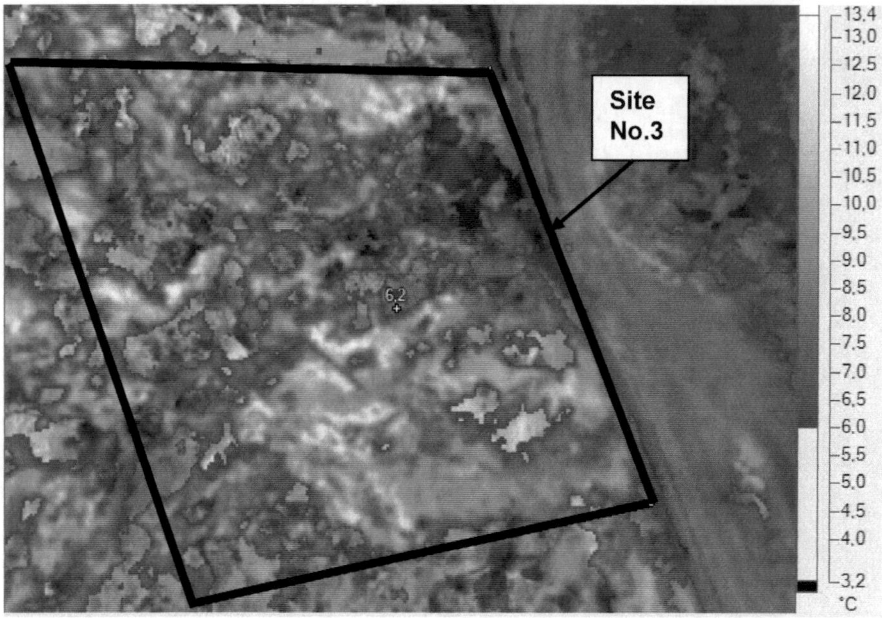

Figure 8. Thermographic data gathered for site No.3.

Table 1. Estimated methane flow rates and main parameters included in the calculation procedure.

Site No.	Surface area[m^2]	Mean T_w [K]	Undisturbed soil temperature $T_{w,ref}$ [K]	Measured wind velocity V [m/s]	Imposed surface emissivity	Heat transfer rate [kW] by equation (2)	Estimated methane flow rate per m^2 [m^3/ h·m^2]
1	350	278.3	276.6	5.0	0.95	17.4	0.0245
2	960	281.5	277.3	5.0	0.95	118.5	0.0607
3	1500	280.5	277.3	5.0	0.95	141.0	0.0462

3. Conclusions

Infrared thermography (IRT) has been applied to the study of biogas leakages from waste disposal sites. This diagnostic tool permits the identification of thermal anomalies on the landfill surface related to the biogas digestion processes. The implementation of this technique, whose applications in landfills documented in the literature are mainly based either on ground or aircraft inspections, consists in the use of unmanned flight vehicles (UAVs) to take infrared (IR) images of the landfill surface. The close-range aerial inspection permits the gathering of data over a large region and at the desired spatial resolution by properly arranging the distance between the UAV and the investigated soil.

A simple calculation procedure, based on the steady-state energy balance between the heat produced by the digestion and the heat dissipated into the atmosphere, capable to provide to a rough estimation of the methane flow rate escaping from the landfill surface, has been presented.

The proposed method has been successfully applied to a case study concerning a large landfill located in Genoa (Italy), from which an amount of biogas of about 50×10^6 m^3 per year is currently extracted. Aerial infrared photographs taken during different days and seasons showed the presence of thermal anomalies associated with biogas digestion over regions along the peripheral boundary of the landfill and revealed for these sites the potential for significant amount (from 2.5 to 6×10^{-2} m^3/ h per m^2 of surface area) of biogas extraction. Despite the limitations of the technique described throughout the paper, this study shows that close-range aerial IRT constitutes a reliable approach for monitoring urban landfills. The presented calculation procedure, based on IRT aerial inspections, is deemed to be a powerful tool to detect the biogas production from a waste disposal site and to provide a preliminary estimation of the methane production potential.

References

[1] Lewis AW, Yuen STS and Smith AJR 2003 Detection of gas leakage from landfills using infrared thermography – applicability and limitations *Waste Management Research* **21** 436–447.

[2] Desideri U, Leonardi D and Proietti S 2007 Application of infrared thermography to study behaviour of biogas captation wells *Proc. of Sardinia 2007, 11th Int. Waste Management and Landfill Symposium (S. Margherita di Pula, Cagliari, Italy).*

[3] Battaglini R, Raco B and Scozzari A 2013 Effective monitoring of landfills: flux measurements and thermography enhance efficiency and reduce environmental impact *Journal of Geophysics and Engineering* **10** 064002.

[4] Belghazal H, Piga C, Loddo F, Stitou El Messari J and Ouazani Touhami A 2013 Geophysical surveys for the characterization of landfills *International Journal of Innovation and Applied Studies* **4** (2) 254–263.

[5] Carlomagno GM and Cardone G 2010 Infrared thermography for convective heat transfer measurements *Experiments in Fluids* **49** 1187–1218.

[6] Tonelli A 2000 Some operative applications of remote sensing *Annali di Geofisica*, **43** (6) 1177–1196.

[7] Stockton GR 2004 Advances in applications and methodology for aerial infrared thermography *Proc. of Conf. SPIE No. 5405, Thermosense XXVI (Orlando, USA).*

[8] Slonecker T, Fisher GB, Aiello DP and Haack B 2010 Visible and infrared remote imaging of hazardous waste: a review *Remote Sensing* **2** 2474–2508.

[9] Beaumont B, Radoux J and Defourny P 2014 Assessment of airborne and spaceborne thermal infrared remote sensing for detecting and characterizing landfills *Proc. of 7th Int. Conf. on Waste Management and the Environment (Ancona, Italy).*

[10] Sartori E 2006 Convection coefficient equations for forced air flow over flat surfaces *Solar Energy* **80** 1063–1071.

[11] Chiarantini L and Coppo P 1995 Infrared monitoring system for urban solid waste landfills: experimental activities for biogas outflow modelling *Proc. of Conf. QIRT 94 –Eurotherm Series 42 (Paris, France)* p. 305–309.

[12] Diot M, le Golvan YM, Bogner J, Burkhalter R, Chanton J, Tregourès A and Schaegis L 2000 Landfill biogas emissions measurements: qualification and quantification of the pathways of emitted methane *Proc. of the Intercontinental Landfill Research Symposium (Luleå, Sweden).*

Modelling of Shaded and Unshaded Shallow-Ground Heat Pump System for a Residential Building Block in a Mediterranean Climate

M Bottarelli[1], **C Yousif**[2]

[1] University of Ferrara, Via Quartieri 8 – 44121 Ferrara, Italy
[2] University of Malta, Barrakki Street, Marsaxlokk, MXK 1531, Malta

Corresponding author: michele.bottarelli@unife.it

Abstract. Heat pumps may be coupled to shallow-ground geothermal fields and used for the purpose of space heating and cooling of buildings. However, quite often it is not possible to locate the geothermal field in cleared grounds, especially in cities where building density is high and land has a high premium. This leads to the possibility of burying the geothermal field under the basement of new building blocks, before construction of the building.
In the present work, the shaded-unshaded arrangement is numerically studied by coupling the software DesignBuilder-EnergyPlus to assess the building's energy requirement with the software FEFLOW to solve the heat transfer equation in porous media. Assuming a standard residential building block, the coupling between the two software is performed by assigning the thermal energy requirement for air conditioning, as calculated by EnergyPlus, to a flat-panel typology of ground heat exchanger simplified in a 2D FEFLOW's domain.
The results show that it is necessary to opt for a dual-source heat pump (air/geothermal) system to ensure that the ground is not frozen or over-heated at peak times and to improve the overall performance of the system.

1. Introduction

The dawn of the EU Energy Performance in Buildings Directive Recast 2010/31/EU has put new demands on new and renovated buildings to achieve zero net-energy buildings by 2021, while at the same time, one is witnessing greater importance being given to energy storage, which in this case is the underground thermal energy storage (UTES). For that scope, ground-coupled heat pumps (GCHPs) are commonly depicted as energy efficient systems for air conditioning [1].

But, whereas the use of GCHPs is widespread in Northern Europe, they are still quite uncommon in Southern Europe. So far, the most common system is the reversible air-source heat pump.

Solar systems can easily satisfy the heating and cooling demands of single storey building blocks, which generally have relatively low energy demand and sufficient roof space area to install solar systems to counter-balance their energy consumption and bring them to near zero-energy status. However, the situation could be completely different in urban areas, where multi-storey buildings are more common due to the high value of land and it would not be profitable for the developer to limit the number of storeys. Furthermore, the available roof space and possibly part of the façade would not be sufficient to offset the total energy demand of the building, even when it is built with the highest energy efficiency standards.

In such cases, the building's footprint could be used for shallow-ground thermal energy storage to be exploited for air conditioning of the building block [2], albeit the fact that the building's footprint does not see the sun, which may reduce the final ground energy balance. Consequently, the use of a traditional GCHP coupled to shallow-ground heat exchangers may not be suitable, whereas a dual-source heat pump (DSHP) combining the air-source with the ground-source could be a good compromise [3,4,5], because it would have the potential of minimizing the size of the ground heat exchanger and utilise the underground thermal energy storage more effectively, while preventing frosting at times of low ground thermal energy [6,7,8].

The present paper reports the numerical analysis of a so-called dual-source heat pump (DSHP) in shaded and unshaded conditions, as a first step for future studies in urban cities.

2. Methodology

In the present contribution, a numerical analysis is carried out for the behaviour of a very shallow ground heat exchanger system coupled with a so-called dual-source heat pump (DSHP), which is able to switch between air and ground heat source/sink, according to the best availability. As a consequence, a DSHP can also help to keep the size of the ground heat exchanger system smaller than that when only the ground-source is used. Here, the ground heat exchanger system was assumed to be buried only a few meters below the slab foundation of a four-storey residential building, located in an urban area and hence surrounded by other building blocks. Therefore, the DSHP has been chosen because the heating and cooling demand per square metre is quite high, as a consequence of building typology and its urban location.

The energy requirement for space heating and cooling was calculated by the software DesignBuilder (DB) coupled with the energy model EnergyPlus, for two different locations (Malta and Ferrara, Italy), which represents a Mediterranean and a continental climate, respectively. Then, the resulting thermal loads in terms of hourly time series were imposed in a 2D finite elements model implemented in FEFLOW software, to evaluate the heat transfer in the ground and to determine the temperature of the working fluid and the energy share of air and ground thermal exploitation.

2.1. DesignBuilder-EnergyPlus Model

DesignBuilder software is one of the most popular graphical interfaces to EnergyPlus software, which is one of the most known energy simulation software tools to model the energy use and thermal loads throughout the building. EnergyPlus is a combination of the best capabilities and features of the previous DOE-2 and BLAST models [9,10].

One of the interesting features of EnergyPlus is that it has taken the approach of organising software tools in modules that can easily work together or separately. Given that EnergyPlus had no visual interface up to a few years ago to allow users to draw and see the building, it was necessary to link it up with a third-party software tool, which is DesignBuilder [11]. DesignBuilder has an extensive library of building materials, which makes the input of components faster. ACAD drawings can be imported into DesignBuilder or otherwise drawings can be created directly into the software itself using 3D elements. Building blocks can also be cut, stretched and partitioned into zones [12].

The building block comprises of semi-basement garages, 3 elevated ground floor apartments, 3 first floor apartments, 3 second floor apartments and 2 top floor penthouses. The building has an average floor area of 100 m² per apartment, except for the top 2 penthouses that have around 125 m² each.

Figure 1 shows the details of the building block as follows:

1. The elevated ground floor apartments have an exposed floor to underground garages.
2. The first floor apartments are fully surrounded by buildings and only have the front, back and central walls exposed to the ambient.

3. The second floor apartments are similar to the first floor apartments but have some more exposed walls on the eastern and western parts as well as part of the ceiling of the front rooms, given that there are terraces on top of them.
4. The penthouses are practically exposed from all sides but have an insulated roof, in accordance with the minimum energy performance requirements as set in the Technical Guidance F for Minimum Energy Requirements in Maltese buildings [13].

The total ground floor area covered by the apartments amounted to 405 m² and this is the area under which the shallow-ground geothermal field is placed, as explained in Section 2.2 below. The total floor area of all the floors is 1621 m², out of which 1,188 m² is conditioned space, while the remainder is the unconditioned space in the common areas and lifts. The set point temperature for the conditioned space was 21 °C for winter and 24 °C for summer. However, the operation of the air-conditioning system was set to match the activity schedule of the occupants. This was set as 07:00-09:00 and 16:00-23:00, which would reflect the lifestyle of a typical working family.

Figure 2 shows the percentage distribution of energy consumption of the main four sectors, namely, space heating, space cooling, lighting and water heating. In this case, only natural ventilation is assumed and hence there are no mechanical fresh air systems. Also, it is assumed that the building has no renewable energy installations. It is clear now that in residential buildings, energy used for space cooling and heating is the bulk amount of energy consumed and one has to focus on reducing it.

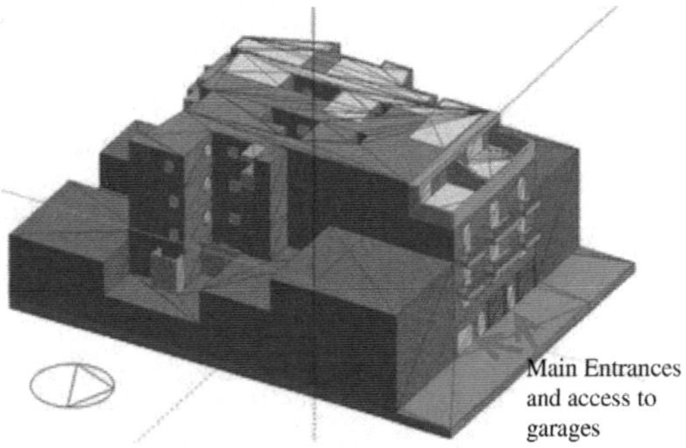

Figure 1. A pictorial demonstration of the building block, with two main entrances and the entrance to the underlying garages.

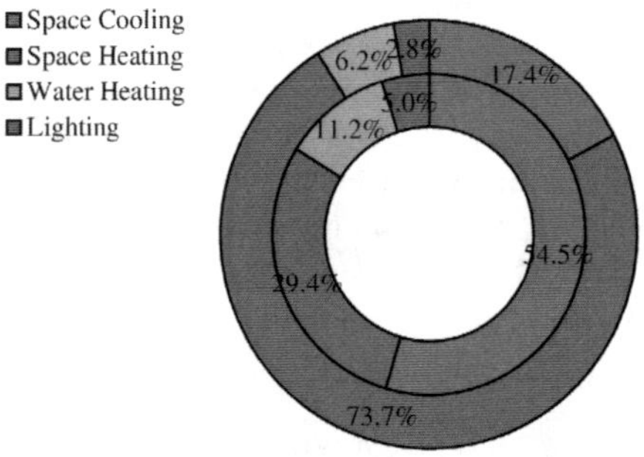

Figure 2. Percentage distribution of energy consumption for Malta (inner doughnut) and Ferrara (outer doughnut).

The building materials for this apartment block are summarized in Table 1. Following the construction of the building block in DesignBuilder, as shown in Figure 1, other input parameters were made according to the specific characteristics of the building. This included the thermal properties of all walls, floors, roofs, windows and doors, as well as all the operational schedules for occupancy and usage of different systems together with the set point temperatures for heating, cooling and water heating and finally the characteristics of all the energy services within the building. One is to note that the construction of this building block does not include any energy efficient measures and the U-values of the building envelope are high. Therefore, this study would give us an idea about the worst case scenario under such circumstances.

Figure 3 shows a comparison between the ambient temperatures in Malta and Ferrara, respectively. It is clear that although the winter temperatures in Ferrara do fall below 0°C, it never drops below 6°C in Malta. At the same time, there are several weeks when the temperatures are quite high in July and August.

A number of scenarios were run in DesignBuilder using the weather data files for Malta and Ferrara. Hourly heating and cooling demands for both models were calculated in DesignBuilder. The weather data for Malta was tailor-made because the default Malta file in DesignBuilder actually refers to Cozzo Spadaro, which is in the nearby island of Sicily. However, it was previously shown that the weather there does not represent that of Malta [14]. The results for the simulations are shown in Figure 4, which are the actual thermal energy requirements for heating and cooling the building. These results served as input values for the design of the dual-source heat pump system. It is to be expected that the heating demand for Ferrara is higher than that for Malta, given that the ambient temperature in winter is much lower there. However, it is interesting to note that the cooling load for Ferrara has reached that in Malta for a significant number of weeks, between June and August. This can be explained by the fact that Ferrara has a continental climate with extremes in weather conditions being experienced in winter and summer, while Malta has a Mediterranean climate and is surrounded by the sea, which moderates temperature swings.

Table 1. Technical description of the building's major envelope elements and materials.

	Actual U-value (W/m^2K)	Description	
External façade	1.513	limestone blocks	180 mm
		air cavity	30 mm
		limestone blocks	180 mm
		lightweight plaster	6 mm
Unconditioned walls	2.040	lightweight plaster	6 mm
and		hollow concrete blocks	230 mm
Internal walls		lightweight plaster	6 mm
Other external walls	2.067	lightweight plaster	6 mm
		hollow concrete blocks	230 mm
		lightweight plaster	6 mm
Floor and ceiling	2.059	lightweight plaster	6 mm
		reinforced concrete	200 mm
		screed	100 mm
		ceramic tiles on floor	12 mm
Main door	3	solid wooden door	35 mm
Windows and doors	6	Single glazing with aluminium frame (no thermal break)	
Façade door	4	Double glazing with 6mm air gap and aluminium frame (no thermal break)	

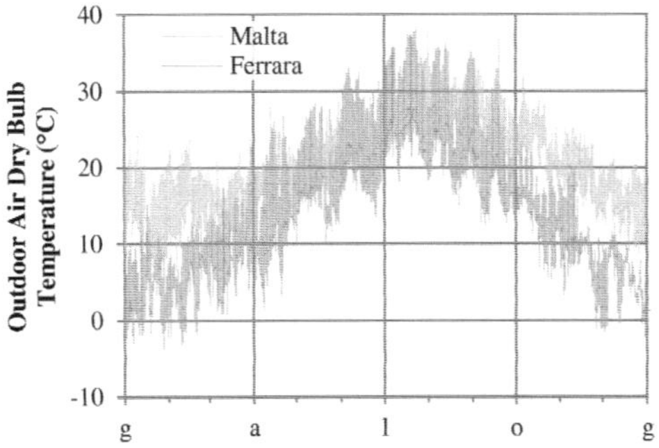

Figure 3. Distribution of external dry bulb temperature in Malta and Ferrara, Italy.

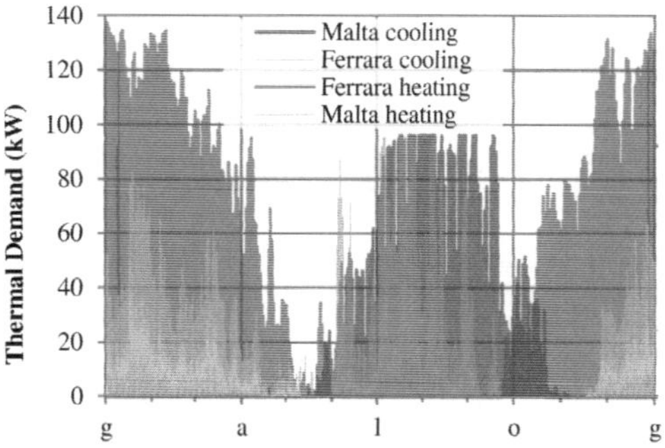

Figure 4. Hourly heating and cooling demand for the building block under the weather climate of Malta and Ferrara, Italy.

2.2. FEFLOW model

FEFLOW is an advanced commercial software that apply the finite element approach for solving flow, heat and mass transfer in porous media, variously saturated and non-homogeneous [15]. In the present study, FEFLOW is used to solve the ground heat transfer through a flat-panel system installed under the building's foundation, and as constrained by the thermal loads resulting from the DesignBuilder analysis. For simplicity, the thermal requirement is here fully applied on the geothermal system, neglecting the reduction driven by the coefficient of performance of the heat pump. As ground heat exchanger, the flat-panel (FP) is innovative and very effective, due to its large surface area that is in contact with the ground to perform the heat transfer process [16]. Its geometry was here implemented into a 2D domain, schematizing a vertical cross-section of heat exchangers and ground laid under the building, as shown in Figure 5. For the purpose of saving simulation time, only half of the building façade was considered (6.75m), thus assuming symmetric the performance.

Flat-panels were considered to be 3m high, vertically installed in a trench 40cm wide and running for the full depth of the plot area of 30 metres. The spanning distance between one trench and another was 1.7m, respectively, according the methodology reported in [17] compared with the very small footprint available. Thermal properties of soil were selected according to [18], taking into account the most widespread shallow lithology of Malta (globigerina limestone) and Ferrara (silty loam). Similarly, the backfill materials were assumed to be of similar material, but partially modified to represent sand (Malta) and sieved loam (Ferrara). All properties are reported in Table 2.

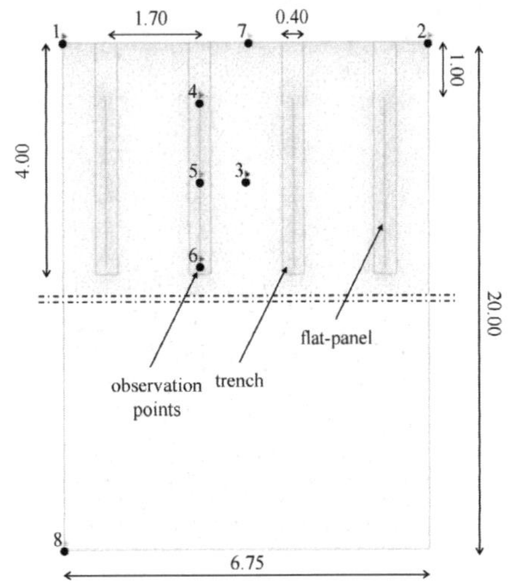

Figure 5. 2D domain and mesh

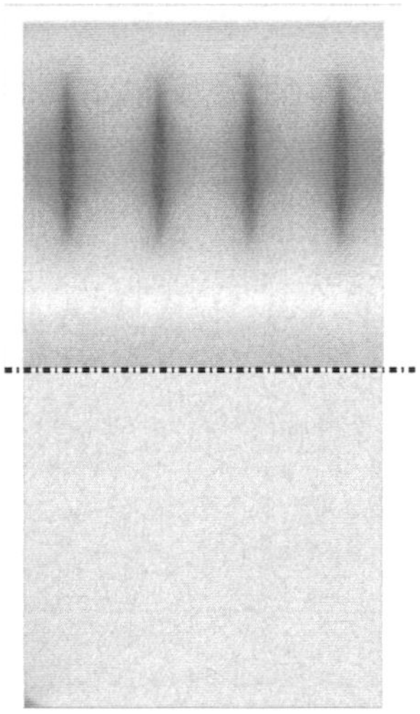

Figure 6. Representative thermal field in winter

Table 2. Ground material characteristics data

Parameter	Malta		Ferrara	
	soil	backfill	soil	backfill
Porosity, [%]	0.1	0.3	0.3	0.3
Density,[kg m^{-3}]	2500	2000	1800	1800
Thermal conductivity, [W m^{-1} K^{-1}]	2.7	2.0	0.8	1.0
Specific heat, [MJ m^{-3} K^{-1}]	2.2	1.6	1.5	1.6
Hydraulic conductivity, [m year^{-1}]	10	3000	10	100

To minimize numerical errors and reduce the computational time, the size of the finite elements was chosen to be fine for the area surrounding the foundation and coarse for the area far from it. The mesh independence of the full final mesh (20,000 finite elements, Figure 5) was checked by doubling the number of elements, without observing significant differences in the resulting temperatures at selected observation nodes.

The vertical sides of the domain were considered adiabatic, because of the symmetry and the adjacent abutting buildings, which were assumed to have the same underground system. Two cases were considered for the top part of the domain, which represents the bottom side of the foundation slab:

- an adiabatic condition, which represents the shaded configuration of the heat exchangers, when installed under the building's footprint, and
- a 1° kind boundary condition; which represents a simplified unshaded condition, similar to a standard installation outdoors. In this case, the hourly time series of the local air temperature (Malta, Ferrara) was assigned, as an approximation of the seasonal weather conditions that only partially affect the heat transfer process on shallow ground heat exchangers, as reported in [16].

The groundwater level was considered 2.0 m deep for the soil surface only for Ferrara, whereas dry condition was assumed for Malta. Finally, the average local air temperature (Malta: 20.3°C; Ferrara: 15.6°C) was assigned to the far bottom side of the building's ground at a depth of 20m.

In order to control the switching between air and ground heat sources/sinks, an user-defined sub-routine was programmed in C^{++} and uploaded as a plug-in module in FEFLOW. The working logic was controlled by the air and average flat-panel temperatures. The average of the panel temperature was taken as the mean of the observation points 4, 5, 6 (see Figure 5). When the flat-panel temperature was energetically more advantageous than the outdoor air temperature, the plug-in sub-routine switches the DSHP to utilize the geothermal resource. The actual energy requirement of the building was then used as a heat flux (boundary condition of 2nd kind). Otherwise, the plug-in triggered the usage of the air source for the DSHP, and in this case, the above boundary condition was removed.

The heat flux ($\dot{h}$) applied at the flat-panels was calculated by dividing the building's thermal requirement ($\dot{H}$) by the full heat transfer surface, taking into account the symmetry of the domain (*2*), the number (n), the length (L) and the height (h) of the flat-panels, as follows:

$$\dot{h} = \frac{\dot{H}}{2 \cdot n \cdot L \cdot h} \tag{1}$$

3. Results

Based on the DesignBuilder simulation outputs for Malta and Ferrara, it is clear that the heating and cooling loads are not balanced for both sites, as shown in Figure 7. Malta requires more cooling than heating, while Ferrara needs more heating than cooling.

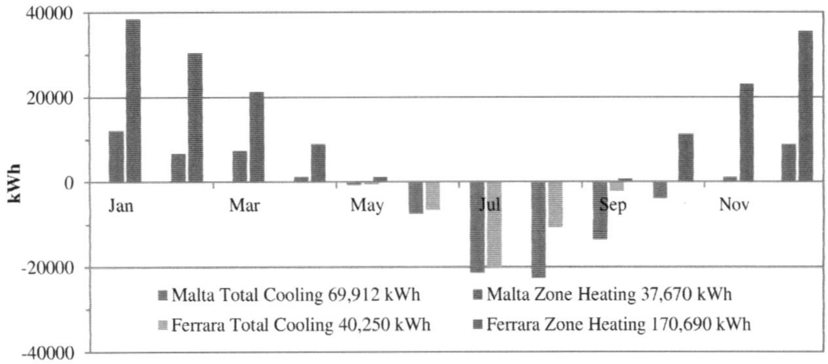

Figure 7. Monthly thermal energy demand for the building block in Malta and Ferrara (cases 1,2).

Four cases were simulated in FEFLOW, taking into account the following heat flux at the flat-panel system:

- the full energy requirement as calculated by DesignBuilder, and applied at the flat-panels (3 m) in *unshaded* conditions taken (*case 1, reference unshaded case*);

- the full energy requirement as calculated by DesignBuilder, and applied at the flat-panels (3m) in the *shade* (*case 2, full shade case*);
- Meeting 50% of the full load energy requirement, and applied at the flat-panels width (3m) in *shaded* conditions (*case 3, shade case with reduced energy load at flat-panel*);
- Meeting 50% of the full load energy requirement, and applied at reduced flat-panels width (1m) in *shaded* conditions (*case 4, shade case with reduced energy load and heat transfer surface at the flat-panel*).

The first case represents a standard installation (*unshaded*), which is taken as the benchmark for comparison purposes. The second case represents the equivalent installation in urban conditions, whereby the flat-panel is buried under the building's footprint (*shaded*). The third case represents the case of reduced energy demand, which could be achieved if the building block was built to higher energy efficiency standards. The fourth case was proposed to evaluate the extent to which the reduction of heat transfer surface would affect the performance of the system, because of a smaller and therefore a cheaper installation (*shaded, reduced energy load, reduced heat transfer surface*). All cases were run for four consecutive years starting from an undisturbed initial condition, until the ground reaches steady temperature trend and thus one would check for any thermal drift in the ground.

The results are summarized in Table 3 and Figure 8 for the cooling season, and Table 4 and Figure 9 for the heating season, where:

- *System average load* is the average thermal power for air conditioning;
- *System max load* is the maximum thermal power for air conditioning;
- *Seasonal energy* is the overall thermal energy for air conditioning;
- *Average specific load FP* is the average heat flux at the flat-panel;
- *Max specific load FP* is the maximum heat flux at the flat-panel;
- *FP energy* is the overall energy exchanged per square metre of flat-panel;
- *FP load energy factor* is the energy share for air conditioning covered by the flat-panel;
- *AIR-FP hour degrees* is hourly seasonal sum of the difference between the temperature at the flat-panel and the air, when the flat-panel is operating;
- *Tmed FP when FP is ON* is the average temperature at the flat-panel when in operation;
- *Tmed AIR when FP is ON* is the air average temperature when the flat-panel is operating;
- *FP time ON* are the operational flat-panel hours;
- *System time ON* are the air conditioning system hours operating;
- *FP share* is the share of the flat-panel operating hours.

For Malta, the share of thermal energy required is 35% for heating and 65% for cooling of the overall energy demand for air conditioning (Tables 3 and 4); for the case of Ferrara, the heating share is 81%, and the cooling share is 19%.

Taking into account the results shown in Figures 8 and 9, Malta shows an enviable mild climate, having on average air temperature between 15°C and 30°C, whereas a harsh continental climate is depicted for Ferrara, where the temperature moves from less than 5°C to more than 30°C. The case 1 is always the worst condition in terms of average working fluid temperature, because it is unable to preserve the thermal energy storage of the summer season for its exploitation during the winter season, and vice versa. On the other hand, the underground thermal energy storage carried out by the adiabatic condition over the soil surface is well depicted by the straight slopes of all other cases, both in spring and in autumn for Malta. For Ferrara, that is clearly only occurring in spring because the low cooling requirement during summer (low operational hours of the heat pump) does not transfer sufficient energy to the ground to be utilized in winter. Moreover, the heating season starts as early as October in Ferrara and this means that the little stored energy from summer would be quickly depleted.

For Ferrara, the winter operating time of the FP system does not change significantly among the four cases (34%, 36%, 38%, 33%), even if the average energy load per square metre of FP is quite different (64, 64, 32, 96 W/m^2). That happens because the energy exploitation of the ground starts early in autumn and the beginning of winter, when the ground temperature is still high. As a consequence, the FP's high performance in heat transfer can be the cause for the fast depletion of the ground thermal storage. After that, the ground-source is only used by the DSHP when hard weather conditions occur (very low ambient air temperature). With regards to this winter behaviour, the case no.4 would seem to be the most profitable, because of the smaller FP surface area installed. During the summer season, case no.4 suffers due to the quick overheating of the ground surrounding the FPs, and therefore, the temperature reaches the same worst conditions as case no.1, within a span of one month only.

Table 3. FEFLOW cooling season results.

COOLING		*Malta*				*Ferrara*			
		Case 1	*Case 2*	*Case 3*	*Case 4*	*Case 1*	*Case 2*	*Case 3*	*Case 4*
System average load	*kW*	32.5	32.5	16.2	16.2	25.7	25.7	12.9	12.9
System max load	*kW*	96.1	96.1	48.1	48.1	95.2	95.2	47.6	47.6
Seasonal energy	*kWh*	69912	69912	34956	34956	40251	40251	20125	20125
FP load energy factor	-	35%	34%	46%	30%	53%	68%	90%	60%
Average specific load FP	*W/m^2*	45.1	45.1	22.5	67.6	35.7	35.7	17.9	53.6
Max specific load FP	*W/m^2*	133.5	133.5	66.7	200.2	132.3	132.3	66.1	198.4
FP energy	*kWh/m^2*	34.1	33.3	22.4	43.6	29.9	38.3	25.2	50.5
AIR-FP hour degrees	*°C*h*	1964	3629	4252	2999	5398	9693	14937	7439
Tmed FP when FP is ON	*°C*	24.0	21.3	21.6	21.7	19.8	16.3	14.1	17.8
Tmed AIR when FP is ON	*°C*	26.4	25.6	25.8	25.6	26.5	26.3	26.2	26.3
FP time ON	*hours*	806	851	1027	763	805	965	1231	881
System time ON	*hours*	2153	2153	2153	2153	1564	1564	1564	1564
FP share	-	37%	40%	48%	35%	51%	62%	79%	56%

Table 4. FEFLOW heating season results.

HEATING		*Malta*				*Ferrara*			
		Case 1	*Case 2*	*Case 3*	*Case 4*	*Case 1*	*Case 2*	*Case 3*	*Case 4*
System average load	*kW*	17.7	17.7	8.8	8.8	46.2	46.2	23.1	23.1
System max load	*kW*	83.1	83.1	41.6	41.6	137.8	137.8	68.9	68.9
Seasonal energy	*kWh*	37670	37670	18835	18835	170690	170690	85345	85345
FP load energy factor	-	57%	57%	76%	47%	34%	36%	38%	33%
Average specific load FP	*W/m^2*	24.6	24.6	12.3	36.9	64.1	64.1	32.1	96.2
Max specific load FP	*W/m^2*	115.5	115.5	57.7	173.2	191.3	191.3	95.7	287.0
FP energy	*kWh/m^2*	29.7	29.7	19.8	36.9	81.7	84.4	45.5	115.7
AIR-FP hour degrees	*°C*h*	2358	4484	7182	3224	3439	4194	4322	3439
Tmed FP when FP is ON	*°C*	15.8	18.5	19.5	17.7	9.0	10.0	9.9	9.6
Tmed AIR when FP is ON	*°C*	13.1	13.5	13.6	13.4	6.3	6.9	6.9	6.8
FP time ON	*hours*	899	886	1205	743	1266	1348	1422	1249
System time ON	*hours*	2129	2129	2129	2129	3698	3698	3698	3698
FP share	-	42%	42%	57%	35%	34%	36%	38%	34%

In Figure 10 and figure 11 are shown the temperatures at the mid-point of the flat-panel (point. no.3), that is representative of the working fluid of the geothermal closed-loop. In comparison with the air temperature when the FP is operating, the average temperature of the working fluid between the unshaded (case 1) and shaded case (case 2) improves by up to 5.0°C for Malta in the heating season, and 3.1 °C for Ferrara. In the cooling season, the improvement is 4.3°C for Malta and 10.0°C for Ferrara. The last high performance in Ferrara is strictly related to the long heating season, which decreases substantially the ground temperature.

Given that no thermal drift occurs in all cases, the energy balance is compensated by the other boundary condition applied in the 2D domain, that is the fixed ground temperature at 20 metres below the ground.

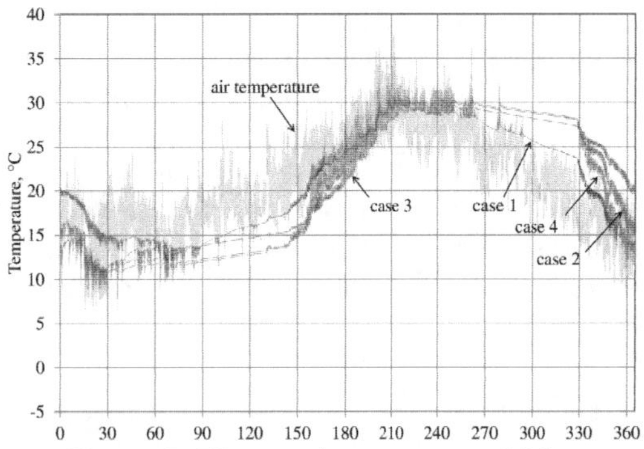

Figure 8: Flat-panel temperature, Malta.

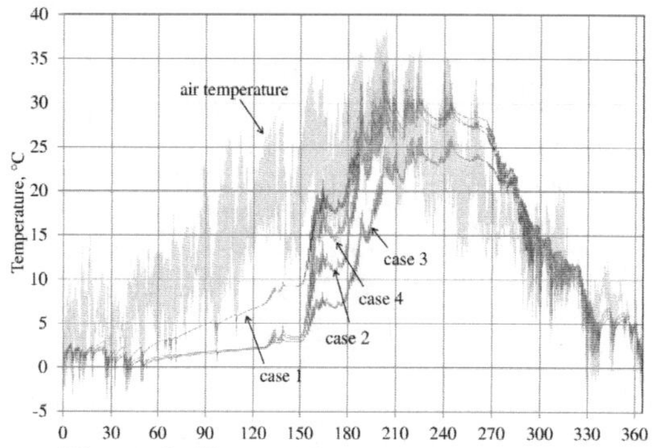

Figure 9: Flat-panel temperature, Ferrara.

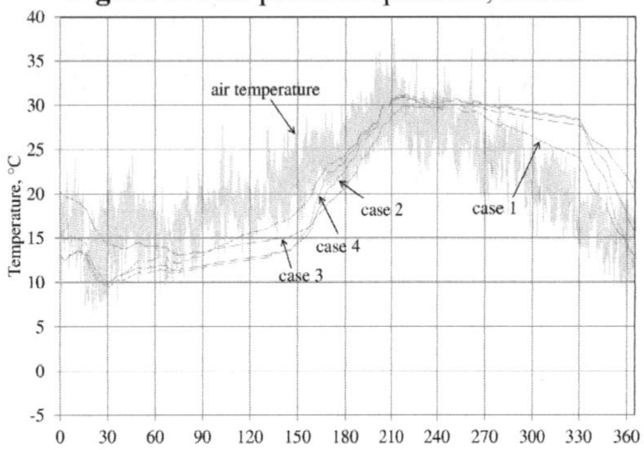

Figure 10. Temperature at mid-point n.3, Malta.

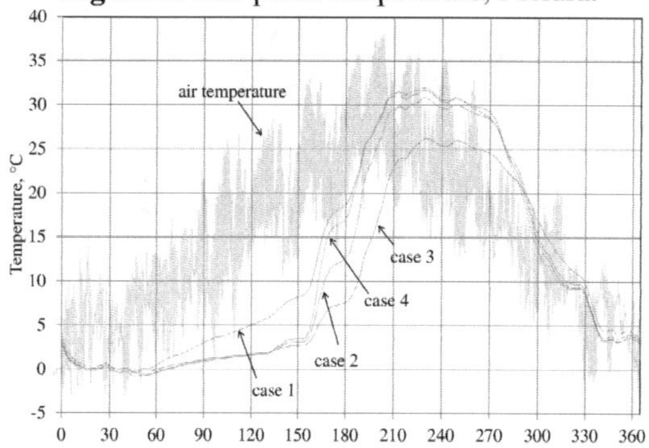

Figure 11. Temperature at mid-point n.3, Ferrara.

4. Conclusions

This paper has attempted to answer the question of whether it is possible to place the shallow ground heat exchangers of a ground-coupled heat pump (GCHP) in the shade, and more specifically under the basement of high rise buildings, without suffering adverse effects such as freezing of the water in the pipes.

The analysis was numerically carried out in two steps. Firstly, the energy requirement for space heating and cooling of a building block was calculated implementing an EnergyPlus model by mean of DesignBuilder; then, the resulting thermal loads were imposed as boundary conditions in FEFLOW, to evaluate the heat transfer carried out by flat-panels in the ground. Two different urban areas were considered to analyse the behaviour according a mild (Malta) and a continental climate (Ferrara, Italy). The building block used purposely a high energy demand for heating and cooling, due to high U-values for walls, windows and floors. As a consequence, preliminary simulations showed that so high thermal loads were not endurable by a common GCHP with shallow ground heat exchangers, and hence it was foreseen a dual-source heat pump (DSHP), which makes use of two thermal energy sources, namely the air and the ground.

With regards of four different scenarios solved, the DSHP was capable of taking up the induced thermal stress in the ground, both in summer and in winter, and to benefit by at least 35% and up to 79% in terms of energy share covered by the geothermal system, according the climate, the energy demand and the sizing of the flat-panels. Moreover, over an extended simulation of four years, the geothermal system suffered no thermal drift in terms of yearly trend, even though the flat-panels were in the shade and the thermal requirement was not balanced over the year.

Therefore, if well controlled, the underground thermal energy storage could be improved and therefore to provide an additional source of renewable energy besides the traditional systems of solar thermal and solar photovoltaic systems.

IOP Conf. Series: Journal of Physics: Conf. Series **796** (2017) 012017 doi:10.1088/1742-6596/796/1/012017

5. References

[1] Hart, D. P. and Couvillion, R., Earth-coupled Heat Transfer. National Water Well Association, Dublin, OH, USA, 1986.

[2] Luo, Z. and Asproudi, C., Subsurface urban heat island and its effect on horizontal ground-source heat pump potential under climate change *Applied Thermal Engineering* **90** 530-537, 2015.

[3] Bottarelli, M., A preliminary testing of a flat panel ground heat exchanger *International Journal of Low-Carbon Technologies* **8** 2 80-87, 2013.

[4] Vocale, P., Morini, G. L. and Spiga, M., Influence of outdoor air conditions on the air source heat pumps performance *Energy Procedia*, **45** 653-662, 2014.

[5] Jang, Y. J., Bae, H. B., Lee, J. S. and Ha, Y. M.,Continuous heating of an air-source heat pump during defrosting and improvement of energy efficiency *Applied Energy* **110** 9-16, 2013.

[6] U.S. Department of Energy (DOE), Energy Savings from Dual-Source Heat Pump Technology. DOE/EE-0220, 2000. http://www.eren.doe.gov/femp

[7] Chiasson, A., A feasibility study of a multi-source hybrid district geothermal heat pump system. *GHC Bullettin* 9-14, 2011.

[8] Wang, H., Liu, X., Feng, G., Kang, Z., Luo, Y., Bai, B. and Chi, L., Simulation analysis of air-ground dual source heat pump operating efficiency *Procedia Engineering* **121** 1413-1419, 2015.

[9] EnergyPlus Documentation, http://nrel.github.io/EnergyPlus

[10] Fumo, N., Mago, P. and Luck, R. Methodology to estimate building energy consumption using EnergyPlus benchmark models, *Energy and Building* **42** 12 2331-2337, December 2010.

[11] Crawley, D. B., *et al*, EnergyPlus: Creating a new-generation building energy simulation program, *Energy and Buildings*, **33** 4 319-331, April 2001.

[12] DesignBuilder software, http://www.designbuilder.co.uk

[13] Building Industry Consultative Council (BICC), Malta, *Technical Guidance F, Conservation of Fuel, Energy and Natural Resources: Minimum Requirements on the Energy Performance of Buildings Regulations 2006*, http://www.bicc.gov.mt/bicc/files_folder/ Cons%20Fuel%20Energy%20Doc%20F%20new.pdf.

[14] Yousif, C., *Assessing Energy Efficiency in Maltese Dwellings*, a Ph.D. Thesis, Universidad de Valladolid, Spain, April 2015.

[15] Diersch, H.-J. G., *FEFLOW Finite Element Subsurface Flow and Transport Simulation System*, Reference Manual, User's Manual and White Papers Vol. I, II, III, IV", Institute for Water Resources Planning and Systems Research, 2005.

[16] Bortoloni, M., Bottarelli, M., Su, Y., A study on the effect of ground surface boundary conditions in modelling shallow ground heat exchangers, *Applied Thermal Engineering*, **111** 1371-1377, 2017

[17] Bortoloni, M., Bottarelli, M., On the sizing of a flat-panel ground heat exchanger, *International Journal of Energy and Environmental Engineering* **6** 2 185-195, 2015.

[18] UNI 11466: Heat pump geothermal systems: design and sizing requirements (2012).

IOP Conf. Series: Journal of Physics: Conf. Series **796** (2017) 012018 doi:10.1088/1742-6596/796/1/012018

A two-dimensional investigation about magnetocaloric regenerator design: parallel plates or packed bed?

C Aprea[1], A Greco[2], A Maiorino[1], C Masselli[1]

[1]DIIN, Università di Salerno, via Giovanni Paolo II 132, 84084 Fisciano, Salerno, Italia
[2]DII, Università degli Studi di Napoli Federico II, P.le Tecchio 80, 80125 Napoli, Italia

Email: cmasselli@unisa.it

Abstract. Magnetic Refrigeration (MR) is a novel refrigeration technique based on eco-friendly solid materials as refrigerants, whom react to the application of magnetic fields, with warming and cooling by magnetocaloric effect. The thermodynamical cycle which best suits the magnetic refrigeration is Active Magnetic Regenerator cycle (AMR). Regenerator is the core of a magnetic refrigerator, since that it plays a dual-role: it operates both as refrigerant and regenerator in an AMR cycle. An AMR cycle consists of two adiabatic stages and two isofield stage. In this paper an investigation is conducted about the magnetocaloric refrigerator design through two-dimensional multiphysics numerical models of two different magnetocaloric regenerators: (1) a packed bed and (2) a parallel plates magnetic regenerators made of gadolinium, operating at room temperature under a 1.5T magnetic field induction. Both models employ water as secondary fluid. The tests were performed with variable fluid flow rate at fixed AMR cycle frequency. The results obtained are presented in terms of temperature span, cooling power, coefficient of performance and mechanical power of the circulation pump, and they indicate under which operating conditions packed bed configuration is to be preferred to parallel plates and vice versa.

1. Introduction

In the last fifteen years the scientific community has oriented more and more interest toward non-vapor compression refrigeration, in order to develop new refrigeration techniques based on solid-state materials [1]. Magnetic refrigeration [2] is the one that most has undergone progressive developments, since it has demonstrated to be an innovative, promising and ecologic technology, based on solid state magnetic materials [3] exhibiting MagnetoCaloric Effect (MCE). Compared to conventional vapor compression systems [4], magnetic refrigeration can be an environment-friendly and efficient technology [5]. Magnetic refrigerant are solid materials, with essentially zero vapor pressure, and therefore they are ecological with no direct Ozone Depletion Potential (ODP) and zero direct Global Warming Potential (GWP) [6].

Specifically MR founds its operation MCE, a physical phenomenon observed in every material with magnetic properties and it consists in an increment of temperature in the magnetic material as a consequence of magnetization under adiabatic conditions. Furthermore, magnetic refrigeration exhibits high potential energetic efficiencies (30-60% of Carnot COP). On the other side MR presents at least two relevant disadvantages: the high costs associated both with the magnetic field generation

[7-8] and magnetic materials [9]. The Active Magnetic Regenerator (AMR) is the core of a magnetic refrigerator system. It is a special kind of thermal regenerator made of magnetic material which works both as a refrigerating and as a heat regenerating medium. The performance of an AMR system strongly depends on the magnetocaloric effect of the magnetic material which the regenerator is made of, on the geometry of the regenerator and on the secondary fluid.

In this paper an investigation is conducted about the magnetocaloric refrigerator design through two-dimensional multiphysics numerical models of two different AMR regenerators. We compare the results of two numerical two-dimensional (2D) multiphysics models [10]: (1) a packed bed and (2) a parallel plates magnetic regenerators made of gadolinium, operating at room temperature under a 1.5T magnetic field induction. Both models employ water as secondary fluid. The tests were performed with variable fluid flow rate, at fixed AMR cycle frequency. The results obtained are presented in terms of temperature span, pressure losses, cooling power, coefficient of performance and they indicate under which operating conditions packed bed configuration is to be preferred to parallel plates and vice versa.

2. Magnetocaloric effect and the AMR cycle

The magnetocaloric effect is a physical phenomenon detected for the first time in iron, by Warburg in 1881[11] as temperature change in the material by varying the intensity of the magnetic field applied, under adiabatic conditions. The entropy of a magnetic solid at constant pressure S(T,H) is a function of both the magnetic field intensity H and the absolute temperature T and it consists of three contributions: magnetic (S_M), lattice (S_{lat}) and electronic (S_{el}) ones:

$$S = S_M(T,H) + S_{lat}(T) + S_{el}(T) \quad (1)$$

In equation (1), the magnetic contribution is function of both H and T, whereas the electronic and lattice contributions only depend on T. Consequently, only the magnetic entropy S_M can be controlled by varying the magnetic-field strength. When the magnetic field is applied to a magnetic material, the magnetic dipoles become oriented according to the direction of the field. If this is done isothermally, it carries to a decrement of the material's magnetic entropy by the isothermal entropy decreasing $(\Delta S_M)_T$. The entropy change can be evaluated as:

$$(\Delta S_M)_T = \int_{H_i}^{H_f} \left(\frac{\partial M}{\partial T}\right)_H dH \quad (2)$$

On the other side, if the magnetization is done adiabatically, the total sample entropy remains constant and the decrease in magnetic entropy is countered by an increase in the lattice and electron entropy. This causes a heating of the material and it takes an increasing of its temperature given by the adiabatic temperature change, ΔT_{ad}. Dually, an adiabatic demagnetization involves the increasing of the material's magnetic entropy, causing a decreasing in lattice vibrations and by that a temperature decrease. The expression for ΔT_{ad} of a magnetic material can be evaluated as:

$$(\Delta T_{ad})_s = -\int_{H_i}^{H_f} \frac{T}{C_H}\left(\frac{\partial M}{\partial T}\right)_H dH \quad (3)$$

where the specific heat can be defined as:

$$C_H = T\left(\frac{\delta s}{\delta T}\right)_H \quad (4)$$

At the Curie temperature a magnetic material exhibits its maximum MCE. There are two types of magnetic phase changes that may occur at the Curie point: First Order Magnetic Transition (FOMT) and Second Order Magnetic Transition (SOMT). A FOMT material shows a discontinuity in the first derivative of the Gibbs free energy, whereas in a SOMT material the first derivative is a continuous function and discontinuity takes place in the second derivative. Therefore, in a FOMT material the magnetization, i.e. the first derivative of the free energy with respect to the applied magnetic field strength, is discontinuous. In a SOMT material the magnetic susceptibility, i.e. the second derivative of the free energy with the field strength, changes discontinuously. Most of the magnetic materials order with a SOMT from a paramagnet to a ferromagnet, ferrimagnet or antiferromagnet [12]. The benchmark material of magnetic refrigeration is considered to be gadolinium (Gd), a rare earth which exhibits a SOMT at $T_{Curie} = 294$ K and it belongs to lanthanide group of elements.

The referring thermodynamical cycle of magnetic refrigeration is the reverse Brayton cycle [13] and, as a matter of fact, in 1982 Barclay introduced an innovative way to follow it up by the introduction of the Active Magnetic Regenerative refrigeration cycle (AMR) [14]. The AMR couples into a single concept what had been before two separate processes: instead of using a different material as a regenerator to recuperate heat from the magnetic material, the AMR concept made use of the refrigerant magnetic material itself. In essence, a temperature gradient is established throughout the AMR and a fluid is used to transfer heat from the cold to the hot end. The working principle of an AMR is presented in Figure 1 where: the dashed line represents the initial temperature profile of the bed in each process, the solid line depicts the final temperature profile of the process.

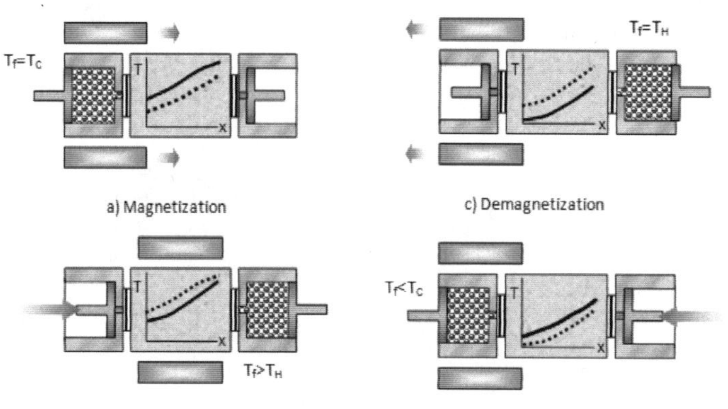

Figure 1. The four processes of an AMR cycle related to an AMR regenerator kept in contact with a cold and a hot heat exchangers.

Assuming that the bed is in steady state condition with the hot heat exchanger at T_H whereas the cold heat exchanger is at T_C, four processes constitute the AMR cycle:

(a) adiabatic magnetization: each particle of the magnetic material which constitute the regenerator, warms up by increasing progressively the intensity of the magnetic field applied, under adiabatic conditions;

(b) isofield cooling: with the magnetic field present at maximum value, the fluid blows from the cold to the hot end of the regenerator. Therefore the regenerator cools down transferring heat to the fluid, whom is expelled in the hot heat exchanger;

(c) adiabatic demagnetization: by decreasing adiabatically the intensity of the magnetic field applied from a maximum until a minimum value, the regenerator cools anymore and a decrement of temperature, equal to $\Delta T_{ad}(T)$ is registered;

(d) isofield heating: while magnetic field has kept to a minimum value, the fluid flows from the hot to the cold end; as a result the fluid, hotter, cools itself by crossing the regenerator and reaching a temperature lower than T_C. At this stage the secondary fluid absorbs heat from the cold heat exchanger at T_C, producing a cooling load.

3. Regenerator geometries and model description

The performances of a magnetic refrigerator AMR based, are mostly influenced by:

- the MCE properties of the magnetic material employed;
- the regenerator geometry and related operating conditions;
- the irreversible heat losses due to non-idealities of the refrigerators.

The criteria to identify the best magnetic material for magnetic refrigeration are reported by Aprea et al. [9,15] and by Franco et al. [16]. The benchmark material is still considered to be gadolinium due to its excellent magnetocaloric properties.

The irreversible heat losses [17] of AMR regenerators have also great influence on the performances of the AMR cycle. Main irreversible heat losses are:

- heat transfer between regenerator and heat transfer fluid;
- pressure drop by flow resistance;
- thermal conduction along the magnetic material
- heat leakage;
- losses due to magnetic hysteresis and eddy currents;
- viscous dissipation in the secondary fluid;
- "dead volume" loss which is due to the heat transfer fluid trapped in the connecting lines between the bed and the heat exchangers.

In the cooling system, the regenerator geometry assumes paramount importance [18-20] for gaining efficient heat transferring between magnetic material and fluid flow, thus minimizing the major loss in the cooling system. Two different regenerators are both widely used: a parallel plate regenerator, which yields lower pressure drops; a packed bed regenerator which can produce a larger temperature span, due to the larger extension of the heat transfer surface.

This paper compares the results of two numerical two-dimensional (2D) multiphysics models: (a) a packed bed and (b) a parallel plates magnetic regenerators made of gadolinium, operating at room temperature under 1.5 T as magnetic field induction. Both models employ water as secondary fluid.

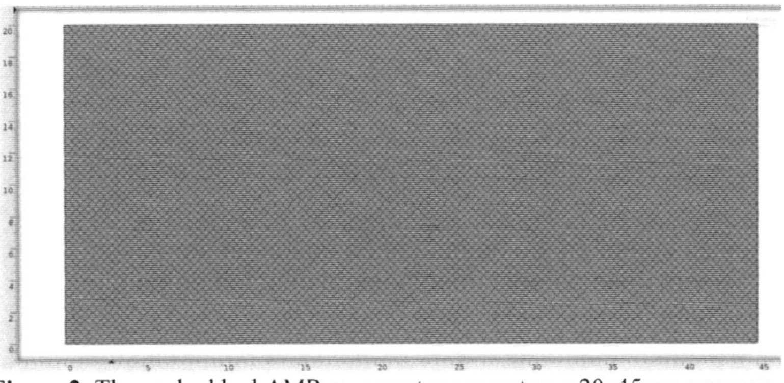

Figure 2. The packed bed AMR regenerator geometry: a 20x45mm wrapper contains 3600 spheres.

The regenerator have a rectangular shape (20x45mm) which acts as a wrapper. The area of the packed bed regenerator is filled with a regular matrix of 3600 circles that constitute the magnetocaloric material, i.e. gadolinium; every circle has a diameter of 0.45mm and the amount of the area occupied by all of the circles is about 60% of the total rectangular area. A group of channels is formed by stacking particles in the regenerator area: the fluid flows through these interstitial channels. In Figure 2 is shown a scheme representing the AMR packed bed regenerator considered.

The area of the parallel plates AMR regenerator contains 27 parallel plates made of gadolinium: every plate has a thickness of 0.5mm. The amount of the area occupied by all of the plates is 60% of the total rectangular area. A group of channels is formed among the plates in the regenerator area, where the secondary fluid flows. Figure 3 reports the parallel plates geometry of the regenerator.

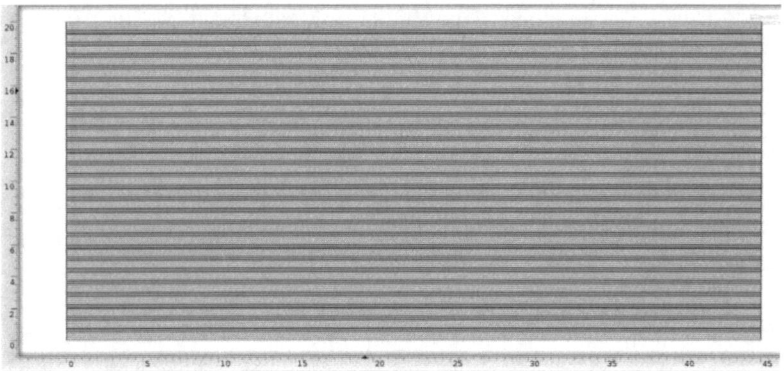

Figure 3. The parallel plates AMR regenerator geometry: a 20x45mm wrapper contains 27 plates and 26 channels.

Both the models takes in account both the fluid flow and heat transfer between the solid and the fluid; the fluid flow in the positive x direction during the isofield cooling process and in the opposite direction during isofield heating. The two models are supported by the same mathematical formulation which describes the AMR cycle, including several distinct groups of equations according to the different processes of the AMR cycle that the regenerator experiences.

The equations that rule the regenerative fluid flow processes, in both directions, are: the Navier-Stokes for the heat transfer fluid and the energy equations for both the fluid and the solid particles. The fluid is assumed incompressible.

The equations during the fluid flow phases are:

$$\begin{cases} \dfrac{\partial u}{\partial x}+\dfrac{\partial v}{\partial y}=0 \\ \dfrac{\partial u}{\partial t}+u\dfrac{\partial u}{\partial x}+v\dfrac{\partial u}{\partial y}=-\dfrac{1}{\rho_f}\dfrac{\partial p}{\partial x}+\upsilon\left(\dfrac{\partial^2 u}{\partial x^2}+\dfrac{\partial^2 u}{\partial y^2}\right) \\ \dfrac{\partial v}{\partial t}+u\dfrac{\partial v}{\partial x}+v\dfrac{\partial v}{\partial y}=-\dfrac{1}{\rho_f}\dfrac{\partial p}{\partial y}+\upsilon\left(\dfrac{\partial^2 v}{\partial x^2}+\dfrac{\partial^2 v}{\partial y^2}\right) \\ \dfrac{\partial T_f}{\partial t}+u\dfrac{\partial T_f}{\partial x}+v\dfrac{\partial T_f}{\partial y}=\dfrac{k_f}{\rho_f C_{fp}}\left(\dfrac{\partial^2 T_f}{\partial x^2}+\dfrac{\partial^2 T_f}{\partial y^2}\right) \\ \dfrac{\partial T_s}{\partial t}=\dfrac{k_s}{\rho_s C_H}\left(\dfrac{\partial^2 T_s}{\partial x^2}+\dfrac{\partial^2 T_s}{\partial y^2}\right) \end{cases} \tag{5}$$

The equations that model magnetization and demagnetization processes are:

$$\begin{cases} \rho_f C_{fp}\dfrac{\partial T_f}{\partial t}=k_f\left(\dfrac{\partial^2 T_f}{\partial x^2}+\dfrac{\partial^2 T_f}{\partial y^2}\right) \\ \rho_s C_H\dfrac{\partial T_s}{\partial t}=k_s\left(\dfrac{\partial^2 T_s}{\partial x^2}+\dfrac{\partial^2 T_s}{\partial y^2}\right)+Q \end{cases} \tag{6}$$

The model takes into account magnetocaloric effect which elevates or reduces the temperature of the solid by the variation of the external magnetic field applied to the regenerator. Hence the MCE temperature variation ΔT_{ad} is converted [21] into a heat source Q:

$$Q = Q(H, T_S) = \frac{\rho_S C_H(H,T_S) \Delta T_{ad}(H,T_S)}{\Delta t} \quad (7)$$

which has the dimensions of a power density and included in the solid energy equation, only for magnetization and demagnetization phases. The term Q is positive during magnetization, negative during demagnetization. Δt is the period of the magnetization/demagnetization process. The coupled equations that govern the AMR cycle, imposed on this model, are solved using Finite Element Method.

The AMR cycle is modeled as four sequential steps. The same time step Δt has been chosen for the resolution during all the four periods of the cycle. The cycle is repeated several times with constant operating frequency until the regenerator reaches steady state operation.

The model has been validated with a Rotary Permanent Magnet Magnetic Regenerator (RPMMR) [10,21,22], developed at the Refrigeration Lab in the University of Salerno in collaboration with the University of Naples Federico II by numerical simulation and experimental tests operating in the same conditions.

4. Numerical simulations and results

Several AMR cycles have been simulated on both the models in order to compare their performances by varying fluid flow rate at fixed AMR frequency. All the results have been obtained working with 1.25 Hz as AMR frequency, in the 288-298 K temperature range and investigating the regenerators behaviors while the secondary fluid velocity changes in 0.0595-0.25 m/s range. The applied magnetic field has a variable intensity from 0 to 1.5 T. The secondary fluid employed is pure water, whereas the magnetocaloric material involved is gadolinium. Cold and hot heat exchangers temperature (T_C=288 K; T_H=298 K) have been chosen to characterize the performance sensitivity of the regenerator in a range which embraces the gadolinium Curie temperature, i.e. 294 K.

The results obtained are presented in terms of temperature span, cooling power, coefficient of performance (COP) and pump power, in order to quantify the pressure losses influence on COPs, for both the regenerator geometries. As a matter of fact the results provide an indication about the operating conditions under which packed bed configuration has to be preferred to parallel plates and vice versa.

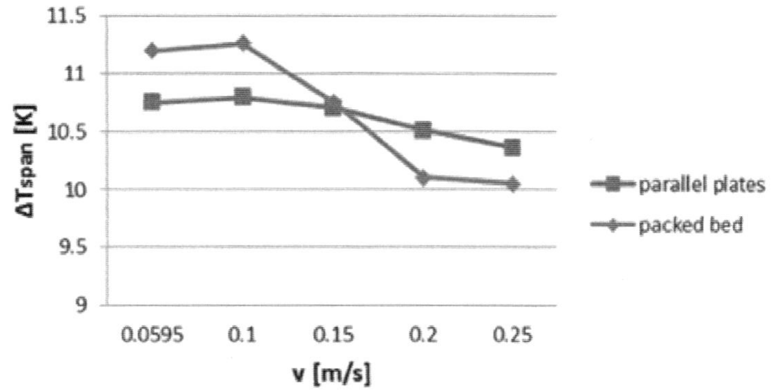

Figure 4. ΔT_{span} evaluated for both the regenerator configuration for variable fluid flow velocity and therefore fluid flow rate.

In Figure 4 it has shown the ΔT_{span} detected for both the regenerator geometries under each fluid flow velocity investigated. Such parameter has been obtained by evaluating the difference between T_H

and the cold side temperature of the secondary fluid averaged in the last process of AMR cycle (fluid flow from hot to cold side of the regenerator), as shown in the following equation:

$$\Delta T_{span} = T_H - \int_{t_M+t_{CF}+t_D}^{t_M+t_{CF}+t_D+t_{HF}} T_f(0,y,t)\,dt \quad (8)$$

Figure 4 reveals that for both the model, the ΔT_{span} decreases with the fluid velocity. Indeed, corresponding to a small fluid mass flow rate, the fluid can be regenerated to reach lower cold side temperature. The maximum ΔT_{span} is reached in correspondence of a fluid velocity of about 0.1 m/s. Furthermore one can observe that although, for low speed, packed bed configuration lets to ΔT_{span} greater than parallel plates ones, for high values of fluid velocity parallel plates geometry ensure larger temperature spans. The latter regenerator configuration seems to exhibit a less sensitive behavior with respect to velocity than the former, in terms of ΔT_{span}.

Figure 5 reports the refrigeration power as a function of fluid velocity, obtained for packed bed and parallel plates AMR regenerator. The refrigeration power has been calculated according to the following equation:

$$\dot{Q}_{ref} = \frac{1}{\tau}\int_{t_M+t_{CF}+t_D}^{t_M+t_{CF}+t_D+t_{HF}} \dot{m}_f C_f\big(T_C - T_f(0,y,t)\big)\,dt \quad (9)$$

Figure 5 clearly shows that the cooling power increases with the fluid flow rate and then decreases reaching a maximum. Indeed,corresponding to a small fluid mass, although the fluid can be regenerated to reach lower exit temperature, a small amount of transfer fluid can produce only a little refrigerating capacity. On the other hand, excessive quantities of transfer fluid perturb the temperature profile of the AMR, decreasing the temperature difference between the fluid and the bed. The bed quickly becomes overwhelmed by the fluid flow and the efficiency of the heat transfer decreases quickly causing a loss in cooling power. From figure 5 one can observe that the points where the geometries confer the maximum refrigeration power differs from each other. As a matter of fact, if the secondary fluid has moved into the regenerator with a velocity of 0.15 m/s, packed bed and parallel plates configurations provide quite the same Q'_{ref}.

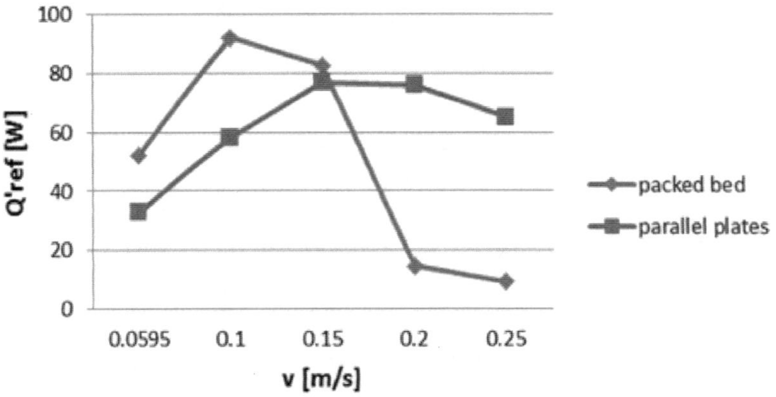

Figure 5. The refrigeration power evaluated for both the regenerator configuration for variable fluid flow velocity and therefore fluid flow rate.

Furthermore, with lower flow speeds, parallel plates shows a drastic reduction of its cooling capacity, higher than packed bed; whereas for values greater than 0.15 m/s, where a slump in packed bed's performances happens, parallel plates responds in a very satisfactory manner, providing refrigerant powers 3 times higher than the former geometry.

To estimate the performance of the two models, the Coefficient of Performance has been introduced as follows:

$$COP = \frac{\dot{Q}_{ref}}{\dot{Q}_H - \dot{Q}_{ref} + \dot{W}_p} \tag{10}$$

whereas Q'$_H$ is the power related to the heat supplied in the environment which has been evaluated according to:

$$\dot{Q}_H = \frac{1}{\tau}\int_{t_M}^{t_M+t_{CF}} \dot{m}_f C_f \left(T_f(L, y, t) - T_H\right) dt \tag{11}$$

W'$_p$ is the mechanical power associated to the circulation pump:

$$\dot{W}_p = \frac{\dot{m}(\Delta p_{CF} + \Delta p_{HF})}{\eta_p \rho_f}(t_{CF} + t_{HF}) \tag{12}$$

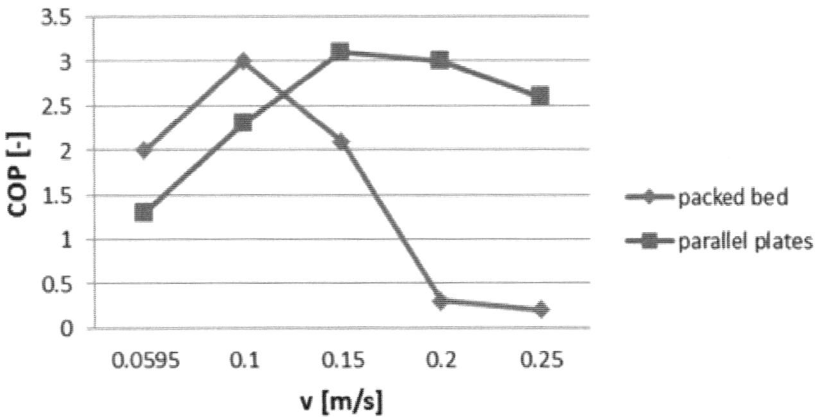

Figure 6. COP estimated for both the geometries for different flow speed.

In Figure 6 the COP evaluated for both the geometries by varying fluid flow rate, are reported. Figure 6 clearly shows that COPs increase until reaching the maximum whom, in both the geometries, assumes the same value (about 3), but the velocity where it is achieved are different. COPs show the same trends of refrigeration power: packed bed is better with low speed, parallel plates are to be preferred for high ones. A packed-bed regenerator is characterized by the best heat transfer surface, with a more marked pressure drop. A flat plates regenerator is characterized by a lower heat transfer surface and lower pressure drops. Therefore another parameter is the responsible of such COP fall in packed bed configuration, for high velocity: pressure losses and, therefore, the mechanical power of the circulation pump.

Figure 7, which exhibits the work of the pump employed to allow the circulation of secondary fluid, reveals a significant reality: despite in the parallel plates configuration the pressure losses are negligible and not influenced by a change of velocity, a packed bed regenerator is strongly affected by pressure losses that seriously limit its operation at high flow speeds.

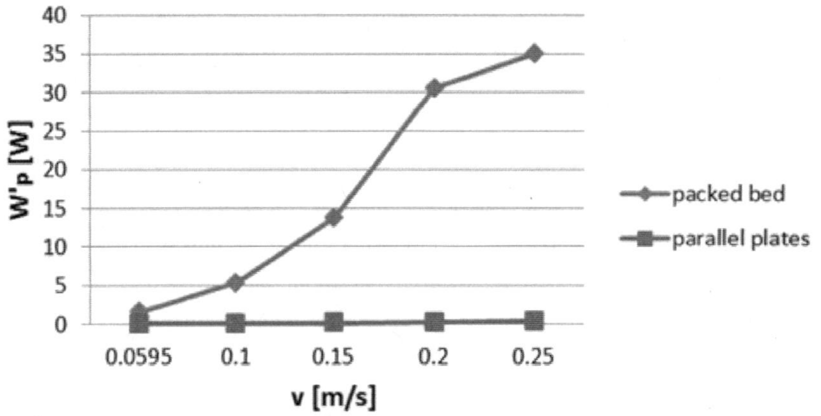

Figure 7. The work of the secondary fluid circulation pump reported versus fluid velocity.

5. Conclusions

In the present paper, two numerical two-dimensional (2D) multiphysics models have been investigated. A packed bed and a parallel plates geometries of magnetic regenerators have been compared. They are either made of gadolinium, working in room temperature range, under 1.5 T as magnetic field induction. A practical model for predicting the performance and efficiency of an AMR refrigerator system for different magnetocaloric materials has been introduced. The results obtained are presented in terms of temperature span, cooling power, coefficient of performance and mechanical power of the circulation pump, when the fluid velocity is varying.

After an accurate analysis of all the above mentioned parameters, a general trend, which describes the performance of the two geometries under a general framework, has been carried out. We can assert that for low values of fluid velocity, packed bed configuration has to be preferred since it offer ΔT_{span}, refrigeration power and COP higher than parallel plates one and therefore a more satisfactory behaviour. For high flow speed the situation is completely upside down: parallel plates geometry provides good results whereas packed bed suffers a significant slump in its performance, caused also by an huge increment of pressure losses whom limits its use.

References

[1] Aprea C, Greco A, Maiorino A and Masselli C 2016 A comparison between different materials in an active electrocaloric regenerative cycle with a 2D numerical model *International Journal of Refrigeration* vol 69 pp 369-382.

[2] Pecharsky V K and Gschneidner K A Jr. 1999 Magnetocaloric effect and magnetic refrigeration *Journal of Magnetism and Magnetic Materials* vol **200** pp 44-56.

[3] Aprea C, Greco A and Maiorino A 2013 The use of the first and of the second order phase magnetic transition alloys for an AMR refrigerator at room temperature: a numerical analysis of the energy performances *Energy conversion and management* vol **70** pp 40-55.

[4] Aprea C, Greco A, Maiorino A, Masselli C and Metallo A 2016 HFO1234yf as a drop-in replacement for R134a in domestic refrigerators: a life cycle climate performance analysis *International Journal of Heat and Technology* vol **34** iss Special Issue 2 pp S212-S218.

[5] Aprea C, Greco A, Maiorino A and Masselli C 2015 Magnetic refrigeration: an eco-friendly technology for the refrigeration at room temperature *Journal of Physics: Conference Series* vol **655** pp 1-13.

[6] Bisio G, Rubatto G and Schiapparelli P 1999 Magnetic systems depending on three or two variables; thermodynamic analysis and some existing and possible applications *Energy conversion and management* vol **40** pp1267-1286.

[7] Aprea C, Greco A, Maiorino A, Mastrullo R and Tura A 2014 Initial experimental results from a rotary permanent magnet magnetic refrigerator *International Journal of Refrigeration* vol **43** pp 111-122.

[8] Aprea C, Greco A, Maiorino A and Masselli C 2016 The energy performances of a rotary permanent magnet magnetic refrigerator *International Journal of Refrigeration* vol **61** pp 1-7.

[9] Aprea C, Greco A, Maiorino A and Masselli C 2015 A comparison between rare earth and transition metals working as magnetic materials in an AMR refrigerator in the room temperature range *Applied Thermal Engineering* vol **91** pp 767-777.

[10] Aprea C, Cardillo G, Greco A, Maiorino A and Masselli C 2015 A comparison between experimental and 2D numerical results of a packed-bed active magnetic regenerator *Applied Thermal Engineering* vol **90** pp 376-383.

[11] Warburg E 1881 Magnetische Untersuchungen. Über einige Wirkungen der Coërcitivkraft *Annalen der Physik* vol **13** pp141-164.

[12] Gschneidner KA jr and Pecharsky VK 2008 Thirty years of near room temperature magnetic cooling: where we are today and future prospects *Intenational Journal of Refrigeration* vol **31** pp 945-961.

[13] Walker G 1983 Claude and Joule-Brayton Systems *The International Cryogenics Monograph Series* chapt *Cryocoolers* ed Springer pp 297-353.

[14] Barclay J A 1982 The theory of an active magnetic regenerative refrigerator *NASA STI/Recon Technical Report* N 83 34087.

[15] Aprea C, Greco A and Maiorino A 2011 A numerical analysis of an Active Magnetic Regenerative Refrigerant system with a multi-layer regenerator *Energy conversion and management* vol **52** pp 97-107.

[16] Franco V, Blázquez J S, Ingale B and Conde A 2012 The magnetocaloric effect and magnetic refrigeration near room temperature: materials and models *Annual Review of Materials Research* vol **42** pp 305-342.

[17] Aprea C, Greco A and Maiorino A 2013 A dimensionless numerical analysis for the optimization of an AMR cycle *International Journal of Energy Research* vol **37** pp 1475-1487.

[18] Aprea C, Greco A and Maiorino A 2011 A numerical analysis of an Active Magnetic Regenerative Cascade system *International Journal of Energy Research* vol **35** pp 177-188.

[19] Aprea C, Greco A and Maiorino A 2012 Modelling an Active Magnetic Refrigeration system: a comparison with different models of incompressible flow through a packed bed *Applied Thermal Engineering* vol **36** pp 296-306.

[20] Dikeos J, Rowe A and Tura A 2006 Numerical analysis of an active magnetic regenerator (AMR) refrigeration cycle *AIP Conference Proceedings*.

[21] Aprea C, Greco A and Maiorino A 2015 GeoThermag: a geothermal magnetic refrigerator *International Journal of refrigeration* vol **59** pp 75-83.

[22] Aprea C, Cardillo G, Greco A, Maiorino A and Masselli C 2016 A rotary permanent magnet magnetic refrigerator based on AMR cycle *Applied Thermal Engineering* vol **101** pp 699-703.

Electrocaloric refrigeration: an innovative, emerging, eco-friendly refrigeration technique

C Aprea[1], A Greco[2], A Maiorino[1], C Masselli[1]

[1]DIIN, Università di Salerno, via Giovanni Paolo II 132, 84084 Fisciano, Salerno, Italia
[2]DII, Università degli Studi di Napoli Federico II, P.le Tecchio 80, 80125 Napoli, Italia

Email: cmasselli@unisa.it

Abstract. Nowadays, the refrigeration is responsible of about 15% of the overall energy consumption all over the world. Actually most of the refrigerant fluids working in vapor compression plants (VCPs) are environmentally harmful, since they presents high GWP (Global Warming Potential), which leads to a substantial warming of both earth surface and atmosphere. Electrocaloric refrigeration (ER) is an innovative, emerging refrigeration technique based on solid state refrigerant that shows a great potential. It fits in the context of environment-friendly refrigeration systems, whom are spreading increasingly to replace VCPs. ER is founded on electrocaloric effect that is a physical phenomenon found in materials with dielectric properties, electrocaloric materials. The thermodynamical cycle that best is addressed to the electrocaloric refrigeration is Active Electrocaloric Regeneration cycle (AER) that consists of two adiabatic and two isofield stages. The core of an electrocaloric refrigerator is the regenerator whom operates both as refrigerant and regenerator in an AER cycle. In this paper, we compare the energetic performance of a commercial R134a refrigeration plant to that of an electrocaloric refrigerator working with an AER cycle. The comparison is performed in term of TEWI index (Total Equivalent Warming Impact) that includes both direct and indirect contributions to global warming.

1. Introduction

Nowadays, the refrigeration is responsible of about 15% of the overall energy consumption all over the world and most modern refrigeration units are based on Vapor Compression Plants (VCP). The traditional refrigerant fluids, i.e. CFCs and HCFCs, have been banned (Montreal 1987) because of their contribution to the disruption of the stratospheric ozone layer (Ozone-Depleting Potential substances ODPs). Human activities have increased the concentration of greenhouse gases in the atmosphere, thus resulting in a substantial warming of both earth surface and atmosphere. The impact of greenhouse gases on global warming is quantified by their GWP (Global Warming Potential). The Kyoto Protocol (1987) sets binding targets for greenhouse gas emissions. National laws and regulations implementing the Kyoto Protocol differ from one another, but they typically call for a phase down of HFC consumption. Therefore actually most of the refrigerant fluids working in vapour compression plants are environmentally harmful. Near all the efforts done [1] in the identification of new refrigerant fluids, environmental friendly, to be employed in VCP, in the last decades the interest of scientific community has oriented itself in studying and developing new refrigeration technologies of low impact in our ecosystem. In a recent past the most promising, eco-friendly, emerging technique seemed to be magnetic refrigeration (MR) [2], whose main innovation consists in employing solid materials as refrigerant, instead of the fluid ones proper of vapour compression. The magnetic refrigerant is a solid, has essentially zero vapour pressure, and therefore it is ecological with no direct ODP and zero GWP [3]. Specifically MR founds its operation on magnetocaloric effect (MCE) [4], a physical phenomenon observed in every material with magnetic properties and it consists in an increment of temperature in the MCE material as a consequence of material's magnetization under adiabatic conditions. Furthermore, magnetic refrigeration exhibits high cooling efficiencies (30-60% of Carnot COP) [5]. On the other side MR presents at least two relevant disadvantages: the high costs associated both with the magnetic field generation [6-8] and magnetic materials [9]. As alternative to MR and therefore to MCE, a similar caloric effect was detected in some materials by the application of an electric field [10-11] and it is called electrocaloric effect (ECE). Electrocaloric refrigeration (ER) is a refrigeration technology based on solid state refrigerants, exhibiting a relevant ECE, that has shown great potential. Even if, as MR, it doesn't presents any ODP or GWP contribution, such technology results more advantageous than MR since it requires high electric fields, easier and cheaper to be generated than magnetic fields proper of the latter. Moreover a 50% of Carnot limit COP could be achieved with electrocaloric refrigeration in small scale applications. The thermodynamical cycle that best is addressed to the electrocaloric refrigeration is Active Electrocaloric Regeneration cycle (AER) that consists of two adiabatic stages (polarization–depolarization) and two isofield stages, corresponding to the heat transfer fluid flows through the regenerator. The Active Electrocaloric Regenerator (AER) is the core of an electrocaloric refrigerator; it is designed as particular thermal regenerator, made of electrocaloric materials and acting both as refrigerant and regenerator in an AER cycle. This paper compares a commercial VCP employing R134a and an electrocaloric refrigerator, under different values of electric field applied, and testing a number of electrocaloric materials as refrigerant, exhibiting several nano-structural differences in order to provide a general framework as detailed as possible, to identify the eco-compatibility of EMs and to evaluate their environmental impact in terms of a greenhouse effect with respect to a vapor compression plant.. The comparison is performed in term of TEWI (Total Equivalent Warming Impact) index that takes into account both direct and indirect contributions to global warming.

2. Electrocaloric effect and the AER cycle

The electrocaloric effect is a physical phenomenon detected in material with dielectrical properties, i.e. electrocaloric materials (EM), which consists in variation of the electrical contribution of the material's entropy as a consequence of an electric field application. The total specific entropy of a particular dielectric material, i.e. an electrocaloric material, consists of three contributions:

$$S = S_{el}(T,E) + S_l(T) + S_e(T) \tag{1}$$

where: S_{el} represents the electric, S_l the lattice and S_e the electron entropy. When the electric field is applied to an electrocaloric material the electric dipoles become oriented according to the direction of the field. If this is done isothermally, it carries to a decrement of the material's electric entropy by the isothermal entropy decreasing ΔS_T. On the other side, if the polarization is done adiabatically, with no thermal transfer toward the ambient, the total sample entropy remains constant and the decrease in electric entropy is countered by an increase in the lattice and electron entropy. This causes a heating of the material and it takes an increasing of its temperature given by the adiabatic temperature change, ΔT_{ad}. Dually, an adiabatic depolarization involves the increasing of the material's electric entropy, causing a decreasing in lattice vibrations and by that a temperature decrease. The expression for ΔT_{ad} of an EM can be evaluated as:

$$(\Delta T_{ad})_s = -\int_{E_i}^{E_f} \frac{T}{c_E}\left(\frac{\partial P}{\partial T}\right)_E dE \tag{2}$$

where the specific heat can be defined as:

$$C_E = T\left(\frac{\delta s}{\delta T}\right)_E \tag{3}$$

In case of isothermal process the entropy change can be evaluated as:

$$(\Delta S_{el})_T = \int_{E_i}^{E_f}\left(\frac{\partial P}{\partial T}\right)_E dE \tag{4}$$

At the Curie temperature, where is located its own electrocaloric phase transition (from ferroelectric state to paraelectric state), an EM shows the peak of ECE, in terms of ΔT_{ad} and ΔS_T. The two possible electrocaloric phase changes that one can observe at the Curie point in electrocaloric materials are the first and the second order. In a first order transition phase the electrocaloric material can achieve a larger ECE than in a second order, due to the jump in the entropy of the material close its Curie temperature. The thermodynamical cycle best fitting to electrocaloric refrigeration is the reverse Brayton cycle [12], which consists of four processes as shown in Figure 1:

- A-B: adiabatic polarization, obtained increasing the electric field intensity from the values E_A to E_B; as a consequence, because of electrocaloric effect, an increasing of EM's temperature, amounting to a $\Delta T_{ad,AB}$ quantity, is registered;
- B-C: heat rejection to an heat sink under a constant electric field E_B;
- C-D: adiabatic depolarization, obtained reducing the electric field intensity from the value E_B to E_A; as a consequence in the electrocaloric material one can observe a decrement of its temperature quantified in $\Delta T_{ad,CD}$;
- D-A: heat absorption from a heat source under constant electric field E_A.

The most used way for heat transferring between the heat source and the heat sinks, in electrocaloric refrigerators, is the employment of a secondary fluid with the purpose to transfer heat from the cold heat sink to the hot heat sink as a consequence of Brayton cycle. In 1982 Barclay introduced [13] an innovative way to consider the reverse Brayton cycle for magnetic refrigeration, through the Active Magnetic Regenerative refrigeration cycle (AMR cycle), concept which has also been extended to electrocaloric refrigeration, through the Active Electrocaloric Regenerative refrigeration cycle, well known as AER cycle. The AER cycle is applied to an AER regenerator made of electrocaloric material whom has the peculiarity to be both refrigerant and regenerator at the same time, whereas a secondary fluids has the function to transfer heat from the cold to the hot end of the regenerator. Substantially every section of the regenerator experiments its own AER cycle, according to the proper working temperature.

Through an AER one can appreciate a larger temperature span across the regenerator. During the entire cycle the electric field varies between a minimum value E_{MIN}, and a maximum, E_{MAX}. The regenerator works between a hot reservoir at T_H and a cold one at T_C. The AER cycle consists of four processes. With reference to Figure 2, during polarization (A) the electric field applied to the regenerator increases its intensity, whereas there the fluid is stationary: as a result an increase of the EM temperature, due to the ECE, is registered. In the cold-to-hot fluid flow process (B), the secondary fluid blows from the cold to the hot end of the regenerator, while the field has kept at a E_{MAX}: the fluid absorbs heat from the regenerator, reaching a temperature higher than T_H and the heat is rejected through a hot heat exchanger. The next process is depolarization (C) where, with no fluid flow, the electric field decreases its intensity until reaching E_{MIN}: as a consequence, the electrocaloric material decreases its temperature accordingly to the ECE. Lastly the fluid flows from the hot heat exchanger to the cold one (D), while the field is at the value E_{MIN}: as a result the fluid, hotter, cools itself by crossing the regenerator, reaching a temperature lower than T_C. At this stage the secondary fluid absorbs heat from the cold heat exchanger at T_C, producing a cooling load.

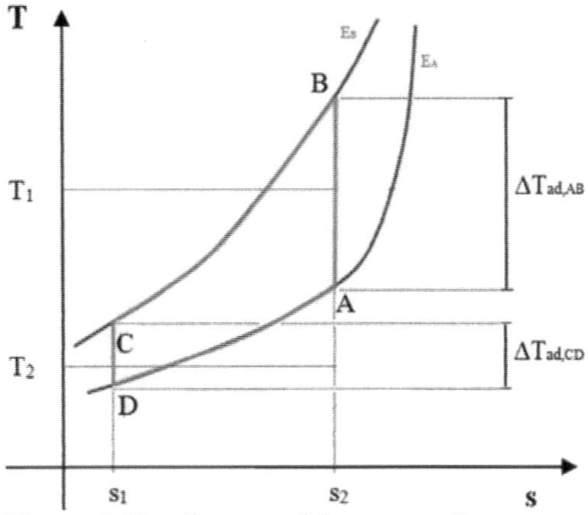

Figure 1. T-s diagram of the reverse Brayton cycle with reference to the ER.

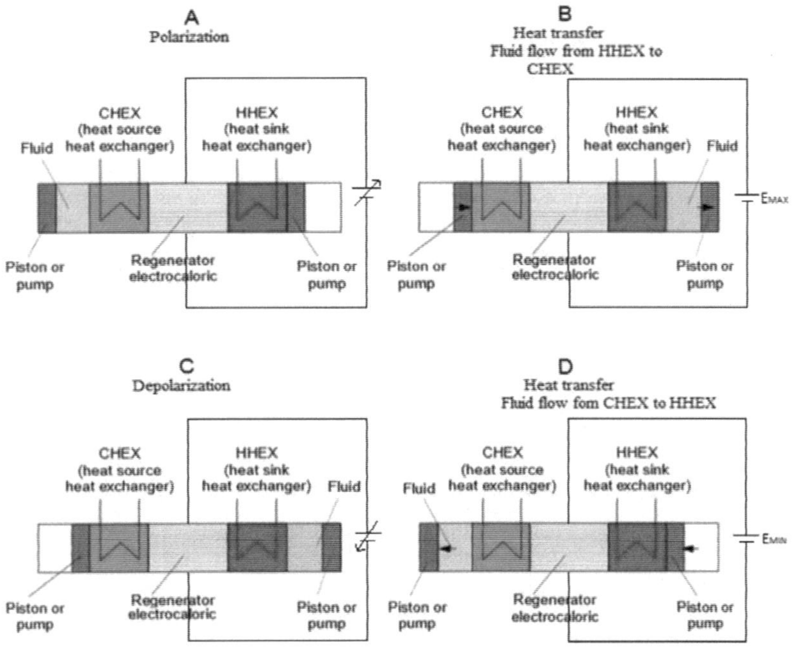

Figure 2. The four processes of an AER cycle related to an AER regenerator kept in contact with a cold and a hot heat exchangers.

3. Electrocaloric materials for AER refrigeration

In the scenario of AER refrigeration at room temperature, electrocaloric materials are classified in three macro-categories: bulk, thin and thick film materials [11]. The use of thin films instead of bulk materials allows the possibility of applying high electric fields with relatively small voltages (higher breakdown electric field) and yielding higher adiabatic temperature variation. In addition, the bulk materials are usually not compatible with the semiconductor integrated circuit technology, which also make them unsuitable in practical applications. A disadvantage of the thin film is that their overall stored energy and heat-sinking capacity are small due to their mass limitations. Another disadvantage of thin films is the fact that they are thermally anchored by substrates, which can reduce the cooling power. Within these two groups EMs can be:

- single crystals (bulk materials);
- ceramics (bulk samples, with thickness higher than 100 μm; thick film (with a thickness greater than 10 μm but lower than 100 μm);
- polymers (thick and thin film, with thickness between 10 μm and 1 μm);

Considering ferroelectric polymers among the most promising electrocaloric materials for cooling application because of their versatility, scalability, weight and cost advantages, in this paper it have been selected some relaxor ferroelectric P(VDF-TrFE-CFE) terpolymer, 59.2/33.6/7.2 mol [14,15] with a film thickness between 4 to 6 μm, non stretched and stretched. In particular, stretching is a process which improves the breakdown field of the considering film. The temperature dependence of ECE of the terpolymer films depends critically on the film preparation conditions. Whereas the uniaxially stretched terpolymer films show pronounced temperature dependence of ECE, the non-stretched films exhibit nearly temperature independent ECE from 5 to 45 °C. Besides, large ECE ($\Delta T>15$ °C) can be induced in both films at 30°C and 150 MV/m. Moreover, in this paper the relaxor ferroelectric terpolymer P(VDF-TrFE-CFE), with a composition of 62.3/29.9/7.8 mol% and a Curie temperature of 26°C, it has been considered as the matrix in which bulky chlorine atoms act as random defects to reduce the ferroelectric domain size and the energy barrier to phase transition; BST ($Ba_xSr_{1-x}TiO_3$) has been selected as the filler. For $Ba_{0.67}Sr_{0.33}TiO_3$ the ferroelectric-paraelectric transition occurs at 27°C. The ECE of the P(VDF-TrFE-CFE)/BST polymer nanocomposities has been significantly enhanced in comparison to those of the polymer matrix and filler, at relatively low electric field [16]. In the group of ceramics it has been considered a thin film of $PbSc_{0.5}Ta_{0.5}O_3$ (PST), a relaxor ceramic which exhibits a second-order-phase transition. Even if PST materials don't show one of the higher ECE in terms of ΔT_{ad} and ΔS_T, they have many advantages like low electrical hysteresis and negligible thermal hysteresis [17]. Another material tested in this work is 0.93PMN-0.07PT, which is made of 93% of the relaxor ceramic $PbMg_{2/3}Nb_{1/3}O_3$ (PMN) and 7% of the ferroelectric perovskite $PbTiO_3$ (PT). 0.93PMN-0.07PT thin film exhibits a giant ECE at its Curie temperature of 27°C [18] but it shows a moderate thermal hysteresis at 18°C [19]. Furthermore, another family of ceramic materials has been studied: the $Pb_{1-3/2x}La_xZr_{0.85}Ti_{0.15}O_3$ (PLZT) antiferroelectric thick films [20]. Such antiferroelectric films have been deposited on a LNO/Si (1 0 0) substrate. A film with a single composition of PLZT 11/85/15 shows a Curie temperature around 111°C. It is possible to use a compositionally graded structure in order to improve many properties of the EM including dielectric properties, saturation polarization and pyroelectric coefficient. Compared to thick films at single composition, the compositionally graded films show higher dielectric constants and larger saturation polarizations. Therefore, also an improved energy storage performance and a giant ECE are observed. It is possible to define an "up-graded" composition thick films by varying the La content from 8 mol % at the substrate interface, to 14 mol% at the top surface; whereas films with the opposite gradient have been called "down-graded" films. The electric cooling efficiency of an electrocaloric material can be evaluated considering the magnitude of $(\Delta S_{el})_T$, ΔT_{ad} and considering two different parameters: RCP(S) and RCP(T). RCP(S) and RCP(T) represent the relative cooling power in isothermal and in adiabatic processes. They are defined by the following equations:

$$RCP(S) = \Delta S_{el_{(\max)}} * \Delta T_{FWHM} \tag{5}$$

$$RCP(T) = \Delta T_{ad_{(\max)}} * \Delta T_{FWHM} \tag{6}$$

where $\Delta S_{el(max)}$ and $\Delta T_{ad(max)}$ are the maximum ΔS_{el} and ΔT_{ad} values at a fixed variation of the electric field and ΔT_{FWHM} is the full-width of temperature corresponding to the half of the maximum.

Table 1 summarizes the characteristics of the presented electrocaloric materials in the room temperature range and it allows a quick comparison between the different electrocaloric materials.

PLZTs show peak values of ΔT_{ad} and ΔS_{el} always higher than that of the other materials: for PLZT single composition, the peak is out of the room temperature range but for the up-graded and down-graded PLZT materials it is within. In the polymer group, P(VDF-TrFE-CFE)/BST67 shows the higher value of ECE. In terms of RCP(T) and RCP(S), P(VDF-TrFE-CFE)/BST67 is the best one but PLZT up-graded also shows acceptable values. P(VDF-TrFE-CFE) has the lower thermal conductivity and therefore the leakage currents under the applied field are negligible (together with the undesired Joule effect).

Table 1. The different electrocaloric materials considered.

Materials	Tc [°C]	ΔE [MV/m]	ΔT_{ad} [°C]	ΔS_{el} [J/kgK]	ρ [kg/m³]	k [W/mK]	C_E [J/kgK]	RCP(T) [K²]	RCP(S) [J/kg]
P(VDF-TrFE-CFE) non stretched	300	70	3.8	25	1800	0.2	1500	355	3993
P(VDF-TrFE-CFE) non stretched	300	100	7	30	1800	0.2	1500	657	5704
P(VDF-TrFE-CFE) Stretched	300	100	8	30	1800	0.2	1500	500	4345
P(VDF-TrFE-CFE)/BST67	311	75	9.2	80	2060	1	1260	1087	10530
P(VDF-TrFE-CFE)/BST71	322	75	9.4	79	2060	1	1260	1038	10000
P(VDF-TrFE-CFE)/BST74	331	75	9.7	78.5	2060	1	1260	1019	4065
P(VDF-TrFE-CFE)/BST77	337	75	9.9	78	2060	1	1260	960	4180
PST	341	77.4	6.2	6.3	8800	1.1	320	657	693
0.93PMN-0.07PT	298	50.9	9	10	8300	1.384	320	125	135
0.93PMN-0.07PT	298	72.3	13	14	8300	1.384	320	160	162
PLZT 11/85/15	111	90	12	10	7900	1.9	330	1320	1200
PLZT upgraded	/	90	28	33	7900	1.9	330	2800	3300
PLZT downgraded	/	90	20	23	7900	1.9	330	1200	1380

4. The TEWI concept

The impact of greenhouse gases on global warming is quantified by their GWP, defined as the mass of CO_2 that would result in the same net impact on global warming as the release of a single unit (kg) of the atmospheric component in question [21]. VCPs produce both a direct and an indirect contribution to global warming. The former depends on the GWP of refrigerant fluids and on the fraction of refrigerant charge released in the atmosphere during operation and maintenance, or not recovered when the system is scrapped. The indirect contribution consists in the so-called energy-related contribution. In fact, a VCP requires electrical energy, produced by a power plant that typically burns a fossil fuel releasing CO_2 to the atmosphere, which is a strong function of its COP, of the power plant efficiency and of the fuel used in the conversion plant that affect the emissions per unit energy converted. The Total Equivalent Warming Impact (TEWI) index takes into account both contributions to global warming of the system. The concept of TEWI was developed to combine the effect of direct refrigerant emission with those due to energy consumption and the related combustion of fossil fuels for the electric energy production. TEWI is the sum of the direct contribution of the greenhouse gases used to make the systems or to operate and the indirect contribution of carbon dioxide emissions, resulting from the energy required to run the systems over their normal lifetimes. The TEWI is calculated as :

$$TEWI = CO_{2,dir} + CO_{2,indir} \quad [\text{kg } CO_2] \qquad (7)$$

$$CO_{2,dir} = RC\left[P_L + \left(\frac{1-P_R}{V}\right)\right]V \cdot GWP \qquad [\text{kg CO}_2] \qquad (8)$$

$$CO_{2,indir} = \alpha \cdot \frac{\dot{Q}_{ref}}{COP} \cdot H \cdot V \qquad [\text{kg CO}_2] \qquad (9)$$

The direct global warming effect of refrigerant fluids, stemming from the absorption they produce of long-wave radiations, depends on their GWP and on the fraction of refrigerant charge released in the atmosphere. The last is mainly due to leakage during the operational plant life time (P_L) and to the residual amounts which, according to the current state of technology, are not recyclable and thus are released to the atmosphere when taking the plant out of operation ($1-P_R$). In the simulation P_L is assumed as 5%, whereas P_R has not been considered. The literature provides some indicative, average levels of CO_2 release per KWh of electrical energy for various countries. For Italy, the value is 0.6 kg CO_2/kWhe. The annual power consumption assumed is 290 kWh per year that corresponds to a commercial medium size wine cooler. R134a is an HFC with zero ODP and a GWP of 1430.

5. Model description

The model presented [22] in this paper is a two-dimensional numerical model of a parallel plates AER regenerator operating at room temperature. The regenerator has a rectangular shape (20x45mm). The area is filled with 54 parallel plates of EM; every plate has a thickness of 0.25mm, in order that it is smaller than the thermal penetration depth δ of every material, where δ is defined as:

$$\delta = \sqrt{\frac{k}{\pi\, f\, \rho\, C_E}} \qquad (10)$$

where f, k, ρ and C_E are the operating frequency and the thermal conductivity, density and specific heat. The amount of the area occupied by all of the plates is 60% of the total rectangular area. A group of channels is formed by stacking plates in the regenerator area, where the secondary fluid flows. The mathematical formulation that describes the AER cycle includes a number of distinct groups of equations according to the different processes of the AER cycle that the regenerator experiences. The equations that rule the regenerative fluid flow processes, in both directions, are: the Navier-Stokes for the fluid flow and the energy equations for both the fluid and the solid particles. With the assumptions that the fluid is incompressible, laminar and the viscous dissipation neglected, the above equations are:

$$\begin{cases} \dfrac{\partial u}{\partial x} + \dfrac{\partial v}{\partial y} = 0 \\ \dfrac{\partial u}{\partial t} + u\dfrac{\partial u}{\partial x} + v\dfrac{\partial u}{\partial y} = -\dfrac{1}{\rho_f}\dfrac{\partial p}{\partial x} + \upsilon\left(\dfrac{\partial^2 u}{\partial x^2} + \dfrac{\partial^2 u}{\partial y^2}\right) \\ \dfrac{\partial v}{\partial t} + u\dfrac{\partial v}{\partial x} + v\dfrac{\partial v}{\partial y} = -\dfrac{1}{\rho_f}\dfrac{\partial p}{\partial y} + \upsilon\left(\dfrac{\partial^2 v}{\partial x^2} + \dfrac{\partial^2 v}{\partial y^2}\right) \\ \dfrac{\partial T_f}{\partial t} + u\dfrac{\partial T_f}{\partial x} + v\dfrac{\partial T_f}{\partial y} = \dfrac{k_f}{\rho_f C_{fp}}\left(\dfrac{\partial^2 T_f}{\partial x^2} + \dfrac{\partial^2 T_f}{\partial y^2}\right) \\ \dfrac{\partial T_s}{\partial t} = \dfrac{k_s}{\rho_s C_{sp}}\left(\dfrac{\partial^2 T_s}{\partial x^2} + \dfrac{\partial^2 T_s}{\partial y^2}\right) \end{cases} \qquad (11)$$

The equations that model the polarization and depolarization processes are:

$$\begin{cases} \rho_f C_{fp}\dfrac{\partial T_f}{\partial t} = k_f\left(\dfrac{\partial^2 T_f}{\partial x^2} + \dfrac{\partial^2 T_f}{\partial y^2}\right) \\ \rho_s C_{sp}\dfrac{\partial T_s}{\partial t} = k_s\left(\dfrac{\partial^2 T_s}{\partial x^2} + \dfrac{\partial^2 T_s}{\partial y^2}\right) + Q \end{cases} \qquad (12)$$

They must take into account the ECE which elevates or lowers the temperature of the solid by the variation of the external field applied to the regenerator. Hence the ECE temperature variation ΔT_{ad} is converted [23] into a heat source Q:

$$Q = Q(E, T_S) = \frac{\rho_s c_{sp}(H,T_s)\Delta T_{ad}(H,T_s)}{\Delta t} \tag{13}$$

which has the dimensions of a power density and included in the solid energy equation, only for polarization and depolarization phases. The term Q is positive during polarization, negative during depolarization. Δt is the period of the polarization/depolarization process. The coupled equations that govern the AER cycle, imposed on this model, are solved using Finite Element Method. The AER cycle is modeled as four sequential steps. The same time step Δt has been chosen for the resolution during all the four periods of the cycle. The cycle is repeated several times with constant operating frequency until the regenerator reaches steady state operation. The refrigeration energy and the energy supplied in the environment are calculated according to the following equations:

$$Q_{ref} = \int_{t_P+t_{CF}+t_D}^{t_P+t_{CF}+t_D+t_{HF}} \dot{m}_f C_f \left(T_C - T_f(0, y, t)\right) dt \tag{14}$$

$$Q_H = \int_{t_P}^{t_P+t_{CF}} \dot{m}_f C_f \left(T_f(L, y, t) - T_H\right) dt \tag{15}$$

The work of the secondary fluid circulation pump is:

$$W_p = \frac{\dot{m}(\Delta p_{CF} + \Delta p_{HF})}{\eta_p \rho_f}(t_{CF} + t_{HF}) \tag{16}$$

The Coefficient of Performance is:

$$\mathrm{COP} = \frac{Q_{ref}}{Q_{rej} - Q_{ref} + W_p} \tag{17}$$

The COP of the vapor compression plant working under the same operating conditions of the above mentioned AER cycle was predicted with the DOE/ORNL Heat Pump model. The VCP used in the simulation is a commercial, small size, R134a refrigerator with a semi-hermetic compressor, air cooled condenser, forced air circulating evaporator [24].

6. Results and discussion

Several AER cycles were simulated with EM, reported in Table 1, employed as refrigerant. In all of the cases, the secondary fluid is pure water. The simulations were performed with fixed AER cycle frequency (1.25 Hz), cold and hot heat exchanger temperature (T_C=292 K, T_C=300 K). Fluid flow rate was varied in the range 0.034÷0.057 kg/s to characterize the performance sensitivity of the regenerator. The results presented were generated for several electric field applied, in order to provide a general framework. The TEWI direct contribution of an AER cycle is zero because of refrigerant solidness, which has essentially zero vapor pressure and GWP. The direct contribution of the vapor compression plant accounts about 10 % of the whole value. The parameter that mostly affects the TEWI indirect contribution is the COP of the plant. Figure 3, which shown the COP of all the materials presented for different fluid flows rates, reveals that upgraded and downgraded thin film of PLZT materials have the highest COPs. Moreover, also the thin film 0.93 PMN-PT, the relaxor ceramic PST and most of P(VDF-TrFE-CFE)/BST polymer nanocomposities exhibit satisfactory performances.

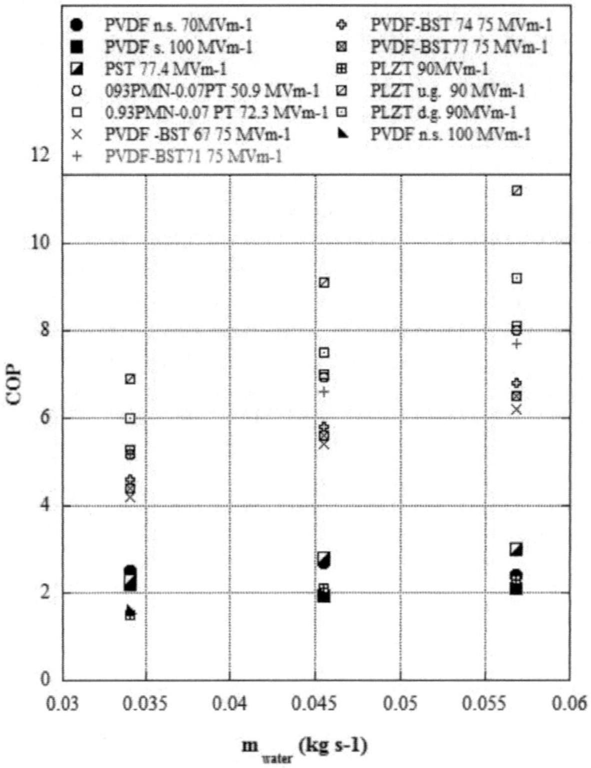

Figure 3. COP of the tested materials vs fluid flow rate.

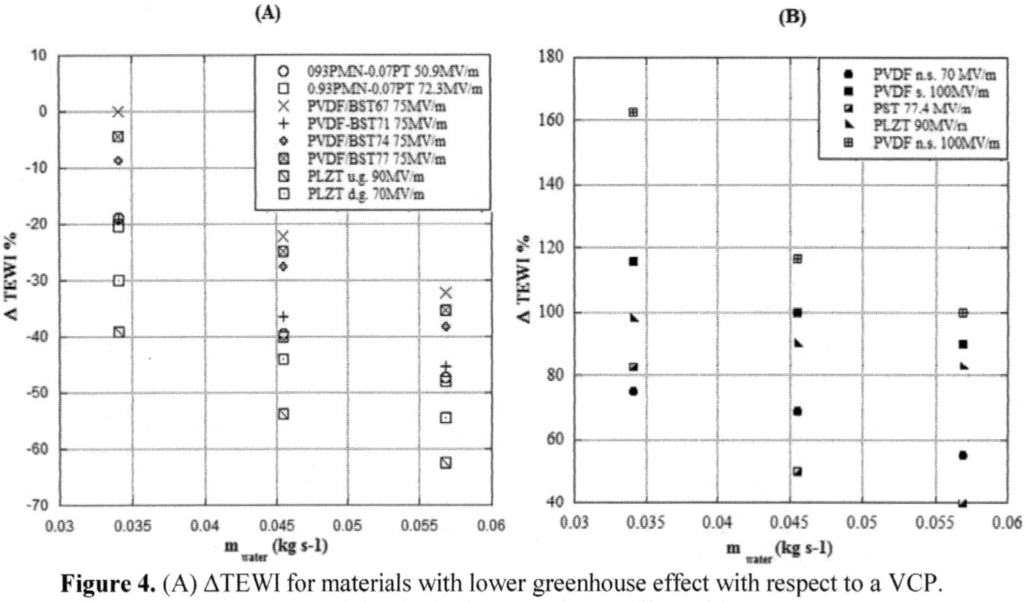

Figure 4. (A) ΔTEWI for materials with lower greenhouse effect with respect to a VCP.
(B) ΔTEWI for materials with higher greenhouse effect with respect to a VCP.

In order to make a comparison ΔTEWI has been defined according to the following equation:

$$\Delta TEWI = (TEWI_{AER} - TEWI_{VC})/TEWI_{VC} \tag{18}$$

In Figure 4 is reported ΔTEWI related to all the detected materials, for different fluid flow rates. Figure 4(A) clearly shows that the AER cycle working with these materials has a lower greenhouse effect with respect to a VCP. In particular with upgraded thin film of PLZT ΔTEWI varies between a minimum of -62.5% to a maximum of -39.1%. Figure 4(B) reports all the materials for which the AER cycle shows an environmental impact higher than the VCP one; materials not suitable for an eco-friendly refrigeration.

7. Conclusions

In the present paper, a practical model for predicting the performance and efficiency of an AER refrigerator system for different electrocaloric materials has been introduced. In particular, the relaxor ferroelectric P(VDF-TrFE-CFE) terpolymer, P(VDF-TrFE-CFE)/BST polymer nanocomposities, the 0.93PMN-0.07PT thin film, thick films of PLZT in single, upgraded and downgraded compositions and thin film of the relaxor ceramic PST have been considered as solid refrigerants. In order to make a comparison, the energetic performances of a vapor compression plant, working in the same operating conditions, have been evaluated by means of a computer program. Attention has been devoted to the evaluation of the greenhouse effect of both AER and vapor compression cycle. The greenhouse effect has been evaluated by means of TEWI index. The simulations clearly show that most of the tested materials has a TEWI index always lower than a VCP. In particular upgraded and downgraded PLZT thin films are the best electrocaloric materials because an AER cycle working with them, shows a greenhouse effect lower than that of a traditional plant of around 50%. Also 0.93PMN-0.07PT and P(VDF-TrFE-CFE)/BST71 reveal promising performances in terms of environmental impact. On the other side the tested thin film of PLZT in single composition, the relaxor ferroelectric P(VDF-TrFE-CFE) terpolymer and PST cannot be considered for refrigeration because they show a contribution in term of greenhouse effect greater than a VCP one. These results indicate that electrocaloric refrigeration is a promising refrigeration technology that could be used in cooler applications. An AER refrigerator can be an eco-friendly technology if it is used with electrocaloric materials that show a significant ECE. From a global point of view (performances and environmental impact), the most promising materials are the upgraded and downgraded PLZT thin films, really performing, whom give a contribution to global warming satisfactorily lower than a vapor compression plant one.

References

[1] Aprea C, Greco A, Maiorino A, Masselli C and Metallo A 2016 HFO1234yf as a drop-in replacement for R134a in domestic refrigerators: a life cycle climate performance analysis *International Journal of Heat and Technology* vol **34** iss Special Issue 2 pp S212-S218.

[2] Aprea C, Greco A and Maiorino A 2014 Magnetic refrigeration: a promising new technology for energy saving *International Journal of Ambient Energy* vol **35** pp 1-20.

[3] Aprea C, Greco A, Maiorino A and Masselli C 2015 Magnetic refrigeration: an eco-friendly technology for the refrigeration at room temperature *Journal of Physics: Conference Series* vol **655** 012026.

[4] Brown GV 1976 Magnetic heat pumping near room temperature *Journal of Applied Physics* vol **47** pp 3673-3680.

[5] Aprea C, Greco A and Maiorino A 2012 Modelling an active magnetic refrigeration system: A comparison with different models of incompressible flow through a packed bed *Applied Thermal Engineering* vol **36** iss 1 pp 296-306

[6] Aprea C, Greco A, Maiorino A, Mastrullo R and Tura A 2014 Initial experimental results from a rotary permanent magnet magnetic refrigerator *International Journal of Refrigeration* vol **43** pp 111-122.

[7] Aprea C, Greco A, Maiorino A and Masselli C 2016 The energy performances of a rotary permanent magnet magnetic refrigerator *International Journal of Refrigeration* vol **61** pp 1-7.

[8] Aprea C, Cardillo G, Greco A, Maiorino A and Masselli C 2016 A rotary permanent magnet magnetic refrigerator based on AMR cycle *Applied Thermal Engineering* vol **101** pp 699-703.

[9] Aprea C, Greco A, Maiorino A and Masselli C 2015 A comparison between rare earth and transition metals working as magnetic materials in an AMR refrigerator in the room temperature range *Applied Thermal Engineering* vol **91** pp 767-777.

[10] Ožbolt M, Kitanovski A, Tušek J and Poredoš A 2014 Electrocaloric refrigeration: thermodynamics, state of the art and future perspectives *International Journal of Refrigeration* vol **40** pp 174-188.

[11] Lu SG and Zhang Q 2009 Electrocaloric materials for solid-state refrigeration *Advanced Materials* vol **21** pp 1983-1987.

[12] Walker G 1983 Claude and Joule-Brayton Systems *The International Cryogenics Monograph Series* chapt *Cryocoolers* ed Springer pp 297-353.

[13] Barclay J A 1982 The theory of an active magnetic regenerative refrigerator *NASA STI/Recon Technical Report* N 83 34087.

[14] Lu S G, Rozic B, Zhang Q M, Kutnjak Z, Pirc R, Lin M, Li X and Gorny L 2010 Comparison of directly and indirectly measured electrocaloric effect in relaxor ferroelectric polymers *Applied Physiscs Letters* vol **97** 201901.

[15] Li X, Qian X S, Lu S G, Cheng J, Fang Z and Zhang Q M 2011 Tunable temperature dependence of electrocaloric effect in ferroelectric relaxor poly(vinylidenefluoride-trifluoroethyleme-chlorofluoroethylene) terpolymer *Applied Physiscs Letters* vol **99** 052907.

[16] Zhang G, Li Q, Gu H, Jiang S, Han K, Gadinski M R, Haque M A, Zhang Q and Wang Q 2015 Ferroelectric polymer nanocomposities for room temperature elcetrocaloric refrigeration *Advanced Materials* vol **27** pp 1450-1454.

[17] Correia T M, Narayan S K, Young J S, Scott J F, Mathur N D, Whatmore R W and Zhang Q 2011 PST thin films for electrocaloric coolers *Journal of Physics D: Applied Physics* vol **44** 165407.

[18] Hamad M A 2013 Room temperature giant electrocalorc properties of relaxor ferroelectric 0.93PMN-0.07 PT thin film *AIP Advances* vol **3** 032115.

[19] Correia T M, Young J S, Whatmore R W, Scott J F, Mathur N D and Zhang Q 2009 Investigation of the electrocaloric effect in a $PbMg_{2/3}Nb_{1/3}O_3$-$PbTiO_3$ *Applied Physics Letters* vol **94** 182904.

[20] Zhao Y, Hao X and Zhang Q 2016 Enhanced energy-storage performance and electrocaloric effect in compositionally graded $Pb_{(1-3/2x)}La_xZr_{0.85}Ti_{0.15}O_3$ *Ceramics International* vol **42** iss 1 part B pp 1679-1687.

[21] Wuebbles D J 1995 Weighing functions for ozone depletion and greenhouse gas effects on climate *Annual Review of Energy and the Environment* vol **20** pp 45-70.

[22] Aprea C, Greco A, Maiorino A and Masselli C 2016 A comparison between different materials in an active electrocaloric regenerative cycle with a 2D numerical model *International Journal of Refrigeration* vol **69** pp 369-382.

[23] Aprea C, Cardillo G, Greco A, Maiorino A and Masselli C 2015 A comparison between experimental and 2D numerical results of a packed-bed active magnetic regenerator *Applied Thermal Eng*ineering vol **90** pp 376-383.

[24] Rice K 2006 DOE/ ORNL Heat Pump model, *Mark VI Version.*

Hourly simulation of a Ground-Coupled Heat Pump system

C Naldi and E Zanchini

University of Bologna, Department of Industrial Engineering, Viale Risorgimento 2, 40136 Bologna, Italy

claudia.naldi2@unibo.it

Abstract. In this paper, we present a MATLAB code for the hourly simulation of a whole Ground-Coupled Heat Pump (GCHP) system, based on the *g-functions* previously obtained by Zanchini and Lazzari. The code applies both to on-off heat pumps and to inverter-driven ones. It is employed to analyse the effects of the inverter and of the total length of the Borehole Heat Exchanger (BHE) field on the mean seasonal COP (SCOP) and on the mean seasonal EER (SEER) of a GCHP system designed for a residential house with 6 apartments in Bologna, North-Center Italy, with dominant heating loads. A BHE field with 3 in line boreholes is considered, with length of each BHE either 75 m or 105 m. The results show that the increase of the BHE length yields a SCOP enhancement of about 7%, while the SEER remains nearly unchanged. The replacement of the on-off heat pump by an inverter-driven one yields a SCOP enhancement of about 30% and a SEER enhancement of about 50%. The results demonstrate the importance of employing inverter-driven heat pumps for GCHP systems.

1. Introduction

Ground-Coupled Heat Pumps (GCHPs) are a very efficient technology for building heating and cooling which reached an important development during the last decades [1].

The design of GCHP systems is usually divided in two parts: the design of the Borehole Heat Exchanger (BHE) field; the choice of the heat pump and the evaluation of its seasonal performance.

Most design methods of BHE fields in the literature are based on the evaluation of the temperature distribution in the borehole field, as a function of time. In these studies, groundwater movement is usually neglected and the ground is considered as an infinite solid medium with constant thermo-physical properties. The problem to be studied is that of conduction in the ground, which is a problem of transient three dimensional heat conduction, for which approximate solutions, either analytical or numerical, are usually employed.

Analytical solutions are available with reference to Infinite Line Source (ILS) models, Infinite Cylindrical Source (ICS) models, and Finite Line Source (FLS) models. In these models, a borehole heat exchanger is considered either as an infinitely long line, or as an infinitely long cylinder, or as a line with finite length, respectively.

The analytical solutions for the ILS model and for the ICS model of a BHE subjected to a constant heat transfer rate per unit length were deduced by Carslaw and Jaeger [2]. The ICS model was employed by Kavanaugh and Rafferty [3] in the design method for BHE fields recommended by ASHRAE [4]. The analytical solution for the FLS model was determined by Claesson and Eskilson [5]. Zeng et al. [6] pointed out that the use of the expression given in Ref. [5], evaluated at the middle of the length of the BHE, yields an overestimation (up to 5%) of the mean temperature field at the BHE surface. The authors recommend to use the value given by that expression when averaged along the BHE length, which is called *g-function*. The *g-functions* are expressions of the time-dependent dimensionless temperature, averaged along the BHE length, due to a uniform and constant heat load which starts at the time instant $t = 0$. Simplified forms of the solution of Ref. [6], which has the form of a double integral, were proposed by Bandos et al. [7] and by Lamarche and Beauchamp [8]. By employing the results obtained in Ref. [7], Fossa [9, 10] proposed simple approximate expressions for the *g-functions* based on the FLS model, which require low computational time and use empirical coefficients, determined through the analysis of different BHE fields.

Numerical simulations of BHE fields can be performed e.g. by means of the software Earth Energy Designer (EED), which is entirely dedicated to borehole heat exchangers and employs own *g-functions*, or through the software TRNSYS, which can perform an energy analysis of the whole building-plant system. However, as pointed out by Fossa and Minchio [11], TRNSYS simulations of BHE fields are not sufficiently accurate in the case of in line fields and/or unbalanced seasonal heat loads. Moreover, TRNSYS does not allow the simulation of heat pumps with inverter.

Recently, very accurate analytical expressions of the *g-functions* for the dynamic simulation of BHE fields have been obtained by Zanchini and Lazzari [12, 13]. These *g-functions*, which are based on the Finite Cylindrical Source (FCS) model, are expressed in the form of polynomial functions of the logarithm of the dimensionless time. Through linear interpolations and superposition of effects, they can be applied for the dynamic simulation of any BHE field, provided that the ratio between length and diameter of each BHE is not lower than 500. In this paper, a code for the hourly simulation of GCHP systems, based on the *g-functions* obtained in Refs. [12, 13], is presented. The code, executable through any programming language and here implemented in MATLAB, applies to mono-compressor on-off and inverter-driven GCHP, used for building heating and/or cooling. Both the heat pump and the coupled BHE field can be simulated for several years. The code is employed to analyze the effects of the inverter and of the total length of the BHE field on the mean seasonal COP (SCOP) and on the mean seasonal EER (SEER) of a GCHP system designed for a residential house with 6 apartments in Bologna (North-Center Italy) with dominant heating loads.

2. Numerical code

2.1. Building and heat pump characterization

In order to characterize the building, the developed code requires as input data the hourly loads E_b during a whole year at the outlet of the generation subsystem (heat pump) for heating and for cooling. The building loads, which can derive from a dynamic simulation, must be set as negative for the heating season and as positive for the cooling season.

In order to characterize a mono-compressor on-off heat pump (ON-OFF HP), the required input data are the values, given by the manufacturer, of heat pump power, COP and EER in heating mode and in cooling mode, for a fixed value of the hot (or cold) water produced, T_w, and for different BHE fluid supply temperatures, T_{fout}. For inverter-driven heat pumps (IDHPs), these input values must be given in correspondence of the maximum, minimum and at least an additional intermediate inverter frequency.

Through interpolations of the manufacturer data, the heat pump power and COP in heating mode and the heat pump power and EER in cooling mode are expressed as second-order polynomial functions of T_{fout}, for each fixed value of T_w. In the case of inverter-driven heat pumps, a family of expressions for the heat pump power and a family of expressions for the heat pump COP (or EER) are obtained for each value of T_w, by varying the inverter frequency between the maximum and minimum value. The heat pump power and COP (or EER) expressions are stopped in correspondence of the cut-off temperature of the related operation mode. In fact, a minimum temperature of the BHE fluid could be fixed in winter (e.g. in order to prevent freezing, if just water is used as BHE fluid) and a maximum temperature in summer.

2.2. Borehole heat exchangers characterization

Regarding the borehole field, the BHE diameter (D), length (L) and thermal resistance per unit length (R_{BHE}) must be set as inputs, as well as the number of boreholes (n_{BHEs}) and their layout pattern and spacing.

The properties of the BHE heat carrier fluid (specific heat capacity at constant pressure, c_{pf}, density, ρ_f, and volumetric flow rate, $\dot{V}_f$) and of the ground (thermal conductivity, k_g, thermal diffusivity, α_g, and undisturbed ground temperature, T_g) are also needed.

The mean dimensionless temperature at the interface between BHE field and ground is evaluated by the code by employing the analytical expressions of the *g-functions* obtained by Zanchini and Lazzari [12, 13], which are given as functions of the logarithm of the dimensionless time, with dimensionless temperature, T^*, and time, t^*, defined as

$$T^* = k_g \frac{T - T_g}{q_0}, \tag{1}$$

$$t^* = \frac{\alpha_g \, t}{D^2}. \tag{2}$$

The term q_0 in equation (1) is a reference thermal power per unit length. The mean value of the building thermal load during January (the month with the highest heating demand), divided by the total length of the boreholes, has been adopted for q_0 in our case study.

The polynomial coefficients of the *g-functions* taken from Refs. [12, 13] have been implemented in the code for several values of the BHE dimensionless length, $L^*=L/D$, and for several values of the dimensionless radial distance from the BHE axis, $r^*=r/D$.

2.3. Calculation of the GCHP system seasonal performance

In order to invoke the proper *g-function* coefficients, the code calculates the values of L^* and r^* required to evaluate the mean temperature at a BHE surface through the superposition of the effects in space. As example, for the case of three in line boreholes with L=105 m, D=0.15 m and a distance between adjacent BHEs of 6 m, the *g-function* coefficients are needed in correspondence of L^* equal to 700 and r^* equal to 0.5 (BHE – ground interface), 40 (central BHE – lateral BHE distance) and 80 (distance between lateral BHEs). If the *g-function* coefficients are not tabulated in correspondence of the required value of L^* (or r^*), the closer lower value and the closer higher value of L^* (or r^*) are considered by the code; then, linear interpolations of the results are employed.

At the beginning of the first hour of the simulated period, the BHE fluid is in thermal equilibrium with the surrounding ground; hence, the BHE fluid mean temperature (T_{fm}) and supply temperature (T_{fout}) are both initialized as equal to the undisturbed ground temperature (T_g).

For the i-th hour of the simulated period, the MATLAB code evaluates the season to which the hour belongs, reads the value of T_{fout} at the beginning of the i-th hour and calculates, through the heat pump second order polynomial functions, the heating (cooling) power that the heat pump is able to deliver and the corresponding COP (EER). For inverter driven heat pumps, a vector for the heat pump power and a vector for the corresponding COP or EER are obtained.

The energy delivered by the heat pump in the i-th hour, $E_{HP}(i)$, is equal to the product of the maximum heat pump capacity and the duration of one hour, t_{hour} (and is set as negative for the heating season), but if $E_{HP}(i)$ turns out higher than $E_b(i)$, it is set equal to $E_b(i)$.

For ON-OFF HPs, the values of the heat pump power and COP or EER in the i-th hour are known from the previous interpolations. The value of the heat pump power for IDHPs can be obtained dividing the module of $E_{HP}(i)$ by t_{hour}. If the result turns out lower than the heat pump power at the minimum inverter frequency, it is set equal to the minimum power (situation corresponding to on-off cycles). The corresponding COP, or EER, of the IDHP is then obtained by applying a second-order polynomial interpolation of the COP, or EER, vector as a function of the heat pump power vector.

The effective hourly values of heat pump COP or EER ($\mathrm{COP}_{eff}(i)$ or $\mathrm{EER}_{eff}(i)$) take into account the heat pump efficiency decay in the case of on-off cycles, which occurs if, during the i-th hour, the heat pump power (at the minimum frequency for IDHPs) is too high compared to the building thermal needs. The values of COP_{eff} or EER_{eff} are evaluated according to the standards EN 14825 [14] and UNI/TS 11300-4 [15], multiplying the obtained COP or EER by the correction factor for on-off condition. The hourly value of the electric energy used by the heat pump is evaluated dividing $E_{HP}(i)$ by COP_{eff}, or EER_{eff}.

The thermal energy exchanged in the i-th hour between the borehole heat exchangers and the ground, $Q(i)$, is evaluated as

$$Q(i) = E_{HP}(i)\left[1 - \frac{1}{\mathrm{COP}(i)}\right] \tag{3}$$

during the heating season, and as

$$Q(i) = E_{HP}(i)\left[1 + \frac{1}{\mathrm{EER}(i)}\right] \tag{4}$$

during the cooling season. $Q(i)$ is negative if $E_{HP}(i)$ is negative, namely if heat is required by the building (extracted from the ground, during winter).

The mean value of the heat flux between BHE and ground per unit BHE length, $q(i)$, is obtained as the ratio between $Q(i)$ and the product of t_{hour} and the total length of the boreholes.

The dimensionless load amplitude of the i-th hour, $A(i)$, is given by the ratio between $q(i)$ and q_0.

At the end of the i-th hour, the dimensionless temperature $T_m^*\big|_{r^*}(i)$, averaged along the BHE length, produced at the dimensionless distance r^* from the BHE axis by a time-dependent dimensionless heat load, with steps of one hour and values given by the coefficients A, is:

$$T_m^*\big|_{r^*}(i) = \sum_{k=1}^{i} A(k)\left\{g\big|_{r^*}\left[(i+1-k)t_{hour}^*\right] - g\big|_{r^*}\left[(i-k)t_{hour}^*\right]\right\}, \tag{5}$$

where the symbol g denotes the *g-functions* and t_{hour}^* is the dimensionless duration of one hour.

By means of equation (5), T_m^* is calculated at r^*=0.5 (BHE-ground interface), and at the dimensionless distances between the BHEs. If the *g-function* coefficients are not tabulated in correspondence of the required value of L^* (or r^*), two hourly values of T_m^* are calculated by the code, in correspondence of the lower and higher closer available values of L^* (or r^*). The actual hourly value of T_m^* for the given L^* (or r^*) is then obtained through linear interpolation.

The mean dimensionless temperature at the surface of a specific borehole of the field is evaluated as the sum of the value of T_m^* produced by the specific BHE at r^*=0.5, and of those produced by the other BHEs of the field at their dimensionless distances from the specific BHE axis (superposition of the effects in space). The mean dimensionless temperature of the BHE field is evaluated as the arithmetic average of the T_m^* values of the different BHEs.

The definition of dimensionless temperature of equation (1) yields the mean temperature of the BHE field at the end of the i-th hour, $T_{m,field}(i)$. In the quasi-stationary approximation, the definition of BHE thermal resistance per unit length, R_{BHE}, yields the BHE fluid mean temperature at the end of the i-th hour, $T_{fm}(i)$:

$$T_{fm}(i) = T_{m,field}(i) + q(i)\, R_{BHE}. \tag{6}$$

The corresponding BHE fluid supply temperature, $T_{fout}(i)$, is obtained by subtracting to $T_{fm}(i)$ half of the difference between the BHE fluid inlet temperature $T_{fin}(i)$ and $T_{fout}(i)$, namely

$$T_{fin}(i) - T_{fout}(i) = \frac{Q(i)}{t_{hour}\, n_{BHEs}\, \dot{V}_f\, \rho_f\, c_{pf}}. \tag{7}$$

The value T_{fout} (i) of the fluid supply temperature at the end of the i-th hour corresponds to that at the beginning of the subsequent ((i+1)-th) hour and is used by the MATLAB code to evaluate the heat pump performance at the subsequent hour, through a for cycle.

The hourly values of T_{fm} during the simulation period allow to check the long-term sustainability of the BHE field in the case of unbalanced building loads.

The seasonal performance of the ground-coupled heat pump during a selected year can be evaluated by means of the Seasonal Coefficient Of Performance (SCOP, ratio between the total energy supplied by the heat pump for heating and the corresponding electric energy used) and of the Seasonal Energy Efficiency Ratio (SEER, ratio between the total energy supplied by the heat pump for cooling and the corresponding electric energy used).

To check the long-term sustainability of a GCHP system, fast simulations of the whole system can be performed for several decades, by employing hourly simulations with the aid of auxiliary monthly simulations. In the case of monthly simulations, monthly building loads are given as inputs and a mathematical model very similar to that presented for the hourly simulation can be used.

3. Characteristics of the case study

A residential building which is undergoing an energy retrofit within the European project HERB (Holistic Energy-efficient Retrofitting of residential Buildings) has been selected to characterize the building heat loads. The house is located in Bologna (North-Center Italy), is composed of three floors with six apartments and has a total heated floor area of 282 m^2. Figure 1 shows the power required by the building P_b at the outlet of the generation subsystem (heat pump), as a function of time, from October 1st to September 30th. The heating season has been set from October to April, included, while the cooling season from May to September, included.

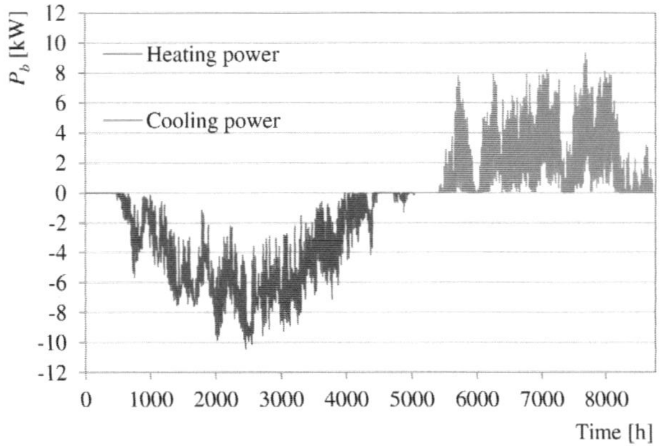

Figure 1. Power required by the building for heating and cooling, from October 1st to September 30th.

In figure 1 heating loads are considered negative, while cooling loads are considered positive, as required by the MATLAB code. Heating loads are dominant; the ratio between the annual thermal energy required by the building for heating and that for cooling is 2.32. The highest magnitudes of the power required by the building are 10.46 kW for heating and 9.32 kW for cooling.

The selected ground-coupled heat pump, used to provide heating and cooling to the building, is an inverter-driven brine-to-water unit. Water is delivered at 40 °C (return temperature 35 °C) during winter and at 7 °C (return temperature 12 °C) during summer. Data of heat pump power, COP and EER have been given by the manufacturer for BHE fluid supply temperatures (T_{fout}) from 5 °C to 18 °C in winter operation and from 15 °C to 35 °C in summer operation, at intervals of 1 °C, in correspondence of inverter frequencies equal to 110 Hz (maximum), 90 Hz, 70 Hz, 50 Hz and 30 Hz (minimum). In heating mode, the values of heat pump power and COP at maximum frequency range from 12.60 kW and 4.41 (T_{fout}=5 °C) to 18.20 kW and 6.40 (T_{fout}=18 °C), respectively. The corresponding values at minimum frequency range from 3.08 kW and 4.82 (T_{fout}=5 °C) to 4.49 kW and 7.37 (T_{fout}=18 °C). In cooling mode, the values of heat pump power and EER at maximum frequency range from 12.10 kW and 4.20 (T_{fout}=35 °C) to 13.80 kW and 6.42 (T_{fout}=15 °C), respectively. The corresponding values at minimum frequency range from 3.01 kW and 4.74 to 3.39 kW and 7.03.

The following ground properties have been set: T_g=14 °C, k_g=1.8 W/(m K), α_g=8.814 × 10^{-7} m²/s. These properties correspond to a soil made of gravel and sand (55%) and clay (45%), partially saturated by groundwater, having a volumetric heat capacity equal to 2.042 × 10^6 J/(m³ K).

The BHE field coupled to the heat pump is composed of three in line double-U boreholes, 6 m spaced, with diameter D=0.15 m and length L either 105 m or 75 m (corresponding dimensionless BHE lengths: L^*=700 and L^*=500, respectively). The BHE pipes, made of high-density polyethylene, have outer diameter 32 mm, internal diameter 26 mm and shank spacing 85 mm. The BHE is sealed with a grout having thermal conductivity 1.6 W/(m K).

The BHE fluid volumetric flow rate, $\dot{V}_f$, is 16 l/min. For the simulations with L=105 m, the BHE fluid is water, whose density ρ_f is 999.25 kg/m³ and specific heat capacity c_{pf} is 4.1896 kJ/(kg K) (properties at 14 °C). The corresponding winter cut-off temperature of the heat pump is 2 °C and the BHE thermal resistance per unit length (obtained through a numerical steady state simulation of the BHE cross section) is 0.0687 (m K)/W. Since simulations with L=75 m would cause too low water temperatures during winter, water is replaced by a mixture of water-glycol (monoethylenglycol 20%), whose ρ_f is 1032 kg/m³ and c_{pf} is 3.89 kJ/(kg K) (properties at 14 °C, from Ref. [16]). The corresponding winter cut-off temperature of the heat pump is -8 °C and the BHE thermal resistance per unit length is 0.0732 (m K)/W.

4. Results

Ten-year hourly simulations of the GCHP system have been performed with the selected IDHP, for the cases of L^*=700 and L^*=500. Figure 2 shows the hourly values of the BHE fluid mean temperature, T_{fm}, obtained during the 10th year of simulation with L^*=700. Each year starts on October 1st and ends on September 30th.

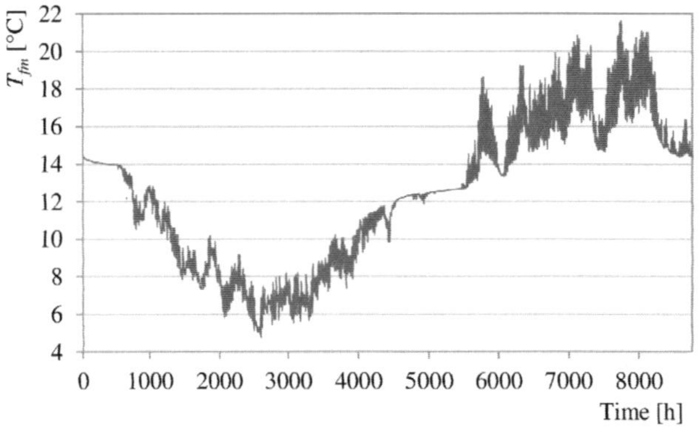

Figure 2. BHE fluid temperature during the 10th year, L^*=700, inverter-driven heat pump.

The first ten points of the curves in figure 3 represent the minimum values of the mean temperature of the BHE fluid ($T_{fm,min}$), reached in each of the 10 years, for the cases of L^*=700 and L^*=500. The other points have been obtained by the aid of monthly simulations, as will be explained in the following.

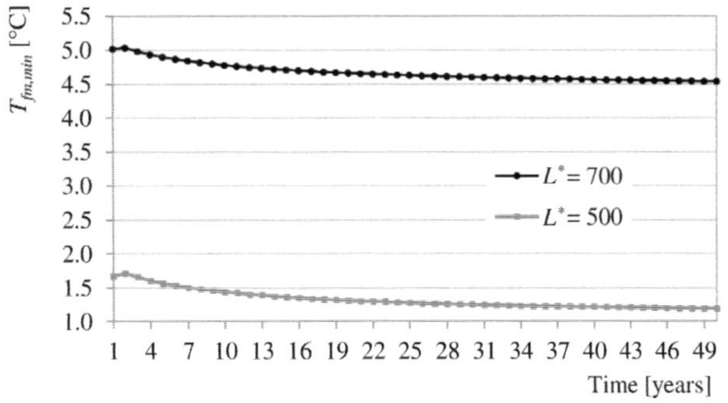

Figure 3. BHE fluid minimum temperatures, IDHP.

An unexpected peak of $T_{fm,min}$ can be seen in figure 3 in correspondence of the second year of simulation. This is due to the fact that the first year starts with the heating season (heat extracted from ground), which, unlike the heating seasons of all the other years, is not preceded by a cooling season (heat injected in ground), that would enhance the fluid temperature. From the second year on, each heating season is preceded by a cooling season and $T_{fm,min}$ has a regular decreasing trend.

Comparing the curves of figure 3, it is evident that, with L^*=500, the BHE fluid reaches lower values of $T_{fm,min}$, compared to the case of L^*=700, mainly due to higher heat loads per unit BHE length.

To analyze the influence of the heat pump inverter on the system seasonal performance during winter and during summer, the simulations are repeated forcing the heat pump to work at its maximum frequency, as a mono-compressor on-off heat pump. Table 1 shows the values of SCOP and of SEER of the ON-OFF HP and of the IDHP, for the 10th year of operation, with both the considered BHE lengths. The performance values of Table 1 do not consider the energy used by the closed loop pump.

Table 1. SCOP and SEER values for the 10th year.

SCOP values	ON-OFF HP	IDHP
L^*=500	3.82	4.97
L^*=700	4.00	5.32
SEER values	ON-OFF HP	IDHP
L^*=500	4.42	6.74
L^*=700	4.45	6.74

From table 1 it can be noticed that the increase of the BHE length yields a SCOP enhancement of about 5% for the ON-OFF HP and of about 7% for the IDHP, while the SEER remains nearly unchanged. Indeed, since cooling loads are low, an increase of the BHE length does not yield a significant SEER improvement. The replacement of the mono-compressor on-off heat pump by an inverter-driven one yields a SCOP increase of about 30% for L^*=500 and of about 33% for L^*=700, together with a SEER enhancement of about 51% for L^*=700 and of about 52% for L^*=500. The most significant improvements on the seasonal performance are obtained by introducing the inverter, whose effect is greater on the summer efficiency (lower building loads, higher number of on-off cycles) than on the winter efficiency (higher building loads, lower number of on-off cycles). Figures 4 and 5 show, respectively, the hourly trend of COP_{eff} and of EER_{eff}, for the period November–January and June–August of the 10th year, with and without inverter, for L^*=700.

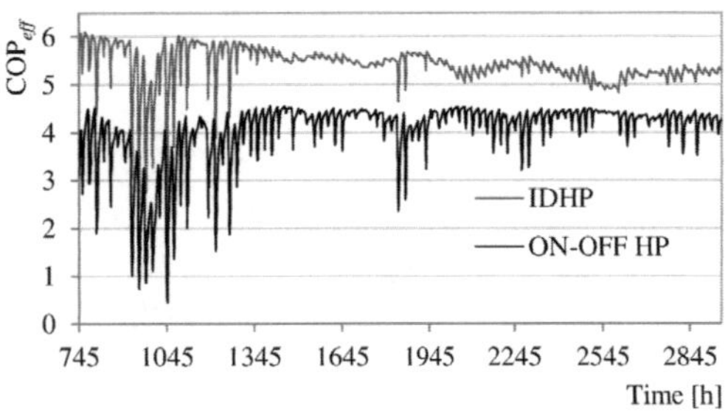

Figure 4. Hourly trend of COP_{eff} from November to January, 10th year, L^{*}=700.

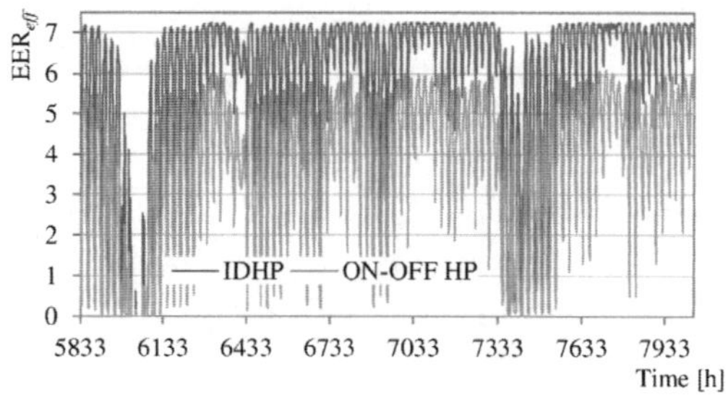

Figure 5. Hourly trend of EER_{eff} from June to August, 10th year, L^{*}=700.

Figures 4 and 5 highlight the better hourly efficiency of the IDHP with respect to the ON-OFF HP, thanks to the possibility of the IDHP of delaying the on-off cycles activation.

A simulation of the GCHP system has been performed for 50 years by means of monthly simulations. Moreover, hourly simulations of the heat pump from the 11th year to the 50th year have been performed by assuming that the hourly values of the BHE fluid mean and supply temperatures during a given year are equal to the corresponding values obtained at the 10th year through hourly simulation, properly translated. The translation constant is the difference between the mean annual value of T_{fm} at the 10th year and that at the given year, both obtained through monthly simulations.

The minimum annual values of T_{fm} so-obtained, from the 11th to the 50th year, are reported in figure 3 for the IDHP with L^{*}=700 or L^{*}=500. From figure 3 it can be noticed that the decrease in $T_{fm,min}$ from the 2nd to the 50th year is less than 0.5 °C for L^{*}=700 and less than 0.6 °C for L^{*}=500. Therefore, in spite of the building unbalanced heat loads, the studied GCHP systems do not reveal long-term sustainability problems.

The accuracy of the assumption of employing monthly simulations for the translation constant has been tested by comparing the difference between the mean annual value of T_{fm} at the 5th year and that at the 10th year, obtained through monthly simulation, with the corresponding difference obtained though hourly simulation. The relative discrepancy between the two differences never exceeds 3.3% in the examined cases. Moreover, the accuracy of the assumption of employing a mean annual value as translation constant has been tested by analyzing the difference between the hourly values of T_{fm} at the 5th year and those at the 10th year, which is plotted in figure 6 for the case of the IDHP with L^{*}=700.

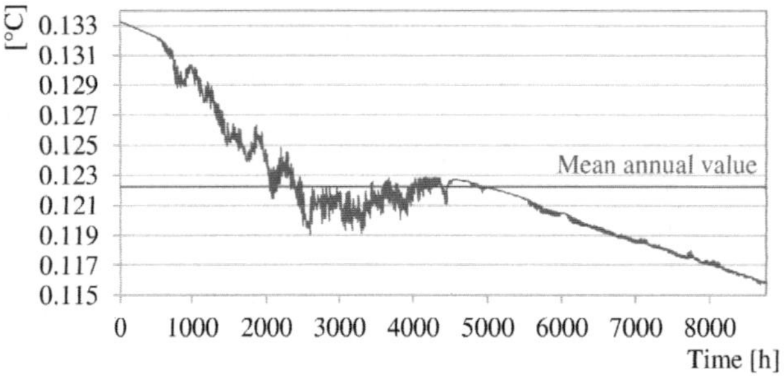

Figure 6. Hourly values of the difference between T_{fm} at the 5th year and T_{fm} at the 10th year, IDHP, L^{*}=700.

The plot of figure 6 shows that this difference is always very close to its mean annual value; the maximum deviation from the mean value is 0.01 °C.

The seasonal performance coefficients obtained with the hourly simulations for the 50th year are reported in table 2.

Table 2. SCOP and SEER values for the 50th year.

SCOP values	ON-OFF HP	IDHP
L^*=500	3.81	4.93
L^*=700	3.98	5.28
SEER values	ON-OFF HP	IDHP
L^*=500	4.43	6.74
L^*=700	4.45	6.74

By comparing the values of table 2 with the corresponding ones of table 1, it is clear that the heat pump seasonal efficiencies remain nearly constant between the 10th and the 50th year.

5. Conclusions

A code for the hourly simulation of Ground-Coupled Heat Pump (GCHP) systems has been developed. The code, which is implemented in MATLAB, employs the *g-functions* obtained by Zanchini and Lazzari [12, 13] and applies to GCHPs with or without inverter, used for building heating and/or cooling. It allows fast hourly simulations for several years and, with the aid of auxiliary monthly simulations, even for several decades of a whole GCHP system, composed by the heat pump and the Borehole Heat Exchanger (BHE) field.

The code has been used to analyze the effects of the inverter and of the total length of the BHE field on the SCOP and SEER of a GCHP system designed for a residential house in Bologna with dominant heating loads. A BHE field with 3 in line boreholes has been considered, with length of each BHE either 75 m or 105 m. The results have shown that the increase of the BHE length yields a SCOP enhancement of about 7%, while the SEER almost does not change. Employing an inverter-driven heat pump instead of an on-off one can yield a SCOP increase of about 30% and a SEER enhancement of about 50%.

The obtained results demonstrate the importance of employing inverter-driven heat pumps for GCHP systems.

References

[1] Lund J W and Boyd T L 2015 Direct Utilization of Geothermal Energy 2015 Worldwide Review *Proc. of World Geothermal Congress 2015* Melbourne, 19–25 April

[2] Carlslaw H S and Jaeger J C 1959 *Conduction of Heat in Solids* (Oxford: Oxford University Press)

[3] Kavanaugh S P and Rafferty K 1997 *Ground-Source Heat Pumps: Design of Geothermal Systems for Commercial and Institutional Buildings* (Atlanta: ASHRAE)

[4] American Society of Heating, Refrigerating and Air-Conditioning Engineers 2007 *ASHRAE Handbook – HVAC Applications* (Atlanta: ASHRAE) chapter 32

[5] Claesson J and Eskilson P 1987 Conductive heat extraction by a deep borehole, analytical studies *Tech. rep. Sweden* (Lund: Lund University)

[6] Zeng H, Diao N and Fang Z 2002 A finite line-source model for boreholes in geothermal heat exchangers *Heat Transfer – Asian Research* **31** (7) 558–67

[7] Bandos T V, Montero A, Fernandez E, Santander J L G, Isidro J M, Perez J, Fernandez de Cordoba P J and Urchueguía J F 2009 Finite line-source model for borehole heat exchangers: effect of vertical temperature variations *Geothermics* **38** 263–70

[8] Lamarche L and Beauchamp B 2007 A new contribution to the finite line-source model for geothermal boreholes *Energ Buildings* **39** 188–98

[9] Fossa M 2011 A fast method for evaluating the performance of complex arrangements of borehole heat exchangers *HVAC&R Res* **17** (6) 948–58

IOP Conf. Series: Journal of Physics: Conf. Series **796** (2017) 012020 doi:10.1088/1742-6596/796/1/012020

[10] Fossa M 2011 The temperature penalty approach to the design of borehole heat exchangers for heat pump applications *Energ Buildings* **43** 1473–9

[11] Fossa M and Minchio F 2013 The effect of borefield geometry and ground thermal load profile on hourly thermal response of geothermal heat pump systems *Energy* **51** 323–9

[12] Zanchini E and Lazzari S 2013 Temperature distribution in a field of long Borehole Heat Exchangers (BHEs) subjected to a monthly averaged heat flux *Energy* **59** 570–80

[13] Zanchini E and Lazzari S 2014 New *g-functions* for the hourly simulation of double U-tube borehole heat exchanger fields *Energy* **70** 444–55

[14] European Standard EN 14825: 2013 Air conditioners, liquid chilling packages and heat pumps, with electrically driven compressors, for space heating and cooling – Testing and rating at part load conditions and calculation of seasonal performance

[15] Italian Standard UNI/TS 11300-4: 2012 Energy performance of buildings – Part 4: Renewable energy and other generation systems for space heating and domestic hot water production

[16] Preisegger E and Flohr F 2010 *VDI Heat Atlas – D4 - Properties of Industrial Heat Transfer Media – D4.1 - Refrigerants: Regulations* (Berlin: Springer Berlin Heidelberg)

CO_2 transcritical refrigeration cycles: potential for exploiting waste heat recovery with variable operating conditions

M Pieve, G Boccardi, L Saraceno, R Trinchieri and G Zummo
ENEA, Via Anguillarese, 301, 00123 Rome, ITALY

E-mail: maurizio.pieve@enea.it

Abstract. An analysis of CO_2 transcritical inverse cycles is presented, for investigating the exploitation potential of the heat rejected from the gas cooler, usually considered as a not recoverable waste heat. The cycle is analyzed in the cooling configuration during the summer season, when there are contemporarily a high temperature of the heat sink, normally over 30 °C, and a high temperature waste heat at the gas cooler, possibly exceeding 100 °C. The latter is usually rejected to the environment, but some useful applications could be analyzed, where a not negligible fraction is somehow recovered. Here a second cooling machine, thermally activated, which adds further cooling capacity to the traditional vapor compression one, is considered. A thermodynamic analysis was performed, with CO_2 evaporation temperature varying between -5 °C and 5 °C, compressor discharge pressure between 74 bar and 150 bar, and gas cooler outlet temperature variable between 30 °C and 45 °C, investigating the influence of the main thermodynamic parameters on the cycle performance by means of the software REFPROP. The depicted model allowed performing a wide sensitivity analysis of all the above mentioned thermodynamic variables, identifying the best operating conditions of the carbon dioxide transcritical inverse cycle, in terms of COP index and waste heat exploitation.

1. Introduction

The use of carbon dioxide as operating fluid in vapour compression refrigeration systems is nowadays receiving a new attention, due to its environmental impact but also due to unexpectedly good performance.

Since the second half of 19th century, such systems were used, at least until 30-40-ties of the past century, when carbon dioxide was slowly replaced by specifically designed synthetic refrigerants; an exhaustive chronicle can be found in many previous studies [1-4]

Over the last decades, the CFC firstly, then the HCFC and finally the HFC refrigerants, which were once expected to be acceptable fluids, were placed in the list of regulated substances due to their impact on climate change, so that a rising interest has been originated in environmentally safe natural working fluids, just like CO_2 is. Carbon dioxide is a very interesting alternative since it is non-flammable, non-toxic (even if it is classified as an asphyxiant gas), inexpensive and widely available, and does not affect the global environment. Carbon dioxide has GWP = 1, but the net global warming impact when used as a technical gas is zero, since the gas is a by-product from industrial production. In addition, CO_2 has excellent thermo-physical properties, leading to good heat transfer, efficient compression and compact system design due to high volumetric capacity.

In the early applications, one of the main problems of CO_2 as a refrigerant fluid was its low critical temperature, but today it can be effectively overcome by operating the system in the transcritical region, with a discharge pressure higher than the critical, which is approximately 73.8 bar with T_{cr} = 30.98 °C.

A transcritical cycle differs from a conventional phase-change vapour compression cycle as for the heat rejection phase in the high temperature cycle section, which is not operated through a condenser, being replaced by a gas cooler, where the CO_2 temperature decreases over a large temperature range. The use of a gas cooler may offer some advantages over the constant temperature vapour condensation phase [1], especially if a heat recovery has to be accounted for. Indeed, when compared with other refrigerants, the CO_2 heat rejection phase may begin at higher temperature levels, possibly exceeding 100 °C. This paper deals with the feasibility of using such waste heat as powering source for a thermally activated cooling machine, as an absorption system, which adds further cooling capacity to the basic cooling cycle.

2. Basic cycle description

A basic cycle configuration is assumed, shown in Figure 1, without any layout complexity, just to fix a reference line which can be rested on for the analysis of more complex cycle arrangements. The analysed cycle is a transcritical CO_2 inverse cycle, thus with a subcritical low pressure and a supercritical high pressure. The evaporating carbon dioxide removes heat from the low temperature heat source, then, at the point 1 (complete vaporization) it is compressed up to the high pressure level, point 2, where the highest pressure and temperature are reached within the cycle. After the compression phase, it enters the gas cooler or any other system of waste heat exploitation, in an ideally isobaric process. The temperature at the exit of gas cooler, point 3, may be as low as the external environment cold sink enables. Then, by entering the throttling valve, the carbon dioxide pressure is reduced down to the low pressure level, where the point 4 states the inlet into the two-phase region, the liquid and the gaseous phases of CO_2 coexisting. The two-phase mixture, point 5, enters the evaporator, being vaporized until the initial conditions.

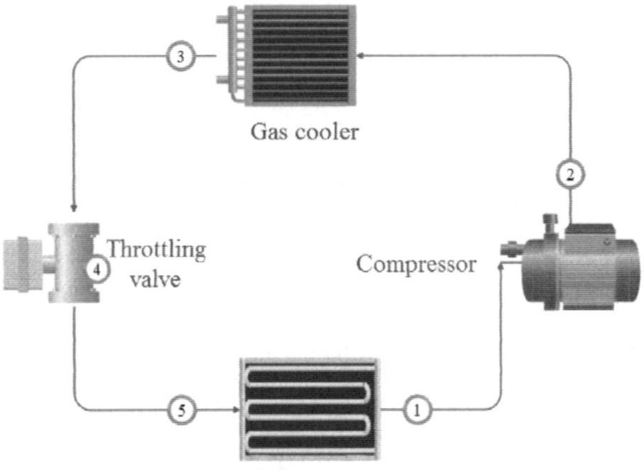

Figure 1. Basic layout of a transcritical CO_2 inverse cycle.

Starting from this basic layout configuration, some usual improvements of the cycle may concern the followings, although they will not be argument of the present study: a double stage compression phase, whose aim is to reduce the compressor power requirement, since the compression curve moves to near isothermal, reducing the power requirement too; expansion devices other than the throttle valve, as for example an ejector or a turbine, which would be able to recover some mechanical work from the expansion of CO_2; the internal heat regeneration, consisting of the heat transfer from the gas cooler outlet stream to the evaporated CO_2 before the latter enters the compressor, so that the quality at the evaporator inlet is lower and the cooling capacity is increased. The last issue would mean also a certain superheat degree at the compressor inlet, which would protect the compressor from unexpected and damaging drops last inside.

Figure 2 shows typical T-s and p-h diagrams for the examined cycle. Some assumptions are made for the analysis: each component is considered in a steady state condition, the kinetic and potential energy variations through it are neglected, as well as the heat and friction losses. As for the compressor isentropic efficiency, its value is set as constant and equal to 0.7, this being the most suitable from a technological point of view [5]. The throttling valve is considered as an isenthalpic device and the inlet conditions at the compressor are those of saturated vapour. Finally, the properties of CO_2 are obtained from REFPROP-NIST software, Ver. 9.0 [6]

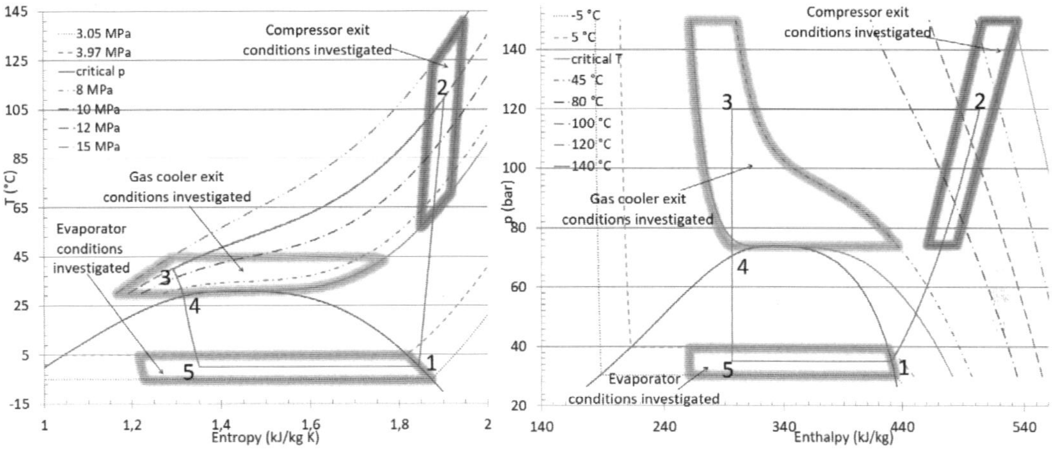

Figure 2. T-s and p-h reference diagrams for a transcritical CO2 inverse cycle.

3. Cycle energy analysis

The Thermodynamics laws can be applied for each component of the simplified cycle, by means of the assumptions previously stated. Each component is considered as a control volume, to which the proper equations apply. This study would investigate, among the others, the effect of some thermodynamic parameters on the COP index, the coefficient of performance of the cycle. Its definition depends on the useful effect the cycle operates for; if a cooling machine is the aim, the COP index means the cooling capacity over the power requirement, i.e.

$$\mathrm{COP} = \frac{\dot{q}_{cool}}{\dot{w}_{net}} \tag{1}$$

where $\dot{q}_{cool}$ is the cooling capacity and $\dot{w}_{net}$ is the mechanical power (usually provided as electric power) required to raise the pressure from the low level to the high value. In the simplest case, referring to the Figure 2

$$\dot{w}_{net} = \dot{w}_{compr} = \dot{m}(h_2 - h_1) \tag{2}$$

with $\dot{m}$ the mass flow rate of CO_2 in the cycle loop and h the specific enthalpy of the fluid. The compressor isentropic efficiency $\eta_{c,is}$ has to be accounted for, which is defined as the ratio of the power required in an ideal reversible (adiabatic and isentropic) compression process to the actual power requirement of the component, as

$$\eta_{c,is} = \dot{w}/\dot{w}_{c,is} \tag{3}$$

where $\dot{w}_{c,is}$ is the power related to the ideal isentropic compression.

If an additional useful effect is considered, besides the cooling capacity, that is the heat recovery potential at the cycle high temperature level, a different index should be used, as some authors proposed [7-13], that is a system COP, named COP_s, defined as the sum of the cooling capacity and the heating capacity over the power requirement:

$$COP_s = \frac{\dot{q}_{cool} + \dot{q}_{heat}}{\dot{W}_{net}} \quad (4)$$

where $\dot{q}_{heat}$ is the sensible heat available at the gas cooler. If all the available heat is recovered, then

$$\dot{q}_{heat} = \dot{m}(h_2 - h_3) \quad (5)$$

In the developed model, a set of equations is to be solved, which links all the thermodynamic states of the cycle. Once the high and the low pressures in the cycle are fixed, as well as the outlet temperature of the gas-cooler, the solving sequence is well defined, if the isentropic efficiency of the compressor is assumed. Table 1 depicts such solving sequence, starting from the compressor inlet conditions, point 1.

Table 1. Set of equations describing the thermodynamic states of the transcritical inverse CO_2 cycle.

State n.	Known variables	Link equation with the other states	Variables to be determined	Notes
1 – compressor inlet	p_1, T_1	$T_1 = T_{sat}(p_1)$	$h_1 = h_1(p_1)$ $s_1 = s_1(p_1)$	
2is – compressor outlet	p_2, $s_{2,is}$	$s_{2,is} = s_1$	$h_{2,is} = h_{2,is}(p_2, s_{2,is})$	Ideal state
2 – compressor outlet	p_2, $\eta_{is,c}$	$h_2 = h_1 + (h_{2,is} - h_1)/\eta_{is,c}$	$T_2 = T_2(p_2, h_2)$ $s_2 = s_2(p_2, h_2)$	Actual state
3 – gas cooler outlet	p_2, T_3	$p_3 = p_2$	$h_3 = h_3(p_3, T_3)$	
4 – two-phase region inlet	h_4	$h_4 = h_3$		
5 – evaporator inlet	p_1, T_1	$h_5 = h_4$	$h_{1,l} = h_{1,l}(p_1)$ $x_5 = (h_5 - h_{1,l})/(h_1 - h_{1,l})$	

A specific application is simulated, that is a cooling machine operating during the summer season, when there are contemporarily a high temperature level for the environmental heat sink, normally over 30 °C, and a high temperature waste heat at the gas cooler, which possibly exceeds 100 °C.

The influence of the following parameters is investigated: the temperature range for the evaporator is set between -5 °C and 5 °C, this being suited for the cooling effect in many applications for environmental air conditioning. As for the discharge pressure at the compressor outlet, its lower limit is set at 74 bar, just above the critical value, and the maximum considered is 150 bar, being currently a very high value from a technological point of view. Finally, the CO_2 temperature at the gas cooler outlet is assumed to vary from 30 °C up to 45 °C, depending of the external sink temperature and the heat exchanger effectiveness; the latter one was not analysed here more in depth. The analysis was performed with no superheat at the compressor inlet.

When dealing with the theoretical COP_s, ideally assuming that all the rejected heat is recovered, the model shows the existence of a maximum, depending on rejection pressure p_{gc} and the gas cooler outlet temperature $T_{out,gc}$; in this regard, an original view can be shown by plotting the COP_s index as iso-curves in a ($T_{out,gc}$, p_{gc}) graph for a fixed evaporator temperature. Figure 3 depicts COP_s ($T_{out,gc}$, p_{gc}) in the ranges investigated by the present study when the evaporator temperature is fixed at 5 °C and the compressor isentropic efficiency at 0.7. Such a diagram shows very concisely the influence of the two variables so that for each couple of values the COP_s is quickly determined. In addition, the iso-COP_s curves allow to identify where the COP_s may increase or decrease faster than elsewhere. In this regard, it can be noted that the highest COP_s values will be get with low pressures at the gas cooler, indeed a small increase of the gas cooler outlet temperature is more detrimental at lower gas cooler pressure than at high values, since a sharply reduction does occur. In other terms, when the iso-COP_s curves are very close each other with respect to a direction, it means a high slope of the COP_s value in that direction. A dividing line of the region where the COP_s slope is very high can be posed as the dotted line in the diagram. It is calculated as the maximum gas cooler outlet temperature which allows a fixed COP_s to be obtained, corresponding also to the optimum gas cooler pressure for a prefixed gas cooler outlet temperature and to the maximum COP_s achievable.

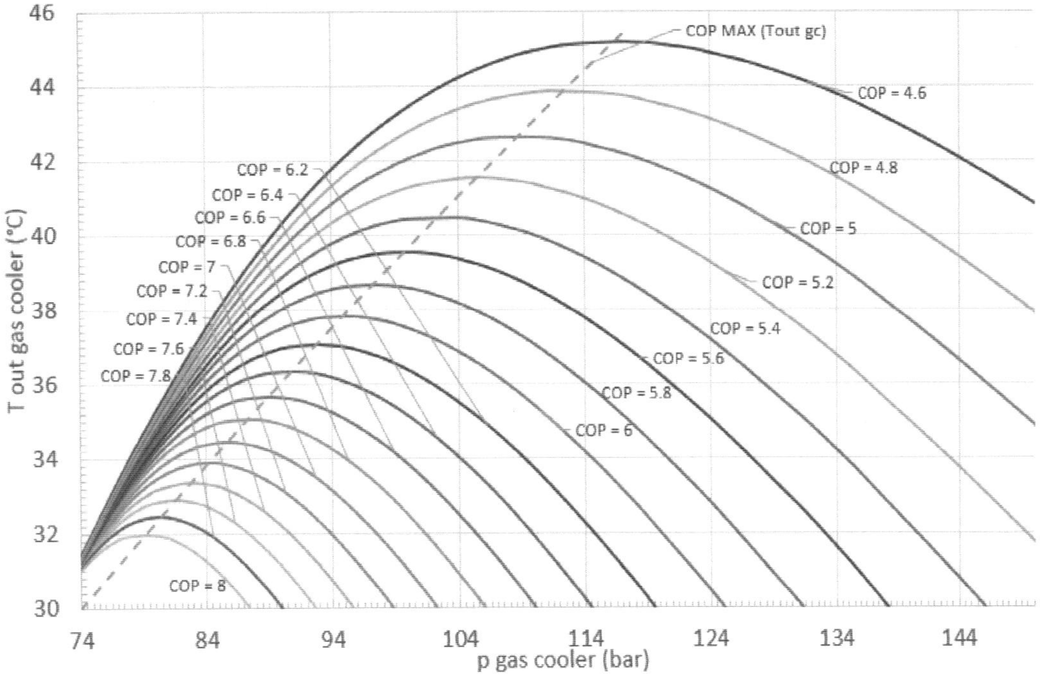

Figure 3. Iso-COP_s (system COP) diagram plot with evaporator temperature at 5 °C and $\eta_{is,c} = 0.7$.

Figure 4 shows the same quantity for an evaporator temperature of -5 °C. As it is fairly predictable, the COP_s is generally lower than that previously shown. For $p_{gc} > 85$ bar, the COP_s reduction is between 18% and 29%. Furthermore, for a fixed gas cooler outlet temperature the maximum COP_s is obtained at higher pressures when the evaporator temperature decreases from 5 °C to -5 °C.

Referring to the results shown in [7], they positively agree with the Figures 3 and 4, reminding the different assumptions as for the internal heat regeneration, supposed with a heat exchanger effectiveness of 60% in [7], while not considered at all in this study.

In addition to the system COP, another important parameter to be considered is the specific cooling and heating capacity, that is the sum of the enthalpy differences across the evaporator and the gas cooler. It is related to the size of the machines, since the cooling and heating capacity are given by the product of the mass flow rate and the corresponding unitary effects. In other terms, if a machine is considered with prefixed cooling and heating loads, a higher value of the total available enthalpy would mean a lower mass flow rate and consequently a lower size of the components. Figure 5 shows as the available total specific enthalpy depends on the gas cooler pressure and the gas cooler outlet temperature.

The total specific capacity increases for raising pressure and lowering gas cooler outlet temperature, however their influence is not linear. Indeed, in the left side of the diagram, a particular behavior of the curves may be noticed, which could be called a knee-curve, where a very steep increase of the available enthalpy occurs, more pronounced with pressure and temperature approaching the critical values. Such a trend, if a stable operation of the machine is the aim, would makes advisable to operate at pressure higher than the knee, so that an unexpected reduction of the pressure or an increase of the temperature can be easily managed without excessively onerous mass flow rate adjustments.

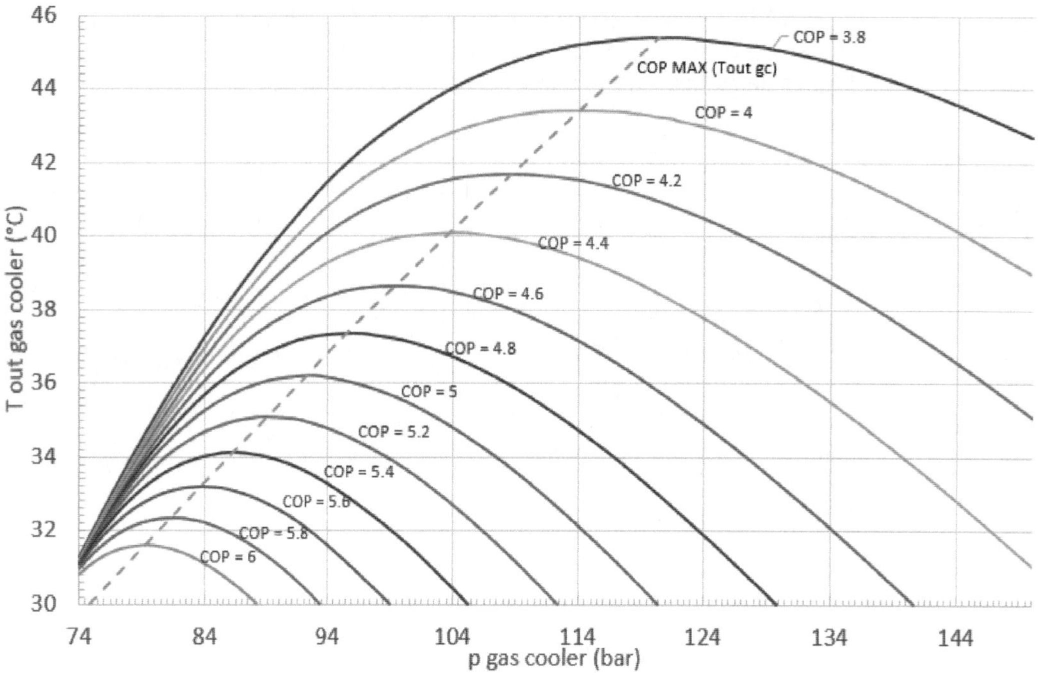

Figure 4. Iso-COP$_s$ (system COP) diagram plot with evaporator temperature at -5 °C and $\eta_{is,c}$ = 0.7.

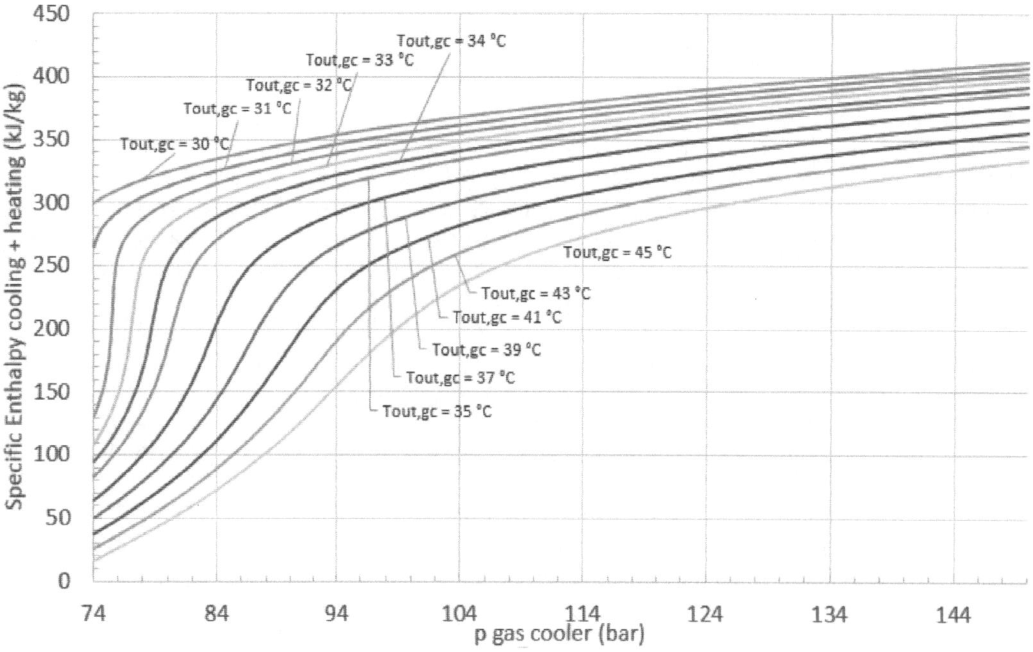

Figure 5. Available total specific enthalpy with T eva = 5 °C and $\eta_{is,c}$ = 0.7.

If the evaporation temperature is fixed at -5 °C, a similar behavior will be noted, shown in Figure 6, with a generalized increase of the available total specific enthalpy, compared with the higher evaporation temperature level. In both cases, at temperatures equal or lower than the critical (30 °C and 31 °C respectively, in the plot), the curves do not undergo such particular feature.

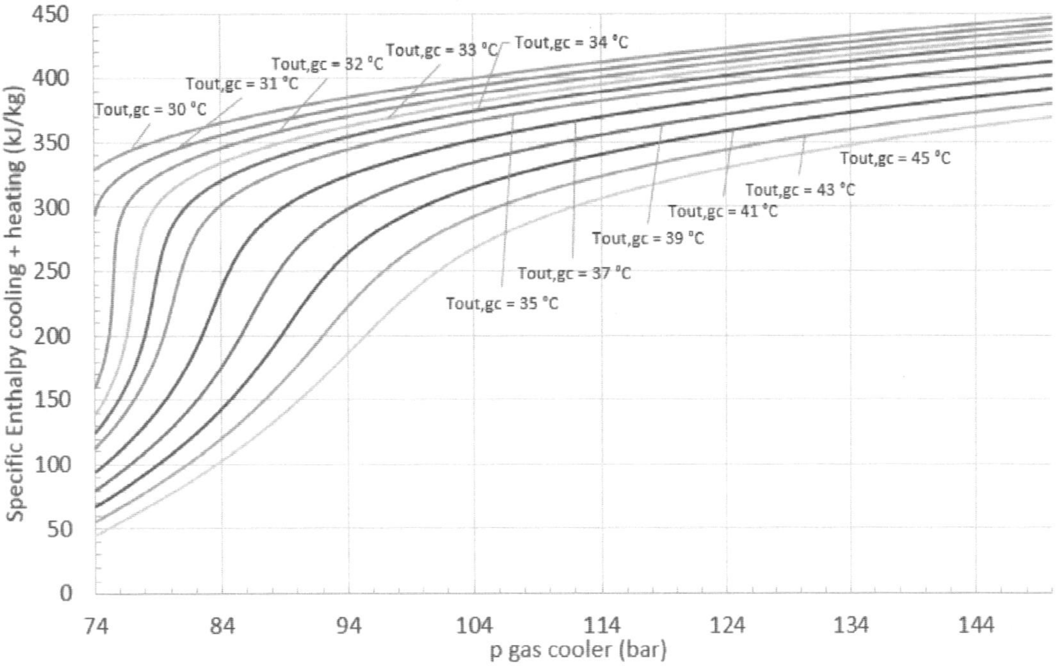

Figure 6. Available total specific enthalpy with T eva = -5 °C and.$\eta_{is,c}$ = 0.7.

4. Additional cooling with absorption chiller coupling

The feasibility of an absorption chiller coupled to the main cooling machine, driven by the exhaust heat rejected at the gas cooler, may be analyzed with some simplifying and conservative assumptions, given the preliminary nature of the present study: the exhaust CO_2 stream is exploited down to 80 °C; the absorption chiller operates with a single effect mode; the COP of the absorption machine, defined as the cooling capacity over the heat power input, is set equal to a mean constant value of 0.6, as if it was independent from the feeding fluid temperature. The performance of such a combined system is evaluated by means of the percentage of cooling COP increase compared with the basic cycle, defined as the total cooling capacity (basic plus absorption cycle) over the compressor power required.

As Figure 7 shows, the additional cooling capacity provides a net positive contribution, which may be very attractive at low evaporation temperatures combined with a high temperature of the heat sink. Quantitatively, if the final heat rejection temperature is 45 °C and the evaporation temperature is -5 °C, a cooling COP increase greater than 30% may be accomplished. Furthermore, a specific trend is shown in the range between 82 and 90 bar, where the cooling COP increase reaches its maximum slope, so that a 20% COP increase can be obtained just at a gas cooler pressure equal to 88 bar. As for a better understanding of the exploitable potential, in the diagram the two investigated evaporation temperatures are depicted, along with the compressor outlet temperatures.

5. Heat recovery specific issues

Whichever the heat recovery system is, an essential tool to understand the relationship between the available heat and its exploitation potential is the T-h diagram, describing the heat recovery profiles.

Figure 8 shows the diagram for an evaporator temperature set at 5 °C and a compressor isentropic efficiency set at 0.7. When the abscissa is 0 kJ/kg, the temperature value indicates the CO_2 temperature at the compressor outlet, so that the influence of the evaporator temperature and the isentropic efficiency are already taken into account. Based on this issue, a minimum value of the high pressure can be evaluated, depending on the heat recovery application. For an absorption cycle heat recovery, as it is the present case, 80 °C is set as the minimum useful value for the heat source at the chiller generator, so that with these cycle parameters the high pressure should be set at least at about 96 bar. When the evaporation temperature is set at the lowest value among those investigated, i.e. -5 °C, better conditions result for the heat recovery capability, as Figure 9 shows, both for the enthalpy difference and for the temperature levels, which are higher. In this case, the minimum value for the high pressure should be fixed at about 82 bar.

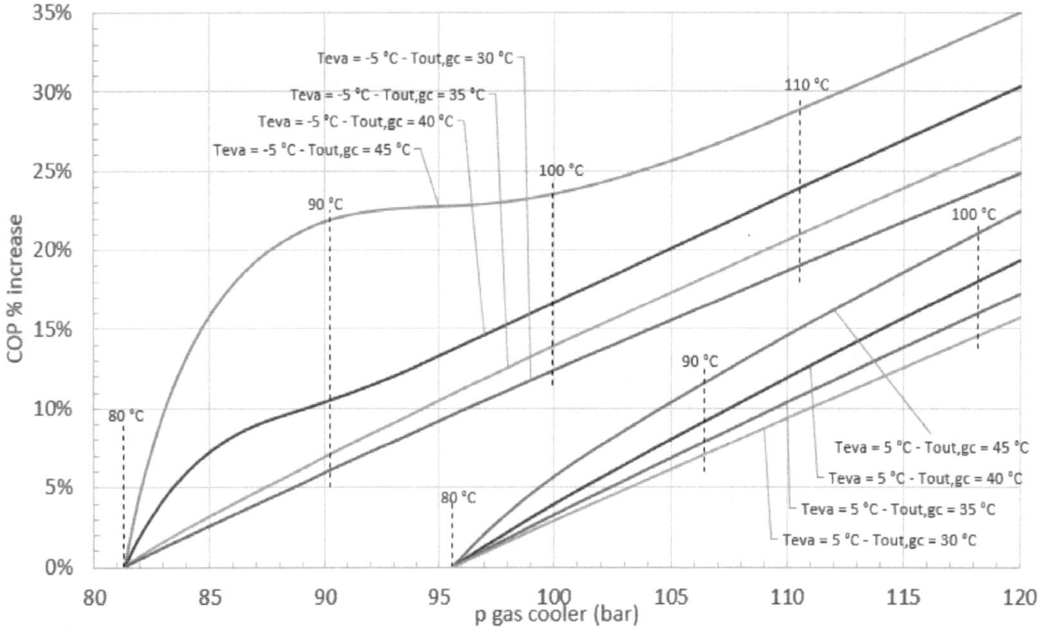

Figure 7. Cooling COP increase for a chiller absorption cycle integration; T out from chiller = 80 °C.

A fundamental issue when dealing with the heat recovery concerns the slope of the depicted curves, which is related to the inverse of the specific heat Cp. When a sharp increase occurs of Cp, especially for pressures close to the critical value, the fluid temperature tends to vary more and more slowly, with noticeable implications as for the heat transfer. The temperature profile of the fluid appointed to the heat recovery will be constrained by the CO_2 one, just as the Second Thermodynamics law states. If the overall waste heat is to be recovered, thus down to the environment temperature, and a rather constant specific heat fluid is used, as the water, its T-h profile would be a straight line that is able to successfully match the CO_2 curves only at the highest pressures. As a lower CO_2 pressure is considered, its behavior deviates more and more from the constant slope trend, so that a higher irreversibility would take place in the heat transfer because of the pinch-point requirements. A viable further option would be to consider many separated system, acting as multiple and variable heat sinks over partial temperature ranges, thus narrower than the entire gas cooler one [14-16]. Such a system would involve a better matching between the cooling and the heating streams even at low gas cooler pressures. On the other hand, it would mean an additional layout complexity, to be carefully evaluated.

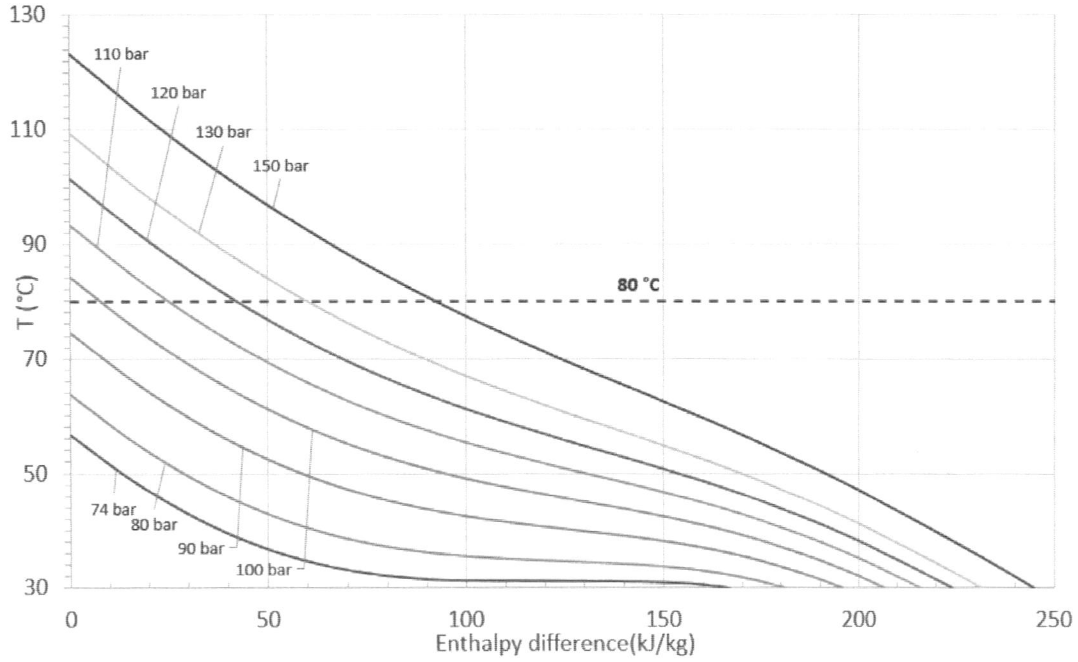

Figure 8. T-h diagram: heat recovery profile when T eva = 5 °C and $\eta_{is,c}$ = 0.7.

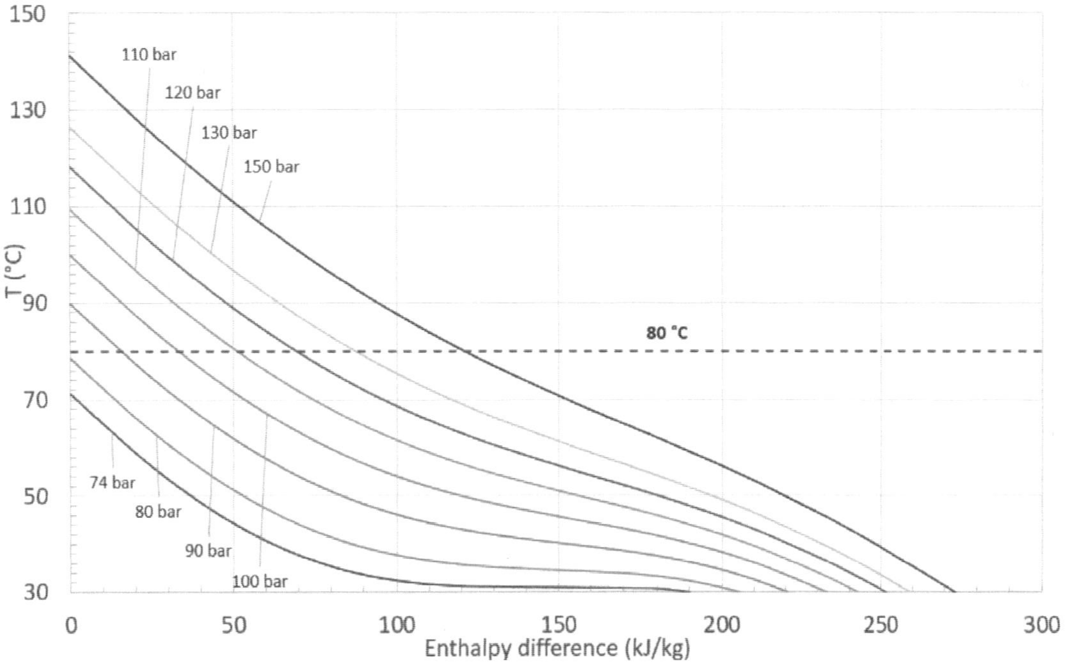

Figure 9. T-h diagram: heat recovery profile when T eva = -5 °C and $\eta_{is,c}$ = 0.7.

6. Conclusions

A preliminary study has been conducted on a basic CO_2 transcritical inverse cycle, in order to evaluate the feasibility of the heat recovery at the gas cooler, where a relatively high-temperature waste heat is available, which usually is rejected to the environment. A general analysis was considered with some reasonable constraints, supposing a cooling machine is operating during summer season for conditioning service, when the environmental heat sink exhibits the worst conditions, assumed between 30 °C and 45 °C. The evaporation temperature was set between -5 °C and 5 °C and the high pressure was varied between 74 bar and 150 bar. As for the theoretically recoverable heat the system COP was calculated, which factors in both the useful effects, cooling and heating. A specific application for the waste heat exploitation was analyzed, that is an absorption chiller machine coupled with the main cycle, so that the latter provides the feeding heat to the former. The total cooling COP, defined as the useful cooling capacity provided by the basic cycle along with the absorption one, has been shown to be capable of exceeding the basic cooling COP alone by more than 30%, being very attractive for the highest temperature values of the heat sink and the lowest evaporation temperatures. Finally, the fundamental key points of the heat transfer have been highlighted related to the influence on the heat recovery capability, namely the T-h profiles matching of the transcritical CO_2 cooling stream to any fluid intended for the heat recovery, and a viable solution has been conceptually outlined.

Acknowledgements

This research was supported and funded by the Electric System Research Program by means of the Program Agreement between ENEA and MISE (Italian Ministry of Economic Development), related to the first Realization Annual Plan (RdS PAR 2015-16).

References

[1] Kim M H, Pettersen J and Bullard C W 2004 *Progr. Energ. Combust. Sci.* **30 (2)** 119-74
[2] Pearson A 2005 *Int. J. Refrig.* **28** 1140-8
[3] Lorentzen G 1994 *Int. J. Refrig.* **17** 292-300
[4] Nekså P, Walnum H T and Hafner A 2010 *Proc. 9th IIR Gustav Lorentzen Conference*, Sydney, 646-59
[5] Dorin SpA 2015, *Dorin Software 2015 V 15.07*
[6] Lemmon E W, Huber M L and McLinden M O 2010 NIST Standard Reference Database 23, Version 9.0
[7] Sarkar J, Bhattacharyya S and Ram Gopal M 2004 *Int. J. Refrig.* **27** 830-8
[8] Sarkar J, Bhattacharyya S and Ram Gopal M 2006 *Int. J. Refrig.* **29** 735-43
[9] Tambovtsev A, Olsommer B and Finckh O 2011 *Proc. 23rd IIR International Congress of Refrigeration*, Prague, 1964-72
[10] Adriansyah W 2004 *En. and Build.* **36** 690-5
[11] Reinholdt L and Madsen C 2010 *Proc. 9th IIR Gustav Lorentzen Conference*, Sydney, 604-12
[12] Funder-Kristensen T, Fösel G and Bjerg P 2013, *5th International Conference on Cryogenics and Refrigeration*, Hangzhou
[13] Colombo I, Maidment G G, Chaer I and Missenden J M 2012 *Int. J. Low C. Tech.* **9** 38-44
[14] Zsebik A, Balikó S and Csata Z 2014 *En. Eng.* **111 (3)** 41-56
[15] Stene J 2005 *Int. J. Refrig.* **28** 1259-65
[16] Heinz A, Martin K, Rieberer R and Kotenko O 2010 *Proc. 9th IIR Gustav Lorentzen Conference*, Sydney, 367-76

Experimental validation of CFD model of thermal fluxes through a multilayer wall

N. Bianco, M. Musto[1] and G. Rotondo

Dipartimento di Ingegneria Industriale, Università degli studi di Napoli Federico II, P.leTecchio 80, 80125, Napoli, Italy

[1]E-mail: marilena.musto@unina.it

Abstract. The evaluation of thermal losses through building envelope is complex due to the presence of different components such as geometrical and structural thermal bridges, multilayer walls, windows and shadow areas. In particular, in presence of material and/or shape discontinuities, the heat flux becomes two-dimensional or three-dimensional and loses its one-dimensionality. Thermal bridges are weak points of the building envelope; the measurement of heat losses through the walls is quite complicated and in particular the detection of bi-tridimensional thermal flux. The integration between the cfd analysis and the experimental study aims the improvement of the measurement technique and the assessment of the dispersion of heat fluxes through multi-layer walls both in the steady and unsteady measurement conditions. The numerical study has been directed for modelling, by means of the commercial numerical code Fluent, of the experimental test section consisting of a multi-layer wall, in which a structural thermal bridge has been inserted in order to make accurate measurements of the three-dimensional heat flux in steady state condition. The study has been developed to validate the CFD model by the experimental set-up, with the aim to investigate possible measurement methodologies of heat fluxes.

1. Introduction

A European directive estimated that buildings are responsible for the 40% of the total energy consumption of the European states [1]. For this reason, the reduction of the heat losses trough the buildings plays an important role on an energy saving policy and for the achievement of the reduction of 20% of energy requirement within 2020; the directive 2012/27/UE [2] establishes a series of objectives to reach. In the past few decades, the growing interest to reduce the energy used in the building sector has led to walls characterized by appropriate thermal properties [3]. In fact, an appropriate construction technique and an appropriate selection of the materials for the walls, can reduce the energy requirement for heating and cooling. The thermal properties of the walls can be evaluated by theoretical models or by stratigraphy. But the experimental values measured in-situ can be very different from those predicted [4, 5]. The discrepancies may be due to imperfection during the construction phase, to aging, to moisture. So, it's necessary to measure the characteristics of the walls in-situ during the normal exercise.

The presence of different building components, such as geometrical and structural thermal bridges, multilayer walls, windows and shadow areas can take to a result not always in compliance to the actual case. In particular, in presence of material and/or shape discontinuities, the heat flux becomes two-dimensional or three-dimensional and loses its one-dimensionality. Thermal bridges are weak points of the building envelope; the measurement of heat losses interesting the walls is quite complicated and in particular the detection of thermal flows bi-tridimensional; they present the greatest analysis complexities. In fact, the thermal bridges determinate heat losses higher than those characterizing a common dispersing surface (i.e., walls without discontinuities). This phenomenon induces uncontrolled thermal losses and hygiene problems, connected to the possible vapor condensation and mold growth. The powerful programs for energy audit of buildings - as DOE-2 [6] and EnergyPlus [7] - do not provide an accurate assessment of thermal bridges and still search and optimization problems require more precise calculation methods. The scientific literature proposes different methods for the analysis of thermal bridges, according to both approaches statistical and numerical. Ceylan and Meyers [8] and Ouyang and Haghighat [9] have implemented the use of Conduction Transfer Function (CTF) for the solution of problems of transient heat transfer. They proposed a new approach to calculate the thermal response factors of a multi-layer wall without finding the roots through calculation of the state transition matrix of a system. Even Stephenson and Mitalas [10] and later Hittle [11] and Hittle and Bishop [12] studied heat transfer through the multi-layer plates. Wang and Chen [13] proposed a stable series expansion approach based on the Routh stability theory declaring that the calculation of TRFs may not be convergent with increase of the terms in expansion polynomials and provided a detailed analysis of the problem.

Afterwards, Renon [14] upgraded the methods developed by Seem [15] and Seem et al. [16], by modelling thermal bridges and showing techniques suitable for transferring the methodology into EnergyPlus [7]. Deque et al. [17, 18] proposed a numerical methodology aimed to characterize thermal bridges, by means of the state-space representation, while Ben Larbi [19] suggested a statistical model for the evaluation of the equivalent thermal transmittance of linear thermal bridges with a development of a library using common geometries. Gao et al. [20] proposed a reduction model of the building heating load, including also the thermal extra-flows due to thermal bridges; their methodology has been, later on, implemented in TRNSYS [21]. Ascione et al. [22], presented a comparison between the results obtained by means of the simplified 1-D models and those obtained by more sophisticated models 2-D or 3-D, in order to point out differences in terms of equivalent conductivity and thermal transmittance, showing that a proper modelling is necessary. In particular, an over estimation of the heat losses, determined by an approximate evaluation, induces higher cost of refurbishing, higher cooling energy requests in summer, minor thermal comfort in naturally ventilated buildings. Asdrubali et al. [23] proposed a methodology to perform a quantitative analysis of some types of thermal bridges, through simple thermography surveys and subsequent analytical processing. From the simple measurement of the air temperature and the analysis of the thermogram, the thermal bridge effect can be estimated as a percentage increase of the homogenous wall thermal transmittance. This term is obtained without further information on the structure of both the thermal bridge and the stratigraphy of the wall. Capozzoli et al. [24] believe that it is necessary to identify an accurate numerical method in order to appreciate the benefits of a proper design. A sensitivity analysis based on an extensive study of the linear thermal transmittance value of many types of thermal bridge, based on the methodology specified in EN ISO 10211, has been carried out. After defining the input design variables and considering a range of variation for each of them for the linear thermal transmittance evaluation, a non-linear regression model has been specifically developed for each analyzed thermal bridge, considering the output values of a numerical code as data set. A method has been performed, based on the obtained results, in order to assess the contribution of each input design variable to the deviation of the linear thermal transmittance for each kind of thermal bridge. Ascione et al. [25] investigated a typical L-shaped thermal bridge and the numerical methodology – compared to the experimental study – reveals very satisfactory results in reliability and accuracy. The percentage gap between calculated and measured heat fluxes varies between −12% and +6%, with an average value close to zero. A satisfactory agreement has been evaluated also with reference to the heat rate through the investigated envelope element.

The integration between the numerical-experimental researches aims the improvement of the measurement technique through knowledge and assessment of the dispersion of heat fluxes in multi-layer wall in the three-dimensional field, both in the steady and unsteady measurement conditions. The study here proposed is aimed to validate the developed CFD model by means of comparisons with experimental measures with the aim of improving methodologies for heat fluxes measurement. The numerical study has been directed to modeling, by means of the commercial numeric code Fluent, an experimental test section consisting of multi-layer wall, in which a structural thermal bridge has been entered in order to make accurate measurements of the three-dimensional heat flux under steady state condition. Globally, the carried out comparisons get very satisfactory results, demonstrating reliability and accuracy of the developed CFD model.

2. Experimental test section

A model of a multilayer wall, composed of two plasterboard panels interleaved by an I-shaped thermal bridge, has been built in order to analyze thermal behavior of the multilayer wall under radiative and convective fluxes. In Figure 1, layout of wall stratigraphy is presented. In particular the stratigraphy is composed of: plasterboard, iron I-shaped thermal bridge and premixed plaster in the middle section and again a plasterboard panel. One of a plasterboard panel was placed in front of an hot panel used as heat source. Dimensions and thermal characteristics are summarized in Table 1. The thermal flux, provided by an electrical resistive panel, will be described with more details in the section 2.2.

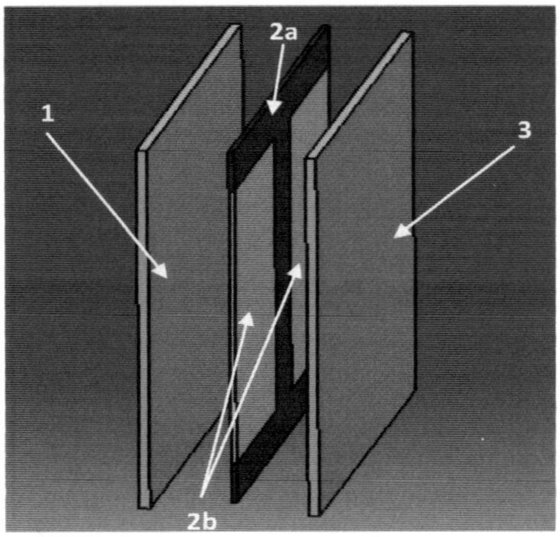

Figure 1. Layout of wall stratigraphy: experimental section

Table 1. Stratigraphy of multi-layer wall: test section

		Dimensions			Thermal characteristics	
	layer materials	height/m	length/m	thickness/m	thermal conductivity/ (W/m K)	c/(J/kg K)
1st layer	Plasterboard	0.490	0.450	0.010	0.500	1138
	Iron thermal bridge (2a)	0.470	0.430	0.005	16.270	502
2nd layer	Premixed plaster x 2 (2b)	0.380	0.190	0.005	0.032	1138
3rd layer	Plasterboard	0.490	0.450	0.010	0.500	1138

2.1. Sensor apparatus

The thin film heat flux sensor used in the experimental tests represents the first stage of the measurement chain. It detects the heat flux that invests a sensible surface by radiation and convection mechanisms. The measurement principle of heat flux sensors is based on measuring the temperature difference across the thermal resistance when a heat flux passes through. Therefore, this kind of sensor in output provides both heat flux and temperature signals. Table 2 summarizes the physical and metrological characteristics of the thin film heat flux sensors attached to the wall and connected to the data logger. The sensors were glued to the surface wall using thermally conductive paste or a thermal pad (100x100 mm^2, 0.5 width and conductivity equal to 5 W(mK)$^{-1}$), in order to ensure a good thermal contact. A better estimation of heat flux can be achieved acquiring a larger number of punctual signals by heat flux sensors, on the examined surface. In order to complete the monitoring campaign, the internal and external air temperatures were measured with a sampling interval equal to 60 s by two PT100 temperature probes, whose metrological characteristics are summarized in Table 3. As shown in Figure 2, they were used 18 sensors, which measure flux and/or temperature.

Table 2. Heat flux sensors: physical and metrological characteristics

Heat flux sensors	
Dimensions	35.1x28.5 mm
Nominal resistance	175 Ω
Nominal sensitivity	2.0 μVW^{-1}m^2
Response time to a step function (63%)	0.60 s
Upper temperature limit	150 °C
Relative uncertainty	±5%
Repeatability	0.6%

Table 3. Temperature probe: metrological characteristics

Sensor type RTD Pt100, Class 1/3B (DIN/IEC751)	
Dimensions	Ø20 x 3 mm
Response time	8 s
Operating range	-50 °C to 125 °C
Uncertainty	± 0.10 °C (at T = 20 °C)

Figure 2. Test section: heat flux sensors attached to external (on the left) and internal (on the right) surfaces

The heat flux sensors are indicated with the numbers from 1 to 9 for the internal surface and from 10 to 18 for the external surface of the test section. Ten k-thermocouple sensors, connected to a data

IOP Conf. Series: Journal of Physics: Conf. Series **796** (2017) 012022 doi:10.1088/1742-6596/796/1/012022

logger, were glued to the same position of the thin film heat flux sensors with the aim to acquire the temperature signal in the same time to obtain better measurement accuracy.

2.2. Heat flux supply

The heat flux supply was composed by the following components (see Figure 3):

- radiative heat source panel;
- multimeter;
- power supply;
- reference resistance (shunt).

In particular, the heat source consists of an upper surface made of a thin copper layer that is able to conduct electric current well, instead the bottom and lateral surfaces are insulated by fiberglass. The copper layer was grooved using a drill press to obtain a series of separated electrical resistors linked in series by electrical bridges. It also was added a transparent plastic film on top of the copper layer to achieve electrical insulation.

Figure 3. Heat flux source (left side); particular of resistance connections (right side)

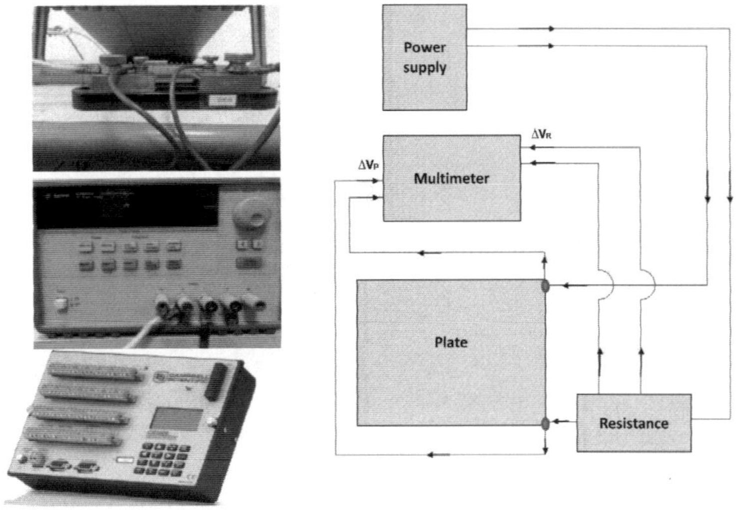

Figure 4. Electrical power supply and data acquisition system for heat flux source panel. Left side: reference electrical resistance, multimeter and data logger. Right side: electrical scheme

This resistance (Figure 3) is linked with an electrical supply (with appropriate safety fuse 5 A) and one of the two power cables (negative pole) is connected in series with a standard reference electric resistance (shunt): $R_C = 0.00066\ \Omega$.

Across the shunt, it was possible to measure the voltage and, consequently, the current circulating in the resistance that is the same circulating in the copper resistance of the heat source plate.
This allows to evaluate the heat flux provided by the heat source panel once measured the voltage difference across the heated panel.

Before every experimental tests run, verification of heat flux provided by source panel was carried out. In Table 4 the comparison between theoretical and experimental results is reported.

Table 4. Validation of heat source panel

Voltage	Current	Power	Theoretical Heat flux	Measured Heat	Percentage
10 V	1.10 A	11 W	60.44 W/m^2	54.85 W/m^2	9.24%
15 V	1.65 A	24.75 W	135.98 W/m^2	123.81 W/m^2	8.95%
20 V	2.20 A	44 W	241.75 W/m^2	219.74 W/m^2	9.10%
		$P_{\text{power supply}} = V \times I$	$\Phi_{board} = \frac{P_{power\ supply}}{A_{board}}$		

2.3. *Experimental test conditions*

The test conditions and the experimental set up are reported in Table 5 and Figure 5, respectively. In particular, in the left side a lateral view of experimental setup can be observed, instead the right side shows a sketch of experimental apparatus with the reference Cartesian axes. The distance between wall and heat source panel is equal to 10.0 cm.

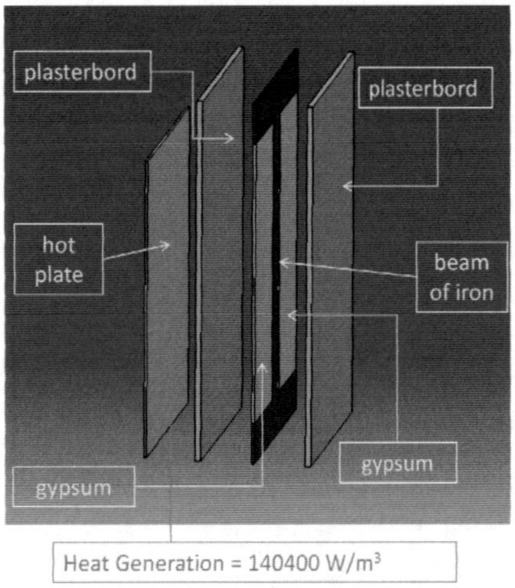

Figure 5. Layout (left side) and experimental apparatus (right side)

Table 5. Test conditions

Measurement duration	60 min
Scan interval (acquisition time of each sample)	5 s
Sampling interval	60 s
Environment temperature	26 °C
Voltage imposed to generate Heat flux	10 V

3. CFD analysis: numerical simulation

In this section, a CFD numerical simulation of the experimental apparatus is presented. The Figure 6 shows the CFD domain. Dimensions and material properties of the wall stratigraphy are the same used in the experimental apparatus (Table 1).

To obtain the temperature and velocity fields, the three-dimensional Navier Stokes equations (Fluent user's guide, 2005) combined with k–ε standard turbulent model in the steady state conditions were numerically solved. To take into account the radiative contribute, the DO model was used. The mesh was realized in Gambit ambient. Due to the mesh size should be neither too sparse, because it would lead to an inefficient analysis of the model, not too dense, because it would require excessive computational time, a sensitivity analysis of the solution on the mesh sizes was carried out. In the computational domain there are areas in which it is interested in having more detailed information, such as area in which there are convective flows. In particular, these convective flows can arise inside the air zone between the copper board and the hot side of the box. In these zones, a boundary layers mesh with the following characteristics was adopted:

- First row (a) = 0.001 m;
- Growth factor (b/a) = 1.3;
- Rows = 7.

The mesh size values in X and Y directions are both equal to 0.01 m, instead in Z direction, it has been adopted different interval spacing, depending on each volume (see Table 6). In Table 7, operative and boundary conditions are summarized.

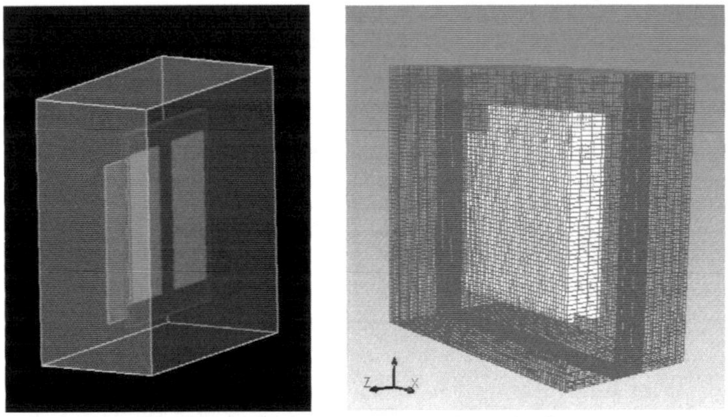

Figure 6. CFD domain and volume discretization

Table 6. Mesh size

Volume Zone	Cell size in Z direction/m
Hot Plate	0.001
Box-Hot (volume between hot plate and wall surface)	0.002
Thermal bridge	0.001
Gypsum	0.001
Air	0.030

IOP Conf. Series: Journal of Physics: Conf. Series **796** (2017) 012022 doi:10.1088/1742-6596/796/1/012022

Table 7. Operative and boundary conditions

Operative Conditions	Pressure=101325 Pa, Temperature=26 °C, g=9.81 m/s
Models	k-ε model, DO (20,20,10,10) radiative model
Materials	Solids: iron, copper, premixed, plasterboard, air
Boundary Conditions	Heat generation imposed on the hot plate: 140400 W/m^3, pressure inlet and outlet conditions imposed on boundary surface of domain

4. Results and Discussion

In this section, both experimental and numerical results are presented and compared to validate the CFD model proposed. The results are given in terms of field and profiles of the total heat flux and temperature in the x, y and z directions. In Figures 7 and 8 heat flux and temperature fields are shown, respectively (on x-y plane). Table 8 summarizes the relative percentage errors between experimental measurements and numerical predictions. It is possible to note that the greatest relative percentage error is equal to 6.1%. Globally, the comparison gets very satisfactory results, demonstrating reliability and accuracy of the developed numerical method.

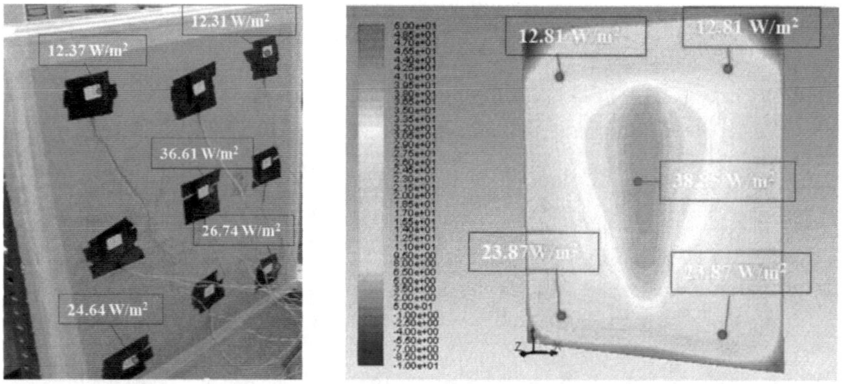

Figure 7. Comparison between experimental and numerical results in terms of heat flux

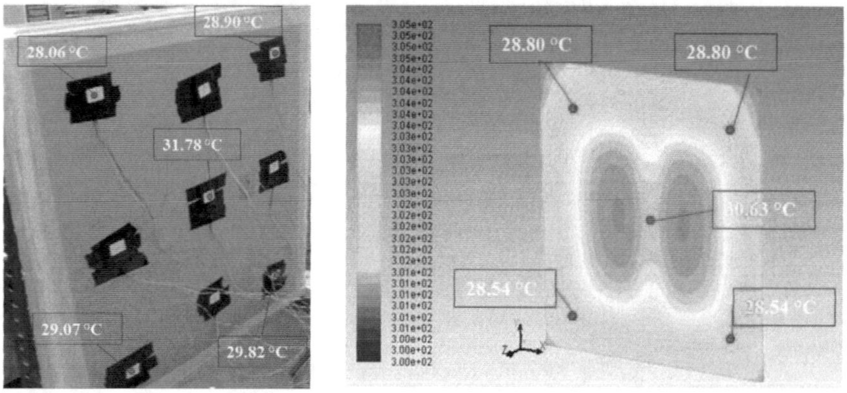

Figure 8. Comparison between experimental and numerical results in terms of temperature

IOP Conf. Series: Journal of Physics: Conf. Series **796** (2017) 012022 doi:10.1088/1742-6596/796/1/012022

Table 8. Comparison between experimental and numerical results: percentage relative error

Sensor position	heat flux	temperature
11	-3.6%	-2.6%
10	-4.1%	0.3%
12	-6.1%	3.6%
14	3.1%	1.8%
13	3.5%	4.3%

The Figure 9 shows the temperature and heat flux inside the wall under investigation. The temperature profiles show that on the vertical symmetry axis the temperature are almost constant, in correspondence of thermal bridge, and decrease from hot surface to cold one; in the same positions, the heat flux decreases rapidly, sign of bi-dimensional thermal field.

5. Conclusion

A numerical analysis for estimating heat flux in a multi-layer wall with a thermal bridges was experimentally validated by setting up an appropriate experimental apparatus. Globally, the comparison gets very satisfactory results, demonstrating reliability and accuracy of the numerical method. The CFD simulation could allow to evaluate, separately, the different heat transfer contributions. Besides, the numerical analysis could be used to design and optimize an experimental setup to measure the different thermodynamic parameters such as the heat transfer coefficient or the thermal transmittance of the wall.

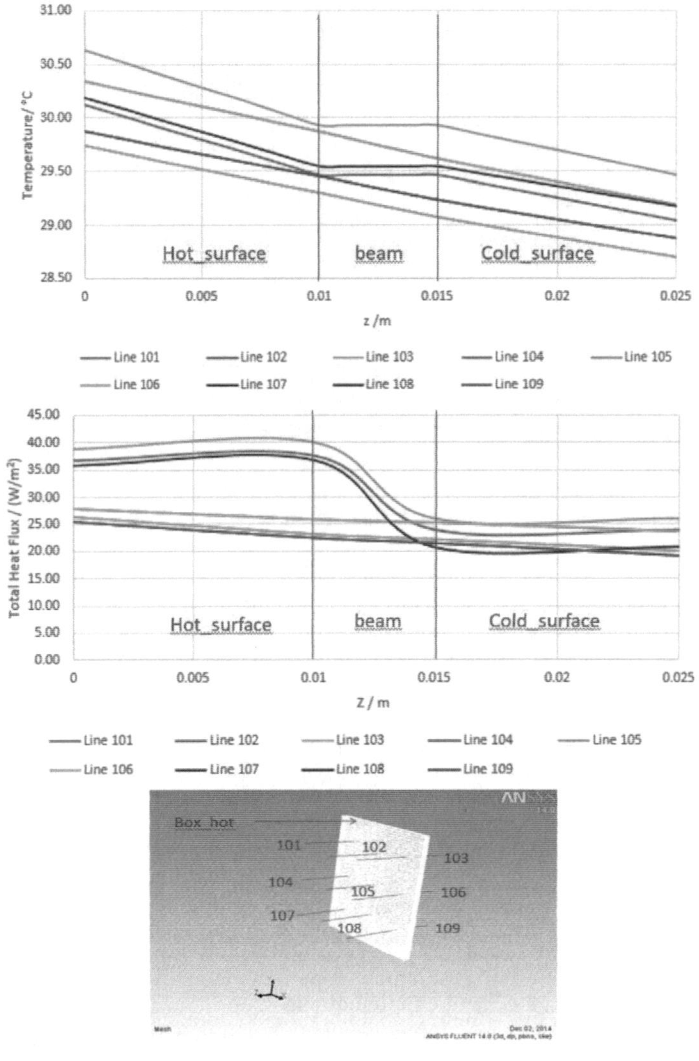

Figure 9. Temperature and heat flux profiles along z axis through the wall

6. References

[1] Directive 2010/31/EU of the European Parliament and of the Council of 19 May 2010 on the energy performance of buildings 2010

[2] Directive 2012/27/EU of the European Parliament and of the Concil of 25 October 2012 on the energy efficiency 2012

[3] Pikas E, Thalfeldt M and Kurnitski J 2014 *Cost optimal and nearly zero energy buildings solutions for office buildings* Energy and Buildings **74** 30-42

[4] Desogus G, Mura S. and Ricciu R 2011 *Comparing different approaches to in situ measurement of building components thermal resistance* Energy and Buildings **12** 2613-2620

[5] Asdrubali F, D'Alessandro F, Baldinelli G and Bianchi F 2014 *Evaluating in situ thermal transmittance of green buildings masonries-A case study* Case Studies in Construction Materials **1** 53-59
[6] Hirsch JJ. DOE, release 2.2. Berkeley (CA): Lawrence Berkeley National Laboratory Simulation Research Group 2005
[7] US Department of Energy. Energy Efficiency and Renewable Energy Office, Building Technology Program, EnergyPlus 2011
[8] Ceylan HT, Meyers GE 1980 *Long-time solutions to heat conduction transient with time-dependent inputs* ASME J Heat Transfer **102** 115–20
[9] Ouyang K, Haghighat F 1991 *A procedure for calculating thermal response factors of multi-layered walls—state space method* Building and Environment **26** 173–77.
[10] Stephenson DG, Mitalas GP 1971 *Calculation of heat conduction transfer functions for multi-layered slabs* ASHRAE Trans 117–26
[11] Hittle DC 1981 *Calculating building heating and cooling loads using the frequency response of multilayered slabs* PhD thesis University of Illinois at Urbana-Champaign
[12] Hittle DC, Bishop R 1983 *An improved root-finding procedure for use in calculating transient heat flow through multilayer slabs* Int J Heat Mass Transfer **26(1)** 1685–93
[13] Shengwei W, Youming C 2002 *A simple procedure for calculating thermal response factors and conduction transfer functions of multilayer walls* Applied Thermal Engineering **22** Issue 3
[14] Renon O 2002 *Thermal bridges modeling in EnergyPlus* Building energy simulation user news **23** (4)
[15] Seem JE 1987 *Modelling of heat transfer in buildings* Ph.D. thesis University of Wisconsin
[16] Seem JE, Klein SA, Beckman WA and Mitchell JW 1989 *Transfer function for efficient calculation of multidimensional transient heat transfer* ASME J Heat Transfer **111** 5–12
[17] Déqué F, Ollivier F, Poblador A 2000 *Grey boxes used to represent buildings with a minimum number of geometric and thermal parameters* Energy Build **31** 29–35.
[18] Déqué F, Ollivier F, Roux JJ 2001 *Effect of 2D modeling thermal bridges on the energy performance of buildings Numerical application on the Matisse apartment* Energy Build **33** 583–87
[19] Ben Larbi A 2005 *Statistical modeling of heat transfer for thermal bridges of buildings* Energy Build **37** 945–51
[20] Gao Y, Roux JJ, Zhao HH and Jiang J 2008 *Dynamical building simulation: A low order model for thermal bridges losses* Energy Build **40** 2236–43.
[21] Klein SA et. al. 2000 TRaNsient SYstem Simulation Program University of Wisconsin: Solar Energy Laboratory
[22] Ascione F, Bianco N, de Rossi F, Turni G and Vanoli GP 2012 *Different methods for the modelling of thermal bridges into energy simulation programs: Comparisons of accuracy for flat heterogeneous roofs in Italian climates* Applied Energy **97** 405–418
[23] Asdrubali F, Baldinelli G and Bianchi F 2012 *A quantitative methodology to evaluate thermal bridges in buildings* Applied Energy **97** 365–73
[24] Capozzoli A, Gorrino A and Corrado V 2013 *A building thermal bridges sensitivity analysis* Applied Energy **107** 229–43
[25] Ascione F, Bianco N, De Masi RF, Mauro GM, Musto M, Vanoli GP 2014 *Experimental validation of a numerical code by thin film heat flux sensors for the resolution of thermal bridges in dynamic conditions* Apply Energy **124** 213–222

IOP Conf. Series: Journal of Physics: Conf. Series **796** (2017) 012023 doi:10.1088/1742-6596/796/1/012023

Summer Thermal Performance of Ventilated Roofs with Tiled Coverings

M Bortoloni[1], **M Bottarelli**[1] **and S Piva**[2]

1 Dipartimento di Architettura, Università di Ferrara, via Quartieri 8, 44121 Ferrara, Italy
2 ENDIF - ENgineering Department In Ferrara, Università di Ferrara, via Saragat 1, 44122 Ferrara, Italy,

Corresponding author E-mail: stefano.piva@unife.it

Abstract. The thermal performance of a ventilated pitched roof with tiled coverings is analysed and compared with unventilated roofs. The analysis is carried out by means of a finite element numerical code, by solving both the fluid and thermal problems in steady-state. A whole one-floor building with a pitched roof is schematized as a 2D computational domain including the air-permeability of tiled covering. Realistic data sets for wind, temperature and solar radiation are used to simulate summer conditions at different times of the day. The results demonstrate that the batten space in pitched roofs is an effective solution for reducing the solar heat gain in summer and thus for achieving better indoor comfort conditions. The efficiency of the ventilation is strictly linked to the external wind conditions and to buoyancy forces occurring due to the heating of the tiles.

1. Introduction

In hot climate reduced building cooling consumption and increased standards of indoor thermal comfort are issues of great relevance. In these regions, the roof plays an important role in reducing the effects of solar radiation when compared to other building elements, due to its extension and exposure to the sun. The common strategies for improving the thermal performance of the building envelope can be summarized in increasing thermal insulation, thermal mass and solar reflectance of the roofs. Other strategies are also available, such as the exploitation of natural convection by adding a ventilated layer below the covering [1]. In cold climates applications of this technology have been studied in depth as strategy to control the vapour [2,3]. Conversely, as will be demonstrated by the following analysis of the literature, in hot climates the performance of ventilated roofs in reducing heat gains is less analysed.

The arrangement of the covering elements over a "batten and counter-batten" system in a traditional pitched roof allows the creation of ventilation ducts under the tiles, Above Sheathing Ventilation (ASV in the following). This volume, where air flows from eaves sections (intake vent) to the ridge, helps to dissipate the excess heat in summer [1]. The efficiency of the ASV is strictly linked to the external wind conditions and to the buoyancy forces induced by the heating of the covering elements. In tiled roofs the air-permeability of the overlapping tiles is an additional and diffused intake/exhaust air-vent system.

Several studies have demonstrated the ASV performance in reducing solar heat gain. Some works followed an experimental approach. In [4] laboratory tests were carried out to evaluate the airflow and the temperature distribution in ASVs, as a function of the power of the solar radiation and the size and shape of the channel. Full-scale models were prepared to run outdoor tests in [5, 6]. In [5] the summer performance of ASVs in combination with a radiant barrier was analysed. A significant performance improvement over a standard ventilated roof was recognized during the daytime. Different sizes of the ventilation duct were compared in [6]. A correlation between solar radiation and ventilation rate, based on experimental data, was discussed in [7].

Other works used numerical models to analyse the mutual relations between air-flows and heat-transfer in ventilated roofs [8-13]. The thermal behaviour of ventilated roofs was analysed with the code Fluent in [8-9] taking into account the airflow due to buoyancy forces only. In [8] the results showed that the heat fluxes can be reduced up to 50% during summer with ventilated roofs. In [9] buoyancy forces are demonstrated to be efficient in producing significant airflows in the channel.

In [10], simulations were carried out to quantify the thermal benefits of a tiled roof over a shingle one, assuming time dependent heat and moisture levels with steady airflow. The estimated benefit was about 14%. A numerical model was developed and validated against experimental data in [11]. It was observed that the airflow induced by the buoyancy forces within the ventilated space reduces the heat flux penetrating into the attic of a 30% in comparison to a direct-nailed roof. In [12] this result was confirmed by means of a different steady state numerical analysis. Anyway, in all these papers the impact of air-permeability in tiled roofs is not investigated and the waterproof covering over the ventilation layer is schematised as a continuous element.

In the present paper a numerical approach is followed to assess the summer thermal performance of a ventilated pitched-roof. The effect on the ventilation efficiency of the air-permeability due to the tile-to-tile opening of a tiled covering is also investigated. Both the cases are compared with the corresponding unventilated roof. The calculations were performed in steady state and repeated for different times of the day and weather conditions.

2. Methods

The commercial finite-element code COMSOL Multiphisycs V4.4 is used for the simulations, to solve the thermo-fluid steady problem in a 2D computational domain. The effect of the buoyancy forces is considered. For the turbulent flow the standard k-ε model is used. For the environmental conditions, data sets based on experimental data are used for wind, solar radiation and temperature.

2.1. Model domain

The numerical simulation of fluid flow and heat transfer in ventilated roofs is a complicated multi-scale problem, on account of the different length scales to be considered. The large scales, pertaining the interactions between wind and building, are strictly connected to the small scales associated with fluid flow within the ASV and the tiny interstices among the overlapping tiles.

In this phase of the research, more attention was paid to the ventilation channel, at the expense of a simplification to the external fluid flow problem. Hence, a 2D domain was modelled as a section of a one floor building of infinite width, thus neglecting the 3D effects of the interaction between building and wind. The one floor building, covered by a pitched roof composed by two slopes with a tilt angle (θ) of 15°, is included in a large external environment 100 m long and 20 m high. Each slope is 5.6 m long; the left one is West-oriented and upwind, whereas the right one is East-oriented and downwind. The scheme of the domain is depicted in figure.1.

In figure 2 are shown a 3D sketch and a cross-section of the ventilated roof. The roof covering is made of Portuguese roof tiles laid on the underlying structure, a batten and counter-batten system. In this phase of the work the latter is not included in the model, although the batten and counter-batten system gives a 3D nature to the thermo-fluid problem. The ventilation is provided by the air flowing into the channel at the eaves section and through the gaps among the overlapped tiles, and partly leaving out at the ridge.

The Portuguese tile shape is characterised by the union of a curved and a flat profile, as shown in

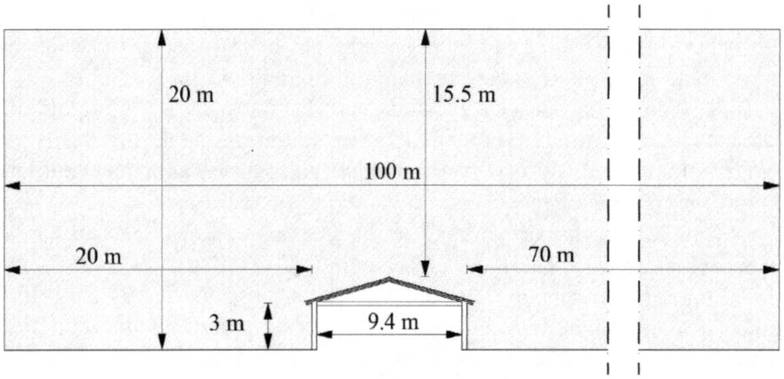

Figure 1. Schematic of the 2D domain.

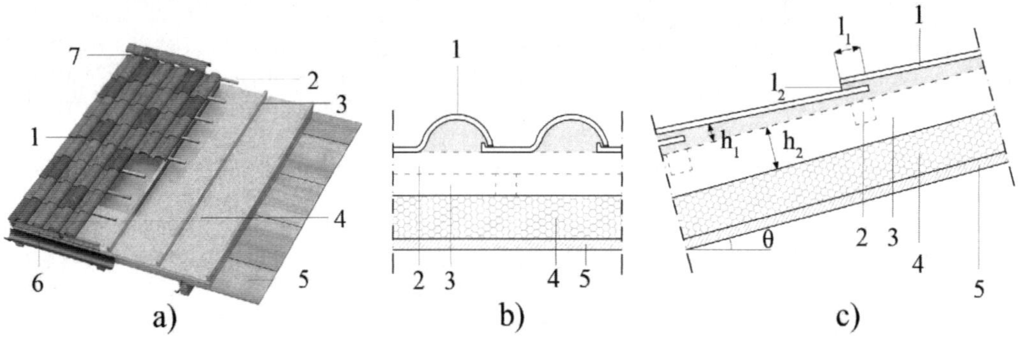

Figure 2. 3D sketch: (a) and cross sections of the ventilated roof (b); 1: tiles; 2: batten; 3: counter-batten; 4: thermal insulation layer; 5: wooden boards; 6) eaves vent comb filler; 7) vented ridge; l_1: length of the tiles overlap; l_2: thickness of the gap between the overlapping tiles; h_1: thickness of the air layer under the tile; h_2: thickness of the ventilation layer; θ: tilt angle.

figure 2b. In the 2D model, the tile shape is simplified in an inclined slab (figure 2c) laid over the batten space with variable height (max 3cm; min 1.5cm), calculated as the average tile height. With regard of the cross sections shown in figure 2, the roof structure is composed of the following elements:

- terracotta tiles, 1, thickness 1 cm;
- tiles overlap, l_1, 5.0 cm long;
- opening gap between the overlapping tiles, l_2, 0.6 cm high.
- air layer h_1 (variable height 1.5-3.0 cm);
- ventilation layer, h_2, (8.0 cm), resulting from the batten (4.0 cm) and counter-batten (4.0 cm) system;
- thermal insulation layer (8.0 cm) of high-density wood fibre;
- wooden board (2.0 cm).

The thermophysical properties are reported in table 1.
Three arrangements of the roof are considered by varying the elements of the base roof structure:

- The ventilated roof (in the following identified as **Vperm**).

Table 1. Thermophysical properties of materials.

Material	Density (kg/m^3)	Thermal conductivity (W/mK)	Specific heat (J/kgK)	Emissivity (-)
Walls	1000	0.800	850	-
Wood fibre insulation	140	0.045	2100	0.90
Wooden board	800	0.120	2100	-
Tiles	1700	0.700	840	0.88

- The non-permeable ventilated roof (**Vimp**), obtained by preventing the airflow through the tiles, so that the ventilation is limited to the air entering at the eaves section.
- The unventilated roof (**Vunv**), where eaves and ridge sections are closed.

Commonly in the numerical models the permeability of the roof is neglected [8-13]; however, when it is considered that in true roofs the gap between the tiles even if small is not avoidable, we take Vperm as the reference one.

The whole mesh is composed of triangular linear elements with 1430000 degrees of freedom. To improve the solution, a greater concentration of elements is imposed near and within the ventilation channels, where the minimum size of the triangular elements is set up to 0.2 mm. For each type of roof different mesh resolutions have been tested in order to produce reliable and mesh-independent results.

2.2. *Case studies*

The analysis is focused on the effects of the wind speed, air temperature and solar radiation on the thermal behaviour of the roof, both in terms of ventilation, heat flux and temperature. The simulations were carried out in steady-state for summer environmental conditions at different time of the day.

A heat flux is applied on the upper surface of the roof tiles, directly exposed to the sun. Based on data available for horizontal surfaces, the solar irradiance on a surface of any orientation and tilt (roof orientation East-West and slope 15°) was calculated according to the method proposed in [14], at hourly scale for the average day of August in Agrigento, a city in southern Italy (37°19' N, 13°30' E). The model is based on daily data of solar radiation. A comparison of the predicted hourly distributions of sun radiation on horizontal surfaces with those measured by the SIAS (Sicilian Information Agro-meteorological Service) at the weather measurement station of Scibica (Agrigento, Italy) shows a very good consistency.

An uniform wind speed profile and an air temperature distribution define the inlet boundary condition in the domain. In the model, wind speed and direction are chosen according to the observations available for the above mentioned location. Wind is supposed entering from the left side (flowing from West to East) of the domain with horizontal direction. The wind speed profile is variable with the altitude, according to the following power law:

$$v = v_0 \cdot \left(\frac{z}{z_0} \right)^{\alpha} \tag{1}$$

where v_0 is the wind speed at the reference height z_0 (10 m), and the parameter α is an empirical exponent depending on the surface roughness, here representative of an urban area and thus equal to 0.3 [15]. The values of wind speed at reference height and air temperature at the inlet were calculated as the average of the hourly mean values measured in August for three years (2012, 2013, and 2014).

An equivalent convective heat transfer coefficient equal to 1.4 $W/(m^2K)$ is applied on the pitched ceiling; it takes into account both the heat transfer from the ceiling surface and the attic and that from the attic to the room (air at 26 °C).

The data sets for wind and solar radiation used in the simulations are reported in Table 2 for different times of the day (8:00; 12:00; 16:00). Two further cases were simulated at 12:00 (Table 2), with reduced reference wind speeds: 2m/s for Case 4 and 1.5 m/s for Case 5.

3. Results and Discussion

For Case 2 of table 2, examples of velocity field around the building for the three different roof typologies (Vperm, Vimp, Vunv) are shown in figure 3.

The velocity fields are similar for the three roof typologies considered; differences are significant only around and within the ASV, according to the different features of the roof. The wind, blowing from left, hits the building and then recirculates at low speed in the proximity of the vertical left wall. The air rising near the wall is then partially diverted into the batten space. The left roof pitch, west-oriented, is directly exposed to the wind; here, behind the eaves section and above the covering, there is a recirculation zone extended up to half of the left slope. The right roof slope, east-oriented, is downwind and affected by a further air recirculation zone due to the wind blowing over the building. Similar observations are suitable for the other cases (Case 1 and Case 3).

The velocity field around eaves and ridge sections is shown in detail in figure 4 for the three types of roof (Vperm, Vimp, Vunv). As for figure 3, the images in figure 4 refer to Case 2 (12:00) of table 2. The two ventilated roofs (Vperm, Vimp) show a different behaviour near the eaves or in the second half of the roof slopes. Both in the Vperm and Vimp cases, the air flowing into the ASV come only from air rising on the left vertical wall. In the case Vperm a part of it flows out through the interstices among tiles. At the inlet, the air velocity is higher in Vperm than in Vimp. In the second half of the upwind slope of case Vperm, the air flows through the tiles into the batten space. In both the Vperm and Vimp cases only a small part of the air flowing in the batten space flows out through the ridge. A significant flow rate continues along the right-side of the roof.

Table 2. Heat flux on the roof tiles, reference wind speed and air temperature.

		Case 1	Case 2	Case 3	Case 4	Case 5
Time	(h)	8:00	12:00	16:00	12:00	12:00
Heat Flux West	(W/m^2)	322	864	675	864	864
Heat Flux East	(W/m^2)	675	864	322	864	864
Wind speed	(m/s)	3.0	4.5	5.0	2.0	1.5

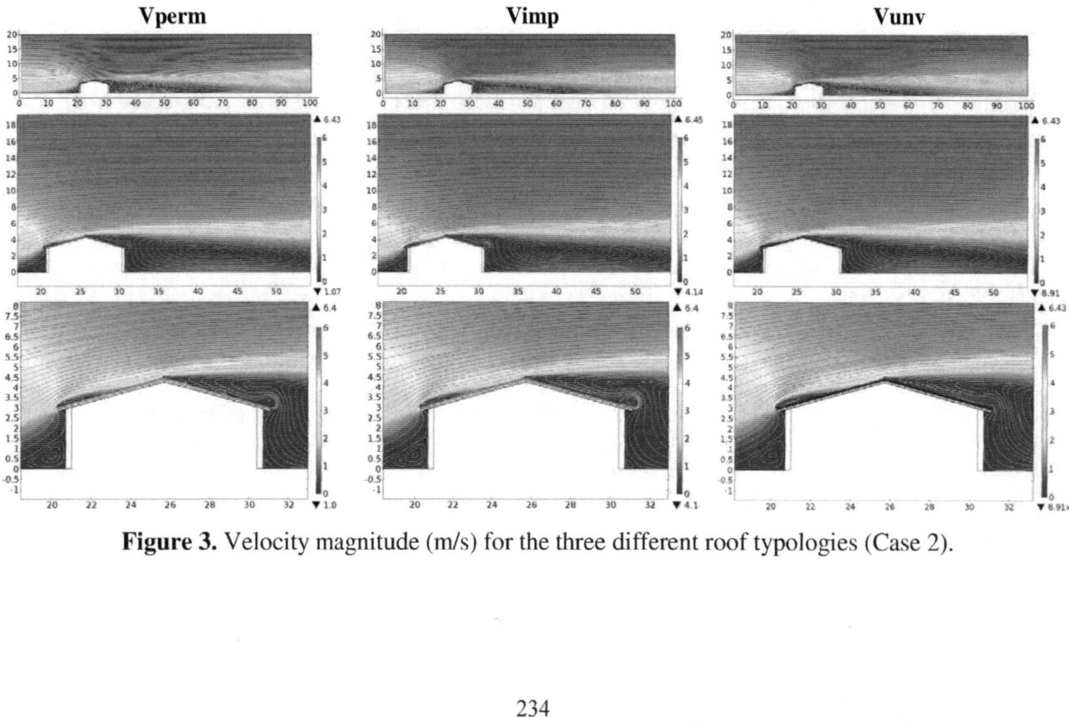

Figure 3. Velocity magnitude (m/s) for the three different roof typologies (Case 2).

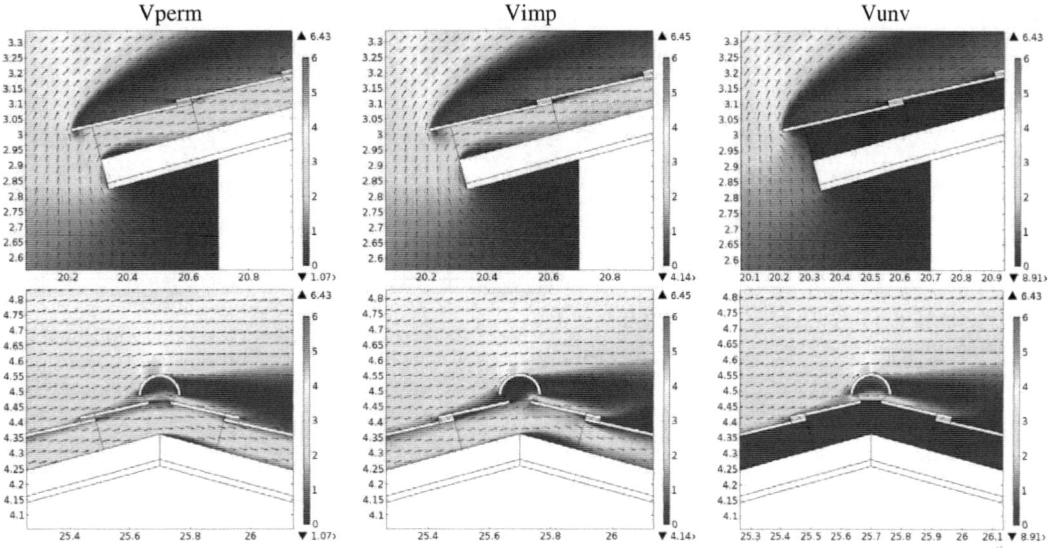

Figure 4. Velocity magnitude (m/s) at the eaves and the ridge sections (Case 2).

The mass flow rate along the batten space (cases Vperm and Vimp) is shown in figure 5 for all the cases codified in table 2. In figure 5, the ridge is 5.4 m far from the eaves section, at the intersection between right and left roof pitch.

In case Vperm the mass flow rate decreases initially due to the air flowing out through the gaps among the tiles. After the midpoint of the left roof pitch the mass flow rate increases due to the airflow passing through the tiles; it does not recover the inlet value in the whole set of simulated cases. The 5% of the air coming from the upwind part of roof goes out at the eaves, while the remaining flows along the right batten space, going out through the tiles and at the eaves section. The mass flow rate is strongly affected by the external wind conditions. In Case 1 (8:00), characterized by a low wind speed, the mass flow rate at the inlet is 0.18 kg/s (for a width of 1 m); it grows up to 0.3 kg/s in Case 2 (16:00), when higher wind speed occurs.

In case Vimp, the mass flow rate is constant, but different, along the two halves of the roof; the air flowing out at the ridge is about 20%.

In case Vperm the mass flow rate varies along the channels due to the permeability of the tiles. For a comparison with the case Vimp we calculated the average mass flow rate along the channel. The difference between Vperm and Vimp is limited (5.8%, 4.2% and 3.3% in Case 1, Case 2 and Case 3, respectively).

Two further cases are simulated to stress the effect of air permeability in ventilated roofs. These are shown in figure 6. For Case 4 (12:00 and a reference wind speed reduced to 2 m/s), the mass flow rate is 17.6% higher in Vperm than in Vimp. Finally, for Case 5 (12:00 and a further reduction of the wind speed at the reference height to 1.5 m/s), in Vperm the airflow direction is inverted and becomes ascending in the last part of the downwind slope. In Vimp the airflow is ascending in the whole right roof pitch. Moreover, in this case the average mass flow rate is 80% higher in Vperm than in Vimp.

In figure 7 is depicted the ratio between the mass flow along the batten space and that calculated at the inlet. For the permeable roof, Vperm, in the upwind channel an initial decrease of 24% is followed by an increase up to 94% of the inlet mass flow rate. Then, a 5% of the air flowing through the ASV flows out at the ridge. Finally, for the three casel 1, 2 and 3, at the eaves section the airflow is reduced to 50%. In the case Vimp, the leakage at the ridge is about 20%.

If the reference wind speed is reduced, Cases 4 and 5, this behaviour changes significantly. In Vperm, the air flowing out at ridge is 11% and 29% for Case 4 and Case 5, respectively. Moreover, the decrease of the mass flow rate is more pronounced in the downwind slope under low wind speed conditions. In Vimp an ascending airflow is observed on the left pitch due to the buoyancy forces. In this case, all the air flowing within both the upwind and downwind channels flows out at the ridge.

A reference mass flow rate can be calculated on the basis of the component of the wind speed projected in the direction of the channel and the area of the eaves section. If we compare the mass flow rate calculated at the inlet with the reference one (table 3), a significant decrease is evident in all cases. In Vperm the inlet mass flow rate is an 80% of the reference one, for all the wind conditions. In Vimp the inlet mass flow rate is a 65% of the reference one. However, a more pronounced reduction to 44.2% is observed in Case 5.

With the aim of evaluating the thermal performance of the roof, temperature and heat fluxes on the surface along the roof are calculated for the different cases and roof typologies.

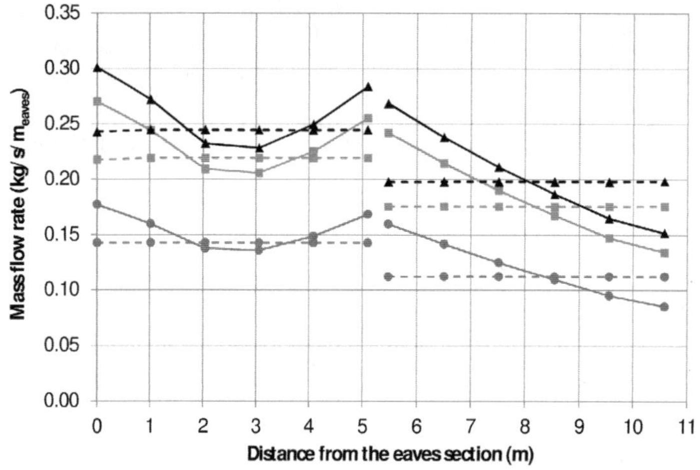

Figure 5. Mass flow rate within the ASV. Continuous line: case Vperm; dotted lines: case Vimp. Symbols:●: Case 1 (8:00); ■: Case 2 (12:00); ▲: Case 3 (16:00).

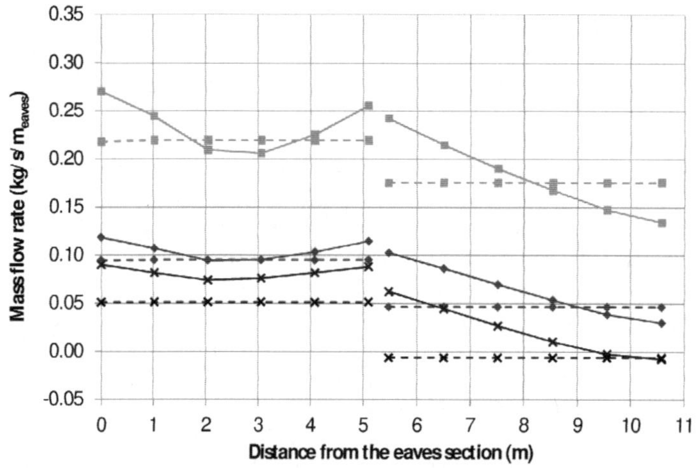

Figure 6. Mass flow rate within the ASV. Continuous line: case Vperm; dotted lines: case Vimp. Symbols: ■: Case 2 (12:00); ♦: Case 4 (12:00); **x**: Case 5 (12:00).

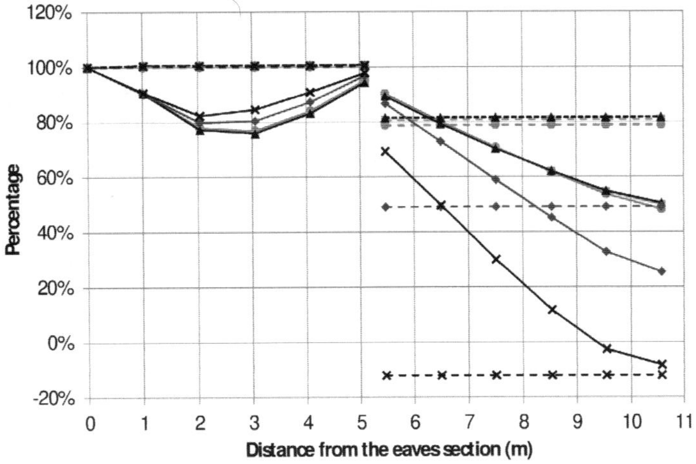

Figure 7. Mass flow rate within the ASV with respect to that at the eaves. Continuous line: case Vperm; dotted lines: case Vimp. Symbols: :●: Case 1 (8:00); ■: Case 2 (12:00); ▲: Case 3 (16:00); ♦: Case 4 (12:00); **x**: Case 5 (12:00)

Table 3 Comparison between reference and inlet mass flow rate.

Case	Mass flow rate (kg/s)					
	Vperm			Vimp		
	Reference	Inlet	Reduction	Reference	Inlet	Reduction
1	0.22	0.18	20.3%	0.22	0.14	35.8%
2	0.33	0.27	19.2%	0.33	0.22	34.9%
3	0.37	0.30	18.8%	0.37	0.24	34.4%
4	0.15	0.12	20.2%	0.15	0.09	36.4%
5	0.12	0.09	22.1%	0.12	0.05	55.8%

In table 4 the average heat fluxes are reported, calculated on the upwind and downwind pitch of the roof and on the lower surface of the ceiling. The air-flow within the ASV reduces the heat fluxes on both roof pitches, regardless of the case considered. In the unventilated roof (Vunv), the heat flux is more than twice of that in Vperm and Vimp. In Case 2 (12:00 and maximum solar radiation), the heat flux calculated on the upwind side is three times higher in Vunv than in Vperm. The heat flux in Vperm and Vimp is always higher downwind than upwind, even for Case 3 (16:00), although the lower solar radiation downwind. In this case, the air is heated up by the tiles while flowing within the upwind roof slope; then heating continues along the batten space in the right-side. On the contrary, for Case 3, in Vunv the heat flux is lower downwind than upwind. The air-permeability is responsible of a significant reduction of the heat fluxes in all the cases. The more the inlet mass flow rate is high, the more Vperm and Vimp show similar values.

In terms of heat fluxes on the ceiling and by taking Vimp as the reference case, the total heat flux (upwind and downwind) in Vperm is about 6.7% and 6% lower than in Vimp, for Case 1 (8:00) and Case 2 (12:00), respectively. For Case 3 (16:00) this difference is reduced to 4% (5.5% upwind, 2.5% downwind), when the average mass flow rate on the downwind roof slope is similar for Vperm and Vimp.

In table 5 the average temperature, calculated on the upwind and downwind pitch of the roof, is reported. The air-flow within the ASV reduces the temperature on both roof pitches, regardless of the case considered. This reduction is higher upwind than downwind, where the mass flow rate decreases. In terms of temperature the effect of the permeability of the roof is not particularly significant.

Table 4. Average heat flux on the ceiling.

Case	Heat flux (W/m^2)								
	Vperm			Vimp			Vunv		
	West	East	Ceiling	West	East	Ceiling	West	East	Ceiling
1	2.5	6.5	4.7	2.7	7.0	5.0	5.8	14.9	10.7
2	3.6	6.3	5.1	3.9	6.7	5.5	10.8	15.8	13.8
3	2.9	3.3	3.2	3.0	3.4	3.3	8.2	6.5	7.6

Table 5. Average temperature on the ceiling.

Case	Temperature (°C)					
	Vperm		Vimp		Vunv	
	West	East	West	East	West	East
1	27.8	30.6	27.9	30.9	30.3	36.6
2	28.5	30.5	28.7	30.7	33.7	37.2
3	28.0	28.4	28.1	28.4	31.8	30.6

4. Concluding Remarks

A numerical approach was followed to evaluate the summer performance of a ventilated pitched-roof in comparison with an unventilated roof, by means of a finite element numerical code, solving the thermo-fluid problem in 2D and under realistic boundary conditions. The performance was evaluated by taking into account the air-permeability of the tiled covering.

In ventilated pitched roofs the mass flow rate within the ASV is strictly linked to the external wind conditions. The upwind eaves section behaves as an air intake and air flows along the whole batten space. As the wind speed decreases, buoyancy forces due to variation in air density have an increasing effect on the ASV behaviour and in our geometry become dominant for 1.5m/s. In this case a reversed flow in the downwind pitch can be observed.

The air permeability of the tiled covering increases the mass flow rate along the ventilation channel. The airflow is higher in the permeable ventilated roof than in the impermeable one. The difference becomes more relevant when the wind speed is low.

The presence of an ASV reduces the heat flux on the ceiling with respect to an unventilated roof, under the considered boundary conditions and with the same thermal insulation. As a consequence, the temperature of the ceiling surface is lower and a better indoor thermal comfort is experienced. In terms of calculated heat flux, the thermal performance of a ventilated roof is enhanced by the air permeability of the tiled covering (-6% in comparison with a non-permeable ventilated roof). This benefit becomes more pronounced (up to -20%) when wind speed is low. This could lead to the consideration that the permeability loses its effectiveness beyond a certain wind speed.

Anyway, the ventilation layer in pitched roofs is effective in reducing the solar heat gain in summer and an improvement of indoor thermal comfort is thus expected.

Finally, ASV could be regarded as an effective solution for a significant energy saving, allowing short payback period of the additional costs.

5. References

[1] Patania F, Gagliano A, Nocera F, Ferlito A and Galesi A 2011 Energy Analysis of Ventilated Roof *Sustainability in Energy and Buildings* eds R J Howlett, L C Jain, S H Lee (Berlin: Springer) pp 15-23

[2] Schunck E, Oster H J, Barthel R and Kießl K 2003 *Roof Construction Manual: Pitched Roofs* (Basel: Birkhauser)

[3] Endriukaityte A, Monstvilas E and Bliudzius R 2005 The impact of climate parameters on air

movement in ventilated roofs air gap *Improvement of Buildings' Structural Quality by New Technologies* eds C Schauer, F Mazzolani, G Huber, G de Matteis, H Trumpf, H Koukkari, J P Jaspart, L Bragan (London: Taylor & Francis) 573-578

[4] Lee S, Park S H, Yeo M S and Kim K W 2009 An experimental study on airflow in the cavity of a ventilated roof *Building and Environment.* **44** 1431-1439.

[5] Dimoudi A, Androutsopoulos A and Lykoudis S 2006 Summer performance of a ventilated roof component *Energy and Buildings* **38** 610-617.

[6] D'Orazio M, Di Perna C, Principi P and Stazi A 2008 Effects of roof tile permeability on the thermal performance of ventilated roofs: Analysis of annual performance *Energy and Buildings* **40** 911-916.

[7] Sandberg M and Moshfegh B 1998 Ventilated-solar roof airflow and heat transfer investigation *Renewable Energy* **15** 287-292.

[8] Gagliano A, Patania F, Nocera F, Ferlito A and Galesi A 2012 Thermal performance of ventilated roofs during summer period *Energy and Buildings* **49** 611-618.

[9] Bianco V, Diana A, Manca O and Nardini S 2016 Thermal Behavior Evaluation of Ventilated Roof under Variable Solar Radiation *Int. J. Heat Techn.* **34** S346-S350.

[10] De With G, Cherry N and Haig J 2009 Thermal Benefits of Tiled Roofs with Above-sheathing Ventilation *Int. J. Building Physics* **33** 171-194.

[11] Miller W, Keyhani M, Stovall T and Youngquist A, 2007 Natural Convection Heat Transfer in Roofs with Above-Sheathing Ventilation *Thermal Performance of the Exterior Envelopes of Buildings: Proc. of ASHRAE THERM X*, Clearwater, FL, Dec. 2007, ASHRAE.

[12] Ciampi M, Leccese F and Tuoni G 2005 Energy analysis of ventilated and microventilated roofs *Solar Energy* **79** 183-192.

[13] Villi G, Pasut W and De Carli M 2009 CFD modelling and thermal performance analysis of a wooden ventilated roof structure *Building simulation* **2** 215-228.

[14] Duffie F and Beckman W, 1991 *Solar engineering of thermal processes* (New York: John Wiley and Sons).

[15] Masters G M 2004 *Renewable and Efficient Electric Power Systems* (New York: John Wiley and Sons).

6. Acknowledgments

The meteorological data used was kindly furnished by S.I.A.S. Servizio Informativo Agrometeorologico Siciliano, viale Regione Siciliana, 4600, 90145 Palermo.

IOP Conf. Series: Journal of Physics: Conf. Series **796** (2017) 012024 doi:10.1088/1742-6596/796/1/012024

The ground surface energy balance in modelling horizontal ground heat exchangers

Marco Bortoloni[1], Michele Bottarelli[1], Yuehong Su[2]

[1] Dipartimento di Architettura, Università di Ferrara, via della Ghiara 36, 44121 Ferrara, Italia
[2] Department of Architecture and Built Environment, The University of Nottingham, Nottingham NG7 2RD, UK

marco.bortoloni@unife.it

Abstract. The performance of horizontal ground heat exchangers (HGHEs) is strongly dependent on climatic conditions, due to the low installation depth. In numerical modelling of HGHEs, the estimation of shallow soil temperature distribution is a key issue, therefore the boundary condition (BC) at the ground surface should be assigned carefully. With this in mind, a model of the energy balance at the ground surface (GSEB), based on weather variables, was developed. The model was tested as the 3rd kind BC at ground surface in modelling HGHEs by means of the FEM code Comsol Multiphysics, solving the unsteady heat transfer problem in a 2D domain. The GSEB model was calibrated and validated with the observed soil temperature at different depths. In addition, the effect on numerical solutions of different BCs, when assigned at the ground surface, was analysed. Three different simulations were carried out applying the GSEB model, the equivalent surface heat flux and temperature as boundary conditions of the 1st, 2nd and 3rd kind, respectively. The results of this study indicate that the use of the GSEB model is a preferable approach to the problem and that the use of the equivalent surface temperature can be considered as a reasonable simplification.

1. Introduction

Nowadays, the reduction of greenhouse gas emissions and the rational use of energy have become a major issue. Worldwide, the great opportunity for energy saving in the buildings sector has been recognised. For the above reasons, recent environmental policies have promoted the energy efficiency of buildings and the spread of renewable energy technologies. Among them, ground-coupled heat pumps (GCHPs) are regarded as a reliable technology and may represent an efficient solution for space heating and cooling, due to their high energy efficiency, if properly designed according to local environmental conditions [1-2].

In GCHPs, a ground heat exchanger is required to thermally couple a heat pump with the ground. The ground heat exchanger usually consists of a piping system installed in vertical boreholes or in shallow diggings. Vertically coupled heat pumps benefit from the relatively stable temperature in the deep ground and uses geothermal energy from the earth whereas a horizontally coupled heat pump uses the seasonal heat storage in shallow soil. As a consequence, the performance of horizontal ground heat exchangers (HGHEs) is strongly affected by the climatic conditions due to the low installation depth. The heat transfer will essentially depend on the temperature difference between the working fluid and the natural ground temperature at the average depth of the HGHE. The seasonal variation of the soil temperature at shallow depth can lead to unfavorable working conditions and, consequently, to an efficiency reduction. Nevertheless, ground thermal drifts are not expected for shallow HGHEs, also after long-term operations, due to the energy balance at the ground surface [3].

The numerical modelling of HGHEs is a difficult task due to the need of an accurate prediction of the shallow soil temperature distribution, which results from the heat transfer occurring at the ground surface according to the local climate. A number of researches have been conducted to predict the performance of HGHEs, following an analytical approach based on the line source theory and cylindrical heat transfer equations [4] or, more recently, by means of numerical models. The annual variation of the soil temperature at different depths can be estimated using a sinusoidal function according to a correlation firstly proposed in [5], under the assumptions that the ground can be considered a semi-infinite solid, and that the surface boundary condition (BC) can be described by a sinusoidal temperature variation. This approach is frequently applied in modelling of HGHEs, were the complex heat transfer phenomena at the ground surface can be reduced to a 1st kind BC. A numerical model for horizontal pipes was developed applying an explicit finite difference scheme in [6], assuming a radially symmetrical temperature profile from inside the pipe, and a far-field BC based on the previously mentioned correlation [5]. A sinusoidal temperature trend was assigned to the ground surface also in [7]. A daily temperature time series was used as BC in [8], as calculated by means of a ground surface energy balance using real weather data. In [9], the surface temperature trend (1st kind BC) was converted to the equivalent heat flux at the ground surface (2nd kind BC), with the aim of considering the variation in the ground surface temperature due to the effect of the heat transfer by HGHE. Despite the long computational time required, numerical studies have been carried out including the mass transfer, in order to take into account the effects of the soil moisture content and migration [10]. Others applied a 3rd kind BC at the ground surface. In [11] only the convective heat flux between air and ground is considered. The external environmental conditions are included in the model by means of energy balance equation validated against experimental data, taking into account the effect of solar radiation, latent and sensible heat transfer in [12]. A earth-air tunnel in a ventilation system has been modelled including the simultaneous heat and moisture heat transfer at the ground surface [13]. On the other hand, a new approach to include the seasonal variation in soil temperature, which has been included as an internal source term instead of comprehensive energy balance at the ground surface, was recently proposed in [14].

The proper assignment of the boundary condition to the ground surface is important in order to carry out an accurate simulation of HGHEs. With this in mind, we proposed a model of the energy balance at the ground surface, which was implemented in the numerical code Comsol Multiphysics as the 3rd kind BC in modelling HGHEs [15]. This approach would allow a more reliable prediction of HGHE performance and moreover, to take into account of the ground surface temperature variation induced by HGHEs. This aspect is not considered with simplified BC (e.g. 1st kind BC). According to [15], this paper provides a further in-depth modelling of the heat transfer processes at the ground surface by improving the convective component of the energy balance by introducing the aerodynamic resistance of the canopy layer. The reliability of the proposed energy balance model has been tested against experimental data. Furthermore, we compared different boundary conditions at the ground surface in modelling HGHEs. The ground surface energy balance model (GSEB), the equivalent surface heat flux and temperature were assigned as boundary conditions of the 1st, 2nd and 3rd kind in three different simulations, respectively.

2. Methods

This study was intended to develop a simplified model of the energy balance at the ground surface, which could be used as a BC in numerical simulation of HGHEs. Moreover, three different approaches to the modelling of the heat transfer at the ground surface were tested (1st, 2nd and 3rd kind BCs) and then compared.

The software COMSOL Multiphysics V5.0, which is based on the finite-element method, was used to solve the 2D unsteady-state heat transfer problem in solids, thus, only conductive heat transport was considered. We developed a model of the energy balance at the ground surface, taking into account of the ground surface properties (albedo and emissivity). A comprehensive weather data set based on experimental data (solar radiation, air temperature, relative humidity, atmospheric pressure and wind speed) was used for simulations. In order to calibrate and test the proposed GSEB model we carried out a preliminary comparison between the calculated and the observed soil temperature at various depths, showing a good agreement.

The GSEB model was applied as the BC at the ground surface in a simulation, which was carried out to analyse the energy performance of a HGHE in heating and cooling. In this study we simulated an innovative horizontal ground heat exchanger, called Flat-Panel, which has been developed recently at the University of Ferrara [16]. Finally, the GSEB model (3rd kind BC), the equivalent heat flux (2nd kind BC) and the equivalent temperature (1st kind BC) on ground surface were applied as boundary condition in three different simulations comparing the numerical solutions in order to identify the proper boundary condition in modelling of HGHEs.

2.1. Model domain

The FEM model was solved for a 2D model domain which comprises a Flat-Panel and a wide part of the surrounding soil (10 m long and 10 m deep). The domain was configured according to the experimental setup at the Department of Architecture of the University of Ferrara, which was equipped with 2 Flat-Panels [16]. The extent of the domain has been selected so that it was large enough to approximate an infinite medium. In the 2D model domain, the Flat-Panel was introduced as 2nd kind boundary condition and modelled as a vertical line, 1 m high and laid within shallow soil, between a depth of 0.9 and 1.9 m, as shown in Fig.1 together with the full mesh. Finally, a symmetric approach was applied to reduce the time required for calculations.

The soil is homogenous and isotropic over the entire domain, in agreement with many approaches to the modelling of HGHEs [17]. Here, the thermos-physical properties of the soil were chosen according to a preliminary field survey which was carried out at the experimental site [18] in order to obtain more reliable simulation. Overall, the soil was found to be a mixture of silt, sand and clay, with a large component of organic matter. In order to solve the energy balance equation at the ground surface, the surface albedo α and emissivity ε (in the infrared) have been considered. The properties of the ground surface were taken for a bare soil according to [19]. The soil properties and that of the ground surface are reported in tab.1.

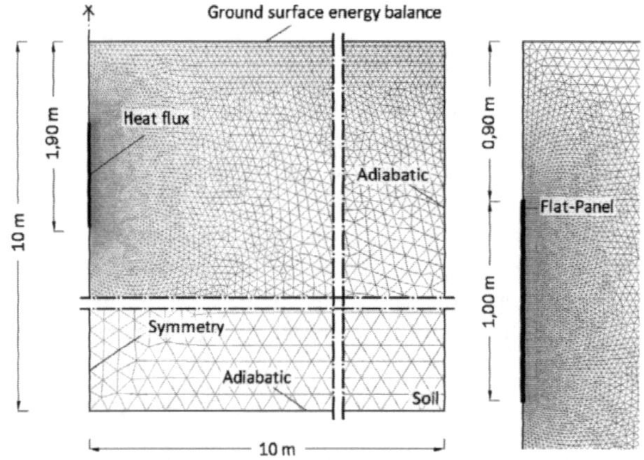

Figure 1. Schematic of the 2D domain, mesh and boundary conditions.

Table 1. Material properties

Thermal conductivity (W/mK)	Density (kg/m^3)	Specific heat (J/kgK)	Surface albedo (-)	Surface emissivity (-)
1.40	1720	2000	0.15	0.95

The heat equation was discretized spatially over the numerical domain by the finite element method using triangular elements with quadratic weighting functions. A higher mesh resolution was set near the Flat-Panel BC and on the top edge of the domain (representing the ground surface) where higher temperature gradients are expected, and coarse in the outer part of the domain. In order to check the grid independence of the solution, a preliminary analysis has been carried out. Different grid refinements have been simulated, progressively increasing the number of elements in the vicinity of the Flat-Panel and of the top boundary, as well in the whole domain. The final mesh is composed of triangular linear with 27200 degrees of freedom.

2.2. Boundary conditions
Thermal boundary conditions were assigned to the outer domain boundaries only. The lower and vertical boundaries in the soil domain were assumed to be adiabatic. On the left vertical side this was due to the condition of symmetry. At the top boundary, representing the ground surface, a 3rd kind boundary condition (BC) has been assigned by means of GSEB model, as detailed in section 2.3. The Flat-Panel was treated as a time-dependent boundary condition by means of a heat flux time series, according to the calculated heating and cooling demand, as described in section 2.4. To run the GSEB model and to determine the HGHE heat flux at hourly scale, a complete set of 2014 weather data of Ferrara, a city in northern Italy, were used in simulations. Finally, in the further analysis, boundary conditions of 1st, 2nd and 3rd kind were imposed at the ground surface in three different simulations. The GSEB model was firstly used as the 3rd kind BC in a preliminary model to assess the equivalent heat flux (2nd kind BC) and the equivalent ground surface temperature (1st kind BC). The initial temperature distribution for simulations was obtained for the 1st day of 2014 from the available soil temperature data.

2.3. The modelling of the energy balance at the ground surface
In the numerical model, the heat is transferred only by conduction between the ground surface and the underlying soil. The temperature in the shallow soil depends on the energy fluxes at the ground surface, so the soil heat flux was defined by the ground surface energy balance equation for the infinitesimally thin surface layer (i.e. the interface between the ground and atmosphere), which can be written in his general form as a function of time and the surface temperature:

$$G(t, T_s) = R - H - LE \tag{1}$$

where:
G = soil heat flux (W/m^2)
R = net radiative energy flux (W/m^2)
H = sensible energy flux (W/m^2)
LE = latent energy flux (W/m^2)

The magnitude of each component is a time-varying function, which depends on weather conditions; therefore, a complete set of weather data is required as an input for equation (1). The GSEB model was developed for grassy surfaces in order to have comparability with the available measurements of soil temperature at different depths. The experimental setup, taken as the reference [16], is covered by a wild meadow which is not subjected to regular watering and mowing. As suggested by [20-21], a detailed modelling of the effect of vegetation would require an additional equation to solve the energy balance in the foliage layer firstly, secondly the energy balance between vegetation and soil. However, these models require the knowledge of several parameters related to the type of vegetation, cutting cycles and irrigation, which are difficult to find. To reduce the computational time, a simplification in modelling the vegetation layer has been introduced by neglecting a separate energy balance equation for vegetation and the underlying soil. The effect of the former one was introduced in each component of the GSEB by means of appropriate correlations.

The radiative component mainly affects the surface temperature, particularly during the summer season. The net radiative flux R (W/m^2) is the sum of the incident shortwave solar radiation into its direct and diffused components (as absorbed and reflected by the surface) and on the longwave radiation received and emitted by the surface. The amount of shortwave and longwave solar radiation reaching the ground surface and the outgoing longwave radiation as well, are reduced by the shading due to the foliage layer. Moreover, the plant canopy absorbs long-wave radiation from the sky and re-radiates to the underlying soil surface. Consequently, a coefficient of shading (s_f=0.50) has been introduced similarly to FASST model [22]. This accounts for the actual surface coverage and for the mutual radiative heat transfer between vegetation and soil, which has not been solved here, contrary to the FASST model. The net radiative energy flux at the ground surface is given by:

$$R = s_f\left((1-\alpha)R_s + R_{ld} - \sigma\varepsilon_s T_s^4\right) \qquad (2)$$

where:
s_f = calibration coefficient of shading
α = surface albedo
R_s = shortwave solar radiation (W/m^2)
R_{ld} = downward longwave solar radiation (W/m^2)
σ = Stefan-Boltzmann constant (W/m^2K^4)
ε_s = surface emissivity
T_s = surface temperature (K)

Aiming to simplify the former GSEB model [15] by eliminating the use of a calibration parameter, the convective heat flux between air and ground surface was calculated taking into account of the wind sheltering by the foliage layer, according to [23]:

$$H = \rho_a c_a \frac{(T_a - T_s)}{r_a} \qquad (3)$$

where:
ρ_a = density of air (1.205 kg/m^3)
c_a = specific heat of air (1005 J/kg/K)
T_s = surface temperature (°C)
T_a = air temperature (°C)
r_a = aerodynamic resistance to heat transfer (s/m)

The aerodynamic resistance, r_a (s/m), is function of wind speed and canopy properties [24]. In this study, a reference a grass cover with a constant height of 0.12 m and a standardised height for wind speed, temperature and humidity of 2 m were assumed. In order to calculate the latent energy flux, the evapotranspiration ET_0 from the vegetated surface was calculated following the FAO Penman-Monteith model that proved to be reliable for different climates and time step [24]. The ET_0 is calculated for a reference grass crop, well irrigated and completely shading the ground. The evapotranspiration was calculated in terms of mass by assuming a constant density of water. A single crop coefficient K_c (0.25) which is usually provided for different agricultural crops and crop growth stages [24]) was introduced in calculating the latent heat flux, in order to take into account the characteristics of the vegetated layer (a not irrigated wild meadow). In our model, variations in vegetation have not been considered due to the objective difficulties in knowing the different growth stages of a wild meadow as well as the watering. In view of this, the latent energy flux at the ground surface is given by:

$$LE = K_c \frac{\rho ET_0 l_h}{3600} \qquad (4)$$

where:
K_c = crop coefficient
ρ = density of water (1.000 kg/m^3)
ET_0 = reference evapotranspiration (l/m^2h)
l_h = latent heat of evaporation (J/kg)

The equation (1) was implemented in Comsol as a model variable. This represents the overall heat flux at the ground surface G (t, Ts), which is a function both of time and surface temperature.

2.3.1. Comparison of simulation results and field data

The reliability of the GSEB model was tested with a preliminary simulation for a whole year. In this simulation, the heat load at the ground heat exchanger was neglected; therefore, the variation in the soil thermal field was only determined by the environmental conditions. A comprehensive data set of 2014 was managed to obtain several hourly scale time series that were needed as input in each term of equation 1. Several weather variables (solar radiation, air temperature, relative humidity, atmospheric pressure and wind speed) were measured by means of a weather station at the Department of Architecture in Ferrara. Measurements of the downward longwave radiation were provided by ARPA-EM, the meteorological service of the Emilia Romagna region. The simulated ground temperature trend was compared with the measured values at different depths (0.1, 0.8, 2.5 m).

The calculated daily average soil temperature is compared against measured values in fig. 2. Three temperature probes have been considered at three different depths (0.1, 0.8, 2.5 m respectively). The analysis of simulated and measured temperature time series showed that the model both over and under-predicts peak daily temperature near the surface, during different periods of the season. Deviations were clear mainly at shallow depths, where fluctuations in temperature are greater. The relationship between measured and simulated temperature was more stable in winter due to the low radiative heat flux and the negligible latent heat flux. In summer the model showed less accuracy, with a maximum difference of 3 °C at a depth of 0.1 m between calculated and measured temperature. The natural growth cycle of the grass covering the ground surface, which was not accounted in the model, may have played a role in varying the energy balance. However, the HGHE is usually installed at a depth between 1 and 2 m, where the daily fluctuations in soil temperature are reduced. The statistical uncertainty was calculated for the entire simulation period. For 2014, it was 0.99, 0.44 and 0.35 °C at a depth of 0.1, 0.8 and 2.5 m, respectively.

Scatter plots of the simulated soil temperatures versus the equivalent measured are shown in fig. 3 to a depth of 0.1 m and 0.8 m. In these graphs, the central line represents the optimum correlation between measured and simulated values, and the two others a span of 2 °C. In both cases the slope of the relationship is close to 1:1, although slight dispersion occurs when the soil temperature is higher than 20 °C.

The heat fluxes related to each component of the GSEB model are shown in fig. 4, for 7 days in winter and summer. Both in winter and in summer, the conductive heat flux in soil (G) shows a strong

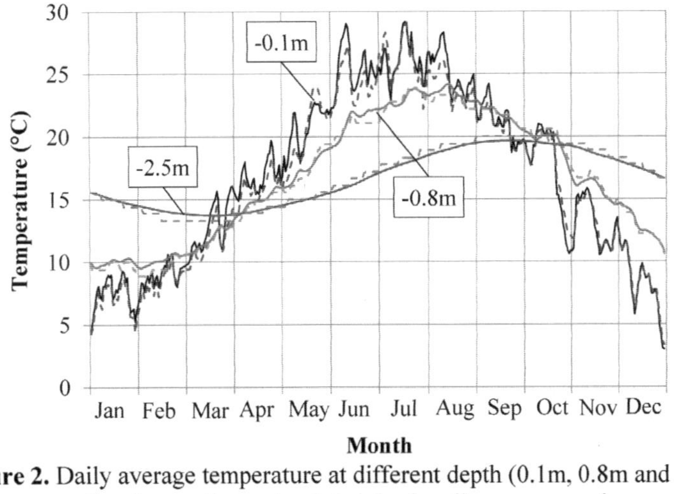

Figure 2. Daily average temperature at different depth (0.1m, 0.8m and 2.5m).
Continuous lines: simulated; broken lines: measured.

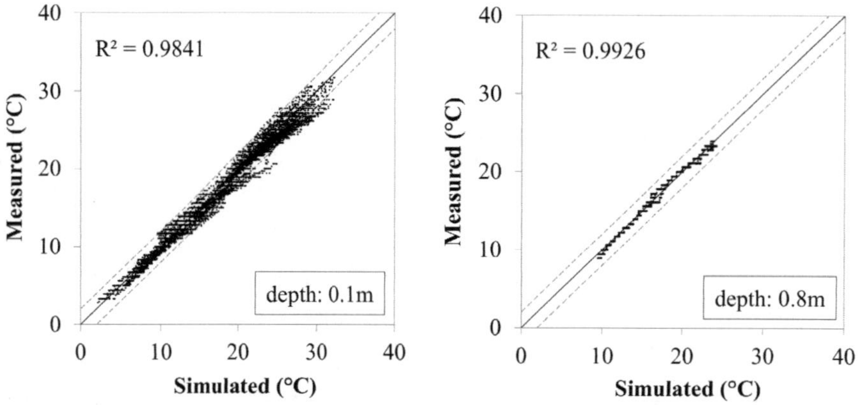

Figure 3. Scatter plot of the simulated and measured hourly temperature.

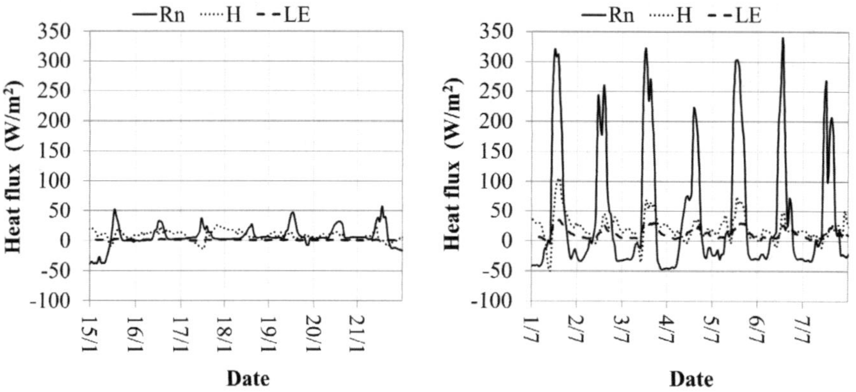

Figure 4. Components of ground surface energy balance during winter and summer.

dependence on the net radiation (R), as a result of the partial shading of the ground surface by the vegetation above. The sheltering effect by the vegetated layer reduces the average convective heat flux (H). The low wind speed, representative of an urban area, furtherly reduces the convective heat flux. H is relatively stable and varies between +100 and -50 W/m^2 depending on the temperature difference between air and surface. Finally, the heat loss from the surface due to the latent heat flux (LE), which is strictly related to the air temperature and soil heat flux, is nearly zero in winter. An increase is observed in summer, with a daily oscillation between 25 and 5 W/m^2 during daytime and night-time, respectively.

2.4. The energy loads for the ground heat exchanger

The energy requirement for space heating and cooling is defined as the amount of energy needed to maintain a constant target value of the building indoor temperature. In GCHPs, the thermal energy is extracted/transferred from/to the ground by means of the ground heat exchanger.

The modelling of the thermal behaviour of a building can be simplified with a lumped parameter approach by reducing the building to a resistance-capacitance model and taking into account the heat transfer through the walls, the thermal inertia of the building and the energy contribution by HVAC systems [25]. A simplified model, based on a lumped resistance-capacitance method, was applied in order to define an hourly energy requirement (both for heating and cooling) to be used in numerical simulation of Flat-Panels [26]. The transient temperature response of the building is determined by formulating an overall energy balance on the entire solid by relating the variation of the internal energy to the rate of heat loss at the surface, according to the difference between the indoor temperature of the building and the outdoor air temperature. In this case the actual ambient air temperature of Ferrara for 2014 was simplified with two sinusoidal trends; on an annual and daily basis, respectively, as shown in fig. 5 together with the hourly heat load profile during winter. A detailed description of the model is reported in [18]. The energy demand has been calculated for a building with an overall transmittance of 0.5 W/m^2K, and a volumetric heat capacity of 100 kJ/m^3. The heating season is supposed from mid-October to April 15th; the cooling season from May 5th to the middle of September. In the model, the heating/cooling system was set to operate to maintain a set point indoor temperature (20 °C in winter and 24 °C in summer). A time schedule was assumed to simulate actual working conditions for a residential building in a mild climate: 4 - 11 AM and 6 - 12 PM during working days, and 6 AM to 12 PM on the weekends. Furthermore, it has been assumed that the heating and cooling power of the system could not exceed 50 W/m^3 of the building when it is turned ON. The daily energy demand (at the Flat-Panel) for both heating and cooling is reported in fig. 6. Under the assumed conditions, the space heating/cooling system worked for a total of 2728 h in heating mode; only 792 h in cooling mode. The calculated energy demand for 2 m^3 of building has been supposed to be the full energy requirement for the geothermal closed loop, therefore the resulting heat flux time series (W/m^2) has been directly applied as the boundary condition at the Flat-Panel in the numerical model. As a result, the maximum heating and cooling heat flux at the ground heat exchanger was 100 W/m^2. On overall terms, during the heating season a total amount of 36.4 kWh (i.e. 72.8 kWh according to the symmetrical domain) for each metre of Flat-Panel have been extracted from soil for 190 days; during cooling season it was 21.8 kWh (43.6 kWh) 145 days.

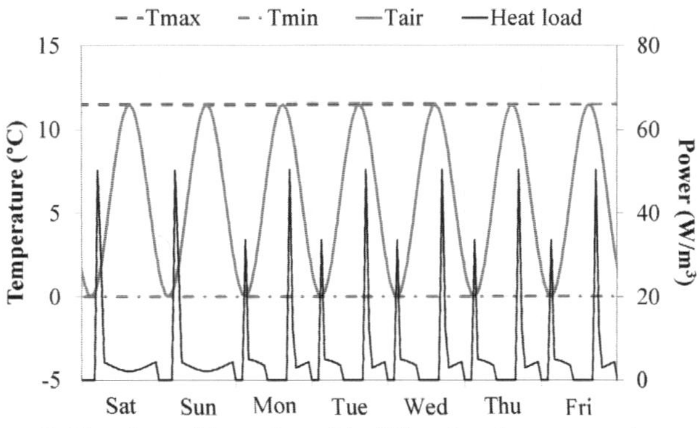

Figure 5. Hourly ambient air and building heating power time series.

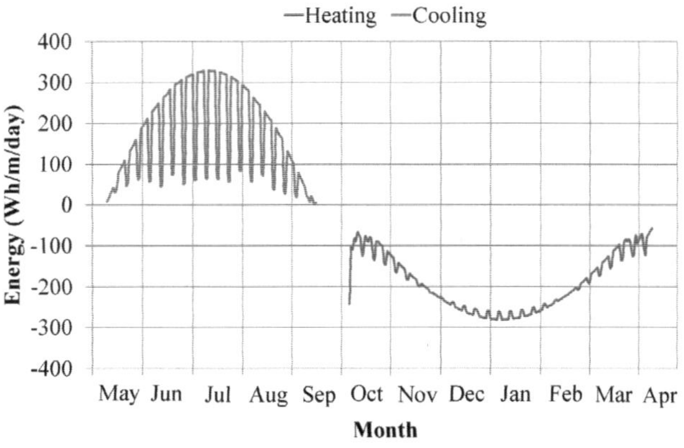

Figure 6. Energy requirements for space heating and cooling at the HGHE.

3. Results and discussion

A comparison between the numerical solution by setting 1st, 2nd BCs and the enhanced GSEB model (3rd kind BC) at the ground surface is here presented, similarly to [15]. A preliminary simulation was run using the GSEB model as the ground surface BC; then the equivalent heat flux and the equivalent surface temperature were obtained. A complete set of weather data of Ferrara for 2014 has been used and the soil temperature profile for the 1st day of 2014 was set as the initial condition. The operation of Flat-Panel was simulated in three new simulations with the proposed GSEB model, the equivalent heat flux and ground surface temperature as BCs, respectively.

The results are compared in terms of the average temperature at the HGHE wall surface which be considered as the temperature at the interface between the ground and the Flat-Panel surface, according to the simplifications and assumptions considered. This temperature is representative of the HGHE performance with the different BCs at the ground surface. Moreover, the temperature in the surrounding soil has been calculated to evaluate the effect of the heat transfer induced by the Flat-Panels.

The resulting time series for each BCs of the daily average temperature at the HGHE wall are shown in fig. 7. In the graph, a whole year is presented, starting in May, when the cooling season begins. The maximum temperature is reached in all three cases in August, according to heat released to the ground. The temperature follows a similar trend in the case where the GSEB and the equivalent temperature are assigned as BC. With the equivalent heat flux, the Flat-Panel surface has a maximum value 1.2 °C higher than the other cases. In the HF case, the thermal drift continues until the beginning of the winter season. The minimum in the temperature trend (at the middle of February in all three cases) is delayed of 45 days in comparison with the temperature at the ground surface and that of the air. When the equivalent heat flux is assigned to the surface (case HF) the temperature drecreases more rapidly. The maximum difference is of 1.2 °C in comparison to the other two cases on a daily average basis. A thermal drift of 1 °C is detected after a year in the case HF, because the equivalent heat flux, which was calculated without the HGHE operating, does not balance the energy demand of HGHE. By contrast, a slight discrepancy is calculated between the case GSEB and T. This results confirms that the use of 1st kind boundary conditions could be considered an acceptable simplification, according to our previous findings.

In fig. 8, a weekly detail of the hourly HGHE operation, when the minimum temperature at the Flat-Panel surface is reached in the heating period. According with the above considerations, temperature values are significantly different in the case of equivalent heat fluxes; therefore the effect of energy extracted or released is over-estimated. In the other two cases the differences, though present, are negligible.

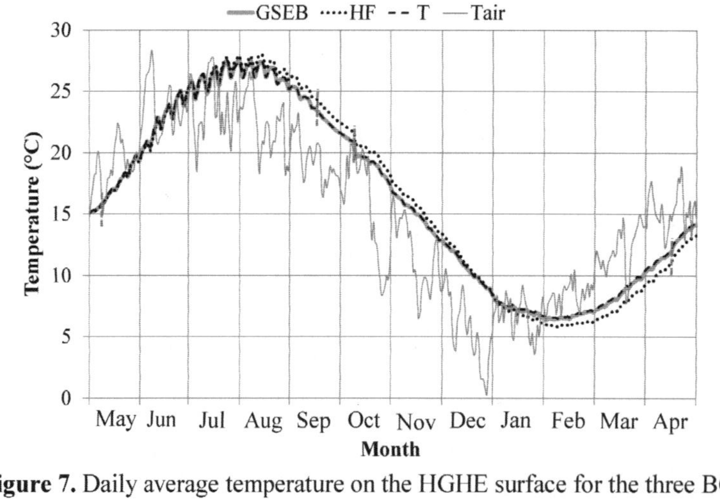

Figure 7. Daily average temperature on the HGHE surface for the three BCs.

Finally, the hourly time series of soil temperature at two different points in the soil domain are shown in fig. 9. A temperature probe is positioned at the average depth of the HGHE (-1.5 m) and 0.5 m in front of it, while the other is above the HGHE, 0.8 m deep in soil. The calculated maximum difference between the case HF and the two other cases is 1.4 °C above the exchanger. Furthermore, a significant thermal drift is maintained over the entire year. By contrast, the soil thermal field is comparable for GSEB and T, similarly to the temperature at the HGHE wall.

4. Concluding remarks

This paper is focused on the improvement of the energy balance at the ground surface model (GSEB) which was implemented in COMSOL Multiphysics to be tested as boundary condition at ground surface. The model requires the ground surface properties and a comprehensive weather dataset for simulations. The results of simulations carried out with the GSEB model were compared with the observed soil temperature data at different depths in 2014, proving to accurately predict the temperature in the soil. Moreover, the effect on numerical solutions of different boundary conditions at the ground surface was analyzed in modelling HGHEs, solving the unsteady heat transfer problem in a 2D computational domain.

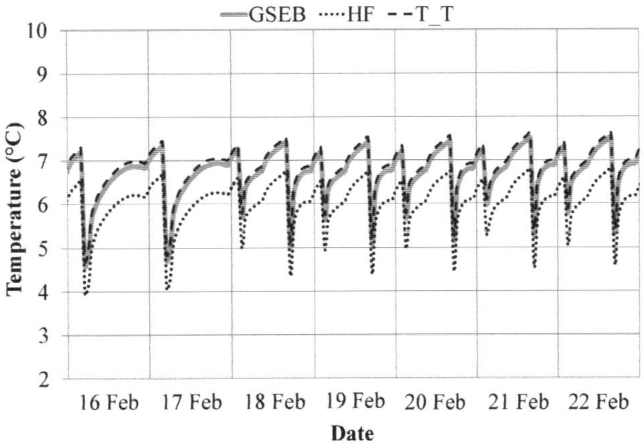

Figure 8. The hourly average temperature at the HGHE surface during a week of operation.

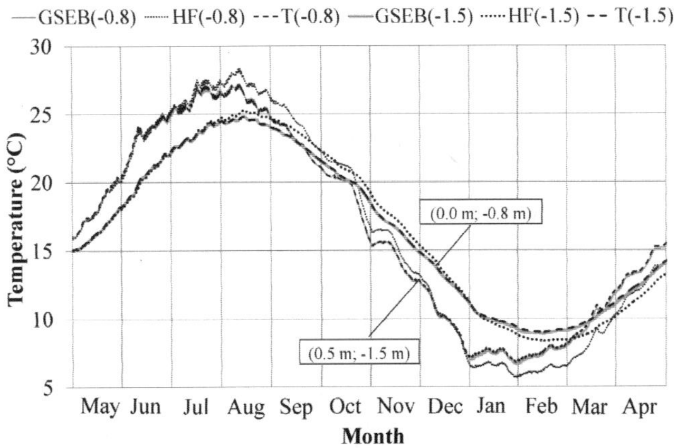

Figure 9. Average temperature in the soil for three boundary conditions.

Simulations were carried out to test the HGHE energy performance in heating and cooling, under the same environmental conditions. The GSEB model, the equivalent heat flux and temperature at the ground surface were used as boundary conditions (BC) of the 1st, 2nd and 3rd kind in simulations.

The solution of the equivalent heat flux at the ground surface is significantly different from the other two cases, and should be considered as a precautionary approach. On the other hand, when a boundary condition of the 1st kind (i.e. equivalent surface temperature) is used, a significant constraint is imposed to the surface temperature, which is not allowed to vary due to the heat transfer by HGHE. However, the correct estimation of the surface temperature is of great importance also in this case, and a preliminary simulation with a GSEB could be required. Consequently, a preferable approach to the problem is the setting of a 3rd kind boundary condition (i.e. the GSEB model), which not affects the calculation time. A correlation between ground surface and air temperature could be developed in the future to determine more quickly a feasible boundary condition in modeling HGHEs.

5. References

[1] Chiasson A D 1999 *Advances in modeling of ground-source heat pump systems* M.Sc. Thesis Oklahoma State University, USA.

[2] Mustaf O A 2008 Ground-source heat pumps systems and applications *Renewable and Sustainable Energy Reviews* **12/2** 344-371

[3] Gan G 2013 Dynamic thermal modelling of horizontal ground-source heat pumps *International Journal of Low-Carbon Technologies* **8** 95-105.

[4] Kavanaugh S P, Rafferty K 1997 *Ground source heat pumps-Design of geothermal systems for commercial and institutional buildings* ASHRAE Applications Handbook.

[5] Kusuda T, Achenbach P 1965 Earth Temperatures and Thermal Diffusivity at Selected Stations in the United States *ASHRAE Transactions* **71** 61-74

[6] Mei V 1986 *Horizontal Ground Coil Heat Exchanger. Theoretical and Experimental Analysis Oak Ridge National Laboratory*

[7] Lee K H, Strand R K 2008 The cooling and heating potential of an earth tube system in building *Energy and Buildings* **40** 486-494

[8] Fujii H, Yamasaki S, Maehara T, Ishikami T, Chou N 2013 Numerical simulation and sensitivity study of double-layer Slinky-coil horizontal ground heat exchangers *Geothermics* **47** 61-68

[9] Bottarelli M, Bortoloni M, Su Y, Yousif C, Aydin A A, Georgiev A 2015 Numerical analysis of a novel ground heat exchanger coupled with phase change materials *Applied Thermal Engineering* **88** 369-375

[10] Piechowsky M 1999 Heat and mass transfer model of a ground heat exchanger: validation and sensitivity analysis *International Journal of Energy Research* **23** 571–588

[11] Kupiec K, Larwa B, Gwadera M 2015 Heat transfer in horizontal ground heat exchangers *Applied Thermal Engineering* **75** 270-276

[12] Demir H, Koyun, Temir G 2009 Heat transfer of horizontal parallel pipe ground heat exchanger and experimental verification *Applied Thermal Engineering* **29** 224-233

[13] Gan G 2014 Dynamic interactions between the ground heat exchanger and environments in earth-air tunnel ventilation of buildings *Energy and Building* **85** 12-22

[14] Sofyan S, Hu E, Kotousov A 2015 A New Approach to Modelling of a Horizontal Geo-Heat Exchanger with an Internal Source Term *Applied Energy* **164** 963-971

[15] Bottarelli M, Bortoloni M, Su Y 2017 A study on the effect of ground surface boundary conditions in modelling shallow ground heat exchangers *Applied Thermal Engineering* **111** 1371–1377

[16] Bottarelli, M 2013 A preliminary testing of a flat panel ground heat exchanger *International Journal of Low-Carbon Technologies* **8(2)** 80-87

[17] Simms R, Haslam S, Craig J 2014 Impact of Soil Hetereogeneity on the Functioning of

Horizontal Ground Heat Exchanger *Geothermics* **50** 35-43
[18] Bortoloni M 2016 *Experimental analysis and numerical simulation of a Flat-Panel ground heat exchanger* Ph.D. Thesis University of Ferrara
[19] Herb W R, Janke B, Mohseni O, Stefan, HG 2008 Ground surface temperature simulation for different land covers *Journal of Hydrology* **356** 327-343
[20] Deardorff J W 1978 Efficient prediction of ground surface temperature and moisture with inclusion of a layer of vegetation *Journal of Geophysical Research* **83** 1889-1903
[21] Best M J 1998 A model to predict surface temperatures *Boundary-Layer Meteorology* **88** 279-306
[22] Frankenstein S, Koenig G 2004 *Fast All-season Soil Strenght (FASST)* Washington, D.C. *U.S Army Corps of Engineers.*
[23] Antonopoulos V 2006 Water Movement and Heat Transfer Simulations in a Soil und Ryegrass *Biosystems Engineering* **95** 127-138
[24] Allen R G, Pereira L S, Raes D, Smith M 1998 *Crop evapotranspiration. Guidelines for computing crop water requirements.* FAO Irrigation and Drainage Paper, 56, Rome: FAO.
[25] Morini G L, Piva S 2007 The simulation of transients in thermal plant. Part I: Mathematical model *Applied Thermal Engineering* **27** 2138-2144
[26] Bortoloni M Bottarelli M 2015 On the sizing of a Flat-Panel ground heat exchanger *International Journal of Energy and Environmental Engineering* **6** 55-63

Acknowledgments
The financial support to this research was provided by the University of Ferrara with funds: "Young Researchers Funds 2014, University of Ferrara".

Methods for the estimation of the energy stored in geothermal reservoirs

Alessandro Franco, Franco Donatini

Department of Energy, Systems, Territory and Constructions Engineering
University of Pisa, Largo Lucio Lazzarino, 56126 PISA

Corresponding author e-mail: alessandro.franco@ing.unipi.it

Abstract. The paper analyze the problem of the estimation of the potential of geothermal reservoirs with the objective of sizing of geothermal power plants in order to reach the goal of a correct matching of renewability and economic related issues. After an analysis of the typical approaches based on the First Order Method diffused in the literature since the late 70s and of the uncertainties connected with the use of simplified approaches, that have determined general overestimation of the geothermal energy stored in the reservoirs and of the size of the plants connected to various geothermal fields, the authors discuss about possible improvements based on the combination of theoretical simplified approaches with experimental data of heat flow, taking into account not only the energy stored, but also the possibility of producing a recharge.

1. Introduction

Geothermal energy is considered a strategic resource for the possibility of producing energy both for direct thermal use and electricity production but its characteristic of renewability are recognized only when a correct exploitation of the reservoirs is carried out. Some areas of the earth are well favoured in terms of the availability of both high temperature and subsurface fluid at relatively shallow depth. These areas are considered "geothermal fields" and "hydrothermal systems". Until now only these areas have been targets for geothermal development for power generation purposes and direct used of thermal energy. The renewability of the resource itself is directly dependent on the type and rate of utilization and the durability of a plant are directly connected to the way the resource is explored and then exploited and to a correct knowledge of a potential assessment, [1, 2].

In the past, geothermal power plants have frequently been built ''as large as possible'' in order to meet economic convenience in a reduced time. But this approach has determined several problems, [3, 4]. In recent times the concept of renewability has been improved introducing the idea of sustainability: this concept include the idea of using geothermal resources using some compensative actions, like cultivation strategies and reinjection. [5, 6]. The proper matching between the reservoir capability and the plants parameters (power size, extraction/reinjection rate) is a critical key point. The main task of potential assessment and sustainable design of a plant is the perspective of resource durability, [7].

Interesting contribution on the definition and evaluation of the sustainability and renewability of the geothermal energy uses are available in the literature, both in some classical contribution, like the one by Muffler and Cataldi, [8] and in more recent papers by Hähnlein et al. [9] and Axelsson [10]. Mainly in case of innovative geothermal utilizations (like for example EGS) the long-term consequences on the environment are not completely known yet. The renewability (and sustainability) reference level can vary, as one can adjust the system size and extraction rate according to an acceptable durability.

In the recent paper by Fox et al. in [11] the renewable capacity of deep systems is assessed and discussed in order to elaborate a rotating utilization strategy. Drilling underground proves that the earth temperature rises with depth at a rate called "geothermal gradient". But the heat content of the earth is so huge, that if it could be utilized, it could cover the entire energy needs for a lot of years and it is practically inexhaustible for human terms. In general heat and suitable temperature are not enough for geothermal exploitation. The availability of fluids, like deep water or saturated steam is also necessary. But the intensive use on some zones caused that some intensive use of geothermal energy caused local cooling down of the reservoir and the extracted flow is not exactly renewed due to heat flow from below, this causing remarkable problems during the production phase.

The worldwide experience in geothermal energy uses demonstrates a lot of cases in which a progressive reduction of production levels after the construction of the power station has been observed so that reinjection has been considered a common practice in geothermal energy uses, [12].

The geothermal potential assessment can be considered a relatively recent field of research. It is well known that, the occurrence of a geothermal system requires a heat source and a working fluid to transfer heat. In general the interest in the use of geothermal energy is connected with the observation of a geothermal anomaly. Considering that the average heat flow in the earth's surface is approximately 60mW/m^2, the occurrence of on anomalous gradient is connected to the relevation of strongest heat flow values, where volcanic activity and presence of fluids and activity transports, the heat flow values overcome the value of 100 mW/m^2 and reach upper levels of the order of 250-1000 mW/m^2 in places characterized by important geothermal activity, like Larderello or Monte Amiata.

The estimation of the heat flow is a good basic element to understand the interest of a geothermal area, but is often not sufficient to clearly understand the capacity of a reservoir. The success of a geothermal project is mainly connected with the dimensions of the reservoir and with a well defined estimation of the energy storage so that a quite long life of the geothermal plant could be obtained.

In the first part of the paper meaningful examples of the techniques and approaches used in the evaluation of the geothermal energy potential are briefly described and discussed, with a particular attention for the medium-low enthalpy resources. Since the preliminary scientific analyses of the potential assessment of geothermal reservoirs it has been stated that a general methodology for every kind of geothermal field does not exist. Although in the literature a lot of different methods and reliable principles have been treated and tested, it seems quite difficult to find a complete multidisciplinary view of the problem, in which the connections between the parameters of plant design, geological and geophysical characterization and reservoir engineering can be clearly evidenced. In general the methods consist in the estimation of the thermal energy content in a reservoir, considering the different contribution from the solid and liquid phases, or from the main underground heat sources in a certain geometric domain. It is now interesting to show some of the main features and characteristics of these methods. Then the authors discuss the perspective of integration of the methodological "First order" methods based on the analysis of simple experimental data coming from geothermal drilling operation, like thermal gradient and local heat flow, for the definition of methods for the definition of sustainable levels of power production.

2. The problems connected with an uncorrected estimation of the energy potential

The excessive exploitation of a geothermal reservoir could determine problems in each kind of geothermal system. The reduction of power in case of vapour dominated geothermal field is well clear but not dramatic. The same problems can be evidenced in case of flash-steam power plants.

In several cases in the industrial practice, fluid extraction reached excessive rates and it has not been possible to maintain productivity in the long-term. The utilization of the Geysers area in California is a well-known example of excessive production [13], or the case of Momotombo geothermal field, analyzed by Porras and Bjornsson [14]. A sustainable perspective in geothermal projects appears to be particularly important in case of small size power plants (below 1 MW of net power produced) based on the utilization of ORC power plants: the high specific costs of the plants and drilling could suggest over sizing and a consequent overexploitation of the reservoir. A temperature reduction of the geothermal source could cause the end of life of the plant because it could be impossible to maintain a correct pinch-point value in the heat exchanger (in Fig. 1 the decrease of the temperature profile of geothermal fluid, caused by a decrease of the source temperature, determines a reduction of pinch point value from the design value *PP1* to the reduced value *PP2*). This problem is important for each kind of ORC, but in particular with reference to advanced heat recovery solutions, like Rankine with superheater and supercritical and for $T_{geo} < 120\text{-}130\ °C$.

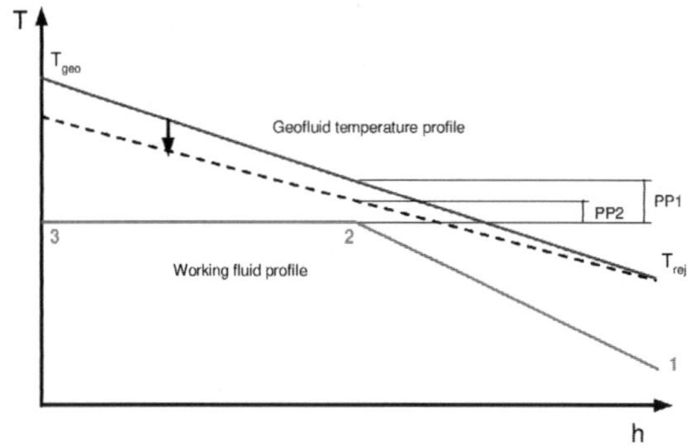

Figure 1. Effects of the temperature decline of the reservoir for the heat exchanger of a binary plant

3. Geothermal potential assessment of hydrothermal reservoirs

Considering the installation of a geothermal power plant the concept of renewability is defined as the ability to maintain the installed power capacity of a plant indefinitely without encountering any resource degradation and without any action; but renewable capacity is often too small for commercial development. In different terms the concept of sustainability has been proposed: this is defined as the ability to economically maintain the installed capacity, over the life cycle of a power plant, by taking practical action (such as, reinjection) in order to compensate for resource degradation (pressure drawdown and/or cooling). [1, 2]. In order to design a plant both with the perspective of renewability and the one of sustainability, the first important step is considered the Geothermal Potential Assessment. The greatest challenges lie in understanding the size and thermal energy of a reservoir as well as the constraints on extracting that thermal energy. A lot of different reliable methods have been used, but a general purpose methodology for every kind of geothermal field doesn't exist [15-16].

Several studies deal with specific geothermal fields worldwide, more than general value surveys. Different instruments and exploration techniques are available today to improve the detail of the information needed to design plants for energy purpose and consequently to estimate the assessment of the reservoir. The geothermal potential of a particular area means mostly the evaluation of temperature (T_{geo}) and pressure (p_{geo}) of the geothermal fluid and of the maximum mass flow rate ($\dot{m}_{geo}$) that can be extracted maintaining the thermal properties of the reservoir for a long time. It can also be defined as the estimation of the energy stored in a reservoir. Resource Estimation Methods can be grouped according two main categories: Methods with no production data and methods integrated with the use of production data. Some of the available methods are enumerated in Table 1.

Table 1. Methods for estimating the energy potential of a geothermal reservoir

Methods without production data	Methods with production data:
• Heat Flow	• Lumped Parameter Models
• Areal Analogy (Power density)	• Decline Curve Analysis
• Volumetric Methods (deterministic & probabilistic)	• Numerical Reservoir Simulation

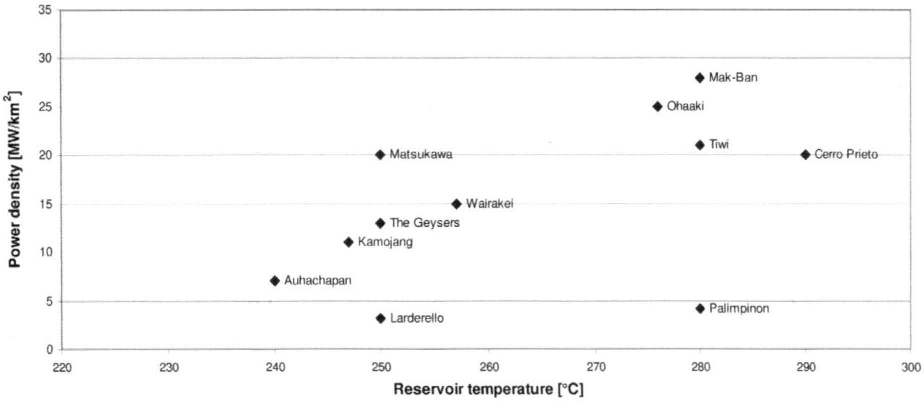

Figure 2. Power density sustained in developed geothermal fields [data available in 17]

Analyzing the data referred to the historical development of geothermal energy use, remarkable differences among the various geothermal fields can be clearly evidenced, as reported in Fig. 2, while the power density sustained in the main geothermal fields, in which industrial utilization of geothermal energy has been developed, is reported. It is possible to observe the differences among fields characterized by the same temperature (see the comparison among Mak Ban, Tiwi and Palimpinon).
One of the most widely and consistently usable methods during all stages is the Volumetric stored heat estimation, early defined in [8]. The basic method involves calculating (usable) heat in place using estimated reservoir volume, rock and fluid characteristics and average temperature, against a reference temperature. Recoverable heat is then estimated by introducing a *Recovery Factor,* which can be seen as fraction of the (usable) heat in place that could be produced feasibly by actual production wells over a reasonable (project) timeframe. Finally a feasible, sustainable plant capacity (MWe) for a given/estimated plant life, conversion efficiency, and power plant availability is defined.

Muffler and Cataldi in [8] identified four methods for assessing geothermal resources: surface heat flux, volume, planar fracture and magmatic heat budget. The total energy available is the fraction extractable and useful of the stored energy in the reservoir at favorable technical and economic conditions. It can be roughly estimated multiplying the total energy stored for an appropriate recovery factor (R), which value is between 0 and 1. The estimation of the recovery factor is not a trivial task, it depends from the site characterization and it can also be a result of estimation according to the individual experience. A brief list of the general results that should be evaluated is here proposed: thermal energy stored (at a certain time) in the reservoir; temperature, pressure and rate of the extracted fluid; chemical composition of the fluid and saltiness (to define the appropriate reinjection temperature T_{rej}); number of wells to be drilled and mutual distances between them; time interval to have an appreciable decrease in the rate and temperature of fluid (or productivity); definition of both the "Base" resource available and of the "Effective" resource, which is useful under favorable and sustainable (economic-environmental-technological) conditions; siting of the reinjection wells (number, mutual distances and interference effects); reinjection strategy (effects of reinjection on productivity); number of wells for compensation. Considering mainly water dominated geothermal field at moderate temperature, the energy potential can be first of all referred to the available temperature of the aquifer (T_R). An equivalent specific energy content of the reservoir (both rocks and fluid), namely $\varepsilon_{(T_R-T_0)}$, can be defined referring to the cooling down to a low temperature level, in this case the environment temperature T_0. Considering the reinjection temperature, T_{rej}, the useful temperature difference is $\Delta T=\left(T_R-T_{rej}\right)$, being $T_{rej} < T_0$ (see Fig. 1). This means that a higher mass flow rate should be extracted for power production, according to the upper limit given by

$$\dot{m}^{*} \cdot c \cdot \left(T_R - T_{rej}\right) \le \dot{m} \cdot \varepsilon_{(T_R - T_0)} \tag{1}$$

The upper limit for the energy potential can be considered as a function of the whole "geothermal system", (as defined in the Introduction and remarked in Fig. 1) as in Eq. (2)

$$\Pi = f\left(geothermal\ system\right) \tag{2}$$

The meaning of the last two equations is that, for given values of maximum energy potential and a well defined value of the temperature difference $T_R - T_{rej}$, a maximum value of mass flow rate (extracted/reinjected) $\dot{m}^*$ can be defined. The value $\dot{m}^*$ clearly depends on a lot of variables (in connection both with natural and technical elements): permeability distribution, hydraulic linking between the production and the reinjection areas, siting of the wells, natural recharge (meteoric water) to the reservoir. So it is clear that the value of ΔT used is inversely proportional to the limit value of the mass flow rate extracted $\dot{m}^*$. If T_{rej} increases, as it can happen in case of rising chemical deposition, the mass flow rate $\dot{m}$ increases, and it could reach excessive values, causing unwanted cooling down of the whole aquifer. Consequently each reservoir presents an optimal combination of mass flow rate extractable and reinjection temperature and a correct design should follow this rule. As it can be seen, this optimization problem involves the whole "geothermal system" (plant, reservoir, environment and their mutual links), so the interdisciplinary approach becomes necessary. The difference $\Delta T=(T_R-T_{rej})$, together with the availability of geofluid and the environment reference temperature (T_0) permits to define the exergy and energy potential of the geothermal field. In evaluation of the commercial potential of a stimulated geothermal reservoir, three must be sufficiently estimated for investment decisions: the available heat content of the reservoir; the optimum thermal extraction rate and the total heat extracted respectively.

4. The first order methods for the definition of the reservoir potential

The thermal extraction rate for a specific reservoir depends on two sets of reservoir characteristics: the heat transfer properties of the reservoir and the flow regime for heat transfer.

The heat transfer properties are determined by the rock-type fracture network which controls the rate of conductive heat transfer to the rock-block surfaces. The energy stored in a geothermal reservoir can be basically determined with the following equation

$$E_R = \rho \cdot c \cdot V \cdot \left(T_R - T_{ref}\right) \tag{3}$$

where c is the volumetric specific heat of the reservoir rock, V is the volume of the reservoir, T_R is the characteristic reservoir temperature, and T_{ref} is a reference temperature.

The thermophysical properties are here referred to the weighted values in terms of mass fraction of rock and groundwater. They can be estimated for example considering that geothermal energy is the heat of the earth included in the rock matrix (for example about 90%) and the pore fluid (for example about 10%). An evaluation of the energy stored based on such a simplified approach determines an overestimation of the potential. Just to understand the problem, considering a volume of the reservoir of 3x3x1 km, typical dimensions of small size reservoirs, like as discussed in [18], a density of 2500 kg/m^3 and a specific heat of 1000 J/kg °C and a temperature difference of 100 °C an energy amount of more than 10^{18} J can be obtained considering some general data about the thermophysical properties ($E_R = 4.5 \cdot 10^{18}$ J). This corresponds to a value of about 108 MTep (about 60% of the total energy used in Italy). Considering the possible installation of a plant, the estimation on the size could be connected to the perspective of maintaining a reduced value of the temperature decrease during the lifetime of the plant, maintaining this value in the range of some degrees (e.g 1°C after 10 years of operation).

Using the same data considered in the previous analysis a potential of more than 2000 MW can be obtained, so that considering a First Law efficiency in the range 0.1-0.15 a power ranging between 200 and 300 MW can be roughly estimated. If a system like the one of Larderello is considered, according to the model described in [19], a dimension of 70x70 km can be considered, so that a volume of about 600 times the one estimated and consequently a power installed of more than 100000 MW could be installed in the area. But this is surely an overestimated value if we consider that at the end of the 70s a running capacity of at least 500 MW caused the first problems.

Similar considerations can be applied to the other important Italian geothermal field: Monte Amiata. In this case, according to Barelli et al. [20], similar conditions of the fluid can be estimated, but a volume of 80 km^3 can be roughly estimated, so that a maximum level of 1200 MW can be estimated. But it is quite clear that such a simplified approach caused an overestimation of the real potential of the reservoirs.

The industrial production based on this kind of estimation has encountered a lot of problems, in connection with industrial exercise of geothermal plants: a real renewable level corresponds to quite low levels of the power, but the interaction with the reservoir (based on reinjection) determines the definition of a sustainable level. Considering this problem, the approach of the geologists has been to reduce the effective value of the energy stored as a function of the real nature of the reservoirs considering its porosity and the real density, and the availability of a geothermal fluid, characterized by the enthalpy value (h_{WH}). The thermal energy that can be extracted at the wellhead is given by

$$E_{WH} = m_{WH} \cdot (h_{WH} - h_{ref}) \tag{4}$$

where m_{WH} accounts for the effective nature of the reservoir

$$m_{WH} = V \cdot (0.5 \cdot \phi_e \cdot X) \cdot \rho_f = V \cdot \phi_R \cdot \rho_f \tag{5}$$

where ϕ_e is the effective reservoir porosity, 0.5 is the fraction of the reservoir that is porous and permeable, X is the degree of saturation, and ρ_f is the fluid density. Hence useful energy is here just the one related to the fluid content of geothermal volume. The ratio between Q_{WH} defined by Eq. (4) and the term defined by Eq. (3) determined a Recovery Factor

$$R_g = \frac{E_{WH}}{E_R} = \frac{V \cdot \phi_R \cdot \rho_f \cdot (h_{WH} - h_{ref})}{\rho \cdot c \cdot V \cdot (T_R - T_{ref})} \tag{6}$$

where ϕ_R, can be considered as a lumped “reservoir porosity”. All of the variables in Eq. (6) are functions of reservoir temperature, pressure and chemical composition, except for ϕ_R, that is the only real unknown variable. According to different analyses reported in the literature, the values of the Recovery factor R_g can vary in general between 0.05 and 0.25 depending on the characteristics of the reservoir. The introduction of the corrections give reasons to a more realistic estimation of the potential of a geothermal reservoir so that reduced values of the energy stored in the reservoir and of the power of a plant using the resource for a well defined period of time (100 years), as reported in Table 2. Analyzing the data of Table 2 it is quite clear the motivation for what before the use of reinjection strategy, the reservoir has been subjected to pressure and temperature decay, causing a reduction of production. The overexploitation of the geothermal reservoir with an increase of the running capacity up to 2200 MW causes an important reduction of the power production during the 90s. The same problems occurred in the Larderello field at the end of the 70s, but the development of reinjection strategy permitted to overcome the problem. The methods discussed before are based on the estimation of the thermal energy stored in the reservoirs. But the reservoir is a closed system with reduced natural recharge of geothermal fluid, while the use of geothermal energy is more directly associated to the fluid. The observation of the reservoirs used for power production shows that pressure and temperature declines continuously with time, particularly in systems that are closed or with small recharge. Production potential is, therefore, often limited by lack of water rather than lack of thermal energy.

Table 2. Properties, recovery factors (R_g) and estimated power production (P) for some reservoirs [16]

Field	**T_R (°C)**	**V (km^3)**	**ϕ_R**	**R_g**	**P (MW$_e$)**
The Geysers	240	150	0.05	0.11	1600
Coso	275	40	0.05	0.08	250
Dixie Valley	220	4/10	0.05/0.13	0.08/0.21	70

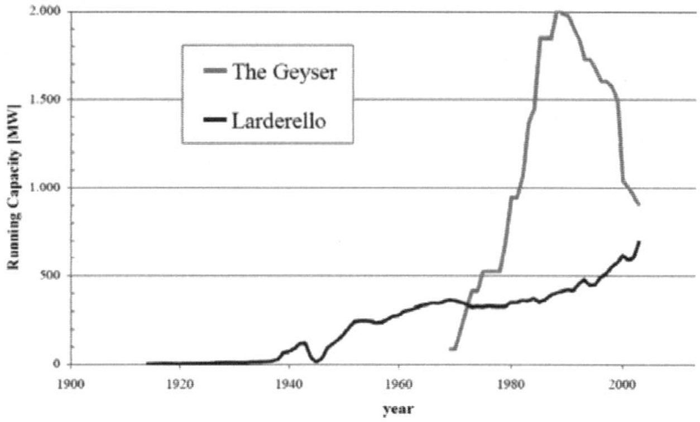

Figure 3. Larderello and The Geysers: a comparison between running capacity [21]

A different first order method consist in the estimation of the thermal energy content in a reservoir, considering the different contribution from the solid and liquid phases, or from the main underground heat sources in a certain geometric domain. In this way it is possible to clearly distinguish the contribution of the rock and of the fluid. Different volumes of rock or fluid are identified, their average temperature is estimated and used to calculate the heat stored, compared to a reference temperature (environment). The thermal energy of a reservoir, E referred to a i-th volume can be calculated from the following equation:

$$E_i = C_{vi} \cdot V_i \cdot (T_i - T_0) \tag{7}$$

where C_v is the volumetric heat capacity ($\rho \cdot c$) of the volume reservoir of volume V_i, T_i is the average temperature of the i-th volume and T_0 a reference temperature (e.g. environment). The volumetric domains containing rock or fluid phase are considered when defining the total porosity ϕ as the ratio between the volume of the empty spaces (V_v) and the total volume V so that:

$$\phi = \frac{V_v}{V} \tag{8}$$

But in general not all the empty spaces in a rock domain are full of liquid or connected and the underground water flow is possible only between the interstitial inter-connected empty volumes, so the effective porosity ϕ_e should be considered. ϕ_e takes into account the effectively interconnected voids, and for this reason it is generally less than ϕ. It is evident that the contribution to the thermal energy content of the reservoir E can be distinguished between the rock (subscript r) and fluid (subscript w):

$$E_i = E_{ir} + E_{iw} \tag{9}$$

$$E_i = (1-\phi_i)\cdot c_{ri}\cdot \rho_i \cdot V_i \cdot (T_i - T_0) + \phi_i \cdot c_w \cdot \rho_w \cdot V_i \cdot (T_i - T_0) \tag{10}$$

In Eq. (10) c_r is the volumetric heat capacity of the rock and c_w the volumetric heat capacity of the fluid (geothermal water), while ρ_i and ρ_w are respectively the density of the i-th rock volume and the density of the water in the pores. The errors made in this case can be then reduced by means of the introduction of a recovery factor R, which can be defined as the ratio:

$$R = \frac{E_S}{E_R} \tag{11}$$

where E_S is the resource which is extractable in the current technological and economic conditions in a sustainable way and E_R is the total resource available in the reservoir. The definition of this factor is similar to indexes used in the oil and gas industry.

Even if with all the corrections introduced with respect to the basic simplified models determines more realistic values of the energy stored a lot of problems are connected to the estimation of the various parameters as largely discussed in the literature. Many factors can be uncertain in the use of the first order methods and those are largely discussed by authors. Among those surely there are: reservoir temperature and rejection temperature, reservoir Area/Size and reservoir thickness are object of specific attention as well as Rock Porosity. But the real problem connected to the use of the first order methods is the definition of the recovery factor. Just to understand the recovery factor varying from 0.05-0.20 in more recently published papers as [1], while values from 0.25 to 0.5 are considered in previously published papers, like [22], only to consider one of them. Considering this particular approach, different evaluation of the power production potential of the Larderello and Mt. Amiata geothermal fields can be estimated, as reported in Figs. 4-5.

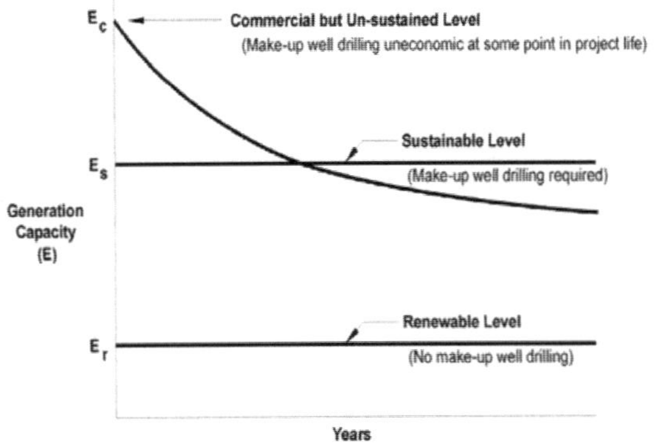

Figure 4. Difference among renewable, sustainable and excessive production connected to geothermal reservoirs: a schematic representation [1].

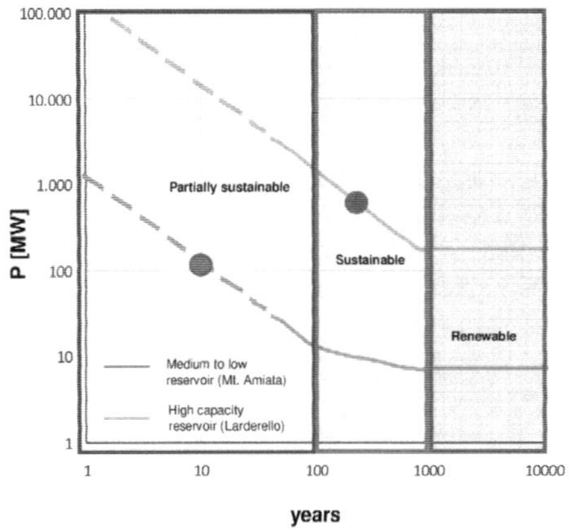

Figure 5. Definition of possible levels of output power for two different Italian geothermal reservoirs.

5. Estimation based on combination of numerical model of the reservoirs combined with geophysical measurement: the importance of natural heat flow

As discussed in the previous section, the "first order" methods, even if useful to have a preliminary evaluation on the energy stored in a geothermal reservoir, have some important intrinsic limitations. The concept of defining a kind of "hot" volume is physically based but important elements, like the permeability of the reservoir are not directly taken into account. The real challenges in the geothermal resource assessment lie in understanding the size and thermal energy of a reservoir as well as the constraints on extracting that thermal energy [16].

Strictly speaking, the power that can be extracted from a geothermal reservoir must be connected not only with the amount of energy stored, but mainly with the heat and momentum transport. So it is surely possible to think to some improvements of the first order methods combining such a kind of approach with some elements that can be derived by experimental data. The natural heat flow appears to be a key element. It involves estimating both convection and conduction components of the natural heat flow from the earth to the surface. The natural heat flow from a geothermal system is the combination of both its advective (e.g., fluid discharge from natural hot springs) and conductive parts and gives a clear indication not only about the energy stored, but also on the thermal recharge of the reservoir. Conductive heat flow, q, can be measured as:

$$\dot{q} = k\frac{dT}{dz} \tag{12}$$

where k is the thermal conductivity of the rock and dT/dz is the vertical temperature gradient. Considering the natural gradient dT/dz = 0.03 °C/m and a typical value of the rock thermal conductivity (k=1.5 – 3 W/mK) a basic value of the conductive heat flow in the range 45-90 mW/m^2! can be obtained. It is clearly possible to understand that the occurrence of heat flow higher than 100 mW/m^2 can be connected with an anomalous value of the thermal gradient as well as the presence of a fluid (water or steam) at a well defined temperature and circulating at a velocity w_f so that the heat flow con be connected to the additional avvective term related to temperature difference between fluid and rocks, ΔT_f and to the velocity of the fluid:

$$\dot{q} = k\frac{dT}{dz} + \rho \cdot c_p \cdot \Delta T_f \cdot w_f \tag{13}$$

If the total heat flux is sensibly higher than 100 mW/m^2, the convective component is remarkable. The term of velocity is directly connected to the hydraulic conductivity k. In general the measured heat flow stands in the range between 20 and 60 mW/m^2. In the oceanic zones the average heat flow is of the order of 100 mW/m^2. In particular cases, the heat flow is anomalously high (e.g. in some places of the Tirrenian coasts 100-450 mW/m^2). The heat flow is certainly better representative of the real energy potential of the reservoir and of the level of natural recharge because according to the description contained in Eq. (13) a direct connection with the fluid flow motion and consequently of the permeability of the shallow reservoir can be considered. The availability of water at a temperature of about 200 °C, considering the typical values for the permeability of shallow reservoirs, in the range between 10^{-9} and 10^{-7} m/s, made reason of additional heat fluxes in the range between 10 to 1000 mW/m^2. Lower are the values of the heat flow rate in case of superheated steam. If a fluid in two phase conditions is present, like in Larderello the heat flow is reduced but it is subject to local remarkable increases in the zones where liquid is available.

New approaches to the geothermal resource assessment involve geophysical techniques enhancements, improvements and public access to the databases, numerical simulation of the fields. Combining the indications furnished by the energy stored in the reservoirs (strictly dependent on the volume rather than on the thermodynamic conditions) and by the heat flow (function of the thermal gradient and of the mass flow rate) it is possible to conduct a better estimation of the energy stored in the reservoir and the power that can be extracted with and without reinjection strategies.

A further refinement is the development of a complete simulation model of the reservoir. One of the authors of the present paper has already tested the aspects of the numerical models and integration with experimental data, [18]. The idea is that physical and hydrogeological analysis of the exploration data can be synthesized in a conceptual model of the reservoir. In this way it is possible to develop a method to pursue the sustainability of a geothermal project defining the production scenario and to meet the right matching between the renewability and the economic profitability of the investment.

6. Conclusions

The geothermal potential assessment is a relatively "young" field of research but it appears to be extremely relevant in order to prevent problems in the future development of energy systems based on the use of geothermal energy resources. In the last years a lot of enhancements have been elaborated and tested by several authors, but at present the problem is already open. The authors have discussed the value of the methods and the application to two Italian geothermal reservoirs, Larderello and Mt. Amiata. Observing the recent history connected to the exploitation of geothermal fields it is possible to conclude that it is not easy to estimate the energy potential of a reservoir, but it can be observed that:

- The idea of a complete renewability is not good for commercial exploitation of geothermal reservoirs: the perspective of sustainability is surely more suitable for giving indications about the and their role in arriving at a decision on the plant capacity which a reservoir can support.
- Even if used for a long time the potential assessment of a geothermal reservoir cannot be related only to the thermal energy content of a reservoir: this gives a general overestimation of the potential while only a part of this energy can be used: the first order methods are not useful.
- The production potential of geothermal fields is often limited by the presence of the fluid rather than of the energy stored; for this reason, the lack of water rather than lack of thermal energy explain a lot of problems connected to the energy production systems.
- The analysis of heat flow measurements could be an important tool for characterizing the potential of shallow geothermal reservoirs, particularly in areas where magmatic heat input does not complicate interpretation of the data that should be correlated to the thermal gradient and to the presence of a fluid at an interesting temperature and of a mass flow rate.

References

[1] Sanyal S K, 2005. Sustainability and renewability of geothermal power capacity, Proceedings of the World Geothermal Congress, Antalya, Turkey.
[2] Rybach L, 2003. Geothermal energy sustainability and environment. *Geothermics* **32**, 463-470.
[3] Stefansson V, 2002. Investment cost for geothermal power plants. *Geothermics* **31**, 263–272.
[4] DiPippo R, 2008. Geothermal power plants: principles, applications, case studies and environmental impact, second ed., Butterworth-Heinemann, New York.
[5] Franco A and Vaccaro M, 2014. A combined energetic and economic approach for the sustainable design of geothermal plants. *Energy Conversion and Management* **87**, 735-745.
[6] Franco A and Vaccaro M, 2012. An integrated "Reservoir-Plant" strategy for a sustainable and efficient use of geothermal resources. *Energy* **37**, 299-310.
[7] Vaccaro M, 2013. Multidisciplinary approach for the sustainable utilization of medium-low temperature geothermal resources, Ph.D. thesis, University of Pisa: http://etd.adm.unipi.it/.
[8] Cataldi R and Muffler P, 1978. Methods for regional assessment of geothermal resources. *Geothermics* **7**, 53–89.
[9] Hähnlein S, Bayer P, Ferguson G and Blum P, 2013. Sustainability and policy for the thermal use of shallow geothermal energy. *Energy Policy* **59**, 914-925.

[10] Axelsson G. 2010. Sustainable geothermal utilization case histories; definitions; research issues and modeling. *Geothermics* **39**, 283-291.

[11] Fox D B, Sutter D, Beckers K F, Lukawski M Z, Koch D L and Anderson B J, 2013. Sustainable heat farming: modelling extraction and recovery indiscretely fractured geothermal reservoirs. *Geothermics* **46**, 42–54.

[12] Kaya E, Zarrouk S J and O'Sullivan M J, 2011. Reinjection in geothermal fields: A review of worldwide experience. *Renewable and Sustainable Energy Reviews* **15**, 47–68.

[13] Khan M A, 2010. The Geysers Geothermal Field, an Injection Success Story, *Proceedings of the World Geothermal Congress*, Bali, Indonesia.

[14] Porras E A and Bjornsson G, 2010. The Momotombo reservoir performance upon 27 Years of Exploitation, *Proceedings of World Geothermal Congress*, Bali, Indonesia.

[15] Mendrinos D, Karytsas C and Georgilakisa P S, 2008. Assessment of geothermal resources for power generation. *Journal of Optoelectronics and Advanced Materials* **10**, 1262-1267.

[16] Williams C F, 2004. Development of revised techniques for assessing geothermal resources, *Proceedings of the Twenty-Ninth Workshop on Geothermal Reservoir Engineering Stanford University*, Stanford, California, SGP-TR-175.

[17] Van Campen B, 2014. Resource assessment, techniques and reporting, Internal Report, The Geothermal Institute University of Auckland Santiago de Chile, 26-29 May 2014.

[18] Franco A and Vaccaro M, 2014. Numerical simulation of geothermal reservoirs for sustainable design of energy plants: a review. *Renewable and Sustainable Energy Reviews* 30, 987-1002.

[19] Romagnoli P, Arias A, Barelli A, Cei M and Casini M, 2010. An updated numerical model of the Larderello–Travale geothermal system, Italy. *Geothermics* **39**, 292–313.

[20] Barelli A, Ceccarelli A, Dini I, Fiordelisi A, Giorgi N and Lovari F, 2010. A review of the Mt. Amiata geothermal system, Italy. *Proceedings of World Geothermal Congress*, Bali, Indonesia.

[21] Bertani R, 2010 Strategic vision for the development of geothermal energy towards 2015, *Proceedings of the Conference of Italian Geological Society*.

[22] Ogena M S and Freeston D H. 1988, Sensitivity analysis of the greater Tongonan field resource assessment, *Proceedings of the 10th New Zealand Geothermal Workshop*, 67-72.

IOP Conf. Series: Journal of Physics: Conf. Series **796** (2017) 012026 doi:10.1088/1742-6596/796/1/012026

Temperature field for radiative tomato peeling

G Cuccurullo and L Giordano
Department of Industrial Engineering, University of Salerno,
Via Giovanni Paolo II 132, 84084 Fisciano (SA), Italy

E-mail: gcuccurullo@unisa.it

Abstract. Nowadays peeling of tomatoes is performed by using steam or lye, which are expensive and polluting techniques, thus sustainable alternatives are searched for dry peeling and, among that, radiative heating seems to be a fairly promising method. This paper aims to speed up the prediction of surface temperatures useful for realizing dry-peeling, thus a 1D-analytical model for the unsteady temperature field in a rotating tomato exposed to a radiative heating source is presented. Since only short times are of interest for the problem at hand, the model involves a semi-infinite slab cooled by convective heat transfer while heated by a pulsating heat source. The model being linear, the solution is derived following the Laplace Transform method. A 3D finite element model of the rotating tomato is introduced as well in order to validate the analytical solution. A satisfactory agreement is attained. Therefore, two different ways to predict the onset of the peeling conditions are available which can be of help for proper design of peeling plants. Particular attention is paid to study surface temperature uniformity, that being a critical parameter for realizing an easy tomato peeling.

1. Introduction

Nowadays, peeling of tomatoes is performed by using steam or lye, which are expensive and polluting techniques. In this connection, interest in radiative heating as means of sustainable peel removal is growing among the scientists, [1-5]. In facts, after a suitable warming up, loosened skin can be easily removed by mechanical pinch rollers. Thus, predicting the thermal response of the tomato during the radiative dry peeling process can lead to good peeling performance and, in addition, it can be of help in designing a suitable radiative heating system, ensuring high quality of peeled tomatoes and energy saving.

A typical configuration realized in industrial plants for tomatoes skin heating adopts a matrix of infrared emitters. The latter illuminates only one side of the target tomatoes which lay on a conveyor system. Since the target is rotating during the transportation, on average, heat is transferred all over the surface of the tomato, [6]. Computational simulation is usually adopted to describe such a process assuming a radiative opaque behaviour for the tomato, [3-4]. On the other hand, several studies are based on radiative volumetric absorption due to the penetration of infrared (IR) radiant energy at a food surface and, consequently, they deal with semi-empirical exponentially decaying models with a finite penetration depth [7-9].

In this context, the novelty herein presented takes into account the effect of the tomato rotating speed both by numerical and analytical point of view. Formerly, a finite element code, i.e. Comsol Multiphysics, is employed to study a 3D model of the rotating tomato; the shape of the target tomato is suitably adjusted by means of a parametric representation, [3], while radiative behaviour is assumed grey and diffuse. A ceramic emitter acting as heat source is conveniently modelled as well. Correspondingly, an analytical model is introduced which involves a semi-infinite slab cooled by convective heat transfer and heated by a pulsating heat source; the latter was described as the positive half wave of a sinusoidal function. The proposed analytical solution shows a pretty good agreement with the corresponding numerical model, until the kernel of the tomato remains at the initial temperature. Thus, an easy to handle model for predicting the onset of the peeling conditions is actually available. However, the numerical solution is required wishing to recover a value for the rotational velocity enabling to get the desired uniformity degree for the surface temperature.

2. Basic equations

In the context of industrial tomato processing, successful peel separation requires the surface temperature to attain values as high as 100°C in a very short warming up period, typically no more than 60 s. Assuming water thermophysical properties apply for tomato, it is easy to check that the penetration depth is confined to really few millimeters under the tomato skin. Such consideration suggests to model the tomato as a semi-infinite body subjected to uniform radiative heating. On the other hand, wishing a more detailed description including spatial disuniformities, a numerical scheme is required which accounts for the irregular geometry of tomatoes.

2.1. Analytical solution

Rapid processing determines only surface peripheral warming up. Thus, the tomato under test is modelled as a purely conductive opaque-to-radiation semi-infinite body with constant thermal properties. Latent heat transfer is neglected as well. Considering that surface temperatures of the infrared heaters are much greater than the target ones, a II-type boundary condition is considered for including radiative IR heating. Tomato rotation around its main axis is accounted for by introducing a sinusoidal pulsating heat source. The latter is described as the positive half wave of a sinusoidal function. In addition, the boundary condition on the wall is supplemented by heat transfer linearly depending on wall temperature, with a constant heat transfer coefficient.

Under such assumptions, the dimensionless energy balance and the related boundary conditions lead to a linear problem:

$$\begin{cases} \dfrac{\partial^2\theta}{\partial\xi^2} = \dfrac{\partial\theta}{\partial\tau} \\ -\left.\dfrac{\partial\theta}{\partial\xi}\right|_{\xi=0} = \sin(\tau)\dfrac{\text{sign}[\sin(\tau)]+1}{2} - Bi\,\theta \\ \theta(\xi\to\infty,\tau)=0 \\ \theta(\xi,\tau=0)=0 \end{cases} \tag{1}$$

where $\theta=(T-T_f)/\Delta T_{ref}$ is the temperature, T_f representing both the surrounding and the initial temperature; the group $\Delta T_{ref} = \dot{q}_0\, x_{ref}/k$ is a reference temperature difference related to the peak amplitude of the wall heat flux, $\dot{q}_0$, and to a reference length $x_{ref} = (\alpha\, t_{ref})^{1/2}$, k and α being the tomato thermal conductivity and diffusivity; the reference time, $t_{ref} = \Omega^{1}$, was chosen such as the dimensionless time resulted $\tau = \Omega\, t$, Ω being the angular velocity of the source. The dimensionless space variable was defined such as $\xi = x/x_{ref}$ and, finally, $Bi = h\, x_{ref}/k$ is the Biot number. The function sign(sin(τ)) is a two-valued function switching each half period between zero or one.

Since the problem is linear, the solution can be sought as the sum of two partial solutions, θ_1 and θ_2, each one affected by a single non-homogeneity arising from the boundary condition at the wall; the

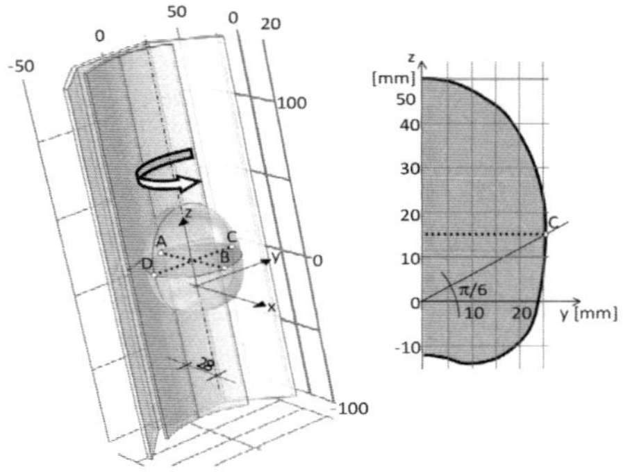

Figure 1. Schematic of the model.

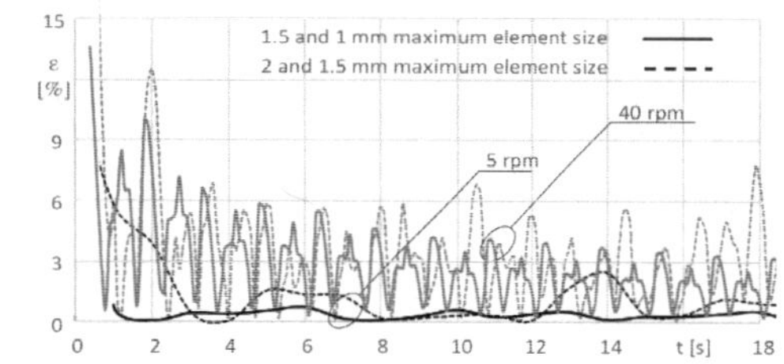

Figure 2. Mesh analysis.

former is due to the heating source and the latter by the cooling surroundings. Then, the two partial solutions have to satisfy two distinct problems derived by the basic one:

$$\begin{cases} \dfrac{\partial^2\theta_1}{\partial\xi^2} = \dfrac{\partial\theta_1}{\partial\tau} \\ \left.\dfrac{\partial\theta_1}{\partial\xi}\right|_{\xi=0} = -\sin(\tau)\dfrac{\text{sign}[\sin(\tau)]+1}{2} \\ \theta_1(\xi\to\infty,\tau)=0 \\ \theta_1(\xi,\tau=0)=0 \end{cases} \qquad \begin{cases} \dfrac{\partial^2\theta_2}{\partial\xi^2} = \dfrac{\partial\theta_2}{\partial\tau} \\ \left.\dfrac{\partial\theta_2}{\partial\xi}\right|_{\xi=0} = Bi\,\theta_2 \\ \theta_2(\xi\to\infty,\tau)=0 \\ \theta_2(\xi,\tau=0)=0 \end{cases} \tag{2}$$

The θ_1-problem is solved in a closed form by the Laplace transform method, while the latter is classically found in literature, see for instance [10]. For such a reason, in what follows, only radiative heat transfer will be dealt with to compare numerical and analytical results without affecting the meaning of the present analysis. After some manipulation, the solution to the first problem is cast as:

$$\theta_1(\xi,\tau) = \int_0^\tau \left(-\frac{e^{-\frac{\xi^2}{4(\tau-y)}}}{\sqrt{\pi}\cdot\sqrt{\tau-y}} \right)\cdot\left(-\sin(y)\cdot\frac{\text{sign}[\sin(y)]+1}{2}\right)\cdot dy = \sum_{k=0}^{n(\tau)-1}(-1)^k z_k(\xi,\tau) \tag{3}$$

with $n = \tau/\pi$, if τ/π is an integer otherwise $n = \text{int}(\tau/\pi)+1$ and

$$\begin{aligned} z_k(\xi,\tau) = &-\frac{1}{2\sqrt{2}}\left(e^{-\frac{\xi}{\sqrt{2}}} F_p(\xi_m)\,\text{erfc}\left(\frac{-\sqrt{2}\tau_r+\xi}{2\sqrt{\tau_r}}\right) + e^{\frac{\xi}{\sqrt{2}}}\text{erfc}\left(\frac{\sqrt{2}\tau_r+\xi}{2\sqrt{\tau_r}}\right) F_m(\xi_p) \right) + \\ &+\frac{1}{\sqrt{\pi}\sqrt{2}} e^{-\frac{\xi^2+2\tau_r^2}{4\tau_r}} \left(F_c\left(\frac{-\sqrt{2}\tau_r+\xi}{2\sqrt{\tau_r}}, \frac{\sqrt{\tau_r}}{\sqrt{2}}\right) F_m(\xi_m) + F_c\left(\frac{\sqrt{2}\tau_r+\xi}{2\sqrt{\tau_r}}, \frac{\sqrt{\tau_r}}{\sqrt{2}}\right) F_p(\xi_p) \right) + \\ &+\frac{1}{\sqrt{\pi}\sqrt{2}} e^{-\frac{\xi^2+2\tau_r^2}{4\tau_r}} \left(F_s\left(\frac{-\sqrt{2}\tau_r+\xi}{2\sqrt{\tau_r}}, -\frac{\sqrt{\tau_r}}{\sqrt{2}}\right) F_p(\xi_m) + F_s\left(\frac{\sqrt{2}\tau_r+\xi}{2\sqrt{\tau_r}}, -\frac{\sqrt{\tau_r}}{\sqrt{2}}\right) F_m(\xi_p) \right) \end{aligned} \tag{4}$$

where $\xi_m = -\tau+\xi/\sqrt{2}$, $\xi_p = \tau+\xi/\sqrt{2}$, $\tau_r = \tau - k\,\pi$; the functions F_c, F_s, F_m and F_p are:

$$F_c(a,b) = \int_0^b u^2 \cos(2au)du \; ; \; F_s(a,b) = \int_0^b u^2 \sin(2au)du \; ; \tag{5}$$

$$F_p(z) = \cos(z) + \sin(z); \; F_m(z) = \sin(z) - \cos(z) \tag{6}$$

The analytical solution was validated in comparison with the corresponding numerical solution obtained by solving a finite element scheme. Moreover, the accuracy of the solution was also verified with reference to the limiting case of a simple sinusoidal source and checking that it fully recovers the first function of the z_k succession.

2.2. Numerical solution

As witnessed by the leading papers by Pan [3-4], the numerical approach allows a quite satisfactory description of the problem at hand, provided that spatial discretization is performed with care as grid dispersion may arise [6]. On the other hand, numerical modelling may be subjected to long execution times, depending on the evaluation of view factors, thus the hemicube approach is suggested, [11]. A schematic representation of the geometry under study is reported in Figure 1. The heating source consists in an infrared emitter coupled to a polished surface for reflecting radiation which illuminates the tomato from one side. The tomato rotates with a prescribed angular velocity around its main axis which is parallel to the IR emitter. Such configuration should possibly resemble a typical industrial setup. The average surface temperature of the emitter is assumed to be 800°C, a value complying to IR emitters standard performance. Tomato is assumed to be a diffuse gray emitter with a surface absorptivity of 95% [3], while air in the enclosure is assumed as a non-participating medium. Latent loss from tomato surface is neglected since it corresponds to less than 2 or 3 % of the total heat loss, due to the shortness of the heating period, [2].

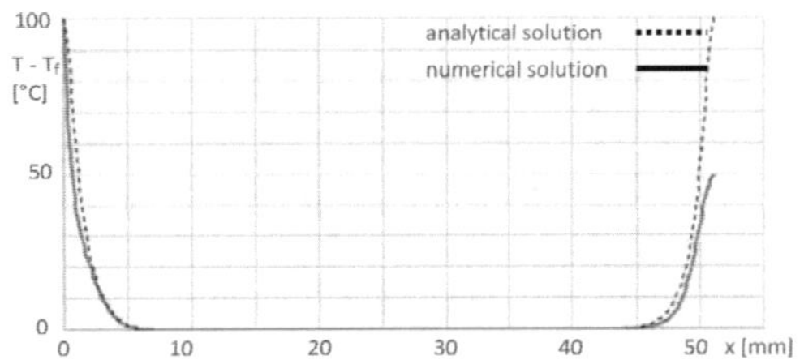

Figure 3a. Temperature profiles along the path AB for 5 after 18 s.

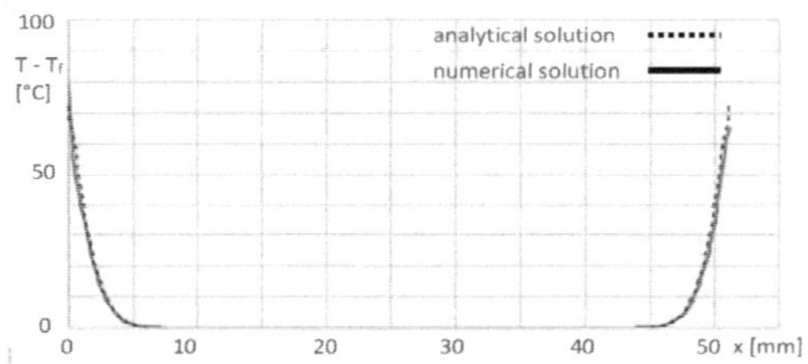

Figure 3b. Temperature profiles along the path AB for 40 rpm after 18 s.

Under the above assumptions, a 3D FEM model was developed by employing the commercial code COMSOL v4.3 [12]. The numerical model determines the unsteady temperature distribution for the rotating tomato by further considering constant flow properties. The computational domain of the tomato was generated based on two-dimensional axisymmetric parametric profile after Li et al., [1]; the related dimensionless shape coefficients were sought as: a = 31 mm, b = 24 mm, c_1 = 0.24, c_2 = 0.38, that identifying a medium size tomato.

The tomato computational domain was discretized using unstructured tetrahedral grid elements. Having in mind to suitably reduce computational burdens, special care was devoted in choosing the mesh size [11]. A mesh convergence study was performed to confirm that results are independent of mesh size during the warming up period, which represents the time needed for the averaged surface temperature to recover the set point value ($T_{objective}$). In practice, such a value is usually selected to be 90°C, which is known to be useful for peeling. The corresponding warming up period turns out to be about 18 seconds. First, a mesh using the fewest, reasonable number of elements was created; the mesh density, expressed in terms of maximum element size, was increased until a satisfactory independence of an objective function by the mesh itself was obtained. The objective function was chosen to be the percentage relative variation of the temperature history, $\varepsilon = (T_{last\text{-}mesh} - T_{previous\text{-}mesh}) / T_{last\text{-}mesh}$, evaluated at the point "C" evidenced in Figure 1. On the above basis, plots of Figure 2 were realized by comparing the objective function evaluated for 2 and 1.5 mm maximum element size, dashed-curves, and for 1.5

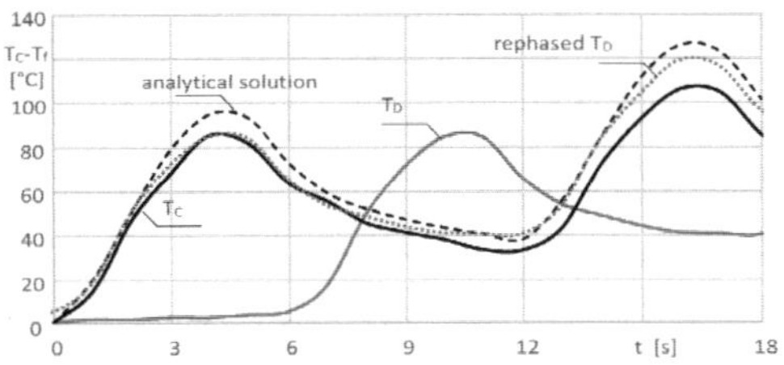

Figure 4a. Temperature – time history of selected points for 5 rpm.

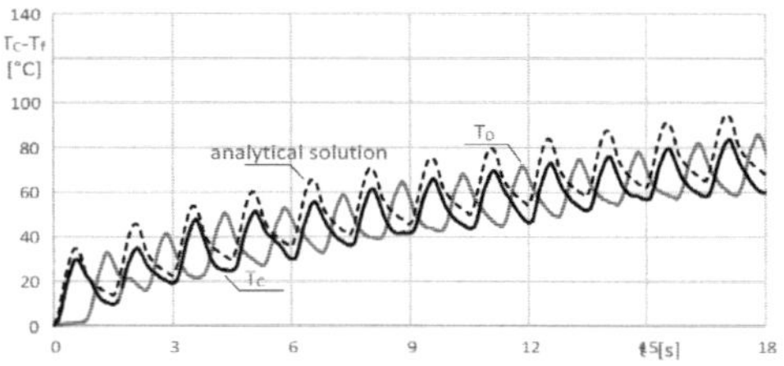

Figure 4b. Temperature – time history of selected points for 40 rpm.

and 1 mm, continuous-curves, for two selected angular velocities, i.e. 5 and 40 rpm. For the lowest angular rotation, results are practically independent of the mesh, while the objective function keeps on fluctuating with double frequency with respect to the source pulsations; in facts, the maximum variations pop up in correspondence of the temperature relative extrema. The same behavior features the highest speed for which, as expected, the objective function attains higher values, even though it becomes regular and contained within the 4% after 9 s for the case related to the finer meshes.

3. Results

3.1. Temperature field

Numerical temperature profiles and the corresponding analytical ones along the path AB during the warming up period are compared in Figure 3 a, b; the points A and B are diametrically opposed and are positioned on the circumference experiencing the maximum wall heat flux, identified by the zenith angle equal to $\pi/6$ in the yz-plane from y-axis. The analytical solution is drawn by assigning to the parameter $\dot{q}_0$ the average value of the wall heat flux over the whole tomato surface as resulting from the numerical solution at the initial time. A first sign about the ability of the semi-infinite model to describe the problem at hand is witnessed by the appropriate estimate for the penetration depth: it turns out to be $(\alpha\, t)^{1/2} = 2$ mm, which is much lower than the tomato average radius after 18 s; moreover, such a value is consistent with the corresponding numerical one, as evident by both plots of Figures 3.

Two occurrences are useful in order to assess that the more the angular velocity increases the more suitable is the semi-infinite model; the former consideration consists in observing that the root mean square errors between the numerical and the corresponding analytical plots turn out to be 7 and 1.5 °C for 5 and 40 rpm, respectively; in addition, the magnitudes of the tangential component of the wall heat flux are contained within 1% and 0.2 % of the normal component for 5 and 40 rpm, respectively. Thus, it can be argued that, actually, unidimensional heat transfer takes place.

The temperature-time histories of the points initially positioned in C and D are reported in Figures 4 for two selected angular velocities, i.e. 5 and 40 rpm, and for both the analytical and numerical solutions. For a fixed speed, the numerically evaluated temperature history of the point C exhibits a steep increase because it moves toward the heating source, while the point D temperature history starts with a zero-slope tangent because its exposition to the source is delayed for one half period. Then, temperatures start to fluctuate periodically that depending on the angular velocity in use while their time-average temperature increases.

However, if point D is returned to being in phase with point C, see the dotted curve in Figure 4a, the two related profiles seem to collapse into a single one within an average relative error of 10.2% in the warming up period. With increasing the angular velocity, the amplitude of temperature fluctuations decreases, Figure 4b, because the wall heat flux is more uniformly distributed along a fixed time-interval, the overall incoming energy being the same. In this case, the average relative error between the two rephased profiles lowers to 5.0 %. This occurrence suggests that profiles individuated at different azimuthal angles lose their individuality as much as the frequency increases, thus enabling the use of the semi-infinite model, dashed line in Figures 4.

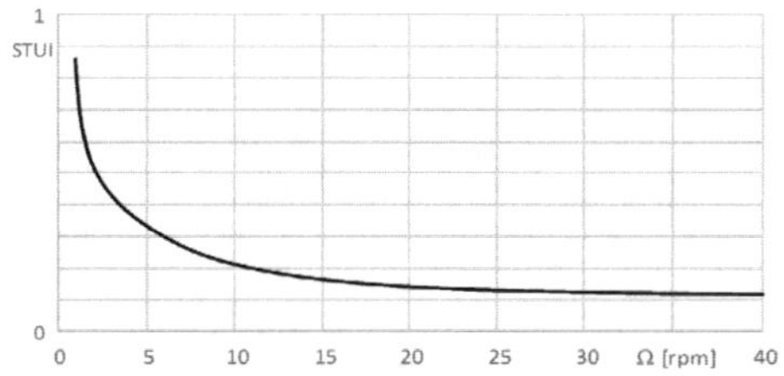

Figure 5. STUI as function of angular velocity after 18 s

It should be emphasized that, for both the rotation speeds, the periodic steady-state response is far from being recovered, thus the sustained solution is not-useful for the specific context at hand.

3.2. Evaluation of surface temperature uniformity

Rapid and uniform surface heating is required to realize an effective IR dry-peeling process, [14-15]. In fact, uneven temperature distributions in tomato surface may determine, in turn, inhomogeneous peeling or product damage. The presence of hot spots can cause deterioration in textural quality and loss in nutritional value of the peeled products, whereas cold spots prevent peel removal. The availability of the numerical scheme allows to study the spatial and temporal temperature uniformity while varying the angular velocity of the sample under test. For such a purpose, a temperature uniformity surface index was introduced which is defined as:

$$STUI = \frac{1}{T_{objective} - T_f} \cdot \sqrt{\frac{\iint \left(T - T_{average}\right)^2 dA}{A}} \tag{7}$$

the area A being the tomato exposed surface.

The STUI turns out to depend on time and angular velocity and is represented in Figure 5 for $t = 18$ s. For low rotation speeds, increasing the angular velocity rapidly improves tomato temperature uniformity whereas the STUI is less sensitive with increasing the angular velocity. From the industrial point of view, such a behavior can suggest how to suitably select a proper angular velocity featuring the tomato conveyance system.

4. Conclusions

Having in mind the typical warming up serving as preparation for radiative tomato peeling, an analytical 1D unsteady heat transfer model was developed. The solution was compared with a 3D model for the problem at hand showing a fair agreement provided that the typical IR dry-peeling conditions are realized. The availability of both analytical and finite element models can be a starting step to gain insight into the rapid surface heating characteristics of IR radiation in the tomato dry-peeling process.

References

[1] Li X, Pan Z, Upadhyaya S K, Atungulu G G & Delwiche M 2011 Three-dimensional geometric modeling of processing tomatoes, *Transactions of the ASABE* **54** 2287–2296.
[2] Li X 2012 *A study of infrared heating technology for tomato peeling: Process characterization and modeling*, Ph.D. Dissertation (Department of Biological and Agricultural Engineering, University of California at Davis, Davis, CA, USA).
[3] Li X, Pan Z 2014 Dry peeling of Tomato by Infrared Radiative Heating: Part I. Model Development, *Food Bioprocess Technol* **7** 1996–2004.
[4] Li X, Pan Z 2014 Dry peeling of tomato by infrared radiative heating: Part II. Model validation and sensitivity analysis, *Food and Bioprocess Technology* **7** 2005–2013.
[5] Rock C, Yang W, Goodrich-Schneider R & Feng H 2011 Conventional and alternative methods for tomato peeling, *Food Engineering Reviews*, **4** 1–15.

[6] Pan Z, Li X, Khir R, El-Mashad H M, Atungulu G G, McHugh T H., Delwiche M 2015 A pilot scale electrical infrared dry-peeling system for tomatoes: Design and performance evaluation, *Biosystems engineering* **137**, 1-8.

[7] Datta A, Ni H 2002 Infrared and hot-air-assisted microwave heating of foods for control of surface moisture, *Journal of Food Engineering*, **51** 355–364.

[8] Prakash B 2012, *Mathematical modeling of moisture movement within a rice kernel during convective and infrared drying*, Ph.D. Dissertation (Department of Biological and Agricultural Engineering, University of California at Davis, Davis, CA, USA).

[9] Trivittayasil V, Tanaka F, Uchino T 2011 Investigation of deactivation of mold conidia by infrared heating in a model-based approach, *Journal of Food Engineering*, **104** 565–570.

[10] Carslaw H S, Jaeger J C 1959 *Conduction of Heat Solids* 2/E (Oxford University Press).

[11] Howell J R, Siegel R, Menguc M P 2010 Thermal radiation heat transfer 5/E (New York: Taylor and Francis).

[12] COMSOL Multiphysics Version 4.3a User Guide, October 2012.

[13] Cuccurullo G, Giordano L and Viccione G 2013 An Analytical approximation for continuous flow microwave heating of liquids, *Advances in Mechanical Engineering*, **2013**

[14] Geedipalli S, Rakesh V, Datta A 2007 Modeling the heating uniformity contributed by a rotating turntable in microwave ovens, *Journal of Food Engineering*, **82** 359–368.

[15] Tanaka F, Verboven P, Scheerlinck N, Morita K, Iwasaki K, Nicola B 2007 Investigation of far infrared radiation heating as an alternative technique for surface decontamination of strawberry, *Journal of Food Engineering*, **79** 445–452.

IOP Conf. Series: Journal of Physics: Conf. Series **796** (2017) 012027 doi:10.1088/1742-6596/796/1/012027

Energy demand hourly simulations and energy saving strategies in greenhouses for the Mediterranean climate

A Priarone[1], M Fossa[1], E Paietta[1] and D Rolando[2]

[1] DIME – Dept. of Mechanical Energy, Management and Transportation Engineering, University of Genoa, Italy, Via Opera Pia 15a, 116145 Genova, Italy
[2] KTH Dept. Energy Technology, Brinnelvagen 68, S-100 44 Stockholm. Sweden

Corresponding Author: a.priarone@unige.it

Abstract. This research has been devoted to the selection of the most favourable plant solutions for ventilation, heating and cooling, thermo-hygrometric control of a greenhouse, in the framework of the energy saving and the environmental protection. The identified plant solutions include shading of glazing surfaces, natural ventilation by means of controlled opening windows, forced convection of external air and forced convection of air treated by the HVAC system for both heating and cooling. The selected solution combines HVAC system to a Ground Coupled Heat Pump (GCHP), which is an innovative renewable technology applied to greenhouse buildings. The energy demand and thermal loads of the greenhouse to fulfil the requested internal design conditions have been evaluated through an hourly numerical simulation, using the Energy Plus (E-plus) software. The overall heat balance of the greenhouse also includes the latent heat exchange due to crop evapotranspiration, accounted through an original iterative calculation procedure that combines the E-plus dynamic simulations and the FAO Penman-Monteith method. The obtained hourly thermal loads have been used to size the borehole field for the geothermal heat pump by using a dedicated GCHP hourly simulation tool.

1. Introduction

One of the main tasks in greenhouse construction is to optimize the conditions for plant growth, generally during the off-season from normal outside field production. Total energy consumption of greenhouse agriculture is steadily increasing because most of the agriculture companies and growers made a shift from unacclimatized to acclimatized greenhouses in order to satisfy the increasing demand of product quality and vegetable production availability all-year-round, which now characterize the European agro-food industry.

During the last 20 years, countries in the Mediterranean climate area have become increasingly competitive producers of greenhouse vegetables and fruit. The covered area in greenhouses in Italy exceeds 42000 hectares, of which 5000 hectares are devoted to crops of vegetables and more than 37000 hectares to floriculture. The greenhouse cultivation represents for the national agricultural system a productive sector of considerable economic importance.

Greenhouse technologies allow cultivating all horticultural species in any region of the world, provided that the greenhouse is properly designed and equipped to control the indoor climate. Over the past few years, researchers have investigated a large number of energy saving techniques for air conditioning of greenhouses and cropping systems by means of studies, simulations and experimentations. Most of the analysed solutions were developed to reduce inside air temperature without compromising the plant crop production, to improve greenhouse insulation, to optimize shape and orientation of the greenhouse (Gupta and Chandra [1], Sethi [2]) or to increase the energy efficiency of the heating and cooling.

Moreover, interest in renewable energies for both heating and cooling has become more intense for greenhouse operators to reduce their energy costs and CO_2 emissions. In fact, despite the large presence of renewable energy sources in several European regions with a significant greenhouse production, the renewable technologies have still a limited application for greenhouse air conditioning. Among these renewable technologies one can mention the solar energy exploitation (Abdel-Ghany and Al-Helal [3], Ozgener and Hepbasli [4]), the photovoltaic applications, the use of solid/residues biomass for heating, the latent heat storage for low temperature heating applications (Tuncbilek et al. [5]) and the geothermal contribution to greenhouse conditioning (Adaro et al. [6], Bakos et al., 1999 [7], Santamouris et al. [8]).

Greenhouse heating is important even in countries with temperate climate, like the Mediterranean region, in order to maximize crop production in terms of quantity and quality and thus to increase the overall productivity of greenhouse. Heating costs not only have a critical influence on the profitability, but in the long-term may also determine the survival of the greenhouse industry. In Italy, the cost of heating currently accounts for approximately 30% on the cost of production in the greenhouse. Apart from the costs problems associated with high-energy consumption, heating is associated with environmental problems through the emission of noxious gases.

An extended review about innovative and renewable heating technologies for greenhouse application has been proposed by Sethi and Sharma [9]. They diffusely illustrate several energy saving solutions as thermal storage, ground-to-air heat exchanger systems, movable insulation/thermal screens, optimal use of the north wall, ground air collectors and aquifer coupled cavity flow heat exchanger systems.

Moreover, Jamal [10] and more recently Sethi and Sharma [11] have proposed a comprehensive description of cooling techniques for greenhouses, including renewable technologies as the roof evaporative cooling, the ground-to-air heat exchanger system and the aquifer coupled cavity flow heat exchanger system.

This paper is related to the Liguria Region research program "Smart Agro-Manufacturing Laboratory (SAM-LAB)". The final goal of the Project was to realize an innovative greenhouse equipped with new materials for the glazed shell and with suitable advanced controls, able to enhance the facility management in the direction of reducing the overall energy needs. The innovative greenhouse has been finally realized at the Regional Agricultural Research Center (Cersaa) located in Albenga (SV). The primary objective of the research was to define solutions for energy efficiency improvement in agriculture and to contribute to the dissemination of methodologies and best practices for the greenhouse sector, even on a large scale and in different territorial and technological contexts.

The present paper illustrates the results of the study with particular attention to the description of the hourly energy demand simulations model built in E-Plus environment and of the sizing method of the BHE (Borehole Heat Exchangers) field coupled to the geothermal heat pump.

2. Numerical dynamic analysis and energy demand hourly simulations: the E-plus model

E-Plus (Energy Plus, U.S. Department of Energy) is a worldwide known open-source software for building hourly energy simulations. The basic input of the model are the meteorological *epw* file containing the hourly data of the site of interest and the model data *idf* file with the details about the geometry, the shell, the selected plant solutions and their control systems.

The analysed greenhouse is located in Albenga, Italy (latitude: 44.3 DD, longitude: 8.47 DD, altitude: 3 m, time fuse: +1 CET) and the meteorological data have been obtained from Meteonorm code.

The greenhouse structure consists in a rectangular aluminum frame with pitched roof. The plan dimensions are 15.3 and 9.9 meters, the eave height is 3.50 m while the roof top height is 5.60 m. The material selected for the greenhouse glazed walls is the low emissivity K-glass type N with a thickness of 4 mm, overall heat transfer coefficient 3.3 W/m^{2}K, normal emissivity equal to 0.05 and a solar gain (g-value) of 71%.

The air conditioning system of the greenhouse provides shadings applied to the glazing surfaces, natural ventilation by controlled opening windows, forced convection of external air and forced convection of air treated by HVAC unit (reversible heat pumps) for both heating and cooling purposes.

In the model the control of each element of the air conditioning system is realized by means of different Objects available in E-Plus (i.e. *Schedules*) with suitable temporal profiles (monthly, daily, hourly or sub-hourly) or on/off specific set points.

The shading is implemented in the model by means of the object *Window Property: Shading Control* that specifies the type, location and controls. The temperature setpoint that triggers the shading is 24°C for the internal air temperature in the greenhouse, for both winter and summer operating mode.

The controlled natural ventilation operates during both winter and summer and it is modeled by means of two different object: the *Zone Infiltration: Design Flow Rate* and the *Zone Ventilation: Wind and Stack Open Area.* Zone infiltration is specified as a design incoming air rate that is modified by temperature difference and wind speed values. This flow rate also fulfils the minimal air change per hour according to a defined time schedule (0.25 during working period, 0.15 during night and holidays). Zone ventilation represents the natural ventilation of external air driven again by wind and stack effects, only when related window openings are activated by their control. The apertures are located near the ground and at the roof top of the greenhouse. The windows opening is triggered by the internal air temperature of the greenhouse and they allow air circulation when this temperature is higher than 24°C during summer or 26°C during winter. An additional control is related to the minimum external temperature at which the natural ventilation starts: external air temperature has to be higher than 16°C and the wind speed lower than 10 m/s.

The forced convection of external air and the forced convection of air treated by HVAC system (heating and cooling) are all modeled by means of the combination of the objects *HVAC Template/Thermostat*, *HVAC Template/Zone/Ideal Loads Air System*, *Design Specification/Outdoor Air*. In particular, the use of external air is driven by particular settings chosen in the field *Outdoor Air Economized Type* that increases the outdoor air flow rate when a cooling load is requested and the outdoor air temperature is lower than the zone (indoor) exhaust air temperature. To cope with heating or cooling loads, the greenhouse inlet air is pre-treated by the HVAC ideal air system object in order to fulfil the following set-points: the temperature of internal greenhouse air must be not higher than 30°C during summer (cooling mode) and not lower than 16°C during winter (heating mode).

Finally the HVAC system is coupled to a Ground Coupled Heat Pump (GCHP) with vertical ground heat exchangers (BHE field). The ground side of the air conditioning system is simulated and designed according to models developed at the University of Genova.

A very significant contribution in greenhouse energy balance is given by the crop evapotranspiration process. To Authors' knowledge no study related to energy consumption in greenhouses considered this particular contribution for evaluating the greenhouse hourly thermal loads (Fabrizio [12]).

3. Evapotranspiration: the model FAO Penman-Monteith

The evapotranspiration consists in the process of transpiration of liquid water thought the leaves and underleaf soil. This water contribution to the greenhouse volume is associated with evaporation and hence with a significant latent heat contribution in the energy balance of the greenhouse control volume.

In this study, the FAO Penman-Monteith model is considered in order to evaluate the hourly reference standard evapotranspiration ET_0 from the greenhouse surface occupied by plants. The real evapotranspiration can be estimated by introducing the effect of the crop by the crop coefficient *Kc*.

The FAO main correlation is the following one:

$$ET_0 = \frac{0.408\left(\frac{dp}{dT}\right)_{sat}(G_n - G_g) + \gamma\frac{37}{T+273}w(p_{sat,T_{iw}} - p_v)}{\left(\frac{dp}{dT}\right)_{sat} + \gamma(1+0.34\,w)} \tag{1}$$

ET_0 standard evapotranspiration [mm/h]

$\left(\frac{dp}{dT}\right)_{sat}$ slope of the saturation curve of vapor [kPa/K]

G_n net irradiation on the vegetation [MJ/(m^{2}h)]
G_g net irradiation on the ground [MJ/(m^{2}h)]
γ psychrometric constant [kPa/K]
w hourly mean wind speed [m/s]
$p_{sat,Thr}$ saturation pressure at hourly mean air temperature T_{hr} [kPa]
p_v hourly mean vapor partial pressure [kPa]

The above model has been conceived for estimating the watering needs of outdoor cultivations and it has been here adapted to the greenhouse conditions by referring to the indoor conditions in terms of temperature, humidity and radiation conditions. In particular, the correlation is employed according to the assumptions here below:

- *solar irradiation:* reduced value (with respect to outdoor conditions) of about 30% to take into account the transmissivity of the glazed surfaces of the greenhouse;
- *air temperature:* is iteratively adjusted from E-plus simulations, starting from an initial condition when no evapotranspiration is present;
- *air humidity:* the first guess value is obtained by the starting simulation in E-Plus with no water gain due to evapotranspiration and then the value is adjusted according to the next iterative simulations of combined Penman-Monteith and E-plus models;
- *wind speed:* a minimum value is assigned (0.6 m/s);
- *rain:* no rain;
- *overcast index:* conversion of the meteorological data (clear sky 0, cloudy sky 10) into the Penman-Monteith scale (clear sky 1, cloudy sky 10).

The Penman-Monteith model is implemented in the E-plus model using the object *Water Use Equipment,* which is a generic object for simulating all water end uses and their evaporation fractions into the zone. The mass flow rate of water into the greenhouse is not constant but it follows the variable evapotranspiration activity of the vegetation as described by the Penman-Monteith model. In particular, the evapotranspiration is greatly influenced by the irradiation and it presents a periodical trend during the cycle day-night and according to the seasonal variation of irradiation during the year (figure 1a).

The hourly values of evapotranspiration have been converted into a three-step daily profile (one profile for each month of the year) in order to be easily managed in a suitable Schedule of the E-plus model (figure 1b). For each daily profile, the first step represents the minimal value of evapotranspiration over the night periods and it is calculated as the monthly average of the minimum daily evapotranspiration value among all days of the month under consideration. The higher step represents the maximum value of evapotranspiration corresponding to the hours with the maximum of solar irradiation and it is estimated as the monthly average value of the maximum daily evapotranspiration among all days of that month. The intermediate step characterizes the remaining daily hours and it is calculated as an average between the two previous values.

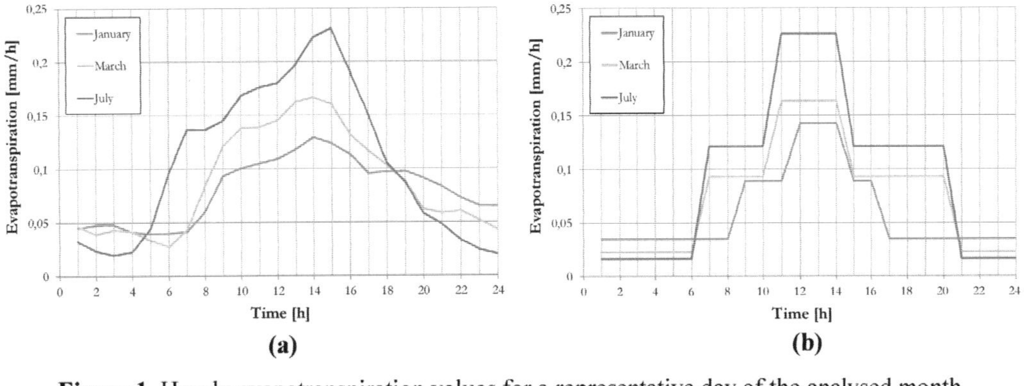

Figure 1. Hourly evapotranspiration values for a representative day of the analysed month, (a) obtained with the Penman-Monteith model and (b) the converted three-step average daily profile.

IOP Conf. Series: Journal of Physics: Conf. Series **796** (2017) 012027 doi:10.1088/1742-6596/796/1/012027

4. Results: the hourly thermal loads

The software E-plus allows to select as hourly output a lot of different parameters.

Among them, the most interesting quantities related to the specific site where the building is built are the *Site Outdoor Air Drybulb Temperature* [°C], the *Site Outdoor Air Relative Humidity* [%] and the *Site Wind Speed* [m/s].

The quantities that evidence the attainment of the design conditions for the zone of interest of the building (in our case a single zone, the greenhouse) are the *Zone Air Temperature* [°C], the *Zone Air Humidity Ratio* [-] and the *Zone Air Relative Humidity* [%].

Figure 2 reports the outdoor and zone temperatures hourly profiles from E-Plus simulations for the month of January and July, respectively. The zone temperatures perfectly fulfil the requested set-points of minimal temperature equal to 16°C during winter and maximal temperature equal to 30°C during summer.

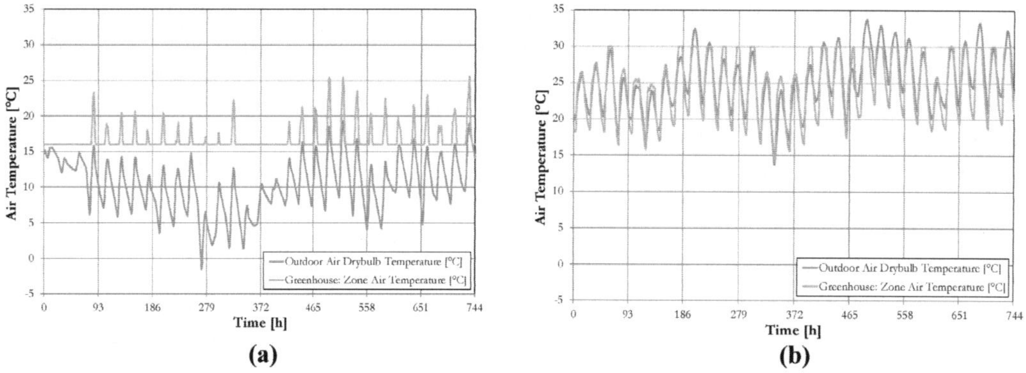

Figure 2. External and indoor (Zone) temperatures profiles from E-Plus simulations for the month of January (a) and July (b).

The main outputs of the objects *Zone Infiltration: Design Flow Rate* and *Zone Ventilation: Wind and Stack Open Area* are the *Zone Infiltration Air Change Rate* [ach] and the *Zone Ventilation Air Change Rate* [ach], respectively. From these outputs one can calculate the Natural Convection Sensible Cooling Rate [MWh] (see table 1 and figure 3).

The more relevant outputs of the object *Water Use Equipment*, strictly linked to the effect of evapotranspiration on hourly thermal loads (latent contribution), are the *Water Use Equipment Total Mass Flow Rate* [kg/s], the *Water Use Equipment Zone Latent Gain Rate* [W] and the *Water Use Equipment Zone Moisture Gain Mass Flow Rate* [kg/s]. From these outputs one deduces the Evapotranspiration heat load [MWh] (see table 1 and figure 3).

The sensible loads for both heating and cooling are distinguished between the loads directly provided by the HVAC system (Supply Air) and those entirely delivered to the zone (Zone). The differences between these two different loads, meaningful only in cooling, represent the contribution of the *Outdoor Air Economized*, i.e. the forced convection of external air. In detail, the significant outputs are the *Zone Ideal Loads Supply Air Sensible Heating Rate* [W], the *Zone Ideal Loads Supply Air Sensible Cooling Rate* [W], the *Zone Ideal Loads Zone Sensible Heating Rate* [W] and the *Zone Ideal Loads Zone Sensible Cooling Rate* [W]. From these outputs one can infer the HVAC Sensible Heating and Cooling Rate [MWh] and the Forced Convection Sensible Cooling Rate [MWh] (see table 1 and figure 3).

Table 1. Energy contribution [MWh] of different elements of the air conditioning system.

Months	HVAC Sensible Heating Rate [MWh]	HVAC Sensible Cooling Rate [MWh]	Forced Convection Sensible Cooling Rate [MWh]	Natural Convection Sensible Cooling Rate [MWh]	Evapotranspiration [MWh]
1	3.8004	0	0	0.0024	0.0151
2	3.1254	0	0	0.0247	0.0439
3	2.0493	0	0.0011	0.2176	0.1769
4	1.4183	0	0	0.4751	0.3653
5	0	0.0004	0.0216	2.0287	0.7480
6	0	0.2971	0.1946	2.5111	1.2237
7	0	1.5003	0.3335	1.5708	1.6563
8	0	1.5181	0.2927	0.7096	1.4616
9	0	0.3289	0.2822	0.3994	1.2052
10	0.6016	0	0.0058	0.3206	0.6997
11	2.2714	0	0	0.0194	0.2268
12	3.7744	0	0	0	0.1557

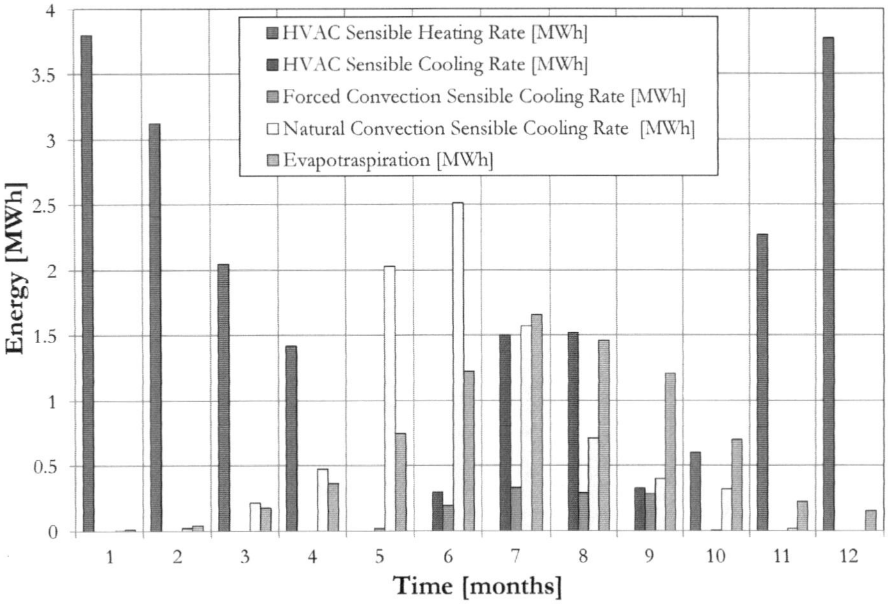

Figure 3. Energy contribution [MWh] of different processes involved in the greenhouse.

From the analysis of table 1 and figure 3 one can deduce the following outcomes.

During the heating season (winter months), the main load is obviously the sensible heating by the HVAC system, coupled with a feeble evapotranspiration activity. However, during the months of March, April and October, a weak cooling, obtained through natural convection, is required during the days with high outdoor temperature and great solar irradiation.

On the contrary, during the cooling season the air conditioning system starts with the natural ventilation, eventually integrated with forced convection of external air. When these actions are not enough or the temperature of the outdoor air become too high and so unfavorable, the system closes the windows and turns on the forced convection of air treated by the HVAC plant.

Finally, the evapotranspiration activity, greatly influenced by the solar irradiation, presents a trend that follows the seasonal variation of irradiation during the year, with a maximum during the month of

June. Furthermore, the evapotranspiration significantly contributes to the refrigeration of the greenhouse and it allows to reduce of energy consumption during the cooling season.

5. Design of the borehole field for geothermal heat pump installation

The exploitation of low enthalpy geothermal resources for air-conditioning represents an important opportunity to reduce the energy consumption and gas emissions even in the air conditioning greenhouse sector. Ground coupled heat pumps (GCHPs) exploit the favourable temperatures of the shallow ground in order to allow high COPs to be obtained. Moreover, in the greenhouse applications the heat pump performance is further enhanced due to the low target temperatures to be pursued at the HP condenser side and to the high target temperatures to be pursued at the evaporator side during the cooling period. In fact, in the current model, the internal greenhouse temperatures to be maintained are 16°C and 30°C during winter and summer, respectively.

Geothermal heat pumps are generally coupled with borehole heat exchangers (BHEs) that play a key role in determining heat pump performance. The borefield design goal is the definition of the best BHE geometry (BHE arrangement, their number and spacing) and the minimum overall length of vertical pipes in order to obtain suitable carrier fluid temperatures from the BHE field able to realize the target seasonal heat pump performance.

Different design criteria are available and all of them rely on a dynamic approach based on the knowledge of the building heat request in time. The simplest and widely employed design method for BHE fields has been developed by Kavanaugh and Rafferty [13], recommended by ASHRAE, and it accounts for three basic heat pulses over 10 years of operations. Monthly time step methods (e.g. EED, Hellström and Sanner [14]) use monthly average heat loads and peak heat loads to model both heating and cooling operations. They are based on the temperature response factor theory (*g*-functions) first introduced by Eskilson [15]. Finally, best analyses are those performed on an hourly time scale: they make use of hourly heat load series to perform a temporal superposition scheme (Bernier et al. [16], Fossa and Paietta [17]).

In the present investigation, the E-plus output data are employed as input for GCHP simulations where the inverse machine COP is iteratively calculated as a function of the variable carrier fluid temperature which in turn depends on the variable temperature of the ground volume surrounding the BHE field.

The selected BHE type is a double-U PE100 pipe with external diameter of 32 mm and length of about 100 m. The perforation is filled by grout with nominal thermal conductivity equal to 1.8 W/m K.

The thermal conductivity of the ground has been evaluated according to the new method of enhanced Thermal Response Test described in [18]. It resulted from measurements equal to about 5 W/m K, a very high value, related to the presence of relevant groundwater movements.

The required length of the BHE field is calculated according to three procedures, namely an hourly and 2 monthly time step calculations that include the use of the EED software (Earth Energy Designer) [14], probably the most widely used design tool for vertical borefield design for GCHP applications. Proprietary code TecGeo (Dalla Pietà and Fossa [19]) is employed to perform monthly simulations with reference to a custom BHE configuration made by 6 BHE arranged in a non-regular U-disposition (see figure 4). Finally the proprietary code MLAA17 [17], which represents a modified version of the Canadian algorithm MLAA [16], is employed for hourly simulations starting from complete E-plus information on greenhouse heating and cooling requirement during the year. The hourly simulation allows the Seasonal Perfomance Factor (SPF or average COP) to be evaluated and it results, for the given HP taken into consideration, equal to about 5.5. EED is employed to validate and check hourly results and the agreement is very good.

All the simulations are carried out along 25 years. The two different approaches (monthly average, i.e. TecGeo, and hourly, i.e. MLAA17) are in close agreement and both able to describe the time varying behaviour of the ground heat exchanger system when responding to the building heat demand as calculated from E-plus simulations. Figure 5 shows the results of a double simulation made by proprietary codes developed at the University of Genova: the hourly values of the fluid temperature obtained by means of MLAA17 seem in very good agreement with the monthly average values calculated using TecGeo.

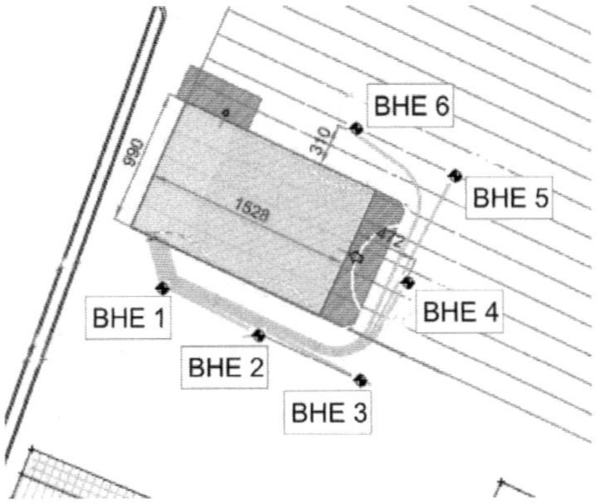

Figure 4. Greenhouse and BHE field layout.

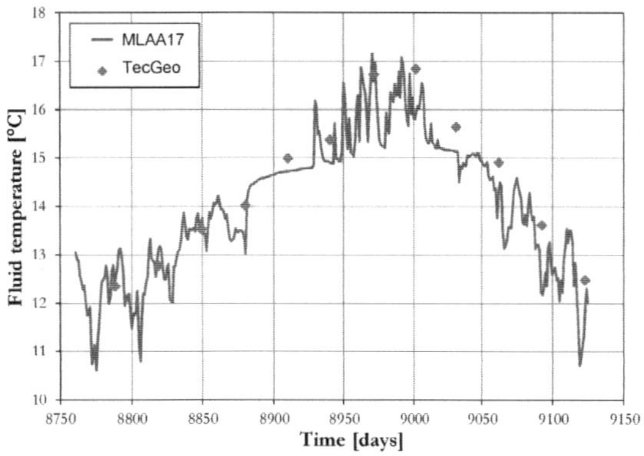

Figure 5. Fluid temperatures from simulations made (25th year) by proprietary codes developed at the University of Genoa, MLAA17 and TecGeo.

6. Conclusions

The paper describes the results obtained in the framework of a Liguria Region research program "Smart Agro-Manufacturing Laboratory (SAM-LAB)". The main goal of the project is to analyse and select air conditioning plant solutions (if possible combined with renewable energies) aimed at improving the energy efficiency in agriculture facilities.

The first part of the study consists in an hourly energy demand simulation of the greenhouse by means of the open-source software Energy Plus (E-plus). The air conditioning system selected for the greenhouse prototype combines shadings applied to the glazing surfaces, natural ventilation through controlled opening windows, forced convection of external air and forced convection of air treated by HVAC unit (reversible heat pumps) for both heating and cooling purposes. In the energy balance of the greenhouse also the crop evapotranspiration is considered as latent heat contribution, and this aspect represents an original contribution of this Research.

From the analysis of the hourly heat loads obtained as outputs of E-plus simulations one can deduce that, during the heating season, the main load is obviously the sensible heating by the HVAC system, coupled with a feeble evapotranspiration activity. During the transitional months (March, April and October) a weak cooling, obtained through the natural convection, is required during the days with high outdoor temperature and great solar irradiation. During the cooling season the system achieves the requested indoor conditions by means of the natural ventilation at the beginning, integrating it with the forced convection of external air or with the forced convection of air treated by the HVAC plant, if necessary.

Finally, the crop evapotranspiration is greatly influenced by the solar irradiation and it presents a trend following its seasonal variation during the year, with a maximum during the month of June. Therefore, the evapotranspiration significantly contributes to the cooling of the greenhouse with a reducing of energy consumption during summer.

The heating and cooling HVAC system couples a reversible heat pump with a borehole heat exchangers (BHEs) field, representing an innovative plant solution when applied to a greenhouse.

To design the Ground Coupled Heat Pump (CGHP), three different design methods are used. Two of them are based on average monthly heat loads combined with seasonal peak (namely EED and TecGeo, the second being a proprietary software developed at Dime, University of Genoa). On the contrary, the third approach uses directly the E-plus hourly heat loads (MLAA17, a proprietary code that represents a modified version of the Canadian algorithm MLAA).

The comparison between results obtained using the different methods reveals a very good agreement and the selected BHE field geometry is composed by 6 BHEs of about 100 m length, arranged in a non-regular U-disposition.

Acknowledgements
This work is included in the set of activities of PAR FAS 2007-2013, in the project SAM LAB.

References

[1] Gupta MJ and Chandra P 2002 Effect of greenhouse design parameters on conservation of energy for greenhouse environmental control *Energy* **27** pp 777–797.
[2] Sethi VP 2009 On the selection of shape and orientarion of a greenhouse: thermal modeling and experimental validation *Solar Energy* **83** pp 21-38.
[3] Abdel-Ghany AM and Al-Helal IM 2011 Solar energy utilization by a greenhouse: general relations, *Renewable Energy* **36** pp 189-196.
[4] Ozgener O and Hepbasli A 2005 A parametrical study on the energetic and exergetic assessment of a solar-assisted vertical ground-source heat pump system used for heating a greenhouse *Building and Environment* **42** pp 11-24.
[5] Tuncbilek K, Sari A, Tarhan S, Ergunes G and Kaygusuz K 2005 Lauric and palmitic acids eutectic mixture as latent heat storage material for low temperature heating applications *Energy* **30** pp 677– 692.
[6] Adaro JA, Galimberti PD, Lema AI, Fasulo A and Barral JR 1999 Geothermal contribution to greenhouse heating *Applied Energy* **64** pp 241-249.
[7] Bakos GC, Fidanidis D and Tsagas NF 1999 Greenhouse heating using geothermal energy *Geothermics* **28** pp 759-765.
[8] Santamouris M, Mihalakakou G, Balaras CA, Lewis JO, Vallindras M and Argiriou A 1996 Energy conservation in greenhouses with buried pipes *Energy* **21** pp 353-360.
[9] Sethi VP and Sharma SK 2008 Survey and evaluation of heating technologies for worldwide agricultural greenhouse applications *Solar Energy* **82** pp 832-859.
[10] Jamal KA 1994 Greenhouse cooling in hot countries *Energy* **19** pp 1187–1192.
[11] Sethi VP and Sharma SK 2007 Survey of cooling technologies for worldwide agricultural greenhouse applications *Solar Energy* **81** pp 1447–1459.
[12] Fabrizio E 2012 Energy reduction measures in agricultural greenhouses heating: Envelope, systems and solar energy collection *Energy and Buildings* **53** pp 57-63.
[13] Kavanaugh SP and Rafferty K 1997 Ground-source heat pumps: design of geothermal systems for commercial and institutional buildings *Atlanta, Georgia: ASHRAE.*
[14] Hellström G and Sanner B 1994 Earth energy designer: software for dimensioning of deep boreholes for heat extraction *Department of Mathematical Physics, Lund University, Sweden.*
[15] Eskilson P 1987 Thermal analysis of heat extraction boreholes *Ph.D. Thesis, Lund University of Technology, Sweden.*
[16] Bernier M, Pinel P, Labib R and Paillot R 2004 A multiple load aggregation algorithm for annual hourly simulations of GCHP Systems *HVAC&R, Research* **10** pp 471-487.
[17] Fossa M and Paietta E 2013 Comparison of multiple load aggregation algorithms for annual hourly simulations of geothermal heat pumps *EGC 2013 Conference* Pisa, Italy.
[18] Fossa M, Priarone A, Rolando D and Berti F 2016 Borefield design and enhanced thermal response test for the Energy Building of the University of Genova *XXXIV National UIT Congress.*
[19] Dalla Pietà D and Fossa M 2006 Modeling and design of borehole heat exchangers for ground-source heat pump applications *XXIV National UIT Congress.*

Cool roofs with high solar reflectance for the welfare of dairy farming animals

G Santunione, A Libbra and A Muscio

Dipartimento di Ingegneria "Enzo Ferrari", Università di Modena e Reggio Emilia, Via Vivarelli 10, 41125 Modena, Italy

Corresponding author: alberto.muscio@unimore.it

Abstract. Ensuring livestock welfare in dairy farming promotes the production capacity of the animals in terms of both quantity and quality. In welfare conditions, the animals can produce at their full potential. For the dairy cattle the most debilitating period of the year is summer, when the stress arising from overheating induces physiological alterations that compromise the animals' productivity. In this study, the summer discomfort of dairy animals is primarily quantified and the production loss is quantified versus the Temperature Humidity Index (THI), which correlates the values of temperature and relative humidity to the thermal stress. In order to reduce or eliminate such thermal stress, it is then proposed to coat the roof of the stables with a paint having high solar reflectance and thermal emittance, that is a cool roof product. This type of roofing solution can considerably limit the overheating of stables caused by solar radiation, thus providing a positive impact on the animals' welfare and improving significantly their productivity in summer.

1. Introduction: thermal comfort of dairy farming animals

Milk production is a delicate process that strongly depends on the animal well-being, which in turn depends on several environmental, physiological and nutritional factors. If the environment provides a situation of comfort, the animals will be able to achieve the most of their productive potential, allowing the maximization of livestock economic results.

The main environmental parameters that determine the microclimate inside a stable are temperature (T) and relative humidity (RH). With regard to temperature, thermal comfort conditions for lactating dairy cows fall between 5°C and 25°C [1,2,3]. Within this range, the physiological rectal temperature is 38.5°C. If the ambient temperature does not fall in the range, the thermal stress increases proportionally to the deviation of temperature from the range limits. The process of temperature control that is put in place by the animals to compensate for the thermal stress is physiologically very fatiguing in terms of energy needs and it reflects directly on productivity. The lack of an ideal microclimate inside the shelter, and the consequent increase in body temperature, involve the activation of self-defense mechanisms such as the decrease in the amount of food ingested, metabolism, respiratory activity and, consequently, productivity. Hyperthermia also alters and worsens the reproductive functions of cells and tissues, constituting one of the main causes of the decrease in the conception rate in summer months [4].

1.1. Temperature Humidity Index (THI)

The importance of thermal comfort of farm animals and, in particular, of dairy cattle has necessitated the identification of conventional parameters of the microclimate inside the stables. The most used among conventional indicators is the Temperature Humidity Index (THI), developed by Thom and Bosen in 1959 [5] for humans, but later adapted for the welfare assessment of lactating cows. It is defined by the formula

$$\text{THI} = \text{DBT} + 0.36 \cdot \text{DPT} + 41.2 \tag{1}$$

where

DBT dry-bulb temperature (°C)
DPT dew point temperature (°C)

Another formula was developed from equation (1) to use ambient temperature and relative humidity instead of the wet bulb temperature [6]:

$$\text{THI} = (1.8 \cdot T + 32) - 0.55 \cdot (1 - RH/100) \cdot (1.8 \cdot T + 32 - 58) \tag{2}$$

where

T ambient temperature (°C)
RH relative humidity (%)

Wiersma [7] divided the THI values obtained from the aforementioned formula under heat stress intervals, ranking them according to their severity (Table 1, Table 2). The studies by Igono and Armstrong [8,9] recognized the value of 72 as the critical threshold of THI to which the beginning of the heat stress of dairy animals is associated and, with this, the decrease in milk production.

Table 1. Values of THI as a function of temperature T and relative humidity RH and progressive levels of stress in dairy animals

T		***RH***																				
°F	°C	0	5	10	15	20	25	30	35	40	45	50	55	60	65	70	75	80	85	90	95	100
75	23.9														72	72	73	73	74	74	75	75
80	26.7							72	72	73	73	74	74	75	76	76	77	78	78	79	79	80
85	29.4			72	72	73	74	75	75	76	77	78	78	79	80	81	81	82	83	84	84	85
90	32.2	72	73	74	75	76	77	78	79	79	80	81	82	83	84	85	86	86	87	88	89	90
95	35.0	75	76	77	78	79	80	81	82	83	84	85	86	87	88	89	90	91	92	93	94	95
100	37.8	77	78	79	80	82	83	84	85	86	87	88	90	91	92	93	94	95	97	98	99	
105	40.6	79	80	82	83	84	86	87	88	89	91	92	93	95	96	97						
110	43.3	81	83	84	86	87	89	90	91	93	95	96	97									
115	46.1	84	85	87	88	90	91	93	95	96	97											
120	48.9	86	88	89	91	93	94	96	98													

Table 2. THI values and stress level

THI values	***Stress level***
THI < 72	absent
72 ≤ THI < 79	mild
79 ≤ THI < 90	high
90 ≤ THI < 99	severe

In summer, the perceived temperature inside the stables turns out to be always higher than that outside due to heat radiated downwards in the far infrared by the roof, which heats up under the effect of solar radiation. This is more extensively explained in section 3. Moreover, the local relative humidity is drastically increased by evaporation of the liquid fraction of manure and the presence of many animals in constrained areas. It was calculated that the stables reach temperature values higher than the external one by at least 3-4°C, and relative humidity values greater than outside by 10%. Considering an average summer temperature of 28°C associated with a mean relative humidity value of 55-60%, as typically recorded by weather stations in the Po Valley, Northern Italy, the THI in the stables easily reaches values corresponding to high levels of stress (>80).

2. Production loss

During the summer period, many farms are affected by a significant decrease in milk production due to thermal stress of cows. Several authors have studied this phenomenon and most of them are concordant in reporting a decline of daily production quantifiable at around 2% for each point of increase above the critical threshold of THI = 72 [10]. Such decline increases in the case of highly productive animals. It was indeed observed that the genetic selection made in the last decades in order to obtain more and more productive animals has helped to make cows very sensitive to overheating since there was not a simultaneous improvement of body thermoregulatory capacity [11,12].

The drop of milk production versus temperature can be estimated by assuming a decline of 2% for each point of THI above 72, as reported in [10]. In this regard, the average value of relative humidity in Modena during summer is around 55% but, as already stated, it can increase by at least 10% inside a stable due to evaporation of the liquid fraction of manure and the respiration rate of animals. Therefore, the THI value and, from that, the drop of milk production, plotted in Figure 1, have been calculated all over the typical range of outside ambient temperature for 65% relative humidity. The outside ambient temperature was assumed to be close to the inside air temperature of the (open) stable and to the temperature perceived by the animals.

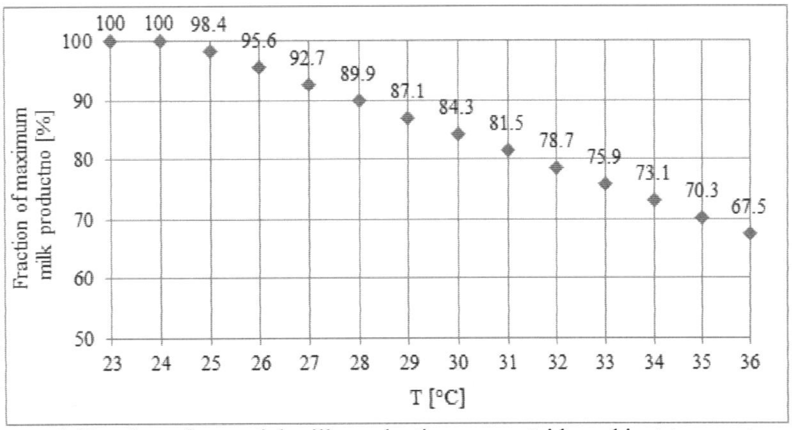

Figure 1. Fraction of potential milk productions vs. outside ambient temperature calculated for a stable located in the province of Modena, Po Valley (inner RH 65%).

The drop in conditions of thermal stress is a consequence of the numerous physiological alterations determined by hyperthermia of the cows. The more the perceived temperature moves away from the thermal comfort threshold (25°C), the more the rectal temperature rises above 38.5°C, compromising various aspects of metabolism and behavior pattern.

2.1. Decrease of food intake

The decrease in milk production due to hyperthermia is primarily related to the decrease in the intake of dry matter [13]. The body of dairy cows has two types of processes available to reject thermal energy: evaporative processes (sweating, tachypnea) and non-evaporative processes (thermal radiation, conduction and convection). Non-evaporative processes become less effective when the ambient temperature increases; on the other hand, evaporative processes are also compromised as a result of an increase in relative humidity. Therefore, during a summer period characterized by high humidity and high temperature, heat dissipation of cows is ineffective in preventing the increase of body temperature. For such reasons, in an attempt to protect themselves from thermal stress, cows enact behaviors to decrease the internal temperature or the heat input resulting from the metabolism, and this happens physiologically when the intake of food is reduced [14].

Kurihara and Shioya [15] reported that the consumption of daily dry matter amounts to an average of about 20 kg for an average daily production of 30 kg of milk, in the absence of stress conditions. This quantity begins to decrease above 25°C. Kurihara and Shioya also showed how food intake varies in relation to the increase in temperature and humidity. For example, if the temperature ranges from 24°C to 28°C, with humidity of 40% (THI 76), the drop in the daily intake of dry substance is about 13%. The problem is also documented by Wheelock et al. [16], which record intake drops of dry matter as high as 30%.

2.2. *Alteration in milk composition*

In addition to the quantitative decrease in milk production, hyperthermia of dairy cows leads to a qualitative deterioration of milk composition. The analyses performed by Sharma et al. on the milk produced by stressed Friesian cows [17] showed differences in the percentages of fat and protein with respect to milk produced in thermal comfort conditions. In particular, as shown in table 3, the fat decreases by more than 5%, and the proteins by almost 9%. Among the proteins involved in the percentage drop there are caseins, fundamental in the curdling process and therefore in production of cheese (especially Parmigiano Reggiano, the most important cheese product of Southern Po Valley). The decrease is also significant for total solids (over 14%) and non-fat solids (almost 7%), which are other indicators of the quality of milk.

Table 3. Percent decrease of components in milk produced by stressed cows

Component	Percent decrease
fats	5.05
proteins	8.91
non-fat solids	14.47
total solids	6.78

2.3. *Fertility decrease*

Thermal stress also contributes to the reduction of fertility typical of the summer period. This involves a considerable economic damage, affecting about 60% of cattle in the world [18]. The most important aspect of summer infertility is its multifactorial nature, since hyperthermia alters and directly worsens cellular functions of various parts and tissues of the reproductive system [19]. Overheating is responsible for the onset of both immediate and delayed effects on the reproductive system. In particular, the quality of the ovarian follicles is compromised and the duration of oestrus is reduced due to the production reduction of the luteinizing hormone (LH) and the secretion of estradiol during the thermal stress period [20]. The reduction of steroidogenic capacity comes from a malfunction of the theca and granulosa cells under stress.

The chronic exposure to heat causes a suppression of the production of progesterone, which leads to a series of negative effects before and after insemination. The low blood levels of this hormone in fact cause an abnormal oocyte maturation, or possibly an early death of the embryo. Moreover, the shortage of progesterone will have a negative impact on the development of the dominant follicle and corpus luteum that subsequently grow up, in addition to being responsible for the alteration of the morphology of the endometrium [21].

All these phenomena lead to a thinning of oestrus, to a decrease in conception rate and increased embryonic mortality. Jordan [22] showed many significant data concerning the effect of climatic factors on fertility: the conception rate falls from 62% to 45% when the rectal temperature rises by 1°C in the 12 hours after insemination; a rectal temperature of 40°C, resulting from an exposure to an external temperature of 32.2°C for 72 hours after insemination, has determined a zero conception rate. The infertility phenomenon due to thermal stress is especially noticeable in high yielding cows. Overheating also affects the bulls, whose sperm produced in stressful conditions is underperforming with respect to the cooler seasons.

2.4. Mortality increase
The mortality of animals is directly related with the THI. The study by Vitali et al. [23], which lasted six years, found that the highest number of deaths among animals occurs in summer. When thermal stress is severe, the loss of water through sweating and rapid breathing contributes to the onset of cardiovascular disorders. Moreover, there are modifications in the fatty acid and glucose metabolism associated with liver dysfunction and oxidative stress. All this increases the probability of death among cows stressed by the heat.

3. Improving thermal comfort of dairy cows
The trends described above, induced by high summer temperatures, have extremely negative repercussions on dairy farms, which would therefore benefit from technical solutions in favor of animal welfare. Typical remedies to lower the average temperature perceived by cattle such as using electric fans, water sprinklers or evaporative cooling systems cause a significant consumption of electricity and also water, moreover they seldom have a decisive effect because they do not intervene on the main cause of thermal discomfort, that is thermal radiation from the overheated ceiling. In fact, if the roof stable is overheated, the ambient temperature to be used for calculation of THI should not be that of the ambient air, but the operative temperature actually perceived inside the stable, about equal to the average of the air temperature and the mean radiant temperature of the building inner surfaces. An effective solution to lower the mean radiant temperature and, with this, the operative temperature is represented by coating the stable roof with solar-reflective products such as paints, membranes, sheaths, etc.

3.1. Cool roofs
Roofing products that contribute to the efficiency of civil engineering structures thanks to surface properties optimized for rejection of solar heat are identified with the technical name of cool roofs. They have been widely exploited in buildings for humans for decades in North America [24,25], more recently also in Europe [26,27] and are extensively considered in policies to counter overheating of both urban areas and dwellings [28,29]. They can of course be used to prevent overheating of roofs also in animal shelters, thus eliminating the infrared thermal radiation from the ceiling. In Figure 2, the thermographic image and the visible picture of the roof of a pig farm near Modena, Northern Italy, are shown. They were both taken in the late summer of 2013 and allow to check, on the right of the thermal image, the surface temperature of a piece of roof coated with solar reflective paint and, on the left, the surface temperature on an uncoated part of the roof. The image reveals how solar radiation heats the uncoated part up to 60°C and above, whereas the part with cool roof paint remains close to ambient temperature.

Figure 2. Thermal image (left) and visible picture (right) acquired in summer of 2013 on a pig shelter roof in the province of Modena, Northern Italy: the roof on the left side, uncoated, has surface temperature in excess of 60°C, while the roof right side, which was coated by a white cool roof paint, remains at a temperature around 30°C.

Below roofs subjected to solar radiation, especially poorly insulated ones, people and animals suffer the so-called "hot head" effect, the strong feeling of discomfort caused by radiation coming from the hot ceiling. Using a calculation model specifically constructed at the Laboratory for Energy Efficiency (EELab) of the University of Modena and Reggio Emilia to evaluate the effect of cool roof of building structures, it was possible to calculate the benefit provided a solar reflective coating (with solar reflectance as high as 85%) applied onto the roof of a typical stable covered with fiber cement panels (with solar reflectance before coating <40%). The calculation revealed that the operative temperature perceived by cows below the uncoated roof can be higher than that of the ambient by more than 3°C, whereas under the cool coated roof it becomes close to that of the ambient. In fact, cool roof products can prevent the onset of the "hot head" effect because they prevent roof overheating and thus thermal radiation from the ceiling.

3.2. Application example of a cool roof on a building

The use of cool coatings on opaque components of the building envelope can have a positive effect, if not decisive, on the problem of summer overheating of buildings and the resulting physical discomfort experienced by people (and animals) housed in them, as well as on the problem of energy and/or water consumption for cooling systems, where present. At the Department of Engineering "Enzo Ferrari" of the University of Modena Enzo Ferrari studies have been conducted on cool roofs for more than a decade [30-34].

A cool roof was installed in 2005 on a building of the Engineering Campus in Modena. This experiment, probably one among the very first cool roofs based on qualified solar-reflective products that have been installed in Italy, made it possible to prove the effectiveness of the technology. The building is a shed without air-conditioning and with relatively low thermal inertia, which always showed an intense overheating in summer and reached internal temperatures often in excess of 40°C. After application of a cool roofing product consisting of a white water-based liquid membrane with waterproofing capacity, a sharp drop of temperature was observed in all measured locations. Coincidentally, the two periods immediately before and after installation of the cool roof were characterized by similar conditions of solar radiation and ambient temperature, so it was possible to immediately verify the effect of the cool roof on the thermal behavior of the shed. In particular, the adoption of the cool roof allowed to reduce the temperature of the ceiling surface, which decreased by 3.7°C on average, as well as the temperature of the indoor air, which decreased by 3.0°C (Figure 3).

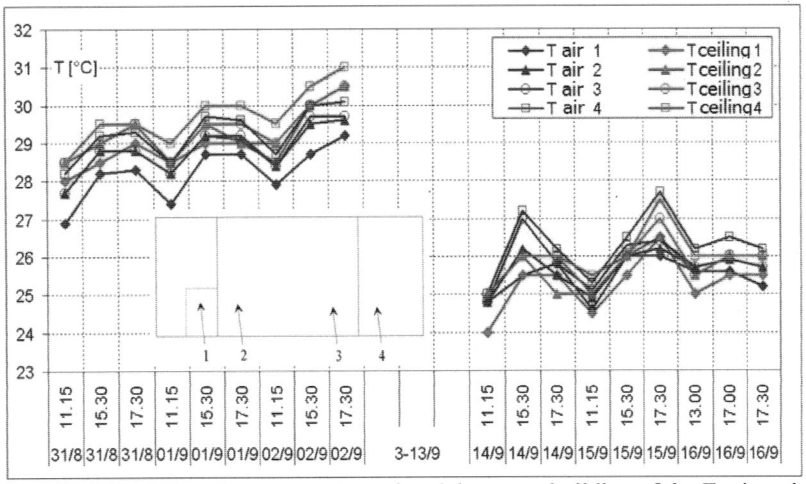

Figure 3. Measured air and ceiling temperatures in a laboratory building of the Engineering Campus in Modena before and after application of a cool roof: the temperature of the ceiling has fallen by 3.7°C on average, that of the air measured at ground level by 3.0°C

3.3. Example of use of a cool roof in farming

In the summer of 2014 a cool roofing product was applied onto the roof of a cattle farm in the province of Cremona, Po Valley, Northern Italy. It was applied on a summer day (July 31st, 2014) with average ambient temperature of 26°C, recorded by a local weather station, and 70% relative humidity measured inside the stable (slightly higher than that considered for Modena in section 2). As previously mentioned, radiative heating is released from the hot ceiling below a roof subjected to solar radiation, thus causing the discomfort of people or animals below. On the other hand, applying a high reflectance material onto the upper roof surface, a high fraction of solar radiation is reflected back to the sky and overheating of roof and ceiling is avoided. The thermographic images in Figure 4 explain the phenomenon.

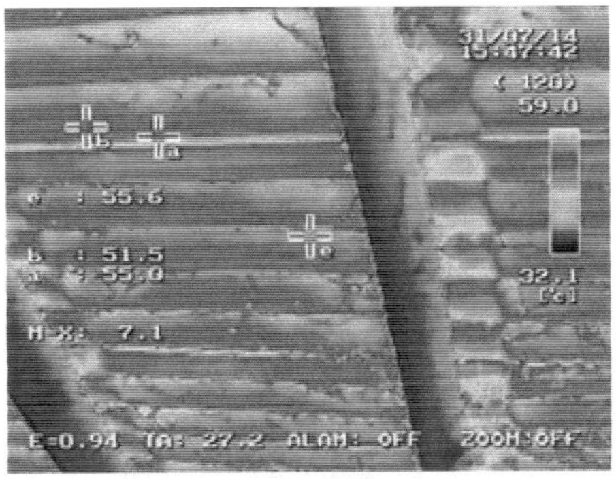

Figure 4a.

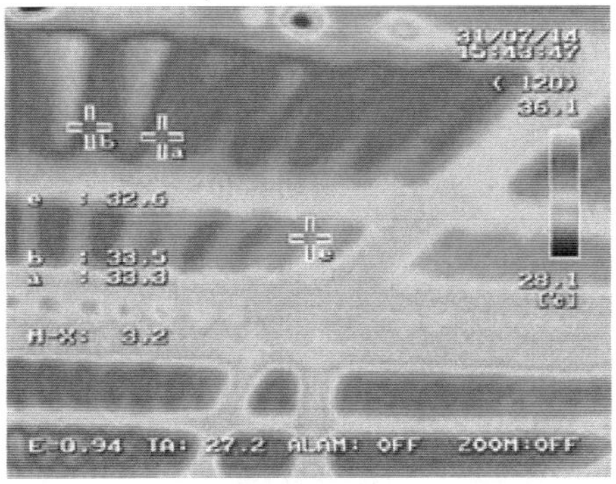

Figure 4b.

Figure 4: Underside roof temperature of a stable in the province of Cremona: a) uncoated roof (average ceiling temperature 55.6°C); b) cool roof (average ceiling temperature 32.6°C). The measurements have been made in the same stable, at the same time of the same day and in the same ambient conditions.

Entering into details, Figure 4a shows a part of the roof with upper surface not coated with solar reflective paint, which reached an average temperature of 55.6°C measured from below. Instead, Figure 4b shows a part with a cool roof coating applied onto the upper surface, with which the ceiling temperature remained at 32.6°C on average. The images were taken in the same stable, and at the same time of the same day, so they clearly show that the solar reflective paint was able to lower the average ceiling temperature by more than 20°C. This phenomenon has a consequence on the operative temperature, that is the temperature perceived by the animals that live below. Since past research has shown [30-34] an average difference of operative temperature between a standard and a cool roof around 3°C, the temperature perceived by the animals below the uncoated and overheated roof was assumed to be 3°C higher than that below the cool roof, which is instead close to the outside ambient temperature. The THI value was then evaluated below both the uncoated roof and the cool roof, considering the measured relative humidity value of 70%. For the ambient temperature in the day of the experiment, a THI value around 80 was calculated due to the increase of the perceived temperature below the uncoated roof, resulting in a potential drop of milk production as high as 16%. Thanks to the application of the cool roof, the perceived temperature was calculated to stay close to the ambient temperature, with a THI value around 75 and a potential production drop of 6%, much less than that in the previously considered condition.

The percentages of potential milk production achieved with and without cool roof were then compared for the considered stable all over the typical range of ambient temperature (Figure 5). Also in this case, the inside air temperature of the (open) stable was assumed to be close to the outside ambient temperature.The difference generally exceeds 10%.

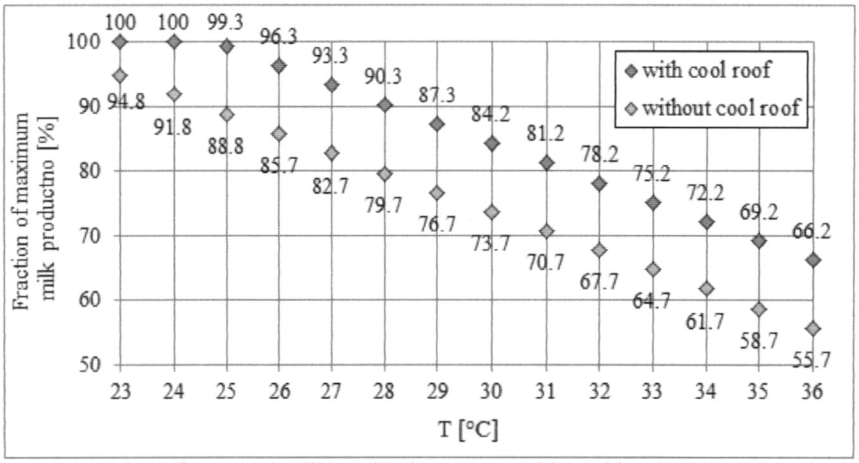

Figure 5. Fraction of potential milk productions vs. outside ambient temperature calculated for a stable in Cremona, Po Valley, considering the temperature perceived by the animals with and without cool roof (inner RH 70%).

3.4. Effects of cool roofs on the corporate balance sheet

The loss of milk production due to thermal stress results in a significant damage to the budget of farms during summer. Therefore, improving animal welfare by improving the microclimate conditions in the stables by means of the application of a cool roof allows to improve the corporate productivity.

In order to translate in economic terms the improvement of productive potential, one can consider the average daily production of a typical stable with about 200 Italian Friesian cows (about 30 kg of milk per cow). Referring to the price of milk for the production of Parmigiano Reggiano, it was calculated that the economic advantage resulting from higher milk production would correspond to about € 350 per day. Considering a surface to be treated of 3000 m^2 and a specific cost of the cool roof coating of 10 €/m^2, the investment could be recovered after 85 cumulative days with average temperature of 26°C or higher. This could amount to one or two calendar years. In such simple economic calculation other benefits from improved animal welfare such as an increased frequency of pregnancies and a reduction of summer mortality were not considered.

4. Concluding remarks

In temperate and hot climates, the microclimate inside animal shelters has an important role for dairy cows. More specifically, summer overheating becomes especially harmful because it is responsible for the onset of thermal stress in the animals. Thermal stress induces a number of physiological alterations that have as direct consequence the decrease of milk production. Indeed, there is a close interdependence between animal welfare and productive or reproductive factors, and the decline in production varies depending on the stress level.

The application of cool roofing materials with high solar reflectance on stable roofs can provide a solution to the problem of overheating of animal shelters in summer due to solar radiation. Such materials reduce the absorption of solar radiation and thus prevent the stable roof to overheat and, consequently, the temperature perceived below to increase, therefore they represent an effective mean to counter the thermal stress of dairy cattle and its negative consequences on the profitability of livestock production.

References

[1] Yeck R G and Stewart R E 1959 A ten-year summary of the psychroenergetic laboratory dairy cattle research at the University of Missouri *Trans. ASAE* **2** 71-77
[2] Ferrari P 2009 Il controllo ambientale nelle stalle *Atti 81ª Fiera Agricola Zootecnica Italiana: Benessere animale: la vacca da latte "dalle parole ai fatti"* (Montichiari, Italy)
[3] Nardone A and Ronchi B. 1999 Lo stress da caldo nelle bovine da latte: benessere e aspetti produttivi *Atti Parliamo di... benessere e allevamento animale* (Fossano, Italy)
[4] Bernabucci U, Lacetera N, Baumgard L, Rhoads R, Ronchi B and Nardone A 2010 Metabolic and hormonal acclimatation to heat stress in domesticated ruminants *Animal* **4** 1167-1183
[5] Thom E C and Bosen J F 1959 The discomfort index *Weatherwise* **12** 57-60
[6] West J 1981 *Effect of Environment on Nutrient Requirements of Domestic Animals* (Washington DC: National Academy Press)
[7] Wiersma F 1990 *Temperature-humidity index table for dairy producer to estimate heat stress for dairy cows* (Tucson: Department of Agricultural Engineering, The University of Arizona)
[8] Igono M O, Bjotvedt G and Sanford-Crane H T 1992 Environmental profile and critical temperature effects on milk production of Holstein cows in desert climate *Int. J. Biometeorology* **36** 77-87
[9] Armstrong D V 1994 Heat stress interaction with shade and cooling *J. Dairy Science* **77** 2044-2050
[10] Bouraoui R, Lahmar M, Majdoub A, Djemali M and Belyea R 2002 The relationship of temperature-humidity index with milk production of dairy cows in a Mediterranean climate *Animal Research* **51** 479-491
[11] Erb H N, Martin S W, Ison N and Swaminathan S 1981 Interrelationship between production and reproduction diseases in Holstein cows. Conditional relationships between production and diseases *J. Dairy Science* **64** 272-281
[12] Nakao T, Moriyoshi M and Kawata K 1992 The effect of postpartum ovarian dysfunction and endometritis on subsequent reproductive performance in high and medium producing dairy cows *Theriogenology* **37** 341-349
[13] West J W 2003 Effects of heat-stress on production in dairy cattle *J. Dairy Science* **86** 2131-2144
[14] Righi F, Quarantelli A, Bonomi A and Renzi M 2004 L'assunzione di sostanza secca nella vacca da latte: parametri correlati, regolazione fisica e chimica *Annuario della Facoltà di Medicina Veterinaria di Parma* **XXIV** 299-315

[15] Kurihara M and Shioya S 2003 *Dairy cattle management in a hot environment* (Taipei: Food and Fertilizer Technology Center for the Asian and Pacific Region) Extension Bulletin 529
[16] Wheelock J B, Rhoads R P, VanBaale M J, Sanders S R and Baumgard L H 2010 Effects of heat stress on energetic metabolism in lactating Holstein cows *J. Dairy Science* **93** 644-655
[17] Sharma A K, Rodriguez L A, Mekonnen G, Wilcox C J, Bachman S K C and Collier R J 1983 Climatological and Genetic Effects on Milk Composition and Yield *J. Dairy Science* **66** 119-127
[18] Calamari L 2003 Condizioni di benessere: gestione degli animali e controllo del microclima *L'ipofertilità nella bovina da latte* (Brescia: Fondazione Iniziative Zooprofilattiche e Zootecniche)
[19] Wolfenson D 2009 Impact of Heat Stress on Production and Fertility of Dairy Cattlen *Proc. Tri-State Dairy Nutrition Conference* (Jerusalem: Department of Animal Sciences, Hebrew University)
[20] Wilson S J, Kirby C J, Koenigsfeld A T, Keisler D H and Lucy M C 1998 Effect of controlled heat stress on ovarian function of dairy cattle: 2. Heifers *J. Dairy Science* **71** 2132-2138
[21] Ahmad N, Schrick F N, Butcher R L and Inskeep E R 1995 Effect of persistent follicles on early embryonic losses in beef cows *Biology of Reproduction* **52** 1129-1135
[22] Jordan E R 2003 Effects of Heat Stress on Reproduction *J. Dairy Science* **86** (E Suppl) E104-E114
[23] Vitali A, Segalini M, Bertocchi U, Nardone A and Lacetera N 2009 Seasonal pattern of mortality and relationships between mortality and temperature-humidity index in dairy cows *Journal of Diary Science* **92** 3781-3790
[24] Taha H, Akbari H, Rosenfeld A and Huang J 1989 Residential cooling loads and the urban heat island-the effects of albedo *Building and Environment* **23** 271-283
[25] Akbari H, Konopacki S and Pomerantz M 1999 Cooling energy savings potential of reflective roofs for residential and commercial buildings in the United States *Energy* **24** 91-407
[26] Synnefa A, Santamouris M and Livada I 2006 A study of the thermal performance of reflective coatings for the urban environment *Solar Energy* **80** 968-981
[27] Santamouris M, Synnefa A, Kolokotsa D, Dimitriou V and Apostolakis K 2008 Passive cooling of the built environment – use of innovative reflective materials to fight heat island and decrease cooling needs *Int. J. Low Carbon Technologies* **3** 71-82
[28] Levinson R, Akbari H, Konopacki S and Bretz S 2005 Inclusion of cool roofs in nonresidential Title 24 prescriptive requirements *Energy Policy* **33** 151-170
[29] Akbari H, Cartalis C, Kolokotsa D, Muscio A, Pisello A L, Rossi F, Santamouris M, Synnefa A, Wong N H and Zinzi M 2016 Local climate change and urban heat island mitigation techniques – the state of the art *J. Civil Engineering and Management* **22** 1-16
[30] Libbra A, Tarozzi L, Muscio A and Corticelli M A 2011 Spectral response data for development of cool coloured tile coverings *Optics and Laser Technology* **43** 394-400
[31] Libbra A, Muscio A, Siligardi C and Tartarini P 2011 Assessment and improvement of the performance of antisolar surfaces and coatings *Progress in Organic Coatings* **72** 73-80
[32] Libbra A, Muscio A and Siligardi C 2013 Energy performance of opaque building elements in summer: Analysis of a simplified calculation method in force in Italy *Energy and Buildings* **64** 384-394
[33] Ferrari C, Gholizadeh Touchaei A, Sleiman M, Libbra A, Siligardi C and Akbari H 2014 Effect of aging processes on solar reflectivity of clay roof tiles *Advances in Building Energy Research* **8** 28-40
[34] Despini F, Ferrari C, Bigi A, Libbra A, Teggi S, Muscio A and Ghermandi G 2016 Correlation between remote sensing data and ground based measurements for solar reflectance retrieving *Energy and Buildings* **114** 227-233

A Further Contribution to the Parametric Analysis of a PCM Energy Storage System

G Casano and S Piva

ENDIF ENgineering Department In Ferrara, Università di Ferrara
via Saragat 1, 44122 Ferrara (I)

Corresponding Author E-mail: stefano.piva@unife.it

Abstract. In recent years heat storage using phase change materials has been also considered in the thermal control of electronic devices. In a recent work we presented some results for a parametric analysis of an energy storage system with a phase change material undergoing a two-levels steady-periodic heat boundary condition, as happens in certain electronic equipments. In particular, a hybrid system composed of a finned surface partially filled with a PCM, was analysed. This solution, which combines both passive (PCM) and active (fins and fans) cooling solutions, is of interest in high power amplifiers characterized by different levels of power dissipation, as is the case of the telecom base station power amplifiers, where the power is proportional to the traffic load. In the present paper we analyze some parameter previously not investigated, in particular the dimensionless transition temperature, for the role played by this parameter in the limitation of the operating temperature reached during the peaks of the power input. The study has provided further useful information for the design of these hybrid cooling systems..

1. Introduction

Heat storage systems using Phase Change Materials (PCMs in the following) offer effective opportunities for managing thermal energy due to the high energy storage density and the isothermal nature of the storage process. Beside interesting applications in many different fields (building envelopes, residential heating and cooling, solar engineering, and spacecraft thermal control applications [1]), in recent years latent heat storage has been also considered in the thermal control of electronic devices, where the heat generated by electronic circuitry must be dissipated to prevent immediate failure and assure long term reliability. In particular for high specific power and/or compactness, the limited capability of the air cooling traditional techniques suggests to benefit from the advantages offered by PCMs, such as: high specific heat, high latent heat, small volume change during phase change, availability at convenient values of melting temperature, non-toxicity, inertness and non-corrosiveness. Furthermore, PCM energy storage systems can be useful to delay the heat release so to reduce the need of heat transfer surface, in particular when the heat dissipation is of periodic nature or suddenly changing.

For a large review of the applications of latent heat storage in electronics some recent papers and volumes are available (Srivatsan *et al.* [2], Ling *et al.* [3], Fleischer [4], Castell and Solé. [5], Casano and Piva [6]).

In some recent papers ([6] and [7]) we presented some results for applications dealing with high power amplifiers characterized by different levels of power dissipation, as is the case of the telecom base station power amplifiers, where the power is proportional to the traffic load. The final goal of this research is the development of a hybrid PCM heat sink for power electronics applications, consisting of a heat sink with part of the channel between the fins filled of a suitable solid-solid PCM. In [7] we analysed experimentally and numerically a TCU consisting of a plane slab of PCM for an on/off heating condition. In [6] we presented some results of a parametric analysis, based on a numerical investigation, for a finned heat sink partially filled of solid-solid PCM. The mathematical model was made dimensionless to allow the characteristic parameters of the application to be evidenced and discussed. However, the large number of independent dimensionless parameters (13) suggested in [6] to concentrate our attention on a limited number of parameters (4), two useful to describe the thermal load and two to characterize the thermal behaviour of the TCU.

In the present paper we add some new results, useful to extend the analysis to some parameter previously not investigated. In particular we concentrate now our attention to the dimensionless transition temperature for the role played by this parameter in the limitation of the operating temperature reached during the peak of the input power. Some new useful information can be now suggested for the design of these hybrid cooling systems.

2. Mathematical Formulation

A sketch of the physical system, consisting of a finned heat transfer surface partially filled of PCM, is shown in figure 1. The heat sink is heated by the electronic system and cooled by a forced flow of air. For the large number of fins and the boundary conditions, when neglecting the longitudinal dependency in the temperature distribution, the analysis can be limited to a 2D portion of the system (figure 2).

The mathematical model representing the physical process is based on the following assumptions:

- PCM is homogeneous and isotropic.
- Thermophysical properties are constant in each phase.
- Phase-change occurs at a single transition temperature between the two solid phases.
- Heat transfer is controlled only by conduction.
- The problem is two-dimensional.

The problem is governed by the Fourier equation, to be solved in the two solid phases of the PCM and in the fin ($i = F$ in the fin; $i = S1$ or $S2$ for the two solid phases of the PCM; $S1$ indicates the area where $T < T_f$, $S2$ that where $T > T_f$).

As the boundary conditions, on the basement a uniform, a time-dependent heat load is applied, characterised by two levels of power. The duty cycle (figure 3) is periodic, with a low power period (q_1 for $0 \leq t \leq t_1$) and a high power period (q_2 for $t_1 < t < P$). The portion of the fin free in air, for

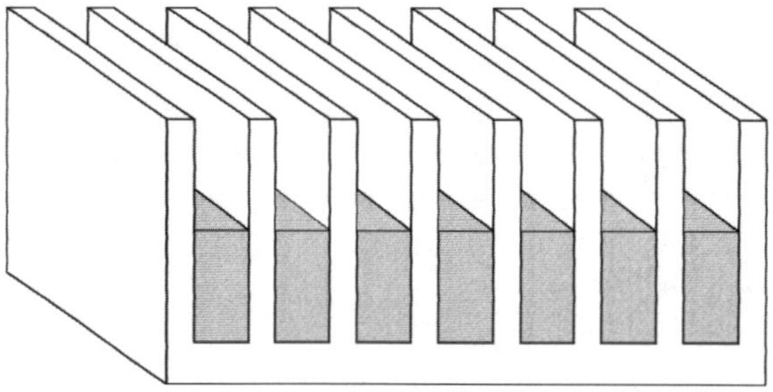

Figure 1. Scheme of the finned heat transfer surface.

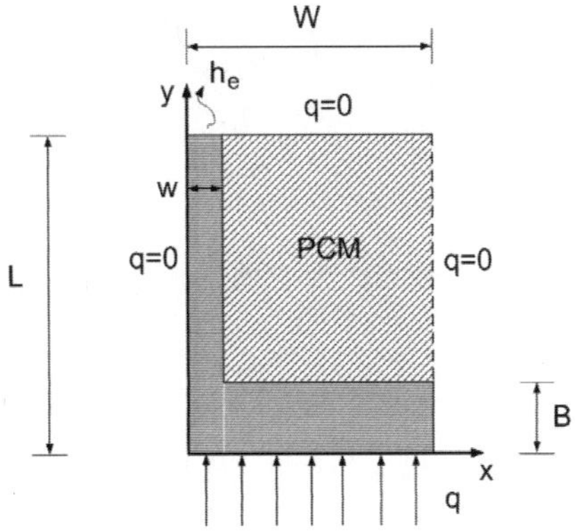

Figure 2. Schematic diagram of the reference domain of the finned heat transfer surface.

simplicity reasons, is substituted by an equivalent heat transfer coefficient, *he**, sized to accommodate the high level heat load. It allows the system to operate during the low power portion of the duty cycle at a value of temperature that is certainly below the transition temperature of the PCM. The remaining surfaces are adiabatic.

The full set of differential equations and boundary conditions is reported in [6]. For reason of brevity here only their dimensionless form is shown. The problem can be thence stated in mathematical form as follows (see the Nomenclature for the definition of the dimensionless quantities):

• Solid phases (for fin, solid 1 and solid 2, respectively):

$$\frac{\partial \theta}{\partial \tau} = Fo_F \mathrm{A}_i \left(\frac{\partial^2 \theta}{\partial \xi^2} + \frac{\partial^2 \theta}{\partial \eta^2} \right) \tag{1}$$

• Solid–solid interface:

$$\theta_{S1} = \theta_{S2} = \theta_f \tag{2}$$

$$\frac{1}{\mathrm{A}_{S1} Fo_F Ste} \frac{\partial \Xi}{\partial \tau} = \left[1 + \left(\frac{\partial \Xi}{\partial \xi} \right)^2 \right] \left(\frac{\partial \theta_{S1}}{\partial \eta} - \frac{\Lambda_{S2}}{\Lambda_{S1}} \frac{\partial \theta_{S2}}{\partial \eta} \right) \tag{3}$$

• Initial condition:

$$\tau = 0\ ,\ 0 \le \xi \le 1 \ \text{ and } \ 0 \le \eta \le 1 \quad \theta(\xi, \eta, \tau = 0) = 0 \tag{4}$$

• Boundary conditions:

$$\xi = 0 \ \text{ and } \ 0 \le \eta \le 1 \quad \frac{\partial \theta}{\partial \xi} = 0 \ \text{ for } \ 0 < \tau \le 1 \tag{5}$$

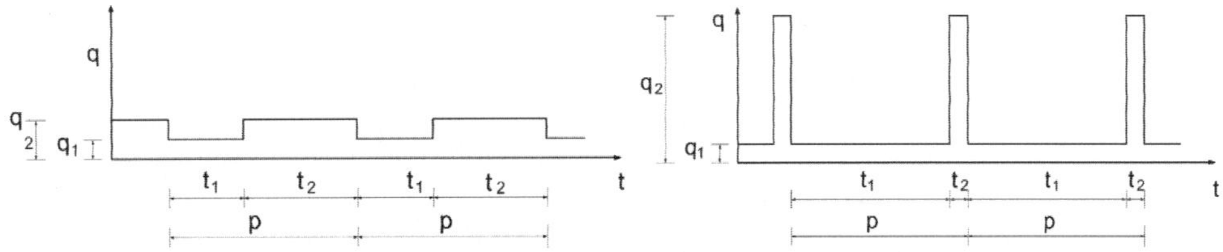

Figure 3. Typical duty cycle of the heating system.

$$\xi = \xi_W \text{ and } 0 \le \eta \le 1 \quad \frac{\partial \theta}{\partial \xi} = 0 \text{ for } 0 < \tau \le 1 \tag{6}$$

$$\eta = 0 \text{ and } 0 \le \xi \le \xi_W \quad \frac{\partial \theta_F}{\partial \xi} = -1 \text{ for } 0 < \tau \le \tau_1 \tag{7}$$

$$\eta = 0 \text{ and } 0 \le \xi \le \xi_W \quad \frac{\partial \theta_F}{\partial \xi} = -\Omega \text{ for } \tau_1 < \tau \le 1 \tag{8}$$

$$\eta = 1 \text{ and } 0 \le \xi \le \xi_w \quad \frac{\partial \theta_F}{\partial \eta} = \frac{Bi}{\xi_W}(\theta_F - \theta_{ai}) \text{ for } 0 < \tau \le 1 \tag{9}$$

$$\eta = 1 \text{ and } \xi_w < \xi \le \xi_W \quad \frac{\partial \theta}{\partial \eta} = 0 \text{ for } 0 < \tau \le 1 \tag{10}$$

$$\xi = \xi_w \text{ and } \eta_B \le \eta \le 1 \quad \frac{\partial \theta_F}{\partial \xi} = \Lambda_i \frac{\partial \theta_i}{\partial \xi} \text{ for } 0 < \tau \le 1 \tag{11}$$

$$\eta = \eta_B \text{ and } \xi_w < \xi \le \xi_W \quad \frac{\partial \theta_F}{\partial \eta} = \Lambda_i \frac{\partial \theta_i}{\partial \eta} \text{ for } 0 < \tau \le 1 \tag{12}$$

The significant dimensionless parameters of the problem are:

$$Fo_F,\ Ste,\ Bi,\ \theta_f,\ \mathrm{A}_{S1},\ \mathrm{A}_{S2},\ \Lambda_{S1},\ \Lambda_{S2},\ \tau_1,\ \Omega,\ \xi_W,\ \xi_w,\ \eta_B.$$

Even though in the mathematical model the dimensionless parameter characterizing the thermal load is τ_1, in the following discussion of the results, we use τ_2 , its complement to 1 ($\tau_2 = 1 - \tau_1$), being this parameter closer to the significance of the heating scheme.

The initial condition, Eq. (4), characterizes an isothermal system at the room temperature. At this time the total energy (sensible and latent) of the system, H_0, is conventionally set $H_0 = 0$.

The time evolution of total energy stored by the system is given by the integral in time of the difference between the incoming and outgoing heat fluxes:

$$\Delta H(t) = \int_0^t \left[q(t)W - \int_0^W h_e \left(T_F(x,L,t) - T_a\right) dx \right] dt \tag{13}$$

where, for the n-th cycle of heating: $q(t) = \begin{cases} q_1 & \text{for} \quad (n-1)P < t \le t_1 + (n-1)P \\ q_2 & \text{for} \quad t_1 + (n-1)P < t \le nP \end{cases}$

The latent energy storage, $\Delta H_L(t)$, is calculated as:

$$\Delta H_L(t) = \rho_{S1} r S_2(t) \tag{14}$$

The sensible energy is calculated as the difference between total and latent energy at the same time.

The reference energy used to make dimensionless the stored energy is the maximum latent energy storable in the system, given by:

$$(\Delta H_L)_{\max} = \rho_{S1} r S_{PCM} \tag{15}$$

where S_{PCM} indicates the surface filled by the PCM ($S_{PCM} = S_1 + S_2$).

The dimensionless latent and total energy are given, respectively, by:

$$\Delta \mathrm{E}_L(\tau) = \frac{\Delta H_L(t)}{(\Delta H_L)_{\max}} \tag{16}$$

$$\Delta\mathrm{E}(\tau)=\frac{\Delta H_L(t)+\Delta H_S(t)}{(\Delta H_L)_{\max}} \tag{17}$$

For the latent and total energy the amplitude of the difference between the maximum and minimum value reached in a period of heating is given, respectively, by:

$$\Delta\mathrm{E}_L=(\Delta\mathrm{E}_L)_{\max}-(\Delta\mathrm{E}_L)_{\min} \tag{18}$$

$$\Delta\mathrm{E}=(\Delta\mathrm{E})_{\max}-(\Delta\mathrm{E})_{\min} \tag{19}$$

A parameter significant in the thermal design of electronics equipment is the mean temperature, and the corresponding dimensionless mean temperature, on the surface receiving the heat flux, given by, respectively:

$$T_m(t)=\frac{1}{W}\int_0^W T(x,0,t)\,dx \tag{20}$$

$$\theta_m(\tau)=\frac{T_m(t)-T_0}{q_1L/\lambda_F} \tag{21}$$

The mean temperature is analyzed also in terms of its average value over a period:

$$(T_m)_{av}=\frac{1}{P}\int_{(n-1)P}^{nP} T_m(t)\,dt \tag{22}$$

$$(\theta_m)_{av}=\frac{(T_m)_{av}-T_0}{q_1L/\lambda_F} \tag{23}$$

For the numerical solution of the problem the Finite Volume Method is used. The numerical procedure and some validation exercises are described in full details in [8]. Here it is relevant to underline that a simplified 2-D approach is followed. Inside each control volume affected by the change of phase, the interface is assumed to advance only in the x direction. The resulting solid/solid transition front assumes thence a stepwise shape. The calculation domain consists of 450 control volumes and the time step is 1 s. The calculations are carried out until the reaching of a steady periodic state.

3. Results and Discussion

In the following discussion, reference will be constantly made to the numerical results produced by Casano and Piva [6] for the same configuration analysed in the present paper.

The problem is highly nonlinear and characterized by a large number of dimensionless parameters. In [6] the discussion was limited to four dimensionless parameters: the Biot number, *Bi*, the power ratio, Ω, the Stefan number, *Ste*, and the time ratio, τ_2. Here we explore the effect of a further dimensionless parameter, the dimensionless temperature of transition, θ_f. For the whole set of simulations we maintained the same dimensionless parameters used in [6]: $\Lambda_{S1}=\Lambda_{S2}=7.411\cdot10^{-4}$, $A_{S1}=1.122\cdot10^{-3}$, $A_{S2}=1.005\cdot10^{-3}$, $\xi_W=4.082\cdot10^{-2}$, $\xi_w=1.020\cdot10^{2}$, $Fo_F=160.2$, $\eta_B=2.041\cdot10^{-2}$. Furthermore, for simplicity we maintained also the following parameters: $T_0 = 30$ °C, $q_1 = 3000$ W/m^2 and $P = 4500$ s.

In figure 4 some results are shown for a fixed *Ste* (*Ste* = 0.012) and a fixed Ω (Ω = 4), for three values of dimensionless transition temperature (θ_f = 9.64, 16.52, 23.41), and for three Biot numbers (*Bi* = 0.059, 0.078, 0.156). The values of the dimensionless amplitude of total, ΔE, and latent, ΔE_L, energy are shown as a function of the dimensionless time ratio, τ_2. The solid line is used for the dimensionless amplitude of the total energy, ΔE, the dashed line for the dimensionless amplitude of the latent energy, ΔE_L.

The dimensionless amplitude of total and latent energy is zero for $\tau_2 = 0$ (steady heating with a low power) and for $\tau_2 = 1$ (steady heating with a high power). Both distributions show a maximum in the interval $0 < \tau_2 < 1$. In general, in Figure 4 it is observed that the maximum values of ΔE_L and ΔE, tend to increase with θ_f for the lowest values of *Bi*, and to decrease for the highest value of *Bi*. For the lowest values of *Bi* ($Bi = 0.059$) the maximum of ΔE tends to a value independent of θ_f; the maximum of ΔE occurs for higher values of τ_2 when increasing θ_f. On the opposite, for the highest values of *Bi* ($Bi = 0.156$) the maximum of ΔE changes a lot with θ_f, while it changes a little with τ_2. Between these two values, for $Bi = 0.078$, the maximum of ΔE changes very little with θ_f, while it changes a lot with τ_2. For the lowest values of *Bi* ($Bi = 0.059$) the maximum of ΔE_L increases with θ_f, with a tendency to become less sensible to the variations of τ_2 for the highest values of θ_f. On the opposite, for the highest values of *Bi* ($Bi = 0.156$) the maximum of ΔE_L decreases with θ_f, whit a progressively lower dependency on τ_2. Between these two values, for $Bi = 0.078$, the maximum of ΔE_L changes moderately with θ_f. In order to obtain large movements of the solid-solid interface, large values of θ_f are needed for limited *Bi*, the opposite for large *Bi*.

For the same values of *Ste* and Ω of figure 4, in figure 5 are shown with a solid and a dashed line, the maximum, $(\theta_m)_{max}$ and the average, $(\theta_m)_{av}$, of the dimensionless mean temperature on the heated surface, respectively, as a function of the dimensionless time τ_2. In general in figure 5 it is observed the same trend already discussed in [6]. The distributions of $(\theta_m)_{max}$ are characterized by decreasing values for increasing values of *Bi*, because a better heat transfer due to larger surfaces of the fins in air or a more intensive convection (high *Bi*), coupled to the same energy input (same Ω), is the reason of a decrease of the temperature. For the whole set of data shown in figure 5, for increasing values of τ_2 a value of $(\theta_m)_{max}$ independent of the value of τ_2 is rapidly reached; this is the value of $(\theta_m)_{max}$ due to the highest power, reached for $\tau_2 = 1$. Only for low values of τ_2 a dependency of $(\theta_m)_{max}$ on this parameter is evident. The distributions of the $(\theta_m)_{max}$ are quite independent of θ_f. Only for low values of τ_2 ($\tau_2 < 0.2 \div 0.4$) it is possible to observe that for the lowest values of *Bi* the asymptotic region of $(\theta_m)_{max}$ shrinks for increasing values of θ_f, the opposite for the highest values of *Bi*. As already recognized in [6], the distributions of $(\theta_m)_{av}$, for increasing values of τ_2, are characterized by a linear increase between the values corresponding to the low and high power steady heating ($\tau_2 = 0$ and $\tau_2 = 1$). This average temperature is clearly related to the dimensionless input energy into the system, as already shown in [6].

In figure 6 some results are shown of the amplitude of total, ΔE, latent ΔE_L, and sensible, $\Delta E - \Delta E_L$, dimensionless energy as a function of the Stefan number, *Ste*, for three values of dimensionless phase transition temperature ($\theta_f = 9.64, 16.52, 23.41$), and for three Biot numbers ($Bi = 0.059, 0.078, 0.156$), for $\tau_2 = 0.6$ and $\Omega = 4$. The solid line is used for the dimensionless amplitude of the total energy, ΔE, the dashed line for the dimensionless amplitude of the latent energy, ΔE_L. For the same values of *Bi* and θ_f of figure 6, in figure 7 is shown with a solid line and a dashed line the maximum, $(\theta_m)_{\max}$ and the average, $(\theta_m)_{av}$, dimensionless mean temperature on the heated surface, as a function of the Stefan number, *Ste*.

In figure 6, for the whole set of calculations, ΔE increases with *Ste* and ΔE_L tends to an asymptotic value independent of *Ste*. The attainment of this asymptotic value depends on *Bi*, *Ste* and θ_f. In figure 6, the effects of θ_f is evident: for increasing values of θ_f, the limiting value $\Delta E_L = 1$ is attained for decreasing values of *Bi*. For the highest values of *Bi* the asymptotic values of ΔE_L is lower than 1 and decreases with θ_f. The dimensionless amplitude of sensible energy (ΔE-ΔE_L) is independent of θ_f; this explains the different slope of the distributions of ΔE: when the latent storage system attains the maximum oscillation, only the sensible energy contributes to the increase of ΔE.

For the same values of τ_2 and Ω of figure 6, in figure 7 are shown with a solid and a dashed line, the maximum, $(\theta_m)_{max}$ and the average, $(\theta_m)_{av}$, of the dimensionless mean temperature on the heated surface, respectively, as a function of the Stefan number *Ste*. In general in figure 7 is observed the same dependency of *Ste* already discussed in [6]. Small reductions of $(\theta_m)_{max}$ are evident for low values of *Ste*; this effect increases with θ_f. In the whole set of data shown in figure 7, $(\theta_m)_{av}$ is the same for the same *Bi*, because this parameter depends only on the dimensionless input energy, here constant ($\Omega = 4$).

IOP Conf. Series: Journal of Physics: Conf. Series **796** (2017) 012029 doi:10.1088/1742-6596/796/1/012029

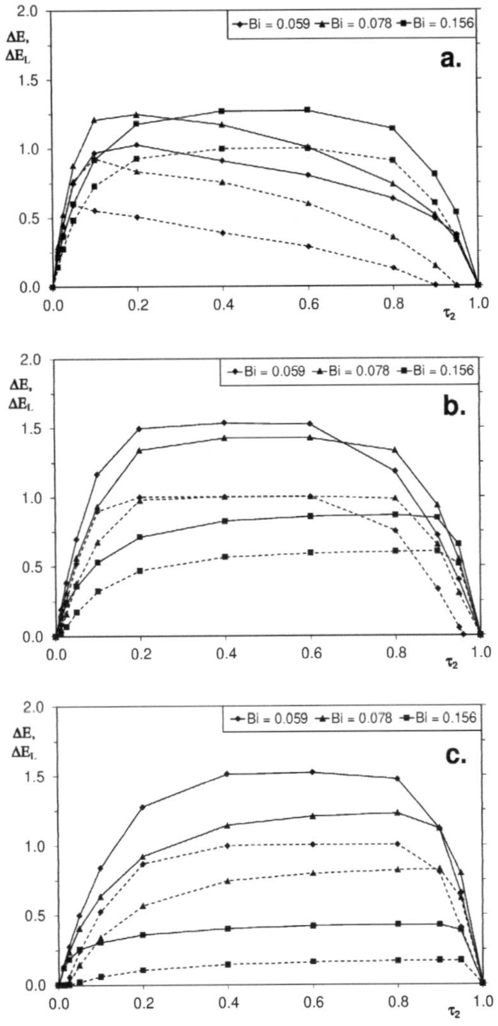

Figure 4. Amplitude of total, ΔE (——), and latent, ΔE_L (– – –), dimensionless energy, for *Ste* = 0.012 and Ω = 4, as a function of τ_2; a.: θ_f = 9.64; b.: θ_f = 16.52; c.: θ_f = 23.41.

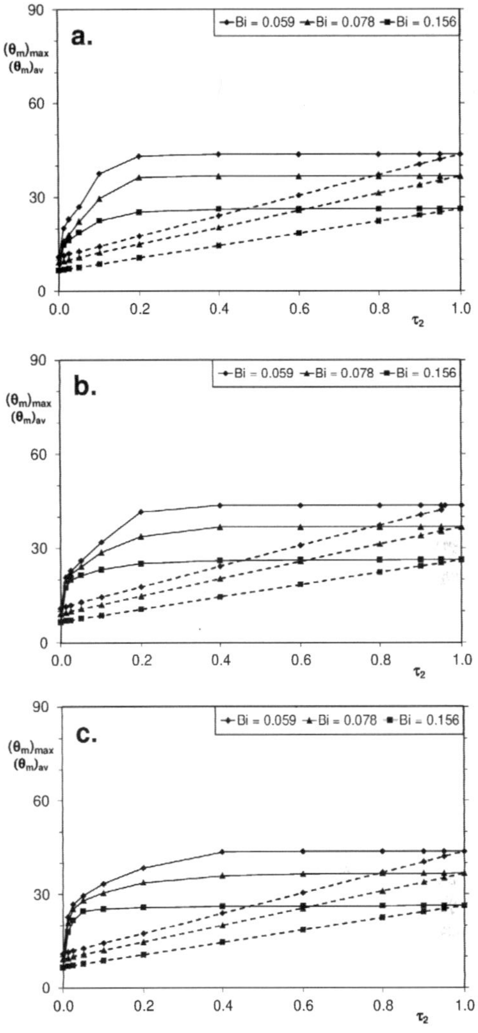

Figure 5. Maximum, $(\theta_m)_{max}$ (——), and average, $(\theta_m)_{av}$ (– – –), dimensionless mean temperature on the heated surface, for *Ste* = 0.012 and Ω = 4, as a function of τ_2; a.: θ_f = 9.64; b.: θ_f = 16.52; c.: θ_f = 23.41.

4. Concluding remarks

A Thermal Control Unit with a PCM energy storage system, designed for the thermal control of electronic devices, has been investigated numerically. This application is characterized by a periodic heating on two levels of power; conduction heat transfer is the prevailing heat transfer process. The main design parameter is the maximum value of average temperature on the heated surface, where the electronic components to be cooled are placed.

The analysis of its energy behaviour shows that the storage of energy varies periodically in time with the two levels of heating.

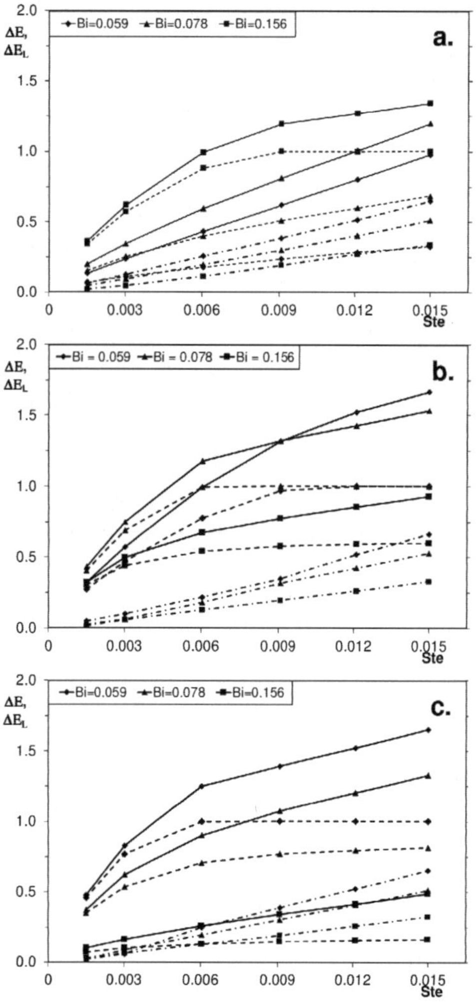

Figure 6. Amplitude of total, ΔE (——), latent, ΔE$_L$ (– – –), and sensible, ΔE-ΔE$_L$ (– · – · –), dimensionless energy, for τ_2 = 0.6 and Ω = 4, as a function of *Ste*; a.: θ_f = 9.64; b.: θ_f = 16.52; c.: θ_f = 23.41.

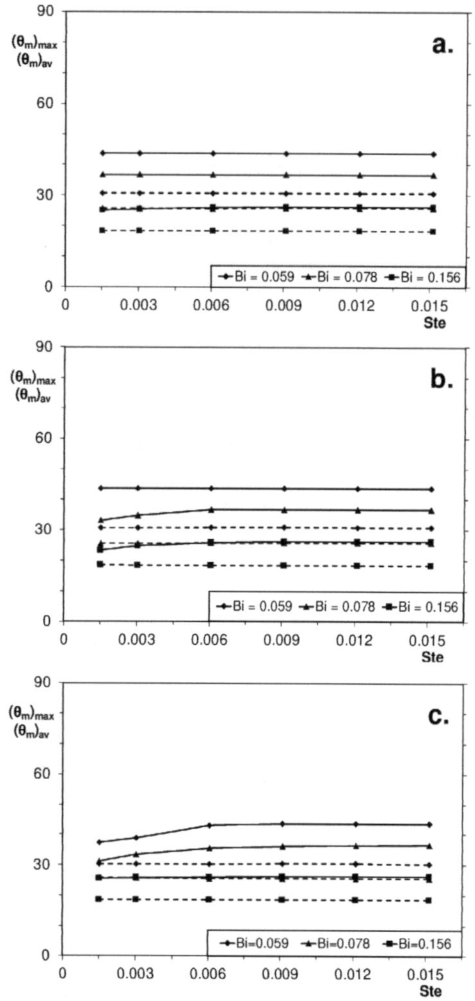

Figure 7. Maximum, $(\theta_m)_{\max}$ (——), and average, $(\theta_m)_{av}$ (– – –), dimensionless mean temperature on the heated surface, for τ_2 = 0.6 and Ω = 4, as a function of *Ste*; a.: θ_f = 9.64; b.: θ_f = 16.52; c.: θ_f = 23.41.

In general it can be observed that a Thermal Control Unit using PCM is suitable to be used when a high power heating acts in short times. During the peak of heating, the phase change can occur; afterwards, it is important that the storage system could release the stored energy in a freezing process. For this hybrid heat storage system, in order to exploit the presence of the PCM, it is important to store and to release an amount of latent energy lower than the maximum latent energy storable.

The dimensionless parameters governing the steady periodic process are thirteen; in a previuos paper [6] we concentrated our attention on two parameters useful to describe the thermal load (Ω and τ_2) and on two more parameters useful to characterize the thermal behaviour of the TCU (*Bi* and *Ste*). In the present paper we explore the effects of the dimensionless temperature of transition, θ_f.

When θ_f is low, a high value of *Bi* helps to control the maximum temperature; the contribution of the PCM to the energy storage can be significant, because a high dissipation to the ambient can sustain the oscillation of the interface position. On the opposite, when θ_f is high, is a low value of *Bi* that helps the complete exploitation of the PCM in terms of oscillations of energy storage. Anyway, in terms of maximum temperature on the heated surface, the behaviour of the Thermal Control Unit seems to be quite independent of θ_f.

A complete exploitation of the PCM requires therefore an accurate design of the system on the base of the parameters affecting the performance of the TCU.

5. References

[1] Sharma A, Tyagi V V, Chen C R and Buddhi D 2009 Review on thermal energy storage with phase change materials and applications *Renewable and Sustainable Energy Reviews* **13** 318-345

[2] Srivatsan S, Anand, J B, Deepak S and Senthil Kumar D 2014 Experimental Analysis of Heat Sinks incorporating Phase Change Materials and Thermal Conductivity Enhancers *Int. J. Engineering Research & Technology (IJERT)* **3** 1197-1204

[3] Ling Z, Zhang Z, Shi G, Fang X, Wang L, Gao X, Fang Y, Xua T, Wang S and Liu X 2014 Review on thermal management systems using phase change materials for electronic components, Li-ion batteries and photovoltaic modules *Renewable and Sustainable Energy Reviews* **31** 427–438

[4] Fleischer A S 2015 *Thermal Energy Storage Using Phase Change Materials. Fundamentals and Applications* (Springer) pp. 8-12

[5] Castell A and Solé C 2015 Design of latent heat storage systems using phase change materials (PCMs) *Advances in Thermal Energy Storage Systems: Methods and Applications* ed. Luisa F. Cabeza (Woodhead Publishing) pp. 298-302

[6] Casano G and Piva S 2015 Parametric Analysis of a PCM Energy Storage System. *Int. J. Heat Technology* **33** 61-68

[7] Casano G and Piva S 2014 Experimental and numerical investigation of a phase change energy storage system *Journal of Physics Conference Series* **501** 012012

[8] Pinelli M, Casano G and Piva S 2000 Solid-liquid phase change heat transfer in a vertical cylinder heated from above *Int. J. Heat and Technology* **18** 61-67

6. Nomenclature

Symbols

B	m	Thickness of the basement
Bi	-	Biot number $(h_e{}^*W)/\lambda_F$
c	J/(kg K)	Specific heat at constant pressure
Fo	-	Fourier number $\alpha_F P/L^2$
$h_e{}^*$	W/m^{2}K	Equivalent heat transfer coefficient
H	J	Enthalpy
L	m	Height of the fin
P	s	Period
q	W/m^2	Heating power
r	J/kg	Latent heat of transition
Ste	-	Stefan number $c_{S1}T_R/r$
t	s	Time
T	K	Temperature
T_R	K	Reference temperature q_1L/λ_F
w	m	Thickness of the fin
W	m	Half width of the fin-PCM system
x	m	Cartesian coordinate
X	m	Solid-solid interface position

IOP Conf. Series: Journal of Physics: Conf. Series **796** (2017) 012029 doi:10.1088/1742-6596/796/1/012029

y	m	Cartesian coordinate
Greek symbols		
α	m^2/s	Thermal diffusivity
A_{Si}	-	Thermal diffusivity ratio α_{Si}/α_F
E	-	Dimensionless energy $\Delta H/(\Delta H_L)_{max}$
η	-	Dimensionless coordinate y/L
θ	-	Dimensionless temperature $(T-T_0)/T_R$
λ	W/(m·K)	Thermal conductivity
Λ_{Si}	-	Thermal conductivity ratio λ_{Si}/λ_F
ξ	-	Dimensionless coordinate x/L
Ξ	-	Dimensionless position of the interface X/L
ρ	kg/m^3	Density
Π	-	Dimensionless input of energy $qt/(\Delta H_L)_{max}$
τ	-	Dimensionless time t/P
τ_i	-	Time ratio t_i/P
Ω	-	Power ratio q_2/q_1
Subscripts		
a		Ambient
av		Average over a period
B		Basement
f		Phase change
F		Fin
m		Mean on the heated surface
max		Maximum
min		Minimum
L		Latent
PCM		Phase change material
S1, S2		Different phases of the solid PCM
w		Half thickness of the fin
W		Half width of the fin-PCM system
0		Initial condition
1, 2		Low and high power

Experimental test of a hot water storage system including a macro-encapsulated phase change material (PCM)

L Mongibello[1], M Atrigna[1], N Bianco[2], M Di Somma[1], G Graditi[1] and N Risi[2]

[1] ENEA – Italian National Agency for New Technologies, Energy and Sustainable Economic Development, Portici R.C. – P.le E. Fermi, 1, 80055 Portici NA, Italy
[2] Dipartimento di Ingegneria Industriale (DII) – Università di Napoli Federico II – P.le Tecchio, 80, 80125 Napoli, Italy

Corresponding author e-mail: luigi.mongibello@enea.it

Abstract. Thermal energy storage systems (TESs) are of fundamental importance for many energetic systems, essentially because they permit a certain degree of decoupling between the heat or cold production and the use of the heat or cold produced. In the last years, many works have analysed the addition of a PCM inside a hot water storage tank, as it can allow a reduction of the size of the storage tank due to the possibility of storing thermal energy as latent heat, and as a consequence its cost and encumbrance. The present work focuses on experimental tests realized by means of an indoor facility in order to analyse the dynamic behaviour of a hot water storage tank including PCM modules during a charging phase. A commercial bio-based PCM has been used for the purpose, with a melting temperature of 58°C. The experimental results relative to the hot water tank including the PCM modules are presented in terms of temporal evolution of the axial temperature profile, heat transfer and stored energy, and are compared with the ones obtained by using only water as energy storage material. Interesting insights, relative to the estimation of the percentage of melted PCM at the end of the experimental test, are presented and discussed.

1. Introduction

By definition, thermal energy storage systems permit to accumulate energy for a later use, thus allowing, for example, a better exploitation of distributed poly-generation systems, which thanks to thermal energy storage systems can produce electricity when it is more economically convenient without or with a limited heat dumping [1,2], or a more efficient use of the thermal energy produced by solar thermal panels by accumulating hot water when it is not required by the thermal utilities for a later use [3,4].

Relatively to the thermal energy storage materials, water is the most used one, essentially because it has practically no cost, a high specific heat, and it is not flammable nor toxic. However, in recent years latent heat thermal energy storage (LHTES), that typically relies on the latent heat relative to the isothermal solid-liquid phase transition of PCMs like paraffins, fatty acids and salt hydrates, has gained great attention, as it can provide much higher energy storage density than the sensible thermal energy storage, and as a consequence it can lead to a consistent reduction of weight and volume of thermal energy storage systems [5-9].

This work analyzes experimentally the dynamical behavior of a hot water storage tank including PCM modules consisting in aluminum bottles of 3 liters partially filled with a commercial bio-based PCM having a melting temperature of 58°C.

In the following, the indoor experimental facility and the PCM modules are described first. Successively, the experimental results relative to the hot water tank including the PCM modules are presented in terms of temporal evolution of the axial temperature profile, heat transfer and stored energy, and they are compared with the ones obtained by using only water as thermal energy storage material.

2. Experimental facility

The experimentation of the hot water storage tank with the PCM has been realized by means of an indoor facility, designed and realized to perform experimental validations of analytical models for the numerical simulation of hot water storage systems including PCMs. Figures 1 and 2 show a sketch and a picture of the experimental facility, respectively.

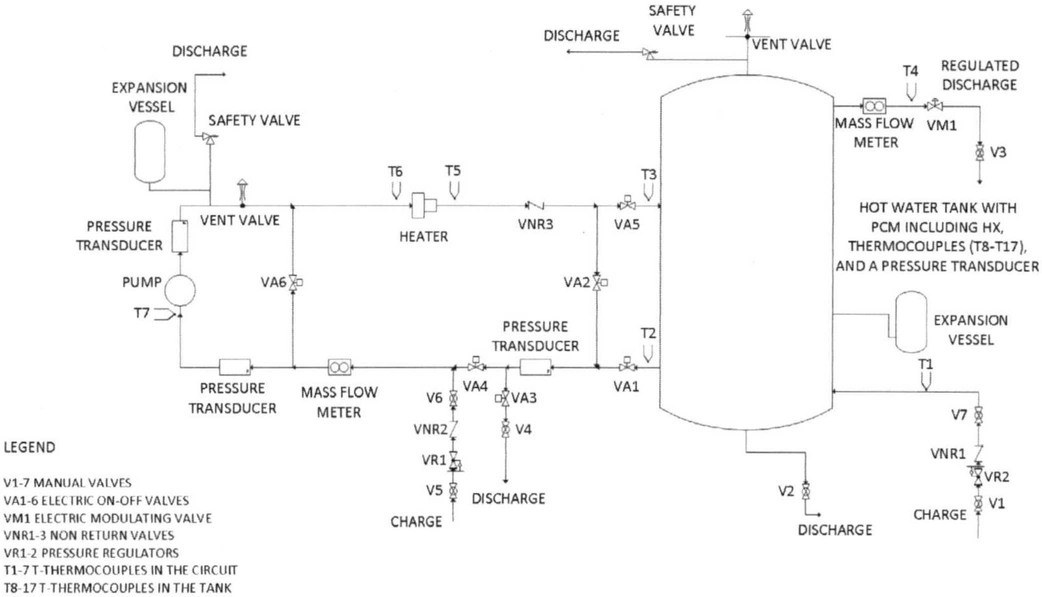

Figure 1. Layout of the experimental facility.

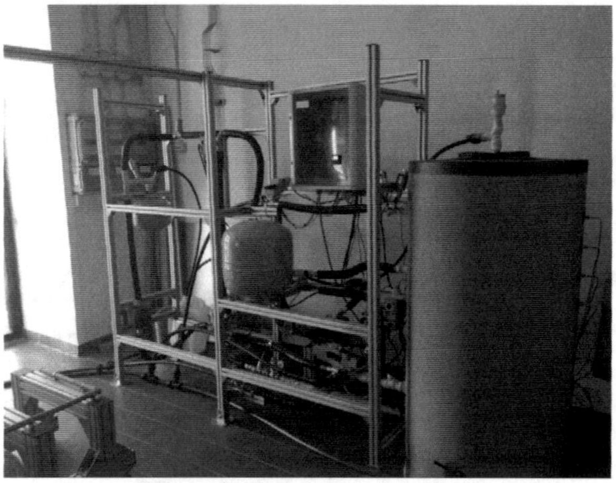

Figure 2. Experimental facility.

The commercial insulated hot water storage tank has a vertical cylindrical shape, a height of 1.27 m, an internal diameter of 0.65 m, and a total capacity of about 420 liters. It is equipped with a 1" coiled-tube heat exchanger having a total heat exchange area of 1.9 m^2, and with 10 type-T thermocouples with the hot junctions equally spaced along the tank axis. Moreover, the hot water tank is connected to an expansion vessel in order to perform tests at different pressure values. The charging of the thermal energy storage tank is realized by means of a fluidic circuit connected to the tank heat exchanger, while the thermal discharging is done by means of an electric modulating valve that allows to regulate the set-point of the water mass flow rate. The heat exchanger inflow and outflow sections are located at about 0.73 m and 0.28 m from the tank bottom, respectively.

Relatively to the charging circuit, the main components are represented by the pump, modulated by an inverter in order to control the mass flow rate through the circuit, and the electric heater providing heating up to 24 kW. A thermoregulator controls the heater in order to regulate the temperature at the tank heat exchanger inflow section. Figures 3 and 4 show the electric heater and the thermoregulator, respectively. The data acquisition and control are made by means of NI modules mounted on NI cRIO-9066 controller. The control panel has been realized by means of the NI software Labview, and all the controlled parameters are regulated using PID controllers.

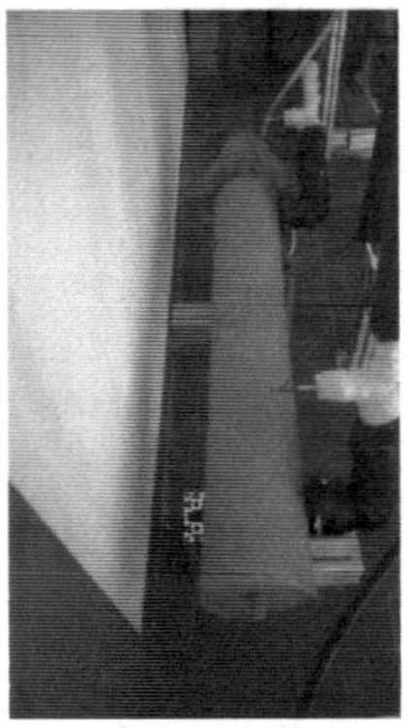

Figure 3. Electric heater.

Figure 4. Thermoregulator.

3. PCM module

Each PCM module consists in an aluminum bottle of 3 liters partially filled with a commercial bio-based PCM, without any additives and without applying any sophisticated technic for the enhancement of heat transfer inside the PCM. Figure 5 and 6 show the aluminum bottle and the PCM, respectively, while table 1 reports the main characteristics of the PCM.

Figure 5. PCM container.

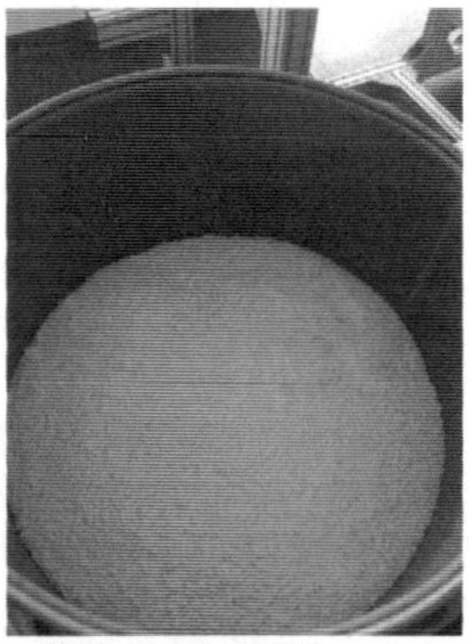

Figure 6. PCM.

Table 1. PCM thermo-physical characteristics.

Property	Value
Melting point	58°C
Latent heat	$225\cdot10^3$ J/kg
Thermal conductivity (liquid)	0.15 W/m/K
Thermal conductivity (solid)	0.25 W/m/K
Density (liquid)	810 kg/m^3
Density (solid)	890 kg/m^3
Specific heat (liquid)	2710 J/kg/K
Specific heat (solid)	2470 J/kg/K

Each bottle contains about 1.7 kg of PCM. Considering that, with the tank water at ambient temperature, the pressure at the tank bottom is regulated at 1.5 barg, and that the tank discharging can be made within a limited range of water mass flow rate, the above quantity of PCM has been calculated so that the module internal pressure remains lower that the one in the tank during both charging and discharging. In such a way, accidental spillage of liquid PCM from the bottle is avoided.

4. Results

In this section the results of two experimental tests, one realized including 14 PCM modules inside the hot water storage tank, and the other one using only water as heat storage material, are reported and compared. In both the cases, all the experimental measures have been recorded with a sample time of 1 s.

Figure 7 shows a picture of the PCM modules inside the tank, which have been positioned in the upper part of the tank through a flange.

Figure 7. PCM modules inside the tank.

The two different thermal energy storage systems have been tested applying the same initial conditions, and the same boundary conditions over the entire experiments duration. Indeed, in both the experimental tests, the energy storage system has been charged for a period of six and half hours, with a constant water mass flow rate through the serpentine of 0.2 kg/s, and with an increment of the water temperature at the serpentine inflow section of 5°C/(30 min), except for the last thirty minutes, in which the temperature at the serpentine inflow has been kept constant and equal to 85°C. In both the cases, the initial temperature of the water at the serpentine entrance was set to 30°C, and about the same initial water temperature profile along the tank axis was established. Furthermore, in both the cases the indoor ambient temperature has been regulated to 23 °C for the entire experiments duration.

In the following, the results relative to the test with water and PCM will be referred to using (w+PCM), while (w) will be used for the ones relative to the case with only water.

4.1. *Temperature axial profiles and accumulated thermal energy*

Figure 8 shows the temporal evolution of the measured water temperature profiles along the tank axis, while figure 9 reports the evolution of the total energy transferred from the serpentine to the tank water, and of the total energy accumulated in the water, from the initial state.

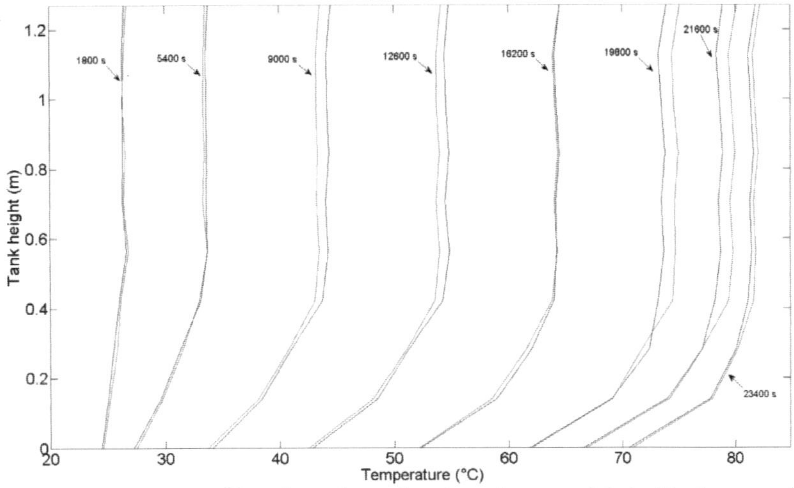

Figure 8. Water temperature profiles along the tank axis: red curves→(w) ; black curves→(w+PCM).

The energy transferred from the serpentine to the tank water has been calculated as the product of the sample time and the difference between the enthalpy flow rate at the serpentine inflow section and the outflow one, while the water internal energy increment has been evaluated using the measured temperatures along the tank axis. As concerns the calculation of the internal energy increment, it has been supposed that the measured temperatures represent the average water temperatures relative to the tank horizontal sections passing through the measuring points. This position, of course, leads to approximated results.

For convenience of discussion, both the experimental tests are divided into two parts. The first part goes from the test start to the instant in which the maximum water temperature in the upper part of the tank with water and PCM reaches the PCM melting temperature (58°C), that occurs after about 13,500 s, and the second one from 13,500 s to the end of the test.

Figure 8 shows that in the first period, due to the lower total heat capacity that characterizes the thermal energy storage tank with water and PCM in this period, the case (w+PCM) experiences a more rapid increase of the water maximum and mean temperatures. In particular, considering the profiles reported in figure 8, it can be seen that the difference between the temperature profile in case (w+PCM) and the one in case (w) increases from the test start to 9,000 s, and then decreases slightly from 9,000 s to 12,600 s. By contrast, in the same period, the case (w) presents a higher heat exchange between the water flowing through the serpentine and the one in the tank, and a higher heat storage in the water, as can be noticed in figure 9. This is explained considering that case (w) presents a slightly higher temperature difference between the water temperature in the serpentine and the one in the tank section where the serpentine is located.

The second period is characterized by the solid-to-liquid phase transition of the PCM, or of part of it. Although the quantity of PCM used is relatively low, the effects of the PCM melting are clear if the differences between the temperature profiles in the two cases from 16,200 s to 21,600 s are considered. Indeed, figure 8 shows that at 16,200 s the two profiles are nearly coincident, and that there is a more rapid increase of the temperatures in the case (w) from 16,200 s to 19,800 s, and also that from 19,800 s to 21,600 s the difference between the profiles remains practically the same. These results are due to the PCM melting, and can be analysed in more detail considering the same arguments used to analyse the first period.

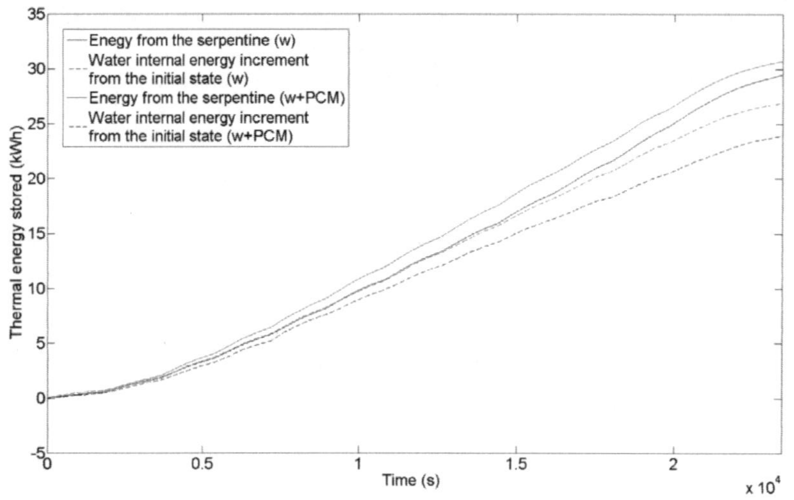

Figure 9. Heat transferred through the serpentine wall, and energy accumulated as sensible heat in the water.

4.2. *Evaluation of the PCM melted fraction*

The most challenging and at the same time intriguing task of the present experimental work has been the estimation of the percentage of melted PCM at the end of the experimental test, namely after six and half hours of charging, using only the above reported experimental data, and without implementing any analytical model for the numerical simulation of the melted fraction of PCM inside the modules. This task has been faced by using two different approaches.

The first one is a qualitative approach, and it is based on the analysis of the temperature profiles in the second period, and, in particular, of the ones at the end of the experimental test. The analysis of the evolution of the temperature profiles in the second period reported before, and the fact that the temperatures profile relative to the case (w+PCM) approaches the one relative to case (w) at 23,400 s suggest that at the end of the experimental test relative to the tank with water and PCM, the energy is accumulated essentially as sensible heat, implying that the PCM is all melted, or most likely, almost all melted. In fact, due to the very low thermal conductivity of the PCM, the characteristic time relative to conduction heat transfer inside the PCM, evaluated using the aluminum bottle radius as characteristic length, is very high and equal to about 11.5 h. So, even if the PCM melting is also influenced by natural convection mechanisms, it is highly probable that the PCM was not completely melted at the end of the test. However, this approach is only qualitative and doesn't allow to calculate the value of the PCM melted fraction.

The second one is an approximated quantitative approach, based on the results reported in figures 8 and 9. Considering the results relative to case (w) in figure 9, namely the two red curves, at each time instant the difference between the continuous curve and the dashed one represents the sum between the heat accumulated in the tank components (walls, insulation, etc.), and the heat losses through the tank external walls. Considering the black curves, which refers to case (w+PCM), at each time instant of the first period the difference between the continuous curve and the dashed one represents the sum between the energy accumulated in the tank components, the heat losses, and the heat accumulated in the PCM modules as sensible heat. In the second period, the difference between the black continuous curve and dashed one also includes the heat accumulated in the PCM as latent heat. Considering that the temporal evolution of the tank water temperature field is very similar in the two cases, as can be seen in figure 8, it has been assumed that, at the end of the tests, the heat accumulated in the tank components and the heat losses in the two cases are equal. So, the sum between heat losses and heat accumulated in the tank components at the end of the tests has been calculated using the last points of the red curves, and then it has been subtracted to the difference between the last points of the black curves. The result represents the sum between the sensible heat accumulated in the solid PCM, the heat accumulated in the PCM as latent heat, and the sensible heat accumulated in the liquid PCM. Finally, neglecting the sensible heat accumulated in the liquid PCM, after calculating the sensible heat accumulated in the solid PCM considering between the initial temperature in the tank and the melting temperature (58°C), it has been possible to calculate the heat accumulated as latent heat, that resulted to be equal to 1.1 kWh. So, since the latent heat relative to all the PCM melted is equal to about 1.5 kWh, the second approach is in good agreement with the first one, suggesting that at the end of the test the PCM was not completely melted.

5. Conclusion

This work reports the results of the experimental analysis of a hot water storage tank including PCM modules consisting in aluminum bottles of 3 liters partially filled with a commercial bio-based PCM having a melting temperature of 58°C. The experimental results relative to the hot water tank including the PCM modules are compared with the ones obtained by using only water as heat storage material. The two different heat storage systems have been charged for a period of six and half hours, applying the same initial conditions, and the same boundary conditions over the entire experiments duration. Results of the present application evidence the effects of the presence of the PCM, and indicate that, at the end of the heat charging, not all the PCM has undergone melting.

References

[1] Di Somma M, Yan B, Bianco N, Luh P B, Graditi G, Mongibello L and Naso V 2016 Multi-objective operation optimization of a Distributed Energy System for a large-scale utility customer *Appl Therm Eng* **101** pp 752-761.

[2] Caliano M, Bianco N, Graditi G and Mongibello L 2015 Economic optimization of a residential micro-CHP system considering different operation strategies *Appl Therm Eng* **101** pp 592-600.

[3] Mongibello L, Bianco N, Caliano M, De Luca A and Graditi G 2015 Transient analysis of a solar domestic hot water system using two different solvers *Energy Procedia* **81** pp 89-99

[4] Mongibello L, Bianco N, Di Somma M, Graditi G and Naso V 2014 Numerical simulation of a solar domestic hot water system *J. Phys.: Conf. Series* **547** 012015

[5] Farid M M, Khudhair A M, Razack S A K and Al-Hallaj S 2004 A review on phase change energy storage: Materials and applications *Energy Conv and Manag* **45** pp 1597-1614.

[6] Zalba B, Marín J M, Cabeza L F and Mehling H Review on thermal energy storage with phase change: Materials, heat transfer analysis and applications 2003 *Appl Therm Eng* **23** pp 251-283.

[7] Sharma A, Tyagi V V, Chen C R and Buddhi D, Review on thermal energy storage with phase change materials and applications 2009 *Renewable and Sustainable Energy Reviews* **13** pp 318-345.

[8] Cabeza L F, Ibanez M, Sole C, Roca J and Nogues M 2006 Experimentation with a water tank including a PCM module *Solar Energy Materials & Solar Cells* **90** pp 1273–1282

[9] Mongibello L, Capezzuto M and Graditi G 2014 Technical and cost analyses of two different heat storage systems for residential micro-CHP plants *Appl Therm Eng* **71** pp 636-642

IOP Conf. Series: Journal of Physics: Conf. Series **796** (2017) 012031 doi:10.1088/1742-6596/796/1/012031

Numerical investigation on thermal behaviors of two-dimensional latent thermal energy storage with PCM and aluminum foam

B Buonomo, D Ercole, O Manca*, S Nardini

Dipartimento di Ingegneria Industriale e dell'Informazione, Università degli Studi della Campania "Luigi Vanvitelli", Via Roma 29, Aversa (CE) 81031, Italy

*Corresponding author: oronzio.manca@unina2.it

Abstract. A numerical investigation on Latent Heat Thermal Energy Storage System (LHTESS) based on a phase change material (PCM) is accomplished. The PCM is a pure paraffin wax with a low thermal conductivity. An aluminum metal foam is employed to enhance the PCM thermal behaviors. The geometry is a vertical shell-and-tube LHTESS made with two concentric aluminum tubes. The internal surface of the hollow cylinder is assumed at a constant temperature above the melting temperature of the PCM to simulate the heat transfer from a hot fluid. The external surface is assumed adiabatic. The phase change of the PCM is modelled with the enthalpy porosity theory while the metal foam is considered as a porous media in Darcy-Forchheimer assumption and the Boussinesq approximation is employed. Local thermal non-equilibrium (LTNE) model is assumed. The results are compared in terms of melting time and temperature fields as a function of time for the charging and discharging phases for different porosities and an assigned pore per inch. Results show that the metal foam improves significantly the heat transfer in the LHTESS giving a faster phase change process with respect to pure PCM, reducing the melting time more than one order of magnitude.

1. Introduction

A big issue in the environment problems is the storage of the thermal energy when it is not required. Hence, in order to store the excess energy, the optimization of a thermal energy storage system (TESS) is necessary to overcome this problem. A thermal energy storage system is a storage device [1] useful to accumulate the thermal energy during the excess of energy and release it when required. There are many types of TESS but the most interesting storage system is a Latent Heat Thermal Energy Storage System (LHTESS). This system is based on the Phase Change Materials (PCMs) that have the peculiarity to change their phase from solid to liquid and vice-versa during the heat transfer maintaining the working temperature nearly constant. The PCM used in this study is the paraffin wax, because it is not corrosive, it is chemically stable, the phase change is reversible and the latent heat is very high [2]. Nevertheless these characteristics are not satisfactory for a TESS because the paraffin wax has a low value of thermal conductivity, making the melting time and the solidification time very long respect to a working cycle of charge and discharge of the system. It is necessary to develop a solution in order to overcome this drawback. In the literature there are many solutions like the addition of highly-conductive nanoparticles [3], the usage of metal foam [4] or finned tubes [5]. Particularly, the open-cell metal foam represents the best solutions to make the LHTESS useful for thermal applications because it has a very high ratio between the area of heat exchange and its volume and the base material has high value of thermal conductivity. Li and Wu [6] numerically studied a 3D model of a shell and tube with PCM for different configuration, Heat Transfer Fluid (HTF) velocity and PCM materials, using the fins to improve the heat transfer. The phase change of the PCM is modelled with the enthalpy porosity method both for charge and discharge phase. The results showed that the extended fins improve the heat transfer by at least 14% respect to the configuration without the fins. The improvement of the PCM with the expanded graphite even reduces the rate of melting by 20%. Atal et al. [7] both experimentally and numerically investigated on a shell-tube arrangement LHTESS with PCM and metal foams. Two different porosities are studied, 95% and 77% and they found that the foam with lower porosity reduces the melting time respect to the higher-porosity foam but in general the metal foam improves the heat transfer respect to the pure PCM. Moreover there is a good agreement between the experimental results and numerical results. A LHTESS for high temperature concentrated solar power plants is numerically studied in [8]. The configuration is a shell-and-tube configuration with a vertical cylindrical tank filled with PCM and various tubes in which the HTF flows from the top to the bottom. The aim of the study is to understand the beneficial effects of the natural convection inside the PCM respect to the pure conductive case. The results show that the melting time is smaller by three hours with respect to the pure conductive case. A similar study to understand the difference between the horizontal configuration and vertical configuration about the effect of the natural convection during the phase change of PCM is made in [9]. Tao and He [10] studied three types of enhancement for a shell-tube configuration and they compare the results. The base case is to fill the system with a pure PCM, the first enhancement is to install fins inside the system, the second and the third are to put a mixed PCM made by different mixture of salts together with fins. The results showed that the only installation of fins reduces the melting time by 36.6% while the second and third enhancement (fins plus mixed PCMs) showed a further reduction of the melting time, respectively by 60.3% and the third 47.7%. Xiao et al [11] experimentally study the thermal behaviour between the PCM paraffin and open-cell metal foam for latent heat storage. Two type of metal foam are employed, copper and nickel with high value of porosity. They found that the copper foam is better than the nickel foam in term of heat exchange, because the effective thermal conductivity is faster for the copper foam than the nickel foam. Zhou et al [12] investigated experimentally the heat transfer in a PCM/metal foam for a device storage system both for charging and discharging process. They found that the metal foam improved the rate of charge and discharge of the system respect to the pure PCM. A two-dimensional LHTESS based on PCM paraffin RT58 with a copper foam is numerically studied using the LTNE model [13]. A similar experimental study was accomplished in [4] where copper foam with the paraffin wax RT58 are employed for storage application.

In this paper a numerical model of the PCM embedded in metal foam is studied. Local thermal non equilibrium model is assumed and the geometry of TESS is a shell and tube configuration with two concentric aluminum tubes. The enthalpy porosity method is employed and the Darcy-Forchheimer law is used for the modelling of the metal foam. The results are shown in term of melting time, temperature and energy storage for charge and discharge phase.

2. Physical model

LHTESS consists in two concentric shell tubes where the PCM with metal foam is enclosed. The inner and outer tube radiuses are respectively 2 mm and 12 mm, as showed in Fig. 1a. The system has a length of 100 mm along the z-direction.

A 2D domain is considered with the axial symmetry along the z-axis as shown in Fig. 1b. A constant temperature (350 K) is assigned to the inner surface to simulate the HTF while the other surfaces are imposed adiabatic. The gravitation acceleration is along the z-axis. The enthalpy-porosity method is employed to model the phase change of the PCM [14]. This model is useful for those materials that melt in a range of temperature, in fact this method defines a region where the solid and the liquid phase are mixed and there is not a clear solid-liquid interface. This region are called mushy region and it is described by a new parameter called “liquid fraction”. The liquid fraction, indicated with Greek letter β, represents

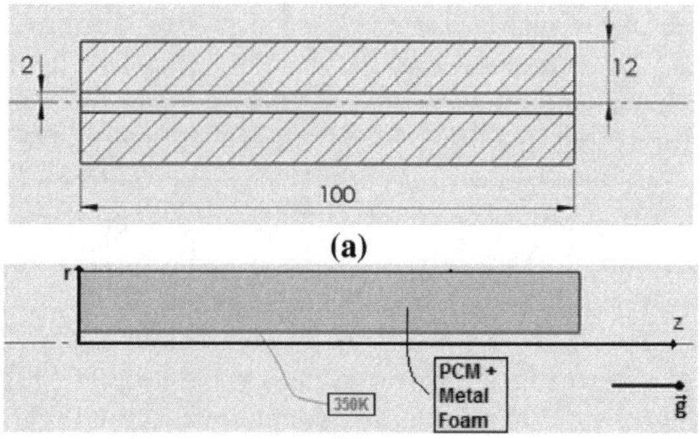

(b)

Figure 1. Schematic configuration of LHTESS (a) and the computational domain (b).

the amount of the liquid phase in a cell of mushy region. It varies from zero to one in the mushy region:

$$\begin{cases} \beta = 0 & for \quad T < T_{solidus} \\ \beta = \dfrac{T - T_{solidus}}{T_{liquidus} - T_{solidus}} & for \quad T_{solidus} < T < T_{liquidus} \\ \beta = 1 & for \quad T > T_{liquidus} \end{cases} \tag{1}$$

T is the local temperature of the cell, $T_{liquidus}$ is the temperature above which the domain is completely liquid and $T_{solidus}$ is the temperature below which it is completely solid. The solid part of the mushy region is modelled by adding a source term in the momentum equation. The metal foam is modelled with the Darcy-Forchheimer assumption, because it behaves as a porous media. To evaluate the heat transfer between the foam and the PCM the Local Thermal Non-Equilibrium (LTNE) model is used. In this model the PCM and the metal foam are not in thermal equilibrium and two temperature functions are present, one for the PCM and the other for the metal foam, therefore two energy equations are necessary to describe the field of the temperature [15]. The buoyancy forces are modelled with the Boussinesq approximation.

The hypotheses in the present work are the following:

1) The computational domain is two-dimensional.
2) The metal foam is isotropic and homogenous.
3) The variation of density in the liquid PCM follows the Boussinesq law.
4) The other properties of PCM are constant.

The governing equations are [15]:

$$\frac{\partial \rho}{\partial t} + \frac{1}{r}\frac{\partial u}{\partial r} + \frac{\partial w}{\partial z} = 0 \tag{2}$$

$$\rho_{PCM}\left(\frac{\partial u}{\partial t} + u\frac{\partial u}{\partial r} + w\frac{\partial u}{\partial z}\right) = -\frac{\partial p}{\partial r} + \mu_{PCM}\left[\frac{1}{r}\frac{\partial}{\partial r}\left(r\frac{\partial u}{\partial r}\right) + \frac{\partial^2 u}{\partial z^2}\right] + S_r \tag{3}$$

$$\rho_{PCM}\left(\frac{\partial w}{\partial t} + u\frac{\partial w}{\partial r} + w\frac{\partial w}{\partial z}\right) = -\frac{\partial p}{\partial z} + \mu_{PCM}\left[\frac{1}{r}\frac{\partial}{\partial r}\left(r\frac{\partial w}{\partial r}\right) + \frac{\partial^2 w}{\partial z^2}\right] + S_z \tag{4}$$

Where u and w are the velocity of the PCM in liquid phase along respectively r and z direction. ρ_{PCM} and μ_{PCM} are the density and viscosity of the PCM. p is the pressure. S_r and S_z are the source terms that taken into account the enthalpy porosity method, the Darcy-Forchheimer and the Boussinesq approximation:

$$S_r = \frac{(1-\beta)^2}{(\beta^3+0.001)^3} A_{mush} u + \frac{\mu_{PCM}}{K} u + \frac{C_f}{\sqrt{K}} \rho u^2 \tag{5}$$

$$S_z = \frac{(1-\beta)^2}{(\beta^3+0.001)^3} A_{mush} w + \frac{\mu_{PCM}}{K} w + \frac{C_f}{\sqrt{K}} \rho w^2 + \rho_{PCM} g \gamma (T_{PCM} - T_0) \tag{6}$$

The first term in equations (5) and (6) models the solid part in the mushy region. In fact when the liquid fraction β is near to zero, this term increase the resistance to flow of the PCM inside the mushy region. The small number (0.001) is used to avoid division by zero; A_{mush} is the mushy zone constant which takes into account the damping of the velocity during the solidification [16]. Its value is set to 10^5 kg/(m^3 s) and it does not affect the rate of melting with the presence of the metal foam. The second and third terms are the Darcy-Forchheimer law, where K is the permeability of the porous media and C_f is inertial drag factor. The last term in equation (6) is the Boussinesq approximation. In local thermal non-equilibrium model the equation of energy for PCM is [15]:

$$\begin{aligned} &\varepsilon \rho_{PCM} c_{PCM} \frac{\partial T_{PCM}}{\partial t} + \varepsilon \rho_{PCM} c_{PCM} \left(u \frac{\partial T_{PCM}}{\partial r} + w \frac{\partial T_{PCM}}{\partial z} \right) = \\ &= k_{eff,PCM} \left(\frac{1}{r} \frac{\partial}{\partial r} \left(r \frac{\partial T_{PCM}}{\partial r} \right) + \frac{\partial^2 T_{PCM}}{\partial z^2} \right) + \alpha_{sf} h_{sf} \left(T_{PCM} - T_{foam} \right) - \varepsilon \rho_{PCM} H_L \frac{\partial \beta}{\partial t} \end{aligned} \tag{7}$$

while for the metal foam is [16]:

$$(1-\varepsilon) \rho_{foam} c_{foam} \frac{\partial T_{foam}}{\partial t} = k_{eff,foam} \left(\frac{1}{r} \frac{\partial}{\partial r} \left(r \frac{\partial T_{foam}}{\partial r} \right) + \frac{\partial^2 T_{foam}}{\partial z^2} \right) + \alpha_{sf} h_{sf} \left(T_{foam} - T_{PCM} \right) \tag{8}$$

Where $k_{eff,PCM}$ and $k_{eff,foam}$ are, respectively, the effective thermal conductivity of the PCM and metal foam and ε is the metal foam porosity. It is hard to predict the effective thermal conductivity in the LTNE model; nevertheless using the relations in [15] it is possible to evaluate the thermal conductivity:

$$k_{eff} = \frac{\sqrt{2}}{2(M_A + M_B + M_C + M_D)} \tag{9}$$

$$M_A = \frac{4\sigma}{\left(2e^2 + \pi\sigma(1-e)\right) k_{foam} + \left(4 - 2e^2 - \pi\sigma(1-e)\right) k_{PCM}} \tag{10}$$

$$M_B = \frac{(e-2\sigma)^2}{(e-2\sigma) e^2 k_{foam} + \left(2e - 4\sigma - (e-2\sigma) e^2\right) k_{PCM}} \tag{11}$$

$$M_C = \frac{\left(\sqrt{2}-2e\right)^2}{2\pi\sigma^2\left(1-2e\sqrt{2}\right)k_{foam} + 2\left(\sqrt{2}-2e-\pi\sigma^2\left(1-2e\sqrt{2}\right)\right)k_{PCM}} \tag{12}$$

$$M_D = \frac{2e}{e^2 k_{foam} + 4 - e^2 k_{PCM}} \tag{13}$$

$$\sigma = \left(\frac{\sqrt{2}\left(2-\left(\frac{5}{8e^3\sqrt{2}}\right)-2\varepsilon\right)}{\pi(3-4e\sqrt{2}-e)}\right)^{1/2} \tag{14}$$

$$e = 0.339 \tag{15}$$

then the effective thermal conductivity of the PCM and metal foam are:

$$k_{eff,PCM} = k_{eff}\big|_{k_{foam}=0} \tag{16}$$

$$k_{eff,foam} = k_{eff}\big|_{k_{PCM}=0} \tag{17}$$

In the equations 7 and 8 the heat exchange between PCM and metal foam occurs by means of the product $\alpha_{sf}h_{sf}$. h_{sf} is the local heat transfer coefficient, α_{sf} is the surface area density of the metal foam. The surface area density represents the total heat exchange area between PCM and metal foam. The relations of Calmidi and Mahajan are employed to evaluate α_{sf} and h_{sf}, [17]:

$$\alpha_{sf} = \frac{3\pi d_f}{\left(0.59 d_p\right)^2}\left(1-e^{\frac{1(1-\varepsilon)}{0.04}}\right) \tag{18}$$

$$h_{sf} = \begin{cases} \left(0.76 Re_d^{0.4} Pr_{pcm}^{0.37}\right)\left(\frac{k_f}{d_f}\right), & 1 \le Re_d \le 40 \\ \left(0.52 Re_d^{0.5} Pr_{pcm}^{0.37}\right)\left(\frac{k_f}{d_f}\right), & 40 \le Re_d \le 1000 \\ \left(0.26 Re_d^{0.6} Pr_{pcm}^{0.37}\right)\left(\frac{k_f}{d_f}\right), & 1000 \le Re_d \le 2\times10^5 \end{cases} \tag{19}$$

where Re_d is the local Reynolds number related to ligament diameter:

$$Re_d = \frac{\rho V d_f}{\mu} \tag{20}$$

d_f is the ligament diameter of the metal foam, d_p is the pore diameter. These physical characteristics are related each other by means of the following relations [17]:

$$\frac{d_f}{d_p} = 1.18\left(\frac{1-\varepsilon}{3\pi}\right)^{1/2}\left(\frac{1}{1-e^{1(1-\varepsilon)/0.04}}\right) \tag{21}$$

$$d_p = \frac{0.0224}{\omega} \tag{22}$$

where ω is the pore density of the metal foam that represents the number of pores across one inch. In this study the pore size is assigned at 20 Pore Per Inch (PPI).

The Darcy-Forchheimer law is the following:

$$\frac{\Delta p}{L} = \frac{\mu}{a}\bar{V} + \rho C_F \bar{V}^2 \tag{23}$$

where Δp is the pressure drop and L is the length of the porous media across which the pressure drop appeared. It describes the behavior of the metal foam. The permeability and drag factor or inertia coefficient values are evaluate by the following, [17]:

$$K = 0.00073(1-\varepsilon)^{-0.224}\left(\frac{d_f}{d_p}\right)^{-1.11} d_p^2 \tag{24}$$

$$C_F = 0.00212(1-\varepsilon)^{-0.132}\left(\frac{d_f}{d_p}\right)^{-1.63} \tag{25}$$

The paraffin wax is the paraffin RT58 of the Rubitherm [18] and the properties are listed in Table 1.

Table 1. Physical properties for the paraffin wax

Physical properties	Paraffin RT58 [18]
Density [kg/m^3]	840
Specific Heat [J/kg K]	2100
Thermal Conductivity [W / m K]	0.2
Dynamic Viscosity [kg/m s]	0.0269
Thermal expansion coefficient [1/K]	0.00011
Melting Heat [J / kg]	180000
Solidus Temperature [K]	321
Liquidus Temperature [K]	335

In the Boussinesq approximation the operating temperature is equals to 321 K.

3. Numerical model

The finite volume method [19] is used for resolving the governing equation (2), (3), (4), (7) and (8) using the commercial code Ansys Fluent 15.0 [20]. For the inner boundary condition, at r = 2 mm T_{PCM} and T_{foam} are equal to 350K.The SIMPLE algorithm is employed for coupling the pressure and the velocity; the green-gauss cell gradient evaluation is used for the spatial discretization l; the second order upwind scheme is used for momentum and energy equation. The convergence is imposed by set the residual to 10^{-5} for continuity and moment equations while 10^{-8} for energy equation. A transient mode is enabled and a temporal independence test is made and the time step size equal to 0.2 s was the best choice for the numerical accuracy. The LTNE has been implemented using a user-defined function for the calculation of the local heat transfer coefficient indicated in equation (19), while the energy equations are automatically solved with a module of FLUENT that overlaps the fluid and solid zone.

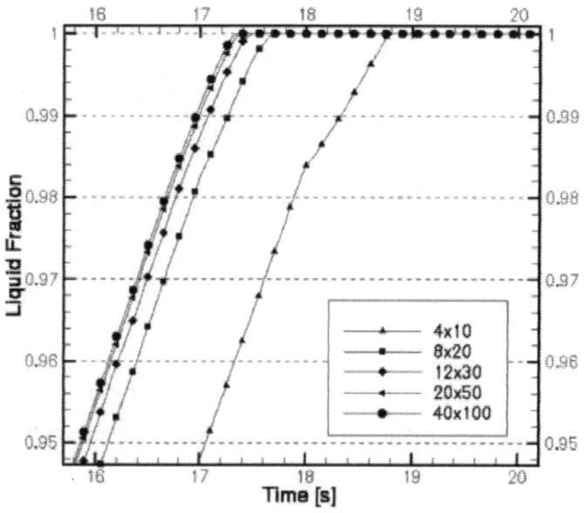

Figure 2 Independent test for the mesh.

The grid independent study is made with five different meshes: 4 x 10, 8 x 20, 12 x 30, 20 x 50 and 40 x 100 nodes and the mesh with 20 x 50 nodes is chosen. In literature there are few experiments about PCM melting in cylinders.

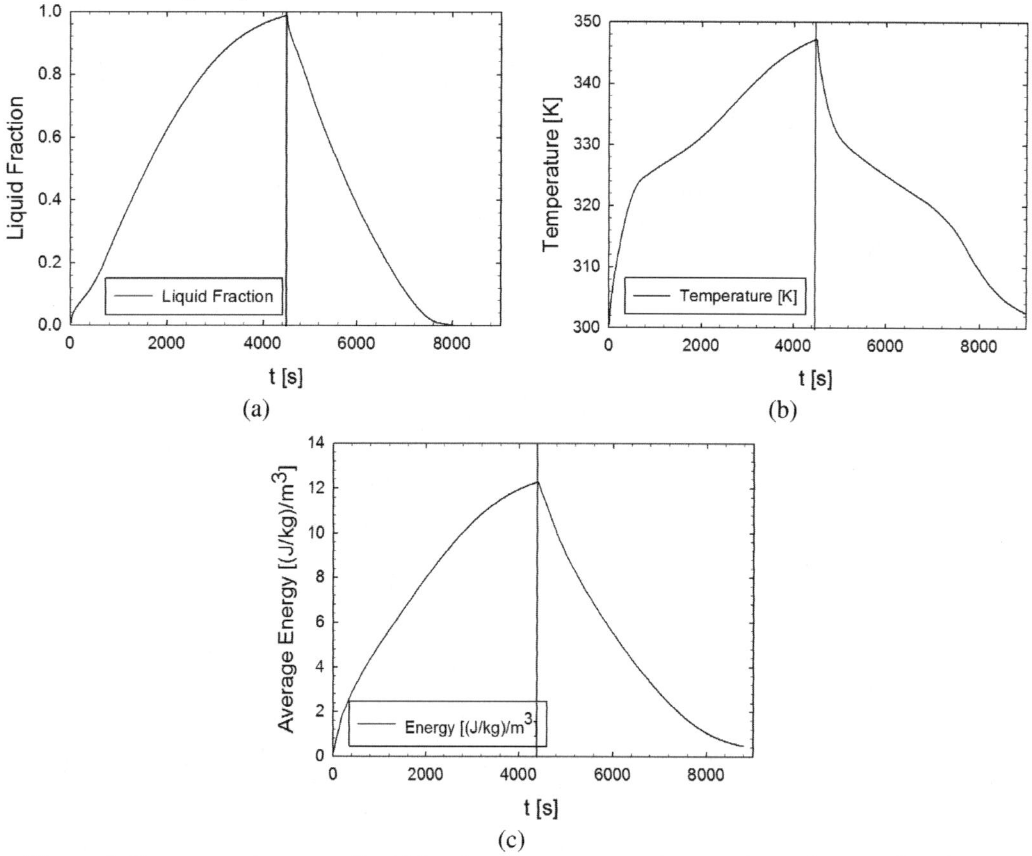

Figure 3 Clean case: a) average liquid fraction b) average temperature c) average specific energy

Nevertheless, the accuracy of the enthalpy-porosity method used in this model is tested using the work of Krishnan et al. [21], where a 2D non-dimensional enclosure is studied. In [21] a two temperature model is implemented with LTNE assumption and various parameters are studied to understand the influence on the PCM melting. Therefore, a comparison between the work [21] and the enthalpy-porosity method of this model is carried out, using the same boundary conditions and properties. The results showed a good agreement. In particular, in different instants of time the evolution of the PCM temperature at the mid-height of the domain is evaluated and the differences between this model and [21] are imperceptible.

4. Results and discussion

The results are expressed for charging and discharging process at different values of porosity, 80% and 90% with an assigned PPI equals to 20. In all cases the LTNE model is employed and for the charging process the temperature of the internal cylinder is assumed to 350 K and for discharging process is 300K. The other surfaces are assumed adiabatic. A comparison in terms of temperature, liquid fraction and specific energy stored is accomplished. Fig. 3 depicted the average liquid fraction (a), average temperature (b) and average energy stored (c) as functions of time for a LHTESS with pure paraffin wax both for charging and discharging process. The rate of melting is very low because of the low value of the thermal conductivity; in fact the charging process is completed at nearly 4500 seconds and the discharging process lasts 4000 seconds. Therefore the discharging process is faster than the charging process in the clean case. Fig. 4 showed a comparison between the pure paraffin wax (CLEAN case) with paraffin inside the metal foam in terms of liquid fraction, temperature and average energy stored.

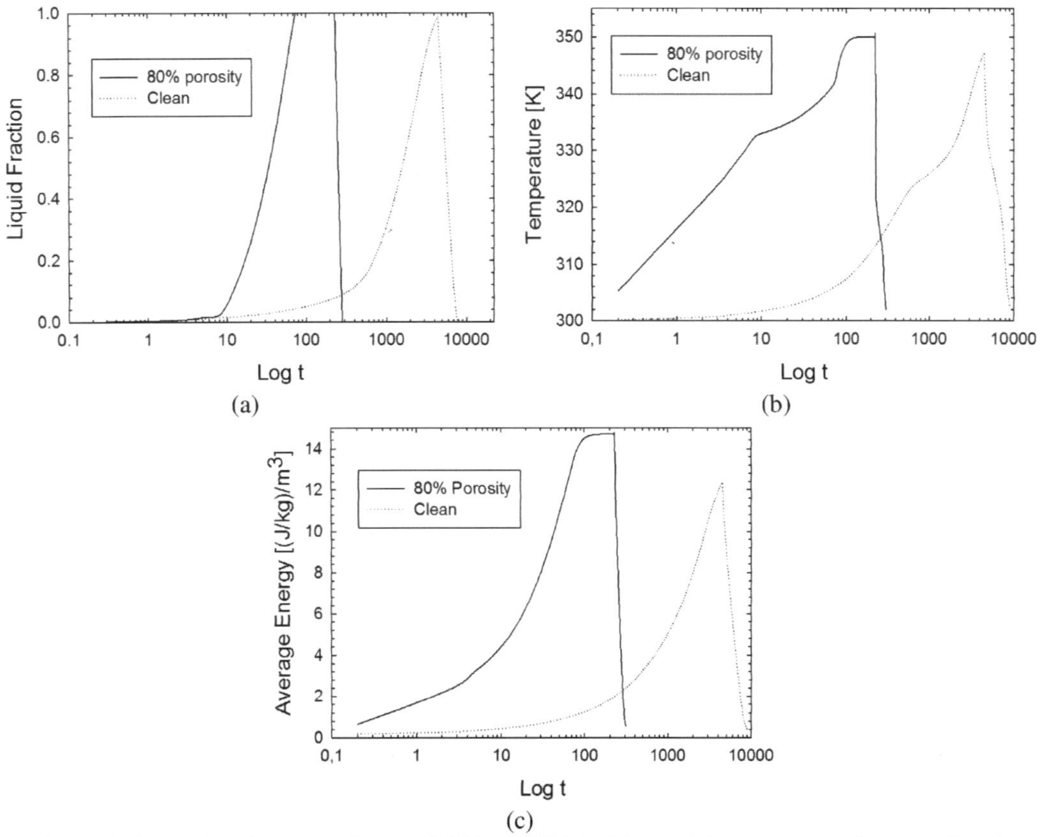

Figure 4 Comparison between the pure PCM and PCM with metal foam in term of average liquid fraction (a), average temperature (b) and energy stored (c) for charging and discharging phase.

IOP Conf. Series: Journal of Physics: Conf. Series **796** (2017) 012031 doi:10.1088/1742-6596/796/1/012031

It is clear that the presence of the metal foam improves significantly the melting rate during the charge phase and the solidification rate during the discharge phase; in fact the scale of the time is logarithmic. Even with the metal foam the discharging process is faster. A further comparison between two metal foams at different porosity value is made in fig 5. the metal foam considered have a porosity equal to 80% and 90%, respectively. The order of magnitude of the rate change phase for charging and discharging process for both porosities is similar but the metal foam with 80% of porosity is faster. Therefore small values of porosity increase the rate of the phenomenon, but the energy stored is lower, as showed in fig. 5c.

5. Conclusions

A numerical investigation on the latent heat thermal energy stored system was accomplished. Both charging and discharging process were performed. The materials employed are the paraffin wax as PCM and the metal aluminum foam. The enthalpy-porosity theory was employed to simulate the phase change process. The metal foam was modelled with the Darcy-Forchheimer law for the fluid flow and the LTNE model for the energy. The results found that the presence of the metal foam improves significantly the thermal performance of the LHTESS; in fact the rate of melting during the charge phase and the rate of solidification during the discharge phase are faster more than one order of magnitude respect to the pure PCM. Moreover, the reduction of the porosity determines an increase of the heat storage and heat release rate.

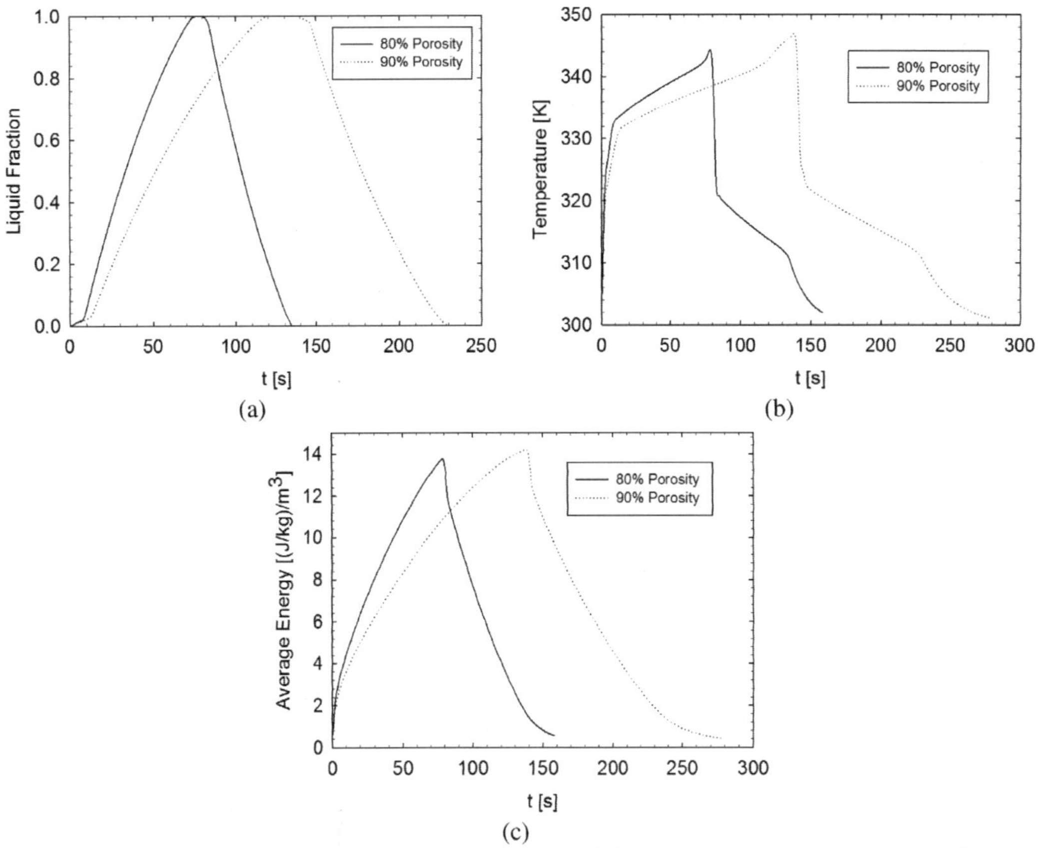

Figure 5 comparison between the two metal foam with different porosity in term of average liquid fraction (a), average temperature (b) and energy stored (c) for charging and discharging phase.

6. References

[1] Antoni G, Marc M, Ingrid M, Ana L, Pablo D, Belén Z and Luisa F C 2010 *Renew. Sust. En. Rev.* **14** 31
[2] Abhat A 1983 *Sol. En.* **30** 313
[3] Khodadadi J M and Hosseinizadeh S F 2007 *Int. Comm. Heat Mass Transf.* **34** 534
[4] Zhao, C Y ,Lu W and TianY 2010 *Sol. En.* **84** 1402
[5] Zhang Y and Faghri A 1996 *J. Enhanced Heat Transf.* **3** 119
[6] Li Z and Wu Z 2015 Sol. En. **122** 382
[7] Atal A, Wang Y, Harsha M and Sengupta S 2016 *Int J. Heat Mass Transf.* **93** 9
[8] Fornanelli F, Camporeal S M, Fortunato B, Torresi M, Oresta P, Magliocchetti L, Miliozzi A and Santo G 2016, Appl. En. **164** 711
[9] Tao Y B and He Y I 2015 *Int. Comm. Heat Mass Transf.* **67** 147
[10] Seddegh S, Wang X and Henderson A D 2016 **93** 348
[11] Xiao X, Zhang P and Li M 2014 *Int. J. Therm. Sci.* **81** 94
[12] Zhou D and Zhao C Y 2011, *Appl. Therm. Eng.* **31** 970
[13] Tian Y and Zhao C Y 2011 *En.* **36** 5539
[14] Voller and V R, Prakash C 1987 *Int. J. Heat Mass Transf.* **30** 1709
[15] Liu Z, Yao Y and Wu H 2013 *Appl. En.* **112** 1222
[16] Al-abidi A, Bin Mat S, Sopian K, Sulaiman M Y and Mohammed A T 2013 *Renew. Sust. En. Rev.* **20** 353
[17] Calmidi and V V, Mahajan R L 2000 *ASME J. Heat Transf.* **122**, 557
[18] Rubitherm GmBH, *www.rubitherm.de.*
[19] Pantankar S V 1980 "Numerical heat transfer and fluid flow", *Hemisphere*, Washington DC.
[20] Ansys Incorporated, Fluent 15.0 User Manual, 2014.
[21] Krishnan S, Murthy J Y and Garimella, S V 2005. *ASME J. Heat Transf. 127* 995

IOP Conf. Series: Journal of Physics: Conf. Series **796** (2017) 012032 doi:10.1088/1742-6596/796/1/012032

Numerical Analysis of a Paraffin/Metal Foam Composite for Thermal Storage

P Di Giorgio*, M Iasiello, A Viglione*, M Mameli*[,1], S Filippeschi*, P Di Marco*, A Andreozzi**, N Bianco****

*DESTEC – University of Pisa – Largo L. Lazzarino 2-56122 – Pisa
**DII, University of Naples - Federico II, P.le Tecchio, 80, 80125 - Naples

[1] E-mail: mauro.mameli@ing.unipi.it

Abstract. In the last decade, the use of Phase Change Materials (PCMs) as passive thermal energy storage has been widely studied both analytically and experimentally. Among the PCMs, paraffins show many advantages, such as having a high latent heat, a low vapour pressure, being chemically inert, stable and non-toxic. But, their thermal conductivity is very low with a high volume change during the melting process. An efficient way to increase their poor thermal conductivity is to couple them with open cells metallic foams. This paper deals with a theoretical analysis of paraffin melting process inside an aluminum foam. A mathematical model is developed by using the volume-averaged governing equations for the porous domain, made up by the PCM embedded into the metal foam. Non-Darcian and buoyancy effects are considered in the momentum equation, while the energy equations are modelled with the Local Thermal Non-Equilibrium (LTNE) approach. The PCM liquefaction is treated with the apparent heat capacity method and the governing equations are solved with a finite-element scheme by COMSOL Multiphysics®. A new method to calculate the coupling coefficients needed for the thermal model has been developed and the results obtained have been validated comparing them to experimental data in literature.

1.Introduction

In the last years, the wide diffusion of technologies that can produce thermal energy from renewable and intermittent sources has increased the need for an efficient TES. Phase-change-materials (PCMs), which release or absorb thermal energy during melting and solidification process, are particularly attractive in this field, since they allow to exchange heat virtually at constant temperature and because of their high thermal energy storage density. However, most PCMs presents the common problem of low thermal conductivity, (around $\mathbf{0.2\ W/mK}$ for most paraffin waxes) that strongly limits the energy charging/discharging rates. Therefore, the introduction of highly conductive materials embedded with PCMs has been already proposed in literature in order to improve the heat exchange process [1]. Metal foams with their low relative density and relatively high thermal conductivity are believed to be a promising material for enhancing heat transfer performance and reducing the charging and discharging periods of PCMs. A significant amount of studies on the thermal conductivity enhancement of PCMs focused on use of porous matrices. Tian and Zhao [2] numerically and experimentally investigated the heat transfer behaviour with a paraffin wax embedded in a copper metallic foam, showing that the addition of the foam can increase the overall heat transfer rate by three to ten times (depending on the metal foam structures and materials) during the melting process. Tong et al [3] inserted an open cell aluminum foam matrix into liquid water and investigated the heat transfer rate during the icing process. Results show that the thermal power exchanged with water during the solidification process is three times higher with the foam. Mancin et al. [4] tested three paraffin waxes with different melting points temperatures (53, 57, 59°C) embedded into a copper metal foam. The experimental results clearly show that the heat transfer capability was improved by a factor of 4 to 8 depending on the metal foam characteristics. Moreover, several mathematical models of the phase change process of a PCM embedded in metallic foam can be found in literature: Alshaer et al. [5] presented a model of the heat exchange in a carbon foam matrix saturated with a PCM. The model was obtained by the Volume Averaging Technique (VAT) based on single-domain energy equation. Mathematical model was validated by comparing its prediction with previous experimental measurements and good agreement was obtained.

Interesting studies with the aim to characterize the foam properties influence on the PCM melting process were carried out by Osama Mesalhy et al. [6] and Lafdi et al [7].In the first one the authors developed a numerical model solving the convection motion of the liquid phase inside a porous matrix, considering Darcy, Brinkman and Forcheimer effects and applying a two energy equations model. The coupling coefficient was estimated simplifying the heat transfer between the foam and the PCM with the heat conduction between two cylindrical layers. In the second paper, the authors experimentally investigated the thermal behaviour of the PCM melting process inside various aluminium foams, finding out that with high porosity and low PPI foams the steady-state regime is reached faster due to the natural convection effect.

The present study aims at developing a robust mathematical model based on the Volumes Of Fluid (VOF) method and the Volume Averaging Technique (VAT) using the non-thermal equilibrium approach, with physically based coefficients and a reasonable computational cost, in order to analyse and predict the phase change process characteristics with different metal foam properties. In particular, a novel approach for the esteem of the interfacial heat transfer coefficient between the solid matrix and the PCM is proposed. The model is qualitatively validated against experimental data in the literature and finally, a simulation in microgravity condition is performed in order to understand what the effect of convection on the overall heat transfer rate is.

2. Model description

2.1. Metal foam models

In order to characterize the heat exchange between metal foam and PCM a 3D model of the foam is necessary. The best foam model in literature was developed by Weaire and Phelan in 1994 [8]. They proposed a new foam element structure, based on eight polyhedrae of equal volume, but different face number. The other foam model widely used in literature is the Kelvin one (see figure 1), that in 1887 proposed a polyhedron with 14 slightly curved faces (eight hexagons and six squares). Because of the easier implementation of Kelvin model, several studies focused on the comparison between the two models: Cunsolo et al. [9] demonstrated that for an air flow there is no significant differences between the two models in case of high porosity foam. For this reason, Kelvin model described and characterized by Iasiello et al. [10] has been chosen. Their work was based on a comparison between simulations made on real foam geometries (obtained with topographical analysis) and ideal Kelvin's ones designed with Surface Evolver. The final obtained open-cell foam element is represented in figure 1.

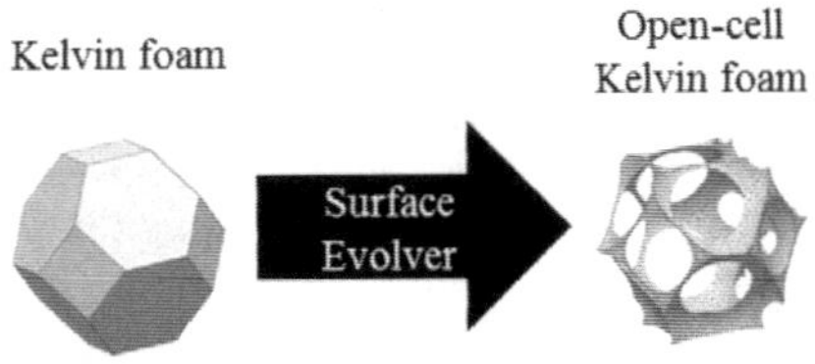

Figure 1. Kelvin model used for metal foam representation modeled with Surface Evolver.

2.2. Modeling approach and basic equations

The model is implemented by using the Volume Averaging Technique (VAT). The variables resolved in the balance equations are averaged on a Representative Volume Element (REV), which can be generally defined as the smallest volume of a porous material for which the continuum assumption is valid [11].

The metal foam-paraffin system is reduced to an equivalent, homogeneous and isotropic porous medium. Two temperature fields are resolved by using a Local Thermal Non-Equilibrium (LTNE) model: one for the metal foam and one for the paraffin.

The numerical model presents the following assumptions:

- The PCM is modeled using the apparent heat capacity method. The temperature dependence of specific heat is described as a continuous function (such as a continuous Dirac Delta function), whose integral is equal to the latent heat. The temperature at which the melting process begins is $T_{start-melt}$, while it ends at $T_{end-melt}$.
- Paraffin viscosity and buoyancy forces are defined trough a step function.
- The liquid PCM is considered incompressible and Newtonian, and subjected to the Boussinesq's approximation.
- The natural convection starts in a REV only when the paraffin is completely melted.
- The changes of density of the paraffin during the whole melting process are neglected. This approximation can be made since the highest impregnation ratio achievable is given by the liquid and solid density ratio. For this reason, only the liquid density is considered. Impregnation ratio is defined as the ratio between the mass of PCM embedded in the metal foam and the theoretic mass.

$$IR = m_{PCM}/(\rho_{PCM}\varepsilon V) \tag{1}$$

The volume averaged equations [11] are:

$$\frac{\partial\langle\rho_{PCM}\rangle}{\partial t} + \nabla\cdot\langle\rho\bar{v}\rangle_{PCM} = 0 \tag{2}$$

$$\frac{\rho}{\varepsilon}\left(\frac{\partial\langle\bar{v}\rangle}{\partial t} + \frac{1}{\varepsilon}\nabla\cdot(\langle\bar{v}\rangle\langle\bar{v}\rangle)\right) = -\nabla\langle p\rangle + \frac{\mu}{\varepsilon}\nabla\cdot(\nabla\langle\bar{v}\rangle) + \rho\beta\bar{g}(\langle T\rangle_{PCM} - T_{start-melt}) - \frac{\mu}{\overline{\overline{K}}}\langle\bar{v}\rangle - \frac{\rho C_F}{\sqrt{\overline{\overline{K}}}}|\langle\bar{v}\rangle|\langle\bar{v}\rangle \tag{3}$$

$$\varepsilon(\rho C_p)_{PCM}\left(\frac{\partial\langle T\rangle_{PCM}}{\partial t} + \bar{v}\cdot\nabla\langle T\rangle_{PCM}\right) = h_v(\langle T\rangle_{foam} - \langle T\rangle_{PCM}) \tag{4}$$

$$(1-\epsilon)(\rho C_p)_{foam}\frac{\partial\langle T\rangle_{foam}}{\partial t} + \mathrm{k}_{\mathrm{eff_{foam}}}\nabla^2\langle T\rangle_{foam} = -h_v(\langle T\rangle_{foam} - \langle T\rangle_{PCM}) \tag{5}$$

Equation (2) is the continuity equation. Equation (3) is the averaged momentum balance equation, where the last two terms on the right side are the Darcy and the Forcheimer terms, respectively. $\overline{\overline{K}}$ is the permeability coefficient, while C_F is Forccheimer coefficient, and they have been determined according to Calmidi [12]. To simulate the phase change in equation (3) and (4), all PCM thermodynamic properties have been defined using a continuous step-function, to improve numerical convergence. In the energy balance equations (4) and (5), $h_v$is the volumetric heat transfer coefficient, and is defined as

$$h_v = h_{sf}a_s \tag{6}$$

where h_{sf} is the Interfacial Heat Transfer Coefficient (IHTC), while a_s is the specific surface area of the metal foam.

2.3. Interfacial heat transfer coefficient (IHTC) determination

The interfacial heat exchange coefficient (h_{sf}) is usually estimated using existing correlation for tubes in cross flow ([13], [14]). These approaches can lead to incorrect results since such correlations have not been obtained for paraffin embedded in metal foam, and do not take into account the different phases in the melting process (i.e. preheating, melting and liquid convective phases).

Another interesting approach was proposed by Mesalhy et al. [6], who assumed the porous matrix geometry to be an intersected mesh of fibres with circular cross section area. The IHTC was then calculated by simplifying the heat exchange between matrix layer and PCM as heat conduction in two cylindrical layers. The correct estimation of this coefficient is important because it deeply affects the phase change process. In order to understand how much the volumetric heat transfer coefficient h_v affects the melting front behaviour, a sensitivity analysis was performed. The results show that there is a threshold value of the volumetric heat transfer coefficient: for higher values, the temperature difference between the metal foam and the PCM is negligible, so a LTE and a LTNE model give basically the same results; for lower values the position of the melting front highly depends on the h_v value. From equation (4) the threshold value depends on:

- the local thermal diffusivity of the PCM;
- the thermal gradients and temperature time derivatives.

In this paper, a new modelling approach for the IHTC calculation is proposed. Three coefficients have to be determined, because three different conditions occur during the phase change process:

- heat exchange between solid PCM and metal foam that occurs when $\langle T\rangle_{foam}$ is lower than $T_{melt-start}$;
- heat exchange between melting PCM and metal foam that ends when $\langle T\rangle_{PCM}$ becomes higher than $T_{melt-end}$;
- heat exchange between liquid PCM and metal foam that begins once the melting process is over.

2.3.1. *Solid-solid and melting IHTCs determination*

The idea in this study is to use the Kelvin 3D model of the foam structure to calculate the IHTC in the conditions listed before. In fact, under the hypothesis of negligible temperature gradient in a single foam cell element, a uniform time dependent temperature is set as boundary condition on the surface of the paraffin cell in contact with the metal foam. On the remaining surface an adiabatic boundary condition is set (symmetry). The heat transfer coefficient per unit of metallic foam surface can be calculated as the time average of:

$$h_{sf} = \frac{1}{t}\int \frac{\dot{Q}_{par-foam}(t)}{S[T_{foam}(t) - \langle T\rangle_{PCM}(t)]}dt \qquad (7)$$

where $\dot{Q}_{par-foam}(t)$is the thermal power at the boundary between the paraffin cell and the metal foam, S is the metal foam surface and $\langle T\rangle_{PCM}$(t) is paraffin cell medium temperature and is defined as:

$$\langle T\rangle_{PCM} = \int_{cell} \frac{T_{PCM}dV}{V_{cell}} \qquad (8)$$

This approach has a high computational cost above all to determine IHTCs during melting conditions because a highly non-linear 3D problem has to be solved. For this reason, the IHTC has been determined using a simplified 1D spherical model, following the same conceptual approach shown before. The time dependent temperature imposed on the boundary is supposed to be the average temperature on the surface of the paraffin sphere: the portion in direct contact with the foam is assumed to be at T_{foam}, while the portion that is not in contact with the foam is supposed to be at $\langle T\rangle_{PCM}$. It follows that T_{foam} can be calculated as

$$T(R,t) = (1 - a_c)\langle T\rangle_{PCM}(t) + a_c T_{foam}(t) \tag{9}$$

$$T_{foam} = \frac{T(R,t) - (1 - a_c)\langle T\rangle_{PCM}(t)}{a_c} \tag{10}$$

where a_c is fraction of the cell external area in direct contact with the foam and is defined as:

$$a_c = \frac{A_{foam}}{4\pi R_{cell}^2} = \frac{V_{cell} a_s}{4\pi R_{cell}^2} \tag{11}$$

h_{sf} has to be defined per area of metal foam surface, so substituting eq. (10) in eq. (7) it follows

$$h_{sf} = \frac{1}{t_f}\int_0^{t_f}\left(\frac{\dot{q}_{par-foam}(t)\cdot a_c}{T(R,t) - \langle T\rangle_{PCM}(t)}dt\right)\left(\frac{1}{a_c}\right) \tag{12}$$

where $T(R,t)$ is the imposed boundary condition of the 1D model and $\dot{q}_{par-foam}$ is the heat flux normalized on the whole paraffin sphere surface; t_f in case of melting heat exchange is the time when the temperature at the cell center becomes higher than $T_{melt-end}$, (i.e. when melting process is concluded) while in case of solid-solid heat exchange has been assumed arbitrarily to be $100s$.It can be verified that h_{sf} shows little dependence from the function imposed as boundary condition ($T(R,t)$), as can be understood from Table 1.

Table 1.$\boldsymbol{h_{sf}}$calculated for various boundary conditions for a 10 PPI 95% porosity cell.

Solid-solid heat exchange Function $T(R,t)$	$h_{sf}[W/(m^2K)]$	Melting heat exchange Function $T(R,t)$	$h_{sf}[W/(m^2K)]$
$298 + 0.02t + 0.005t^2$	337.71	$300 + 0.01t$	335.84
$298 + 0.1t$	314.95	$300 + 0.01t + 0.0002t^2$	365.47
$298 + 2\log(t+1)$	286.67	$300 + 2\log(t+1) + 0.01t$	290.02

Notice that h_{sf} does not vary significantly between melting and solid–solid conditions. For this reason, this coefficient will be assumed to be constant in both phases.

The 1D approach was numerically validated for the simplest case of solid-solid heat exchange that occurs before melting process starts. The results obtained with the 1D model are compared with those obtained with a 3D transient model of the heat conduction in a single cell. Table 2 reports the IHTCs obtained for the 85% porosity foam at various PPI with both 3D and 1D model.

Table 2.IHTC comparison for various PPI for an 85% porosity metal foam calculated with the 1D and the 3D model.

Porosity ε	PPI	Cell diameter [m]	$h_{sf}[W/(m_{foam}^2K)]$ 3D model	$h_{sf}[W/(m_{foam}^2K)]$ 1D model
	10	0,00623	271	257
85%	20	0,00312	550	495
	30	0,00156	1077	1048

2.3.2. *Convective IHTC determination*

When the paraffin wax is completely melted, natural convection can occur. To determine the heat transfer coefficient in this condition, the magnitude of the velocity field induced has to be estimated. To do so, the volume averaged model was solved with a first try IHTC to determine the order of magnitude of the velocity field. In order to determine the interfacial heat transfer coefficient during the liquid-solid heat exchange, a 3D stationary model was solved (see figure 2).

Table 3. IHTC during natural convection for various velocity inlet conditions.

	$\varepsilon = 0.94;\ 10\ PPI$	$\varepsilon = 0.94;\ 20\ PPI$	$\varepsilon = 0.94; 0\ PPI$
u_o[m/s]	$h_{sf}[W/(m^2K)]$	$h_{sf}[W/(m^2K)]$	$h_{sf}[W/(m^2K)]$
$1 \cdot 10^{-5}$	293.4	596.7	1194.3
$5 \cdot 10^{-5}$	281.4	568.6	1138.4
$1 \cdot 10^{-4}$	340.4	690.5	1367.1
$5 \cdot 10^{-4}$	344.6	704.4	1453.8

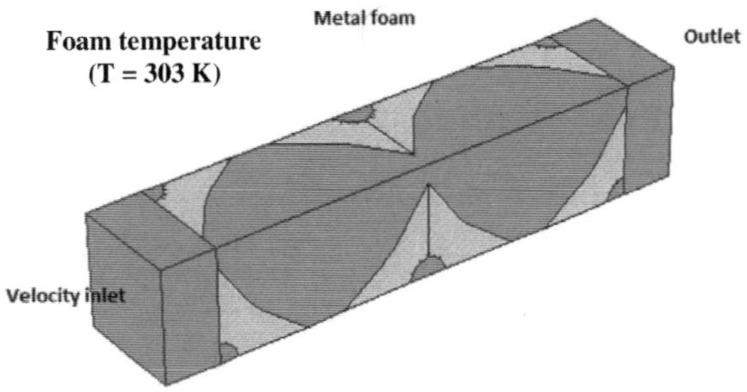

Figure 2. Schematic representation of the 3D stationary for the IHTC calculation during natural convection. The different boundary conditions for both fluid-dynamic and thermal model are reported.

Mass, momentum and energy equations are solved. For the momentum equation, the velocity found with the first iteration is used as inlet boundary condition for the simulation. No slip condition is used at the solid/fluid interface. At the outlet, the atmospheric pressure is assumed. For the energy equation, a uniform ambient temperature is assumed at the fluid inlet, while a uniform temperature value of 303 K is employed at the solid fluid interface. Symmetry condition is used on the other boundary surfaces.

The finite element commercial code COMSOL Multiphysics is employed to solve the governing equations and boundary conditions. In Table 3 are reported the IHTCs calculated for velocity inlet conditions close to the one calculated after the first iteration.

3. Model validation and results

In order to validate the model developed, the experimental data by Lafdi et al. [7] have been considered. The system was heated from the left side and cooled by a heat sink on the right side using an aluminum plate chilled by flowing water. The validation is only qualitative and not quantitative, because of the uncertainty on the boundary conditions on the heater side and on the heat sink side. In fact, it is not possible to estimate the contact resistance between heater plate and metal foam that can play an important role during the dynamic of the melting process, and the thermal power absorbed by the heat sink. Moreover, the experimental characterization of the apparent specific heat during phase change is not reported in the article. Therefore, it was assumed that the specific heat-temperature function during phase change has to be a continuous Dirac Delta function.

Table 4.IHTC during and coupling coefficients during melting and solid-solid heat exchange.

		$\varepsilon = 0.94$			
	h_{sf}		$a_s[m^2/m^3]$	$h_{sf}a_s[W/m^3]$	$h_{sf}a_s[W/m^3]$
PPI	Solid-solid	Melting		Solid-solid	Melting
10	314.7	368.2	511.24	$1.61 \cdot 10^5$	$1.88 \cdot 10^5$
20	626.2	678.5	1022.48	$6.40 \cdot 10^5$	$6.94 \cdot 10^5$
40	1388	1353	2044.97	$2.84 \cdot 10^6$	$2.77 \cdot 10^6$

The chosen comparison parameters are melting front shape after 80 minutes (4800 s) and the heater plate temperature evolution. For a further model validation, we are planning our own experimental campaign, where metal foam and heated plate will be brazed, in order to reduce as much as possible, the contact resistance. Table 4 shows the IHTCs and the coupling coefficient that have to be inserted in equations (3) and (4) for the melting and solid-solid heat exchange.

Table 5. Metal foam effective properties.

Metal foam effective properties $\varepsilon = \mathbf{0.94}$	10 PPI	20 PPI	40 PPI
Effective thermal conductivity $[W/mK]$	3.67		
Density $[kg/m^3]$	162		
Forchheimer coefficient	0.0153		
Foam permeability $[1/m^2]$	$1.037 \cdot 10^{-7}$	$2.583 \cdot 10^{-8}$	$6.458 \cdot 10^{-9}$

Comparing the results shown in table 4 with those in Table 3, it is possible to assume a constant IHTC for all the heat exchange phases for each case, since the values obtained are similar and differ about 10%. This assumption does not affect the calculation precision but reduces the calculation time, because IHTC is now a constant value instead of a temperature depending function. The system is heated with a constant and uniform heat flux (28.9 W). The thermal power was supposed to be completely delivered to the metal foam, since its effective conductivity is much higher than paraffin one (see table 3 in [7]). Foam effective thermal conductivity was calculated according to (12), as suggested by the supplier (ERG Aerospace) [15].

$$k_{eff} = \frac{1-\varepsilon}{3} k_{Al} \tag{13}$$

The heat sink was simulated imposing the initial temperature at the corresponding boundary for all the modelled cases. The experimental data shows that the mean temperature on the heater plate is generally lower for lower PPI foams. This trend is in good agreement with the experimental data, as can be seen comparing figures 3a and 3b.

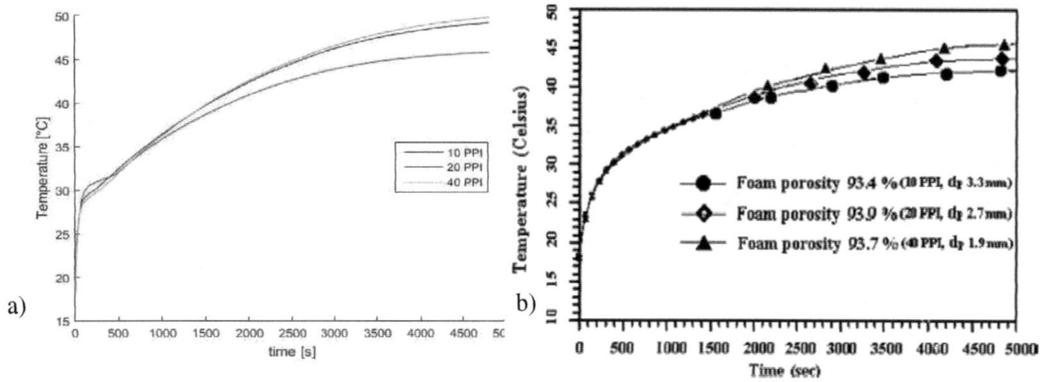

Figure 3. Comparison between experimental and mathematical results of heater temperature.

Moreover, with the 10 PPI foam a steady state condition is achieved before the other two cases in both experimental and model results. In fact, even if the coupling coefficient is lower for lower PPI foams, the permeability factor increases as well as the heat transferred because of natural convection.

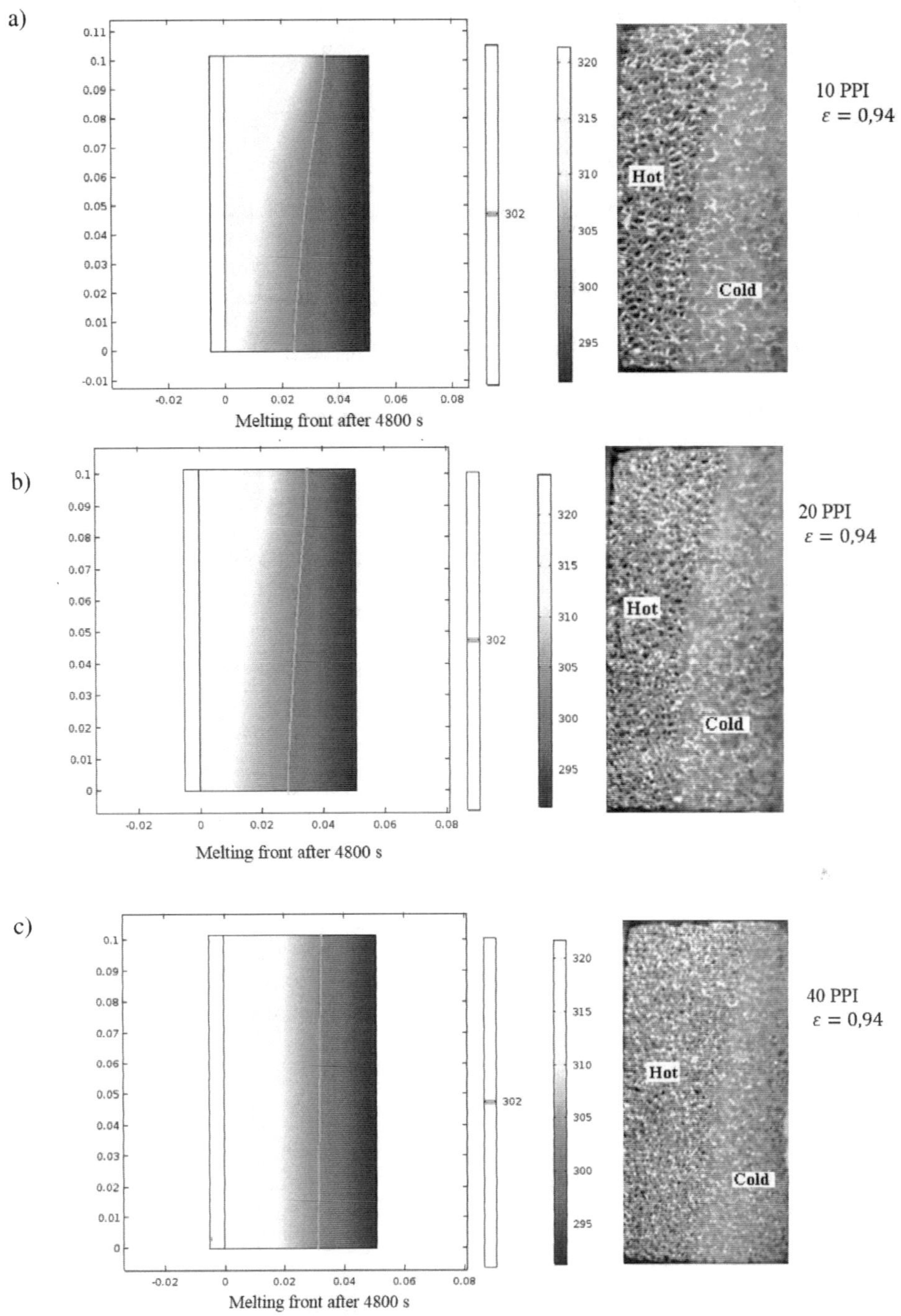

Figure 4.Comparison between experimental and model melting front after 4800 s. The grey line identifies the melting front in the model results.

Analysing the 10 PPI foam temperature contour, there are areas where paraffin is hotter that metal foam. The importance of convective heat transfer can be also seen from the melting front shape that is much more irregular for the 10 PPI foam. Figure 4 shows a comparison between the model and the experimental melting front for the 10 PPI, 20 PPI and 40 PPI foams. The higher IHTC typical of higher

PPI foams seems to have little influence on heater temperature in the cases analysed. This happens because of the low thermal flux that has to be dissipated. Notice that in order to have a further validation on the IHTCs calculated, a higher thermal power should be furnished to the system.

4. Conclusions

A numerical model based on solving the volume averaged conservation equations for mass, momentum and energy with phase change (melting) is developed in order to study the effect of adding a high thermal conductivity matrix on the performance of PCM energy storage. The convection motion of the liquid phase inside the porous matrix is solved considering the Darcy, Brinkman and Forcheimer effects. The local thermal non-equilibrium assumption is considered due to the large differences in thermal properties between the solid matrix and PCM by applying a two-energy equation model. A novel model to calculate the interfacial heat transfer coefficient (IHTC) is proposed by simulating heat exchange between the PCM and the Kelvin ideal metal foam. For the solid-solid and the melting heat exchange phase the 3D transient model was simplified with a 1D transient model to reduce computational cost. This approach was numerically validated since the results obtained with the 1D model are very close to the 3D model ones in the case of solid-solid heat exchange. For the liquid phase a 3D stationary model was solved for different velocity inlet conditions. The interfacial heat transfer condition is similar for the three cases analysed, so a constant value has been chosen for the simulation. The results of the averaged equations showed good agreement with the experimental results found in literature as for the melting front shape and heater temperature evolution, but an experimental setup will be built in order to further validate the model developed. With the coupling coefficient calculated, a microgravity condition model was built and the results were compared with the ones previously obtained.

Nomenclature

a_s	Specific foam area per unit of volume (1/m)
a_c	Paraffin cell area fraction in contact with foam (m^2_{foam}/m^2_{cell})
C_F	Forcheimer coefficient
C_p	Specific Heat (W/kg·K)
d_c	Cell diameter (m)
d_p	Pore diameter (m)
g	Gravity acceleration (m/s^2)
h_{sf}	Interfacial heat transfer coefficient (W/m^2·K)
h_v	Volumetric heat transfer coefficient (W/m^3·K)
k	Thermal conductivity (W/m·K)
$\overline{\overline{K}}$	Porous medium permeability (1/m^2)
L	Latent heat of fusion (J/kg)
m	Mass (kg)
p	Pressure (Pa)
$\dot{q}_{par-foam}$	Heat flux on the paraffin cell boundary (W/m^2)
$Q_{par-foam}$	Thermal power exchanged between paraffin and metal foam (W)
S	Surface (m^2)
T_{PCM}	PCM temperature (K)
$T_{start-melt}$	Initial melting temperature (K)
$T_{end-melt}$	Final melting temperature (K)
T_{foam}	Foam temperature (K)
T(R,t)	Temperature imposed for h_{sf} determination (K)
t	Time (s)
t_f	Time of simulation for h_{sf} determination (s)
$\bar{v}$	Velocity vector (m/s)
V	Volume (m^3)
V_{cell}	Metal foam cell volume (m^3)

Abbreviations

IHTC	Interfacial Heat Transfer Coefficient
PPI	Pore Per Inch
LTNE	Local Non Thermal Equilibrium
LTE	Local thermal equilibrium
VAT	Volume Averaging Technique
REV	Representative Volume Element
IR	Impregnation Ratio
PCM	Phase Change Material
TES	Thermal Energy Storage

Greek symbols

ε	Porosity
μ	Dynamic viscosity (kg/m^2·s)
ρ	Density (kg/m^3)
β	Volume expansion coefficient (1/K)

Other symbols

x, y, z	Coordinates (m)
<>	Volume average

References

[1] Fan L and Khodadadi J M 2011 Thermal conductivity enhancement of phase change materials for thermal energy storage: A review *Renewable and Sustainable Energy Reviews* **15 (1)** 24-46.

[2] Tian Y and Zhao CY 2011 A numerical investigation of heat transfer in phase change materials (PCMs) embedded in porous metals *Energy* **36 (9)** 5539–5546.

[3] Tong X Khan J A and Amin M R 1995 Enhancement of solidification heat transfer by inserting metal-matrix into phase change material Proceedings of the 1995 *ASME International Mechanical Engineering Congress & Exposition San Francisco, CA, USA 1 1-8.*

[4] Mancin S Diani A Doretti L Hooman K and Rossetto L 2014 Experimental analysis of phase change phenomenon of paraffin waxes embedded in copper foams *International Journal of Thermal Sciences* **90** 79-89.

[5] Alshaer W G Nada SA Rady M A Le Bot C and Del Barrio EP2014 Numerical investigations of using carbon foam/PCM/Nano carbon tubes composites in thermal management of electronic equipment *Energy Conversion and Management* **89** 873–884.

[6] Mesalhy O Lafdi K Elgafy A and Bowman K (2005), Numerical study for enhancing the thermal conductivity of phase change material (PCM) storage using high thermal conductivity porous matrix *Energy Conversion and Management* **46(6)** 847-867.

[7] Lafdi K Mesalhy O and Shaikh S 2007 Experimental study on the influence of foam porosity and pore size on the melting of phase change materials *Journal of Applied Physics* **102 (8)** paper number 083549.

[8] Weaire D and Phelan R 1994 A counter-example to Kelvin's conjecture on minimal surfaces *Philosophical Magazine Letters* 69107–110.

[9] Cunsolo S Iasiello M Oliviero M Bianco N Chiu W K and Naso V 2016 Lord Kelvin and Weaire–Phelan Foam Models: Heat Transfer and Pressure Drop *Journal of Heat Transfer* **138 (2)** paper number 022601.

[10] Iasiello M Cunsolo S Oliviero M Harris W M Bianco N Chiu W K and Naso V 2014 Numerical Analysis of Heat Transfer and Pressure Drop in Metal Foams for Different Morphological Models *Journal of Heat Transfer* **136 (11)** paper number 112601.

[11] Whitaker S 1999 The method of volume averaging (Norwell, MA, USA: Kluwer Academic Publishers).

[12] Calmidi V 1998 Transport phenomena in high porosity metal foams Ph.D thesis: University of Colorado.

[13] Liu Z Yao Y and Wu H 2013 Numerical modelling for solid–liquid phase change phenomena in porous media: Shell-and-tube type latent heat thermal energy storage *Applied Energy* **112** 1222–1232.

[14] Li W Q Qu Z G He Y L and Tao W Q 2012 Experimental and numerical studies on melting phase change heat transfer in open-cell metallic foams filled with paraffin *Applied Thermal Engineering* **37** 1-9.

[15] http://www.ergaerospace.com/Thermal-transfer.html.

The role of the saver in a HPLCs heat transfer process: a transient analysis

M. Lorenzini [1], **P. Valdiserri** [2], **M. Pagnoni** [3]

[1] DIN – Dipartimento di Ingegneria Industriale
Via Fontanelle 40, 47121 Forlì, Italy
[2] CIRI Edilizia e Costruzioni – Centro Interdipartimentale per la Ricerca Industriale
Via del Lazzaretto 15/5, 40131 Bologna, Italy
[3] Pagnoni Impianti s.r.l.
Strada per Cascina Restelli 26-28, 20886 Aicurzio (MB), Italy

Corresponding author email: marco.lorenzini@unibo.it

Abstract. This work describes the experimental validation of a mathematical model implemented in Matlab/Simulink®, which is used to investigate energy consumption in the production process of high-pressure laminate composites (HPLCs). First, experimental onsite temperature measurements are used to determine parameters for the core component, i.e. the package, consisting of several layers of different materials. These otherwise hard to assess quantities are the generated power due to exothermal reactions in the cross-linking phase, and the piecewise approximation for the variable heat capacity of the impregnated paper over the heating phase of the process. The capability of the whole model (including the hot and cold water storage tanks) to replicate the behaviour of the actual production process has been verified. The validated model has been employed to investigate the influence of a saver tank for the spent hot water to be employed in the early stage of the heating phase of the process, with the load and discharge phases lasting three or five minutes each. Computations demonstrate that significant savings in energy (up to 34) can be obtained.

1. Introduction

High-pressure laminate composites (HPLCs or HPLs), also known by the trademark name of Formica™ [1], have a wide range of applications in furniture and as building material. HPLs are normally made of a decorative layer of impregnated melamine paper and a sturdier, thicker core (Kraft paper impregnated with resins) for structural purposes. Research and development in HPLs has so far focused mainly on their mechanical characteristics, in particular on their stiffness and strength, both tensile and flexural [2], and on the way in which high-pressure curing affects these properties [3]. In the last decade, owing to the constant drive to reduce production costs, attention has also been devoted to the production process of impregnated paper (with melamine formaldehyde (MF) and urea formaldehyde (UF) resins) as the surface quality of decorative low-pressure laminates is strongly affected by this [4]. Contrary to other materials employed for similar uses produced at lower pressures, HPLs require thermal cycling for resins to reticulate.

In particular, the paper is pressed between two plates (at about 7-10 MPa) and heated to a given temperature (between 160 and 180 °C depending on the type of material), where it stays for some time (about fifteen minutes as a rule), then cooled back to ambient temperature while still being pressed in order to avoid buckling problems when the press plates are released.

A single press accommodates a number of openings (ten to twelve in standard realizations), in each of which the components of the so-called package are laid (figure 1). The package consists of a platen (carbon steel) working as tray, a number of Kraft paper sheets (called blanket or cushion, to even the heat flux distribution and compensate possible unevenness of the surface which often occur in the production).

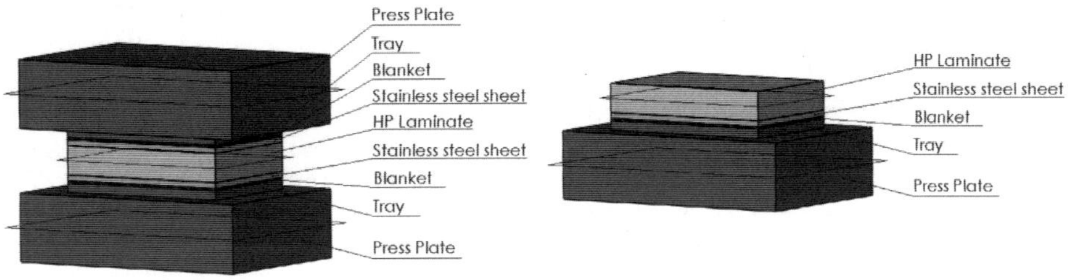

Figure 1 – Sketch of the package (left) and of the half which is simulated (right).

Above them, a modular assembly is laid; this consists of a stainless steel sheet with high surface finish (to give the decorative paper the proper surface appearance and pattern) a sheet of decorative paper, one of support material (consisting of several Kraft paper sheets), a detach pad (that also constitutes the symmetry layer for the assembly) to avoid the HPL sticking to the one immediately above, another support-decorative pair and another stainless steel. Depending on the size of the openings and on the thickness of the HPL, the modular assembly may be repeated a number of times for each opening (from one to ten). The package is topped by another series of Kraft paper sheets and by a lid to make the package symmetrical. Heating and cooling of the package is achieved through channels machined in the press plates, through which the heat transfer fluid (usually subcooled water at high pressure) flows. Heating and cooling strategies are often dictated by economic and layout constraints, and may include one or several storage tanks for the spent fluid in order to employ part of the waste heat.

High-pressure lamination is over half a century old, yet the process is still managed empirically and with massive resort to rules of thumb and workshop experience. This attitude is now starting to be changed due to more pressing energy saving demands, which call for a reduction in process inefficiencies. In the case of HPLs, little exists in the open literature, regarding the characterization of the production process and possible ways to reduce its energy consumption without significantly increasing production times.

In a previous work [5] a Simulink® [6] model of the thermal cycle through which HPLs are produced was employed to study the influence on temperature distribution in the laminate, in particular in the heating phase, of the thermal conductivity and thickness of the different components of the package: it was concluded that the thermal conductivities of the tray and stainless steel sheet are of little significance in modifying the temperature distribution in the laminate, while that of the blanket is decisive: if it is low, the temperature over the thickness of the laminate tends to be more uniform, and *vice versa*. Moreover, reducing the blanket's thickness high temperatures in the laminate are achieved in shorter times but the gradients increase as well, to values which may lead to local burns in the HPL due to overheating.

This work is a continuation of the previous investigation, moving from the behavior of the package to that of the whole process loop. In order to validate the model previously obtained of the press station experimental measurements of the temperature history of the process were carried out during an actual production run. Once the validity of the model was proven and some of its parameters fine-tuned, a storage tank for spent hot water was added and its influence on the process in terms of energy savings and increase in process times investigated

2. Model validation for the package and for the complete circuit

2.1. Model parameters to be fine-tuned

The model which was employed to investigate the influence of the different components of the package on the temperature distribution along the thickness of the laminates was described in detail in [5]. Here, only the part relevant to validation, namely the laminate within the package, is discussed, namely the package consisting of the tray, the blanket, the stainless steel sheet and half of the HPL. Every component is divided into several layers for each of which the temperature is computed.
The number of layers into which the package is divided depends on the thickness of every component, the thermal conductivity and thermal capacity of the material. In the present case, the blanket is divided into four layers, whilst the half of the HPL has eight layers. For each component of the package and for each layer a time-dependent equation is obtained from the energy balance [7] for the layer and discretized using finite differences [8]. For the inner nodes, the equations have the form customary for non-stationary conduction with internal generation (due to evaporation of the moisture in the soaked paper), so at node m

$$\frac{kA\left(T_{m-1}-T_m\right)}{\Delta x}+\frac{kA\left(T_{m+1}-T_m\right)}{\Delta x}-hL\Delta x\left(T_m-t_{\mathrm{A}}\right)+\dot{q}'''A\Delta x=\rho A\Delta x c\frac{dT_m}{d\tau} \qquad (1)$$

where:

- k ($\mathrm{W\cdot m^{-1}K^{-1}}$) is the thermal conductivity of the element ;
- Δx (m) is the thickness of the layer
- A ($\mathrm{m^2}$), L (m) are the area and the perimeter of the layer;
- T_m , T_{m-1} and T_{m+1} (K) are the temperature of the considered layer, and the ones of the two adjacent layers;
- h ($\mathrm{W\cdot m^{-2}K^{-1}}$) is the convective heat transfer coefficient (obtained from [5]) with the surrounding air at temperature t_A (K);
- $\dot{q}'''$ ($\mathrm{W\cdot m^{-3}}$) is the thermal power generated inside the layer due to exothermal reactions during the cross-linking phase;
- ρc ($\mathrm{J\cdot m^{-3}K^{-1}}$) is the volumetric thermal capacity of the material.

At the interface of adjacent layers continuity of the heat flux is imposed, while the contribution due to contact resistance was neglected, as it would not be significant for the present calculations and also considering the high pressures that keep the materials together.

In Eq.(1), $\dot{q}'''$ is not known *a priori*, so some experimental data are needed to determine it. The specific heat capacity c for the laminate is also unknown, and varies depending on the stage of the process: it starts at rather high values (due to moisture and impregnating agents in the paper), and progressively drops as the laminate is heated, to reach a minimum in the end phase of the cross-linking and remains constant during the successive cooling step. In order to have an estimate of the two quantities discussed above, temperatures were measured experimentally at a plant during batch operations.

2.2. Experimental measurements and package model validation

A short description of the plant where measurements were carried out is now given: this is relevant to the validation procedure, as layouts may differ significantly depending on the production site. The plates are heated with hot water at constant temperature from a storage tank, until cross-link temperature, T_k, is reached (in this case T_k=122 °C), then warm water is shut off and cold water from another storage tank is forced through the press plates. The cooled water is recirculated to the storage tank, where it is heated up again, while during the cooling phase the hotter water out is sent to a storage tank for reuse, till its temperature drops below a threshold value: below this, the fluid is circulated directly to the cooling tower and returned to ambient temperature. In order to validate the model, temperature measurements of the ambient air, inlet and outlet water, blanket and laminates were carried out by means of K-type, calibrated thermocouples with a total uncertainty of ±0.5 K. Data were acquired every 10 seconds with a $5^{\frac{1}{2}}$-digits multimeter and recorded to file. Temperatures for the blanket were measured at several layers, around 0.15 m from the border of the plate on both short and long sides, while for the laminate three layers were chosen, the temperature being measured on both the long and short sides for each of them. The values obtained for the water outlet temperature coincide with that of the plate, as was demonstrated previously [5] and were therefore employed to represent the temperature history of the latter. The temperatures of the central layer of the laminate were averaged and employed to obtain values of heat generation due to cross-linking (in the interval 80-122 °C in the experiment). The results of measurements are reported versus time τ in figure 2, where error bars are not shown as they would be tiny and just hamper the readability of the plots. Plate temperature (light blue) exhibits the behaviour typical of lumped-capacitance bodies, while the laminate has a strong lagging in temperature (orange).

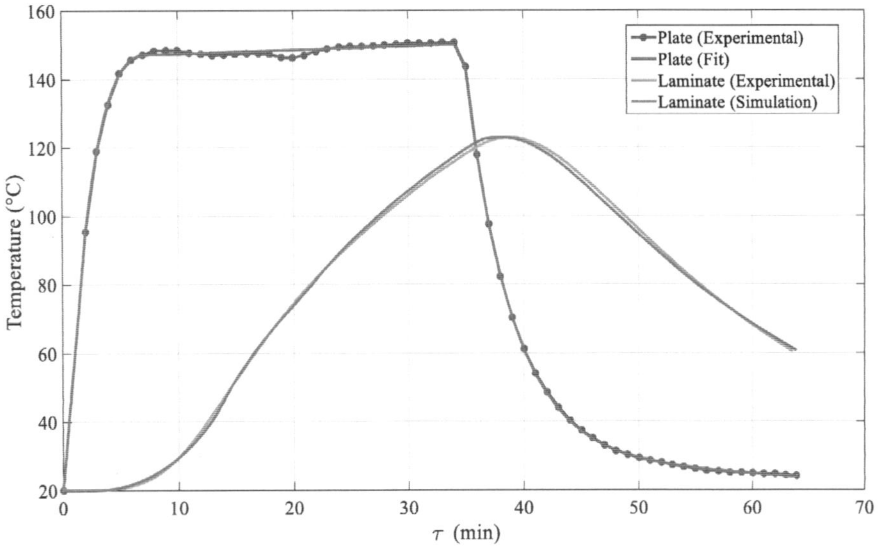

Figure 2 – Experimental temperature data and approximating temperature curve.

Experimental plate temperatures are used to obtain a functional approximation for the plate temperature (continuous red line in figure 2), which is used as input temperature for a reduced model consisting of the package only: this allows the determination of cross-linking heat generation and of specific heat capacity. The former is $\dot{q}''' = 22\mathrm{kW}\cdot\mathrm{m}^{-3}$, while the trend of c in the heating phase is shown in figure 3: it must be remarked that once c reaches its minimum value, its heat capacity becomes constant and remains so over the whole cooling step.

With the values just determined, the simulated temperature evolution within the laminate is shown again in figure 2: a very good agreement with the experimental data can be appreciated. It is also to be noticed that the maximum temperature in the laminate is reached almost five minutes after the hot water has been shut off.

2.3. Validation of the complete model

In order to test the whole Simulink model, which comprises the hot and cold water storage tanks, the heat generation due to cross-linking and the thermal capacity just determined have been employed as parameters for the package, and a simulation of the batch process has been run. In order to replicate the behavior of the actual production process, which involves feeding hot water to the plates at almost constant temperature, the capacity of the storage tank has been increased accordingly.

The results of the simulation are shown in figure 4: again, the agreement between numerical and experimental data is quite satisfactory.

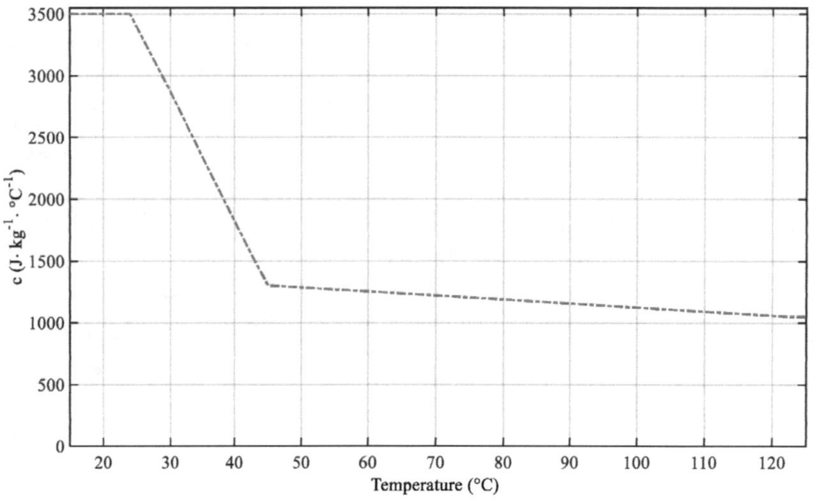

Figure 3 – Change of specific heat capacity of the laminate with temperature in the heating phase.

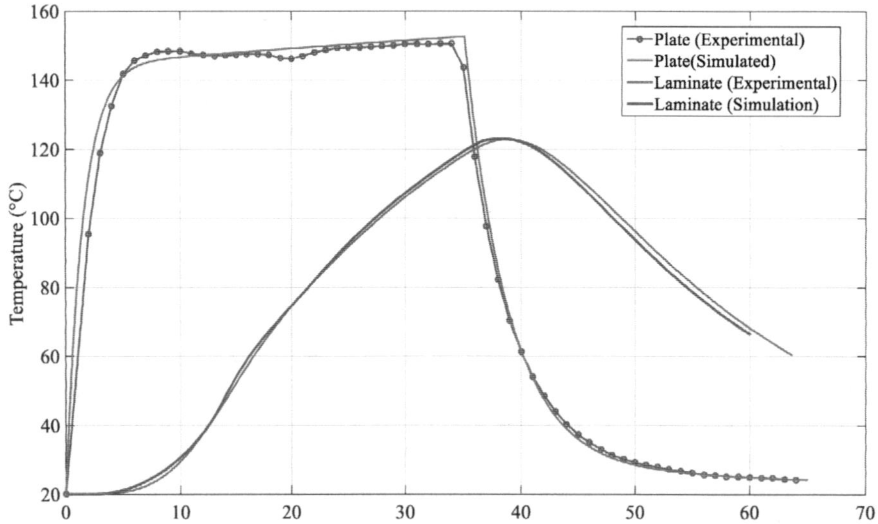

Figure 4 – Comparison between experimental data and simulation of the whole plant.

3. Auxiliary hot water storage tank

To decrease energy consumption, a tank (saver) to store the spent hot water from the plates during the cooling phase is added to the model. The saver has a storage volume V=10 m^3, makes the hot water available for heating the plates in the initial phase of next press run. As a consequence of this, the times needed to complete the process increases because of the lower temperature of the hot water supplied to the plates. As an example, simulations have been run keeping all parameters unchanged and imposing two equal values for the load and discharge times of the saver, namely τ_s=3 min and τ_s=5 min. Simulations have been run for a number of cycles such that the system gives the plates about the same energy during the heating phase as it receives from them during cooling. Results in terms of temperature versus time for both the plates and the laminates at the fourth cycle and τ_s=5 min are shown in figure 5.

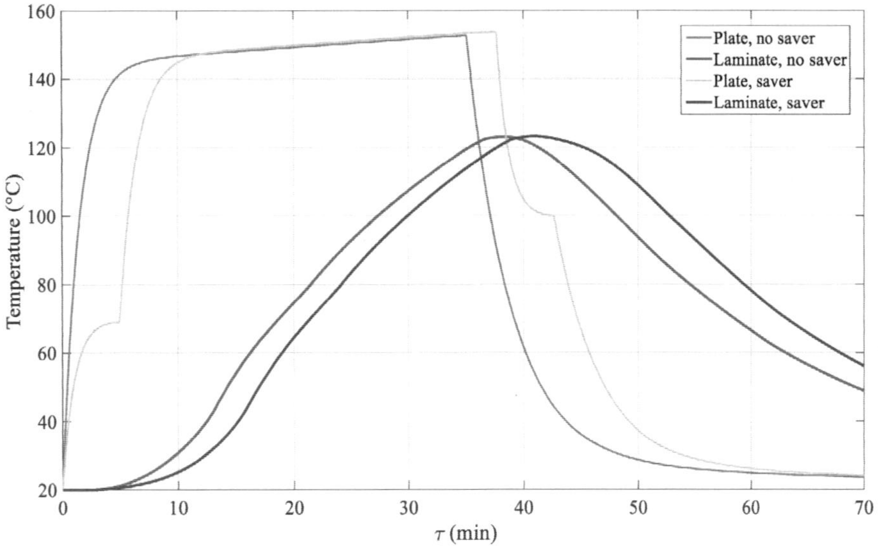

Figure 5 – Temperature profiles with and without saver, τ_s=5 min.

Comparison of the temperature profiles of the plate with and without saver evidences that how the use of the storage tank delays the process, which lasts a few minutes (between two and four) longer. The time when the saver ceases to be used is clearly visible both for the heating and the cooling steps. When the temperature curve flattens (this meaning that the plates are roughly at the same temperature as the water fed to them), hot water must be supplied at a higher temperature, i.e. from the hot water tank, and temperature steeply increases. During cooling, when the water from the saver has been warmed to the temperature of the plates, heat removal becomes ineffective, and colder water must be supplied, i.e. from the cold water tank. Again, the temperature drops sharply and reaches its steady-state value. The sharp change experienced by temperature when the saver is no longer used is not seen in the laminate: this is due to the lagging behaviour of the heated mass, but an increase in the time needed to reach a certain temperature is visible – from two to four minutes for the conditions investigated in this work – and affects production times adversely.

The advantage of using a saver lies in the reduction of energy consumption, as shown in table 1, which refers to τ_s=3 minutes, and table 2, which is for τ_s=5 minutes. Each table bears in the top row the cycle from the start of the production process to which the data in columns refer. During the first run, the saver is assumed to be at ambient temperature, so that it cannot be used in the first heating of the plates, but only after the hot water flowing from the plates during the cooling process has been made available. During cooling, hot water is stored in the saver, which makes the average water temperature increase, which corresponds to recovery of some thermal energy from the hot stream, before this is sent to the cooling towers.

As can be seen from table 1, this recovery is around 45% of the total energy needed. At the start of the second cycle, the hot water in the saver, which has a temperature of roughly 70°C, is used to heat the plates up for three minutes, thus supplying about 18% of the whole heating load and decreasing the temperature in the tank. At the start of the cooling phase the process starts again, with an increase in the percentage of the heating load covered by the water from the saver, which is 29% in this case.

Table 1. Energy performance of saver, τ_s=3 min

		I	II	III	IV	V	VI	VII
Saver (Heating Phase)								
Initial temperature	(°C)	--	70.0	85.9	93.1	96.4	97.9	98.6
Final Temperature	(°C)	--	53.4	64.0	68.9	71.1	72.1	72.5
Energy recovered	(MJ)	0	692	914	1016	1062	1084	1092
Percentage recovery	(-)	0%	18%	24%	27%	28%	29%	29%
Saver (Cooling Phase)								
Initial temperature	(°C)	30.0	53.4	64.0	68.9	71.1	72.1	72.5
Final Temperature	(°C)	70.0	85.9	93.1	96.4	97.9	98.6	98.7
Energy recovered	(MJ)	1670	1360	1218	1154	1126	1112	1106
Percentage recovery	(-)	45%	36%	32%	31%	30%	30%	29%
Energy supplied through boiler	(MJ)	3748	3056	2836	2736	2690	2670	2660

Table 2. Energy performance of saver, τ_s=5 min

		I	II	III	IV
Saver (Heating Phase)					
Initial temperature	(°C)	--	72.4	88.1	97.8
Final Temperature	(°C)	--	53.9	64.0	70.3
Energy recovered	(MJ)	0	774	1009	1154
Percentage recovery	(-)	0%	21%	27%	31%
Saver (Cooling Phase)					
Initial temperature	(°C)	30.0	53.9	64.0	70.3
Final Temperature	(°C)	72.4	88.1	97.8	98.9
Energy recovered	(MJ)	492	1435	1290.4	1202
Percentage recovery	(-)	13%	38%	34%	32%
Energy supplied through boiler	(MJ)	3748	2972	2740	2596

If the water in the saver is used for a longer time, as is the case for Table 2, which corresponds to τ_s=5 minutes, the percentage of energy recovered is slightly higher, around 31%; as a drawback, the time needed to complete a cycle increases to about four minutes as compared to when no saver is employed.

4. Conclusions

This work describes the experimental validation of a Simulink model used to investigate the production process of high-pressure laminate composites. For the core component, i.e. the package, consisting of several layers of different materials, experimental temperature measurements have allowed the determination of two otherwise hard-to-compute quantities, namely the generated power due to exothermal reactions in the cross-linking phase, and the piecewise approximation for the variable heat capacity of the impregnated paper over the heating phase of the process. The parameters thus determined have been used to check the capability of the whole model (including the hot and cold water storage tanks) to replicate the behaviour of the actual production process. The validated model has been employed to investigate the influence of a storage tank for the spent hot water (a so-called saver) to be employed in the early stage of the heating phase of the process, with the load and discharge phases lasting three or five minutes each. Computations demonstrate that the process times increase between two and four minutes, with average savings slightly higher for higher saver operation times (τ_s=5 min) and averaging 34 % of the total energy consumption.

References

[1] Formica Corporation 2012 *Fabrication Advice* (Auckland: New Zealand)
[2] Žaludek M, Rusnáková S, Fojtl L and Rusnak V 2014, "Experimental study of mechanical behavior of HPL composites", EAN 2014 – *52nd International Conference on Experimental Stress Analysis*.
[3] Martins J M, Almeida M L, Coelho C M, Ferra J and Carvalho, L H 2015 A new methodology to evaluate the cure of resin-impregnated paper for HPL *J. Adhesion* **91** 10-11 792-800
[4] Roberts R J and Evans P D 2005 Effects of manufacturing variables on surface quality and distribution of melamine formaldehyde resin in paper laminates *Composites* A **36** 1 95-104
[5] Valdiserri P, Lorenzini M, Pagnoni M 2013 A transient analysis of heat transfer process in HPL production *Proc. 31st UIT Heat Transfer Conference* 651-9
[6] http://www.mathworks.it/products/simulink/
[7] Bird RB Stewart WE and Lightfoot EN 2002 *Transport Phenomena 2nd ed* (New York: Wiley)
[8] Nellis G and Klein S 2008 *Heat Transfer* (Cambridge: Cambridge University Press)

Energy performance of a ventilation system for a block of apartments with a ground source heat pump as generation system

Michael Lucchi, Marco Lorenzini, Paolo Valdiserri

Alma Mater Studiorum – Università di Bologna
DIN Dipartimento di Ingegneria Industriale – Bologna, Italy

Corresponding author email: paolo.valdiserri@unibo.it

Abstract. This work presents a numerical simulation of the annual performance of two different systems: a traditional one composed by a gas boiler-chiller pair and one consisting of a ground source heat pump (GSHP) both coupled to two thermal storage tanks. The systems serve a bloc of flats located in northern Italy and are assessed over a typical weather year, covering both the heating and cooling seasons. The air handling unit (AHU) coupled with the GSHP exhibits excellent characteristics in terms of temperature control, and has high performance parameters (EER and COP), which make conduction costs about 30% lower than those estimated for the traditional plant.

1. Introduction

Building energy consumption amounts to almost 40% of Europe's yearly energy needs and contributes to around 36% of CO_2 emissions within the European Union. In order to reduce these figures, EU member States are required to implement energy efficiency measures for buildings in compliance with the Energy Performance of Buildings Directive [1]-[3].

Different strategies can be pursued to achieve the objective of reducing the fossil primary energy consumption, such as increasing the insulation thickness for the building envelope, using more efficient HVAC systems and exploiting renewable energies [4].

With the increase of insulation thickness for the walls and better airtight window glasses and frames, a growing concern is represented by proper ambient ventilation to ensure high standards of air quality. Owing to the decrease in the overall transmittance of the building envelope, thermal losses due to ventilation constitute a growing share of the total heating/cooling demands, between 20 and 50% depending on building insulation and compactness, air change rate, indoor heat sources, temperature and relative humidity set points and outdoor climatic conditions [5]. Heat Recovery (HRV) from exhaust air in buildings represents a widespread energy saving strategy, which allows to transfer energy between exhaust air and fresh supply air by means of air-to-air heat exchangers integrated within the air handling unit (AHU). Savings in primary energy due to HRV can be significant, depending on the type of air supply system, heat exchanger configuration and power needed to operate the heat recovery unit [6].

Regarding renewable energy, in 2009 the European Parliament recognizes aero-thermal, geothermal and hydrothermal energy as renewable energy source (RES) [7]. The employing of ground-coupled heat pump (GCHP) systems for heating and cooling of new and existing buildings satisfies both the directive

on renewable energy and the reduction of primary energy consumption needs. GCHP can reach higher energy efficiency than those of an air source heat pump system because the soil can provide higher temperature for heating and lower temperature for cooling than air does [8].
The issue of energy saving in HVAC is of particular interest for buildings in locations whose weather conditions on a typical year provide for significant heating and cooling loads, making it necessary to operate the air conditioning system almost continuously over the whole year, switching from heating mode to cooling mode depending on the season. This is the case in Italy for the river plain area in the North, especially its eastern part, where winters are cold and summers are very warm and humid, as for Bologna [9].
Traditionally, AHUs for the maintenance of the thermal comfort in buildings consist of heating coils fed with water heated in a gas boiler and cooling coils fed with water chilled in an air-cooled chiller (when intermediate fluid configuration is preferred instead of one with direct expansion); moreover HRV systems are increasingly integrated within the AHU for a better overall efficiency[10].
The aim of this work is to compare the heating and cooling energy performance of a traditional AHU serving a block of flats in Bologna with those of an AHU employed for the same task but coupled with a ground-source heat pump for the production of both hot and cold water. Both systems analyzed make use of sensible heat recovery systems for the preheating or precooling of supply air, depending on the operation mode, and thermal energy storage tanks in order to uncouple hot and cold production systems from sudden load variations within the building.
A numerical investigation method has been adopted, carrying out simulations over a typical year exploiting the huge potential of TRNSYS, a lumped-parameter simulation program for energy system transient analysis [11].
The temperature control inside the building has been analyzed in terms of displacement from the defined set point; besides, the temperature oscillations inside the heat storage tanks, the consequent cycling and the performance of GSHP in terms of COP and EER have been investigated. Finally, the annual energy consumption and operational costs for the two technical solutions are evaluated and compared.

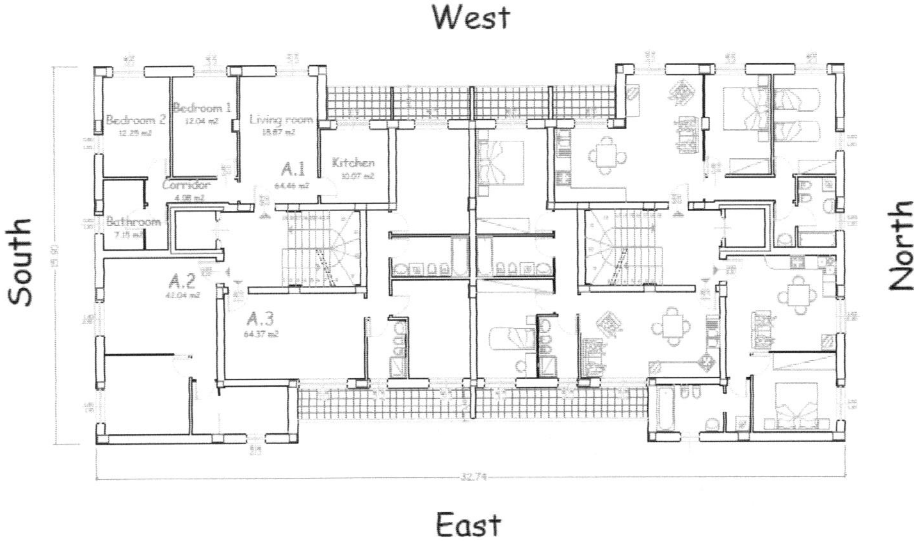

Figure 1. Floor drawing of the block of flats.

IOP Conf. Series: Journal of Physics: Conf. Series **796** (2017) 012034 doi:10.1088/1742-6596/796/1/012034

2. Simulation

The building chosen for the simulation consists of a block of flats located in Bologna. Its structure is typical of the last decade of the 20th century for this kind of constructions in the area. It is a three-storied building with six flats per floor of three types which differ in the net surface area (A_1=64.46 m^2, A_2=42.04 m^2 and A_3=63.37 m^2 respectively) and with the East-West direction acting as a mirror line. The total net surface of the flats is A_{nf}=1025 m^2, whereas the total area, including walls, stairs and lifts is A_{tf}=1369 m^2. The net volume of the building to be conditioned is V_n=2768 m^3, and the envelope through which heat is exchanged with the ambient is A_n=1933 m^2. Fig.1 reports a floor drawing of the block of flats.

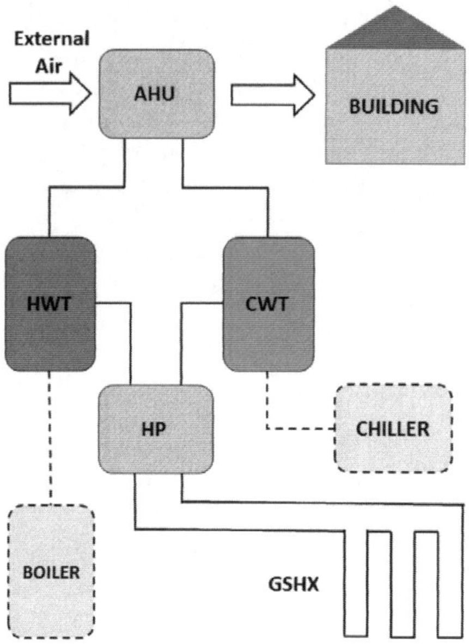

Figure 2. Sketch of the HVAC systems.

The two HVAC systems investigated serve the whole building and share some common components, namely the storage tanks for hot (HWT) and chilled (CWT) water, as it emerges in Fig.2, where a sketch of the HVAC system in the two configurations has been reported. Each flat is served by a ventilation system supplying treated air to the load by means of an independent air handling unit (AHU), which draws the thermal power needed from the HWT or from the CWT depending on the time of the year. The HRV within the AHU exchanges sensible heat only (i.e. no latent heat due to condensation/evaporation of moisture in the air is recovered) and its efficiency can be calculated as:

$$\varepsilon_S = \frac{T_{PR}-T_o}{T_e-T_o} \tag{1}$$

ε_S is the sensible heat efficiency, T_{PR} is the supply air temperature after the heat recovery process, while T_o and T_e are respectively the outdoor and the exhaust air temperatures. In this study, a fixed-plate cross-flow heat recovery system has been considered and a value of 0.7 has been adopted for the sensible efficiency, according to indications given in [9] and assuming a medium-to-high efficiency heat exchanger.

The water tanks act as buffer systems, to dampen temperature oscillations due to sudden changes in outdoor conditions and to cover peak loads that might occur. The tanks are modeled as stratified, with a volume capacity V_t=4 m^3 each. Depending on the choice of the HVAC system, the HWT and CWT are fed separately by a gas boiler and an air-cooled chiller, or together by a water-to-water heat pump (HP),

which exchanges thermal energy with the ground by means of a ground source heat exchanger (GSHX), where a mixture of water and glycol circulates. On the basis of the design indications given in [12], the geothermal field consists of 10 vertical probes, whose buried length is 100 m; the total surface area occupied by the geothermal field is about 490 m^2. The nominal thermal power the gas boiler can supply is P_b=56 kW with an overall efficiency η_b=0.85. Referring to an existing unit, the chiller can supply a cooling power P_c=42.6 kW with a rated energy efficiency ratio EER_c=3.15, considering water inlet and outlet temperatures of 12°C and 7°C at the evaporator and air temperature of 35°C at the condenser; in the simulation the data of the existing unit were employed in order to establish EER values as a function of the outdoor air temperature, as reported in Fig.3, where EER curves are plotted for different values of the evaporator outlet temperature.
The ground source heat pump was also modeled to replicate an actual machine. The nominal cooling power is P_{hph} = 45 kW with nominal EER=4.10, when the temperatures of the fluid at the evaporator are 12/7°C and the temperatures of the fluid at the condenser are 30/35°C; the nominal heating power is P_{hpc} = 53 kW with a coefficient of performance COP = 4.10, when the temperatures of the fluid at the evaporator are 10/7°C and the temperatures of the fluid at the condenser are 40/45°C . Actual EER and COP curves were deduced from builder's data to better replicate the machine's actual behavior.
The temperature set points of water feeding the heat exchangers within the AHU are 40°C and 11°C for heating coil and cooling coil respectively, assuming a dead band of ±2°C for temperature control purpose.

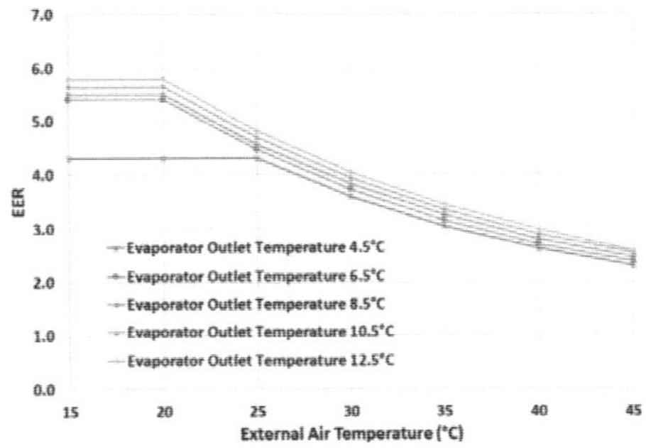

Figure 3. EER curves for the chiller.

Fig.4 shows EER curves versus condenser inlet temperature for different values of the evaporator inlet temperature, while COP values versus evaporator inlet temperature for fixed condenser inlet temperatures are plotted in Fig. 5 for the working ranges of interest.

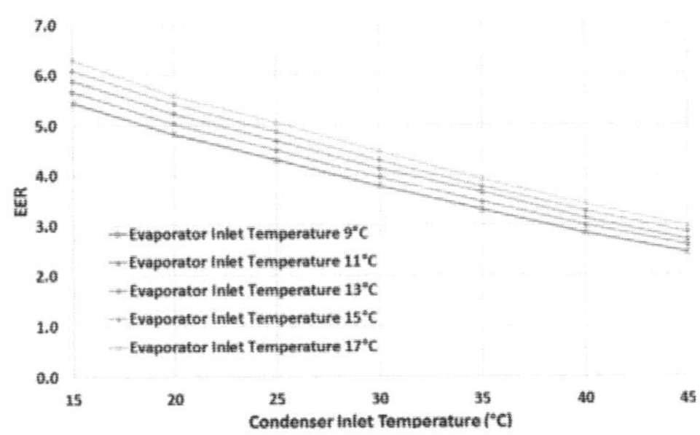

Figure 4. EER curves for the heat pump.

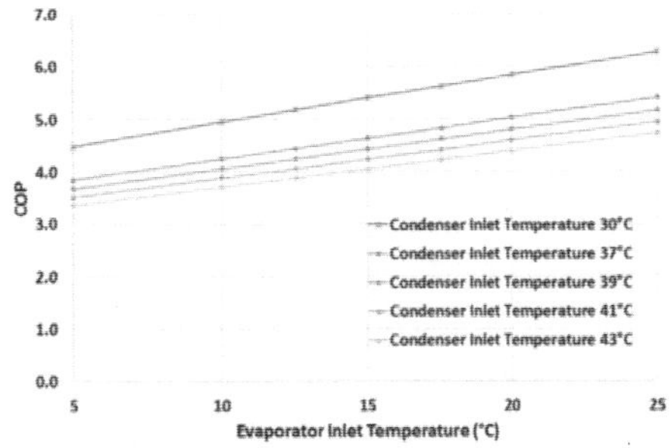

Figure 5. COP curves for the heat pump.

The layout of the building-HVAC system for the case of the ground-source heat pump is shown in Fig.6.

The AHU is not shown in Figure 6 as a single block: its components are represented instead, namely the heating and cooling coils, the heat recovery unit, the fan coils and the exhaust air mixer. The ground source heat pump is also shown separately from the borehole heat exchanger loop. The control system to keep temperature in all rooms at their set values is not shown in the picture. A similar setup was employed for the traditional system with gas boiler and chiller, not shown here.
All building details were modelled for the purpose of thermal analysis by determining their U-values, a summary of which is reported in Table 1 together with the corresponding thickness.

Table 1. U-values and thickness for different structures.

	Thickness (m)	U-value ($W m^{-2} K^{-1}$)
Outer Walls	0.36	0.45
Inner Walls	0.24	0.54
Ground Floor	0.50	0.24
Inner Ceilings	0.45	0.75
Windows	-	2.83

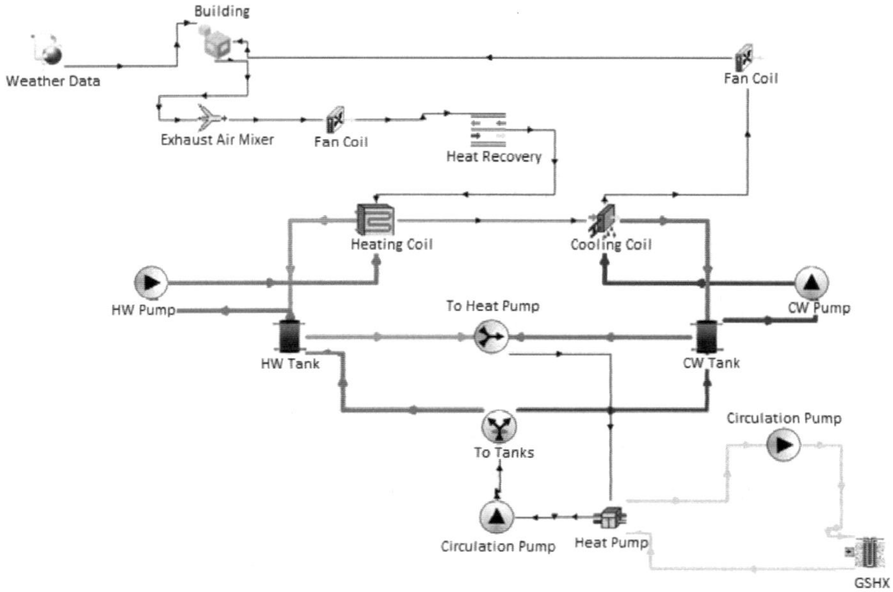

Figure 6. Layout of the heat pump system and load.

In the Bologna region, regulations concerning the heating season define it as the period from October 15th to April 15th, while the cooling season is the remaining part of the year. During the heating season, room temperature is set at $T_{set,h}$=20°C, and heating operates from 6:00 am to 12:00 pm. In the remaining hours, heating is switched on only when temperature in any one room drops below T_l=18°C. During the cooling season, the set point temperature is $T_{set,c}$=26°C, and the system operates in the same hours as in winter, but it is never switched on between 0:00 am and 6:00 am. Over the whole year, whenever outdoor temperature is between 20°C and 26°C, the system operates in free cooling mode. Weather data for the city of Bologna were obtained from the Meteonorm database.

IOP Conf. Series: Journal of Physics: Conf. Series **796** (2017) 012034 doi:10.1088/1742-6596/796/1/012034

For both system configurations, simulations were run at time steps of 5 minutes over the whole year, spanning both the heating and cooling seasons. Several data were collected as output and the most significant shall be discussed in the following section.

3. Analysis of the results

Temperature control for the rooms was more than satisfactory, as exemplified in Fig.7. The outdoor and bedroom temperatures over a 24-hours period are plotted for two days in winter and two in summer. Two of the chosen days are representative of the average temperature of the period (November 2nd and June 30th for winter and summer respectively) while the remaining two are those when outdoor temperatures reached their extreme values (January 13th and August 18th respectively): for all cases temperature conditions in the bedroom were quite stable, also owing to the capacitive effects of the storage tank. Even when outdoor temperature dropped below the design value of -5°C (night hours of Jan 13th, where it was -6°C on the average) conditions in the room never fell below 18°C.
The other rooms had a similar behavior. In Fig.7 it is also possible to notice how bedroom temperatures displayed a smoother trend during days of average conditions while they had a slightly oscillating behavior during the days of extreme conditions.

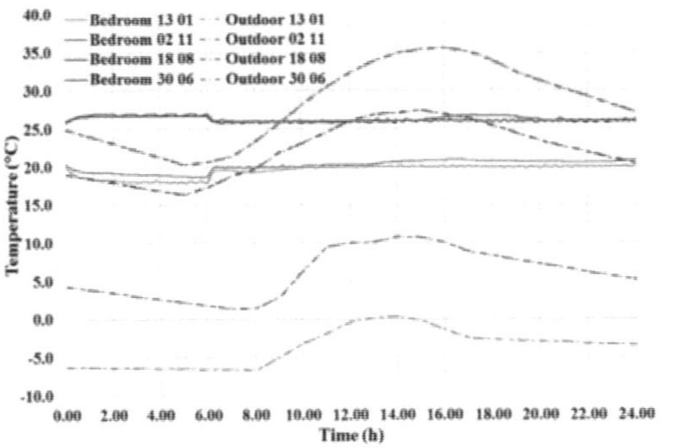

Figure 7. Outdoor and room temperature for different days of the year.

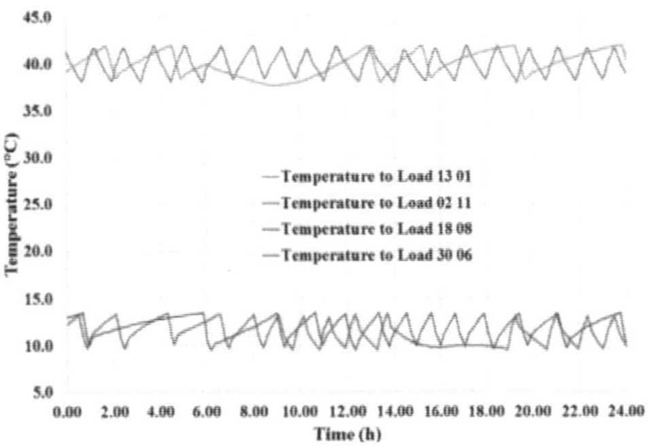

Figure 8. Temperature of water from storage tank to load.

To assess the behavior of the storage tank, the temperature of the water fed to the load (building) was analyzed during the same days as for temperature control of the thermal zones, as shown in Fig.8. In this case, temperature related to winter days are higher (as thermal energy must be supplied to the building) than those occurring in summer days (when the thermal load must be removed from the building). During the days representative of the period, the water temperature oscillates regularly between 40±2°C in winter and 11±2°C in summer, which indicates how the tank is working properly as thermal energy buffer. During days with harsher temperature conditions, the amplitude of temperature oscillations is still around ± 2 °C, but frequency decreases, especially in winter, due to the larger energy demands of the load. The trend of temperature in the early hours of Jan 13th, as the set-point increases to 20°C (at 6:00) differs from that of the representative winter day (Nov. 2nd), because of the increased energy demands of the building. The same happens in the summer on Aug 18th , but in the afternoon hours, when outer temperature reaches its maximum.
The performance of the heat pump is analyzed for the representative days in winter and summer. Figure 9 shows the power to load, EER and outdoor temperature for June 30th. It is evident how the thermal energy removed from the load follows the trend of the outdoor temperature: as the latter increase so does the former. Moreover, when the heat pump operates (which corresponds to a decrease of the water temperature to load, with reference to Fig.8) there is an increase in the amount of thermal energy removed. The EER of the heat pump during operating times are rather high, averaging around 5.0. A similar behavior occurs during winter operation, as shown in Fig.10.

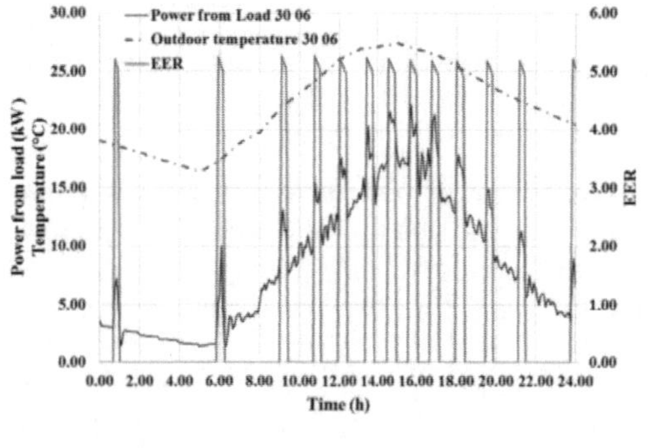

Figure 9. Summer performance of heat pump.

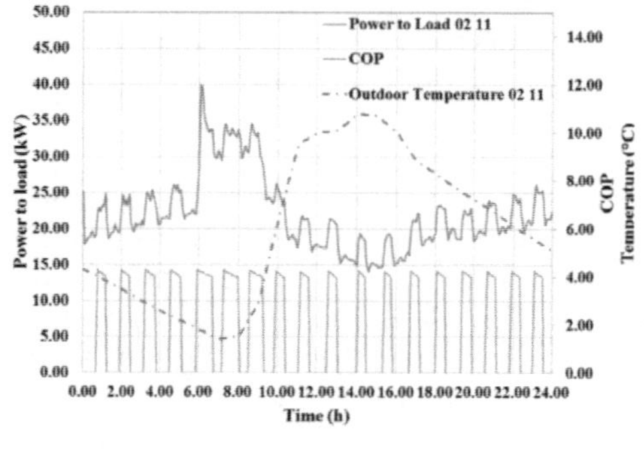

Figure 10. Winter performance of heat pump.

This time the power to load and the outdoor temperature have opposite trends: when the temperature outside decreases the thermal energy which needs to be supplied to the building increases. In the same way, while the heat pump operates, the power supplied to load increases due to the increase in the temperature of the water from the storage tank to the load, as shown in Fig.7.

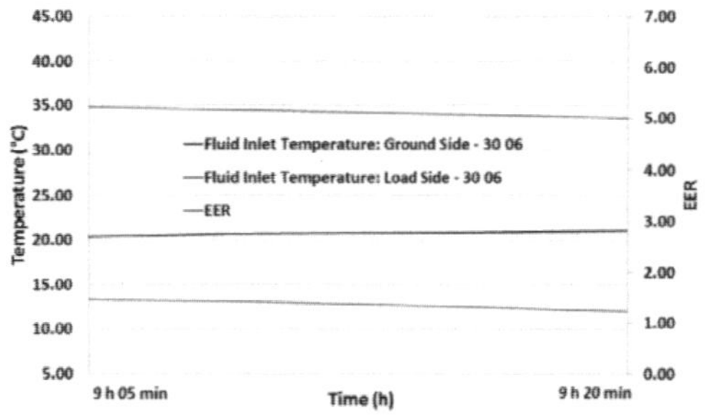

Figure 11. EER in a representative summer day as function of working fluids inlet temperatures. Particular of an operating period.

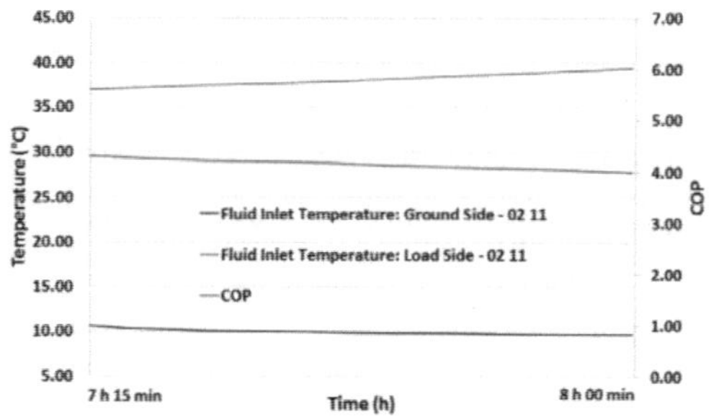

Figure 12. COP in a representative winter day as function of working fluids inlet temperatures. Particular of an operating period.

The values of the coefficient of performance are lower than those obtained for the EER, but they are still satisfactory. This difference is largely due to the smaller temperature difference at which the ground-source heat pump operates during the summer months as opposed to the operating conditions in winter, as can be seen in Fig.11 and Fig.12, where the EER and COP trends are shown for different fluid inlet temperatures at the two ends of the heat pump for two representative days, focusing on two operating periods. At the ground side, the fluid inlet temperature is about 20°C in summer and 10°C in winter, with a soil temperature (not shown here) which is about 15°C throughout the year.

The above results show that the geothermal plant and storage tanks are correctly designed to cover the thermal demands of the block of flats over the entire year. The ground source heat pump operates with a thermal reservoir, the soil, which is on the average at a far more stable and higher temperature than ambient air (which is employed in air-to-air and air-to-water heat pumps).

A qualitative analysis on the possibility of soil thermal depletion has also been carried out: during the heating season, the thermal energy subtracted from the ground was estimated in E_{g_win} =64.416 MWh, while during the cooling season the thermal energy supplied amounts to E_{g_sum}=43.768 MWh.

Neglecting solar radiation at ground surface and conduction from surrounding soil, a ground temperature decrease is to be expected.
The seasonal electricity consumption of the GSHP is L_{el_win} = 22.334 MWh in winter and L_{el_sum} = 7.324 MWh in the summer, which yield a seasonal coefficient of performance SCOP=3.88 and a seasonal energy efficiency ratio SEER = 4.97. The high values achieved are due to the comparatively high and constant temperature of the outside heat source/sink (the ground) with respect to ambient air.

Table 2. Electricity and natural gas yearly consumption and costs.

	Traditional	GSHP	Δ	Δ %
Annual electricity consumption (kWh)	37410	58809	+ 21399	+57
Annual natural gas consumption (Sm^3)	11417	0	- 11417	-100
Total annual cost (€)	16615	11762	- 4853	-29
Average annual cost per flat (€)	923	653	- 270	-29

The operating costs for the ground-source heat pump have been compared to those of the traditional heating and cooling solution which employs a gas boiler and an air-cooled chiller for winter and summer conditions respectively. The results are shown in Table 2.
The total energy demand is obviously the same for both systems, but the GSHP uses electrical energy only, whereas the boiler-chiller pair employs natural gas to operate in the winter months. The amount of electrical energy in kWh for the chiller and GSHP and the amount of natural gas for the boiler over a year are shown in the second and third row from the top, with the difference reported as absolute and percent values in the last two columns right. The last two rows show the total yearly costs and the yearly costs per flat. The price of electricity and natural gas for Italy have been estimated to be 0.20 € $(kWh)^{-1}$ and 0.80 € $(Sm^3)^{-1}$ respectively, taxes and duties included. From the results above it turns out that the use of an AHU coupled with a GSHP instead of a traditional plant brings yearly savings in operating costs of about 30%; however, the cost analysis carried out is not sufficient to justify the installation of such a system, because it must be accompanied with a detailed economic analysis of the initial investment.

4. Conclusions

In the present work a ground source heat pump system for heating and cooling a block of flats located in northern Italy (city of Bologna) has been numerically simulated. The energy consumption results have been compared with the ones obtained by employing a traditional system composed by a gas boiler for winter heating coupled with an air-cooled chiller for summer cooling. The simulations show that the yearly primary energy and the operating costs for the ground source heat pump system are lower than the ones for the traditional boiler-chiller combination system. Moreover, the GSHP systems satisfy the requirement of the European directive on the employing of renewable energy technology for space heating and cooling.

References

[1] Directive, 2002/91/EC of the European Parliament and of the Council of 16 December 2002 on the energy performance of buildings.
[2] Directive 2010/31/EU of the European Parliament and of the Council of 19 May 2010 on the energy performance of buildings.
[3] European Commission, 2020 Vision: Saving energy, 2011, web access at https://www.energy.eu/publications/saving-energy-2011.pdf.
[4] De Boeck, L., Verbeke, S., Audenaert, A., De Mesmaeker, L., Improving the energy performance of residential buildings: A literature review, *Renewable and Sustainable Energy Reviews* **52** (2015) 960-975.

[5] Valdiserri, P., Study of the energy performance of a retrofitting office, Building Simulation Applications *BSA 2015, 2nd IBPSA-Italy Conference* Bozen-Bolzano, 4th–6th February 2015.
[6] Mardiana-Idayu, A., Riffat, S.B., Review on heat recovery technologies for building applications, *Renewable and Sustainable Energy Reviews* **16** (2012) 1241-1255.
[7] Directive 2009/28/EC of the European Parliament and of the Council of 23 April 2009 on the promotion of the use of energy from renewable sources
[8] Sarbu, I., Sebarchievici, C., General review of ground-source heat pump systems for heating and cooling of buildings, *Energy and Buildings* **70** (2014) 441–454.
[9] Roulet, C., *Ventilation and Air Flow in Building*, 2008.
[10] *ASHRAE Handbook: Heating, Ventilating and Air Conditioning Applications*, 2011
[11] TRNSYS 17, A transient system simulation program user manual, Solar Energy Laboratory, University of Wisconsin-Madison, USA, 2012.
[12] Eicker, U., *Energy Efficient Buildings with Solar and Geothermal Resources*, 2014.

Comparative CFD simulations of a hydrogen fire scenario

M Nobili and G Caruso
Department of Astronautical, Electrical and Energy Engineering – Nuclear Section
Sapienza, Università di Roma – C.so Vittorio Emanuele II, 244 – 00186 Roma, Italy

Email: matteo.nobili@uniroma1.it

Abstract. Hydrogen leakage and fire ignition and propagation are safety concerns in several industrial plants. In a nuclear fusion power plants the separation of hydrogen and tritium takes place in different steps, among which one or more electrolyzers are foreseen. A fire scenario could take place in case of leakage of hydrogen. In such cases, it is important to prevent the spreading of the fire to adjacent rooms and, at the same time, to withstand the pressure load on walls, to avoid radioactivity release in the surrounding environment. A preliminary study has been carried out with the aim of comparing CFD tools for fire scenario simulations involving hydrogen release. Results have been obtained comparing two codes: ANSYS Fluent© and FDS. The two codes have been compared both for hydrogen dispersion and hydrogen fire in a confined environment. The first scenario is aimed to obtaining of volume fraction 3D maps for the evaluation of the different diffusion/transport models. In the second scenario, characterized by a double-ended guillotine break, the fire is supposed to be ignited at the same time of the impact. Simulations have been carried out for the first 60 seconds. Hydrogen concentration, temperature and pressure fields are compared and discussed.

1. Introduction

This paper summarises the calculations carried out to compare two different computer codes, Fire Dynamics Simulator (FDS) and ANSYS Fluent in the simulation of the release and the ignition of hydrogen after the break of a pressurised pipeline.

Fire Dynamics Simulator [1] is a computational fluid dynamics model of fire-driven fluid flow. The software solves numerically a form of the Navier-Stokes equations appropriate for low-speed, thermally-driven flow, with the emphasis on smoke and heat transport from fires. The core algorithm is an explicit predictor-corrector scheme, second order accurate in space and time. Turbulence is treated by means of Smagorinsky form of Large Eddy Simulation. FDS uses, in its default configuration, a single step chemical reaction whose products are tracked via a two-parameters mixture fraction model. By default, two components of the mixture fraction are explicitly computed, the mass fraction of the unburnt fuel and the mass fraction of the burnt fuel. The code solves the radiative heat transfer using the radiation transport equation for grey rays with a technique similar to finite volumes method for convective transport.

ANSYS Fluent is a well-known commercial CFD code [2] and its main characteristics will not be reported here due to the large amount of selectable tools, schemes and models.

The simulations assume the rupture of a hydrogen pressurized line and the release of the gas in a closed environment in which several electric equipment can lead to safety issue for fires and explosions. It has been assumed that the room is provided with an operating HVAC system and the first 60 seconds after the break have been analysed.

2. Release conditions and geometry description

The hydrogen leakage takes place without the intervention of safety system for at least one minute. The accident is postulated to occur in a pressurized line with a nominal flow rate of 150 Nm^3/h. In normal operation P-T conditions in the rupture section are 7 bar and 40°C.

The release rate in the first milliseconds has been evaluated according to [3], considering the system as a vessel with a hole in the wall. The simplification is safety oriented because it overestimates the release, neglecting the pressure drops of the circuit. The calculated initial release rate is 0.555 kg/s and it drops to the nominal flow rate after 220 milliseconds.

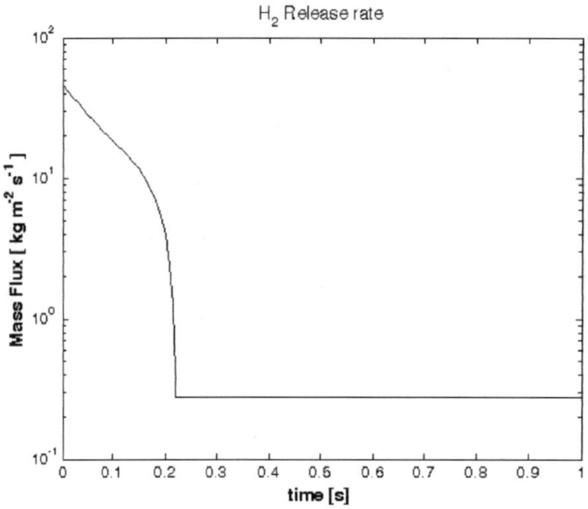

Figure 1. Mass Flow Inlet profile

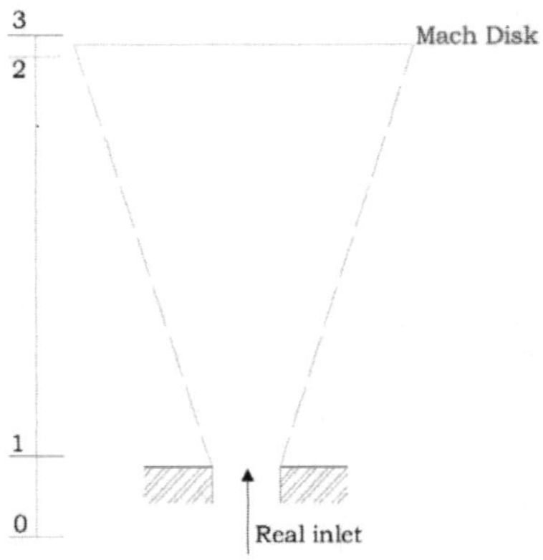

Figure 2. Schematic representation of the single Mach disk model approach

Being the line pressurized at 7 bar, the accidental break results in a critical flow of hydrogen. An ideal single Mach disk model [4] was applied to obtain the inlet dimensions in subsonic conditions. The chocked flow at the real inlet expands isentropically to a location where a single Mach disk forms. All the hydrogen is assumed to flow through the disk and, due to the isentropic expansion, the gas is in supersonic condition just upstream of the Mach disk but it is in subsonic state, but still compressible, at the downstream location. Across the Mach disk the Shapiro's normal shock relations are valid and they allow to calculate the flow conditions and the Mach disk dimensions.

Starting from the stagnation condition in section 0, the following relations allow to calculate the new inlet dimension.

$$P_{1t} = P_0 \left(\frac{2}{\gamma+1} \right)^{\frac{\gamma}{\gamma-1}} \quad P_{2t} = P_{1t} \quad P_2 = P_{2t} \left(1 + \frac{\gamma-1}{2M_2^2} \right)^{-\frac{\gamma}{\gamma-1}} \quad P_3 = P_{3t} \left(1 + \frac{\gamma-1}{2M_3^2} \right)^{-\frac{\gamma}{\gamma-1}}$$

$$T_{1t} = T_0 \left(\frac{2}{\gamma+1} \right) \quad T_{2t} = T_{1t} \quad T_2 = \frac{T_{20}}{1 + \frac{\gamma-1}{2M_2^2}} \quad M_3 = \left(\frac{\frac{M_2^2+2}{\gamma-1}}{\frac{2\gamma}{\gamma-1} M_2^2 - 1} \right)^{0.5} \tag{1}$$

Where the subscript *t* indicates the *total* quantities. The position of the disk cannot be determined by this model, but it is reasonable to consider this virtual inlet in the position of the real one, with a negligible error. FDS imposes to use square surfaces and with this limitation the virtual inlet has the dimension of 0.11×0.11 m^2, while the real pipe has an internal diameter of 0.04 m.

2.1. Computational grids

The assumed dimensions of the room are 5 m (width), 6 m (length) and 3 m (height). Grid size, for both FDS and ANSYS Fluent, has been differentiated according to preliminary considerations on velocity and temperature gradients. In Table 1 the principal features of both grids are summarized.

Table 1. Mesh main features.

	Minimum size	Maximum size	Elements number
FDS	1 cm × 1 cm × 2.5 cm	10 cm × 10 cm × 10 cm	926640
ANSYS Fluent	1 cm × 1 cm × 2.5 cm	10 cm × 10 cm × 10 cm	1569921

In figure 3 and 4 the two computational grids and their details are shown.

Figures 3a, 3b and 3c show FDS mesh and its details and it can be seen that with this code, the domain is made up of rectilinear volumes. Limitations on elements shape is compensated by a greater flexibility on mesh alignment. ANSYS Fluent mesh was obtained with Gambit 2.4 and it is made of both hexahedral and tetrahedral mesh elements. Gambit does not have the possibility to switch from a finer regular mesh to a coarser regular mesh, so the transition needs a mix of hexahedral and tetrahedral elements. This different approach motivates the difference in the elements total number. To maintain the same grid size in the inlet zone, the transition to the outer zone requires a larger number of elements in the Gambit mesh.

3. Results

The codes comparison has been performed in two different situations:

1) Release and dispersion of hydrogen from the break in the pressurized line, without any trigger;
2) Release and early trigger of hydrogen leading to a fire scenario.

In both cases the initial condition of temperature and pressure in the room is 25°C and 1 atm. Wall, both parietal and equipment, are treated as adiabatic. The HVAC system was maintained operative during both the simulations, the inlet and outlet air velocity was 2 m/s.

3.1. Hydrogen release

A hydrogen release from a tank in a closed environment may lead to dangerous concentrations that can cause fast deflagration or even detonation, jeopardizing the integrity of the structure.

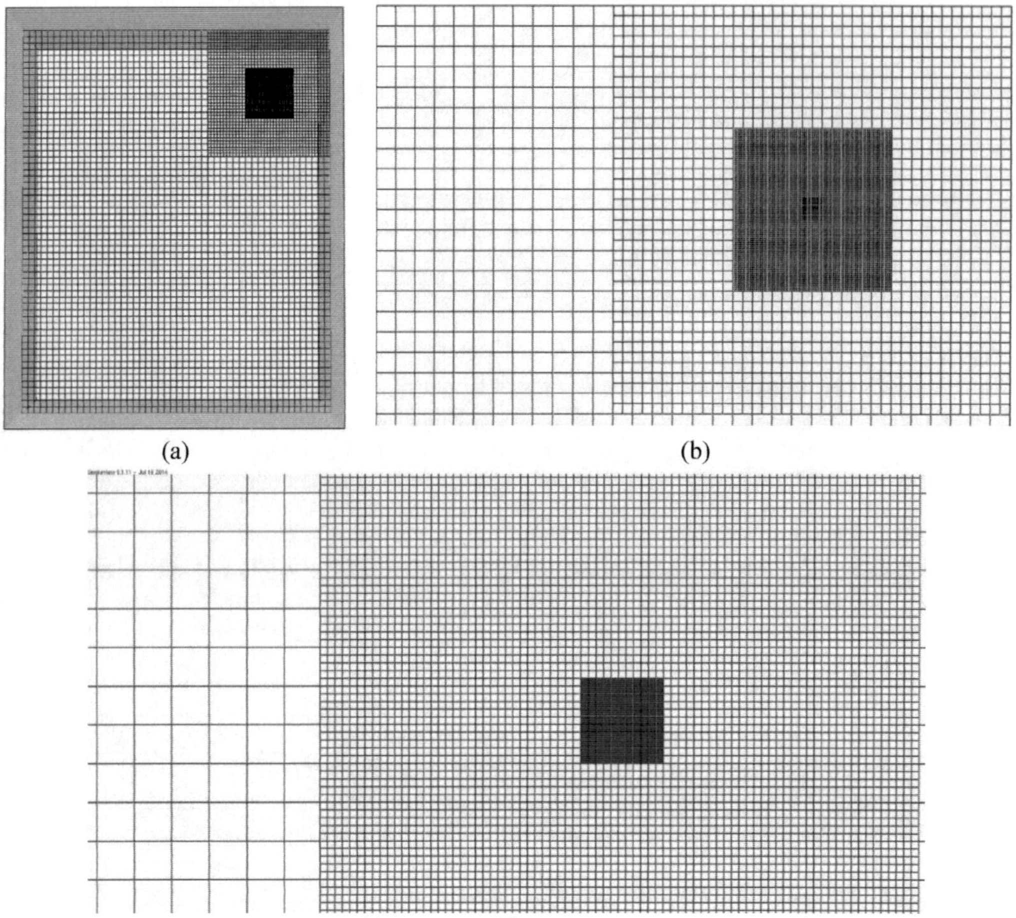

(a) (b) (c)

Figure 3. (a) FDS rectangular grid (b) detail of the intermediate zone (c) Finer zone and inlet mesh

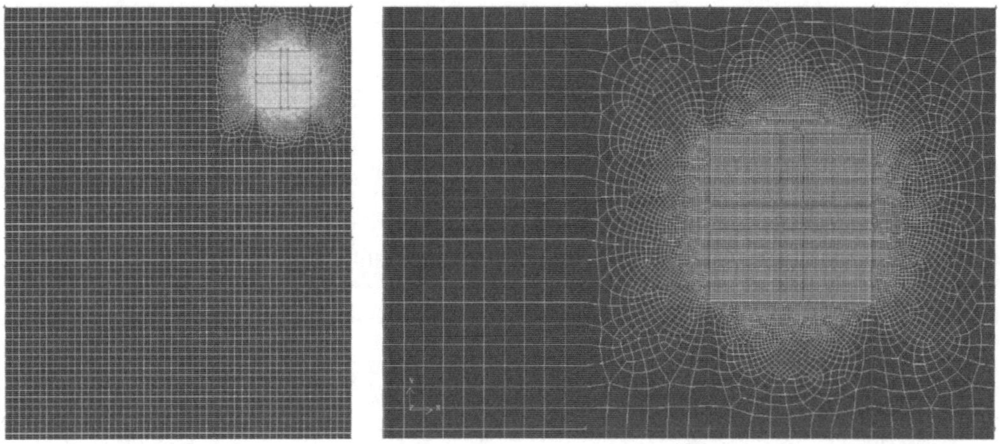

(a) (b)

Figure 4. (a) ANSYS Fluent grid (b) detail of the intermediate and inlet mesh

In some cases, such as nuclear industry, this is a main safety issues due to the potential uncontrolled release of radionuclides.
At 1 bar the minimum concentration necessary to support the combustion, *lower explosive limit LEL*, is about 4.19% in volume and the correspondent *upper explosive limit UEL* is about 74.6% [5].
For this reason, the present CFD tools comparison is focused on the distribution of hydrogen in the room and in particular on the differences in volume concentration of the H_2-air mixture.

Five different parameters are presented to compare the two simulation tools:

1) Total amount of hydrogen in the room and within *LEL* and *UEL;*
2) Volume fraction time trend in 3 control points above the break;
3) Volume fraction along the vertical axis above the rupture, at the end of pressurized discharge (0.22 s);
4) Contours of volume fraction on a vertical plane at 0.1, 1.0 and 10.0 seconds.

3.1.1. Hydrogen total mass and hydrogen mass within flammability limits
Hydrogen flows within the room with the same mass flux in both calculations. The HVAC system removes part of this hydrogen through an outlet fan located at the centre of the ceiling. The comparison of the two calculations shows a greater ventilation effectiveness in the ANSYS Fluent calculations. It carries more hydrogen with the "fresh" air flow, and, initially, it keeps the flammable fraction below the value calculated by FDS. While maintaining a lower amount of fuel in the room, ANSYS Fluent calculation after few seconds foresees a higher mass within the flammability limits.

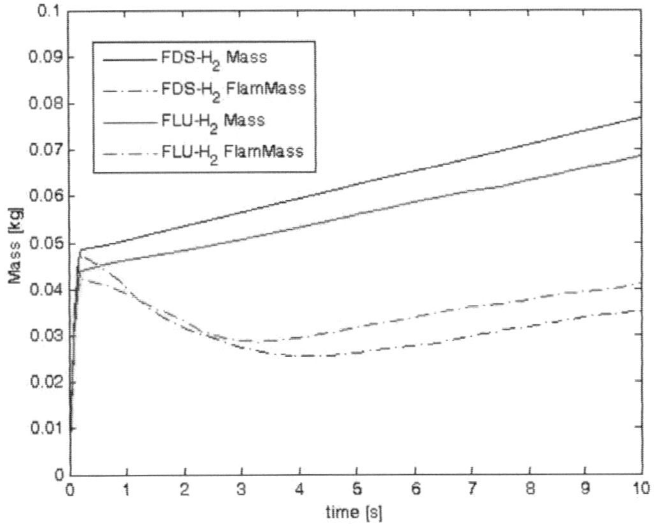

Figure 5. H_2 total mass and H_2 mass within flammability limits

3.1.2. H_2 volume fraction
Time trends of hydrogen volume fraction have been plotted for three different point in the vertical direction, at 1, 2 and 3 meters above the break. Due to its turbulence treatment, the results of FDS calculations are more unstable, but the main trend is replicated by ANSYS Fluent analysis. Higher differences are in correspondence of the inversion in the flammable mass trend. Fluent overestimates the volume fraction at 1 meter along the axis in terms of both peak height and width. The gap between the two calculation for the lower control point is anyway significant and non-negligible. In case of the higher control points 2 and 3 instead, the results obtained with RANS calculation are in accordance with FDS results.

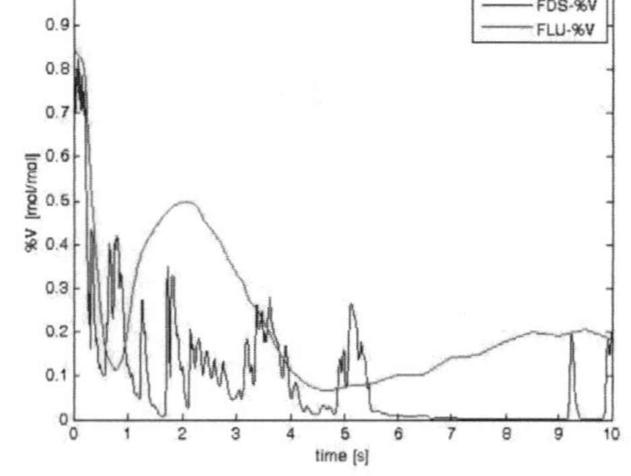

(a)

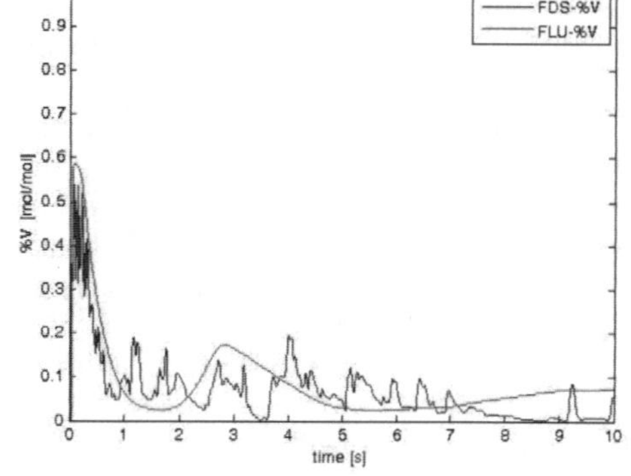

(b)

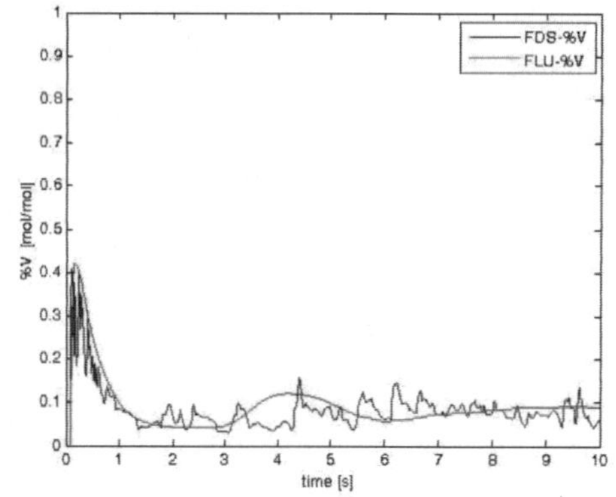

(c)

Figure 6. (a) H_2 %vol vs time at 1m along the inlet axis, (b) at 2m, (c) at 3m

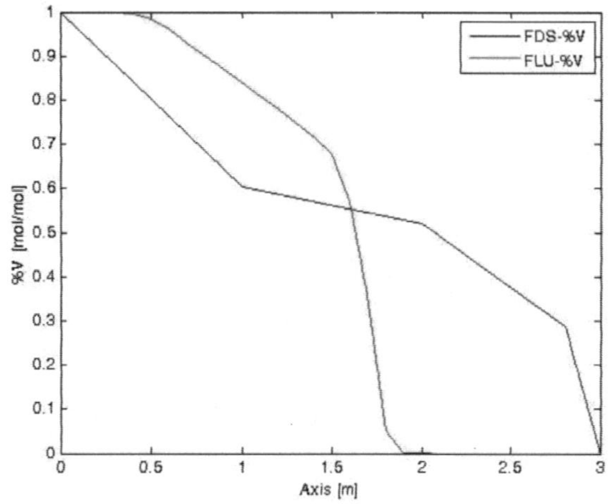

Figure 7. H_2 %vol along the inlet axis at 0.22s (end of pressurized phase)

Figure 7 shows the comparison of the hydrogen volume fraction calculated along the vertical line in correspondence of the break, at the end of the pressurized release phase. At this time the hydrogen cloud shows an opposite behaviour in the two simulation. ANSYS Fluent results show a denser cloud in the lower zone and a rapid decrease at about 1.5 meters. FDS calculations lead to an almost linear behaviour, the volume fraction varies smoother between its maximum at the break level and its minimum at ceiling.

3.1.3. H_2 volume fraction contour

The contours of the hydrogen volume fraction are shown at 0.1s, 1s and 10s on a XZ plane cutting the inlet section (Fig. 8). ANSYS Fluent results are not affected by diffusion near the break for the pressurized release phase, but the dimension of the hydrogen cloud appears overestimated. The shape of the cloud is similar to a typical jet-shape. The obtained contours are comparable with the experimental Rayleigh scattering images of unignited hydrogen leak provided in [6].

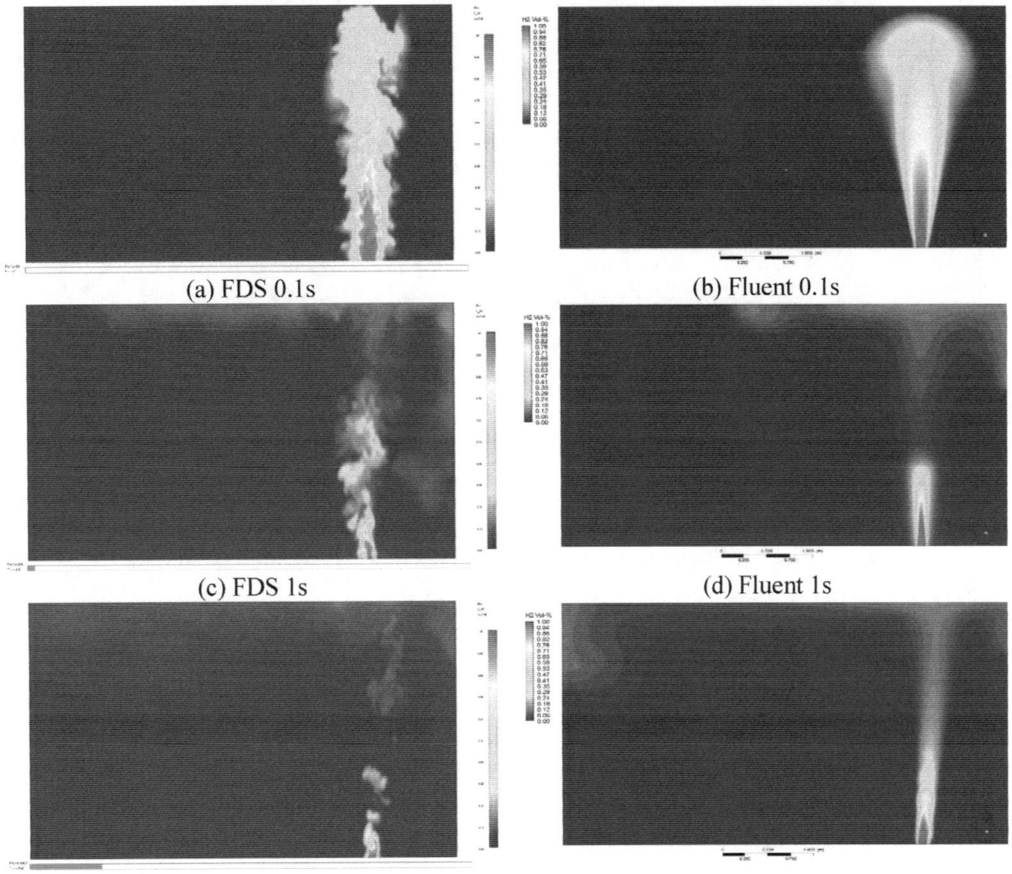

(a) FDS 0.1s (b) Fluent 0.1s

(c) FDS 1s (d) Fluent 1s

(e) FDS 10s (f) Fluent 10s

Figure 8. H_2 %vol contours in a vertical plane

3.2. Fire scenario

In this second comparison calculation, a triggering of fire at the beginning of the release is assumed to occur. Results have been compared with respect to the:

1) Maximum gas temperature in the room;
2) Temperature distribution along the inlet axis;
3) Temperature contours.

3.2.1. Maximum gas temperature vs Time

The results from ANSYS Fluent calculation show a higher peak at the beginning of the transient and, for the first 10 seconds, they overestimate the temperature with respect the FDS results. After an initial instability, the temperature trend calculated with Fluent is smoother, while FDS profile is susceptible of oscillations due to adopted LES turbulence formulation. The difference is constantly about 150°C, as shown in Fig. 9.

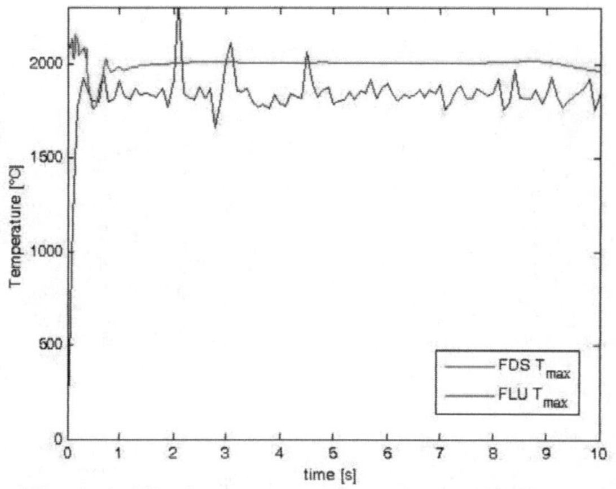

Figure 9. Maximum temperature in the fire flames

3.2.2. Temperature along the inlet axis for different times and temperature contours

The following figures show different temperature distributions along the inlet axis. Figure 10a and figure 10b refer to 0.1 and 0.22 seconds, respectively. During the pressurized discharge phase the temperature difference in the higher part of the flame is quite large, about 1000 °C. The punctual nature of these data sets increases the difference due to velocity fluctuations. The comparison for t=1s (Fig. 10c) shows a very different axial distribution. FDS results lead to an almost flat profile, describing a situation in which seems to be the absence of combustion in the central zone. ANSYS Fluent results show instead a different profile, with higher value in the central segment. It seems that in FDS simulation, at t=1s, the hydrogen cloud is spread in the room and the higher temperature peak are localized in different zones. At t=10s (Fig. 10d) the distributions along the axis is similar, higher temperature are reached around the break and near the ceiling, where higher gas concentrations are achieved.

These behaviours can be seen in the contour maps in Fig. 11.

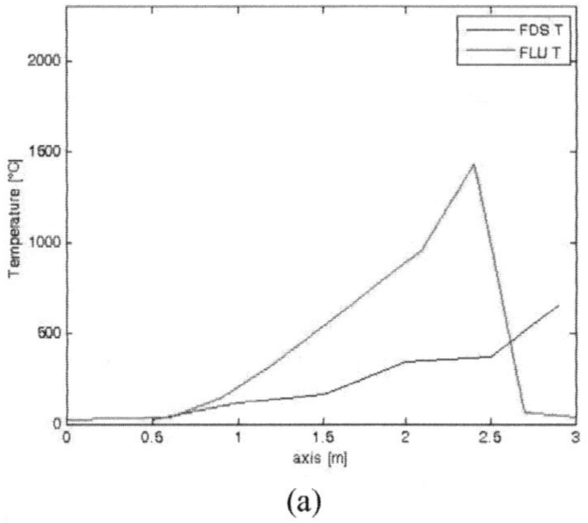

(a)

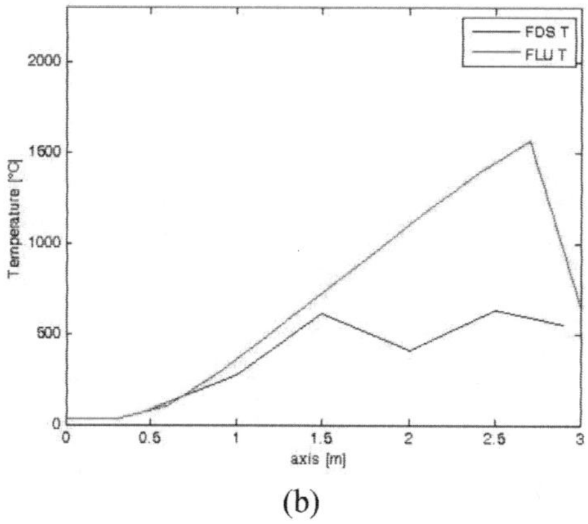

(b)

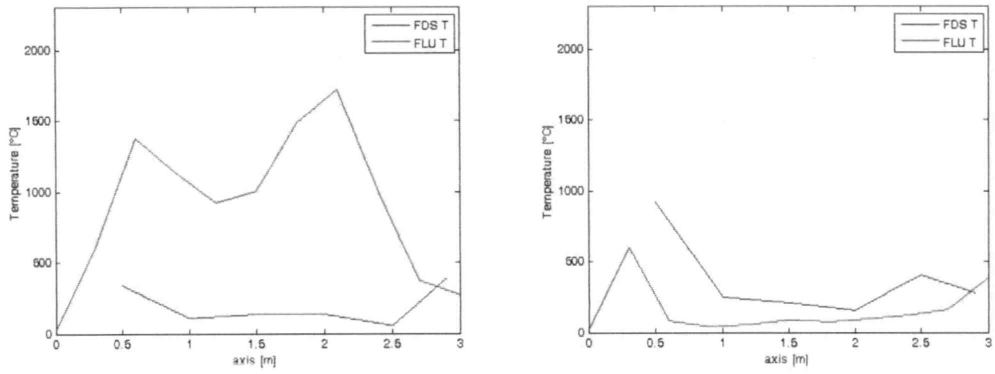

(c) (d)

Figure 10. Gas Temperature along inlet axis – (a) 0.1s (b) 0.22s (c) 1s (d) 10s

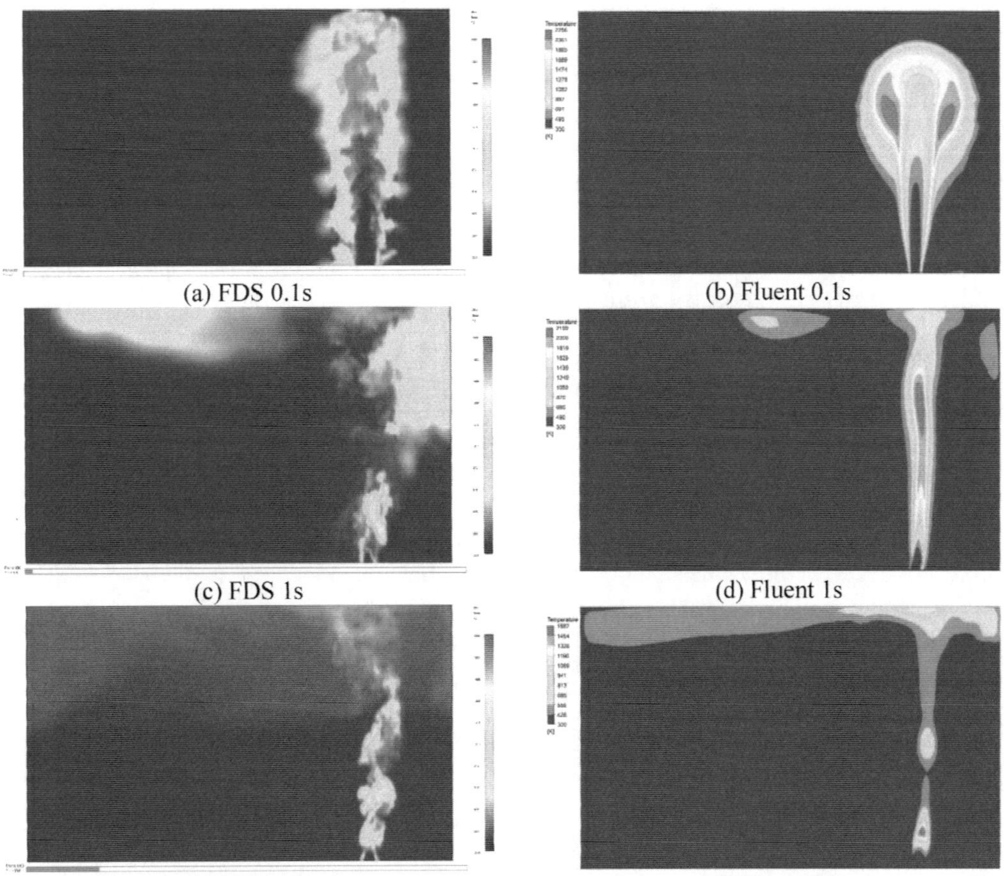

(a) FDS 0.1s (b) Fluent 0.1s

(c) FDS 1s (d) Fluent 1s

(e) FDS 10s (f) Fluent 10s

Figure 11. Temperature contours

4. Conclusions

FDS models and methods have been validated in the past over a large number of experimental results and test cases involving fire scenarios [7,8]. Even if a safety margin has to be taken into account due to unavoidable errors linked to numerical solutions, it has been proven to be a suitable code for safety analyses of fire accident. Its low flexibility and its high simplicity in geometry definition and mesh generation result in a higher computational speed, which allow to execute LES calculation of complex unstable phenomena in a reasonable time. On the other side, combustion models and LES turbulence treatment in ANSYS Fluent require a greater computational effort, which is mainly due to its adaptability. RANS models for turbulence are not, theoretically, suitable for highly unstable phenomena description and simulation, due to the averaging methods which tend to flatten the profiles and the fluctuations. In the present comparison these aspects have been clearly identified but, at the same time, a good match in some results have been obtained. Anyway, a deeper comparative analysis is needed and validation over experimental results for particular cases with clear boundary conditions are due, but in a first approximation the results from RANS based numerical simulation of fire accident can be considered quite reliable.

5. References

[1] Fire Dynamics Simulator *User's Guide*
[2] ANSYS Fluent 15.0 *User's Guide*
[3] van den Bosch C J H et al., *Methods for the calculation of Physical Effects Due to releases of hazardous materials (liquids and gases) – PGS 2 (CPR 14E), TNO "Yellow Book"*
[4] Birch A D, Hughes D J and Waffield F 1986 *Combust. Sc. Tech.* **52** 161-171
[5] Cohen N 1992 *Flammablity and Explosion limits of H_2 and H_2/CO: A Literature Review*
[6] Houf W and Schefer R 2008 *Int. J. Hydrogen Energy* **33** 1435-1444
[7] Fire Dynamics Simulator *Technical Reference Guide*
[8] Floyd J 2006 *Siting Requirements for Hydrogen Supplies Serving Fuel Cells in Non-combustible Enclosures*

Transient Heat Conduction in a Wall Exposed to a Fire: an Analytic Approach

G Casano and S Piva

ENDIF ENgineering Department In Ferrara, Università di Ferrara
via Saragat 1, 44122 Ferrara

Corresponding Author E-mail: stefano.piva@unife.it

Abstract. Fire safety engineering requires a detailed understanding of fire behaviour and of its effects on structures and people. Currently, advanced numerical codes for the prediction of the fire behaviour are available. However, they often require heavy calculations and long times. In this context analytical solutions can be useful for a fast analysis of simplified schematizations. It allows a more effective final utilization of the advanced fire codes. In this contribution, for a separation wall exposed to a fire the temperature history is analysed of the unexposed wall surface. Due to the limitations of the model, the temperature in the fire room changes stepwise, with a final value typical of a post-flashover condition. Nevertheless, with an appropriate choice of the heat transfer coefficient, the thermal action on the surface exposed to fire becomes that due to a fire following the standard temperature-time curve. The solution is then obtained by applying the separation of variables to the heat conduction equation. The problem is made dimensionless and the results are analysed in order to validate their significance. This simplified model allows to obtain useful information on the magnitude of the temperature reached.

1. Introduction

The capability of a reliable prediction of the thermal behaviour of a wall exposed to a fire is a relevant topic in Fire Safety Engineering for different reasons: structural resistance, evacuation and safety of the people, protection of stored materials. The result is that the "procedure of determining the temperature development in members on the basis of the thermal actions and the thermal material properties" (EN 1991-1-2 [1]), synthetically defined as the temperature analysis, is a much studied subject.

For the prediction of the temperature history in load bearing or non-load bearing walls, many models, characterized by different levels of complexity, are available in the literature. For concrete walls (EN 1992-1-2 [2]) and for masonry walls (EN 1996-1-2 [3]), "the calculation methods for thermal response shall be based on the acknowledged principles and assumptions of the theory of heat transfer" and "shall include the consideration of (…) the relevant thermal actions specified in EN 1991-1-2" and "the temperature dependent thermal properties of the materials". However, "the influence of moisture content and of migration of the moisture (…) may conservatively be neglected" and in concrete "the temperature profile in a reinforced concrete element may be assessed omitting the presence of reinforcement". We can conclude that in the current standards for the prediction of the temperature history in a concrete or masonry wall, the solution of the Fourier equation with convective and radiative boundary conditions is regarded as appropriate. For complex geometries, for non linear problems (e.g. thermal properties variable with temperature) or for multiple physical processes (e.g. combustion coupled to heat transfer) different numerical codes are commercially available, while analytical solutions are usually not easily found.

At the same time, when the fire resistance of building components and constructions has to be declared, the building elements are classified in different classes, or combinations of those, by specifying different performance criteria. Normally the fire resistance classification is expressed as a time limit in minutes (15, 30, 45, 60, 90, 120, 180, 240, or 360) which shows the time the performance criteria is fulfilled during a standardized fire test. Specific criteria for the classes are: R Load bearing capacity, E Integrity, I Insulation. By following EN 1992-1-2 [2], "where compartmentation is required, the elements forming the boundaries of the fire compartment, including joints, shall be designed and constructed in such a way that they maintain their separating function during the relevant fire exposure. This shall ensure, where relevant, that:

- integrity failure does not occur (…)
- insulation failure does not occur (…)
- thermal radiation from the unexposed side is limited."

In the present paper we concentrate our attention on insulation (Criterion "I" in the following), that is the ability of a separating element of a building construction when exposed to fire on one side, to restrict the temperature rise of the unexposed face below specified levels. In particular "Criterion "I" may be assumed to be satisfied where the average temperature rise over the whole of the non-exposed surface is limited to 140 K, and the maximum temperature rise at any point of that surface does not exceed 180 K." (EN 1992-1-2 [2]). This temperature rise can be measured or, as previously discussed, predicted. Here, in order to verify the criterion "I", the prediction of the time-space temperature distribution in a separating wall is obtained analytically. A large review of the analytical formulations used for the temperature prediction for walls exposed to fire is given by Wang and Tan [4].

The analytical approach necessarily limits the results for several reasons:

1) The boundary conditions are necessarily simple: imposed temperature, heat flux, third type [5]. In order to simulate an increase of temperature on the heated surface similar to that given by the standard temperature-time curve (EN 1991-1-2 [1]), the choice has fallen on a boundary condition of the third type with a stepwise change of temperature. In this case with an appropriate choice of the heat transfer coefficient it is possible to obtain the desired temperature history on the face of the wall exposed to fire.

2) The geometry are simple, mainly reduced to single layer wall of infinite extension. However, this is a largely diffused typology for the separating elements of load bearing or non-load bearing walls in fire compartments.

Nevertheless, we want to demonstrate that analytical models are important in analyzing conduction phenomena in fire systems. For this aim we investigated the thermal performance of a wall exposed to a fire. The goal is to estimate the time needed to the wall to attain the limiting value of temperature on the face unexposed to fire when the wall is used as the separation from a fire room.

2. Model

The problem is schematically shown in Fig. 1. We want to model the typical laboratory installation for testing the fire resistance of non load bearing walls, as required by EN 1363-1 [6] and EN 1364-1 [7]. A constant properties, single layer, wall is exposed to a fire on one face. On the opposite side of the wall the ambient temperature is constant. The temperature in the fire room changes stepwise, with a final value typical of a post-flashover condition. The boundary conditions on the walls are of the third type. The thickness of the wall is thin with respect of its height and width so that the wall can be considered of unlimited extension and the heat flow one-dimensional in the direction of the thickness.

The solution is obtained by applying the separation of variables to the heat conduction equation, as suggested by Carslaw and Jaeger [8] and Ozisik [9].

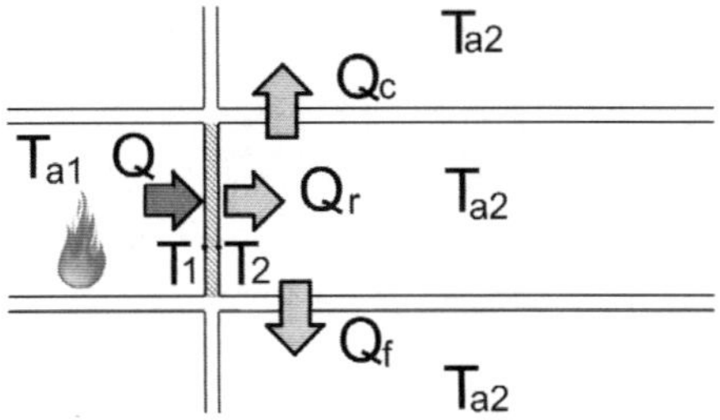

Figure 1. Schematic of the problem.

The problem is stated mathematically as follows:

Conservation of energy: $$\frac{\partial T}{\partial t} = \alpha \frac{\partial^2 T}{\partial x^2} \tag{1}$$

Boundary conditions:

$x = 0$ $$h_1 \left[T_{a1} - T(x = 0,t)\right] = -\lambda \frac{\partial T}{\partial x}(x = 0,t) \tag{2}$$

$x = L$ $$-\lambda \frac{\partial T}{\partial x}(x = L,t) = h_2 \left[T(x = L,t) - T_{a2}\right] \tag{3}$$

Initial condition:

$t = 0$ $$T(x,t = 0) = T_{a2} \tag{4}$$

For the solution of the problem the separation of variables can be applied. So the solution of Eqs. (1-4) can be decomposed as:

$$T(x,t) = u(x) + v(x,t) \tag{5}$$

When substituting Eq. (5) in Eqs. (1-4), it follows that the problem is now given by the solution of the following differential equations:

$$0 = \frac{d^2 u(x)}{dx^2} \tag{6}$$

$$\frac{\partial v(x,t)}{\partial t} = \alpha \frac{\partial^2 v(x,t)}{\partial x^2} \tag{7}$$

where Eq. (6) deals with the Steady state component of the solution and Eq. (7) with its Time dependent component.

The Steady state component of the solution is coupled to the following boundary conditions:

$x = 0$ $$h_1 \left[T_{a1} - u(x = 0)\right] = -\lambda \frac{du}{dx}(x = 0) \tag{8}$$

$x = L$ $$-\lambda \frac{du}{dx}(x = L) = h_2 \left[u(x = L) - T_{a2}\right] \tag{9}$$

Equations (6) and (8-9) are then expressed in dimensionless form, according to the following dimensionless variables:

$$U = \frac{T_{a1} - u(x)}{T_{a1} - T_{a2}} \qquad \xi = \frac{x}{L}$$

Equation (6) and related boundary conditions, Eqs. (8-9), become:

$$0 = \frac{d^2 U}{d\xi^2} \tag{10}$$

$\xi = 0$ $$\left.\frac{dU}{d\xi}\right|_{\xi=0} = Bi_1 U(\xi = 0) \tag{11}$$

$\xi = 1$ $$\left.\frac{dU}{d\xi}\right|_{\xi=1} = Bi_2\left(1 - U(\xi = 1)\right) \tag{12}$$

The dimensionless parameters Bi_1 and Bi_2, are given by:

$$Bi_1 = \frac{h_1 L}{\lambda} \qquad Bi_2 = \frac{h_2 L}{\lambda}$$

The solution of Eqs. (10-12) is given by:

$$U = A + B\xi \tag{13}$$

where:

$$A = \frac{Bi_2}{Bi_1 + Bi_2 + Bi_1 Bi_2} \tag{14}$$

$$B = \frac{Bi_1 Bi_2}{Bi_1 + Bi_2 + Bi_1 Bi_2} \tag{15}$$

The Time dependent component of the solution (Eq. 7) is coupled to the following boundary conditions:

$x = 0$ $$-h_1 v(x = 0, t) = -\lambda \frac{\partial v(x = 0, t)}{\partial x} \tag{16}$$

$x = L$ $$-\lambda \frac{\partial v(x = L, t)}{\partial x} = h_2 v(x = L, t) \tag{17}$$

By coupling Eq. (4) and Eq. (5) we obtain the initial condition, given by:

$t = 0$ $$v(x, t = 0) = T_{a2} - u(x) \tag{18}$$

Equations (7) and (16-18) are then expressed in dimensionless form, according to the following dimensionless variables:

$$V = \frac{v(x,t)}{T_{a1} - T_{a2}} \qquad \tau = \frac{t}{L^2/\alpha} \qquad \xi = \frac{x}{L}$$

Equation (7) and related boundary and initial conditions, Eqs. (16-18), become:

$$\frac{\partial V}{\partial \tau} = \frac{\partial^2 V}{\partial \xi^2} \tag{19}$$

$\xi = 0$ $$\left.\frac{\partial V}{\partial \xi}\right|_{\xi=0} = Bi_1 V(0, \tau) \tag{20}$$

$\xi = 1$ $$\left.\frac{\partial V}{\partial \xi}\right|_{\xi=1} = -Bi_2 V(1, \tau) \tag{21}$$

$\tau = 0$ $$V(\xi, \tau = 0) = -1 + U(\xi) \tag{22}$$

Following the method of separation of variables, the function $V(\xi, \tau)$ can be expressed as:

$$V(\xi, \tau) = \sum_{n=1}^{+\infty} A_n X_n(\xi) \Theta_n(\tau) \tag{23}$$

$X_n(\xi)$ and $\Theta_n(\tau)$ are solution of:

$$\frac{\partial \Theta_n}{\partial \tau} = -\beta_n^2 \Theta_n \tag{24}$$

$$\frac{\partial^2 X_n}{\partial \xi^2} = -\beta_n^2 X_n \tag{25}$$

A general solution of Eq.(24) can be expressed in terms of the exponential function, always tending to zero for $t \rightarrow +\infty$:

$$\Theta_n = \Theta_{0,n}\, e^{(-\beta_n^2 \tau)} \tag{26}$$

A general solution of Eq.(25) can be expressed in terms of trigonometric functions:

$$X_n = C_{1,n} \sin(\beta_n \xi) + C_{2,n} \cos(\beta_n \xi) \tag{27}$$

By substituting Eqs. (26) and (27) in Eq. (23), it follows:

$$V_n(\xi, \tau) = A_n \Theta_{0,n} \left[C_{1,n} \sin(\beta_n \xi) + C_{2,n} \cos(\beta_n \xi) \right] e^{(-\beta_n^2 \tau)} \tag{28}$$

The boundary conditions (Eqs. (20) and (21) give a relationship between $C_{1,n}$ and $C_{2,n}$:

$$C_{1,n} = \frac{Bi_1}{\beta_n} C_{2,n} \tag{29}$$

$$\left[Bi_2 \sin(\beta_n) + \beta_n \cos(\beta_n) \right] C_{1,n} + \left[-\beta_n \sin(\beta_n) + Bi_2 \cos(\beta_n) \right] C_{2,n} = 0 \tag{30}$$

The solution of this system of equations gives:

$$\frac{\sin(\beta_n)}{\cos(\beta_n)} = \frac{\beta_n (Bi_1 + Bi_2)}{\beta_n^2 - Bi_2 Bi_1} \tag{31}$$

It can be easily proved that, after substituting Eq. (29) in Eq. (28), the functions $V_n(\xi, \tau)$ satisfy Eqs. (19-22):

$$V_n(\xi, \tau) = \frac{C'_{2,n}}{\beta_n} \left[Bi_1 \sin(\beta_n \xi) + \beta_n \cos(\beta_n \xi) \right] e^{(-\beta_n^2 \tau)} \tag{32}$$

More generally:

$$V(\xi, \tau) = \sum_{n=1}^{+\infty} \frac{C'_{2,n}}{\beta_n} \left[Bi_1 \sin(\beta_n \xi) + \beta_n \cos(\beta_n \xi) \right] e^{(-\beta_n^2 \tau)} \tag{33}$$

or, more simply:

$$V(\xi, \tau) = \sum_{n=1}^{+\infty} C'_{2,n} X_n(\xi) e^{(-\beta_n^2 \tau)} \tag{34}$$

where:

$$X_n = \frac{1}{\beta_n} \left[Bi_1 \sin(\beta_n \xi) + \beta_n \cos(\beta_n \xi) \right] \tag{35}$$

It can be easily proved that the functions $X_n(\xi)$, with the boundary conditions, Eqs.(26) and (27), guarantee:

$$\int_{\xi=0}^{\xi=1} X_n(\xi) X_m(\xi) d\xi \begin{cases} = 0, & if\ n \neq m \\ \neq 0, & if\ n = m \end{cases} \tag{36}$$

The admissible values of the constant of separation β_n are the roots of the transcendental equation Eq.(31). The search of these roots can be easily executed by means of a numerical procedure.

The constants $C_{2,n}$ that satisfy the boundary condition Eq. (20) are determined by means of the following procedure. From Eqs. (22) and(34) it follows:

$$\sum_{n=1}^{\infty} C'_{2,n} X_n(\xi) = F(\xi) \tag{37}$$

By multiplying both sides of Eq.(37) by X_m and integrating from $\xi = 0$ to $\xi = 1$, on the basis of the property given by Eq. (36), it follows:

$$C'_{2,n} = \frac{\int_0^1 F(\xi) X_n(\xi) d\xi}{\int_0^1 X_n^2(\xi) d\xi} \tag{38}$$

By calculating the integrals in Eq. (37), it follows:

$$\int_0^1 F(\xi) X_n(\xi) d\xi = \frac{1}{\beta_n}\left[\frac{A Bi_1 - B}{\beta_n} + \left(A + B + \frac{Bi_1 B}{\beta_n^2}\right)\sin(\beta_n) + \frac{B-(A+B)Bi_1}{\beta_n}\cos(\beta_n)\right] \tag{39}$$

$$\int_0^1 X_n^2(\xi) d\xi = \frac{Bi_2\left(Bi_1^2+\beta_n^2\right)+\left(Bi_2^2+\beta_n^2\right)\left(Bi_1+Bi_1^2+\beta_n^2\right)}{2\beta_n^2\left(Bi_2^2+\beta_n^2\right)} \tag{40}$$

From Eqs. (33, 38, 39, 40) it follows:

$$V(\xi,\tau) = 2\sum_{n=1}^{+\infty} \frac{\left(Bi_2^2+\beta_n^2\right)\left[\frac{ABi_1 - B}{\beta_n} + \left(A+B+\frac{Bi_1 B}{\beta_n}\right)\sin(\beta_n) + \frac{B-(A+B)Bi_1}{\beta_n}\cos(\beta_n)\right]}{\left(Bi_1^2+\beta_n^2\right)\left[Bi_2 + \left(Bi_2^2+\beta_n^2\right)\right] + Bi_1\left(Bi_2^2+\beta_n^2\right)} \cdot \left[Bi_1 \sin(\beta_n\xi) + \beta_n \cos(\beta_n\xi)\right] e^{\left(-\beta_n^2\tau\right)} \tag{41}$$

Finally, the temperature distribution is given by:

$$T(x,t) = T_{a1} - (T_{a1} - T_{a2})(U - V) \tag{42}$$

3. Validation

In order to guarantee the reliability of the numerical code, two validation exercises are reported.

An important step for the analytical solution is the search of the eigenvalues of the problem, given by Eq. (31). The search of these roots is executed by means of an iterative numerical procedure. In Grigull and Erk [10] are reported the first 8 eigenvalues for a single layer wall of constant properties, where the heating scheme is similar to that proposed in the present paper. The initial condition is an assigned distribution of temperature. On the two sides of the wall we see the same heat transfer coefficient and air temperature. In Table 1 the eigenvalues calculated with the our procedure are compared with those calculated by Grigull and Erk [10].The comparison is carried out for $Bi = 0.2$. It is evident that the agreement between the results is excellent.

As the second validation exercise, a 5 mm infinite steel sheet which forms the boundary between two compartments, both initially at a temperature of 20°C, is analyzed. The temperature of the air in one of the compartments is suddenly increased to 100°C, as proposed by Drysdale [11]. The

Table 1. Comparison between eigenvalues calculated and available in the literature, for the same problem.

Eigenvalue N.	Present work	Grigull and Erk [10]
1	0.3111	0.311
2	1.6320	1.632
3	3.1731	3.173
4	4.7335	4.734
5	6.2991	6.299
6	7.8667	7.867
7	9.4354	9.435
8	11.0047	11.005

convection heat transfer coefficient is h = 20 W/m^{2}K; the thermal properties are: ρ = 7850 kg/m^3, c_P = 460 J/kgK, λ = 45.8 W/mK. The analytical and reference solutions are compared in Fig. 2. The comparison shows a very good agreement.

4. Results and Discussion

The problem now considered is schematically shown in Fig. 1. In particular, as an example of application of the analytical solution derived in Par.2, in Fig. 3 we report the results for a concrete wall characterized by a thickness of 100 mm.

The thermal and physical properties of concrete with calcareous aggregates are those specified by the Standard EN 1992-1-2 [2]. However, in the latter the data of specific heat, thermal conductivity and density are given as a function of temperature in the range $20 \leq T \leq$. 1200 °C; in our calculations we used average values in the temperature range 20 ÷ 800 °C, precisely: ρ = 2197.2 kg/m^3, c_p = 1047.4 Jkg^{-1}K^{-1} and λ = 0.927 Wm^{-1}K^{-1}. The choice of the temperature interval derives from some preliminary analyses for thicknesses in the range $60 \leq L \leq 175$ mm. For the typical thermal loading of these walls we found minimum and maximum temperature values always inside the selected temperature range.

The procedure to define the boundary conditions are different for the two faces. On the surface not exposed to fire the coefficient of heat transfer by convection and radiation is h (x = L) = 9 Wm^{-2}K^{-1} and the ambient temperature T_{a2} = 20 °C. On the surface exposed to fire we used the procedure proposed by Buchanan and Rao Munukutla [12]; they suggested that "the fire-exposed face of the wall is assumed to be slightly cooler than the furnace temperature, by a factor which varies from 0.6 at time zero to 0.96 after one hour of exposure, from recent tests" [13]. Based on the latter results and on the standard temperature-time curve suggested by EN 1991-1-2 [1], incidentally in accordance with ISO 834-1 [14], we obtained for the temperature of the fire-exposed face a logarithmic time distribution:

$$T(x=0,t) = 15.873 + 200.8\ln(t) \tag{43}$$

were T is in °C and t in min. The coefficient of heat transfer by convection and radiation was then chosen so to obtain, for the face exposed to fire following the ISO / EN Standard temperature-time curve, the best approximation to Eq. (43). For the case shown in Fig. 3, the result is h (x = 0) = 27.1 Wm^{-2}K^{-1}. In Figure 3 the whole set of temperature distributions used to determine h (x = 0) are also reported. The calculated temperature on the face exposed to fire (continuous line) is a little big higher than that of Eq. (43) (line with short hatching) in the first hour, then tends to follow with a good accuracy the reference temperature.

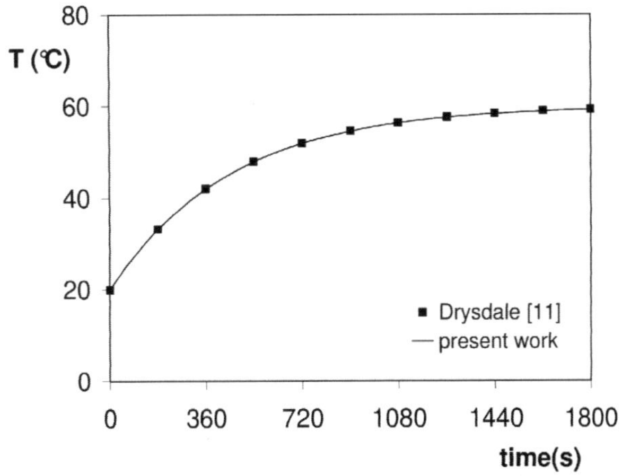

Figure 2. Validation exercise: comparison between analytical (——) and reference (■, [11]) solutions.

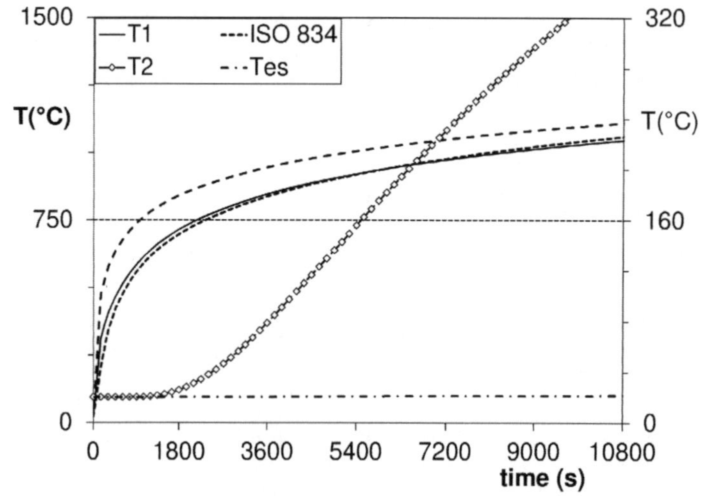

Figure 3. Temperature versus time for a concrete wall (L = 100 mm).

In Figure 3 the temperature evolution on the non fire-exposed face shows a trend qualitatively in full accordance with those shown in the experimental investigations reported in [13]. The face is undisturbed during the initial time, then tends to exhibit a fast increase of the surface temperature. The fire resistance time of this wall, defined as the time required for the unexposed surface temperature to rise 160°C (an increase of 140 °C over the ambient temperature 20 °C), is 91.5 min.

The code based on the analytical solution proposed in Par. 2 has been used to predict the behaviour to fire, in terms of temperature trends on the unexposed surface, for non-load bearing concrete walls of different thickness, taking as the reference the limiting values proposed by EN 1992-1-2 [2]. The temperature rise of the unexposed face is the parameter used to assess the ability of a separating element of a building construction to classify the Insulation "I", when exposed to fire on one side. By following EN 1992-1-2 [2], Criterion "I" may be assumed to be satisfied "where the average temperature rise over the whole of the non-exposed surface is limited to 140 K, and the maximum temperature rise at any point of that surface does not exceed 180 K".

Figure 4 shows on the abscissa the wall thickness and on the ordinate axis the time required to reach the limiting temperature set by EN 1992-1-2 [2] on the surface not exposed to the fire. The Standard EN 1992-1-2 [2] provides the class wall as a function of thickness by means of tables. The range of thicknesses examined is between a minimum of 60 mm and maximum of 175 mm. The data provided by the Standard [2] are given in the form of a bar graph. Differently the results obtained from the analytical solution are provided by means of symbols interpolated by a continuous line of parabolic trend. The two distributions are in good agreement; as expected, the calculated values are always greater, even if though only slightly, than the values proposed by the standard for the different classes of wall. This agreement is more pronounced for small and medium-thick walls, the opposite for very thick walls. Necessarily, the comparison can be only qualitative, because nothing is known about the nature of the material used for the collection of the data shown in EN 1992-1-2 [2]. Nevertheless, the good agreement between two trends guarantees the reliability of the results obtained with the analytical method proposed in Par. 2, at least for a simple layer wall.

5. Concluding Remarks

Interesting analysis and characterization of conduction phenomena in structures exposed to fire can be obtained with analytical models, even if limitations of geometry, properties and boundary conditions are necessarily evidenced.

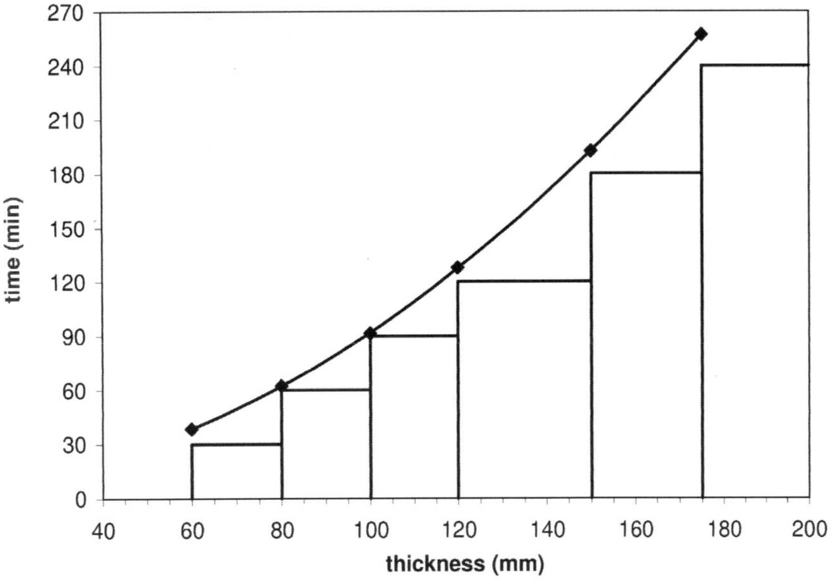

Figure 4. Criterion "I": comparison between calculated (—♦—) and reference (bars) values for non load bearing concrete walls.

The analytical approach proposed here for the study of the performance of a wall separating a room with a fire from a different compartment, has provided useful results. With this analytic solution we are able to easily calculate the time for which the temperature of the surface unexposed to the fire reaches the limiting values given by the current standards to characterize its performance.

Trends and temperature history are those expected for these geometries and materials. A comparison of standard and predicted values for the Criterion "I", indicating the insulation performance of the walls, shows a good agreement; thus the significance of the methodology for a simple prediction of "I" is demonstrated.

6. References

[1] EN 1991-1-2:2009 Eurocode 1: Actions on structures - Part 1-2: General actions - Actions on structures exposed to fire.
[2] EN 1992-1-2:2004 Eurocode 2: Design of concrete structures - Part 1-2: General rules - Structural fire design.
[3] EN 1996-1-2:2005 Eurocode 6: Design of masonry structures. General rules. Structural fire design.
[4] Wan Z H and Tan K H 2007 Temperature prediction for multi-dimensional domains in standard fire, *Commun. Numer. Meth. Engng.* **23** 1035-55.
[5] Ezekoye O A 2016 Conduction of heat in solids *SFPE Handbook of Fire Protection Engineering* ed M J Harley 5th Edition. (New York Springer) chapter 2 pp 25-52.
[6] EN 1363-1:2012 Fire resistance tests Part 1: General Requirements.
[7] EN 1364-1:2015 Fire resistance tests for non-loadbearing elements. – Part 1: Walls.
[8] Carslaw H S and Jaeger J C 1986 *Conduction of Heat in Solids*, second ed. (London Oxford University Press).
[9] Ozisik M N 1993 *Heat Conduction*, second ed. (New York John Wiley & Sons).
[10] Gröber H and Erk S 1961 *Fundamentals of Heat Transfer* (McGraw-Hill).
[11] Drysdale D 2011 *An Introduction to fire dynamics* 3rd ed (New York John Wiley & Sons).

[12] Buchanan A H and Rao Munukutla V 1991 Fire Resistance of load-Bearing Reinforced Concrete Walls, *Fire Safety Science Proc. of the Third Int. Symposium*, pp. 771-780. doi:10.3801/IAFSS.FSS.3-771.

[13] Woodside A, de Ruiter J J and Wade C A 1991 Fire resistance of New Zealand concretes, SR34 Building Research Association of New Zealand.

[14] ISO 834-1:2015 Fire-resistance tests -- Elements of building construction -- Part 1: General requirements.

Critical heat flux in a multi-minichannel heat sink. Effect of the heated length-on-diameter ratio

R MASTRULLO[1], A W MAURO[1,2] and L VISCITO[1]

[1]Department of Industrial Engineering, Federico II University of Naples, P.le Tecchio 80, 80125, Naples (Italy)

[2]e-mail: alfonsowilliam.mauro@unina.it

Abstract. This paper exhibits saturated CHF experimental values obtained with R134a and R1234yf, working at saturation temperatures from 25 °C up to 65 °C (i.e. reduced pressures from 0.16, 0.20 and up to 0.46, 0.54, respectively). The mass flux was let to vary from 150 up to 350 kg/m^2 s. All tests were performed with an aluminum multi-minichannel heat sink, made up of seven rectangular ducts, each of them 2 mm wide, 1 mm high and 35 mm long. Two heated lengths of 25 and 35 mm were structured, in order to study two different L_h/D_{eq} ratios. The results show that critical heat flux is enhanced with increasing the mass flux and decreasing the saturation temperature. A greater L_h/D_{eq} ratio leads instead to lower CHF values.

1. Introduction

The ongoing progresses in technology have made possible the use of densely packed integrated circuits (ICs) [1]-[2], laser mirrors and concentrated solar cells [3]. For the correct functioning of these devices, the heat must be dissipated within very small areas, thus leading to the necessity of removing high heat fluxes that exceed the air and the liquid forced convection cooling power.

Two-phase heat exchange devices represent valuable resources to replace the single-phase cooling methods. Microchannels and minichannels, by exploiting the boiling mechanism of synthetic refrigerants, are well suited for this task and provide large heat transfer surface area per unit length [4].

The operative limit of such system is represented by the critical heat flux (CHF), referring to the situation in which the heated wall loses contact with the liquid phase, yielding to a sudden drop of the heat transfer coefficient and therefore a sharp rise of the wall temperature, which may exceed the melting point of the channel material [5]. The capability to determine CHF is therefore of great importance, since a heat sink should be designed with a considerable amount of safety relative to the maximum heat flux dissipation that can be handled or the minimum refrigerant flow rate.

Different geometries, fluids and operating conditions have already been presented in scientific literature [6], although still limited sources cope with new low-GWP refrigerants at medium-to-high saturation temperatures. Furthermore, most of the studies are related to single channel geometries [7]; even if the basic CHF phenomenon in rectangular minichannels could be similar to that in round tubes, the existence of non-heated walls and possible mal-distribution [8] may lead to significant discrepancies among the experimental results available in literature. It is also known that the CHF is related to the channels heated length L_h and its impact has been usually measured with the use of the heated length over the equivalent diameter of the channel L_h/D as dimensionless parameter. Generally,

CHF decreases with increasing L_h/D and some researchers [9]-[10] have reported threshold values of L_h/D beyond those it does not affect the critical condition anymore.

Wojtan et al. [11] investigated different conditions of boiling R134a and R245fa to obtain the saturated critical heat flux in 0.5 and 0.8 mm internal diameter tubes with two different heated lengths. They found that with the extension of the heated length from 20 to 70 mm, the drop of the CHF value was very intense (up to 40 W/cm^2).

Tanaka et al. [7] collected existing data and developed a CHF correlation for thin rectangular channels applicable to a wide range of heated lengths. The authors found that CHF was highly affected by the L_h/D parameter, especially for low mass flux condition, in which CHF dropped more than 350% passing from $L_h/D = 50$ to 179.

Wu and Li [10] also took experimental CHF values from existing databases, observing that CHF greatly decreased with L_h/D, for small values of the heated length. At a threshold of $L_h/D = 150$, it presented a negligible effect on the saturated CHF.

In this paper, some experimental saturated CHF results are presented in an aluminum multi-minichannel heat sink for two different working fluids, R134a and its low-GWP substitute R1234yf. With the same equivalent diameter for each channel of 1.33 mm, two different test section set ups have been arranged in order to achieve different heated lengths of 25 and 35 mm. The resulting L_h/D ratio is then 18.75 and 26.25, respectively, and its effect is analyzed in the form of diagrams and boiling curves. All tests present a saturation temperature range of 25-65 °C and a mass velocity range of 150-350 kg/m^2 s.

2. Experimental test rig and measurements uncertainties

The test facility has been assembled in the Refrigeration Laboratory, at the Industrial Engineering Department, Università degli Studi di Napoli Federico II. The whole set-up is made up of two different closed loops. In the main refrigerant loop, the working fluid conditions in terms of test section inlet saturation temperatures, sub-cooling and mass flow rate, are set and monitored. The latter is a water closed loop which is thermally controlled by a thermostatic bath. A schematic of the entire test rig is displayed in Fig. 1.

2.1. Refrigerant loop and water loop description

The main loop is the black line depicted in Fig. 1. The working fluid flows is driven by means of a magnetic gear pump through a Coriolis mass flow meter up to the test section, in which a Pt100 resistance thermometer and a piezoelectric pressure transducer measure the inlet thermodynamic conditions. The pressure drop across the test section is measured with a differential pressure transducer and another Pt100 resistance thermometer is placed at the outlet section. A manually-controlled throttling valve is employed in order to adjust the system pressure and the mass flow rate during the experiments. The refrigerant in two-phase conditions is condensed with a plate heat exchanger and then flows into a liquid receiver. The working fluid is finally sub-cooled by means of a double pipe heat exchanger before the pump suction head which closes the loop.

The magnetic gear pump may elaborate volumetric flow rates within a range of 1.3-2.5 dm^3/min, changing its rotating speed via inverter from 1650 rpm to 3400 rpm. In order to achieve very low mass flow rates flowing into the test section, the refrigerant loop is equipped with a by-pass circuit, manually controlled, which is able to recirculate the refrigerant surplus directly into the liquid receiver.

The secondary loop provides demineralized water to the sub-cooler and the condenser. Both the double pipe and the plate heat exchangers may be excluded from the water flow by employing two ball-cock valves. The desired water temperature is obtained with a remote-controlled external thermostatic bath. The secondary fluid specific volume variations are restricted thanks to an expansion vessel.

Both the main and the secondary loop are provided with several Pt100 resistance thermometers to monitor the thermodynamic conditions in the entire test rig (see Fig. 1).

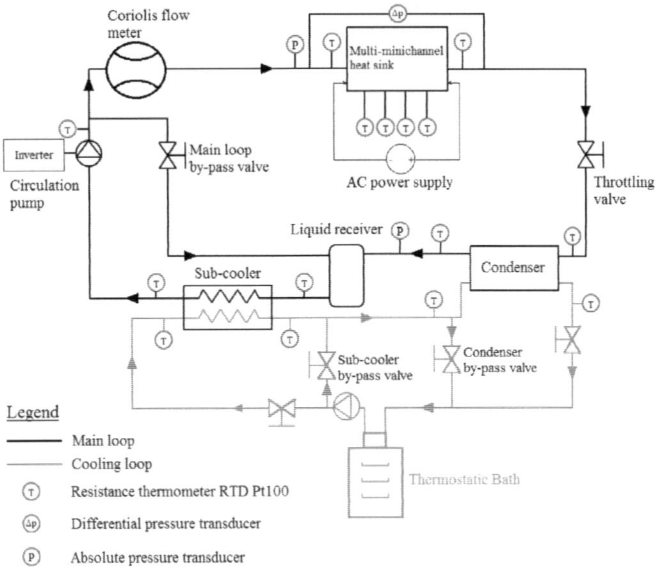

Figure 1. Sketch of the test facility

2.2. Test section

The test section used for all the experiments presented in this paper consists of an aluminum multi-minichannel heat sink, horizontally placed in the test rig. Seven channels are carved in the aluminum block, each of them 35 mm long, 2 mm wide and 1 mm high (entailing an equivalent diameter of 1.33 mm), and two manifolds at the inlet and the outlet are provided in order to prevent flow maldistribution. The structure is sealed with an aluminum cogged cover and a rubber gasket. Four Pt100 cylindrical resistance thermometers are equidistantly placed along the channels for the measurement of the wall temperature. The actual distance between the RTDs placement and the channels wall is s = 2.5 mm. The heat is supplied from the bottom thanks to a slot carved in the aluminum main block, by means of AC power supply and a solid state relay which is able to give up to 400 V.

Two different heated lengths are used in this paper: in the first set of experiments, the heat is provided with a ceramic heater directly put in the slot underneath the test section. The heating element is able to supply up to 697 W (at 25 °C) over a base area of 2.50x2.50 cm^2.In this case, the heated length is $L_h = 25$ mm and the L_h/D ratio (with D the equivalent diameter of each minichannel) is equal to 18.75.

For the second set of experiments, the heat is provided with a cartridge ceramic heater, accommodated in a copper block, whose pyramidal edge is put in the slot underneath the aluminum test section. According to the manufacturer, the cartridge heater is able to provide up to 3000 W (at 600 °C and 400 V). In this case, the slot carved in the aluminum block has been modified in order to obtain a heated length equal to the channels length ($L_h = 35$ mm and $L_h/D = 26.25$). The copper block geometry has been chosen by simulating the temperature field with different dedicated software. The pyramidal shape allows to preserve the perpendicular direction of the heat flux in respect to the minichannels cross section.

The whole test rig is insulated with an appropriate layer of synthetic rubber, with a thermal conductivity less than 0.042 W/(m K) within the range of operating conditions investigated. Particular

IOP Conf. Series: Journal of Physics: Conf. Series **796** (2017) 012037 doi:10.1088/1742-6596/796/1/012037

attention has been paid to the test section: in the first set of experiments, the aluminum block has been covered with several layers of synthetic rubber, whilst in the case of the cartridge heater, a considerable amount of mineral wool (0.038 W/m K) has been used to cope with the higher temperatures of the copper block.

Both the L_h/D test section arrangements have been tested in liquid single phase in order to verify the correct insulation. The imposed heat has been compared to the heat absorbed by the sub-cooled liquid and the heat losses have been then calculated.

The adiabaticity tests for the first test section have shown that the heat losses could be considered negligible. In this case, the effective heat $\dot{Q}_{eff}$ has been considered equal to the electrical measurement.

The adiabaticity tests for the second test section arrangement ($L_h/D = 26.25$, with the presence of the copper block) have led to calculated heat losses approximately equal to 5-6 %, at any heat imposed. It was found that the heat losses were mainly dependent on the difference between the copper element temperature and the ambient temperature. Therefore, a simple equation for the evaluation of the heat losses has been implemented and used for the calculation of the effective heat:

$$\dot{Q}_{loss} = 1.478 \cdot \left(T_{copper} - T_{amb}\right)^{0.5706} \quad (1)$$

Fig. 2 displays a schematic picture of the test section with its geometrical characteristics. Fig. 3 shows the test section corresponding to the higher L_h/D ratio with the copper block exposed before the insulation process.

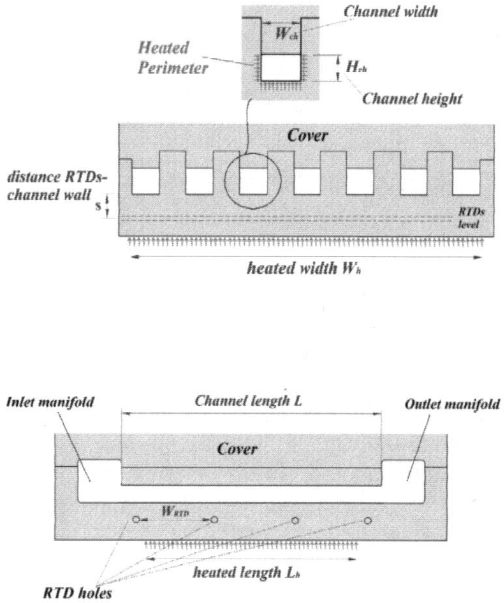

Feature	Value	Symbol
Number of channels	7	N
Channels width	2 mm	W_{ch}
Channels height	1 mm	H_{ch}
Channels length	35 mm	L
Equivalent diameter: $\frac{4 \cdot W_{ch} H_{ch}}{2 \cdot (W_{ch}+H_{ch})}$	1.33 mm	D
Heated width	25 mm	W_h
Heated lengths	25/35 mm	L_h,
Distance RTD-wall	2.5 mm	s
Distance between RTDs	10 mm	W_{RTD}

Figure 2. Geometrical details of the test section

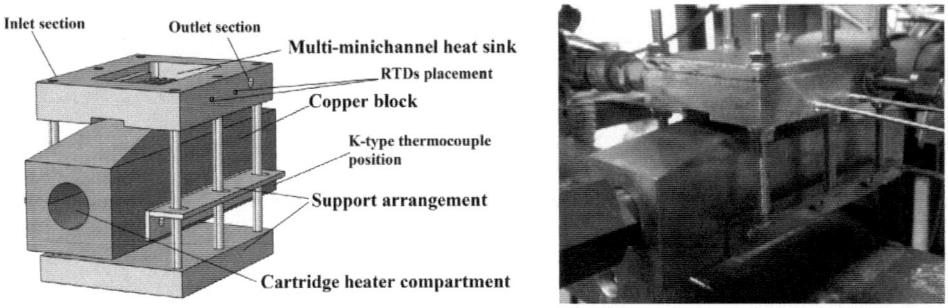

Figure 3. Schematic and real set-up of the test section with the copper block and the cartridge heater, corresponding to the highest L_h/D arrangement

2.3. Measurement uncertainties

The temperature measurements at the inlet and the outlet of the test section, as well as in the rest of the test rig, are obtained with 4-wire Pt100 resistance thermometers placed outside the pipe walls with a nano-aluminum thermal compound. According to the manufacturer, their maximum uncertainty should not exceed 0.180 °C. The four Pt100 cylindrical RTDs for the measurement of the wall temperature carry instead a lower overall uncertainty of 0.154 °C.

Two absolute pressure transducers (0-50 bar) measure the absolute pressure at the inlet of the test section and right before the liquid receiver. By taking into account the non-linearity and the repeatability effects, the resulting uncertainty is 0.3% of the read value.

The differential pressure transducer (0-60 kPa) gives a span error of 0.75% at full scale, resulting in an absolute uncertainty of 0.45 kPa.

The Coriolis mass flow meter has been calibrated up to the 2% of the full scale (2.3 g/s). The maximum uncertainty was found to be 1% of the measurement and this value is used for the whole range of our experiments.

The electrical power measurements have been taken with a digital wattmeter (0-4.00 kW), whose uncertainty is claimed to be 1% of the reading.

Thanks to the manufacturers information, the B type uncertainty has been composed with the standard deviation of the measured parameters in the recording time. Finally, the expanded uncertainty of the measured parameters has been evaluated by considering a coverage factor $k = 2$, thus guaranteeing a confidence level of 95%. The expanded uncertainty of all the derived parameters is then evaluated by means of the law of propagation of errors.

All the experiments carried-out for this paper present an uncertainty of the CHF inferior to 3 %. The operating parameters evaluated at the critical conditions in terms of mass velocity G and saturation temperature T_{sat} have an uncertainty always inferior to 16 % and 3.5 %, respectively. The higher uncertainty of the mass flux is due to fluctuations of the mass flow rate approaching the critical condition, thus resulting into a higher standard deviation.

3. Method

3.1. Experimental procedure

The purpose of each test is to describe a complete boiling curve from the onset of the boiling region up to the critical condition. All the boiling curves were therefore obtained in steady state conditions, meaning that the desired inlet pressure (i.e. saturation temperature), inlet sub-cooling and mass velocity were set and kept constant throughout the test procedure.

The system pressure was fixed by setting the desired temperature of the thermostatic bath, while the mass flow rate was obtained with a specific inverter frequency for the circulation pump.

Significant adjustments of the mass flow rate and system pressure could be achieved by manually turning the main loop by-pass valve.

Small alterations of the inlet temperature and therefore of the inlet sub-cooling were possible with the aid of the sub-cooler by-pass circuit.

Once all the desired parameters were fixed, for both the test sections the heat was supplied in small increments by changing the voltage applied to the ceramic elements within the range 0-230 V.

All the parameters were monitored and controlled with Labview software and Arduino One controller.

As regards the first test section, the heater temperature was constantly monitored with a K-type thermocouple integrated into the heater itself and the power supply was automatically shut off when it reached the limit of 120 °C not to irreparably damage the test section. The integrity of the second test section was instead ensured by two K-type thermocouples inserted in the copper structure. In this case, the power supply was expected to be cut off in case the copper block reached the threshold of 300 °C.

The system was considered stabilized when the relative uncertainty of the main parameters such as the mass flux, saturation temperature and heat imposed were inferior to 10%, 2% and 3%, respectively, in the recording time of 2 minutes with a recording frequency of 1 Hz. For each point of the boiling curve, the nominal value was assigned to the sample average value.

The following data reduction process was implemented in MATLAB software [12], while the calculation of all the thermodynamic properties was carried out with the software REFPROP 9.0, developed by Nist [13].

3.2. *Data reduction*

The mass flux inside the test section is evaluated from the measured mass flow rate:

$$G = \frac{\dot{m}}{W_{ch} \cdot H_{ch} \cdot N} \tag{2}$$

where N, W_{ch} and H_{ch} are the number of minichannels, the channels width and height, respectively, and $\dot{m}$ is the measured mass flow rate.

The inlet saturation temperature and enthalpy are evaluated with the inlet temperature and the inlet pressure and the aid of the software REFPROP 9.0. [13].

The actual wall temperature is calculated by considering 1-D heat conduction inside the aluminum test section. A preliminary analysis with dedicated calculation software has in fact confirmed that the axial conduction in the aluminum block is negligible if compared to the main flux in the vertical direction.

$$T_{w,i} = T_{RTD,i} - \frac{\dot{q}_b \cdot s}{k_{al}} \tag{3}$$

In the above equation, $T_{RTD,i}$ is the measured temperature of the ith thermocouple, k_{al} is the aluminum thermal conductivity (fixed at 240 W/m K) and $\dot{q}_b$ is the base heat flux, evaluated as the ratio of the effective heat power (see equation (1)) over the base heated area, which is 2.50x2.50 cm^2 and 2.50x3.50 cm^2 for the first and the second test section, respectively.

The actual wall heat flux is instead evaluated with the same procedure used by Park and Thome [14], by considering the heated perimeter and a fin efficiency η.

$$\dot{q}_w = \frac{\dot{Q}_{eff}}{N \cdot L_h \cdot (W_{ch} + 2H_{ch} \cdot \eta)} \tag{4}$$

The value of the fin efficiency is firstly guessed equal to 0.90, then it is evaluated with an iterative procedure until it reaches a fixed value, according to the following equations [15], in which the HTC is the mean heat transfer coefficient obtained at the highest wall superheat and W_{fin} is the fin width:

$$\eta = \frac{\tanh(m \cdot H_{ch})}{m \cdot H_{ch}} \tag{5}$$

$$m^2 = \frac{HTC \cdot 2(W_{fin} + L_h)}{k_{al} \cdot W_{fin} \cdot L_h} \tag{6}$$

$$HTC = \frac{\dot{q}_W}{\Delta T_{w,max}} \tag{7}$$

The outlet enthalpy and quality are finally obtained with an energy balance on the test section. The outlet pressure is calculated by subtracting the pressure drop from the inlet pressure value.

$$h_{out} = h_{in} + \frac{\dot{Q}_{eff}}{\dot{m}} \tag{8}$$

3.3. CHF detection method

The critical condition usually refers to the point of the boiling curve in which a sharp rise of the wall superheat is detected without significant increment of the imposed heat flux. In this case, the liquid phase in contact with the heated wall is replaced by the vapor, which is not able to handle the thermal load due to its lower heat transfer efficiency.

In case of multi-minichannel systems, other researchers [14]-[16]-[17] have found that the sharp rise of the wall temperature is instead substituted by a gentler change in the boiling curves slope. This behavior is probably due to the heat redistribution aided by conduction phenomena, which is more evident when increasing the thickness of the solid substrate and/or the number of minichannels [18]. Furthermore, the mass velocity also has a mitigating effect on the critical condition [14]-[19].

Fig. 4 displays four different experimental boiling curves taken with R134a at T_{sat} = 45 °C with two different mass velocities and with the two L_h/D available. The wall superheat on the x-axis is the maximum value among the four RTDs placed along the channels. Fig. 4(a) refers to the minimum mass flux investigated of 150 kg/m^{2}s, while Fig. 4(b) refers to the maximum mass flux tested of 350 kg/m^{2}s. In both figures, the shape of the curves significantly changes with L_h/D, indicating an effect of the solid copper substrate. The change in the boiling curve slope is also more abrupt in the first case of G = 150 kg/m^{2}s, confirming the effect of the mass velocity on the critical values.

There is therefore the need to identify an objective method to detect the CHF condition. Following the same approach of Mauro et al. [16], the critical heat flux has been defined as the heat flux in which the boiling curve decreases its slope below a chosen threshold of 1.0 W/cm^2 K.

Pragmatically, the slope of the boiling curve presents an irregular trend and it is not always monotonically decreasing. The search of CHF begins only when the slope (evaluated with spline interpolation from the experimental data) is always below a threshold value of 1.5 W/cm^2 K. Fig. 5 depicts the slope of the boiling curve obtained with refrigerant R134a at T_{sat} = 65 °C and G = 300 kg/m^2 s. The CHF research begins from the green line, in which the slope is always inferior to the threshold of 1.5 W/cm^2 K. The actual CHF is found when the curve first reaches the imposed threshold of 1.0 W/cm^2 K. This point is displayed with a red star in the diagram in Fig. 5. Either in case this method does not find the CHF or the critical wall superheat is higher than 25 K, the CHF value is conventionally defined as the wall heat flux corresponding to a wall superheat of 25 K.

In this procedure, the overall CHF uncertainty should take into account the uncertainty related to the boiling curve slope, which is affected by the wall superheat excursion. However, since the CHF phenomenon is obtained with sharp increases of the wall temperature for small heat flux variations, the uncertainty of the measured temperature is not of primary importance and the CHF overall uncertainty has been calculated by only taking into account the electrical measurement.

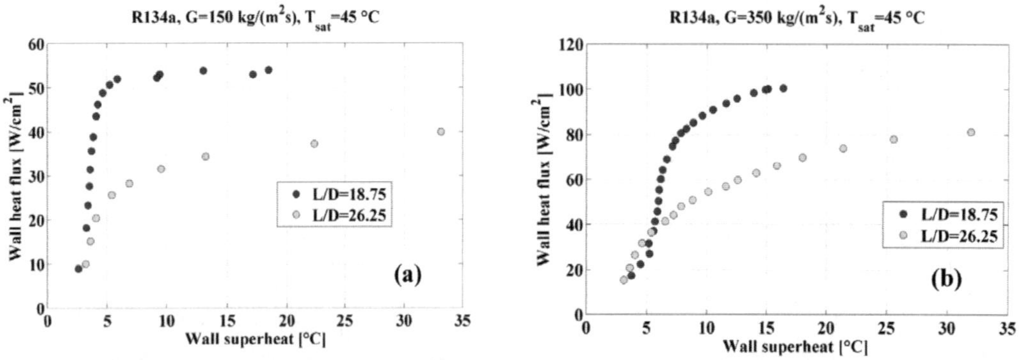

Figure 4. Boiling curves obtained with refrigerant R134a at T_{sat} = 45 °C. Effect of L_h/D ratio at: (a) G = 150 kg/m² s and (b) G = 350 kg/m² s.

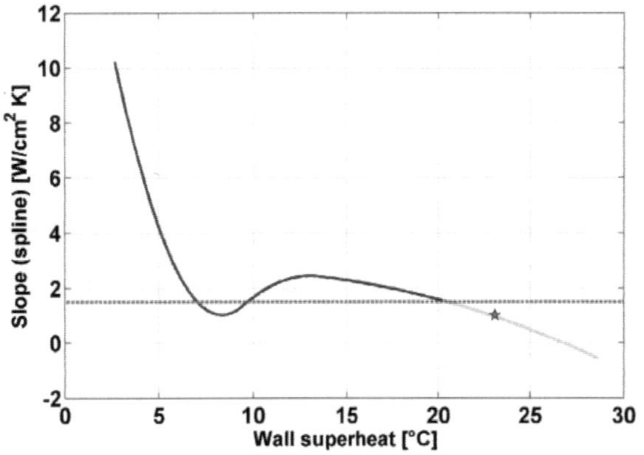

Figure 5. CHF detection method applied to R134a at G = 300 kg/m² s and T_{sat} = 65 °C. The red dashed line identifies the search threshold of 1.5 W/cm² K, and the actual CHF value is obtained at a slope of 1.0 W/cm² K, highlighted with the red star

4. Experimental results

The main CHF tests have been performed with refrigerant R134a and its low-GWP substitute R1234yf, within a saturation temperature range of 25-65 °C and a mass flux range of 150-350 kg/m² s. Each test has been repeated for two different L_h/D ratios of 18.75 and 26.25, respectively. The effect of all these parameters on both fluids is exposed in Fig. 6(a)-(b), in which the experimental CHFs and their expanded uncertainty are displayed against the saturation temperature. Each marker and color refer to different mass velocities and the two different L_h/D ratio values are highlighted with solid and dashed line styles.

It is quite evident that the effect of the saturation temperature on the detected critical heat fluxes is almost negligible. Regarding refrigerant R134a (Fig. 6(a)), the CHF values remain nearly the same regardless the mass flux or the L_h/D ratio. The highest decrement is recorded for G = 300 kg/m² s and L_h/D = 26.25, with the CHF passing from 71.9 to 62.3 W/cm² when the saturation temperature rises from 25 to 65 °C. For refrigerant R1234yf (Fig. 6(b)), the CHF values remain nearly the same at any mass flux for the lowest L_h/D ratio = 18.75, whereas the descending trend is more evident for the highest L_h/D ratio of 26.25: at G = 352 kg/m² s, the CHF reduces of approximately 20% (from 71.0 to 56.6 W/cm²) at increasing saturation temperature.

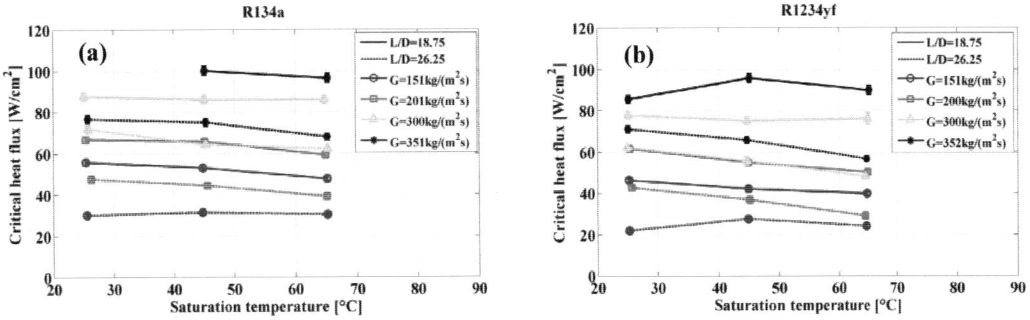

Figure 6. Experimental CHF values with their expanded uncertainty band as a function of the saturation temperature. Each color and marker are referred to a particular mass velocity. The solid and dashed lines refer to L_h/D ratios of 18.75 and 26.25, respectively. The examined fluids are: (a) R134a and (b) R1234yf.

The parametric effect of the working fluid, the L_h/D ratio and the mass flux is depicted in Fig. 7, in which the experimental CHF values with their expanded uncertainty have been plotted versus the mass velocity, for both fluids and L_h/D ratios investigated. The saturation temperature has been instead fixed to 65 °C.As expected, the mass flux has a great effect on the critical values, since CHF increases significantly (up to 130%) passing from 150 to 350 kg/m^2 s, assuming almost a linear trend, for both fluids and L_h/D ratios tested.

Between the two fluids, R134a shows better cooling performances than R1234yf: at the same conditions, the CHF for the new low-GWP fluid is roughly the 20% lower than that of R134a.

A different L_h/D ratio obtained by changing the heated length L_h in the two test sections used leads to significantly different CHF values. For both refrigerants, at $L_h/D = 26.25$ the CHF results are lower and the gap is larger for higher mass velocities. Similar results have been obtained in other works available in open literature [7]-[10]-[11]. Specifically, for refrigerant R134a, at G = 150 kg/m^2 s, CHF passes from 47.0 to 31.0 W/cm^2, while at G = 350 kg/m^2 s, it goes from 91.0 to 58.0 W/cm^2, with a reduction of 36%. For refrigerant R1234yf, at G = 150 kg/m^2 s CHF passes from 40.5 to 22.0 W/cm^2, while at G = 350 kg/m^2 s, it goes from 90.2 to 58.0 W/cm^2, with a reduction of 36 % as well. Similar trends are observed for different saturation temperatures.

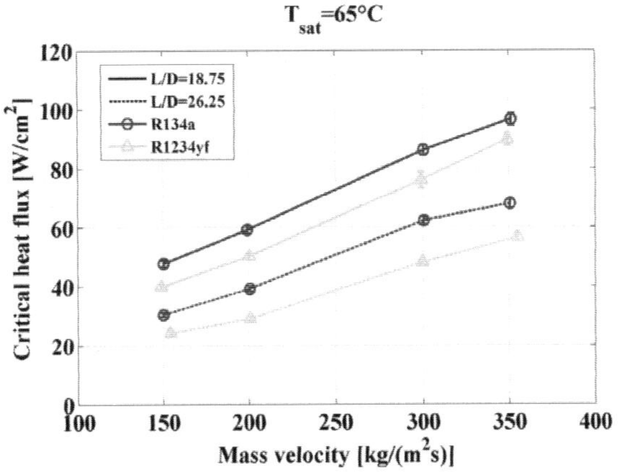

Figure 7. Experimental CHF values with their expanded uncertainty band as a function of the mass velocity for refrigerants R134a and R1234yf with two different L_h/D ratios of 18.75 and 26.25. The diagram refers to a saturation temperature of 65 °C.

5. Conclusions

The saturated critical heat flux (CHF) in a multi-minichannel heat sink at different operating conditions has been studied in this paper. The effect of the mass velocity, saturation temperature and working fluid has been first investigated. Two test section set-up arrangements have been structured in order to achieve two different heated lengths and therefore two different L_h/D ratios. The effect of this parameter has also been studied.

The experimental results have shown that R134a has better cooling performances that those of its substitute R1234yf, with a CHF from 10 to 20 % higher when evaluated at the same thermodynamic conditions.

The effect of the saturation temperature is almost negligible for both fluids and all the mass velocities, but it is amplified in case of the highest L_h/D ratio of 26.25, for which the CHF reduction is roughly the 20 % when passing from 25 to 65 °C

CHF increases linearly with mass velocity, regardless the fluid used and the saturation temperature. The trends do not change with the two different L_h/D ratios.

The heated length has a strong influence on the experimental CHF results. By increasing the L_h/D ratio, the CHF significantly reduces of about 35 % for both fluids in all the operating conditions tested. Slightly greater differences are found for higher mass velocities.

Although the obtained results are in agreement with other works available in open literature, the effect of the L_h/D ratio has still to be fully clarified in the available correlations and further research with new L_h/D ratios, fluids and thermodynamic conditions would be of use to increase the experimental database.

References

[1] United States Semiconductor Industry Association (SIA), European Semiconductor Industry Association (ESIA), The Japan Electronics and Information Technology Industries Association (JEITA), The Korean Semiconductor Industry Association (KSIA), The Taiwan Semiconductor Industry Association (TSIA) 2009 *Int Technol Roadmap Semicond.* http://www.itrs.net/Links

[2] Mudawar I2001 Assessment of high-heat-flux thermal management schemes. *IEEE Trans Compon Packag Technol* **24** 122-141

[3] Royne A, Dey CJ and Mills DR 2005 Cooling of photovoltaic cells under concentrated illumination: a critical review. *Solar Energy Materials and Solar Cells* **84** 451-483

[4] Palm B 2001 Heat transfer in microchannels *Microscale Thermophys Eng* **5** 155-175

[5] Bergles AE and Kandlikar SG 2005 On the nature of critical heat flux in microchannels *J Heat Transf* **127** 101-107

[6] Ribatski G 2013 A critical overview on the recent literature concerning flow boiling and two-phase flows inside micro-scale channels *Experimental heat transfer: a Journal of thermal energy generation, transport, storage, and conversion* **26:2-3** 198-246

[7] Tanaka F, Hibiki T and Mishima K 2009 Correlation for flow boilingcritical heat flux in thin rectangular channels *Journal of heat transfer* **131** 121003-7

[8] Saha S K and Celata G P 2016 Instability in flow boiling in microchannels *Springer briefs in applied science and technology, Thermal engineering and applied science* **8884** DOI 10.1007/978-3-319-23431-1

[9] Boyd R D 1985 Subcooled flow boiling critical heat flux (CHF) and its application to fusion energy components. Part II: a review of micro-convective, experimental and correlation aspects *Fusion Technology* **7** 31-52

[10] Wu Z and Li W 2011 A new predictive tool for saturated critical heat fluxin micro/mini-channels: effect of the heated length-to-diameter ratio International Journal of Heat and Mass Transfer **54** 2880-2889

[11] Wojtan L, Revellin R and Thome J R 2006 Investigation of saturated critical heat flux in a

single, uniformly heated microchannel *Experimental Thermal and Fluid Science* **30** 765-774.

[12] MATLAB release. *Natick, Massachussets, United States, The MathWorks, Inc.*

[13] Lemmon E W, Mc Linden M O and Huber M L 2009 REFPROP *NIST Standard Reference Database* **23**, *Version 9.0*

[14] Park J E and Thome J R 2010 Critical heat flux in multi-microchannel copper elements with low pressure refrigerants *Int. J. of Heat and Mass Transfer* **53** 110-122

[15] Baehr HD and Stephan K 2006 Heat and mass transfer *Springer 2nd edition*

[16] Mauro A W and Thome J R, Toto D and Vanoli G P 2010 Saturated critical heat flux in a multi-microchannel heat sink fed by a split flow system *Experimental Thermal and Fluid Science* **34** 81-92

[17] Koşar A and Peles Y 2006 Critical heat flux of R-123 in silicon-based microchannels. *Journal of Heat Transfer* **129** 844-851

[18] Saenen T and Thome J R 2016 Dynamic numerical microchannel evaporator model to investigate parallel channel instabilities *Journal of Electronic Packaging* **138** 010901-13

[19] Mastrullo R, Mauro A W, Thome J R, Vanoli G P and Viscito L 2016 Critical heat flux: performance of R1234yf, R1234ze and R134a in an aluminum multi-minichannel heat sink at high saturation temperatures *Int. J. Thermal Sciences* **106** 1-17

Numerical Simulation of Thin Film Breakup on Nonwettable Surfaces

Nicola Suzzi[1] **and Giulio Croce**[1]

[1]DPIA - Dipartimento Politecnico di Ingegneria e Architettura - Universitá di Udine - Via delle Scienze - 33100 - Udine (UD) - Italy

E-mail: `giulio.croce@uniud.it`

Abstract. When a continuous film flows on a *nonwettable* substrate surface, it may break up, with the consequent formation of a dry-patch. The actual shape of the resulting water layer is of great interest in several engineering applications, from in-flight icing simulation to finned dehumidifier behavior modeling. Here, a 2D numerical solver for the prediction of film flow behavior is presented. The effect of the contact line is introduced via the disjoining pressure terms, and both gravity and shear are included in the formulation.
The code is validated with literature experimental data for the case of a stationary dry-patch on an inclined plane. Detailed numerical results are compared with literature simplified model prediction. Numerical simulation are then performed in order to predict the threshold value of the film thickness allowing for film breakup and to analyze the dependence of the dynamic contact angle on film velocity and position along the contact line. Those informations will be useful in order to efficiently predict more complex configuration involving multiple breakups on arbitrarily curved substrate surfaces (as those involved in in-flight icing phenomena on aircraft).

1. Introduction

A thin layer of liquid can spread over a solid substrate as a collection of tiny droplets, as an ensemble of rivulets or as a continuous film. The details of this transitions have an impact on different engineering problems: as an example, in-flight icing phenomenon on aircraft surface is driven by the evolution of the water layer fed by a stream of super-cooled droplets from the cloud; depending on the cloud water content and the fluid flow field and temperature, the liquid layer may take one of the above mentioned forms and influence the consequent ice accretion and its effect on aircraft aerodynamics. Also fogging and defogging of a car windshield or heat transfer on dehumidifier fin surfaces are affected by condensation and evaporation phenomena with droplets growth, coalescence and motion over solid substrates.

The early studies reported in open literature followed an analytical/integral approach: a first paper by Hartley and Murgatroyd [1] considered the separation of a laminar film flowing vertically under gravity, proposing a critical breakup threshold for the liquid flow rate based on a force balance at the apex of the dry-patch. In [2], the authors modeled the transition between still droplets and moving ones via a Lagrangian approach, simulating the process of birth, growth, coalescence of the individual droplets and detecting their motion by a force balance. A different approach is the straightforward numerical solution of the two dimensional lubrication equation. Witelski and Bowen [3] proposed an efficient implicit scheme for the numerical integration of the lubrication equation, via an ADI factorization. The authors considered only fully wetted surfaces.

Schwartz and Eley [4] introduced the disjoining pressure term in order to model the substrate wettability and the development of dry-patches. The disjoining pressure definition was adjusted to match the static contact angle. The modeling of the contact angle via disjoining pressure term was later validated by Zhao and Marshall [5], who also investigated the finger instability phenomenon, induced by surface heterogeneity. Sellier [6] considered the perturbation induced by a 3D obstacle on the one dimensional flow of a film flowing down a vertical plane, focusing on the liquid distribution around the obstacle and the influence of the obstacle shape, rather than on the dry-patch dynamics.

The aim of this paper is the development of an efficient solver for the governing lubrication equation suitable as a first step for the analysis of complex, real world configurations like the one encountered, as an example, in in-flight icing configuration, involving the evolution and interaction of several rivulets, film and dry-patches. Thus, a general, fully implicit, 2D solver is described, commenting on its CPU time efficiency. An upwind discretization is chosen, rather than the central schemes used in [3–8], in order to be able to model also the convection dominated flows where the shear stress is the main driving force (as done, for continuous films, in [9]). The code is first validated with experimental evidences [10, 11] and then applied to the analysis of dry-patch stability dynamics. In particular, the maximum mass flow rate allowing for the growth of a dry-patch is computed from the fully 2D model. Furthermore, the evolution of the contact angle along the whole 2D profile of the patch is presented and commented, while in previous literature references the analysis was limited to the region close to the patch apex.

2. Mathematical model

Consider a thin film of height h flowing along an inclined plane under the action of both gravity and shear. Let α be the inclination of the plane with respect to the horizontal and $\boldsymbol{\tau_a}$ the shear applied by an external gas flow. The continuity equation offers:

$$\frac{\partial h}{\partial t} = -\nabla \cdot \boldsymbol{Q} = -\nabla \cdot (\boldsymbol{u}\, h) \tag{1}$$

For low Reynolds numbers, $\mathrm{Re} \leq 1$, the lubrication theory allows to write the average film velocity as a function of the shear stress and of the gradient of pressure p:

$$\boldsymbol{u} = -\frac{\nabla p}{3\,\mu}\, h^2 + \frac{\boldsymbol{\tau_a}}{2\,\mu}\, h \tag{2}$$

Substituting Eq. (2) in Eq. (1) leads to

$$\frac{\partial h}{\partial t} = -\nabla \cdot \left(-\frac{\nabla p}{3\,\mu}\, h^3 + \frac{\boldsymbol{\tau_a}}{2\,\mu}\, h^2 \right) \tag{3}$$

The pressure field of the liquid is given by Schwartz and Eley [4] as

$$p = \rho\, g \left(h\, \cos\alpha - x\, \sin\alpha \right) - \sigma\, \nabla^2 h - \Pi \tag{4}$$

where

- $\rho\, g\; (h\, \cos\alpha - x\, \sin\alpha)$ represents the contribution of gravity pressure, x increasing in the downhill direction of the plane;
- the capillary pressure $\sigma\, \nabla^2 h$ is due to the curvature of the liquid-gas interface;

- the disjoining pressure Π, related to the intermolecular forces between liquid and solid substrate, is given by

$$\Pi = B \left[\left(\frac{h_{wet}}{h} \right)^n - \left(\frac{h_{wet}}{h} \right)^m \right] \tag{5}$$

  where h_{wet} is the height of the precursor film and the parameters m, n, B have to be defined according to the constraints $1 < m < n$, $B > 0$.

In particular, writing the equilibrium of forces for a liquid volume near the contact line, Schwartz and Eley [4] obtained the following expression of the coefficient B,

$$B = \frac{1}{h_{wet}} \frac{(n-1)(m-1)}{(n-m)} \sigma\,(1 - \cos\theta_e) \tag{6}$$

where θ_e is the equilibrium contact angle.

Zhao and Marshall [5] demonstrated consistency of Eq. (6), but they observed that values of the precursor film height h_{wet} much larger than the physical one (which is related to the range of intermolecular forces) are usually chosen for numerical simulations, leading to inaccurate results. Hence they introduced the correction factor f,

$$f = 1 + 6.069 \left(\frac{h_{wet}}{h_\infty} \right) + 161.7 \left(\frac{h_{wet}}{h_\infty} \right)^2 - 1547 \left(\frac{h_{wet}}{h_\infty} \right)^3 + 5890 \left(\frac{h_{wet}}{h_\infty} \right)^4 \tag{7}$$

which multiplies the coefficient B.

Defining the non-dimensional quantities

$$\bar{h} = \frac{h}{h_\infty}\ ;\ \bar{\boldsymbol{x}} = \frac{\boldsymbol{x}}{h_\infty}\ ;\ \bar{t} = \frac{t}{(h_\infty/u_\infty)}\ ;\ \bar{\tau}_a = \frac{\tau_a}{(\sigma/h_\infty)}\ ;\ \bar{p} = \frac{p}{(\sigma/h_\infty)}\ ;\ \bar{\Pi} = \frac{\Pi}{(\sigma/h_\infty)} \tag{8}$$

and introducing Bond and Capillary numbers, Eqs. (3) and (4) can be rearranged in non-dimensional form

$$\frac{\partial \bar{h}}{\partial \bar{t}} = -\frac{1}{\mathrm{Ca}} \nabla \cdot \bar{\boldsymbol{Q}} = -\frac{1}{\mathrm{Ca}} \nabla \cdot \left(-\frac{\nabla \bar{p}}{3} \bar{h}^3 + \frac{\bar{\boldsymbol{\tau}}_{\boldsymbol{a}}}{2} \bar{h}^2 \right) \tag{9}$$

$$\bar{p} = \mathrm{Bo} \left(\bar{h} \cos\alpha - \bar{x} \sin\alpha \right) - \nabla^2 \bar{h} - \bar{\Pi} \tag{10}$$

The fourth order, non-linear PDEs, Eqs. (9) and (10), are solved numerically on a 2D domain on the inclined plane surface. The special case of gravity and shear acting in the same direction is considered. Thus, the average velocity of the undisturbed film is given by:

$$u_\infty = \frac{\rho g \sin\alpha}{3\mu} {h_\infty}^2 + \frac{\tau_a}{2\mu} h_\infty \tag{11}$$

3. Spatial discretization

The 2D spatial domain $L_x \times L_y$ is discretized using an uniform, orthogonal grid composed by $n_{el} = n_i \times n_j$ elements of dimensions Δx and Δy. The convective term $\nabla \cdot \boldsymbol{Q}$ of Eq. (1) is discretized via a first order upwind scheme. Hence, the flux $\boldsymbol{Q} = \boldsymbol{u}\,h$ is decomposed into two contributions. With reference to x direction, $Qx^+ = u^+ h$ refers to the the wave propagation with positive velocity, while $Qx^- = u^- h$ refers to the wave propagation with negative velocity.

Using finite volume notation and referring to non-dimensional quantities, the volume integral of the divergence for the i,j−th element can be expressed as an integral over its boundary surface $\Omega_{i,j}$, with $\hat{\boldsymbol{n}}$ outward unit normal vector:

$$\frac{1}{V_{i,j}} \int_{V_{i,j}} \nabla \cdot \bar{\boldsymbol{Q}} = \frac{1}{V_{i,j}} \int_{\Omega_{i,j}} \bar{\boldsymbol{Q}} \cdot \hat{\boldsymbol{n}}\, d\Omega = \frac{\bar{Q}x_{i+\frac{1}{2},j} - \bar{Q}x_{i-\frac{1}{2},j}}{\Delta \bar{x}} + \frac{\bar{Q}y_{i,j+\frac{1}{2}} - \bar{Q}y_{i,j-\frac{1}{2}}}{\Delta \bar{y}} \tag{12}$$

The fluxes that cross the faces of the $i, j-$th element in the x direction can be calculated as:

$$\bar{Q}x_{i-\frac{1}{2},j} = \frac{1}{2}\left(\bar{Q}x^{+}{}_{i-1,j} + |\bar{Q}x^{+}{}_{i-1,j}|\right) + \frac{1}{2}\left(\bar{Q}x^{-}{}_{i,j} - |\bar{Q}x^{-}{}_{i,j}|\right) \tag{13}$$

$$\bar{Q}x^{+}{}_{i,j} = \left(\bar{u}^{+}\,\bar{h}\right)_{i,j} = \left(-\frac{1}{3}\frac{\bar{p}_{i+1,j} - \bar{p}_{i,j}}{\Delta\bar{x}}\,\bar{h}^{2}_{i,j} + \frac{\bar{\tau}_a}{2}\,\bar{h}_{i,j}\right)\bar{h}_{i,j} \tag{14}$$

$$\bar{Q}x^{-}{}_{i,j} = \left(\bar{u}^{-}\,\bar{h}\right)_{i,j} = \left(-\frac{1}{3}\frac{\bar{p}_{i,j} - \bar{p}_{i-1,j}}{\Delta\bar{x}}\,\bar{h}^{2}_{i,j} + \frac{\bar{\tau}_a}{2}\,\bar{h}_{i,j}\right)\bar{h}_{i,j} \tag{15}$$

The expressions of the fluxes $\bar{Q}x^{+}{}_{i,j}$ and $\bar{Q}x^{-}{}_{i,j}$ ensure that the pressure gradient on each face is always computed with a centered difference:

$$\left(\frac{\partial\bar{p}}{\partial\bar{x}}\right)_{i\pm\frac{1}{2},j} = \frac{\bar{p}_{(i\pm\frac{1}{2})+\frac{1}{2},j} - \bar{p}_{(i\pm\frac{1}{2})-\frac{1}{2},j}}{\Delta\bar{x}} \tag{16}$$

The pressure field, Eq. (10), is calculated from the values of the film distribution $\bar{h}$ and substituted in Eqs. (14) and (15). The capillary pressure $\nabla^2\bar{h}$, the only contribution depending on the spatial derivates of $\bar{h}$, is implemented via a centered scheme, leading to

$$\left(\frac{\partial^2\bar{h}}{\partial\bar{x}^2}\right)_{i,j} = \frac{\bar{h}_{i-1,j} + \bar{h}_{i+1,j} - 2\,\bar{h}_{i,j}}{\Delta\bar{x}^2} \tag{17}$$

for the x direction. Combining centered scheme to calculate the capillary pressure and upwind to calculate the fluxes through the element faces, it is easy to demonstrate that the higher order derivates of $\bar{h}$ are treated with a centered scheme.

4. Time discretization

Von Neumann analysis for an explicit Euler discretization of the linearized PDE

$$\frac{\partial\bar{h}}{\partial\bar{t}} = -\frac{1}{3\,\mathrm{Ca}}\frac{\partial^4\bar{h}}{\partial\bar{x}^4} \tag{18}$$

gives the following stability constraint for the time step:

$$\Delta\bar{t} \le \frac{3}{8}\,\mathrm{Ca}\,\Delta\bar{x}^4 \tag{19}$$

Eq. (19) clearly shows that an explicit scheme to solve Eqs. (9) and (10) is computationally too expensive if good resolution ($\Delta\bar{x} < 1$) is required. The same conclusion is reported by several authors [3–5, 12]. Thus, an implicit Euler scheme is chosen. A straightforward implementation would lead to a sparse system of non-linear algebraic equations. However, on the structured grid used in the present work, the computational time can be greatly reduced following an alternate direction implicit (ADI) scheme. Witelski and Bowen [3] considered the numerical resolution of high-order non-linear diffusion equations and suggested to treat explicitely the mixed derivates of $\bar{h}$ when applying the ADI approximate factorization.

Thus the film flux is decomposed into two components isolating the mixed derivates of $\bar{h}$, which derive from the gradient of the capillary pressure, in the expression of $\bar{\boldsymbol{Q}}$:

$$\bar{\boldsymbol{Q}} = \bar{\boldsymbol{F}} + \frac{\bar{h}^3}{3}\begin{Bmatrix} \dfrac{\partial}{\partial\bar{x}}\left(\dfrac{\partial^2\bar{h}}{\partial\bar{y}^2}\right) \\ \dfrac{\partial}{\partial\bar{y}}\left(\dfrac{\partial^2\bar{h}}{\partial\bar{x}^2}\right) \end{Bmatrix} \tag{20}$$

Linearizing $\bar{\boldsymbol{F}}$ by means of the Jacobian matrix and applying the approximate factorization of [3], the sparse, non-linear system of algebraic equations can be decomposed into two pentadiagonal, linear systems to solve together.

It is worth to point out that only first-order accuracy in time is obtainable from two-step schemes (as Eulero and Crank-Nicolson) when mixed derivates are involved in ADI factorization. The time step $\Delta\bar{t}$ is dinamically adjusted as,

$$\Delta\bar{t} = \frac{\kappa}{\max\left\{\frac{|\bar{h}_{i,j}^{n+1}-\bar{h}_{i,j}^{n}|}{\bar{h}_{i,j}^{n+1}}\right\}}\,\Delta\bar{t} \tag{21}$$

where κ is maximum allowed increment $\Delta h/h$ (usually $\kappa \leq 0.02$).

5. Boundary and initial conditions

At the inlet boundary ($x = 0$), the undisturbed film flow per unit length $\dot{V} = u_\infty\, h_\infty$ is supposed to enter the domain, giving $\bar{\boldsymbol{Q}}_{\boldsymbol{in}} \cdot \hat{\boldsymbol{n}} = 1$; remembering Eq.(4), this condition implies that the inlet pressure equals the one of the undisturbed film, $\bar{p}_\infty = \text{Bo}\cos\alpha + \bar{\Pi}_\infty$. At the outlet boundary ($x = L_x$), zero flux entering the domain and zero pressure gradient, $\bar{\boldsymbol{Q}}_{\boldsymbol{in}} \cdot \hat{\boldsymbol{n}} = 0$ and $\nabla\bar{p}\cdot\hat{\boldsymbol{n}} = 0$ respectively, are imposed. At the lateral boundaries, simmetry conditions $\bar{\boldsymbol{Q}} \cdot \hat{\boldsymbol{n}} = 0$ and $\nabla\bar{p}\cdot\hat{\boldsymbol{n}} = 0$ are imposed.

The onset of the dry-patch is induced by a discontinuity in the initial wetting properties of the substrate, rather than introducing a physical obstacle as in [6]. In particular, the film flow is iniatialized with a nonwettable small square patch, imposing: zero flux through the nonwettable patch boundaries, $\bar{\boldsymbol{Q}} \cdot \hat{\boldsymbol{n}} = 0$; fixed film height $\bar{h}_{i,j} = \bar{h}_{wet}$ inside the nonwettable patch.

Two sets of different initial conditions can be chosen:

- dry condition $\Rightarrow \bar{h}_{i,j} = \begin{cases} 1 \text{ if } x_{i,j} < 0.1\,L_x \\ \frac{h_{wet}}{h_\infty} \text{ if } x_{i,j} > 0.1\,L_x \end{cases}$
- wet condition $\Rightarrow \bar{h}_{i,j} = 1$.

The wet condition describes the flow of a film, initially undisturbed, perturbed by the obstacle; the dry condition allows to study the contact line motion and possibly the phenomena related to finger instability.

6. Validation

We validate the code simulating the experimental configurations of [10, 11] and comparing the dry-patch shape with the prediction of the empirical model of Podgorski et al. [10], which validity is confirmed by its perfect agreement with several experimental data [10, 11]. Silicon oil, $\rho = 950\ (\text{kg/m}^3)$, $\mu = 0.019\ (\text{Pa s})$ and $\sigma = 0.021\ (\text{N/m})$, and equilibrium contact angle equal to $\theta_e = 20°$ are chosen for the following simulations.

Starting from an undisturbed film flow on the inclined plane, we apply the nonwettable small patch defined in the previous paragraph in order to perturb the flow and eventually generate a dry-patch. The choice of the obstacle size L_o follows Sellier [6]:

$$\frac{L_o}{h_\infty} = (\text{Bo}\,\sin\alpha)^{-\frac{1}{3}} \tag{22}$$

Each computation is identified by the inclination α of the plane with respect to the horizontal, the magnitude of the shear τ_a and the undisturbed film flow per unit length $\dot{V}$; τ_a is set to zero for the following simulations. Grid independence analysis suggested the use of a non-dimensional spatial discretization step in the range $[0.025 - 1]$, and the precursor film heigth h_{wet} is set at

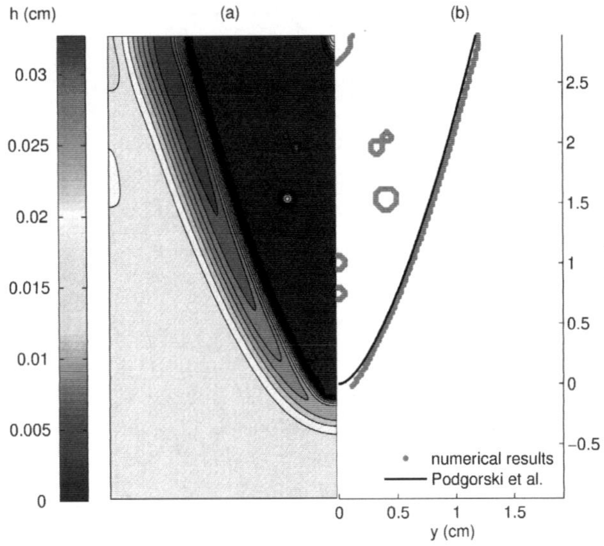

Figure 1. Computed film height (a); dry-patch validation (b). $\dot{V} = 5.9 \times 10^{-3}\,\text{cm}^2/\text{s}$, $\alpha = 30°$.

5% of h_∞, following [4, 5]. The choice of disjoning pressure parameters $(m, n) = (2, 3)$ is driven by stability considerations [4, 5].

The numerical result of figure 1(a) shows the stationary distribution $h(x, y)$ of the film height. Confirming both experimental and analytical evidences [10, 11, 13], a capillary ridge with higher film thickness and flow rate appears near the contact line of the resulting dry-patch. Moreover, some residual drops, that are slowly draining, can be observed. Figure 1(b) compares numerical results with the dry-patch model given in [10]: it can be noticed that the numerical dry-patch fits exactly the empirical model.

7. Results and discussion

The implicit code was applied to the dry-patch stability analysis, estimating the maximum critical mass flow allowing for its persistence. Following the procedure of experimental tests in [10], the undisturbed flow is first perturbed in order to generate a dry-patch, as in the previous paragraph. Once the stationary condition is reached, the obstacle is removed: if the dry-patch is stable (i.e. the contact line doesn't move and a new stationary condition is immediatly reached) the imposed flow rate $\dot{V}$ is lower then the critical one; if the contact line starts advancing or the dry-patch closes up, $\dot{V}$ is higher than $\dot{V}_{cr}$.

Running the code for increasing value of the film flow rate until the critical condition is reached, a set of points describing the critical flow rate dependence on the inclination of the plane was obtained. This allowed us to compare the numerical results with an empirical model, which simplifies the determination of the critical flow rate. This model derives from the assumption that a capillary ridge, where the liquid flows along the contact line direction, surrounds the dry-patch. Following [10, 14], mass, momentum and energy balances are written for the ridge volume surrounding the dry-patch apex. Following the procedure described in detail in [14] we obtain:

$$1 - \cos\theta_e = \mathrm{We} + \mathrm{Bo}\left(\cos\alpha + \frac{S_r}{h_\infty{}^2}\sin\alpha\right) \tag{23}$$

$$\frac{p}{(\sigma/h_\infty)} = \frac{\mathrm{We}}{2} + \mathrm{Bo}\left(\cos\alpha + \frac{1}{2}\frac{L_r}{h_\infty}\sin\alpha\right) \tag{24}$$

where S_r and L_r are the ridge transverse section area and the ridge width, normal to the contact line, respectively. The mean pressure inside the ridge is given by the capillary pressure:

$$p = -\frac{1}{L_r}\int_0^{L_r} \sigma \frac{d^2h}{dl^2}\,dl = -\frac{\sigma}{L_r}\left[\frac{dh}{dl}\right]_0^{L_r} = \frac{\sigma}{L_r}\left[\tan\Theta\right]_0^{\theta_e} = \sigma\frac{\tan\theta_e}{L_r} \tag{25}$$

Introducing the non-dimensional parameter $\gamma = L_r\,h_\infty/S_r$ in order to describe the ridge shape, calculating Weber number and Bond number with respect to the inlet film flow rate $\dot{V} = u_\infty\,h_\infty$, the critical flow rate $\dot{V}_{cr}$ can be calculated solving Eqs. (23), (24) and (25). Figure 2 shows

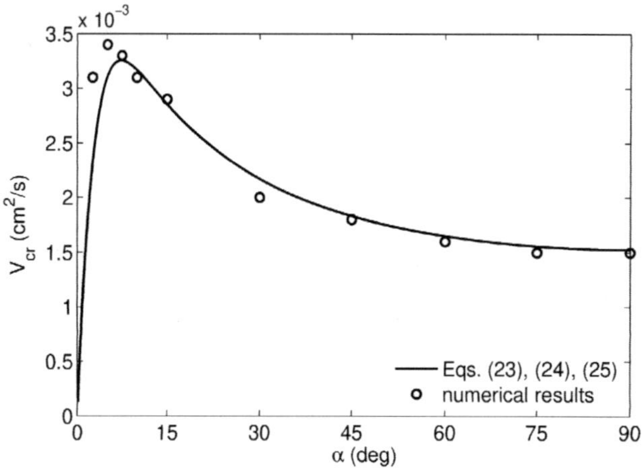

Figure 2. Critical flow rate for $\theta_e = 20°$.

the critical flow rate as a function of the plane inclination for $\theta_e = 20°$ deriving from numerical simulations (circles), fitted by the simplified model (continuous line). The empirical model described above best fits the numerical results when $\gamma = 0.921$.

The dynamics of dry-patch re-wetting was also studied: if the imposed flow rate exceeds the critical value the dry-patch is unstable and will be someway shed away once the nonwettable patch is removed. Figure 3(a) shows the dry-patch contact line at different time step for a high value of the film flow rate, $\dot{V} >> \dot{V}_{cr}$. The dry-patch first adjusts its shape after the obstacle removal and then advances with a constant velocity, due to gravity force, rather than closing up. In figure 3(b) the position of dry-patch apex is traced as a function of time, confirming that a characteristic draining velocity, lower then the undisturbed film velocity u_∞, is reached. From other simulations, which results are not shown due to space limitation, it was observed that, when the flow rate is close to the critical value, $\dot{V} \simeq \dot{V}_{cr}$, a characteristic draining velocity is still reached, confirming the dry-patch motion along the inclined plane, but the dry-patch also tends to close up with a lower velocity, perpendicular to the contact line itself, due to capillary effect.

Finally, the consistence of the disjoining pressure model for describing the dynamics near the contact line is also analyzed. The value of the equilibrium θ_e, in fact, is used to define the

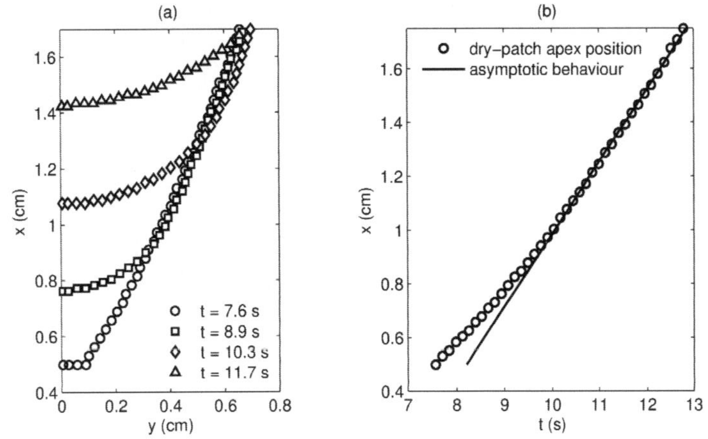

Figure 3. Contact line at different time (a); dry-patch apex position as a function of time (b). $\dot{V} = 7.5 \times 10^{-3}\,\text{cm}^2/\text{s}$, $\alpha = 30°$.

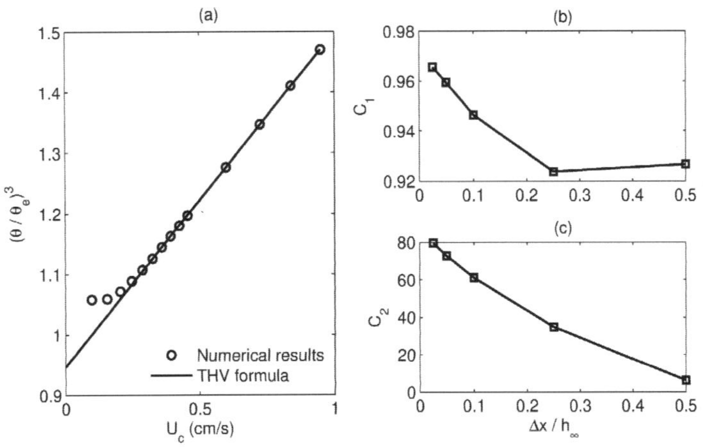

Figure 4. Dynamic contact angle for 1D film flow as a function of contact line velocity, $\Delta x = 0.1\,h_\infty$ (a); influence of discretization accuracy on regression coefficients of Eq.(26) (b).

disjoining pressure from Eq. (6). However, the actual θ is calculated from the slope of the film thickness distribution at the contact line, i.e. from the local modulus of the gradient of h. Zhao and Marshall [5] validated Eqs. (5) and (6) in the case of one dimensional film flow advancing along an inclined plane, concluding that the calculated dynamic contact angle respects the modified Tanner–Hoffman–Voinov formula,

$$\theta^3 = \theta_e^3 \left(C_1 + C_2 \frac{\mu\, u_c}{\sigma} \right) \; ; \; \frac{u_c}{u_\infty} = \frac{1 - (h_{wet}/h_\infty)^3}{1 - (h_{wet}/h_\infty)} \tag{26}$$

where θ and θ_e are the dynamic and static contact angles respectively, u_c is the contact line velocity, which depends on the undisturbed film velocity and the precursor film thickness, and C_1, C_2 are constants (C_1 depending on the spatial discretization accuracy, which influences the calculated value of θ).

Simulations were run considering the 1D problem (a film flowing over an inclined plane initially dry) with $\theta_e = 20°$ and $\alpha = 30°$. Figure 4(a) shows the dynamic contact angle as a function of the contact line velocity u_c: the 1D numerical results (circles) are fitted by Eq. (26); in order to guarantee a good resolution near the contact line, it was chosen a spatial discretization step equal to $\Delta x/h_\infty = 0.1$. The effect of a change in the spatial discretization accuracy on the evaluation of θ is shown in figure 4(b). It can be noticed that the regression coefficient C_1 of Eq. (26) tends to 1 (i.e. the condition $\theta = \theta_e$ if $u_c = 0$ is respected) for high discretization accuracy ($\Delta x \to 0$).

The dynamic contact angle is also studyed in the 2D case: a dry-patch is generated perturbing the undisturbed flow by an obstacle, which is then removed allowing for the contact line movement. Simulations were run for different values of the imposed flow rate $\dot{V}$: a moving dry-

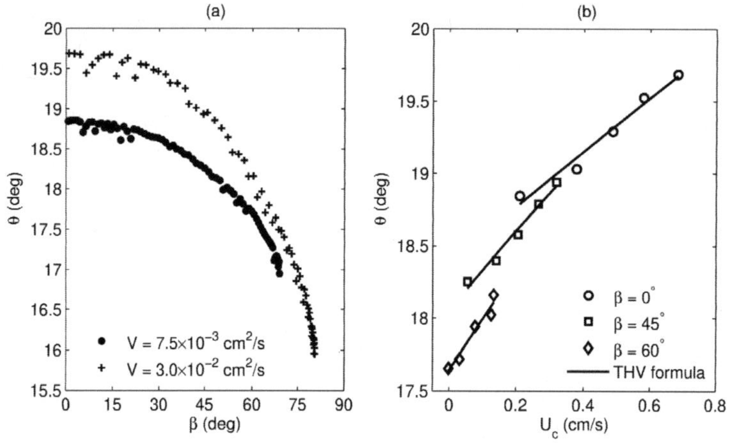

Figure 5. Dynamic contact angle as a function of the contact line inclination for two different flow rates (a); dependence of the contact angle on contact line velocity, for $\dot{V} = 0.75, 1.5, 2.0, 3.0 \times 10^{-2}\,\mathrm{cm^2/s}$ (b).

patch is captured and the contact angle calculated. Figure 5(a) demonstrates that the dynamic contact angle strongly depends on the position, here identified by the angle β between the normal to the contact line and the x direction ($\beta = 0$ at the patch apex). The contact angle θ decreases for increasing value of β. This trend is confirmed by the relation $\tan\theta \propto x^{-\frac{3}{4}}$, obtained via post-processing of the results of [13], where the authors solved analitically a simplified version of Eqs. (3) and (4). Two different shapes of θ as a function of β, see figure 5(a), were obtained for two different flow rates: the discrepancy between the two curves is due to the additional dependence of θ from the contact line velocity. In figure 5(b) three values of β are chosen and the contact angle θ is plotted, for each β, against the contact line velocity, showing that Eq. (26) is still valid. The contact line velocity is calculated numerically, tracing shape and position of the contact line when marching in time, under the assumption that it always moves along its normal direction.

Comparing the performances of the implicit code with the explicit one on the same test case of figure 1 reveals a gain of about $10^2 - 10^3$ on Δt and a corresponding strong reduction in terms of simulation time, of the order of 10^{-2}. Moreover, time step of about 10^6 the Neumann limit, Eq. (19), are reached running the ADI code with spatial discretization step of about h_{wet}.

8. Conclusion

An implicit code was implemented in order to predict the distribution of a gravity driven film over an inclined plane under the assumpions typical of lubrication theory. The code was first validated generating a dry-patch by the implementation of an obstacle and comparing its shape with the one predicted by a literature empirical model of Podgorski et al. [10].

The dry-patch stability analysis allowed to define a critical flow rate, above which the dry-patch moves under the action of gravity. Such threshold value is of obvious importance for a number of practical implementations including heat transfer and film flows. As confirmed also by experimental evidences [10, 11], the model of Hartley and Murgatroyd [1] is not able to predict whenever a dry-patch is stable or not. A novel symplified model for the critical mass flow rate, useful as a quick and easy global threshold value for practical applications, was thus described and validated against numerical predictions.

Once the disjoining pressure model for describing the surface wettability had been validated for a 1D film flow, the dynamic contact angle along the contact line of a moving dry-patch was analyzed: the contact angle decreases when the liquid flows along the contact line tangential direction, in agreement with previous analytical evidences.

The model allows for the analysis of shear driven film and the numerical solution of lubrication equations can be applied to more complex problems, including multiple dry-patch generation, dynamics of drops growth and coalescence or problems involving condensation and evaporation. Moreover, practical problems, such as in-flight icing phenomena or fogging and defogging simulations on car windshield, involve complex geometries, which may require unstructured grids for an easier description of the physical domain. However, even in the present structured grid form, the code may provide a useful insight on the physical mechanisms driving the evolution of a discontinuous film, as well as important quantitative indication on proper threshold values for possible film breakups and transitions.

Nomenclature

B	disjoining coefficient ($\mathrm{N\,m^{-2}}$)
$C_{1,2}$	THV formula coefficients
g	acceleration of gravity ($\mathrm{m\,s^{-2}}$)
h	film thickness (m)
L	length (m)
m, n	disjoining exponents
p	pressure ($\mathrm{N\,m^{-2}}$)
$\boldsymbol{Q}$	mass flow per unit length ($\mathrm{m^2\,s^{-1}}$)
S	section area ($\mathrm{m^2}$)
t	time (s)
$\boldsymbol{u}$	flow velocity vector ($\mathrm{m\,s^{-1}}$)
$\dot{V}$	undisturbed film flow rate ($\mathrm{m^2\,s^{-1}}$)
x, y	Cartesian coordinates (m)

Greek symbols:

α	inclination of the plate (deg)
β	angle between the normal to the contact line and the plate downhill direction (deg)
γ	aspect ratio
θ	contact angle (deg)
μ	dynamic viscosity (Pa s)
Π	disjoining pressure ($\mathrm{N\,m^{-2}}$)
ρ	density ($\mathrm{kg\,m^{-3}}$)
σ	surface tension ($\mathrm{N\,m^{-1}}$)
$\boldsymbol{\tau}$	shear stress ($\mathrm{N\,m^{-2}}$)

Dimensionless numbers:

$\mathrm{Bo} = \rho\, g\, {h_\infty}^2/\sigma$	Bond number
$\mathrm{Ca} = \mu\, u_\infty/\sigma$	Capillary number
$\mathrm{Re} = \rho\, u_\infty\, h_\infty/\mu$	Reynolds number
$\mathrm{We} = \rho\, {u_\infty}^2\, h_\infty/\sigma$	Weber number

Subscripts and superscripts:

$\bar{\ }$	dimensionless
∞	undisturbed film
a	external gas flow
cr	critic
e	equilibrium
i, j	number of the grid element
n	number of the time step
o	obstacle
r	capillary ridge
wet	precursor film

IOP Conf. Series: Journal of Physics: Conf. Series **796** (2017) 012038 doi:10.1088/1742-6596/796/1/012038

References

[1] Hartley D E and Murgatroyd W 1964 *Int Journal of Heat and Mass Transfer* **7** 1003
[2] De Candido E, Croce G and D'Agaro P 2012 *Heat Transfer Engineering* **33**(13) 1130
[3] Witelski T P and Bowen M 2003 *Applied Numerical Mahtematics* **45** 331
[4] Schwartz L W and Eley R R 1998 *Journal of Colloid and Interface Science* **202** 173
[5] Zhao Y and Marshall J S 2006 *Journal of Fluid Mechanics* **559** 355
[6] Sellier M 2015 *Int Journal of Multiphase Flow* **71** 66
[7] Weidner D E, Schwartz L W and Eres M H 1997 *Journal of Colloid and Interface Science* **187** 243
[8] Mata M R and Bertozzi A L 2011 *Journal of Computational Physics* **230** 6334
[9] Bourgault Y, Habashi W G, Dompierre J and Baruzzi G S *International Journal for Numerical Methods in Fluids* **29** 429
[10] Podgorski T, Flesselles J M and Limat L 1999 *Physics of Fluids* **11** 845
[11] Rio E and Limat L 2006 *Physics of Fluids* **18** 032102
[12] Diez J A and Kondic L 2002 *Journal of Computational Physics* **183** 274
[13] Wilson S K, Duffy B R and Davis S H 2001 *Euro Journal of Applied Mathematics* **12** 233
[14] Suzzi N, Croce G and D'Agaro P 2015 *Journal of Physics: Conference Series* **655** 012041

Experimental characterization of a silicone oil-in-water droplet generator based on a micro T-junction

B Rostami, B Pulvirenti, G Puccetti and G L Morini

DIN – Alma Mater Studiorum Università di Bologna, Laboratorio di Microfluidica, Via del Lazzaretto 15/5, Bologna 40131, Italy

Corresponding Author Email: behnam.rostami2@unibo.it

Abstract. This paper deals with the emulsion of two immiscible fluids in a micro T-junction. An opposed-flow micro T-junction obtained by means of square microchannels (with a side of 300 μm) fabricated in a pure fused glass chip has been used for the formation of silicone oil-in-water (O/W) droplets. The experimental results have been obtained by considering both pure deionized water and a mixture of deionized water and surfactant (Tween 20) as the continuous flow. The results shown in this paper highlight that the presence of surfactant, also in very small concentrations, is able to change drastically the flow patterns of the two-phase flow generated by the T-junction. Concentration in weight of Tween 20 between 1 and 2% in the continuous flow is able to promote highly monodispersed emulsions with low polydispersity, especially for low flow rate ratios between the dispersed and continuous phase flows. On the contrary, by avoiding the use of surfactant, a stratified flow is obtained. The experimental results obtained in this work have been used in order to link the depth ratio of the stratified flow and the non-dimensional length of the plugs in droplet-based flow to the flow rate ratio between the dispersed and continuous flows.

1. Introduction

Micro T-junctions can be employed in order to obtain stable emulsions characterized by droplets with regular dimensions of the orders of tens/hundreds of microns. Microdroplets are now widely used in DNA and blood analysis [1], chemical reactions [2], food processing [3] and drug discovery [4]. Stable emulsions are characterized by droplets (dispersed or droplet forming phase) generated in the liquid core (continuous phase). The most studied emulsions in literature are those obtained by mixing water and oil; micro T-junctions and flow-focusing microdevices are the most preferred droplet generators due to their ease in fabrication and their capability to generate stable emulsions with droplet sizes easily controlled by means of the flow rates of the continuous and dispersed phase imposed at the inlet of the device. Thorsen *et al.* [5] were the first researchers who proposed the use of a T-junction into a microdevice for the generation of droplets [5, 6] as well as bubbles [7].

About the modelling of these devices, Garstecki *et al.* [8] observed that the squeezing mechanism is the main mechanism of the droplet formation and they proposed a simple physical model for the analysis of squeezing which is actually the basis of the developed correlations for these devices. Steegmans *et al.* [9] and Xu *et al.* [10] concentrated their attention on the physical mechanisms which control the droplet breakup. Guillot and Colin [11] and Tan *et al.* [12] focused their analysis on the influence of the continuous and dispersed phase flow rates on the control of the flow patterns

IOP Conf. Series: Journal of Physics: Conf. Series **796** (2017) 012039 doi:10.1088/1742-6596/796/1/012039

generated. Their conclusion is that the flow rate ratio α between the dispersed and continuous phase ($\alpha=Q_d/Q_c$) is the main responsible of the transition between different flow patterns.

Series of analysis have been focussed on the influence of the junction geometry on the characteristics of the two-phase flows; Das and Das [13], Raj *et al.* [14] and Steegmans *et al.* [15] studied the emulsion production by means of Y-junction by changing the angle between the inlet and main channel while Yeom and Lee [16] studied λ-junction. However, by observing the literature, the cross-flow and perpendicular T-junction geometries are the most studied configurations for the generation of emulsions [17]. The T-junction can be used under two operative configurations: the cross-flow and the opposed-flow configurations. Qian and Lawal [18] suggested using the opposed-flow configuration when gas bubbles are generated in a liquid continuous flow. Recently, Shui *et al.* [19] tested the opposed-flow configuration also for droplet generation by using both O/W and W/O droplets.

In this paper an experimental campaign focused on the analysis of the generation of stable silicone oil-in-water droplets in a T-junction under opposed-flow configuration is described. Firstly, the two-phase flow generated by means of a micro T-junction while no surfactant is added to the continuous phase is analyzed in a wide range of volumetric flow rates of oil (from 5 to 100 ml/h) and water (from 5 to 120 ml/h). Secondly, the effect of the presence of a small concentration of surfactant (=2 wt%) in the continuous phase is analyzed by observing the variation of the flow pattern due to the presence of the surfactant for each couple of volumetric flow rates (water from 1 to 25 and oil from 0.1 to 13 ml/h) imposed at the inlet of the T-junction.

2. Experimental apparatus

The experimental test rig shown in figure 1 is composed of an inverted microscope (1, *Nikon Eclipse TE2000-U*) with a double system of illumination obtained by means of a Mercury lamp (2a, *Nikon C-SHG1*) which is used to illuminate the T-junction from the bottom and a COB LED lamp (2b, *100W, 9000 Lumens*) which is able to illuminate the microdevice from the top. An air immersion lens with a numerical aperture NA=0.25 and magnification M=10x (3, *Nikon, CFI DS 10X*) is used to zoom the T-junction; a high-speed camera (4, *Olympus I-speed II*) is connected to the inverted microscope in order to be able to acquire a series of images with a maximum frame rate of 5000 fps. A LCD monitor (5) is used for a run-time visualization of the flow within the microdevice. In order to impose the silicone oil and the water volumetric flow rates at the inlet of the junction, two independent syringe pumps (6, *Harvard Apparatus PHD4400 Programmable* and 7, *Cole-Parmer Version Hills*) are employed.
By considering the declared accuracy of the syringe pumps, the uncertainty on the imposed volumetric flow rate Q_c and Q_d is equal to ±0.5% and ±0.35%, respectively.

A commercial micro T-junction (*Translume Co.*) has been chosen in this study; it is obtained by means of two perpendicular square microchannels having a side of 300 μm; the microchannels are microfabricated on a fused silica glass chip. The top surfaces of the microchannels are polished (optical polishing); on the contrary, the sidewalls and the bottom walls (translucent floor) of the channels are characterized by a random roughness of the order of 1 micron. The top surfaces show hydrophilic behaviour; on the contrary the other walls might be considered slightly hydrophobic.

All the square microchannels have the same width and depth, denoted as w_c, h_c for the continuous phase and w_d, h_d for the dispersed phase channel. The main characteristics useful to control the droplet generation within the T-junction are: the flow rate Q associated to the continuous (Q_c) and dispersed phase (Q_d), the main physical properties of the fluids involved (density ρ, dynamic viscosity μ) for both continuous and dispersed phase fluids and the interfacial tension σ between the liquids. The Buckingham-Pi theorem shows that the fluid dynamic behaviour of the T-junction is a function of the following dimensionless quantities: $\Lambda=w_d/w_c$, $L_c=h_c/w_c$, $L_d=h_d/w_d$, $\alpha=Q_d/Q_c$, $\lambda=\mu_d/\mu_c$, named width ratio, aspect ratio of the continuous and dispersed phase, flow rate ratio and viscosity ratio, respectively. The Capillary number ($Ca=Q\mu/\sigma wh$) and Reynolds number ($Re=\rho Q/\mu h$) associated to the dispersed and continuous phase. In this case, since the T-junction is made up of square microchannels, all the geometrical ratios Λ, L_c and L_d, are equal to unity.

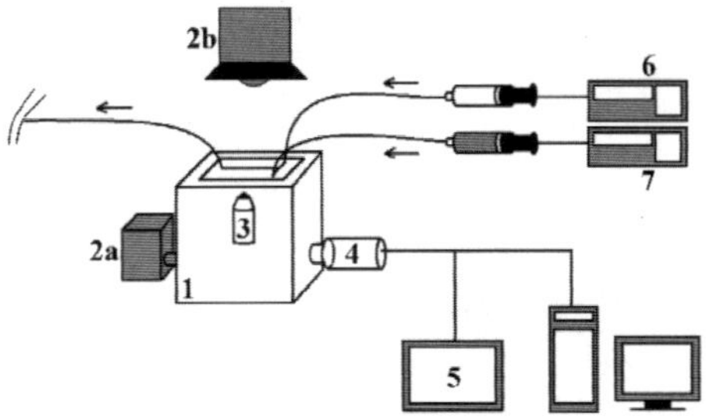

Figure 1. The experimental setup

The T-junction has been tested in this work under opposed-flow configuration. In figure 2a sketch of the T-junction with the indication of the position of the inlets of the dispersed and of the continuous phase is given: the continuous and dispersed phase meet at the junction head-to-head.

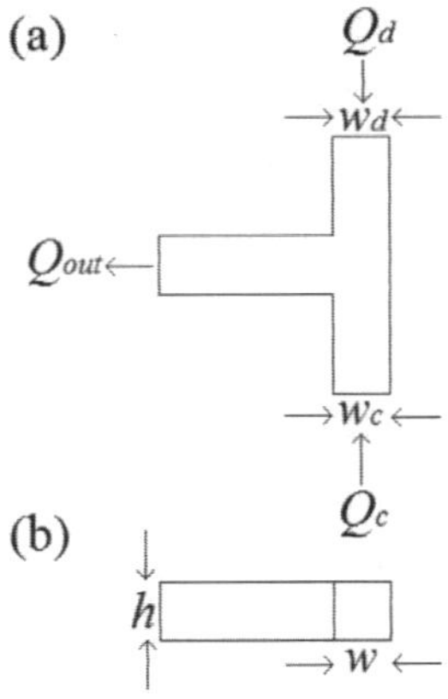

Figure 2. Sketch of the top (a) and side view (b) of the opposed-flow T-junction tested in this work

The silicone oil (Sigma Aldrich, viscosity 20 cSt, density 0.95 g/ml @ 25 °C) has been employed as the dispersed phase in the tests and the deionized water (viscosity 0.89 cSt, density 0.998 g/ml @ 25 °C) is considered as the continuous phase. Tween 20 (Sigma Aldrich, density 1.095 g/ml @ 25 °C) is used in a series of experimental tests as surfactant added to the deionized water in small concentrations (<2 wt%) in order to help the generation of stable oil-in-water emulsions (O/W droplets). Tween 20 (miscible in water) is a non-ionic detergent widely used in biochemical applications. The Hydrophilic-Lipophilic Balance (HLB) value of a surfactant is generally used to determine if the surfactant is hydrophobic or hydrophilic. Emulsifiers with high HLB values enhance the production of oil-in-water; on the contrary, surfactants with low HLB values give water-in-oil (W/O) emulsions [19]. Tween 20 is a hydrophilic surfactant having HLB=16.7 and it can be proficiently used for O/W emulsions.

3. Results

3.1. The effect of surfactant

Since the surface effects are dominant at microscale, small concentrations of surfactants added to the continuous phase can strongly enhance the droplet generation in microdevices as demonstrated by many papers [19, 20]. The surfactant adsorption in liquid-liquid systems acts dynamically in the droplet formation and strongly influences the values assumed by the interfacial tension σ. If σ is reduced, the mixing energy decreases and, as a consequence, the droplet generation becomes easier.

The presence of surfactant added to the continuous phase is more important when a T-junction under opposite-flow configuration is tested; in this configuration, in fact, the mixing tends to generate stratified flow instead of droplets at the exit of the junction, so the presence of a surfactant can help to create stable emulsions, as demonstrated in [19].

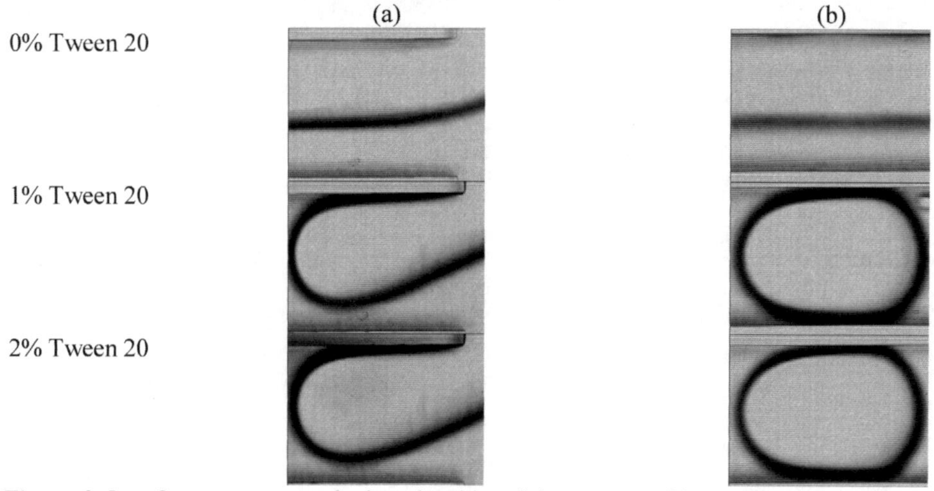

Figure 3. Interface geometry at the junction (a) and downstream (b) as a function of the concentration of Tween 20 added to the water when Q_c=25 ml/h and Q_d=5 ml/h (α=0.2)

Figure 3 shows the effect of Tween 20 on the silicone oil droplet formation in water. A volumetric flow rate of water equal to 25 ml/h has been imposed at the entrance (bottom-top direction in figure 3) against a volumetric flow rate of 5 ml/h of silicone oil (top-bottom direction in figure 3). The images are recorded by means of the speed camera and the evolution of the interface is reconstructed by means of a Matlab code using the Image Toolbox functions [21]. Without surfactant, the streams are not able to generate a droplet regime but they originate a stratified flow downstream of the junction.

Now, the same test has been repeated by adding Tween 20 to the water flow; since the critical micelle concentration (CMC), defined as the surfactant concentration above which micelles are spontaneously formed, of Tween 20 in water is equal to 0.0074% (w/v), values of concentration larger than 1% are more than sufficient in order to reduce to its minimum value the interfacial tension σ.

For this reason, two values of concentration of Tween 20 in water have been used during the tests: 1 wt% and 2 wt%. The largest value has been used in order to verify if the minimum value of the interfacial tension has been reached when the lowest concentration value is adopted.

By adding Tween 20 in water with a concentration of 1 wt%, the geometry of the interface at the junction drastically changes due to the reduction of σ; now, the largest flow rate of water (with respect to the silicone oil flow rate) is able to squeeze the silicone oil against the edge of the T-junction by generating the detachment of a regular oil droplet, as evidenced by the image taken downstream the T junction (figure 3b). In this condition a stable O/W emulsion is generated.

The results shown in figure 3 demonstrate that the increase of concentration between 1 to 2% is not able to produce any change on the geometry of the interface at the junction (which means the droplet formation mechanism is the same) and on the features of the oil droplets obtained downstream.

3.2. Water without surfactant

As evidenced by figure 3, without surfactant the flow regime after the T-junction is typically a stratified regime in which two parallel streams of silicone oil (on the top in figure 3) and water (on the bottom in figure 3) fill the cross section of the channel in laminar regime. Twenty three different experimental runs have been carried out by varying the inlet flow rates in a wide range from 5 up to 120 ml/h. For each value assumed by the flow rate ratio α^{-1} ($=Q_c/Q_d$), the ratio between the depth of the water (H_c) and the oil (H_d) streams has been calculated. For a fixed viscosity ratio (in this case

μ_c/μ_d=0.0467) and fixed geometry and flow configuration of the T-junction, the depth ratio H_c/H_d associated to the stratified flow, easily obtained by the image post-processing as shown in figure 4a, is a function of α. The uncertainty on H_c and H_d is linked to the thickness of the interface which depends on the illumination system. In this case the typical thickness of the interface is of the order of 14 pixels which corresponds to an uncertainty of 19 μm. The trend of the depth ratio as a function of flow rate ratio is shown in figure 4b.

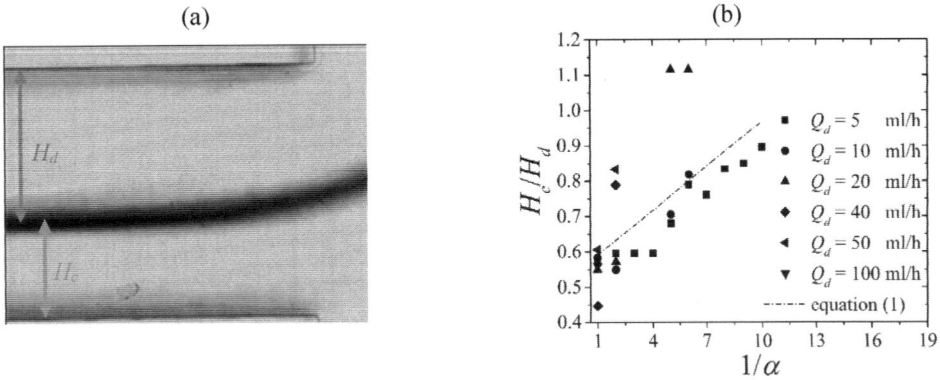

Figure 4. A sketch with H_c and H_d (a) and the trend of the depth ratio H_c/H_d for a stratified flow of water and silicone oil as a function of the flow rate ratio α in an opposite-flow T-junction (b)

The results show that, especially for low oil flow rates (Q_d<10 ml/h), the depth ratio tends to show a linear dependence on the flow rate ratio α. For large oil flow rates all the results tend to collapse to the same value in correspondence of the minimum value of α tested.

The experimental values of H_c/H_d, scale with the correlation:

$$H_c/H_d = 0.549 + (0.042/\alpha) \tag{1}$$

which is represented in figure 4b by means of a dash-dot line. It is interesting to note that the viscosity ratio μ_c/μ_d in this case is equal to 0.0467, approximately similar to the slope of the linear correlation. In this sense, the experimental results suggest that the expression of the correlation can be generalized as follows:

$$H_c/H_d = a + (b/\lambda\alpha) \tag{2}$$

where λ is the viscosity ratio (μ_d/μ_c). In this way the role of the viscosity ratio on the value of the layer depths of the stratified flow is highlighted. About the coefficients a and b introduced in equation (2), they can be seen as a function of the T-junction geometry and the width ratio $\Lambda=w_d/w_c$. More experimental tests with T-junctions having different geometries with different working fluids are needed in order to determine the right value of these coefficients. The depth ratio is strongly influenced by the flow rate of the fluid having lower viscosity (i.e. water in this work); on the contrary, variations of the flow rate linked to the fluid with higher viscosity (i.e. silicone oil) make marginal variations of the depth ratio.

When the flow rate ratio α is increased, the oil droplets start to be generated. For a fixed silicone oil flow rate, a critical value of the water flow rate there exists above which the droplet generation begins. For a fixed oil flow rate of Q_d=5 ml/h the droplet generation starts only when Q_c>80 ml/h; for Q_d=10 ml/h, the droplet generation starts for Q_c>60 ml/h. The critical flow rate ratio in correspondence of which the droplet generation becomes possible seems to have a non-monotonic trend as evidenced by the flow pattern map shown in figure 5. Maybe it is anticipated to see the droplet generation when

IOP Conf. Series: Journal of Physics: Conf. Series **796** (2017) 012039 doi:10.1088/1742-6596/796/1/012039

Q_c>45 ml/h for Q_d=20 ml/h, but it seems that with the increase in Q_d, the same trend cannot be seen due to the power of the oil flow which does not allow to be pinched by the continuous flow.

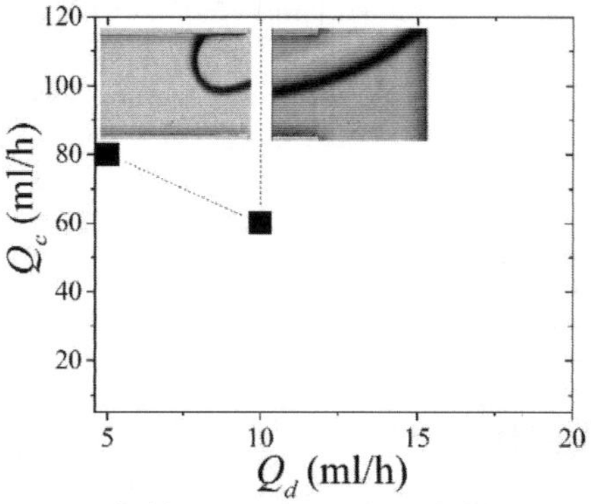

Figure 5. Flow pattern map for water (without surfactant) and silicone oil mixtures as a function of the imposed dispersed (silicone oil) and continuous (water) flow rates

3.3. Water with surfactant

As explained above, the use of Tween 20 added to water changes the flow patterns generated with the T-junction and the droplet formation is greatly enhanced. In this section, all the experimental results have been obtained by adding to water Tween 20 with a concentration equal to 2% in weight.

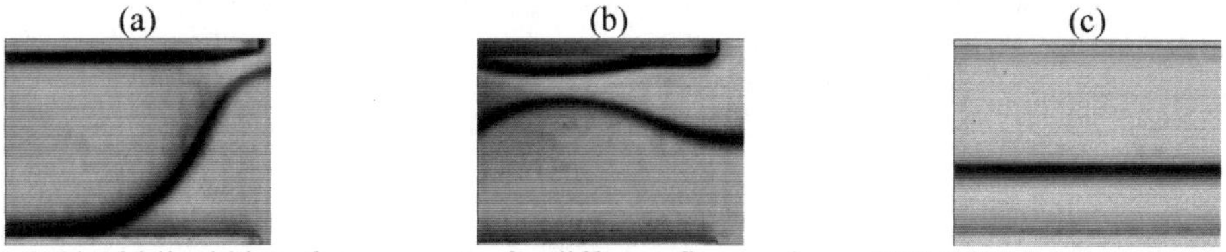

Figure 6. Liquid-liquid interface geometry for different flow regimes: DTJ (a), DC (b) and SF (c)

The water flow rate has varied between 1 to 25 ml/h while the silicone oil flow rate is changed from 0.1 to 13 ml/h. In this case, stratified flow is observed only for very high flow rates of silicone oil. Working with a wide range of flow rates both for dispersed and continuous phase enables the observation of different flow patterns. Three different flow regimes have been identified, namely droplet generation at T-junction (DTJ), downstream of the main channel (DC) and stratified flow (SF). The transition from one flow pattern to another one occurs when the imposed flow rates at the inlets are changed. Figure 7 shows the typical trend of the liquid-liquid interface for the three regimes defined above. In DTJ regime the detachment of the droplet occurs at the edge of the T-junction (see figure 6a), in DC regime the detachment of the droplet occurs downstream (see figure 6b) an in SF regime no droplets are observed (see figure 6c). For low values of α (<2), the droplet generation at the T-junction (DTJ) is mainly due to the squeezing mechanism; the silicone oil is squeezed by water against the edge of the T-junction by generating the detachment of a regular oil droplet. On the contrary, when α is increased the droplet generation occurs downstream of the channel (DC). When the oil flow rate is increased a stratified flow occurs (SF). In figure 7 the three different flow patterns are shown as a function of α and the water flow rate (Q_c). It is clear that in lower water flow rates, the larger values of (α) are needed in order to obtain the passage from one pattern to another one. By increasing α, the detachment point of the droplet moves downstream along the main channel (DC) and the thread formed before detachment increases in size and an elongated filament is observed in the main channel. Stratified flow (SF) is observed only for large values of α.

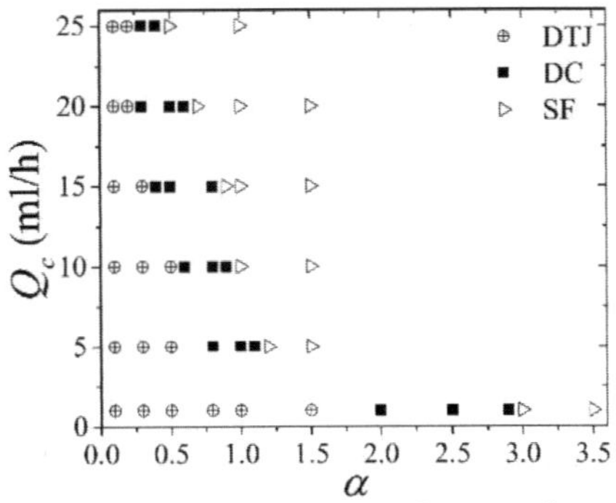

Figure 7. Flow pattern map for a T-junction in opposite-flow configuration as a function of flow rate ratio α (silicone oil-water with a concentration of 2 wt% of Tween 20)

The non-dimensional length of the generated droplets, defined as the ratio between the droplet axial length and the channel width ($\bar{L}=L/w$), is shown in figure 8 as a function of α for both DTJ and DC regimes. The dimensionless length of the droplets tends to increase when α increases. Figure 8 confirms that an increase in Q_c or a decrease in Q_d will produce smaller droplets both in DTJ and DC regimes. When Q_c is increased, the droplets are pinched off earlier and the plug size is reduced. On the other hand, in DTJ regime, an increase in α will result in longer droplets; the experimental data seem to put in evidence that the droplet volume is influenced by the imposed flow rates of both fluids and not only by α [10, 20, 22].

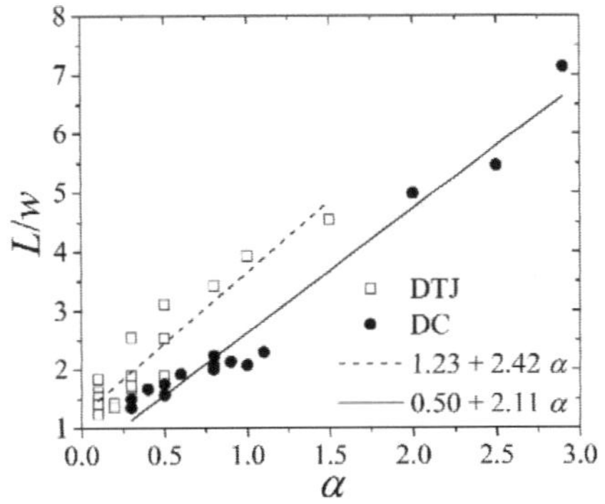

Figure 8. Non-dimensional length of the droplets as a function of flow rate ratio α for different continuous flow rates (empty symbols: DTJ regime; full symbols DC regime)

The increase in α (by increasing Q_d) does not affect the droplet volume significantly; this result indicates that the dispersed flow rate influences the droplet volume through the formation time and it becomes important when low continuous flow rates are imposed at the inlet of the junction. The breakup mechanism in DC regime is similar to the mechanism responsible of the droplet formation in DTJ regime but in DC regime the outlet channel is never completely blocked and the shear force plays a more important role in droplet detachment. By increasing α, the shear force becomes the most important force which governs the fluid flow.

The typical values of polydispersity, defined as the standard deviation of the droplet size distribution divided by the mean droplet size, are reported in table 1. NA means that no experimental test has been carried out under these conditions. For low Q_c, highly monodispersed droplets are generated especially in DTJ regime; it can be observed that the typical values of polydispersity are very low. On the contrary, polydispersity increases in DC regime and very different droplets in size may be obtained especially before the transition from DC to SF regime. Generally, by increasing Q_c

and/or α, the polydispersity increases. $\bar{L}$ obtained in DTJ regime as a function of α shows a linear trend of the data in agreement with the relationship proposed by many authors [22] for squeezing ($\bar{L}=\varepsilon+\omega\alpha$) where ε and ω are two fitting constants which depend on the channel geometry [10].

Table 1. The polydispersity of $\bar{L}$ as a function of the imposed values of α and Q_c (Q_c in ml/h)

	Q_c					
α	1	5	10	15	20	25
0.1	0.0233	0.0128	0.0199	0.0121	0.0204	0.0245
0.2	NA	NA	NA	NA	0.0177	0.0230
0.3	0.0258	0.0093	0.0124	0.0097	0.0160	0.0444
0.4	NA	NA	NA	NA	NA	**0.1164**
0.5	0.0308	0.0239	0.0112	0.0209	0.0383	NA
0.6	NA	NA	NA	NA	**0.0972**	NA
0.7	NA	NA	NA	NA	NA	NA
0.8	0.0389	0.0215	0.0249	**0.1416**	NA	NA
0.9	NA	NA	0.0354	NA	NA	NA
1.0	0.0466	0.0246	NA	NA	NA	NA
1.1	NA	0.0244	NA	NA	NA	NA
1.5	0.0706	NA	NA	NA	NA	NA
2.0	**0.1209**	NA	NA	NA	NA	NA
2.5	**0.3638**	NA	NA	NA	NA	NA
2.9	**0.4616**	NA	NA	NA	NA	NA

Figure 9 shows how the fitting coefficients ε and ω have been deducted by using the experimental data; a value of 1.23 for ε and 2.42 for ω has been determined. Similar values have been suggested by other authors for different T-junctions under different flow configurations. The experimental results obey to the squeezing law but it seems that both flow rates (continuous and dispersed) and not only the ratio α influence the droplet length. In figure 9 the Y-bar indicates the droplet polydispersity.

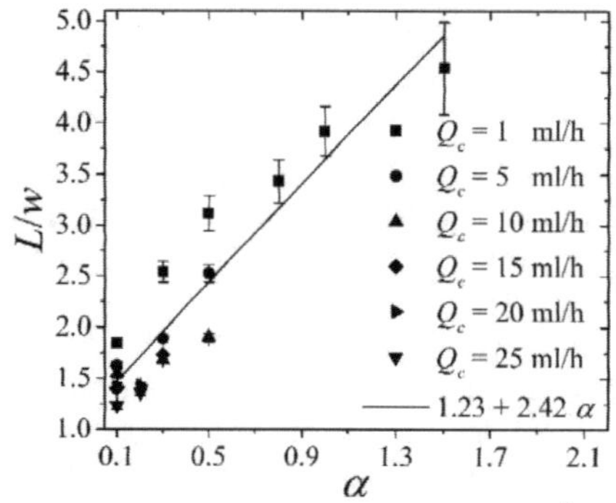

Figure 9. Non-dimensional length of the droplets as a function of α for different continuous flow rates (DTJ regime)

In figure 10 the non-dimensional length of the droplets is given as a function of the imposed continuous flow rate for fixed α. The trend shown in figure 10a is in agreement with previous work [23]; as proposed by Chiarello *et al.* [20], if the droplet non-dimensional length is normalized with $\alpha^{0.25}$ (see figure 10b), the experimental data obtained with different values of α show a more evident correlation. In addition, from figure 10 it is evident that the cases obtained by imposing a continuous flow rate (Q_c) less than 2 ml/h are characterized by a large value of polydispersity (see also table 1).

These results seem to confirm, as highlighted by other researchers for different flow configurations of the T-junction [10, 20, 22], that $\bar{L}$ in DTJ regime cannot be considered as a function of the flow rate ratio α only but it shows a power-law dependence on the continuous Capillary number, which is proportional to the continuous flow rate.

By considering $\bar{L}$ in DTJ regime both the flow rate ratio and the continuous flow rate and adopting the form of the correlations proposed by Xu *et al.* [10] and Chiarello *et al.* [20], the experimental results obtained in this work by using the T-junction under opposite flow configuration suggest to use β=0.25, in agreement with [20] and m=−0.2 in agreement with [10] for a correlation in form of $k\alpha^{\beta}Q_c^{m}$ while k=4 to obtain a good tuning with the experimental data.

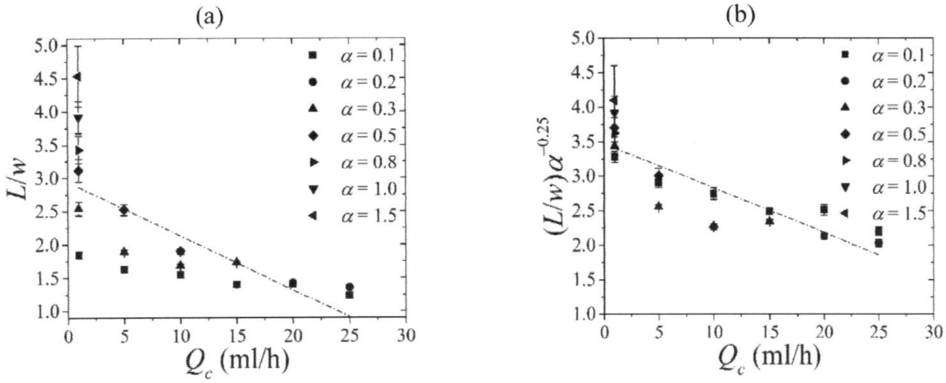

Figure 10. Non-dimensional length (a) and normalized non-dimensional length (b) of the droplets as a function of continuous flow rate for different flow rate ratios for DTJ regime

Figure 11 confirms that, for α<0.5, the power-law dependence of $\bar{L}$ on the continuous Capillary number (flow rate) and α is able to reproduce with an acceptable accuracy the trend of the experimental data obtained in this work.

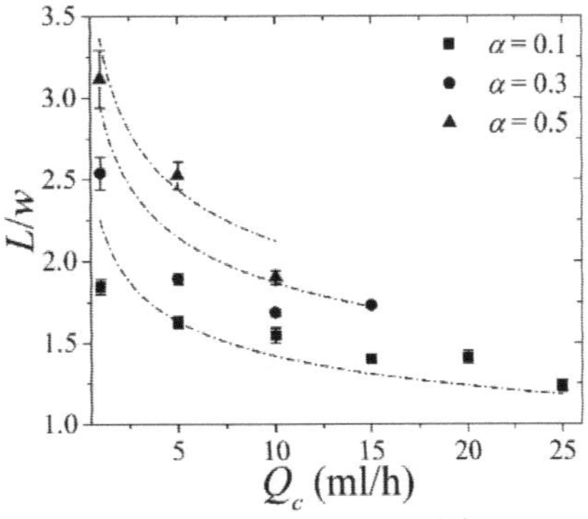

Figure 11. Non-dimensional length of the droplets compared for α<0.5 with the power-law correlation $k\alpha^{\beta}Q_c^{m}$ (DTJ regime)

4. Conclusion

In this work the production of silicone oil-in-water (O/W) droplets within a micro T-junction under opposite-flow configuration is experimentally analyzed. Firstly, the role played by the presence of small concentrations (<2 wt%) of surfactant (Tween 20) in the continuous phase on the droplet generation has been studied. It has been clearly demonstrated that the use of Tween 20 in water greatly enhances the production of silicone oil plugs; on the contrary, without the use of Tween 20 the main flow regime which can be obtained in a T-junction under opposite flow configuration is the stratified regime when the dispersed flow rate is lower than 100 ml/h and the continuous flow rate is lower than 120 ml/h. In absence of surfactant, an analysis of the depth ratio of water and oil in stratified regime has been made for a wide range of water and silicone oil flow rates; a simple correlation has been proposed in order to predict the value of the depth ratio as a function of the flow rate ratio (α) of the dispersed and the continuous phase. When Tween 20 is added to water three different flow regimes can be detected experimentally, which are namely DTJ (droplet formed at the T-junction), DC (droplet formed at the downstream of the channel) and SF (stratified flow) regimes. The experimental data confirms that the non-dimensional length of the oil plugs depends not only on the flow rate ratio α but also on the continuous Capillary number (flow rate). It has been demonstrated that the correlations proposed by Xu *et al.* [10] and Chiarello *et al.* [20] for the prediction of the non-dimensional length of the plugs still hold for a T-junction under opposite flow configuration.

References

[1] Xiang X, Chen L, Zhuang Q G, Ji X H and He Z K 2012 *Biosens. Bioelectron.* **32** 43-49.
[2] Song H, Chen D L and Ismagilov R F 2006 *Angew. Chem. Int. Edit.* **45** 7336-7356.
[3] Schemberg J, Grodrian A, Romer R, Gastrock G and Lemke K 2009 *Eng. Life. Sci.* **9** 391-397.
[4] Dittrich P S and Manz A 2006 *Nat. Rev. Drug Discov.* **5** 210-218.
[5] Thorsen T, Roberts R W, Arnold F H and Quake S R 2001 *Phys. Rev. Lett.* **86** 4163-4166.
[6] Tice J D, Song H, Lyon A D and Ismagilov R F 2003 *Langmuir* **19** 9127–9133.
[7] Gunther A, Khan S A, Thalmann M, Trachsel F and Jensen K F 2004 *Lab. Chip.* **4** 278–286.
[8] Garstecki P, Fuerstman M J, Stone H A and Whitesides G M 2006 *Lab. Chip.* **6** 437-446.
[9] Steegmans M, Schron C and Boom R 2009 *Chem. Eng. Sci.* **64** 3042-3050.
[10] Xu J, Li S, Tan J and Luo G 2008 *Microfluid. Nanofluid.* **5** 711-717.
[11] Guillot P and Colin A 2005 *Phys. Rev. E.* **72** 066301.
[12] Tan J, Xu J, Li S and Luo G 2008 *Chem. Eng. J.* **136** 306-311.
[13] Das A K and Das P K 2010 *Langmuir* **26** 15883-15894.
[14] Raj R, Mathur N and Buwa V V 2010 *Ind. Eng. Chem. Res.* **49** 10606-10614.
[15] Steegmans M L J , Schroen K and Boom R M 2010b *AICHE J.* **56** 1946-1949.
[16] Yeom S and Lee S Y 2011b *Exp. Therm. Fluid Sci.* **35** 387-394.
[17] Xu J H, Luo G S, Li S W and Chen G G 2006c *Lab. Chip.* **6** 131-136.
[18] Qian D and Lawal A 2006 *Chem. Eng. Sci.* **61** 7609-7625.
[19] Shui L L, van den Berg A and Eijkel J C T 2009 *Lab. Chip.* **9** 795-801.
[20] Chiarello E, Derzsi L, Pierno M, Mistura G and Piccin E 2015 *Micromachines.* **6** 1825-1835.
[21] Pulvirenti B, Rostami B, Puccetti G and Morini G L 2015 *J. Phys.: Conf. Ser.* **655** 012028.
[22] Liu H and Zhang Y 2011 *Phys. Fluids.* **23** 082101.
[23] Christopher G F, Noharuddin N N, Taylor J A and Anna S L 2008 *Phys. Rev. E.* **78** 036317.

Motive flow calculation through ejectors for transcritical CO_2 heat pumps. Comparison between new experimental data and predictive methods

G Boccardi[1], G Lillo[2], R Mastrullo[2], A W Mauro[2], M Pieve[1] and R Trinchieri[1]
[1] Energy Department Production, Conversion and Use of Energy Division, Laboratory of Development of Chemical Processes and Thermofluidynamic,
ENEA, C.R.Casaccia - Via Anguillarese 301, 00123 Roma (Italy)
[2] Department of Industrial Engineering, Federico II University of Naples, p.le Tecchio, 80 - 80125, Napoli (Italy)

E-mail: gino.boccardi@enea.it

Abstract. The revival of CO_2 as refrigerant is due to new restrictions in the use of current refrigerants in developed countries, as consequence of environmental policy agreements. An optimal design of each part is necessary to overcome the possible penalty in performance, and the use of ejectors instead of throttling valves can improve the performance. Especially for applications as CO_2 HPs for space heating, the use of ejectors has been little investigated. The data collected in a cooperation project between ENEA (C.R. Casaccia) and Federico II University of Naples have been used to experimentally characterize several ejectors in terms of motive mass flow rate, both in transcritical CO_2 conditions and not. A statistical comparison is presented in order to assess the reliability of predictive methods available in the open literature for choked flow conditions.

1. Introduction

In order to contribute to the reduction of the global warming phenomenon, the latest F-Gas Regulation [1] will limit the use of high GWP refrigerant fluids in different sectors. Carbon dioxide is a very interesting alternative since it is non-flammable, nontoxic, inexpensive and eco-friendly, having GWP=1. Neska et al. [2] showed high energy efficiency in the sanitary hot water production using CO_2 as refrigerant, although the energy performance strongly decreases when the gas-cooler water inlet temperature increases. In this case, lower performances are mainly due to the throttle losses in the expansion process.

However, it is possible to improve the performance of the overall system using an ejector system as substitute of a common expansion valve. Minetto et al. [3], through a numerical simulation, highlight that only the use of an ejector system can improve carbon dioxide heat pump performance leading to COP values similar to R410A system. Elbel et al. [4] and Lucas and Koehler [5] show COP and exergy efficiency improvements up to 17% using an ejector system in transcritical refrigeration cycle. Furthermore, Lawrence and Elbel [6] pointed out that the performance of the ejector strictly affects the ejector cycle performance.

Lucas and Koehler [5] experimental study shows a maximum in ejector efficiency with respect to high-side pressure. Maximum ejector efficiencies of 22% have been measured. Furthermore, the ejector efficiency decreases with increasing gas cooler outlet temperature and decreasing evaporation pressure. The same study shows that the driving mass flow rate increases with increasing high-side pressure and decreases with increasing gas cooler outlet temperature at fixed high-side pressure. Lee et al. [7] and Liu et al. [8] experimental studies show that an optimum design configuration exists, in terms of throat diameter, which maximizes the performance of the system, for fixed operating conditions.

A fundamental variable for the ejector sizing is the motive (or driving) mass flow rate; its theoretical calculation is particularly complicated because:

- there are often transcritical conditions at the ejector inlet;
- there are always sonic velocity conditions at the throat, so the driving mass flow rate is independent of the downstream conditions and depends only on the nozzle inlet conditions, Lucas et al. [9];
- there is often two phase flow in the throat.

The processing of the experimental tests allowed to obtain several data on flow rate conditions through two ejectors of different diameter.

In this paper, experimental data have been compared to the predictions of mass flow rate through throttling devices obtained by three methods: the first one is a recent method proposed by Lucas et al. [9]; the second one is based on an HEM developed from Leung ([10] and [11]); the last one is the HNE-DS method, proposed by Diener et al. [12] and used in ISO/DIS4126 [13], for the two-phase mass flow rate calculation in chocked conditions through nozzles and safety valves.

2. Prediction Methods

The accurate prediction of the two-phase mass flow rate is a quite tricky issue, due to incomplete knowledge of the complex thermal-fluid dynamic phenomena occurring between the two phases. In particular, the following issues should take into account:

- the close interaction between vapour quality and sharp changes in pressure;
- possible thermodynamic non-equilibrium (metastable conditions);
- the potential different velocity of the two phases;
- the sound velocity in two-phase flow that can change quickly. Its calculation is still a matter of discussion.

The equation for calculating the dischargeable mass flow rate "G_r" through a throttling device having a geometric seat area "A" is:

$$G_r = k_d\, G_t\, A \tag{1}$$

where G_t is the theoretical flux in an ideal (isentropic) nozzle and k_d is the two-phase "discharge coefficient". The Homogeneous Models calculates G_t assuming that the two-phase mixture is homogeneous and liquid and vapour phases run at the same velocity inside the valve. Consequently, all the physical parameters are defined via their averages values weighted on the vapour quality. The Homogeneous models can be divided into two main groups, the "HEM", Homogeneous Equilibrium Models, and "HNE", Homogeneous Non-Equilibrium Models, depending on whether the thermodynamic equilibrium at the throttling device outlet is assumed or not.

HEM method estimates G_t (Leung [10] and [11]) by the following equation:

$$G_t = \frac{\sqrt{-\omega \ln(\eta) - (\omega - 1)(1 - \eta)}}{\omega \cdot (1/\eta - 1) + 1} \cdot \sqrt{\frac{2 p_0}{v_0}} \tag{2}$$

where η is ratio between the outlet and inlet pressures p_{out}/p_0 and ω is a characteristic parameter, function of the inlet conditions only:

$$\omega = \frac{x_0 (v_{g,0} - v_{l,0})}{v_0} + \frac{c_{pl,0} T_0 p_0}{v_0} \cdot \left(\frac{v_{g,0} - v_{l,0}}{\Delta h_{v,0}} \right)^2 \tag{3}$$

Here the first term denotes the fluid compressibility at inlet conditions whilst the second one expresses the compressibility depending on the evaporation due to depressurisation. In choked conditions, in eq. (2) the pressure ratio η is replaced by the choked pressure ratio $\eta_c = p_c/p_0$ calculated by

$$\eta_c^2 + (\omega^2 - 2\omega)\cdot(1-\eta_c)^2 + 2\omega^2 \cdot \ln \eta_c + 2\omega^2 \cdot (1-\eta_c) = 0 \tag{4}$$

This model has limits in the case of short orifice, (Fletcher [14], Fisher et al. [15]) even if they refer mainly to tests on tubes and nozzle.

The ISO method [13], derived by the previous HEM method, introduces a parameter "N" related to the boiling delay and depending on η_c and x_0. This parameter, that multiplies the second term of eq. (3), is calculated by:

$$N = \left(x_0 + c_{pl,0} T_0 p_0 \cdot \frac{v_{g,0} - v_{l,0}}{\Delta h^2_{v,0}} \ln(1/\eta_c) \right)^a \tag{5}$$

N may assume values in the range; for high values the method tends to the HEM (for N=1 the equations become the same) whilst for low values the non equilibrium hypothesis prevails. The exponent *a* depends on the nozzle type and is 0.6 for short nozzle and orifices. N appears in the η calculation too and, therefore, the method requires an iterative procedure of calculation. Both methods can also be used for single phase flows but not in transcritical conditions.

Moreover, the ISO method is recommended when the inlet pressure is less than 50% of the thermodynamic critical pressure of the fluid. This suggestion depends on the fact that, in this pressure range, rapid changes in the physical properties of the fluids may occur which may cause errors on the calculation method. In the case of HEM, this recommendation is less strict because the method of calculation is different.

The CO_2 mass flow rate through an ejector can also be estimated by means of a method proposed by Lucas et al [9], obtained using the driving mass flow rates measured in an experimental campaign [16] on a multi-ejector heat pump using R744 as refrigerant.

In this method η_c, is calculated by

$$\eta_c = \frac{p_{th}}{p_{in}} = 0.0871942 \cdot \frac{p_{in}}{p_{cr}}^{0.9519907} \cdot \frac{\rho_{in}}{\rho_{cr}}^{2.348013} + 0.39387 \tag{6}$$

where p_{th} corresponds to the choked pressure.

Supposing an isentropic expansion in the converging driving nozzle, the mass vapour quality, the density and the enthalpy in the throat are calculated using the inlet entropy of the driving nozzle and the pressure at the throat obtained by (6).

With this method, the driving mass flow rate through an ejector is estimated by equations

$$u_{th} = \sqrt{2 \cdot (h_{in} - h_{th})} \tag{7}$$

$$G_t = A_{th} \cdot \rho_{th} \cdot u_{th} \tag{8}$$

This method can be used to predict the driving mass flow rate for every test conditions because it does not depend on the parameter like ω or on a rapid change in the physical proprieties of fluids.

3. Experimental setup

3.1 Test facility

The experimental data used in this paper were obtained testing a 30 kW CO_2 air-water heat pump equipped with a multi-ejector expansion pack as throttling device. The tests were performed in the experimental facility "Calorimetro Enea" at ENEA (Casaccia) research center, a climatic chamber suited to test air-to-water reversible heat pumps with thermal capacity up to 50 kW, according to UNI EN 14511/2011 [17].

The experimental set consists of an alternative semi-hermetic compressor (CP) driven by an inverter, a plate heat exchanger (GC), a finned coil (EVAP), a plate internal heat exchanger (IHE), an electronic valve (EEV) and a multi-ejector expansion pack (EJEC). Figure 1 shows the multi-ejector CO_2 system including four different ejector geometries, with throat diameters from 0.7 mm to 2.0 mm.

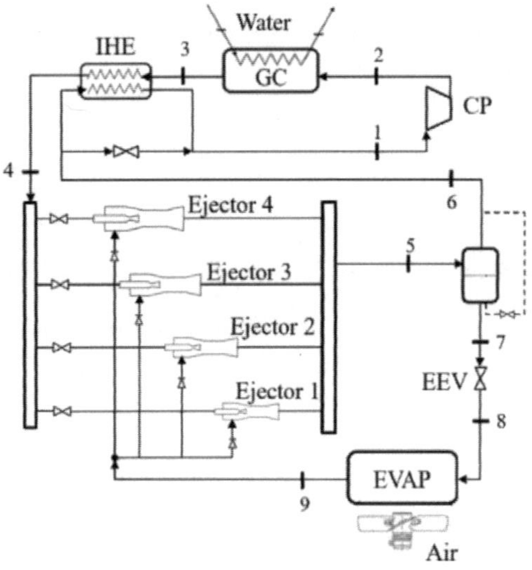

Figure 1. Layout of multi ejector CO_2 system

The heat pump control system can activate each ejector independently, depending on boundary conditions, with 16 different configurations; an ejector schematic is shown in figure 2. Temperature and pressure sensors are installed at the inlet and outlet of each component. K-type and J-type thermocouples were placed to measure the temperatures and piezoelectric transmitters to measure the pressure. The water volumetric flow rate is measured by an electromagnetic transmitter. The electrical power is measured by a wattmeter. All measurement instruments are characterized by high accuracy, according to [17].

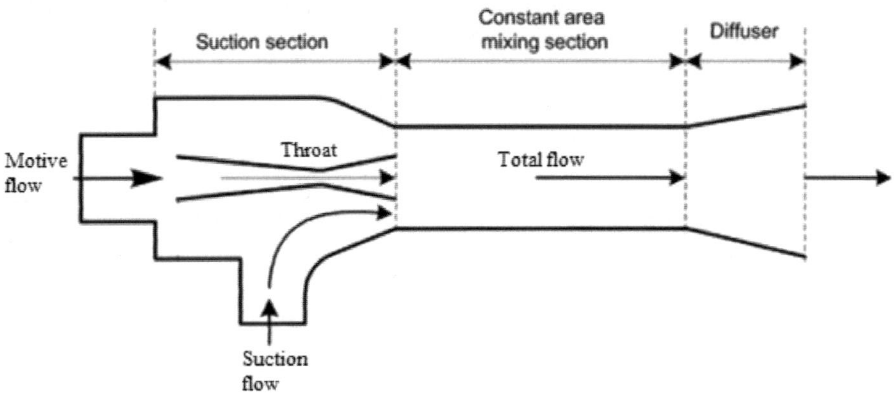

Figure 2. Schematic of the ejector system

Table 1 summarizes the instrument specifications and their uncertainties, as indicated by the manufactures.

Table 1. Measurement instrument and calibrated uncertainties.

Measurement	Range/Unit	Uncertainty
Temperature (K-type)	0/150 °C	± 1.1 K
Temperature (J-type)	-40/80 °C	± 1.1 K
Pressure	0-60/0-100/0-160 bar	0.08%
Water volumetric flow rate	0/200 l/min	0.02% of reading
Electrical power	0/25 kW	Precision class 0.5

3.2 Experimental procedure

All the tests were run according to [17]. The boundary conditions in terms of air temperature, relative humidity, water temperature and water mass flow rate were fixed and kept constant during the tests. Also, it was possible to set the outlet evaporator and inlet compressor superheating. Furthermore, it was possible to set manually the chosen ejectors configuration; for some test situations, two ejectors worked one by one. The trends of the thermodynamic measured variables were monitored via a software designed specifically. The main thermodynamic parameters were recorded and processed using Matlab software. The instabilities generated by the control system regulation are limited according to [17].

The motive mass flow rate has been calculated with an energy balance at the gas-cooler (the condenser of a CO_2 heat pump) with the equation:

$$G_{MF} = \frac{Q_{GC}}{h_{IN,GC} - h_{OUT,GC}} \tag{9}$$

The effect of variables uncertainties on the uncertainty of this calculation has been evaluated according to equation:

$$u_c(G_{PF}) = \sqrt{\left[\left(\frac{\partial G_{PF}}{\partial \dot{Q}_{GC}}\right)^2 \cdot u_c{}^2(\dot{Q}_{GC})\right] + \left[\left(\frac{\partial G_{PF}}{\partial \Delta h_{GC}}\right)^2 \cdot u_c{}^2(\Delta h_{GC})\right]} \tag{10}$$

and, for each test, it has been lower than ±7%.

4 Results and discussion

Table 2 shows the available experimental data subdivided by the inlet pressure range and the throat diameter.

Table 2. Experimental test subdivision

Throat ϕ [mm]		1.41	2
Not transcritical pressure conditions	$p<p_{cr}$	7	8
Transcritical pressure conditions	$p\geq p_{cr}$	18	12
Total		25	20

In NTr conditions, we have used Lucas correlations, HEM and ISO methods for calculating driving mass flow rate; with $p>p_{cr}$ we only able to use the Lucas correlations. For a proper comparison among the three methods, consistently to the isentropic expansion hypothesis assumed by Lucas, in (1) we set k=1.

In order to evaluate the prediction reliability of every model, a parameter r, defined as

$$r = \frac{G_t}{G_r} \tag{11}$$

(the ratio between the mass flow rate of the model and the actual mass flow rate), has been introduced; of course, the more r is close to 1, the more the prediction is good.

The figure 3 depicts r as a function of p_{in}; the different markers denote the predictions of three methods. The parameter r is better for the Lucas correlations and the values are almost constant with the p_{in}. HEM presents a similar trend with a r difference of almost 0.1 less. ISO predictions increase with p_{in} decreases; for the lowest p_{in}, r is the better.

Figure 4 shows p_c as a function of p_{in}. The p_c values are very important for mass flow rate prediction because it is present both in eq. (2) as η and in the Lucas correlations. In Figure 4 it is possible also to observe as Δp (difference between p_{in} and p_c) changes with p_c decreasing for the three methods: HEM and ISO show always Δp values between 10-15 bar, whilst Lucas correlations predict p_c values almost constant.

When p_{in} is lower than 60 bar, HEM and ISO p_c predictions are comparable with Lucas ones: in figure 3, for these p_{in} range, also r values are close each other. In particular, in the case of lowest p_{in}, corresponding to ISO and HEM p_c values lower than Lucas, ISO has the best r.

The first two figures indicate how the HEM method is good while ISO method gives acceptable results only with $p_{in}< 60$ bar, with a trend towards better performance with decreasing pressure. This behavior is consistent with the suggestion found in [13] even if p_{in} is far greater than to the suggested value (<50% p_{cr}, corresponding to < 36.88 bar for CO_2).

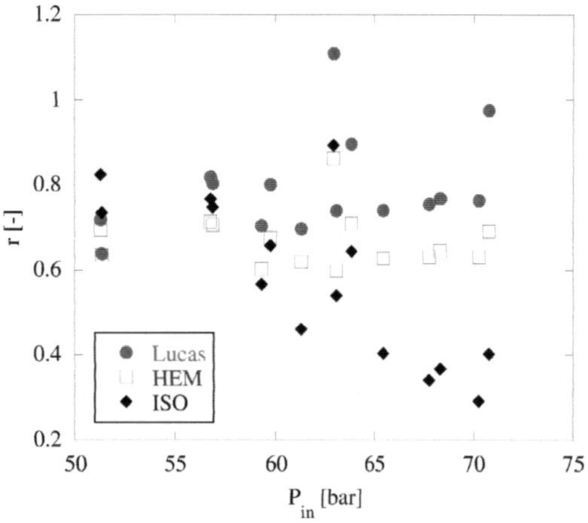

Figure 3. r calculated by Lucas, HEM, ISO methods vs. p_{in}

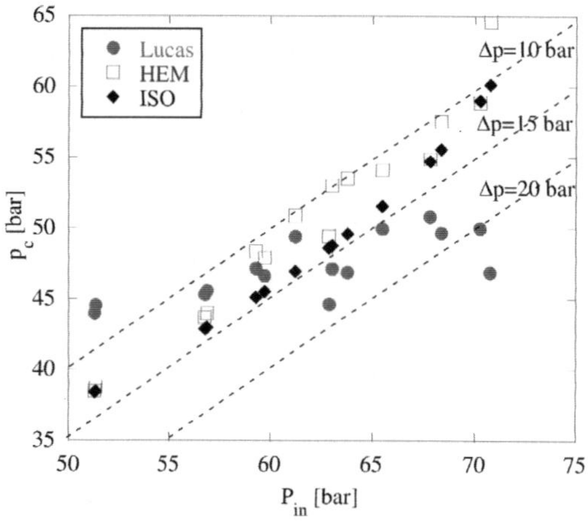

Figure 4. p_c calculated by Lucas, HEM, ISO methods vs. p_{in}

Hereafter we compare only HEM and LUCAS predictions. The vapour quality is an important parameter for evaluating the performance of a model in case of two-phase flow (§. 2). The development of the calculations with the correlations of Lucas involves x definition at the throat; even the HEM method estimates x, but it does this at inlet conditions and more generally, such as state of the fluid. Figure 5 depicts r as function of x calculated by Lucas; for the same points, the HEM predictions on CO_2 thermodynamic state are reported. Lucas correlations calculate x values until 0.18, and *r* increasing with x. For HEM, the flow is often subcooled, with *r* values almost constant, or saturated; in these cases, *r* increases with x.

By figure 6 we try to evaluate the difference between Lucas and HEM predictions as function of difference of the calculated p_c. The data indicate that p_c value is fundamental for method performance; the difference between *r* values decreases with p_c values difference.

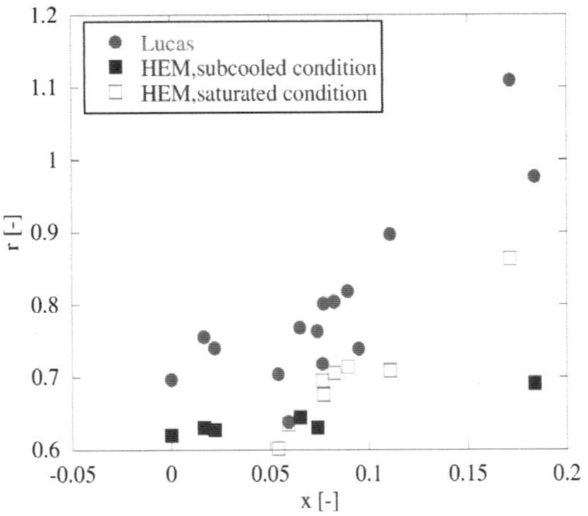

Figure 5. *r* calculated by Lucas and HEM methods vs. x

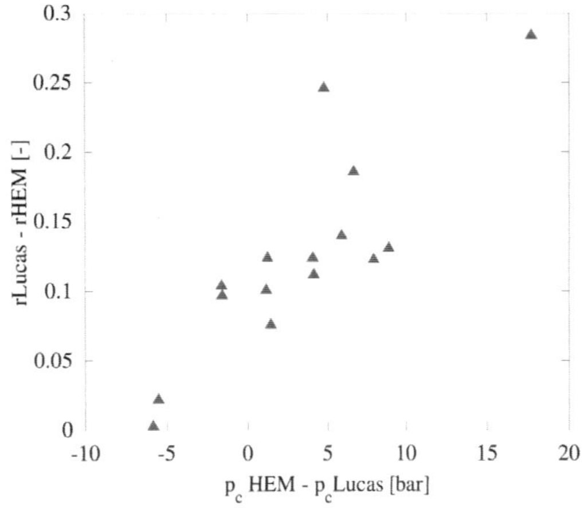

Figure 6. Difference *r* vs. difference p_c calculated by Lucas and HEM

In transcritical conditions, if we segregate the data by the throat diameters, it is possible to observe that the Lucas predictions depend on p_{in} and the predictions are as better as the throat is larger (figure 7); the *r* values are on average better than the not transcritical tests. In figure 8, it is interesting to notice the Lucas prediction behaviour in the whole range of inlet test pressure. Even when compared with HEM predictions (not transcritical tests), predictions by Lucas method do not show an evident dependence on p_{in}. This behaviour seems to be inconsistent with what previously suggested in (§.2) for the sonic conditions.

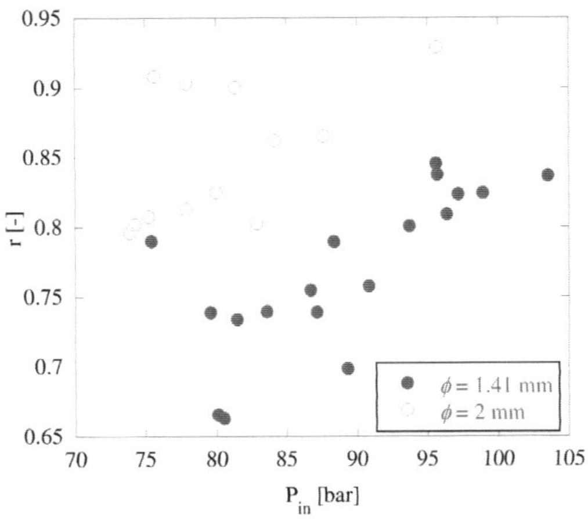

Figure 7. *r* calculated by Lucas method vs p_{in}

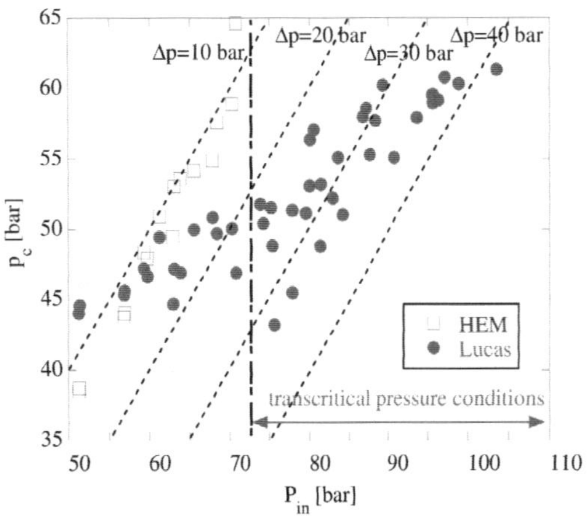

Figure 8. p_c calculated by Lucas and HEM methods vs. p_{in}

Figure 9 depicts *r* as function of x; we are not able to observe an *r* dependence on x. Instead, it is evident that tests with $\phi = 2$ mm have better predictions, as figure 7 shows.

In Figure 10, we evaluate the geometry influence (two different diameters) on the predictions of HEM models (NTr only) and Lucas as a function of the specific mass flow rate.

In the case of HEM predictions, we do not observe evident difference between the two diameters; instead, *r* Lucas values are better for 2 mm diameter orifice.

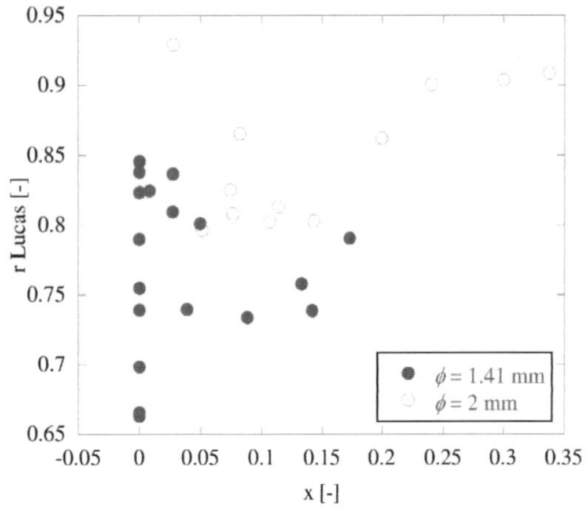

Figure 9. *r* calculated by Lucas method vs. x

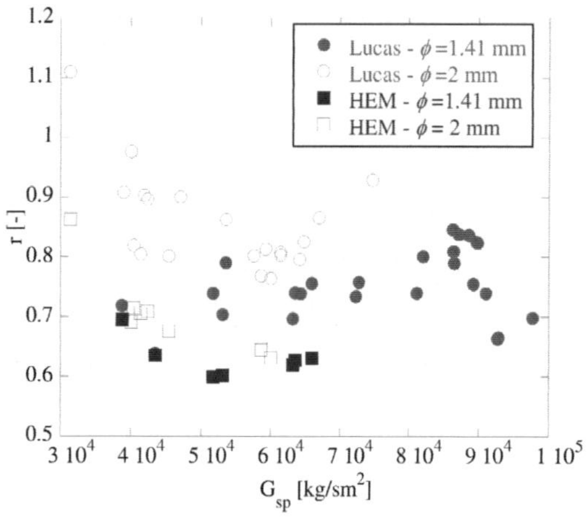

Figure 10. *r* calculated by Lucas and HEM methods vs. G_{sp}, parameter ϕ_{th}

5 Conclusions

In order to improve the knowledge about ejector performance, experimental tests were carried-out at the laboratory DTE-PCU-SPCT of the ENEA (Casaccia) research center. A complete heat pump system with multi-ejector pack was tested in a climatic chamber. Subcritical and transcritical motive nozzle inlet conditions, with high side pressure range between 50 bar and 105 bar, were investigated. Therefore, new experimental data in terms of motive nozzle mass flow rate were obtained. The results were elaborated using two existing prediction methods (HEM and ISO) and Lucas correlation [9], in order to find the best methods to evaluate the motive nozzle mass flow rate. We observed that:

- the calculation methods available for choked mass flow rate (HEM, ISO) cannot be used in transcritical conditions;
- Lucas correlations are useful at every pressure inlet conditions but, in same situations, its prediction reliability seems to be conditioned by the original database [9,16] (e.g. different performances for different geometries);
- HEM can be used in NTr condition and its performances are comparable with Lucas predictions
- ISO appears unusable for p>55 bar; under this pressure, it might operate better than the other methods.

Therefore, further studies and experimental tests are necessary to:

- ➢ Improve Lucas method with new experimental data
- ➢ Evaluate the possibility of application of HEM and ISO methods, both improving performance in not transcritical conditions and trying to expand the deployment in the transcritical range.

Acknowledgements

This work presents results related to the activities carried out with financing funds of the Electricity System Research (Italian Ministry for Economic Development, RdS - PAR 2015-16)

References

[1] Regulation (EU) No 517/2014 of the European Parliament and the Council of 16 April 2014 fluorinated greenhouse gases and repealing Regulation (EC) No 842/2006. *Off. J. Union*

[2] Neksa P, Rekstad H, Zakeri GR and Schiefloe PA 1998 CO_2-heat pump water heater: characteristics, system design experimental results. *Int. J. Refrigeration* **21** 172–79.

[3] Minetto S, Cecchinato L, Brignoli R, Marinetti S and Rossetti A. 2016 Water-side reversible CO_2 heat pump for residential application *Int. J. Refrigeration* **63**: 237-50.

[4] Elbel S and Hrnjak P 2008 Experimental validation of a prototype ejector designed to reduce throttling losses encountered in transcritical R744 system operation. *Int. J. Refrigeration* **31(3)** 411–22.

[5] Lucas C, Koehler J. 2008 Experimental investigation of the COP improvement of a refrigeration cycle by use of an ejector *Int. J. Refrigeration* **35 (6)** 1595-1603.

[6] Lawrence N and Elbel S. 2015 Analysis of two-phase ejector performance metrics and comparison of R134a and CO_2 ejector performance *Science and Technology for the Built Environment* **21(5)** 515-25.

[7] Lee J S and Kim M S 2011 Experimental study on the improvement of CO 2 air conditioning system performance using an ejector. *Intern. J of Refrigeration* **34(7)**, 1614-1625.

[8] Liu F, Groll E A and Ren J 2016 Comprehensive experimental performance analysis of an ejector expansion transcritical CO_2 system. *Applied Thermal Engineering* **98** 1061–1069

[9] Lucas C, Koehler J, Schroeder A, Tischendorf C, 2013 Experimentally validated CO_2 ejector operation characteristic used in a numerical investigation of ejector cycle *Int. J. Refrigeration* **36** (3), 881e891. http://dx.doi.org/10.1016/j. ijrefrig.2012.10.035.

[10] Leung J C 1986 A generalized correlation for one-component Homogeneus Equilibrium flashing choked flow *AIChE Journal* **32** 1622

[11] Leung J C 1995 The omega method for discharge rate evaluation. *International Symposium on Runaway Reactions and Pressure Relief Design* Boston 367-393.

[12] Diener R, Schmidt J, 2004 Sizing of Throttling Device for Gas/Liquid Two-Phase Flow Part 1: Safety Valves *Process Safety Progress* **23**, No.4 335-344.

[13] ISO/DIS4126-10 2008 Safety devices for protection against excessive pressure – Part 10: Sizing of safety valves for gas/liquid two-phase flow.

[14] Fletcher B 1984 Flashing flow through orifices and pipes *CEP* 76-81

[15] Fisher H G, Forrest H S, Grossel S S, Huff J E, Muller A R, Noronha J A, Shaw D A, Tilley B J 1992 Emergency relief system design using DIERS technology. *The DIERS Project Manual*, AIChE, New York

[16] Lucas C, Koehler J, 2012 Experimental investigation of the COP improvement of a refrigeration cycle by use of an ejector *Int. J. Refrigeration* **35** (6) 1595 1603.

[17] UNI. 2011. Norma UNI EN 14511. Condizionatori, refrigeratori di liquido e pompe di calore con compressore elettrico per il riscaldamento e il raffrescamento degli ambienti. Parte 3: Metodi di prova.

Nomenclature

A	area	(m^2)	***greek letters***			
C	specific heat capacity	(J/kg K)	ϕ	diameter	(m)
G	mass flow rate	(kg/h)	ω	compressible flow parameter	(-)
h	specific enthalpy	(J/kg)	ρ	density	(kg/m^3)
K	discharge coefficient	(-)	$\Delta h_{v,0}$	latent heat of vaporization	(J/kg)
P	pressure	(bar)	η	pressure ratio	(-)
Q_{GC}	heating capacity	(kW)			

IOP Conf. Series: Journal of Physics: Conf. Series **796** (2017) 012040 doi:10.1088/1742-6596/796/1/012040

T	temperature	(°C)
u_c	uncertainty	(-)
u	velocity	(m/s)
v	specific volume	(m^3/kg)
x	mass vapour quality	(-)
r	prediction reliability	(-)

Abbreviations

NTr	Not transcritic pressure conditions
Tr	Transcritic pressure conditions
HEM	Homogeneous Equilibrium Model
HNE	Homogeneous Non Equilibrium Model
ISO	ISO/DIS4126-10 sizing method

subscripts

0	stagnation inlet conditions
c	choked (sonic) flow condition
cr	thermodynamic critical conditions
g	gas
GC	gas-cooler
in	inlet conditions
l	relative to the liquid phase
MF	motive mass flow
out	outlet conditions
r	actual
sp	specific
t	theoretical
th	throat conditions

Gas holdup and flow regime transition in spider-sparger bubble column: effect of liquid phase properties

Giorgio Besagni*, Fabio Inzoli*, Giorgia De Guido, Laura A. Pellegrini****

* Politecnico di Milano, Department of Energy, Via Lambruschini 4a, 20156, Milano
**Politecnico di Milano, Dipartimento di Chimica, Materiali e Ingegneria Chimica "G. Natta", Piazza Leonardo da Vinci 32, 20133, Milano

E-mail: giorgio.besagni@polimi.it

Abstract. This paper discusses the effects of the liquid velocity and the liquid phase properties on the gas holdup and the flow regime transition in a large-diameter and large-scale counter-current two-phase bubble column. In particular, we compared and analysed the experimental data obtained in our previous experimental studies. The bubble column is 5.3 m in height, has an inner diameter of 0.24 m, it was operated with gas superficial velocities in the range of 0.004–0.20 m/s and, in the counter-current mode, the liquid was recirculated up to a superficial velocity of -0.09 m/s. Air was used as the dispersed phase and various fluids (tap water, aqueous solutions of sodium chloride, ethanol and monoethylene glycol) were employed as liquid phases. The experimental dataset consist in gas holdup measurements and was used to investigate the global fluid dynamics and the flow regime transition between the homogeneous flow regime and the transition flow regime. We found that the liquid velocity and the liquid phase properties significantly affect the gas holdup and the flow regime transition. In this respect, a possible relationship (based on the lift force) between the flow regime transition and the gas holdup was proposed.

1. Introduction

Two-phase bubble columns are equipment used for bringing one or several gases into contact with a liquid phase. They are built in several forms, but the simplest configuration consists in a vertical cylinder with no internals, in which the gas enters at the bottom through a gas distributor. The liquid phase may be supplied in batch mode or it may be led in either co-currently or counter-currently to the upward gas stream. This type of contacting devices has found many applications in the chemical, petrochemical and biochemical industries thanks to a number of advantages they provide in both design and operation (i.e., the lack of any mechanically operated parts, low energy input requirements and reasonable prices). Despite the simple column arrangement, the interactions between the phases inside the reactor are extremely complex, making their design and scale-up very difficult. Indeed, although a large amount of research is ongoing, the correct prediction of the fluid dynamics inside bubble columns is still hardly possible without experimentation. In particular, the global and local fluid dynamics in bubble columns can be described by global as well as local flow properties—i.e., the gas holdup (ε_G) and bubble size distribution (BSD). The gas holdup is a dimensionless parameter defined as the volume of the gas phase divided by the total volume. It determines the residence time and, in combination with the BSD, the interfacial area for the rate of interfacial mass transfer. The

global and local flow properties are related to the prevailing flow regime, which can be distinguished (in large-diameter bubble columns) in (i) the homogeneous flow regime, (ii) the transition flow regime and (iii) the heterogeneous flow regime [1]. The homogeneous flow regime—associated with small superficial gas velocities, U_G—is referred as the flow regime where only "*non-coalescence-induced*" bubbles exist (e.g. as detected by the gas disengagement technique, see ref. [2]). The homogeneous flow regime can be distinguished into "*pure-homogeneous*" (or "*mono-dispersed homogeneous*") flow regime and "*pseudo-homogeneous*" (or "*poly-dispersed homogeneous*" or "gas maldistribution") flow regime. The transition from the homogeneous flow regime toward the heterogeneous flow regime is a gradual process in which a transition flow regime occurs. The transition flow regime is identified by the appearance of the "*coalescence-induced*" bubbles and is characterized by large flow macro-structures with large eddies and a widened bubble size distribution due to the onset of bubble coalescence. At high gas superficial velocities, a fully heterogeneous flow regime is reached [3]; it is associated with high coalescence and breakage rates and a wide variety of bubble sizes. It is worth noting that, in large-diameter bubble columns, the slug flow regime may not exist because of the Rayleigh–Taylor instabilities [4]. The transitions between the three prevailing flow regimes depend on (i) the operation mode, (ii) the design parameters and (iii) the gas/liquid phases of the bubble column. For example, using a sparger that produces mainly very small bubbles the homogeneous flow regime is stabilized [5], whereas the "*mono-dispersed homogeneous flow regime*" may not exist if large bubbles are aerated [2] up to a "*pure heterogeneous flow regime*" from the beginning [6].
The many relationships between the bubble column fluid dynamic parameters and the various variables characterizing the system make it difficult to find general correlations for the precise design of bubble columns (i.e., the correct estimation of the gas holdup) [7]. Indeed, the large variation of gas holdup values presented in the literature leads to the development of a large number of correlations for the gas holdup. To this end, we have set up a large-scale and large-diameter bubble column to study the global and local bubble column fluid dynamics and, thus, to provide rational basis for bubble column design and scale-up [2]. The bubble column has a d_c = 0.24 m inner diameter, H_c = 5.3 m height (H_c/d_c > 20). The diameter of the column, its height and the sparger openings were chosen considering the well-known scale up criteria: generally, a diameter greater than 0.15 m, an aspect ratio larger than H_0/d_c >5 (H_0 is the liquid free level) and sparger openings larger than d_0 > 1 mm guarantee results that could be used for scaling-up [8]. In our previous papers the influence of the column and sparger design [2, 9], of the operation modes [2, 9, 10, 11, 12] and of sodium chloride (as an electrolyte) concentration [13], ethanol concentration [14] and monoethylene glycol concentration [15] over the bubble column fluid dynamics was studied. This paper summarizes and compares the previous experimental results and further discusses this topic. In particular, the relationship between the flow regime transition and the gas holdup are discussed. The present work represents the first step of a larger research framework focused on establishing a large dataset of gas holdup data to develop general gas holdup correlations. This paper is structured as follows. In Section 2 the experimental setup and dataset is described, in Section 3 the experimental results are presented and are summarized and compared in Section 4. Finally, conclusions are drawn.

2. The experimental setup and the methods

2.1. Gas holdup measurements

The experimental facility (Figure 1a) is a non-pressurized vertical pipe made of Plexiglas® with d_c = 0.24 m and H_c = 5.3 m. A pressure reducer controls the pressure upstream from the rotameters (1) and (2), used to measure the gas flow rate (accuracy ± 2% f.s.v., E5-2600/h, manufactured by ASA, Italy). A pump, controlled by a bypass valve, provides water recirculation, and a rotameter (3) measures the liquid flowrate (accuracy ± 1.5% f.s.v., G6-3100/39, manufactured by ASA, Italy). The values of gas density (used to compute the superficial gas velocity) are based upon the operating conditions existing at the column mid-point (computed by using the ideal gas law). The gas distributor, is a spider-sparger distributor (Figure 2a and 2b) with hole diameters d_o = 2 - 4 mm (Figure 2b). The spider sparger, has six arms made of 0.012 m diameter stainless steel tubes soldered to the center cylinder of the sparger. The sparger was installed with the six holes located on the side of each arm facing upward. These holes are distributed with an increasing diameter moving toward the column wall (Figure 2a). The gas and liquid temperatures were checked and maintained constant at room temperature during all the experiments (22 ± 1 °C). It is worth noting that the diameter of the column, its height and the sparger openings were chosen considering the scale-up criteria listed in the introduction [8].

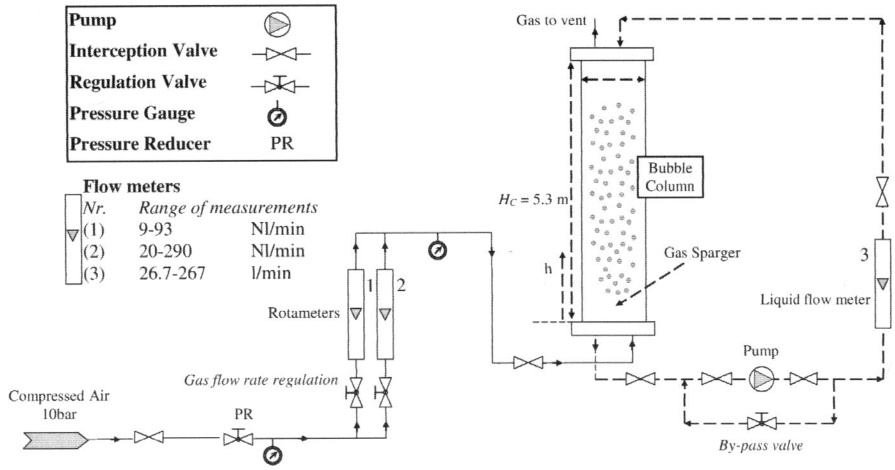

Figure 1. The experimental facility.

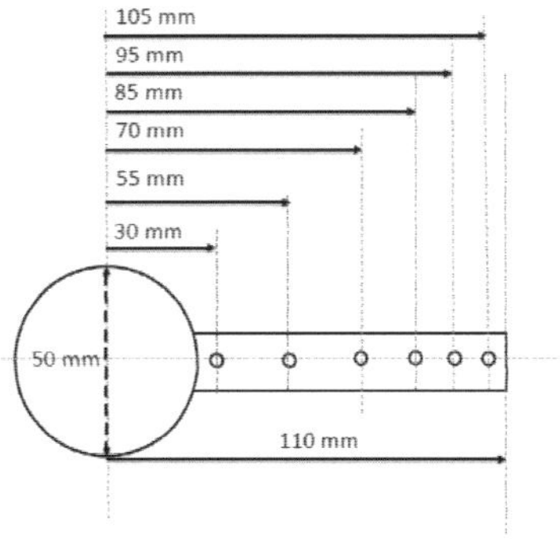

(a) Spider sparger: view from top **(b) Spider sparger: distribution of the openings**

Figure 2. The spider sparger.

In all the runs, filtered air was used as the dispersed phase and various fluids (tap water, aqueous solutions of sodium chloride, a water-ethanol mixture and solutions of water-monoethylene glycol of different concentrations) were employed as the liquid phase to study the influence of the liquid properties on the bubble column fluid dynamics. In such a case, a liquid of a known composition was charged to the column and the gas flow rate adjusted to the desired value. The liquid phases were chosen to study the influence of organic (ethanol) and inorganic (sodium chloride) active agents, and viscous solutions:

- aqueous solution of ethanol (EtOH): 0.05%wt;

- aqueous solution of sodium chloride (NaCl): between 0.1 and 1%wt (corresponding to a concentration ratio 0.17 < c/c_t < 1.17, Table 1; where c is the NaCl molar concertation and c_t = 0.145 l/mol is the critical value for the coalescence inhibition, as reported in the literature [16] and discussed in [13]).
- aqueous solution of monoethylene glycol (MEG): 11 concentrations between 0.05 and 80%wt; to investigate the "*dual effect*" of the viscosity as described by Besagni et al. [15]. The properties of the solutions tested are summarized in Table 2. The properties for pure water have been taken from those available in the literature [17-19]. As for the physical properties of water-MEG mixtures, the work by Sun and Teja [20] has been taken into account for evaluating their densities and viscosities as a function of the mass fraction of monoethylene glycol, of the temperature and of the properties of the two components constituting the binary system. As for the surface tension of aqueous monoethylene glycol solutions at 298.15 K, the correlation found in the product guide provided by MEGlobalTM [21] has been used.

Table 1. NaCl concentrations tested.

***c* [mol/l]**	0	0.02	0.07	0.12	0.145[1]	0.170
c/c_t [-]	0	0.14	0.48	0.84	1	1.17

[1] *Critical concentration, c_t = 0.145 mol/l [16]*

Table 2. Liquid phases tested (properties evaluated at T = 25°C, p = 101325 Pa).

$c_{MEG,wt}$ [%]	ρ_L [kg/m3]	μ_L [mPa*s]	σ [mN/m]
0	997.086	0.8903	0.0715
0.05	997.158	0.8917	0.0715
0.1	997.229	0.8928	0.0715
0.5	997.801	0.9019	0.0713
0.75	998.159	0.9077	0.0712
1	998.516	0.9135	0.0711
5	1004.208	1.0106	0.0696
8	1008.443	1.0894	0.0685
10	1011.249	1.1450	0.0677
40	1051.150	2.4287	0.0583
80	1094.801	7.9655	0.0502

2.2. Gas holdup measurements

Measurements of the bed expansion allowed the evaluation of the gas holdup ε_G. The procedure involves measuring the location (height) of liquid free surface when air flows in the column. The gas holdup is, then, obtained using the following relation:

$$\varepsilon_G = \frac{(H_D - H_0)}{H_D} \tag{1}$$

Where H_D and H_0 are the heights (measured from the sparger) of the free surface after and before aeration, respectively. The free surface before aeration is H_0 = 3.0 m.

2.3. Flow regime transition analysis

Although the flow transition from the homogeneous to the transition flow regime does not happen instantaneously [1], the definition of an approximate transition point is helpful for modelling the hydrodynamic of bubble columns. In this study, we employ two methods from the literature for investigating the flow regime transition: (i) the swarm velocity and (ii) the drift flux method.

2.3.1. Swarm velocity method

The swarm velocity method was proposed by Zuber and Findlay [22] and is based on the evaluation of the swarm velocity, U_{swarm}:

$$U_{swarm} = U_G / \varepsilon_G \tag{2}$$

In this method, the swarm velocity is plotted against the superficial gas velocity: U_{swarm} is constant in the homogeneous flow regime and starts to increase - at a certain transition superficial velocity, U_{trans} - as the system enters the transition/heterogeneous flow regime. The appearance of the first large bubble is responsible for the increase in the swarm velocity and is an indication of flow regime transition. This method was employed also by Krishna et al. [23], Letzel et al. [24] and Gourich et al. [25], Ribeiro and Mewes [11] and Besagni et al. [2, 9-15].

2.3.2. Wallis plot method

The drift-flux method was proposed by Wallis [26] and was widely used in the literature [2, 12-15, 26, 27]. This method is based on the drift flux (that represents the gas flux through a surface moving with the speed of the two-phase mixture) and it is experimentally obtained as follows:

$$J_T = U_G\left(1-\varepsilon_G\right) \tag{3}$$

Theoretically, the drift flux is written in terms of a parameter, U_b, whose dependence upon ε_G varies with the prevailing flow regime:

$$J_E = U_b\left(1-\varepsilon_G\right) \tag{4}$$

The idea is to employ a model for U_b valid for the homogeneous flow regime and to plot J_E and J_T in the same graph as a function of ε_G. In the homogeneous flow regime J_E is equal to J_T and the transition point is, thus, defined when:

$$J_T \neq J_E \tag{5}$$

The evaluation of U_b is a matter of discussion in the literature, as different models were proposed and applied. In this study, we follow the approach of Krishna et al. [28], which is based on the empirical model of Richardson and Zaki [29]:

$$U_b = u_\infty\left(1-\varepsilon_G\right)^{n-1} \tag{6}$$

where n is fluid-dependent ($n \cong 2$ for water) and u_∞ - the terminal velocity of an isolated bubble - should be fitted with the aid of the experimental data in the determination of the flow regime transition point. Combining Eq. (4) and Eq. (6) results:

$$J_E = u_\infty \varepsilon_G\left(1-\varepsilon_G\right)^n \tag{7}$$

3. The gas holdup data

3.1. Air-water

Figure 3 displays the gas holdup measurements for the air-water system. At low superficial gas velocities, the relationship between the gas holdup and the superficial gas velocity is linear, followed by a change in tendency at a transition superficial gas velocity, U_{trans}. The linear trend corresponds to the homogeneous flow regime and the change in tendency is due to flow regime transition toward the transition flow regime. Indeed, above the transition velocity, "*coalescence-induced-bubbles*" begin to appear, and the bubble coalescence increases the average rise velocity and reduces the gas residence time in the column, thus reducing the gas holdup versus gas velocity slope. It is worth noting that the homogeneous flow regime, in the present bubble column, is characterized by poly-dispersed BSDs and, therefore, it is classified as a "*pseudo-homogeneous*" flow regime [2]. The shape of the gas holdup curve is the one typically found for similar sparger geometries: the sparger used in this study has "*large*" holes (d_0 > 1 mm) and, as expected, no peak can be observed in the gas holdup curve [3].

Upon increasing the liquid flowrate, a faster increase in the gas holdup is observed at low U_G, and the transition point also moves toward lower superficial gas velocities. This change is explained by the effect of the liquid flow, which slows down the rise of the bubbles, leading to higher gas holdup: the more compact arrangement of the bubbles leads to an earlier flow regime transition. Our results prove that the counter-current mode has an influence on the gas holdup, as widely discussed in [2], and in agreement with the literature [30-33]. The reader may refer to our previous paper for a comprehensive discussion concerning the hydrodynamic of the system [2].

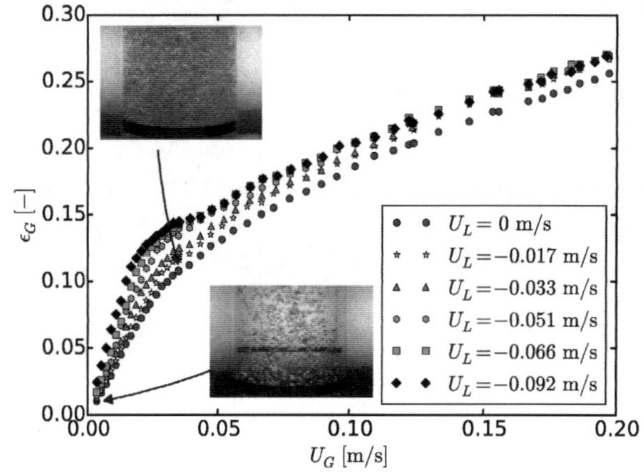

Figure 3. Gas holdup – air-water (influence of the counter-current mode).

3.2. Air-water-NaCl

Most electrolytes inhibit bubble coalescence in water [34-38] and, in this respect, a key concept is the transition concentration, c_t, defined as the concentration above which bubble coalescence is drastically reduced [37, 38]. In the case of NaCl, the threshold is c_t = 0.145 mol/l [9]. Depending on the concentration of the electrolyte, we may define a "*coalescent flow regime*" ($c/c_t \leq 1$) and a "*non-coalescent flow regime*" ($c/c_t > 1$); in the present study, five NaCl concentrations ($0 \leq c/c_t \leq 1.17$, in both the "*coalescent flow regime*" and the "non-*coalescent flow regime*") were tested: Figure 4 displays the gas holdup measurements for the air-water-NaCl system.

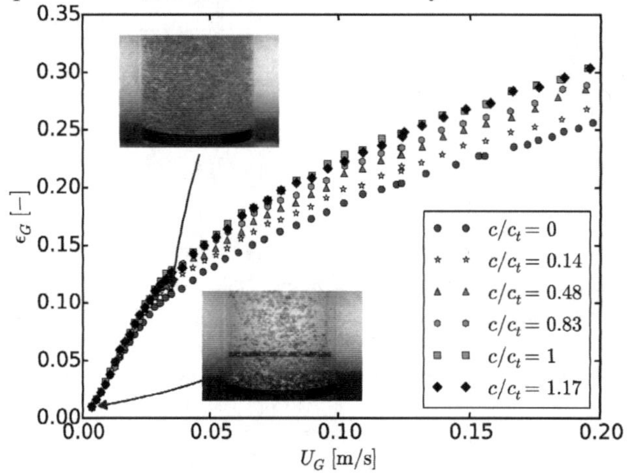

Figure 4. Gas holdup – air-water-NaCl.

The addition of NaCl, up to the critical concentration, increases continuously the gas holdup and stabilizes the homogeneous flow regime due to the coalescence inhibition effect [16, 39]. Above the critical concentration, there is no remarkable difference in the gas holdup and flow regime transition. An interesting aspect is the non-linearity of the electrolytes effect upon the gas holdup. The curve for c/c_t = 0.14 and 0.48 is already shifted to considerably higher ε_G values in comparison to the curve related to $c/c_t = 0$, while the relative distance between the curves associated with c/c_t = 0.48, 0.83 and 1 is considerably lower. The literature agrees that the increase in gas holdup is a consequence of the homogeneous flow regime stabilization, which is further discussed in Section 4. Further details on the global and local flow properties were presented by Besagni and Inzoli [13].

3.3. Air-water-EtOH

The alcohol molecules are composed of hydrophilic and hydrophobic parts that are adsorbed at the interface when dissolved in water, causing the coalescence suppression [40]. In the present case, the addition of ethanol stabilizes the homogeneous flow regime, due to the coalescence inhibition effect of alcoholic solutions, and, as a consequence (Section 4), increases the gas holdup (Figure 5). Moreover, the addition of ethanol changes the bubble shape and size distributions: we have observed an increased number of small and spherical bubbles (Figure 6b) [14]. It is worth noting that the difference between the two gas holdup curves increases with increasing U_G. Further details on the global flow properties and the bubble size distributions were presented by Besagni et al. [14].

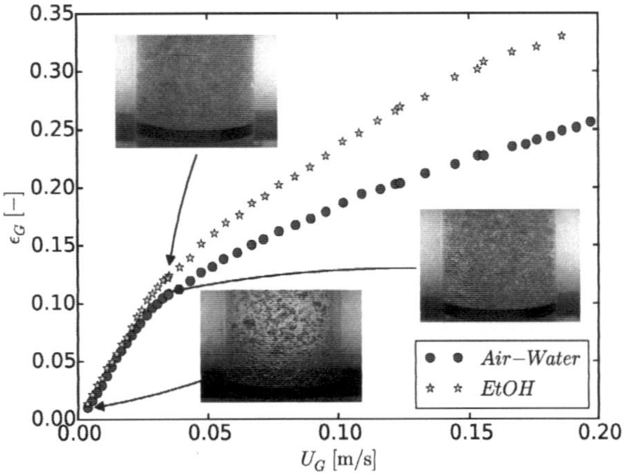

Figure 5. Gas holdup – air-water-EtOH.

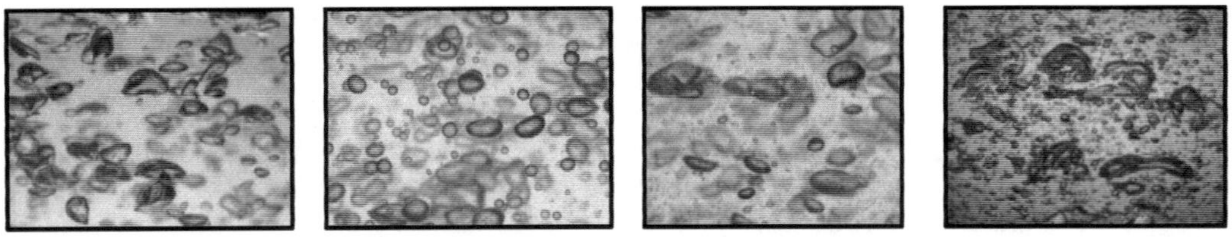

(a) Air-water **(b) Air-water-EtOH** **(c) Air-water-MEG 5%** **(d) Air-water-MEG 80%**

Figure 6. Influence of the liquid phase properties on the bubble shapes.

3.4. Air-water-MEG

Figure 7 displays the gas holdup measurements for the air-water-MEG system. The gas holdup continuously increases by increasing the MEG concentration up to $c_{MEG,wt}$ = 5% (μ_L = 1.01 mPa·s, Figure 7a), along with the contribution of small bubbles (Figure 6c). On the other hand, if the concentration is further increased from $c_{MEG,wt}$ = 5% (μ_L = 1.01 mPa·s) to $c_{MEG,wt}$ = 80% (μ_L = 7.97 mPa·s), the gas holdup decreases (Figure 7b). For this last concentration, the gas holdup curve lies even below that obtained for pure water: indeed, increasing the viscosity, the tendency to coalescence prevails, creating large cap-bubbles (Figure 6d) rising the column at a higher velocity, thus reducing the gas holdup. A similar behaviour was observed in the early study of Wilkinson et al. [41] and in more recent studies [42]. In addition, at high viscosity and high U_G "*coalescence-induced*" bubbles, rising the column, were observed. An interesting discussion concerning the contribution of the "*coalescence-induced*" bubbles to the gas holdup structure in highly viscous liquid phases was proposed by Yang et al. [43]. Further details on the global flow properties and the bubble size distributions were presented by Besagni et al. [15].

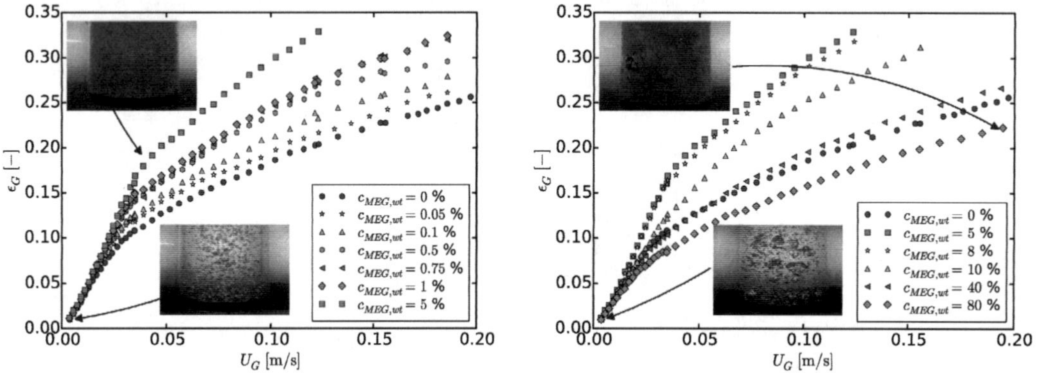

(a) Low viscosities **(b) Moderate/high viscosities**

Figure 7. Gas holdup – air-water-MEG.

4. The flow regime transition

The flow regime transition points were investigated by the methods presented in Section 2.3. The values of the transitional gas velocities (U_{trans}) and transitional gas holdups (ε_{trans}) are in agreement between the two methods and, following the proposal of Ribeiro and Mewes [39] and Besagni and Inzoli [2], the transition points were evaluated as the mean of the two values. The results of the analysis are presented in Figure 8 and are summarized below:

- **air-water-NaCl.** The homogeneous flow regimes is stabilized while increasing the electrolyte concentration till the critical concentration;
- **air-water-EtOH.** The homogeneous flow regimes is stabilized by the addition of EtOH;
- **air-water-MEG.** The addition of MEG stabilizes or destabilizes the homogeneous flow regime depending on the MEG concentration: the low viscosities stabilizes the homogeneous flow regime, whereas moderate/high viscosities destabilize the homogeneous flow regime.

Regardless of the liquid phase considered, it is interesting to observe the relationship between the homogeneous flow regime stabilization (or destabilization) and the gas holdup increase (or decreasing): if the homogeneous flow regime is stabilized/destabilized, the gas holdup increases/decreases. It has not escaped our notice that the flow regime transition is related to the prevailing BSDs of the system. The detailed discussion of the BSDs is beyond the scope of this paper and the interested reader should refer to the experimental results presented in refs. [14, 15].

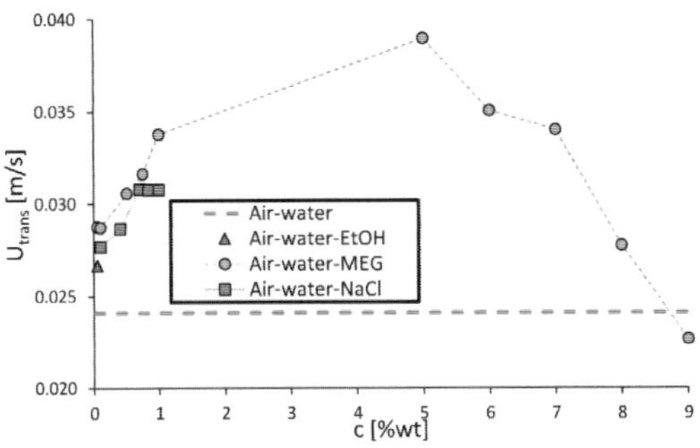

(a) Influence of liquid phase properties on U_{trans}

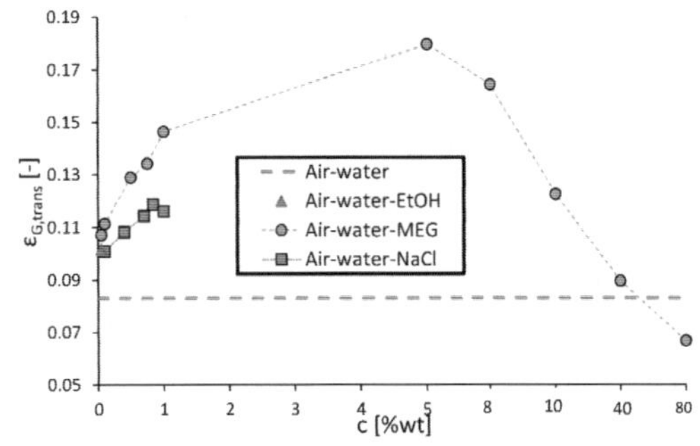

(b) Influence of liquid phase properties on $\varepsilon_{G,trans}$

Figure 8. The flow regime transitions: influence of the liquid phase properties.

The experimental results suggest that a change in the liquid phase properties mainly affect the bubble interfacial properties, thus resulting in change in the bubble size distributions (i.e., the balance between coalescence and break-up) and, finally, changing the gas holdup curve. This concept was firstly proposed by Besagni [44] and has been further discussed by Besagni et al. [14-15]. The above-mentioned relationship between the bubble scale and the laboratory-reactor scale may be governed by the lift force. In particular, the lift force induces the changes in the bubble size distributions and affects the motion of the small and large bubbles [2, 15]. In this respect, the lift force affects the bubble column fluid dynamics and the flow regime transitions. The larger bubbles, having a negative lift coefficient, move toward the center of the pipes, thus, promoting the "*coalescence-induced*" bubble and the flow regime transition. On the contrary, the small bubbles, having positive lift coefficient, stabilize the homogeneous flow regime. This concept has been widely discussed and verified in our previous paper [15], with respect to the MEG system. Of course, this concept should be verified in future studies with respect to other systems (i.e., air water and active agent systems).

5. Conclusions

We have experimentally investigated the influence of liquid phase properties on gas holdup and flow regime transition in a large-scale counter-current bubble column. The bubble column is 5.3 m in height, has an inner diameter of 0.24 m, it was operated with gas superficial velocities in the range of 0.004–0.20 m/s and, in the counter-current mode, the liquid was recirculated up to a superficial velocity of -0.09 m/s. Air was used as the dispersed phase and various fluids were employed as liquid phases. The experimental dataset consist in gas holdup measurements and was used to investigate the global fluid dynamics and the flow regime transition between the homogeneous and the transition flow regimes. We found that the liquid velocity and the liquid phase properties significantly affect the gas holdup and the flow regime transition. The experimental results suggest that a change in the liquid phase properties affect the bubble interfacial properties, thus resulting in change in the bubble size distribution and, finally, changing the gas holdup curve. A stabilization/destabilization of the homogeneous flow regime results in an increase/decrease of the gas holdup increase (or decreasing). The relationship between the bubble scale and the laboratory-reactor scale may be governed by the lift force. Further studies should focus on the lift force as the link between the bubble scale and the laboratory-reactor scale; in this respect, extensive experimental campaigns to obtain information on bubble sizes and shapes should be performed. Possibly, the lift force may be included, in future studies, in gas holdup correlations to link the local and the global scales.

6. Acknowledgment

The authors wish to express their thanks to Itelcond srl and to Ing. Luca Primavesi, in particular, for having supplied monoethylene glycol for the experimental campaign.

References

[1] Nedeltchev S 2015 *Chem. Eng. Sci.* **137** 436-44
[2] Besagni G and Inzoli F 2016 *Chem. Eng. Sci.* **146** 259–290
[3] Sharaf S, Zednikova M, Ruzicka M C and Azzopardi B J 2015 *Chem. Eng. J.* **288** 489–504
[4] Kataoka I and Ishii M 1987 *Int. J. Heat Mass Tran.* **30** 1927-39
[5] Mudde R F, Harteveld W K and van den Akker H E A 2009 *Ind. Eng. Chem. Res.* **48** 148-159
[6] Ruzicka M C, Drahoš J, Fialová M and Thomas N H 2001 *Chem. Eng. Sci.* **56** 6117-24
[7] Shah Y T, Kelkar B G, Godbole S P and Deckwer W D 1982 *AICHE J.* **28** 353-379
[8] Shaikh A. and Al-Dahhan M. 2013. *Ind. Eng. Chem. Res.* **52** 8091-8108
[9] Besagni G and Inzoli F 2016 *Chem. Eng. Sci.* **145** 162–180
[10] Besagni G, Guédon G and Inzoli F 2014 *J. Phys Conf Ser* **547** 012024
[11] Besagni G, Guédon R G and Inzoli F 2016 *ASME J. Fluids Eng* **138** 011302-15
[12] Besagni G and Inzoli F 2016 *Ext. Therm. Fluid. Sci.* **74** 27–48
[13] Besagni G and Inzoli F 2015 *J. Phys Conf Ser* **655** 012039
[14] Besagni G, Inzoli F, De Guido G and Pellegrini A M 2016 *Chem. Eng. Res. Des.* **112** 1-15
[15] Besagni G, Inzoli F, De Guido G and Pellegrini A M 2016 *Chem. Eng. Sci.* **158** 509-538
[16] Zahradnik J, Fialova M, Rruvzivc k, M., Drahovs, J., Kavstanek F and Thomas N H 1997 *Chem. Eng. Sci.* **52** 3811-26
[17] Kestin J, Sokolov M and Wakeham W A 1978 *J. Phys. Chem. Ref. Data* **7** 941-948
[18] Vargaftik N, Volkov B and Voljak L 1983 *J. Phys. Chem. Ref. Data* **12** 817-820
[19] Perry R H and Green D W 2008 *Perry's chemical engineers' handbook* (New York)
[20] Sun T and Teja A S, K 2003 *J. Chem. Eng. Data* **48** 198-202
[21] MEGlobalTM 2015 *Ethylene Glycol. Product Guide*
[22] Zuber N and Findlay J A 1965 *J. Heat Transf* **87** 453-68
[23] Krishna R, Wilkinson P M and Van Dierendonck L L 1991 *Chem. Eng. Sci.* **46** 2491-6
[24] Letzel H M, Schouten J C, van den Bleek C M and Krishna R 1997 *Chem. Eng. Sci.* **52** 3733-9
[25] Gourich B, Vial C, Essadki A H, Allam F, Belhaj Soulami M and Ziyad M 2006 *Chem. Eng. Process.* **45** 214-23
[26] Wallis G B 1969 *One-dimensional two-phase flow* (New York)
[27] Passos A D, Voulgaropoulos V P, Paras S V and Mouza A A 2015 *Chem. Eng. Res. Des.* **95** 93-104
[28] Krishna R, Ellenberger J and Maretto C 1999 *Int. Commun. Heat Mass* **26** 467-75
[29] Richardson J F and Zaki W N 1997 *Chem. Eng. Res. Des.***75**, Supplement S82-S100
[30] Otake T, Tone S and Shinohara K 1982 *J. Chem. Eng. Jpn.* 338–340
[31] Baawain M S, El-Din M G and Smith D W 2007 *Ozone: Sci. Eng.* **29** 343–352
[32] Biń A K, Duczmal B and Machniewski P 2001 *Chem. Eng. Sci.* **56** 6233–6240
[33] Jin H, Yang S, He G, Wang M and Williams R A 2010 *J. Chem. Technol. Biot.* **85** 1278–1283
[34] Craig V S J, Ninham B W and Pashley R M 1993 *J. Phys. Chem-US* **97** 10192-7
[35] Deschenes L A, Barrett J, Muller L J, Fourkas J T and Mohanty U 1998 *J. Phys. Chem B* **102** 5115-9
[36] Keitel G and Onken U 1982 *Chem. Eng. Sci.* **37** 1635-8
[37] Lessard R R and Zieminski S A 1971 *Ind. Eng. Chem. Fund.* **10** 260-9
[38] Marrucci G and Nicodemo L 1967 *Chem. Eng. Sci.* **22** 1257-65
[39] Ribeiro Jr C P and Mewes D 2007 *Chem. Eng. Sci.* **62** 4501-9
[40] Albijanić B, Havran V, Petrović D L, Durić M and Tekić M N 2007 *AIChE J.* **53** 2897-2904
[41] Wilkinson P M, Spek A P and van Dierendonck L L 1992 *AIChE J.* **38** 544-54
[42] Rabha S, Schubert M and Hampel U 2014 *AIChE J.* **60** 3079-3090
[43] Yang J H, Yang J I, Kim H J, Chun d H, Lee H T and Jung H 2010 *Chem Eng Process.* **49** 1044-1050
[44] Besagni G 2016 *Bubble column fluid dynamics: experimental and numerical investigations* Ph.D. Thesis, Politecnico di Milano

Foam evolution in a processed liquid solution

E Salerno[1], **P Levoni**[1,2], **G S Barozzi**[1] **and A Malfatto**[3]

[1] DIEF- Dipartimento di Ingegneria "Enzo Ferrari", Università di Modena e Reggio Emilia, Via Vivarelli 10 - 41125 Modena (Italy)
[2] MIMESIS fluid dynamics & energy, Strada Contrada 309 - 41126 Modena (Italy)
[3] Sidel S.p.A., Via La Spezia 241/A - 43125 Parma (Italy)

Email: elisabetta.salerno@unimore.it

Abstract. Foam formation in a carbonated solution undergoing pouring and decompression is investigated with the use of high-speed imaging. Operational conditions similar to those encountered in industrial bottling processes are applied to inspect the mechanisms that control the foaming behaviour in practical filling applications. The evolution of the foam column during pressure release is analysed in quantitative terms by extracting the foam thickness from the images. The bubble dynamics inside the solution, and the destabilization processes on the foam column are seen to have a paramount effect on the observed foam evolution trend. The contributions to foam formation given by the nuclei entrained in the bulk liquid and by the bubble-generating sites on the container walls are finally distinguished and discussed.

1. Introduction

Foams frequently arise in industrial applications involving filling and pouring of liquids, such as food processing, casting of molten metals, and pharmaceuticals and chemicals manufacturing. In those processes, the formation of foam is an unwanted effect, since it retards the production process and can compromise the product quality. Consequently, finding suitable strategies able to inhibit the inception of foam is of primary importance in those fields.

It is now well known that in practical situations the presence of pre-formed gaseous nuclei is an essential factor for a solution to foam [1]. These nuclei can be as small to escape optical detection [2], and may be introduced in the liquid by several mechanisms, including bubble entrainment by liquid-liquid impacts [3–6], and bubble entrapment along the container walls during filling [7, 8]. The growth of nuclei to macroscopic sizes may be activated by pressure reduction or temperature increase, which can drive the system to a supersaturation state [1]. The turbulent dynamics generated inside the container can also trigger the growing of nuclei by cavitation [9]. Once they reach a critical size, bubbles migrate towards the free surface of the liquid where their aggregation gives rise to a column of foam. On the other hand, foam will decay spontaneously over time, its deterioration being dictated by the kinetics of four processes: disproportionation, gas diffusion, drainage and coalescence [10–12].

The mechanisms governing foam growth and stability, as well as their influencing parameters, have been extensively studied in different research areas for many years. However, most of the published studies consider one single phenomenon at a time, and are usually performed under simplified conditions. For these reasons, the exact incidence of the various processes on the production of foam is still quite unclear; this is especially true in the context of the filling processes, where a variety of mechanical and fluid dynamical events are interrelated. The present research aims at clarifying for the first time the contributions given by the totality of the foaming mechanisms to foam formation in a carbonated liquid, as a consequence of the pressure-filling process and the subsequent decompression phase. For this purpose, a series of experiments was carried out on a carbonated, optically transparent commercial soft drink, poured into a bottle with the use of a testing filling machine. High-speed imaging was adopted to inspect the phenomena occurring inside and above the solution, which was processed under realistic conditions of pressure and flow rate. The comparison between the images obtained on the foam column and inside the solution allowed the identification of the main phenomena at the basis of foam development. The recorded image sequences were also analysed in quantitative terms and the trend of the foam level during decompression was reconstructed.

2. Experimental procedures

2.1. Experimental set-up

In order to reproduce the fluid dynamical conditions encountered in practical filling applications, a testing filling machine for the bottling of carbonated beverages was employed in the experimentation. The machine transfers the liquid product from a pressurised reservoir to the container at controlled conditions of pressure and flow rate – their values being those typically used in the beverage industry. The container used for the tests was a brand-new 1.5 l plastic bottle. Since a swirl-type valve was used for filling, the bottle was chosen of cylindrical body, having an almost constant diameter (d = 88 mm), not ribbed and with a smoothly curved shoulder, to prevent accidental detachments of the annular filling jet from the container walls. The working fluid was carbonated on site to 4.3 volumes of CO_2, and maintained at the fixed temperature of 17.5 °C.

Experiments were performed with the following filling procedure: first, the internal pressure of the bottle was equated to the reservoir pressure, set at 4.4 bar gauge. This was achieved by introducing carbon dioxide in the bottle while venting to the atmosphere the air contained. Next, the liquid was injected into the bottle at a specific constant flow rate. The bottle was filled only partially, i.e. to just 1 l, to prevent foam overflows and allow an accurate measurement of the foam level. After the filling was completed, a short settling period was waited and then the pressure was gradually released. The decompression was operated in two steps, by consecutively opening a snift valve connected to ambient air. The opening time for second snift (Δt_2) was three times the opening time of the first snift (Δt_1), and equalled the intermediate rest period (Δt_{rest}) set between the two decompressions. The actual duration of each step cannot be provided since protected by non-disclosure agreement. At the end of each filling sequence, the bottle was emptied out, rinsed and dripped.

High-speed imaging was used for the recording of foam and bubble dynamics during decompression. The imaging system was arranged as sketched in figure 1.a. Two NanoSense MkIII CMOS cameras from Dantec Dynamics were positioned on one side of the filling machine for the simultaneous observation of the flow phenomena in distinct regions of the bottle. Two different configurations were implemented. In the first configuration (configuration A), one camera was zoomed in the bottle headspace to capture foam evolution, while the other one monitored the rising bubbles close to liquid/foam interface. The cameras were fitted with a zoom lens (AF Nikkor 28-80 mm) and a macro lens (Micro Nikkor 60 mm f/2.8), respectively. In the second configuration (configuration B), both the cameras were focused inside the liquid at different heights. In this case, they were equipped with macro lens to inspect bubble dynamics in the whole filled volume at the maximum achievable spatial resolution. Magnifications up to 100 μm/pixel were obtained. Image distortions in the liquid due to the bottle cylindrical geometry were minimized by encasing the bottle into a square-sided Plexiglas container. The optical compensation box, shown in figure 1.b, was filled with water for the matching of the refractive indexes.

Image sequences were recorded at 200 fps, which was the maximum rate compatible with the duration of the experiments. Acquisitions started at the beginning of the decompression sequence, triggered by a signal transmitted by the filling machine to the cameras. The microcontroller Arduino Uno was employed for the signal handling. A maximum synchronization error of 6.6 μs was measured, including the contribution of the cables. Three 27.2 W/m LED tubes were used as the light sources in backlight configuration. The tubes were positioned close to the test container to provide intense illumination. A uniform distribution of light was obtained by interposing a light diffuser between the LED sources and the compensation box.

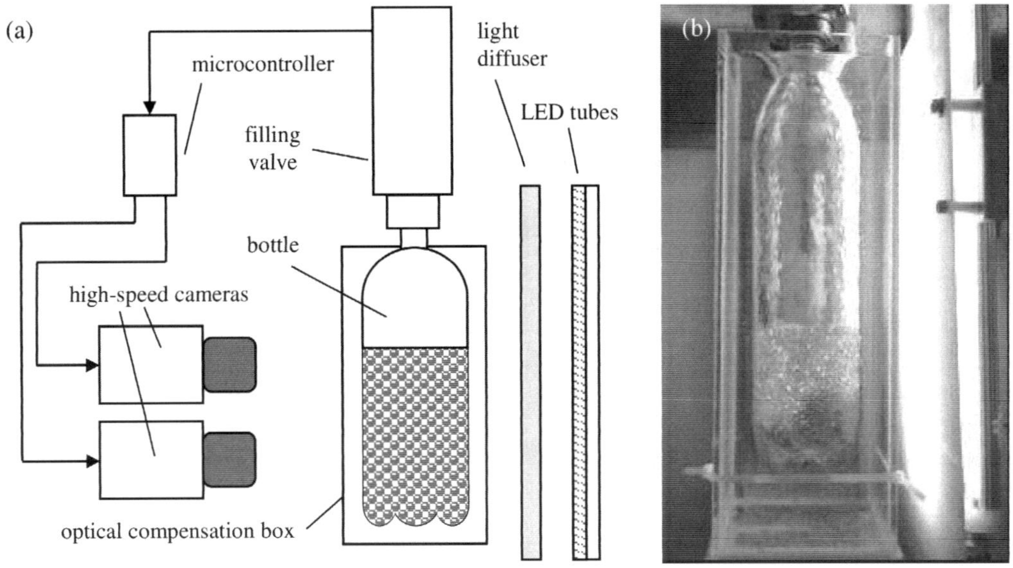

Figure 1. Schematic of the imaging system in side view (a), and snapshot of the bottle encased in the optical compensation box during filling (b).

2.2. *Data reduction*

The extrapolation of qualitative and quantitative information on the foam evolution and bubble behaviour was performed by processing the acquired images with the software ImageJ (provided by NIH, USA). Images collected inside the liquid were examined by direct visual inspection. No automated detection techniques could be applied because of the very high bubble concentration. The size of the bubbles was evaluated by approximating the nuclei with ellipsoids, from whose volume the equivalent spherical diameter was calculated. Subpixel accuracy could be obtained with this method.

For the estimate of the foam thickness, a procedure based on the extraction of intensity profiles was implemented. Images acquired in the bottle headspace were first filtered to suppress noise, and then reduced to a region with an almost uniform background. The profiles of the pixel intensities were drawn for the frames of interest, by plotting the horizontally averaged grey value along the vertical direction of the selected area. Each curve was analysed to identify the position of the upper and lower foam interface according to pre-determined mathematical criteria. The thickness of the foam was thus derived, with a maximum error of 9%.

3. Results and discussion

To allow the comparison between acquisitions recorded in different configurations, a preliminary repeatability analysis was carried out. The test was repeated twenty times with the cameras set in configuration A. The observed foam behaviour was qualitatively the same in all the cases; moreover, the maximum foam thickness was obtained at the same time instant with an error of 3%, its measured values showing a variation of 5%. Hence, the process can be considered reproducible.

The experiment was then performed again in both configurations. The foaming phenomena visualized inside and above the liquid during decompression are detailed in the following section.

Next, the trend of the foam level as reconstructed from a whole recorded sequence is presented and commented.

3.1. Foam and bubble dynamics

The inspected foam and bubble dynamics were divided into five main stages, here described in chronological order in terms of the dimensionless parameter $t' = t / \Delta t_1$. Exemplifying image sequences are provided, reporting i) in the first column, foam evolution as captured by the camera focused in the bottle headspace in configuration A, and ii) in the second and third column, what observed for the same time instant in the upper and lower portions of the liquid, respectively, with the cameras set in configuration B. The beginning of the first decompression step is taken as the reference time.

3.1.1. Early growth. At the beginning of the decompression stage, foam has not developed yet. As shown in figure 2.a, just a collar of bubbles is occasionally present, formed by the nuclei introduced by the impinging filling jet and already risen to the liquid free surface. Further clustering is prevented by the tail-end of the jet which is still descending along the bottle walls. The liquid bulk is populated by a large amount of bubbles having different size and distribution (figures 2.b and 2.c). They all originate from entrainment processes taking place during filling. Bubble diameters ranging from 260 μm, which is the minimum perceivable size, to 4.83 mm have been measured. The biggest bubbles are found in the upper portion of the fluid, in rapid ascension driven by buoyancy. Below this region, bubble mean diameter gradually decreases downwards. Sub-millimetric nuclei prevail in the bottom of the bottle: such small bubbles are more affected by the liquid flow field, and are easily carried down deep into the fluid where they are retained even for long times.

Within the first four tenths of the first decompression step, all the entrained bubbles having diameters larger than 2 mm before decompression reach the free surface. Their aggregation gives rise to the first layer of foam, displayed in figure 2.d. Meanwhile, the gaseous nuclei gradually increase in size due to pressure reduction. Inside the liquid, this results in some alterations of the motion of the bubbles: as the buoyancy force increases, the nuclei which are already pointing upwards are accelerated towards the free surface, while those entangled in the fluid flow become able to overcome drag. This second effect is obtained when a critical diameter is attained, which depends on the residual agitation level of the liquid. At this stage of the process, the critical diameter is on the order of the millimetre: a sudden change in direction has been observed for bubbles with diameter of 1.27 ± 0.12 mm in the upper portion of the fluid (figure 2.e) and of 1.11 ± 0.08 mm in the lower portion (figure 2.f). The slight difference between the two regions is symptomatic of a fluid velocity field which is still non-uniform.

Bubble enlargement produces, as an additional consequence, the appearance of free nuclei that before were too small to be detected. Furthermore, at nearly $t' = 0.4$ also gaseous cavities entrapped on the bottle internal walls become visible. Hence, the liquid seems pervaded by an increasingly greater amount of bubbles, as can be observed by scrolling down the images in the second and third column of figure 2. Such a trend continues in the next few instants ($0.4 < t' < 0.8$). In this time-interval no more bubbles come into view, but a few nucleation sites on the walls activate and start to release bubbles of some hundreds of microns in diameter, with frequencies as high as 24 bubbles per second.

The population of nuclei in the liquid undergoes changes also in size distribution. Since bigger bubbles rise more rapidly, the lower portion of the bottle soon remains occupied by nuclei of homogeneously small size, whose diameter grows in time up to nearly 0.6 mm. The upper portion of the bottle houses a more varied population, bubbles arriving at the liquid-foam interface with diameters of even 2.3 mm. Fed by the continuous arrivals of bubbles, the foam column keeps growing (figure 2.g). During expansion, deterioration processes begin to affect the top layers of the foam: bubble burst occurs when an unstable size is reached, as a consequence of the pressure decrease or due to coalescence between adjacent nuclei.

Figure 2. *Early growth.* Scale bar 10 mm.

3.1.2. Cloud formation. At about t' = 0.8, bubbles moving in the liquid bulk experience a global upward deviation of their trajectories. Visually, such a behaviour translates into the formation of a cloud of bubbles departing from the bottom of the bottle (figures 3.b and 3.c). The phenomenon is dictated by the smallest entrained nuclei left in the liquid that change direction simultaneously once they get the critical size. Relative to the previous stage, the current measured value of the critical diameter is lower, i.e. of 0.69 ± 0.11 mm, and identical in the whole fluid, suggesting that the flow field has attenuated and levelled out. Besides the nuclei responsible of its generation, the also cloud incorporates i) entrained bubbles, typically of bigger size, already moving towards the free surface, ii) bubbles previously released from the sites on the walls, iii) entrained bubbles of size even smaller than the critical one, dragged into the wake of the rising bubbles, and, iv) bubbles torn away from the solid surface by the cloud passage.

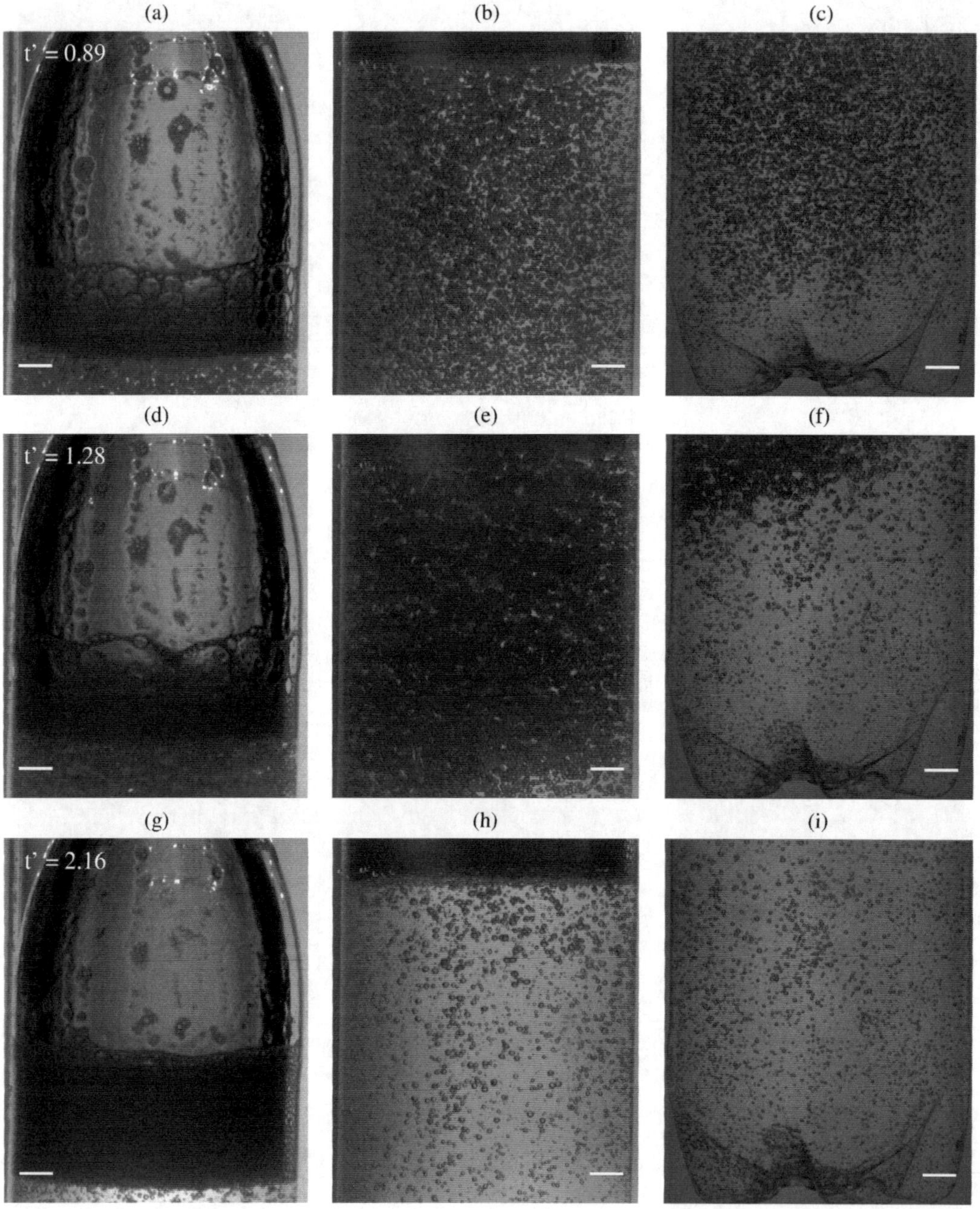

Figure 3. *Cloud formation.* Scale bar 10 mm.

While ascending, the nuclei continue to enlarge and the cloud thickens. Since the path of the nuclei is more and more obstructed, their speed is slowed down and it takes about 1.4 Δt_l for the cloud to rise completely. Hence, the process completes at t' = 2.2 (third row of figure 3), that is, well after the end of the first decompression step, extending over the first third of the intermediate rest period. Images shown in the first column of figure 3 demonstrate that cloud ascent promotes a significant expansion of the foam column. However, foam growth is not constant. On the one hand, it is interrupted by coalescence and burst of the bubbles composing the upper (and older) layers of the foam, like the ones appearing in figure 3.d. On the contrary, the new layers formed by the bubbles carried by the cloud are more compact and homogeneous, and so more stable. On the other hand, the interface between the liquid and foam is not fixed: initially the developing gaseous phase inside the liquid makes the interface shift upwards. As the cloud evolves, the rising nuclei start to crowd at progressively lower levels, and the interface is brought back to nearly the starting height.

3.1.3. First collapse. With the cloud passage, the liquid empties out of all the nuclei introduced by entrainment. Nucleation sites already identified on the bottle walls continue to release bubbles in the remaining part of the intermediate rest period (2.2 < t' < 4.0), but with progressively slower rates. Further sites between the ones visible in figures 3.h and 3.i become active in this stage. These, in general, produce a few bubbles, whose size and frequency of release are highly influenced by interactions with other nuclei. Due to the scarce supply of new bubbles, the foam behaviour is dominated by decay. The homogeneous structure of the foam hinders its immediate deterioration. With passing time, drainage and coalescence induce structural changes in the whole foam column, involving fusion, deformation and rearrangement of bubbles, as can be observed by comparing figures 3.g and 4.a. As a result, bubble skins get thinner and weaker, ultimately causing the rupture of the top layers of the foam.

3.1.4. Second growth and collapse. The second decompression step, operated at t' = 4.0, is instantly perceived as a sudden but moderate swelling of the bubbles in the foam. The foam structure is then further destabilized and subsequently exposed to a faster decay. Indeed, figures 5.a and 5.d show that the foam height drops, ultimately assuming, at t' = 5.3, levels similar to those observed in the very early stages of the process (figure 2.d). The new pressure decrease also affects the bubbles inside the liquid, including those bound to the bottle walls. As can be seen from the images in figures 5.b and 5.c, as compared with figures 4.b and 4.c, the bubble diameter increases rapidly, up to 30% in the first three tenths of the second snift.

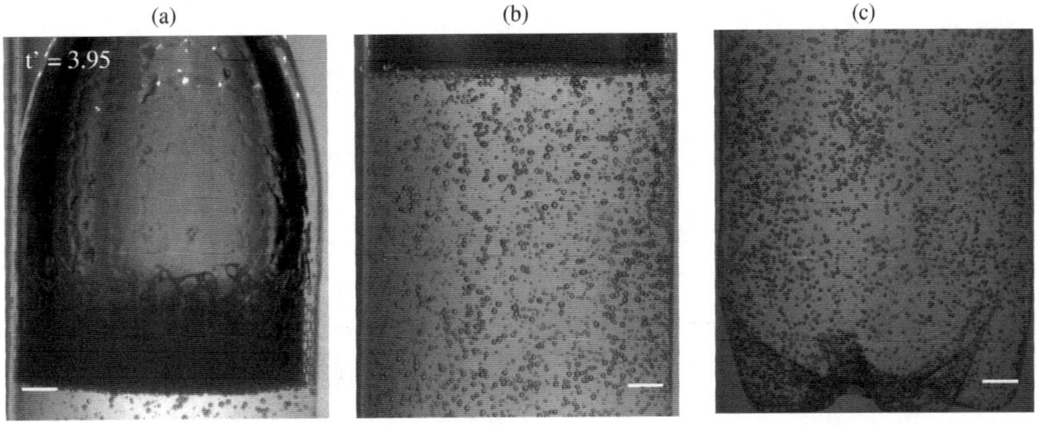

Figure 4. *First collapse.* Scale bar 10 mm.

The persistence of a supersaturation state in the liquid promotes the growth of the nuclei also in the following instants. Consequently, more bubbles become able to overcome adhesion and detach from the solid surface. Large bubbles from the bottom of the bottle stimulate, in turn, the release of other nuclei when bumping them during ascension. In addition, nucleation sites activated during the first snift recover their original frequencies of release. All those factors produce a gradual increment in the rate of bubbles arriving at the liquid-foam interface (figure 5.b). Such a tendency culminates at approximately t' = 4.8, after which no apparent variations in the characteristics of the emerging bubbles are observed.

3.1.5. Final equilibrium. When t' > 5.5 the foam height stops reducing. Any decay of the existing layers is compensated by the massive arrival of new bubbles, as large as those responsible of the generation of the first layer of foam. Hence the foam column is progressively deprived of residual bubbles from the first decompression step, which are substituted by freshly originated nuclei of uniform size. This process contributes to create an equilibrium state which lasts till the end of the process, as proved by figure 6. From figure 6.c it is also evident that despite the augmenting buoyancy force and the collisions with rising nuclei, clusters of bubbles are held on the bottle walls. These may give an additional delayed contribution to the foam, as they can be activated by mechanical agitation once the bottle is detached from the filling machine.

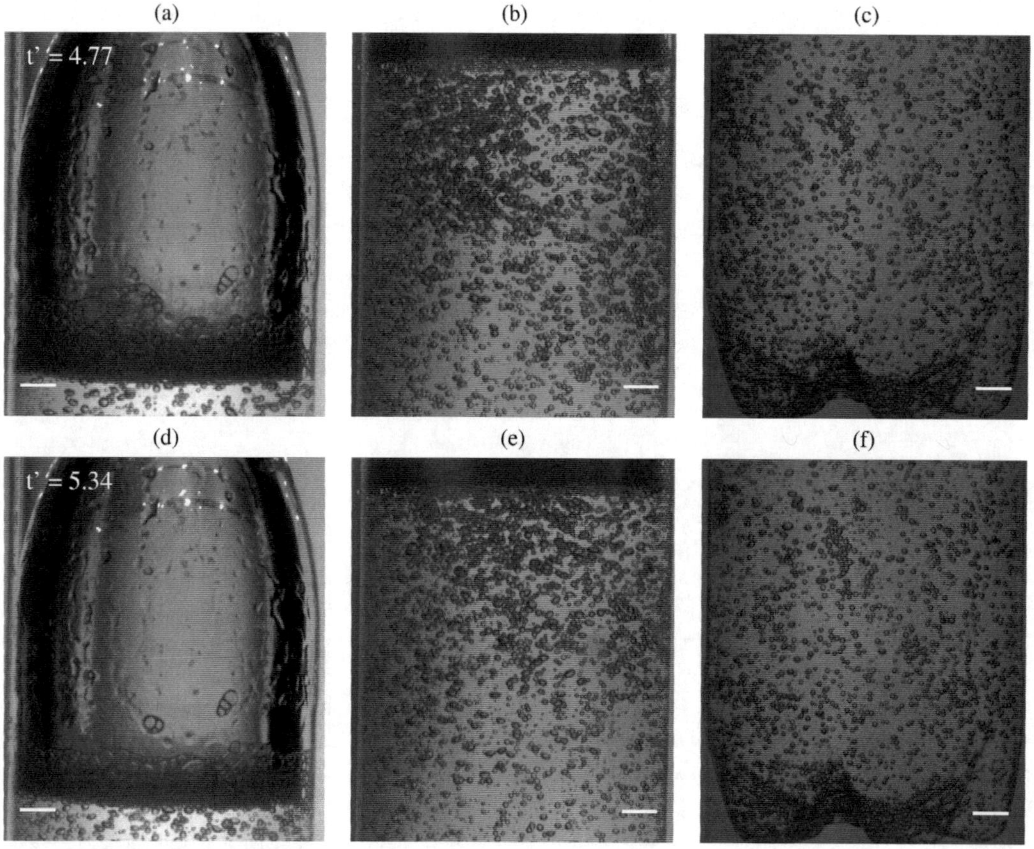

Figure 5. *Second growth and collapse.* Scale bar 10 mm.

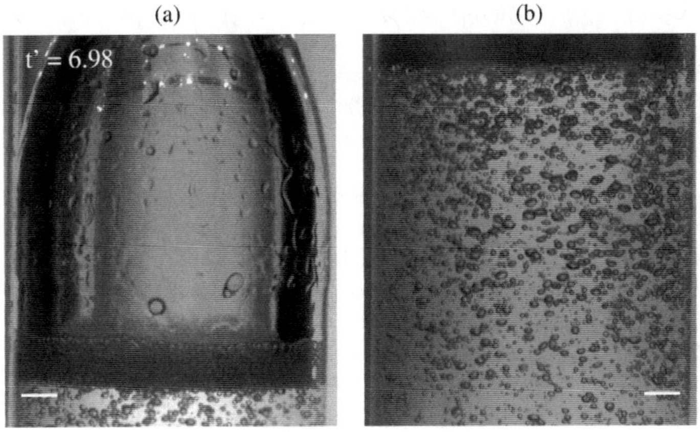

Figure 6. *Final equilibrium.* Scale bar 10 mm.

3.2. Plot of the foam trend

To understand how the above foaming dynamics reflect on the amount of foam produced, the image sequence relative to configuration A was analysed in quantitative terms. The foam thickness was extracted from 86 frames using the procedure described in section 2.2. The measurements were intensified when rapid changes in the foam column were observed, so as to reconstruct the whole trend with the best accuracy. The resulting graph of the foam thickness versus time is displayed in figure 7. Boundaries between the different stages of the decompression sequence are also indicated.

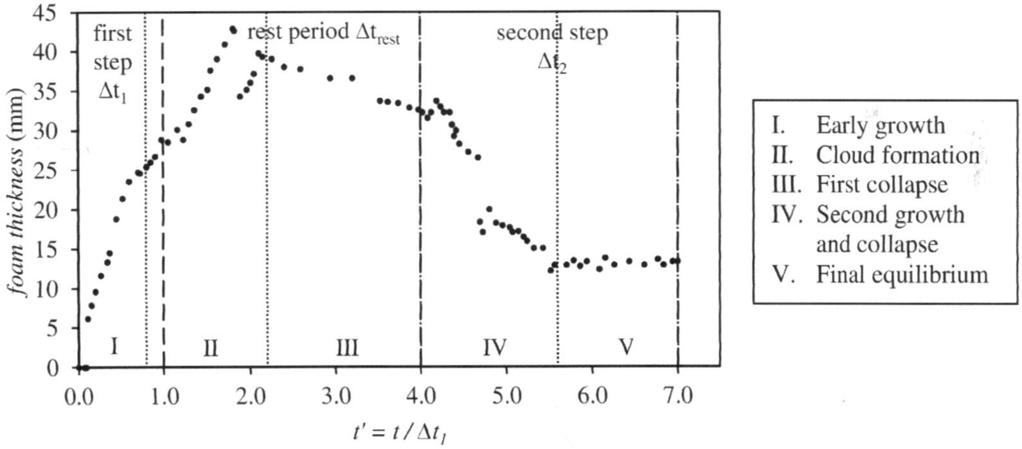

Figure 7. Graph of the foam level versus time. Dashed and dotted lines represent the boundaries between i) the decompression steps and ii) the identified stages of foaming dynamics, respectively.

The graph shows that the foam level grows rapidly during the first half of the first decompression step, fostered by the rise of the largest entrained bubbles. The growth is not immediate, but a short latency time has to be waited to allow the aggregation of the initial layer of foam. Later, the foam continues to expand with a slower rate. The change in slope should be attributed to variations occurring in the population of the ascending nuclei, such as their mean diameter and rising velocity. Nevertheless, the subsequent generation and development of the cloud, as reported in section 3.1.2, produce a further increase in the foam thickness. Bumps are present in this segment of the curve; these are caused by the occasional burst of bubbles on the top of the column. Though the cloud has not completely risen, at t' = 1.8 the foam reaches its maximum thickness, i.e. 43 mm, corresponding to the 26 % of the filled volume of liquid. The subsequent rupture of a large bubble originated by fusion from those visible in figure 3.d produces a sudden drop in the foam height. Next, the level is partially restored by the tail nuclei of the cloud as they eventually reach the liquid-foam interface.

With the cloud arrival (t' = 2.2), the production of foam induced by the first decompression step ceases. On the contrary, the foam level starts to decrease despite the incoming of some residual bubbles from the bottle walls. Initially (2.2 < t' < 3.2), the decay is almost imperceptible, impeded by the compact structure of the foam. Only a few bubbles burst, their size being too small to produce a significant reduction of the column. Approaching the end of the intermediate rest period, the number and size of the exploding bubbles gradually increase, being promoted by the advancing deterioration processes. However, this translates into a lowering of just 4 mm of the foam height. A more drastic collapse of the foam column is observed after the snift valve is opened again: following the modest peak at t' = 4.2, to be attributed exclusively to a swelling of the existing layers, the thickness of the foam reduces of 18 mm within an interval of 11 tenths. The decrease is not continuous but occurs in steps, as it is related to simultaneous ruptures of multiple large bubbles.

The last portion of the curve is characterized by the convergence of the foam thickness to a constant value, i.e. 13 mm. This can be considered as the equilibrium value in the balance between the further deterioration of foam, on the one hand, and the renewed production of bubbles, on the other one, under the conditions achieved with the second decompression step.

4. Conclusions

The foaming phenomena occurring in a carbonated beverage during bottling have been examined with the use of high-speed imaging. It is ascertained that with the adopted decompression sequence, nuclei entrained in the bulk liquid during filling play a major role in the thickening of the foam. Changes occurring in the population of the rising bubbles reflect on the rate of growth of the foam level. In particular, the development of a cloud provided by a global deviation of bubble trajectories determines the achievement of the maximum thickness. Other factors participate in defining the initial growth of the foam, including the conditions at the free surface as decompression initiates, and the onset of foam destabilization processes. Those last phenomena strictly depend on the structure of the foam and influence its subsequent evolution. The end of the process is dominated by the competition between the deterioration of the existing layers and the release of bubbles from sites located on the bottle walls, which allow the establishment of an equilibrium height of the foam.

References

[1] Jones F S, Evans G M and Galvin K P 1999 *Adv. Colloid Interface Sci.* **80** 27–50
[2] Harvey E N, Barnes D K, McElroy W D, Whiteley A H, Pease D C and Cooper K W 1944 *J. Cell. Comp. Physiol.* **24** 1–22
[3] Biń A K 1993 *Chem. Eng. Sci.* **48** 3585–630
[4] Qu X, Goharzadeh A, Khezzar L and Molki A 2013 *Exp. Therm. Fluid Sci.* **44** 51–61
[5] Ray B, Biswas G and Sharma A 2015 *J. Fluid Mech.* **768** 492–523
[6] Salerno E, Levoni P and Barozzi G S 2015 *J. Phys.: Conf. Ser.* **655** (2015) 012036
[7] Bankoff S G 1958 *AIChE J.* **4** 24–26
[8] Chen P, Wu W and Jones B G 2009 *Nucl. Eng. Des.* **239** 2035–41
[9] Brennen C E 2014 *Cavitation and bubble dynamics* (New York: Cambridge University Press)
[10] Pugh R J 1996 *Adv. Colloid Interface Sci.* **64** 67–142
[11] Malysa K and Lunkenheimer K 2008 *Curr. Opin. Colloid In.* **13** 150–62
[12] Shokribousjein Z, Deckers S M, Gebruers K, Lorgouilloux Y, Baggerman G, Verachtert H, Delcour J A, Etienne P, Rock J, Michiels C and Derdelinckx G 2011 *Cerevisia* **35** 85–101

IOP Conf. Series: Journal of Physics: Conf. Series **796** (2017) 012043 doi:10.1088/1742-6596/796/1/012043

An Innovative Enhanced Wall to Reduce the Energy Demand in Buildings

F Fantozzi[1], S Filipeschi[1], M Mameli[1], S Nesi[1], G Cillari[1], M B H Mantelli[2], F H Milanez[2]

[1]Department of Energy, Systems, Land and Construction Engineering - University of Pisa, Largo L. Lazzarino, 56122, Pisa, Italy
[2] Dr., Mechanical Engineering Department, Federal University of Santa Catarina UFSC, Brazil
Corresponding author: Sauro Filippeschi- s.filippeschi@ing.unipi.it

Abstract. Energy saving in buildings is one of most important issues for European countries. Although in the last years many studies have been carried out in order to reach the zero-consumption house the energy rate due to passive solar heating could be further enhanced. This paper proposes a method for increasing the energy rate absorbed by opaque walls by using a two phase loop thermosyphon connecting the internal and the external façade of a prefabricated house wall. The evaporator zone is embedded into the outside facade and the condenser is indoor placed to heat the domestic environment. The thermosyphon has been preliminary designed and implanted into a wall for a prefabricated house in Italy. An original dynamic thermal model of the building equipped with the thermosyphon wall allowed the evolution of the indoor temperature over time and the energy saving rates. The transient behaviour of the building has been simulated during the winter period by using the EnergyPlus™ software. The annual saving on the heating energy is higher than 50% in the case of a low consumption building.

1. Introduction

Energy saving in buildings is one of the most important issues in European countries. There are several strategies to reduce the energy demand in buildings, such as ventilation, thermal insulation, passive heating and cooling plants. One of the most effective methods is indeed the use of passive solar energy, which is directly supplied to buildings through the windows and through the opaque walls. An interesting research trend in this topic at the moment is the use of "smart envelopes" that are able to exploit the solar energy for the production of both active and passive heating.

This paper proposes an innovative method to increase the energy rate coming from the solar opaque elements by using a loop two phase thermosyphon which directly connects the internal and the external house wall surfaces. Zhang et al. [1] presented a paper on a loop-heat-pipe based solar thermal facade (LHP-STF) installed in a reference residential building. They studied the technical evaluation and the economic advantages of applying this solution in three typical European climates, including North Europe (Stockholm), West Europe (London) and South Europe (Madrid). The research results indicated that the LHP-STF could contribute to the hot water heat load throughout the year, with substantially reduced heating load in the winter, and yet a slight increased cooling load in the summer. However it is well known that when the gravity field can be exploited, the two-phase loop thermosyphons perform better and they are cheaper than capillary loops [2] [3] and therefore they seem to be more attractive for this kind of applications.

As a first attempt to recover energy from unglazed wall, a new concept with heat pipes or thermosyphons integrated into the facade of a building has been presented by Sun et al. in 2014 [4] and tested by Zhang et al. in 2015 [5] for Chinese climates.

The basic idea consists of a two phase device, named microgravity heat pipe, which presents the evaporator implanted in the outdoor plaster and the condensing section implanted in the inner surface. The main advantage of this solution is the two-phase device efficient unidirectional heat transfer characteristics given by natural convection. The device, indeed, works like a thermal diode that transfers heat directly into the building during winter if the temperature difference between indoor and outdoor is positive and, at the same time, prevents the heat losses in the opposite case. Sun et al in 2014 [4] propose to insert an intelligent manual valve to stop the natural circulation inside the two-phase device. This allows for preventing building heating in summer and consequently increasing the demand on cooling energy.

In 2015, Zhang et al. [5] tested a Wall Implanted Heat Pipe WIHP made of an evaporator section built with 24 capillary pipes (ID=2.7 mm, 600 mm long) implanted into a mortar plaster and a similar sized condenser implanted into the internal plastering layer. This device has been filled with R600a. They have tested a 1720x1720 mm^2 surface facing south at different outside temperature, with an internal air temperature of 18°C. They found out that heat transfer transmittance value of the wall with WIHP ranges from 0.76 to 0.987 W/m^2 K as the outside surface temperature increases from 24 °C to 42 °C. They affirm that this solution decreases the annual energy heating demand by approximately 15% in Jinan climate.

In 2016 Fantozzi et al. [6] proposed a totally new concept two-phase device to enhance the heat transfer performance in prefabricated houses with metallic case. The new device has been modeled and applied to a low consumption house sited in Pisa. The paper shows that the Wall Thermosyphon (WT) is able to enhance the building thermal comfort by keeping the indoor temperature close to the thermal comfort standard for most of the day. The energy saving rates were very high and overcomes rates of 23% of the heating demand during the winter period in the case of a low consumption building.

However, this preliminary analysis presented in [6] is very basic and relative to the specific house described in the paper. In order to understand the real impact of this device on the energy saving rates in Italy a large analysis must be made taking into account more general aspects. The role of the exposition and the surface extension, the impact of the ratio between the heated volume and the WT surface on the global saving rates, the amount of the energy that must rejected by the systems and other important issues has been analyzed in this paper.

2. The Wall Thermosyphon

2.1. The Working principle

The actual device [6] aims at enhancing the performances of the device proposed by Sun et al. [4]. A loop thermosiphon has been embedded into a wall of prefabricated house as schematically shown in Figure 1. A loop thermosyphon consists of an evaporator (hot source), a condenser (cold source) and two adiabatic arms [7]. In the proposed solution the evaporator is brazed or welded to a metallic façade. This solution can be applied to several prefabricated houses because the use of metallic façades has found a large diffusion in modern architecture. As described in [6] the metallic plate is irradiated by the sun and the loop thermosyphon directly transfers the heat inside the room through the condenser. As the metallic plate temperatures overcomes the inside air temperature the loop thermosyphon transfer heat into the room. This device cannot be switched off as the inside temperature overcomes the standard comfort temperatures. In this case a large amount of heat should be rejected outside through a hollow space inside the internal façade where a finned condenser tube is placed (Figure 1). This cavity is connected both to indoor and outdoor through four opening and closing air vents. The working principle scheme is shown in [6]: as the inside temperature is too high the outdoor vents are opened and the indoor vents are closed and as the indoor temperature is too low, the indoor vents are opened and the outdoor vents are closed.

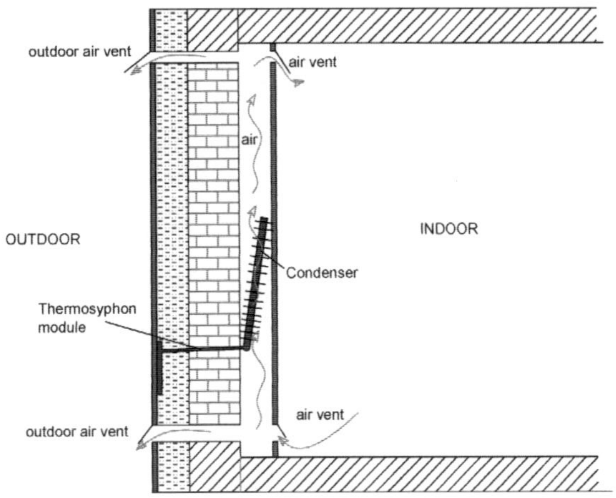

Figure 1. Wall Thermosyphon working principle scheme.

The wall loop thermosyphon designed for this application is similar to the devices described by Milanez and Mantelli [8] and it is shown in Figure 2. The evaporator consists of two vertical aluminum pipes (ID=12 mm) connected with two manifolds (ID=32 mm). The condenser is made of two horizontal manifolds and 5 tilted pipes equipped with 20 rectangular fins 10 mm interspaced. The working fluid is the refrigerant R141. All the dimensions of the loop thermosyphon are shown in Fig. 2.

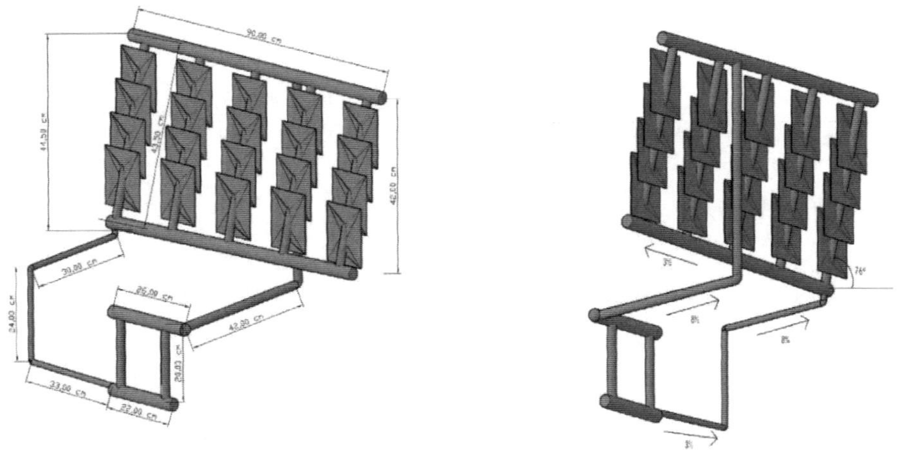

Figure 2. Loop Thermosyphon dimensions.

2.2 The reference building

The paper proposes to apply this innovative device in a prefabricated wall of a single floor reference house shown in Fig. 3. The reference house has been realized with high efficiency solutions and could be labeled as "low consumption house" in the Italian standard certification labeling procedure with an energy demand lower than 2000 kWh per year.

IOP Conf. Series: Journal of Physics: Conf. Series **796** (2017) 012043 doi:10.1088/1742-6596/796/1/012043

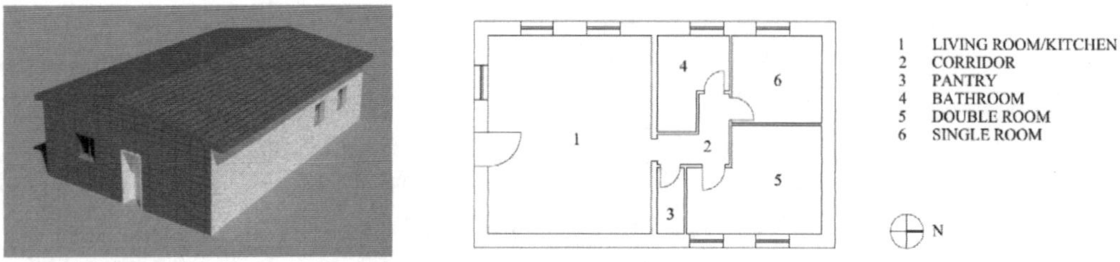

Figure 3. Sample Built House, Rendering and Reference House Map.

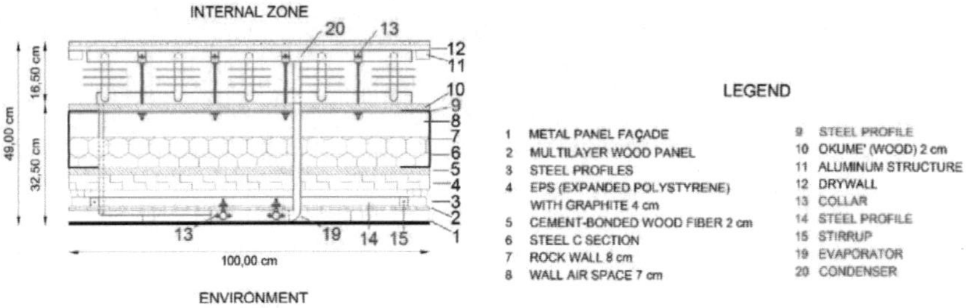

Figure 4. Vertical wall with the embedded wall thermosyphon.

Such low consumption rates have been reached by using highly insulated walls. The heat transfer transmittances of all the elements are reported in Table 1 and their surfaces are reported in Table 2.

Table 1. Heat Transfer Transmittances

Envelopes elements	Total Heat Transfer Transmittance H [W/m^2 K]
Outside Wall	0,21
Inside Wall	0,39
Roof	0,27
Window	1,59
Floor	0,5
Door	1,18

Table 2. Surfaces of envelope elements case 1

Envelopes elements	Dimension 1 [m]	Dimension 2 [m]	Surface [m^2]	Window surface [m^2]	Door Surface [m^2]
Floor	7	12	84	-	-
Roof	7	12	84	-	-
South Wall	7	3	21	1,08	2,4
Nord Wall	7	3	21	-	-
Est Wall	12	3	36	2,16	-
West Wall	12	3	36	4,32	-

Being the reference building a prefabricated house, the walls are previously made in a factory and assembled together on site to realize the building. The walls are therefore made of several modules 1 m wide and 3 m high. Each module is equipped with 3 wall thermosyphons as shown in [6].

3. Simulation tools and dynamic building analysis

3.1 Dynamic simulation building criteria

The whole prefabricated house has been simulated during its transient behavior by using the software EnergyPlus™ based on a finite difference approach. The implemented algorithm uses an implicit finite difference scheme which is also coupled with an enthalpy-temperature function able to account for the phase change energy phenomena. The building simulated in EnergyPlus™ has no internal gains (people, lights or electric equipment) except ventilation gain with a flow rate of 0,35 V/h (a slightly higher value than that imposed by the Italian legislation for the people comfort standard that is 0,3 V/h) uniformly distributed during a day.

The wall thermosyphon passive plant has been simulated as a dynamic external heating load affected by the outdoor façade temperature and the indoor air temperature. This load hourly changes along the winter period as well as the solar heat flux and the temperature changes. The simulation criterion is detailed in [6] and here briefly summarized. The heat power directly transferred by the WT is positive as the difference between the outdoor façade temperature and the indoor air temperature is positive and it is equal to:

$$q = \frac{T_{SW} - T_{IA}}{R_{wth}} \tag{1}$$

where R_{wth} is the WT thermal resistance of the wall module.

$$R_{wth} = R_{th} + R_{air} \tag{2}$$

where R_{th} is the thermal resistance of the two-phase devices, R_{wall} is the thermal resistance of the wall and R_{air} is the thermal resistance of the free convection heat transfers at the condensing section. The thermal resistance and the thermal capacitance of the loop thermosiphon are negligible in comparison with the wall. The thermal resistance of the thermosyphon (R_{th}) has been estimated with the model described by Milanez and Mantelli [8] and it is considered constantly equal to 0.03 m^2KW^{-1}. If the thermal resistance of the finned condenser section is equal to 0.05 m^2KW^{-1} the thermal resistance of the device is therefore 0.08 m^2KW^{-1}. The total power that the WT is able to transfer inside the building for each hour is given by the heat flux coming from Eq. (s) multiplied for the relative surface. The power transferred by the WT in an hour has been inserted into the EnergyPlus™ platform as a simple convective heat load. The WT energy rate is therefore highly affected by the climate conditions. The energy rates have been calculated by referring to a standard day for each month.

3.2 Monthly Standard Day selection

The energy rates saved during a whole year have been hourly calculated starting for a standard day each month. The energy rates of each day has been therefore multiplied by the number of days of each month comprised in the standard winter season (1st November-15th April for Pisa city). The software EnergyPlus™ uses weather data coming from more than 2100 locations in the world. The weather data are arranged by World Meteorological Organization region and different Country: "Pisa-S. Giusto" is the climate database used in these simulations. These data have been collected from meteorological measuring stations for a year without any filters and post processing activity. They show a wide spreading. In order to limit the error due to the spreading of the weather data a standard day has been defined for each month starting from the original weather database. The standard day selection has been made by comparing the EnergyPlusTM weather database with those coming from the standard UNI EN

10349-1:2016 which reports the average value of several climatic parameters for a standard day of each month. Only the parameters affecting the WT performances have been considered for the selection:

- Site Outdoor Air Dry-bulb Temperature (T_{air} [°C])
- Incident Solar Radiation Rate per Area on South Wall (W_{south} [W/m^2])
- Incident Solar Radiation Rate per Area on East-West Wall ($W_{east\text{-}west}$ [W/m^2])

Table 3. Discrepancy between the EnergyPlus™ database and the standard UNI EN 10349-1:2016

Selected Standard day	Relative Error T_{air} [%]	Relative Error W_{south} [%]	Relative Error $W_{east\text{-}west}$ [%]	Average Error [%]
29/01	32,39	5,92	17,61	18,64
6/02	6,96	7,99	39,13	18,03
29/03	10,21	0,12	13,58	7,97
26/04	6,95	18,15	12,28	6,41
20/11	3,94	16,63	36,75	19,11
15/12	9,28	15,32	34,58	19,73

The relative error between the average value of each single parameter listed above and the daily average value shown in UNI EN 10349-1:2016 has been calculated for every day in a month. The selection of the standard climatic parameter is an essential element of most building energy efficiency programs, however there is no widely accepted scientific technique for its delineation. A wide review of this technique is shown in [9]. In this paper the standard day is the day that minimize the average error calculated among the relative errors of each single parameter listed above. Table 3 reports the selected day and shows the minimum average error join with the relative errors of each single parameter.

Table 3 shows that the relative errors of each single parameter are quite high even for the selected day. The highest errors are relative to the parameter $W_{east\text{-}west}$ going from 12,28% (April) up to 39,16 % (February). The relative errors of T_{air} are low and they are generally lower than 10% for all the warming period except for January.

Note that the parameter $W_{east\text{-}west}$ has however a minor impact on the WT performance of the other two parameters because it is generally lower than W_{south} and the eat west surface have been consider only in a simulation scenario.

3.3 Building simulation cases

The wall thermosyphon has been applied to the building described above. The main aim of this paper is to evaluate the energy rated saved as the wall thermosyphons are applied to the reference building. Previous simulations [6] showed that a building can have very high energy saving rates (ranging from 20% up to 40%) for warming period at Pisa. These energy saving percentages have been obtained without any thermal control systems applied to the building and the indoor air temperature can reach temperatures higher than those recommended to maintain the thermal comfort.

In order to maintain the indoor thermal comfort, the temperature control system above described must be activated. The thermosyphon cannot be switched off so it continuously transfers heat indoor, the control system decides if the heat must be used or rejected outdoor. The temperature control system proposed in this paper is totally passive. Furtherly it has been assumed that the thermal control system is able to naturally remove any amount of energy. The efficiency of this system will be experimentally verified in future. In order to evaluate the impact of the wall thermosyphon on the energy demand in a building is therefore fundamental calculated both the energy saving rates and the energy rates that must be rejected outdoor. The energy rates are highly affected by both the orientation and the surface cover by the thermosyphon for this reason the building has been simulated in three different significant configurations:

- Case 0: building without any WT
- Case 1: building with WT embedded into the south wall (S=7x3 m^2)
- Case 2: building 90° clockwise rotation with WT embedded into the south wall (S=12x3 m^2)
- Case 3: Case0 with WT embedded into the south, east and west walls (S=31x3 m^2)

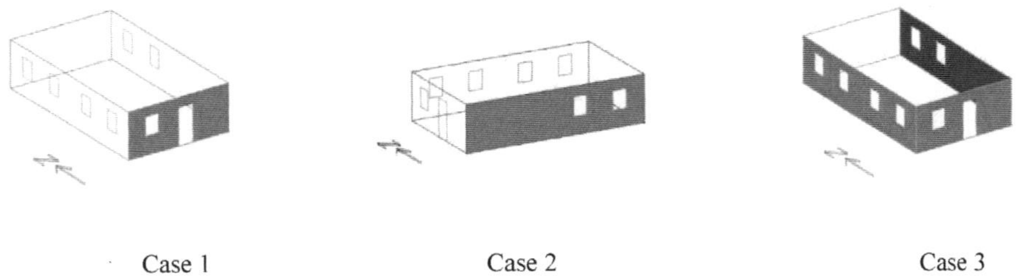

Figure 5: Different façade orientations and different WT surfaces studied in the simulations

4. Results and discussions

4.1 Temperature evolution along a standard day for each month

The WT thermosyphon module functioning can be well understood by observing the evolution of the indoor air temperatures along the standard day for each month. Figures 6 and 7 show the hourly evolution of the indoor air temperature T_{IA} along the selected day of February and November, respectively. Both the figures show the indoor air temperature with and without the wall thermosyphon embedded in the wall and the minimum and the maximum air temperature value recommended to maintain the indoor thermal comfort (20 and 22 °C). The indoor air temperature in the case of WT embedded shown in Figures 6 and 7 are both referred to the case 1. The effect of the wall thermosyphon is evident in both the months.

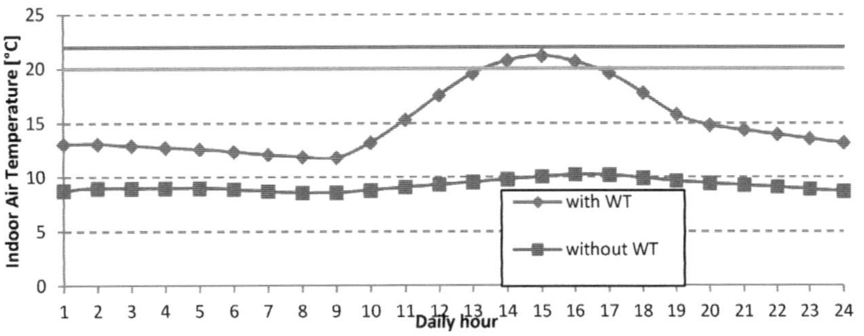

Figure 6. Daily evolution of the indoor air temperatures for the selected day February in case 1 simulation with and without thermosyphons

In February the indoor air remains constantly under the maximum standard comfort temperature threshold and no energy rate must be rejected from the building. In November (Figure 7) the indoor air temperature overcomes the maximum temperature threshold from 12.30 to 15.30 p.m if none thermal control system is active. In this time the heat transferred by the thermosyphon must be removed from the building and outdoor rejected. The hourly temperature evolution with the temperature control systems active for the standard day of November is shown in Figure 7.

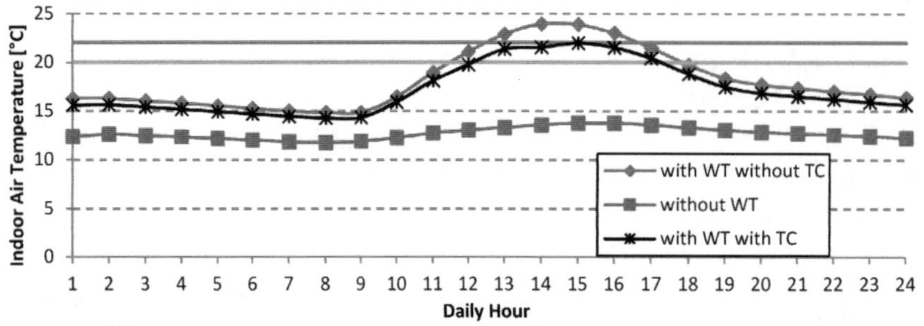

Figure 7. Daily evolution of the indoor air temperatures for the selected day of November in case 1 simulation with and without wall thermosyphon (WT) e Thermal Control System (TC)

4.2 Energy saving

The energy saving rates have been calculated with the temperature control system activated. Figure 8 shows the monthly energy heating demands with the thermosyphon is embedded into the wall compared with those without Thermosyphon for case 1, case 2 and case 3. The wall thermosyphon module allows high energy saving rates by decreasing the energy demand down to 50-55 % even if the reference building is a low consumption house (heating energy annual demand lower than 20 kWh/m^2). Figure 8 shows that the energy heating demand slightly decreases as the WT module surface faced at south increases (case 2). The energy demand furtherly decrease as the WT module surface is extended at east and west facades. The energy saved percentage with respect the case 0 (without WT) is 52.11% for case 1 and 60.33 % for case 2.

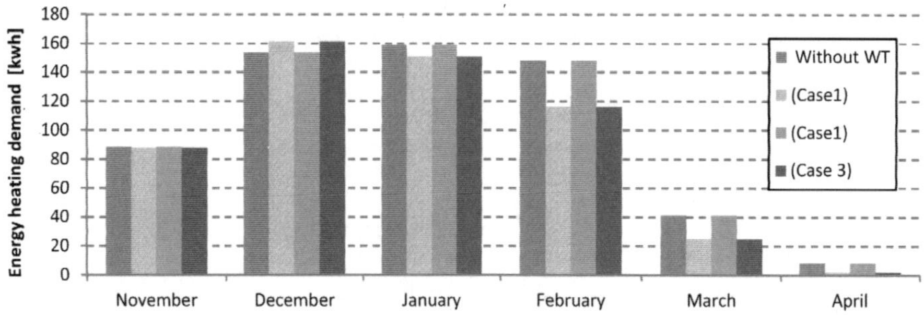

Figure 8. Monthly energy heating demand for the house with or without the wall thermosyphon system for cases 1, 2 e 3.

This behavior is due to the effect of the thermal comfort control system that limits the energy saving rates because rejects the most of heat transferred by the thermosyphon along a day (Figure 7). However even if the WT surface (Case 3) grows up to 93 m^2 in comparison with 21 m^2 (Case 1) the energy demands slightly decreases during an year. Even if the most of the surface are east and west faced, the surface enhancement is rather relevant. In this case the energy saved percentages in a whole year grows up to 64.25%. The energy saving is not the only parameter that allows the efficiency characterization of this module. The heat in excess must be rejected because it could make the indoor temperature higher than the thermal comfort limitation. For each month the energy rates that the thermosyphon transfers inside the room (transferred energy)has been evaluated as well as the energy rates that must be outside rejected (rejected energy). These energies have been compared with the energy that WT transfers inside the rooms without any Temperature Control system (TC).

These energy rates are monthly shown in Figg. 9, 10 and 11 for case 1, case 2 and case 3, respectively. Note that the energy rate transferred into the building without any temperature control system (TC) is lower than the sum of the energy really transferred into the house and the energy outdoor rejected. This behavior is due to thermal control system. If none temperature control system is activated the indoor temperature freely raises and the heat transferred by the WT decreases as the inside temperature decreases. On the other hand if the indoor temperature is constantly maintained at 22°C the heat losses are constant and the heat transferred by the WT is lower.

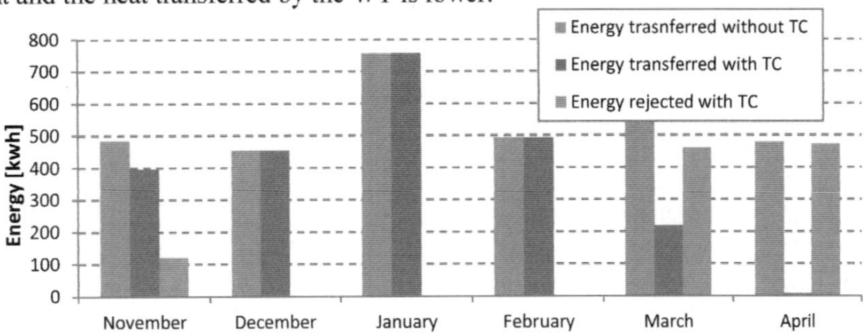

Figure 9. Monthly energy rates for the house with or without the temperatures control system for case 1 simulation.

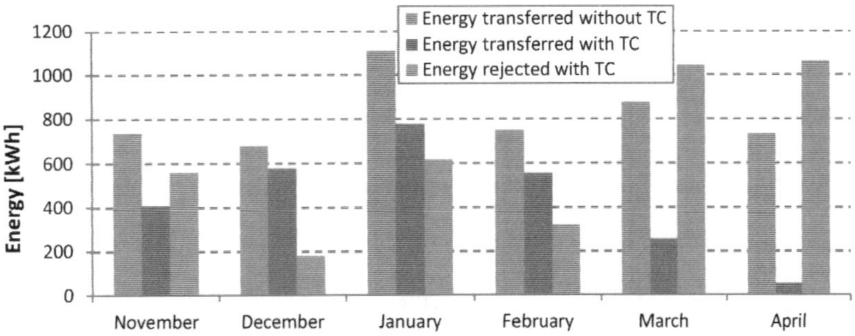

Figure 10. Monthly energy rates for the house with or without the temperatures control system for case 2 simulation.

Figures 9, 10 and 11 furtherly show that as the WT surface increases the heat transferred into the buildings increases as well as the heat outdoor rejected. The systems works so efficiently that the energy rates that must be removed increases as the surface south faced increases (case 2 – Fig. 11).

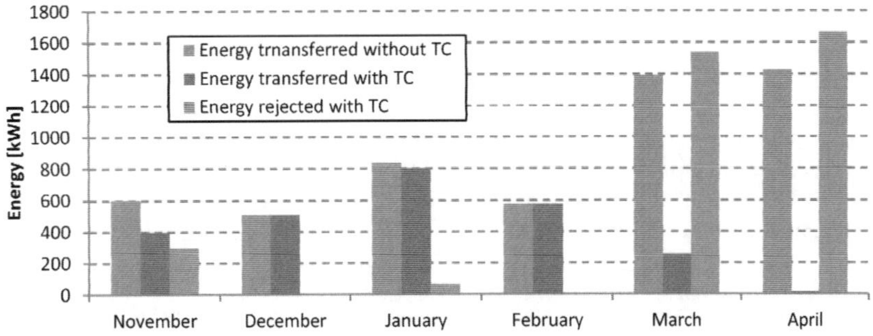

Figure 11. Monthly energy rates for the house with or without the temperatures control system for case 3 simulation.

Figure 10 shows that for the case 2 the energy must rejected from the building during all the warming period even in January and December. In April the rates that must be removed only for 15 days is more than 400 kWh. It is really high and if the TC system no efficiently works a real discomfort could be really observed. The problem becomes huge for case 3 simulation, as shown in Fig. 11. The energy rates that must be removed by the TC system are higher than 1000 kWh for March and April and the temperature inside the TC system could reach a really high value. This paper shows that the efficiency of the TC system is crucial for the Wall thermosyphon systems because it transfers so efficiently the sun radiation inside the building that the discomfort condition could be quickly reached.

If the efficiency of the TC systems is known, each building could be optimized by calculating the WT surface in order to conjugate the maximum high saving rate performance with a good thermal comfort standard inside the rooms.

5. Conclusions

This paper presents an innovative systems that reduce the energy heating demand in the prefabricated houses in Italy. The system is totally passive and consists of a loop thermosyphon that is embedded inside a wall of a building. The wall thermosyphon has a low inertia and quickly and efficiently transfers the heat irradiated by the sun inside the building. The wall thermosyphon works so efficiently that in the case of a low consumption prefabricated house (heating energy demand lower than 20 kWh/m2) the energy saving rates are higher than 50% in all the cases simulated. Unfortunately, the paper shows that energy that must be rejected during the warmer months could reach very large rates that could create discomfort conditions inside the building. If the WT thermosyphon cover the surface south, east and west faced, for example, the energy saving percentage is 62%, but the amount of energy that must be removed is higher than 3400 kWh per year (40 kWh/m2 year). This paper shows that the efficiency of the TC system is crucial for the Wall thermosiphon systems because it transfer so efficiently the sun radiation inside the building that the discomfort condition could be quickly reached.

6. References

[1] Zhang X, Shen J, Adkins D, Yang T, Tang L, Zhao X, He W, Xu P, Liu C, Luo H 2015 The early design stage for building renovation with a novel loop-heat-pipe based solar thermal facade (LHP-STF) heat pump water heating system: Techno-economic analysis in three European climates, *Energy Conversion and Management*, **106** 964–986.

[2] Filippeschi S 2011 Comparison between miniature periodic two-phase thermosyphons and miniature LHP applied to electronic cooling equipment, *Applied Thermal Engineering* **31** 795-802.

[3] Jafari D, Franco A, Filippeschi S, Di Marco P 2016 Two-phase closed thermosyphons: A review of studies and solar applications, *Renewable and Sustainable Energy Reviews* **53** 575–593

[4] Sun Z, Zhang Z, Duan C 2015 The applicability of the wall implanted with heat pipes in winter of China, *Energy and Buildings* **104** 36–46.

[5] Zhang Z, Sun Z, Duan C 2014 A new type of passive solar energy utilization technology—The wall implanted with heat pipes, *Energy and Buildings* **84** 111–116.

[6] Fantozzi F, Filippeschi S, Mameli M, Mantelli M B H Milanez F H 2016 How a wall thermosyphon can enhance the energy savings in a prefabricated house in Italy, *Joint 18th IHPC and 12th IHPS conference* Jeju South Korea June 12-16.

[7] Franco A., Filippeschi S. 2010 Experimental analysis of Closed Loop Two Phase Thermosyphon (CLTPT) for energy systems, *Exp. Thermal and Fluid Science* **51** 302-311.

[8] Milanez F H, Mantelli M B H, 2010 Heat transfer limit due to Pressure drop of a two-phase loop Thermosyphon, *Heat Pipe Science and Technology, An International Journal* **1(3)** 237–250 .

[9] Walsh A, Cóstola D, Chebel Labaki L 2017 Review of methods for climatic zoning for building energy efficiency programs, *Building and Environment*, **112**, 337-350.

[10] Caruso G Fantozzi F Leccese F 2013 Optimal theoretical building form to minimize direct solar irradiation *Solar Energy* **97** 128-137.

U-PHOS Project: Development of a Large Diameter Pulsating Heat Pipe Experiment on board REXUS 22

P Nannipieri[1], M Anichini[1], L Barsocchi[1], G Becatti[1], L Buoni[1], F Celi[1], A Catarsi[1], P Di Giorgio[1], P Fattibene[1] E Ferrato[1], P Guardati[1], E Mancini[1], G Meoni[1] F Nesti[1], S Piacquadio[1], E Pratelli[1], L Quadrelli[1], A S Viglione[1], F Zanaboni[1], M Mameli[1*], F Baronti[1], L Fanucci[1], S Marcuccio[1], C Bartoli[1], P Di Marco[1], N Bianco[2], M Marengo[3], S Filippeschi[1].

[1] University of Pisa, Largo Lucio Lazzarino 2, 56122 Pisa, Italy.
[2] University of Naples Federico II, Corso Umberto I, 40, 80138, Naples, Italy.
[3] University of Brighton, Lewes Rd, Brighton BN2 4AT, United Kingdom.

*Corresponding Author Email: mauro.mameli@ing.unipi.it

Abstract. U-PHOS Project aims at analysing and characterising the behaviour of a large diameter Pulsating Heat Pipe (PHP) on board REXUS 22 sounding rocket. A PHP is a passive thermal control device where the heat is efficiently transported by means of the self-sustained oscillatory fluid motion driven by the phase change phenomena. Since, in milli-gravity conditions, buoyancy forces become less intense, the PHP diameter may be increased still maintaining the slug/plug typical flow pattern. Consequently, the PHP heat power capability may be increased too. U-PHOS aims at proving that a large diameter PHP effectively works in milli-g conditions by characterizing its thermal response during a sounding rocket flight. The actual PHP tube is made of aluminum (3 mm inner diameter, filled with FC-72), heated at the evaporator by a compact electrical resistance, cooled at the condenser by a Phase Change Material (PCM) embedded in a metallic foam. The tube wall temperatures are recorded by means of Fibre Bragg Grating (FBG) sensors; the local fluid pressure is acquired by means of a pressure transducer. The present work intends to report the actual status of the project, focusing in particular on the experiment improvements with respect to the previous campaign.

Keywords: Large Diameter Pulsating Heat Pipe, Sounding Rocket, Milli-gravity.

1 Introduction

Passive systems such as heat pipes are becoming the most popular choice for high heat power dissipation in electronics. The Pulsating Heat Pipe (PHP), patented by Akachi [1], [2], is a relatively new member of the Heat Pipe family and consists of a serpentine capillary tube, folded in many turns and closed end-to-end in a loop. The tube is first evacuated and then partially filled with a refrigerant fluid which naturally resides in the forms of liquid slugs and vapor plugs. The oscillatory motion of the slugs and plugs due to phase change phenomena allows transportation of heat. In order to obtain the desired liquid slugs/vapor plugs flow regime in ground conditions, the internal tube diameter must be smaller than a critical value (static Bond criterion).

$$d_{cr} = 2\sqrt{\frac{\sigma}{g(\rho_l - \rho_v)}}$$

The fluid motion is indeed activated when the system is heated up in the evaporator section: the vapor bubbles expands and push the liquid slugs towards the condenser zone, where the heat is released to the cold source and the condensation process occurs. From the theoretical point of view, just by considering the static Bond criterion, it seems that in reduced gravity conditions, the inner tube diameter may be increased. Actually, Gu et al [3] and Mameli et al [4] asserted that viscous and inertial effects always play a significant role in the definition of the flow pattern within the device that cannot be discounted. Not surprisingly, considering the new dynamic confinement criterion based on Weber and Garimella numbers, it seems that a two-phase loop may still works as a PHP in reduced gravity conditions with an inner tube diameter larger than the one evaluated on static ground conditions. In order to assess the validity of the previous assertion, it is necessary to conduct experiments in a reduced gravity environment, to decouple completely the inertial effects from buoyancy. Based on the above consideration, Mangini et al. [5] tested a PHP with an inner diameter of 3mm larger than the critical diameter calculated on static Earth gravity condition (1.7mm for FC-72 at 20°C) both on ground and on milli-gravity condition, experienced during ESA 61st Parabolic flight campaign. On ground, the device works as a closed loop two-phase thermosiphon when gravity assisted, and as a pure conductive medium when placed horizontally. When gravity level decreases, a sudden transition of the flow pattern from stratified to slug flow appears, and the device is able to work as a PHP.

The reduced gravity environment experienced during the parabolic flight is not long enough (20s) to reach an operational steady state. In order to obtain a longer period of milli-g, PHOS experiment tested a large diameter PHP on board REXUS 18 sounding rocket [6,7,8]. PHOS did not experience the reduced gravity environment due to a failure of the rocket de-spin system, thus, despite the test rig was perfectly operating, no significant results have been obtained for the characterization of a PHP, except on ground condition. U-PHOS is born from the necessity to perform a second test of the PHP in a long enough milli-g environment, because of the failure of the previous one. The aim of this work is to illustrate the objectives and the test-cell improvements with respect to the PHOS project and the expected results coming from the next REXUS 22 sounding rocket campaign. The experiment is going to be launched in March 2017, on board of REXUS 22 rocket, with an expected milli-gravity environment lasting almost 90s. The inner diameter is the same of PHOS experiment (3mm), with a new, more compact design very similar to the experiment that will be tested on the International Space Station. The secondary goal of the project is to use a different cooling system consisting of a phase change material (PCM), i.e. paraffin wax, embedded with a metal foam.

2 Experimental set-up

The test-cell of the experiment consists of a closed-loop PHP, a heat source which provides the thermal power to the evaporator section, and a composite metal foam/PCM heat sink.

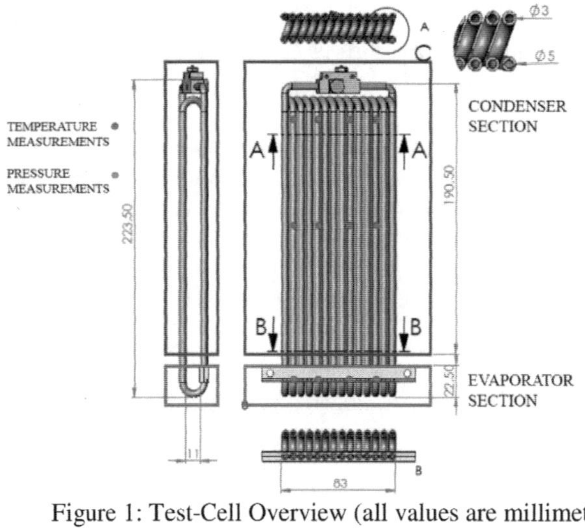

Figure 1: Test-Cell Overview (all values are millimetres).

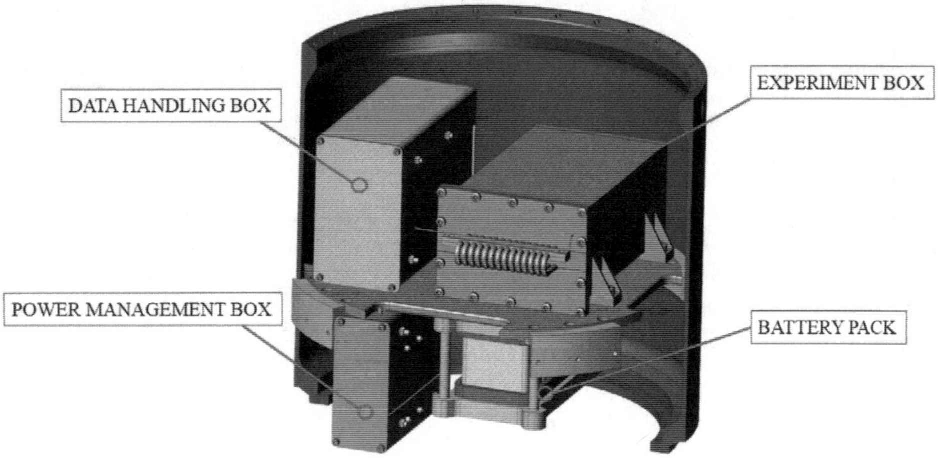

Figure 2 Experiment Assembly.

The condenser section of the PHP and the PCM are contained inside an airtight box (named experiment box), in order to avoid any hazardous leakage of PCM. The evaporator section of the PHP emerges from the experiment box and hosts the thermo-heater elements.

The test-cell and the peripheral facilities are mounted inside a cylindrical module, which constitutes the REXUS 22 rocket casing. The peripheral facilities include the battery pack, in order to provide power to the electrical heater and the experiment; the data handling unit, which contains the electronic boards for the management of the data on board; and the power management unit, including the electronics boards able to control the power to the experiment. The test-cell drawings and the architecture of the entire experiment are shown in figure 1 and figure 2.

2.1 Test-cell

The PHP is made of annealed aluminum (6063) tube with inner diameter of 3 mm and outer diameter of 5 mm. The tube is folded in a single loop with fourteen turns at the evaporator, arranged in a staggered configuration, as illustrated in Figure 3. The tube is closed in a loop by a T-junction, with two additional ports: one used for the emptying and filling procedures, the other one hosting a pressure transducer (Kulite®, XCQ-093, 1.7 bar A, max error: 0.05% FSO).

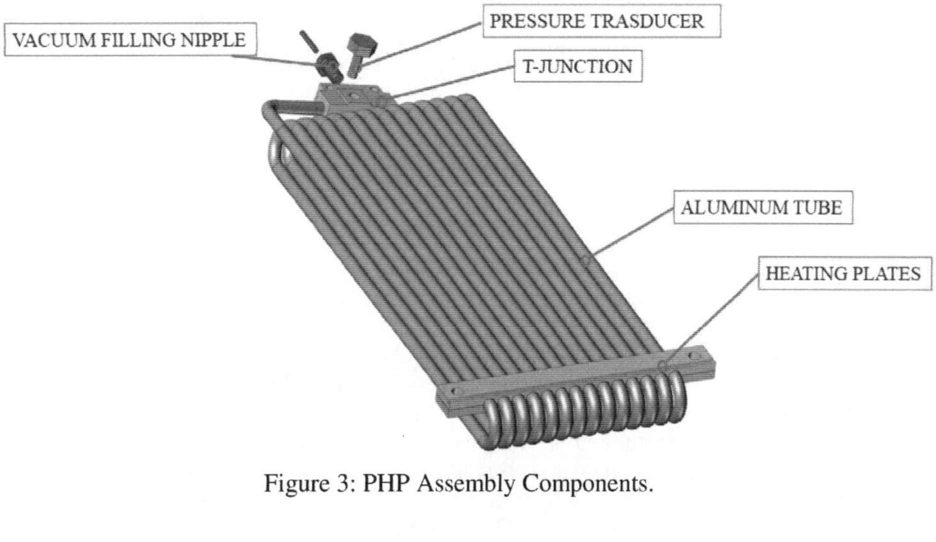

Figure 3: PHP Assembly Components.

The PHP is first evacuated by a two-stage turbo pump (Agilent® DS102, TwissTorr 84S), then is partially filled with a working fluid, FC-72, with a volumetric ratio of 0.5 ± 0.025 and sealed. The design of the test-cell follows the criteria already chosen for PHOS Project, nevertheless the configuration has been upgraded with respect to the previous PHOS Project [6,7,8] in order to fulfil the space constraints required by the International Space Station Thermal Platform and allow a better recirculation of the fluid.

2.1.1 *Evaporator Section*

The PHP evaporator section consists of a series of two ceramic heaters (plates) connected in parallel. In PHOS experiment the heating element was realized by means of a heating cable wrapped around the tube, but the thermal contact was not optimized. The new heaters, custom made by INNOVACERA® on our design, are placed on a flat aluminium surface brazed on the PHP. The nominal resistance is 18 Ω at 25°C. The heating elements are powered up to 200W (corresponding to a radial wall to fluid local heat flux of 12.6 W/cm^2) by 8 battery cells (SAFT® MP176065). The ceramic plates are in contact with the aluminium tube by means of an aluminium slab (heating plates in figure 3) glued on it. For safety reasons, the maximum temperature is fixed at 150°C, controlled by an electro-thermal switch.

2.1.2 *Condenser Section*

In order to be independent from the cooling section thermal inertia, a stable temperature should be maintained in the condenser section. Heat dissipation at constant temperature can be obtained using a phase change material in close contact with the condenser section. The material chosen is Octadecane paraffin wax. Its characteristics are shown in table 1. The paraffin wax should absorb heat via latent heat of fusion, and keep the condenser at constant temperature. However, as the thermal conductivity of the paraffin wax is small, after a certain amount of time, the melting front stops. Hence, the liquid paraffin surrounding the tubes is heated up (sensible heating), leading to an increase of the condenser temperature. This phenomenon is even more evident in milligravity condition due to the absence of convection. In order to avoid this effect, that was evident during the PHOS experiment, the paraffin thermal conductivity is enhanced by embedding it inside a metallic foam. Furthermore, since the metallic foam is highly porous, the overall mass of the PCM/ metallic foam assessment does not increase too much.

Table 1: Paraffin Wax Characteristics

Density	814 (solid), 774 (liquid) [*kg/m^3*]
Specific Heat	2160 [*J/kg.K*]
Melting Point	28.0 [*°C*]
Boiling Point	317.9 [*°C*]
Heat of Fusion	244 [*kJ/kg*]
Thermal Conductivity	0.15 [*W/mK*]

Several studies in literature calculate an overall thermal conductivity ten times higher than the paraffin [9, 10], by using copper foams. Metallic foams are characterized by two parameters: porosity (the ratio between the foam weight and the weight of the same solid volume composed of the same material) and pore density (the number of pores in a linear inch PPI). From literature [11] it seems that the foam pore density influences the convective heat transfer, so it should not be important in a milli gravity environment, while the porosity affects the melting process: higher values of the foam density mean higher thermal conductivity and, consequently, lower temperature variations at the condenser. For this reason, the best solution seems to be a relatively high density foam (88% porosity). Further tests are planned in the near future in order to correctly characterize the behavior of the embedded paraffin with the foam.

2.1.3 *Temperature Measurements by optic fibers*
The measurement is made by an SBI (Sensor Bragg Interrogator), consisting in an optical fiber which is doped in several locations obtaining small embedded sensors. A laser light is injected in the fiber and different wavelength spectra are reflected by the sensors depending on the sensor temperature. The SBI is indeed equipped with receptors able to measure the reflected wavelengths and thus to calculate the temperatures for all the sensor locations. The chosen configuration consists in 2 fibres each one with 12 sensors on them. Temperatures can be acquired with up to 1 KHz sampling frequency, with an accuracy of ±1 °C, in a range between 0 and +150°C. It is worth noting the great advantages resulting from this configuration in terms of weight, reduced complexity of cabling and measure accuracy. The position of the FBG sensors on the PHP is schematically shown in Figure 4. Sensors position is symmetrical on both the PHP sides.

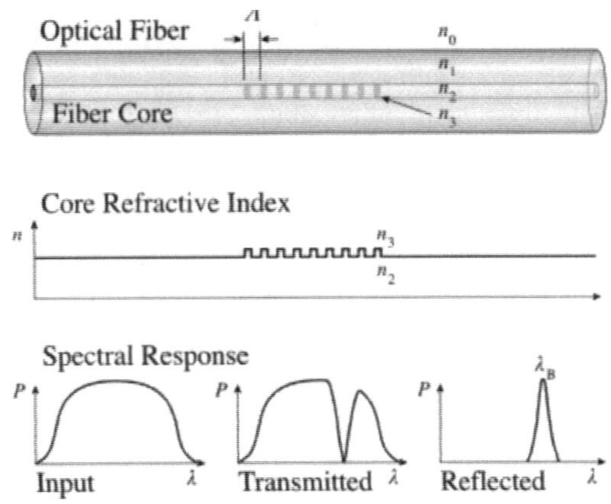

Figure 4: Optical fiber basic working principles.

As shown in figure 5, the experiment will be turned on fifteen minutes before launch. The main operation that will be conducted during the pre-launch phase is a preliminary heating of the PHP, done in order to keep the PCM (Phase Change Material) at a temperature close to its melting point. This operation is fundamental, since PHOS Project [6,7,8] highlighted that the paraffin temperature was too low at the beginning of the experiment to perform its function. Moreover, the ambient temperature at launch is strongly varying and frequently very low since the launch will take place at Kiruna, Sweden, on March, with typical environment temperatures varying from -30°C to 10°C. At lift-off the experiment supplies a different power from the one supplied in the pre-heating phase. It supplies 200W to the PHP, instead the pre-heating power is about 40W. The experiment will also start acquiring and saving (and sending to ground) temperature data, pressure and supplied power. Taken t = 0s as the lift-off time, after almost 65s a rocket despin (yo-yo) system will be activated, in order to obtain milli-g conditions. The apogee will be reached at almost t150s. At t+180s the experiment will be turned off and the on-board electronics will continue sending data on ground for redundancy.

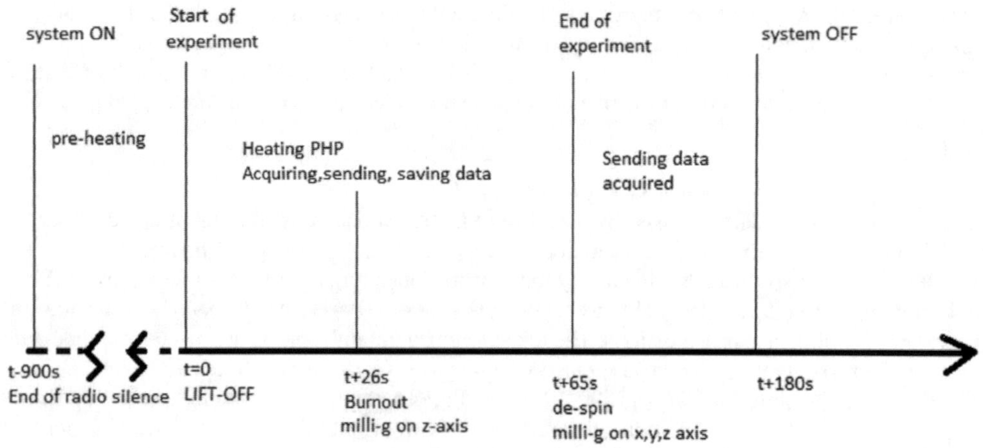

Figure 5: Experiment Timeline.

2.2 *Electronic system*

The objective of the electronic system is to provide the heating power necessary for the experiment, to acquire data and to communicate with the ground station. These functions are onto 4 independent sub-modules: an Actuation Module, an Interface Module, a Main Control Unit (MCU) and a Power Distribution Unit (PDU) shown in figure 6.

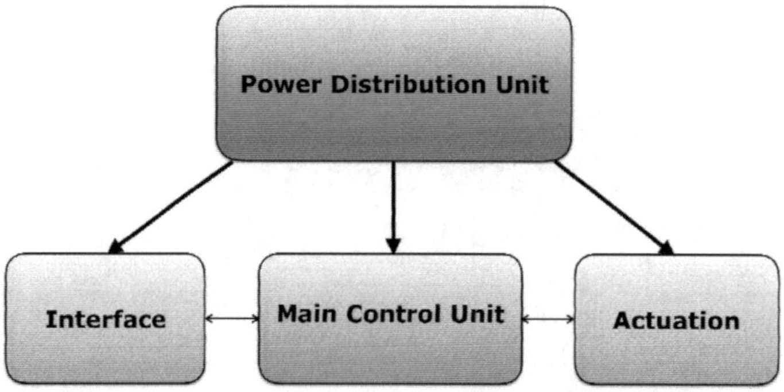

Figure 6: Electronic System Overview

The actuation module controls a heating system during the experiment, to provide the necessary heating power to it. It does it in two different phases: the pre-heating phase, where the system has to maintain the PCM temperature close to its melting point, and the heating phase, where a constant power shall be delivered to the PHP. This functionality has been implemented combining a Digital Controller and a PWM (Pulse Width Modulation) Driver, together with a Sensing Module to acquire the delivered power. The PWM signal passes through a gate driver and connects heaters to the power supply, controlling the power delivered with the duty cycle of the PWM Signal. Note that the delivered power is low pass filtered by the thermal pole of the heater, resulting in a mediated output thermal power. The interface module has the role to host a secondary sensor acquisition, which will not be used for the system control. The module will acquire internal pressure values of the PHP and three axis acceleration values. The communication will be established with the REXUS Service Module and, through it, with the ground station. The link is created through opto-isolators and the communication protocol is RS-422. The Main Control Unit (MCU) is a commercial board from Embedded ARM (TS7200®): it is a programmable board hosting a LINUX® distribution with on board memory and interface features such as ADC channels and RS-485 port. The board will be responsible for the timeline and the control of the experiment. Moreover, the board interacts with the data acquisition board responsible for temperatures acquisition (SBI). The Power distribution module hosts several DC-DC converters in order to guarantee the appropriate voltage to all the electronic components. One of the DC-DC converters (DCM24AP480T320A50 VICor®) provides the heating power to the heating module.

3 Expected results

With the results coming from the REXUS campaign it will be possible to analyse and to understand the thermal-fluid dynamics of the device response at varying acceleration, in time. Temperature and pressure inside the PHP are expected to show a different oscillating behaviour after the occurrence of milli-gravity. From a scientific point of view this experiment could provide fundamental data in order to improve the prediction capability of numerical models in milli-g environment and help to understand the physical phenomena that characterize the start-up at low-g conditions. Moreover, it could be possible to analyse the performance of PCM with a metallic foam and improve the numerical modelling of this complex phenomenon.

3.1 Ground tests

In order to compare U-PHOS results with the PHOS results on ground, the experiment is going to be tested both in vertical and horizontal position. In vertical position, the device operates as a two-phase thermosyphon. The liquid phase accumulates close to the evaporator section and boils continuously. The generated vapor rises towards the condenser section and, sometimes, drags up fluid batches (liquid and vapor) in the so-called "bubble lift" mode.

In horizontal position, due to the fluid stratification: the liquid phase resides in the lower tube region and the vapour phase on the higher one. Still, in the new configuration, a small gravity head will be present, and the device will work once again as a loop thermosiphon. Since the gravity head is smaller than the vertical case, the performance is expected to be lower too.

3.2 Flight test

The flight test that will be performed during REXUS 22 launch aims to confirm the predicted behaviour of the PHP obtained during the parabolic flights. As previously asserted, a PHP has been tested in milli-g environment during the 61st ESA Parabolic Flight Campaign [5]. Each parabolic manoeuvre is divided in 3 parts: 20 s at 1.8g (hyper-gravity) followed by 22 s at 0.01g (milli-gravity) and 20-25 s at 1.8g (hyper-gravity). Figure 7 shows the temporal evolution of the pressure inside the PHP during one parabolic sequence.

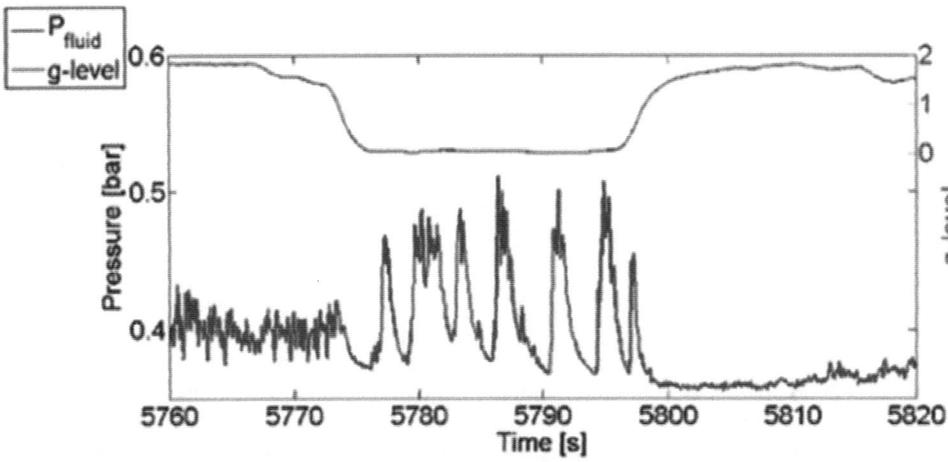

Figure 7: Fluid Pressure inside PHP and Gravity Field [5].

As illustrated in figure 7, during milli-gravity condition pressure fluctuation inside the PHP in recorded. This fluctuation is due to fluid motion inside the tube, mainly due to the slug and plug flow pattern, as the gravity vector components are almost zero during the milli-g condition.

No significant data was obtained from the PHOS experiment because of a malfunction of the de-spin system. We expect that the U-PHOS zero gravity results will be similar to the one in figure 7: hopefully the pressure fluctuations will be present during the whole micro-gravity period (90s) which is longer than the one achievable in parabolic flights.

3.3 Data analysis plan

Data will be post-processed and analyzed in order to detect evaporator and condenser external wall temperature evolutions, fluid pressure variation frequency and amplitude, fluid temperature variation frequency and amplitude. All data recorded will be plotted against time and in particular:

- Tube wall temperature in the evaporator and condenser zone will be plotted together with the acceleration and heat power input.
- Local fluid pressure in the condenser section will be plotted together with the acceleration and power heat input, in order to estimate the time response to low-g environment of the system.
- FFT analysis on the pressure signal will be carried out in order to verify if different thermal/hydraulic behaviours are characterised by a dominant frequency.
- Melting front of the paraffin will be detected over time acquiring temperatures by means of the FBG sensors. Experimental data will be compared with the ones previously obtained on ground and the ones coming from the numerical model of the PCM melting phenomenon.

From the acquired data it will be possible to evaluate the equivalent thermal resistance and the thermal performances of the device in milli-g environment.

4 Conclusions

The objectives and test cell of U-PHOS Project have been explained in details, with reference to its thermal, mechanical and electronic design. U-PHOS Project aims to characterize the behaviour of a large diameter Pulsating Heat Pipe (PHP), a promising device for thermal management in space application.

The experiment is going to be tested in milli-gravity condition, on board REXUS 22 sounding rocket. The test-cell of the experiment consists of an annealed aluminium tube, of internal diameter of 3 mm, bended in 14 turns and closed end-to-end in a loop. The test-cell is equipped with ceramic heaters as heating source, and paraffin wax embedded with metal foam as cooling section. Temperature and pressure data are collected during the whole experiment both on ground and flight condition. In particular temperature data are detected using fibre brag grating sensors inserted inside an optic fibre and in contact with the external wall of the PHP.

The experiment is an upgrade of a previous experiment (PHOS) that was launched on board REXUS 18 sounding rocket. Enhancement of the heat sink thermal conductivity and a better accuracy, reliability and integrability of the temperature sensors, are the main innovations with respect to PHOS. Important results will be obtained for the characterization of the device in low-gravity environment. At the present time the experiment is under critical design review. The launch is scheduled for March 2017.

Acknowledgements

The present work has been carried out as part of the REXUS BEXUS program, supported by European space agency (ESA), German Aerospace Research Establishment (DLR) and Swedish National Space Board (SNSB). The ESA MAP INWIP project for funding the project, the International Scientific Team on Pulsating Heat Pipes, the school of engineering and AAVID Thermalloy, especially Marco La Foresta and Lorenzo Caporale, for the crucial help in the test cell manufacturing. We also would like to thank Alessandro Signorini and INFIBRA Technology for the great help. The authors would like to thank the PHOS team for the support and encouragement to continue their work.

References

[1] Akachi H 1990 *US Pantent* 4.921.041 Structure of a heat pipe.

[2] Akachi H 1993 *US Patent* 5.219.020 Structure of a micro heat pipe

[3] Gu J, Kawaji M and Futamata R 2004 Effects of gravity on the performance of pulsating heat pipes *J. Thermophys. Heat Transfer* **18** 371-378.

[4] Mameli M, Araneo L, Filippeschi S, Marelli L, Testa R and Marengo M 2014 Thermal response of a closed loop pulsating heat pipe under a varying gravity force *Int. J. Therm. Sci.* **80** 11-22.

[5] Mangini D, Mameli M, Georgoulas A, Araneo L, Filippeschi S and Marengo M 2015 A pulsating heat pipe for space applications: ground and microgravity experiments *Int. J. Therm. Sci.* **95** 53-63.

[6] Creatini F, Guidi GM, Belfi F, Cicero G, Fioriti D, Di Prizio D, Piacquadio S, Becatti G, Orlandini G, Frigerio A, Fontanesi S, Nannipieri P, Rognini M, Morganti N, Filippeschi S, Di Marco P, Fanucci L, Baronti F, Manzoni M, Mameli M, Marengo M, 2015 Pulsating Heat Pipe Only for Space: Results of the REXUS 18 Sounding Rocket Campaign *XXXIII UIT Congress* L'Aquila, Italy, Journal of Physics: Conference Series 655 (2015) 012042 doi:10.1088/1742-6596/655/1/012042.

[7] Creatini F, Guidi GM, Belfi F, Cicero G, Fioriti D, Di Prizio D, Piacquadio S, Becatti G, Orlandini G, Frigerio A, Fontanesi S, Nannipieri P, Rognini M, Morganti N, Filippeschi S, Di Marco P, Fanucci L, Baronti F, Manzoni M, Mameli M, Marengo M, 2015 Thermal Response of a Pulsating Heat Pipe on Board the Rexus 18 Sounding Rocket: PHOS Experiment Chronicles *UK Heat Transfer Conference*, Edinburgh UK.

[8] Creatini F, Guidi GM, Belfi F, Cicero G, Fioriti D, Di Prizio D, Piacquadio S, Becatti G, Orlandini G, Frigerio A, Fontanesi S, Nannipieri P, Rognini M, Morganti N, Filippeschi S, Di Marco P, Fanucci L, Baronti F, Manzoni M, Mameli M, Marengo M, 2015 Phos Experiment: Thermal Response of A Large Diameter Pulsating Heat Pipe On Board Rexus 18 Rocket 2015 *22nd ESA PAC Symposium* Tromso, Norway.

[9] Xiao X, Zhang P, Li M 2013 Preparation and thermal characterization of paraffin/metal foam composite phase change material *Applied Energy* **112** 1357-1366.

[10] Xiao X, Zhang P, Li M 2014 Effective thermal conductivity of open-cell metal foams impregnated with pure paraffin for latent heat storage *Int. J. of Thermal Sciences* **81** 94-105.

[11] Lafdi K, Mesalhy O, Shaikh S 2007 Experimental study on the influence of foam porosity and pore size on the melting of phase change materials *J. of Applied Physics* **102** 083549.

IOP Conf. Series: Journal of Physics: Conf. Series **796** (2017) 012045 doi:10.1088/1742-6596/796/1/012045

Asymmetry during a horizontal annular flow in a micro-channel: optical measurements and effect of dimensionless numbers

C CAPO [1], T LAYSSAC [2], S LIPS [2], A W MAURO [1,3] and R REVELLIN [2]

[1] Federico II University of Naples, Department of Industrial Engineering, P. le Tecchio, 80, 80125, Naples (Italy)

[2]CETHIL UMR5008, Université de Lyon, CNRS, INSA-Lyon, F-69621, Villeurbanne (France)

[3] e-mail: alfonsowilliam.mauro@unina.it

Abstract. New applications of HFC refrigerants in organic Rankine cycles at high saturation temperatures and the wider use of CO_2 for air-conditioning have pushed research to the characterization of two-phase heat transfer at medium/high reduced pressures and have pointed out the effect of these operating conditions on asymmetric distribution of refrigerant around tube perimeter and its indirect effect on heat transfer. Currently there is a lack of data about asymmetric distribution of liquid film at the wall, especially for refrigerants and micro-channels. In order to have a physical evidence of this asymmetry also for micro-channels and approach to a relationship between this phenomenon and dimensionless parameters, new data are here presented. The asymmetric annular flow of the refrigerant R245fa inside a horizontal, round 2.95 mm inner diameter channel is studied with pictures captured by a high speed video camera. The experimental results here presented were obtained at saturation temperatures equal to 20 °C and 40 °C at low mass velocities (50, 100 and 200 kg $m^{-2}s^{-1}$) to asymmetric distribution, enriching the database presented in previous studies. The new dimensionless parameter, eccentricity, has been related to the dimensionless groups: Froude and Bond numbers, and Martinelli parameter, showing the mutual correlation among them.

1. Introduction

Two-phase flows are encountered in many industrial applications related to heat transfer process or to fluid transport such as evaporators and condensers for the refrigeration and air-conditioning sectors, steam generators of nuclear plants or ORC cycles, and pipelines for petrochemical transportation.

Among two-phase flows, the annular two-phase flow is the most desired due to the intense heat transfer induced by the vapor core draining the thin liquid film at the wall. The knowledge of the relation between heat transfer and operating parameters is of primary importance for an accurate design in order to get the desired value of the heat transfer coefficient to balance the heat transfer resistances at each fluid side, being important to not exceed with the heat transfer intensity only at one side, since this would not be useful for the global thermal resistance, while penalizing considerably the pressure drops and the energy consumptions.

Several models have been developed to describe the behavior of an annular flow in literature, and it is well established that several concurrent factors affect the heat transfer and pressure gradients [6, 7]; they are related to the geometry (horizontal or vertical configuration, diameter and surface aspect), thermophysical properties, liquid film thickness, liquid and vapor velocities, and liquid droplet entrainment.

The knowledge of the liquid film thickness at the wall is of primary importance to characterize an annular two-phase flow. In facts, simple models proposed in literature have successfully correlated the Nusselt number and the shear stress at the wall to the liquid film thickness [for example: 8, 9], which is directly related to the void fraction and the fraction of liquid entrained in the vapor core. Most of the models available in literature are based on the hypothesis of the symmetry (uniformity) of the liquid film distribution around the perimeter.

Applications of refrigerants to ORC cycles recovering waste heat and also a lot of plants working with fluids evaporating at a reduced saturation pressure close to 0.5 or higher are appearing. The operating conditions of these applications are outside the range of applicability of methods currently available. In facts, recent experimental studies have investigated this new range of operating conditions and have pointed-out the effect of gravity on heat transfer coefficients variation around the perimeter related to asymmetric liquid distribution at the wall [1, 2, 3] for horizontal flows.

Several works in literature deal with the determination of the liquid film thickness during annular flows.

Luninski *et al.* [10] measured the film thickness of air-water mixtures at different positions around the tube periphery, in channels with diameters ranging from 8.15 to 12.5 mm.

Ong and Thome [11] proposed a top/bottom liquid film thickness comparison for refrigerants R134a, R236fa and R245fa during flow boiling in small channels of 1.03, 2.20 and 3.04 mm diameter.

Other experiments were performed by Farias *et al.* [12], Fukano and Osaka [13], Laurinat and Hanratty [14], Masala *et al.*[15].

Cioncolini and Thome [16], Hurlburt and Newell [17] and Schubring and Shedd [18] proposed to predict quantitatively the evolution of eccentricity mainly for slightly stratified air-water flows.

In the studies of Cioncolini and Thome [16] and Schubring and Shedd [18], the eccentricity is defined as the ratio between top film thickness and bottom film thickness as shown by Eq. (1):

$$ecc_{ratio} = \frac{t_{top}}{t_{bottom}} \tag{1}$$

where t_{top} and t_{bottom} are respectively top and bottom film thicknesses.

Cioncolini and Thome [16] proposed a correlation for predicting the eccentricity ecc_{ratio}:

$$ecc_{ratio} = \frac{0.0789 Fr_v^{1.90}}{1+0.0789 Fr_v^{1.90}};\ Fr_v > 1 \tag{2}$$

with

$$Fr_v = xFr_{vo};\ Fr_{vo} = \frac{\frac{G}{\sqrt{\rho_{vap}}}}{\sqrt{g(\rho_{liq} - \rho_{vap})D}} \tag{3}$$

The correlation proposed by Schubring and Shedd [18] is defined as:

$$ecc_{ratio} = 1 - \exp(-0.63 Fr_{ss}) \tag{4}$$

$$Fr_{ss} = \frac{xG}{\rho_{liq}(g t_{mean})^{0.5}} \tag{5}$$

with t_{mean} the mean film thickness calculated with the following empirical correlation:

$$\frac{t_{mean}}{D} = \frac{4.7}{x}\left(\frac{\rho_{vap}}{\rho_{liq}}\right)^{\frac{1}{3}}\left(\frac{GD}{\mu_{liq}}\right)^{-\frac{2}{3}} \tag{6}$$

Hurlburt and Newell [17] proposed a correlation to evaluate the grade of asymmetry at the bottom of the tube with respect to the mean film thickness on the circumference of the tube as follows:

$$ecc_{mean} = \frac{t_{mean}}{t_{bottom}} \tag{7}$$

The correlation proposed by Hurlburt and Newell [17] is:

$$ecc_{mean} = 0.2 + 0.7\left(1 - e^{-\frac{1}{75}\left(\left(\frac{x}{1-x}\right)^{0.5} Fr_{hn} - 20\right)}\right) \tag{8}$$

valid for $\left(\frac{x}{1-x}\right)^{0.5} Fr_{hn} > 20$ and with $Fr_{hn} = \frac{xG}{\rho_{vap}(gD)^{0.5}}$

These predictive methods are based on data mostly related to air-to-water flows at ambient temperature, with large diameters (they range between 8.8 and 98 mm) and large excursions of the mass velocity; hence, they are calibrated in a large range of Froude numbers; on the other side these methods have been calibrated on a limited range of reduced pressures (mainly low reduced pressures) and they have not a good agreement with experiments at low mass fluxes or medium reduced pressures, as demonstrated in the assessment by Layssac *et al.* [5].

The work by Donniacuo *et al.* [3] showed experimental results during the evaporation of R245fa in a 2.95 mm horizontal tube. The operating conditions in which the experiments were performed are: mass velocities equal to 300 and 400 kg $m^{-2}s^{-1}$ and saturation temperatures of 40, 60, 80 and 100°C. They experimentally showed that there is a combined effect of the Froude and Bond numbers on the stratification; the interrelationship being much evident at high reduced pressure or low mass fluxes. Recently, Layssac et al [3] have proposed a correlation to account for the effects on the asymmetry both of the mass velocity, the diameter and of the thermodynamic properties.

The correlation proposed has the following form:

$$ecc_{diff} = \frac{2.7Fr_{vo}^{-1.2}\chi}{1+2.7Fr_{vo}^{-1.2}\chi} \; for Fr_{vo}^{-1.2}\chi > 0.01 \tag{9}$$

$$ecc_{diff} \approx 0 \; for Fr_{vo}^{-1.2}\chi \leq 0.01$$

where ecc_{diff} is defined as in the work by Donniacuo *et al.* [3] according to which the flow stratification is evaluated by considering the relative position of vapour core center compared to the flow center divided by internal radius as asymmetry parameter, whose strictly equivalent formula is:

$$ecc_{diff} = \frac{t_{bottom} - t_{top}}{D} \tag{10}$$

where D is the internal diameter, t_{bottom} and t_{top} respectively the bottom and the top film thicknesses. The limit cases of this definition correspond to a centered flow for an eccentricity of 0 and a full stratified flow with a bottom film thickness tending to the internal diameter when eccentricity is 1.

Equation (9) included the Martinelli parameter, related to the case of a turbulent-turbulent two-phase flow, defined as:

$$\chi = \left(\frac{1-x}{x}\right)^{0.9}\left(\frac{\rho_{vap}}{\rho_{liq}}\right)^{0.5}\left(\frac{\mu_{liq}}{\mu_{vap}}\right)^{0.1} \tag{11}$$

The Martinelli parameter is able to include the effect of the thermodynamic properties, which is important to have a predictive model able to capture predictions when there is a large variation of the reduced pressure. In facts, the paper [3] showed that this method is much more improved also when compared to a database of experiments including a large variation of the Froude number and of the thermodynamic properties (by means of the Martinelli parameter).

Layssac *et al.* [3] proposed a correlation, with no Bond number, Eq. (9), due to its weak influence in the statistics, since it ranged only from 11 to 18.

The aim of this work is to expand the range of operating conditions investigated previously to have a larger database, especially in those operating ranges that enlarge the effects of asymmetry and of the Bond number. In particular, the experiments were performed at mass flux velocities (G) from 50 to 300 kg $m^{-2}s^{-1}$ and saturation temperatures (T_{sat}) of 20, 30, 40, 50° C, with Bond numbers from 8 to 10.

2. Experimental test facility, optical measurement technique and data reduction

The experimental test facility is well described in the work by Charnay *et al.* [1]. The same test bench was used to adjust the experimental conditions in terms of mass flow rate, vapor quality and saturation pressure at the inlet of the glass tube, placed at the end of the test section presented in Fig. 1(a).

The optical method used to measure the liquid film thickness is the same as that described in the paper presented by Donniacuo *et al.* [3]. A high-speed video camera system (Photron Fastcam SA3120KM2) enables the flow visualization by recording image sequences. The light system used for the visualization is provided through an adjustable light behind the visualization glass tube. The optical arrangement is a backlit imaging arrangement, resulting in a light gradient at the liquid–vapor interface due to shadowgraph effect. A high-speed camera system and a light behind the tube enable the visualization of the flow by recording frame sequences. The camera enables to get the local liquid film thickness with a mean resolution of 200 pixels.mm^{-1}.

In each condition of temperature and mass velocity, 4 series of 1363 frames are taken, corresponding to a total time of 2.7 s. The frame size is of 1024*1024 pixels and each pixel has a grayscale value ranging from 0 to 255. A MATLAB program has been conceived by Donniacuo *et al.* [3] which is able to get local film thickness for annular flows. An example of an image and the processing of the top and bottom thicknesses versus time is reported in Figure 1(b).

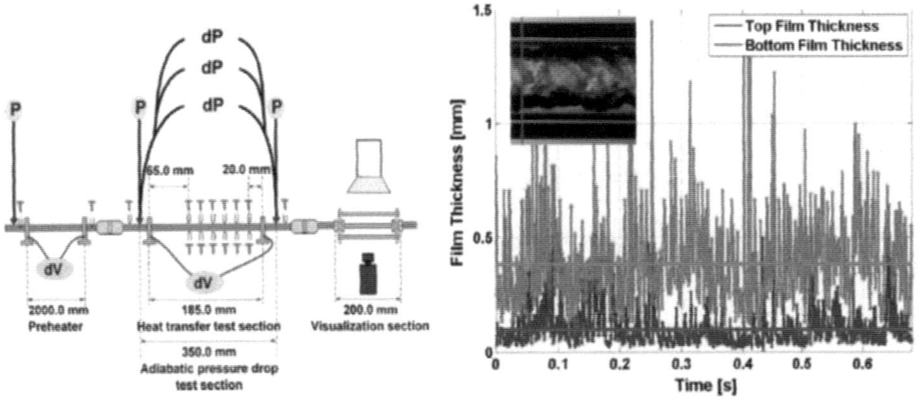

Figure 1. a) Schematic of the test facility. b) Local and average film thicknesses at saturation temperature of 80 °C, mass velocity of 300 kg m^{-2} s^{-1} and vapor quality of 0.35.

This method, based on grayscale analysis, has been improved to be applied to intermittent flows where single-phase profiles have to be detected by Layssac *et al.* [5], also permitting to sensibly reduce the uncertainty on interface detection for annular parts. Since the direct measurement suffers of the refraction effects through the glass of the tube, a correction factor was introduced. In case of intermittent flow the average value of the thickness is calculated only while there is a two-phase flow.

The uncertainty on the apparent film thickness is ± 2 pixels. At each operating condition, the top and bottom liquid film thicknesses were calculated as the average value from 5452 recorded frames. The uncertainty of each measurement is calculated, taking into account several sources of uncertainty: the inner and outer diameter dimensions, the variation of liquid R245fa refractive index with the temperature and the limitations of image resolution due to finite pixel dimensions. The combined uncertainty for film thickness measurement was evaluated as following:

$$\delta t_{real} = 2\sqrt{\left(\frac{\delta t_{app}}{EF.Sc}\right)^2 + \left(\frac{t_{app}\delta_{EF}}{-Sc.EF^2}\right)^2 + \left(\frac{t_{app}\delta_{Sc}}{-EF.Sc^2}\right)^2} \tag{12}$$

with t_{app} the apparent film thickness, t_{real} the real film thickness EF the enlargement factor due to optical deformation, and Sc the scaling factor. In the present study, δt_{app}= 2 pixels, δ_{EF} = 0.047, δ_{Sc} = 0.84 mm/pixel for an outer diameter of 909 pixels. As a conclusion, δt_{real} for the top film thickness ranges between 0.02 mm and 0.03 mm depending on the experimental conditions.

3. Experimental results

This paper presents experimental results during an adiabatic two-phase flow of the fluid R245fa in a smooth tube with an internal diameter of 2.95 mm, varying the mass flux, G, in the range 50-300 $kgm^{-2}s^{-1}$ and the saturation temperature from 20 and 50 °C, in the whole range of vapour qualities.

Figure 2 (a)-(b) reports an example of the measured dimensionless top and bottom thicknesses as a function of the vapour quality at a constant mass flux, varying the Bond number from 8 to 10, where the Bond number is defined as:

$$Bd = \frac{g\,(\rho_{liq}-\rho_{vap})\,D^2}{\sigma} \tag{13}$$

including the effect of the variation of the saturation pressure, through the variation of the liquid and vapour densities with respect to the surface tension.

Consistently to the results of Donniacuo *et al.* [3] it is evident how the saturation pressure affects the dimensionless thickness, especially at the bottom of the tube and for low vapour qualities. In particular, increasing the saturation pressures the thickness at the bottom increasing for the same mass flux and vapour quality. This effect is present also at the top of the tube where the variation is less relevant.

In this work to evaluate the stratification a new parameter, S, the symmetry, is defined. S is the complement to 1 of the eccentricity, already defined by Eq. (10), as presented in the next equation

$$S = 1 - \frac{t_{bottom}-t_{top}}{D} \tag{14}$$

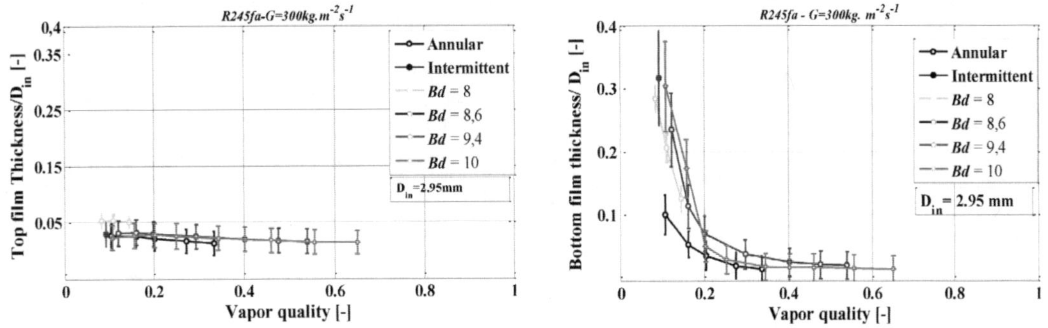

Figure 2. Dimensionless liquid film thickness at the wall for the fluid R245fa, an internal diameter of 2.95mm, a mass flux of 300 kg m^{-2}s^{-1}, varying the Bond number from 8.0 to 10.0 (i.e. the saturation temperature from 20°C to 50°C). a) Measurements at the top of the tube. b) Measurements at the bottom of the tube.

When symmetry is equal to 1 there is no stratification, since the eccentricity is zero. On the contrary, zero is the expected value of the symmetry in case of higher stratification. This new parameter enables to organize the experimental results as shown in Fig. 3(a)-(b) which displays that symmetry tends to 1 when the vapour quality increases. The limit case of zero, geometrically corresponding to a bottom film thickness tending to the top side of the tube is not reached in the present case. First bubbles appearing with nucleation, the symmetry parameter was evaluated to 0.35. The symmetry parameter is thus more pertinent than the eccentricity as it enables to highlight the behaviour of centred and almost centred flows, while keeping the same mathematical limits.

During the experiments two different kinds of flow patterns were observed: annular flow and intermittent flow where the intermittent is considered as the combination of bubbly flow, bubbly-slug flow and slug flow.

At a fixed value of G, equal 200 kg m^{-2}s^{-1}, by fixing also the value of the vapour quality, Fig. 3(a) shows the decline of symmetry when the Bond number is increasing from 8 to 10 and the vapour only Froude number is decreasing from 12 to 7.5. The definition of the vapour only Froude number used in this work is the one presented in Eq.(3)

The obtained trend of the symmetry, when these two dimensionless numbers are varying, is justified by the growth of $T_{sat.}$

Taking into account that the Bond number is the ratio between gravity and surface tension forces meanwhile the Fr_{vo} number is the ratio between inertia and gravity forces it is possible to resume that higher saturation temperature provides a higher Bond number and a lower Fr_{vo} number. This means higher stratification due to fact that the importance of the gravity forces is greater if compared to the surface tension and inertial forces.

Fig. 3(b) deals with the evolution of the symmetry at a Bond number equal to 8.6.with Fr_{vo} number varying from 2.6 to 15. A fixed value of Bond number implies that the saturation temperature is constant: the thermodynamic properties of the fluid do not change and consequently the buoyancy and surface tension forces are the same. Fr_{vo} number increases because of the growth of G and the only force rising is the inertial force. It is possible to see that when the Fr_{vo} number increases, symmetry increases that means lower stratification.

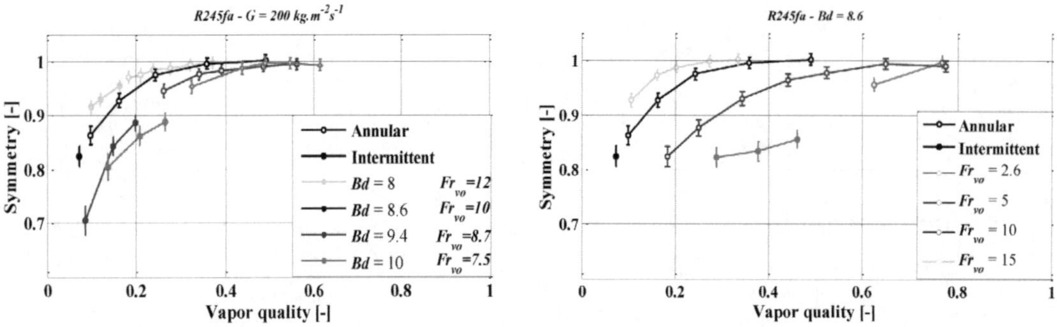

Figure 3. Plot of the experimental values of the symmetry parameter as defined in Eq. (14) as a function of the vapor quality, for the fluid R245fa. a) at G=200 kg m^{-2}s^{-1}, varying the Bond number from 8.0 to 10; b)at Bond number equal to 8.6 and varying Fr_{vo} from 2.6 to 15.

The complement to 1 of the correlation proposed by Layssac *et al.* [5], introduced before in Eq.(9), provides a method for the prediction of the symmetry, as presented in the next equation. This method has an additional benefit to keep all the data points and to avoid an arbitrary limit for the value of $Fr_{vo}^{-1.2}\chi$ of 0.01 as presented in Layssac *et al.* [5]. The new correlation is:

$$S_{predicted} = 1 - ecc_{diff} = 1 - \frac{2.7 Fr_{vo}^{-1.2}\chi}{1+2.7 Fr_{vo}^{-1.2}\chi} \tag{15}$$

For a statistical analysis, the mean percentage error, MPE, and the mean average percentage error, MAPE, are considered. They are defined as follows:

$$MPE = \frac{1}{n}\sum_{i=1}^{n} \frac{S_{meas,i} - S_{pred,i}}{S_{meas,i}} \tag{16}$$

$$MAPE = \frac{1}{n}\sum_{i=1}^{n} \left| \frac{S_{meas,i} - S_{pred,i}}{S_{meas,i}} \right| \tag{17}$$

where n is the sample size, S_{meas} the experimental value of the symmetry and S_{pred} the predicted symmetry. MPE and MAPE obtained for the present data are respectively of -0.1% and 4.0%.

The comparisons of the measured values of the symmetry with the predictions of the new model, Eq. (14) are presented in Fig.4a) where it is possible to see that 94% of the available data points fall within ±10% error band.

Globally, it appears the present data points dispersion is slightly higher than in the case of Layssac *et al.* [5] experimental data while the mean predictions in both cases do not overestimate or underestimate the experimental data. The larger dispersion in present case can be related to the experimental conditions. Actually, the range of temperatures enables to reach a Bond number of 8, compared with the Layssac *et al.* [5] data which ranges from 11 to 18. This increases relative importance of surface tension effects on the two-phase flow, compared with buoyancy and inertia. Furthermore, the surface tension tends to replace inertia in mechanisms to centre the flow due to low mass velocity (50 kg m^{-2}s^{-1}). Consequently, the new experimental data would not be well predicted for low mass velocity and low temperature since the geometry of the flow would be more conducted by surface tension compared with Layssac *et al.* [5] database.

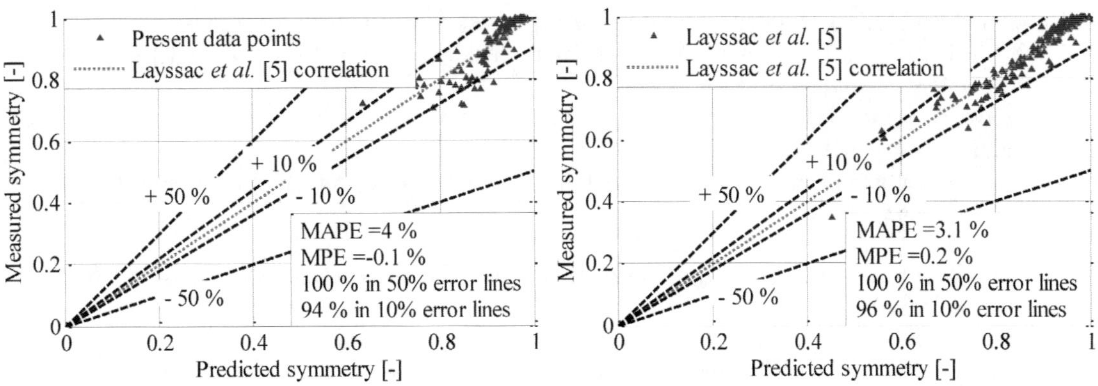

Figure 4. Plot of the statistical comparison between a) the new experimental results and the predictive method from equation (15); b) Layssac *et al.* [5] experimental results and the predictive method from equation (15).

4. Conclusions

New data points have been obtained on Charnay et *al.* [1] test facility. These points correspond to lower temperatures and lower mass velocities compared to the work led by Layssac *et al.* [5].

The general trend of two-phase flow symmetry was still well predicted by the analysis of dimensionless numbers such as vapour only Froude number and Bond number. Introduction of a new symmetry parameter enabled to easily analyze all types of flows, especially centred flows and then to avoid arbitrary limit for the field of application of the correlation of Layssac *et al.* [5].

The comparison between present experimental data and correlations has led to a larger dispersion with respect to the previous work by Layssac *et al.* [5], especially for not centred flows. This may be due to the higher relative importance of surface tension in the present conditions. Thus, it is planned to take into account these points, as well as bibliography data, in the optimization algorithm developed and described by Layssac *et al.* [5], in order to calibrate the correlation in a wider range of operating conditions. This would enable to highlight the contribution of surface tension in the correlation. Furthermore, the surface tension effect could be further analysed by reducing the internal diameter of the test section and by measuring the side film thicknesses.

References

[1] Charnay R, Revellin R and Bonjour J 2015 Flow boiling heat transfer in minichannels at high saturation temperatures: Part I – Experimental investigation and analysis of the heat transfer mechanisms *International Journal of Heat and Mass Transfer* **87** 636-652

[2] Mastrullo R, Mauro A W, Thome J R, Toto D and Vanoli G P 2012 Flow pattern maps for convective boiling of CO2 and R410A in a horizontal smooth tube: Experiments and new correlations analyzing the effect of the reduced pressure *International Journal of Heat and Mass Transfer* **55** 1519-1528

[3] Donniacuo A, Charnay R, Mastrullo R, Mauro A W and Revellin R 2015 Film thickness measurements for annular flow in minichannels: Description of the optical technique and experimental results *Experimental Thermal and Fluid Science* **69** 73-85

[4] Mauro A W, Cioncolini A, Thome J R and Mastrullo R 2014 Asymmetric annular flow in horizontal circular macro-channels: Basic modeling of liquid film distribution and heat transfer around the tube perimeter in convective boiling *International Journal of Heat and Mass Transfer* **77** 897-905.

[5] Layssac T Lips S and Revellin R 2016 Prediction of stratification during intermittent and annular flows of R245fa in an horizontal mini-channel *ICMF-2016 – 9th International Conference on Multiphase Flow* May 22nd –27th 2016 Firenze Italy.J.C.

[6] Wallis G B 1969 *One-dimensional two-phase flow* McGraw-Hill New York.

[7] Hewitt G F and Hall-Taylor N S 1970 *Annular two-phase flow* Pergamon Press, Oxford.
[8] Asali J C, Hanratty T J and Andreussi P 1985Interfacial drag and film height for vertical annular flow *AIChE Journal***31** (6) 895-902.
[9] Henstock W H and Hanratty T J 1976 The interfacial drag and the height of the wall layer in annular flows *AIChE Journal***22** 990-1000.
[10] Luninski Y, Barnea D and Taitel Y 1983 Film thickness in horizontal annular flow *Can. J. Chem. Eng.***61** 621–626.
[11] Ong C L and Thome J R 2011 Macro-to-microchannel transition in two-phase flow: Part 1 – Two-phase flow patterns and film thickness measurements *Experimental Thermal and Fluid Science* **35** 37–47.
[12] Farias P S C , Martins F J W A , Sampaio L E B , Serfaty R and Azevedo L F A 2010 Liquid film characterization of horizontal, annular, two-phase, gas-liquid flow using time-resolved LASER-induced fluorescence *15th Int. Symp. on Applications of Laser Techniques to Fluid Mechanics* 5th-08th July Lisbon Portugal.
[13] Fukano T and Ousaka A1989 Prediction of the circumferential distribution of film thickness in horizontal and near-horizontal gas–liquid annular flows *Int. J. Multiphase Flow* **15** 403–419.
[14] Laurinat J E, Hanratty T J and Jepson W P 1985Film thickness distribution for gas-liquid annular flow in a horizontal pipe *PhysicoChem. Hydrodynam* **6** 179-195.
[15] Masala T Harvel G and Chan J S Separated two-phase flow regime parameter measurement by a high speed ultrasonic pulse-echo system *Review of Scientific Instruments* **78**.
[16] Cioncolini A and Thome J R 2013 Liquid film circumferential asymmetry prediction in horizontal annular two-phase flow *Int. J. Multiphase Flow* **51** 44-54.
[17] Hurlburt E T and Newell T A 1997 Prediction of the circumferential film thickness distribution in horizontal annular gas-liquid flow *Air conditioning and refrigeration center Report ACRC TR-111* University of Illinois Urbana USA.
[18] Schubring D and Shedd T A 2009 Critical friction factor modelling of horizontal annular base film thickness *Int. J. Multiphase Flow* **35** 389-397.

Nomenclature

Bd	Bond number	(-)
D	diameter	(m)
ecc	eccentricity	(-)
EF	enlargement factor	(-)
Fr	Froude number	(-)
g	acceleration of gravity	($m\ s^{-2}$)
G	mass flux velocity	($kg\ m^{-2}s^{-1}$)
MAPE	mean average percentage error	(%)
MPE	mean percentage error	(%)
S	symmetry	(-)
Sc	scaling factor	($pixel\ mm^{-1}$)
t	film thicknesses	(m)
T	temperature	(°C)
x	vapor quality	(-)

greek letters

μ	dynamic viscosity	($kg\ s^{-1}\ m^{-1}$)
ρ	density	($kg\ m^{-3}$)
σ	surface tension	($N\ m^{-1}$)
χ	Martinelli parameter	(-)

Abbreviations

ORC	organic Rankine cycle

subscripts

app	apparent
bottom	bottom side
diff	difference
EF	enlargement factor
hn	Hurlburt and Newell method
In	internal
liq	liquid phase
mean	mean value
pred	predicted value
sat	saturation
Sc	scaling factor
meas	measured value
ss	Schubring and Shedd method
top	top side
vo	as the cross section was filled only by the vapor
vap	vapour phase
v	related to the vapor

Preliminary numerical studies of an experimental facility for heat removal in natural circulation

C Bertani, M De Salve, M Caramello, N Falcone, A Bersano and B Panella

Energy Department, Politecnico di Torino, Corso Duca degli Abruzzi, 24, 10129 Torino, Italy

E-mail: marco.caramello@polito.it

Abstract. In recent years particular attention has been dedicated to passive safety systems for heat removal in nuclear power plants. Passive safety systems can achieve a high level of safety, as they carry out their mission relying solely on physical principles like natural circulation, without any need of operators or energy sources. To qualify these systems and components experimental activities are necessary to study and to understand the governing physical phenomena. The present paper shows the design of an experimental facility to be installed in the laboratories of the Energy Department of Politecnico di Torino. The facility is inspired by the decay heat removal system for ALFRED reactor and comprehends a heated bayonet tube and a heat sink for the heat removal (a heat exchanger inside a pool). The thermal power is in the order of 1 kW. A RELAP5-3D model of the facility has been developed and sensitivity analyses were performed to highlight the geometry of the heat exchanger, the final heat sink, and the mass of water inside the loop. The results of this phase serve to understand the physical limits of the facility, to demonstrate a preliminary feasibility and to optimize the geometry for the desired operating conditions.

1. Introduction

The safe operation of nuclear reactors has always been a topic of great interest both in the scientific and social environments. In the last few years existing power plants underwent through an extensive series of assessments and adaptations to integrate lessons learnt from recent accidents [1] while new reactor projects have been object of careful analyses to comply with the "safety by designs" principles [2]. In the framework of the systems for decay heat removal, a promising road to enhance the reactor safety is to use passive safety systems [3], especially in the light of the experience feedback given by generation III+ reactors. Such systems are designed to operate independently by relying solely on physical principles without the need of operators intervention. Passive safety systems have been tested and qualified for the operation in light-water reactors [4], but, as the components used may be different their application to liquid metal reactors has still to be confirmed. The Advanced Lead Fast Reactor European Demonstrator (ALFRED) [5] is a lead cooled fast reactor that makes use of bayonet heat exchangers to remove the heat from the primary system [6]. The heat exchanger geometry has some specific characteristics that may affect the natural circulation, such as the flow path and the regenerative heat transfer. It is therefore important to perform experiments for experience acquisition and code validation.

In the Energy Department of Politecnico di Torino many activities are carried out to study natural circulation in passive safety systems with bayonet heat exchangers. An experimental facility has been designed and is currently in the procurement and installation phase. The facility consists of a 1 meter heated bayonet tube and a heat exchanger inside a pool for final heat sink. The maximum power of the facility is in the order of 1 kW. In parallel to the experimental activity, a RELAP5-3D model of the facility has been developed to support the design by means of sensitivity studies and to perform pre-test calculations. The present paper describes the geometry and the phenomena to be studied in the experimental facility as well as the model and the preliminary sensitivity studies developed by means of the system code RELAP5-3D for design support. The results obtained suggest that the facility should be able to represent most of the relevant physical phenomena for the safety system and therefore may provide important informations for the development of these components.

2. Facility description

Figure 1 shows a drawing of the facility. Heat is generated on the bayonet heat exchanger at the bottom, where water enters from the downcomer pipe (see figure 2), reverses its direction and upflows in the annular region created between the coaxial pipes. At the outlet, the fluid exits transversely through a T-junction and crosses the vertical riser before reaching the heat sink. In this part, the fluid exchanges heat with water at ambient conditions housed in an annular pool around the pipe before entering again the bayonet from the top.

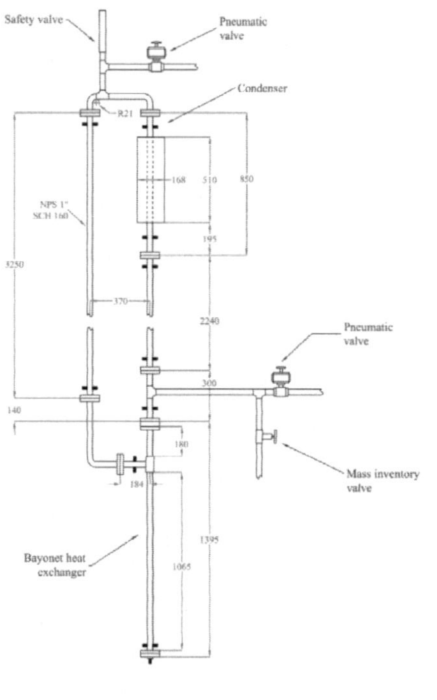

Figure 1. Drawing of the facility.

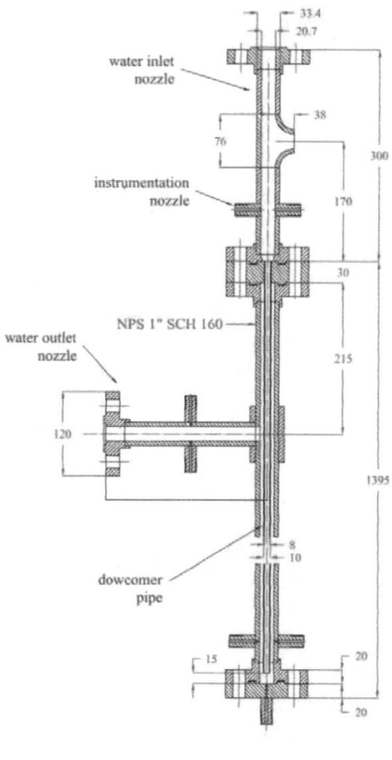

Figure 2. Drawing of the bayonet heat exchanger.

The overall motion of the fluid will be driven by buoyancy. The heat source increases the enthalpy of the fluid while the heat sink does the contrary, and the density difference drives the flow.

The facility is equipped with pressure transducers and thermocouples to monitor the condition of the fluid in several positions. Additional lines have been added to fill the facility with water, vent noncondensables and perform cold tests. These lines can be connected to an open circuit with a pump to characterize the pressure drops in forced circulation. A safety relief valve is set on the top of the facility to prevent accidents deriving from overpressure transients. The different regions of the system are coupled by means of flanges for fast inspection and maintenance.

Heat is generated by means of electrical wires wrapped around the outermost tube of the bayonet heat exchanger. Considering the equipment available on the market and the surface area of the tube the power input should be of the order of 1.5 kW.

At this stage, the facility has been installed in the laboratory and cold tests are undergoing to characterize the hydraulic resistance of the loop.

3. RELAP5-3D Model

Validation of design, technical choices and sensitivity studies on the behavior of the facility have been supported by means of the commercial software RELAP5-3D©, that is a highly generic code for simulation of thermal-hydraulic problems in both nuclear and nonnuclear systems involving mixtures of vapor, liquid, noncondensable gases, and volatile solute [7]. Figure 3 shows a sketch of the Nodalization created for the study. The model comprises both the facility and the final heat sink.

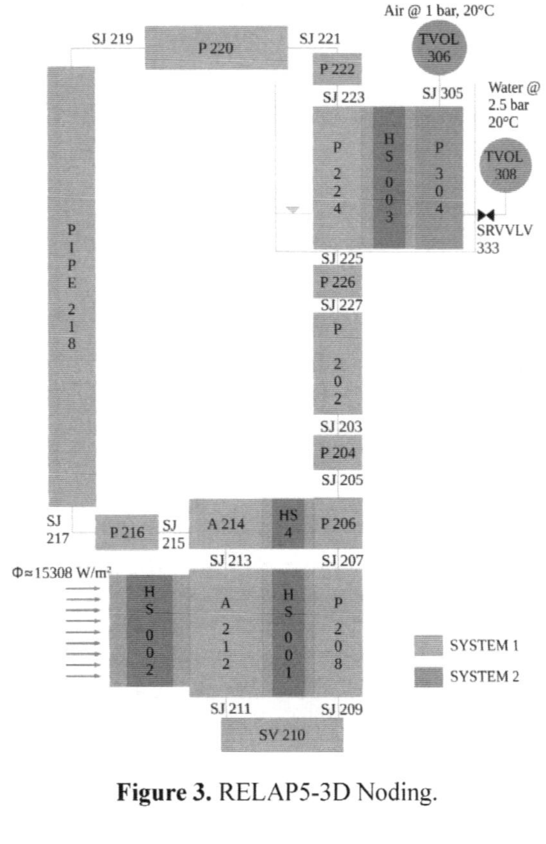

Figure 3. RELAP5-3D Noding.

The bayonet's downcomer is described by means of pipes 206 and 208. Single volume 210 models the volume of water in the inversion region. The bayonet's riser is modelled by means of annulus 212. Heat structures 1 and 2 models respectively (1) the regenerative heat transfer between the inner tube and the outer tube and (2) the outermost tube over which an imposed heat flux on the outside simulates the heating wires. Heat structure 4 models the heat transfer between the outlet of the bayonet and the uppermost region of the inner tube. The condenser is modelled by means of pipe 224 which transfers heat to the pool - modelled as pipe 304- through heat structure 3. Servo valve 333 provides water to the pool in order to keep the level to a user's defined set-point.

4. Results and discussion

This section presents the results obtained by means of the system code RELAP5-3D©. In the first part we have hypothesized an initial condition for the circuit and the water pool and have observed how the system reaches a steady state operation in natural circulation. In the second part, we modified the initial mass inventory in the circuit and the water level set point in the pool to understand how these two quantities affect the steady state operation. We have defined a custom void fraction expressing the volume fraction of steam at the initial pressure of 1 bar. It is the ones' complement of the volume occupied by the liquid phase. This parameter allows the filling level of the facility to be understood (when this value is 0, the loop is full of liquid water).

4.1. Startup procedure

The circuit is supposed to start in saturation condition at the pressure of 1 bar and the air in the facility is supposed to be completely vented out. The initial void fraction of the circuit is 50% (liquid water occupies an eight of more or less 2.7 meters while saturated steam is on the top). The timing of the events is as follows:

t ∈ [0-50] s: power source is off and water pool is empty
t ∈ (50-110] s: power source is turned on (ramp between 0 and 1.5 kW)
t ∈ (110-170] s: power is kept constant and pool is empty
t ∈ (170-230] s: water at 30 °C and 1 bar is injected in the final heat sink to form a level of 14 cm
t ∈ (230-6000] s: power input and water level in the pool are kept constant.

The objective of the simulation is to understand if the system can reach a stable operating condition under natural circulation.

As the power input is turned on the pressure starts to increase and the water level in the loop decreases. Heat is removed by means of a pool boiling mechanism and the steam produced inside the bayonet reaches the top of the loop passing through the water level with a slug flow pattern. When the water is injected in the final heat sink natural circulation triggers and the flow rate stabilizes around a value of more or less 13 g/s. When the water in the pool reaches saturation and the power removed equalizes the power input the pressure stabilizes around 17.5 bar. The circuit then works in saturated conditions, apart from the downcomer channel below the final heat sink where water is below saturation of about 6 °C. The pressure reached by the circuit is such that the temperature difference between the loop and the pool can evacuate the power input with the natural circulation flow rate. Figures 4 to 8 show the behavior of some quantities during the transient.

When the water in the final heat sink starts to boil (around t=1100 s), a power jump in the heat removed from the final heat sink can be observed (figure 4), while pressure behavior in the loop experiences a sudden variation before stabilizing (figure 5).

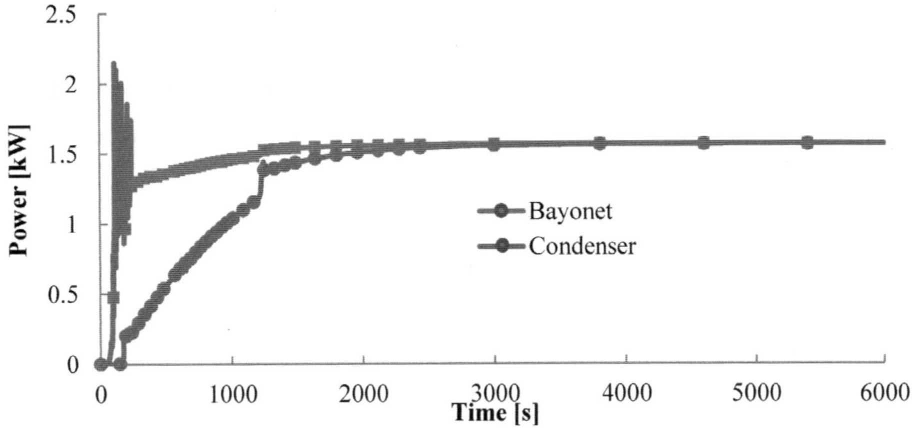

Figure 4. Power exchanged in the bayonet and the pool.

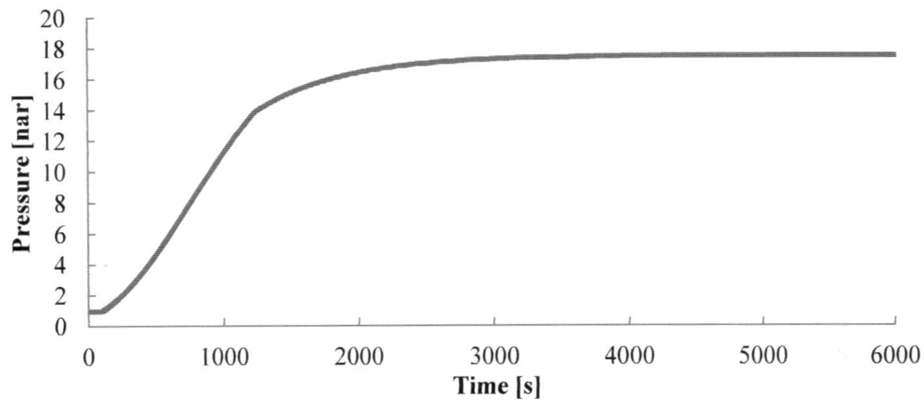

Figure 5. Pressure of the circuit.

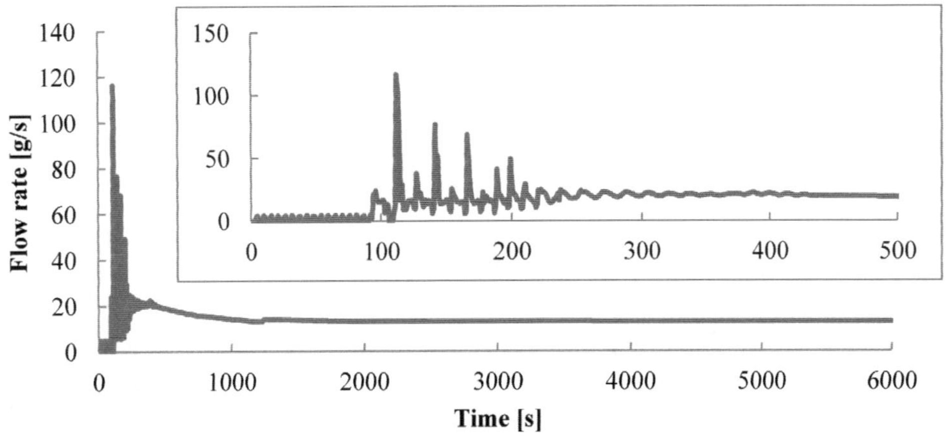

Figure 6. Flow rate at the outlet of the bayonet.

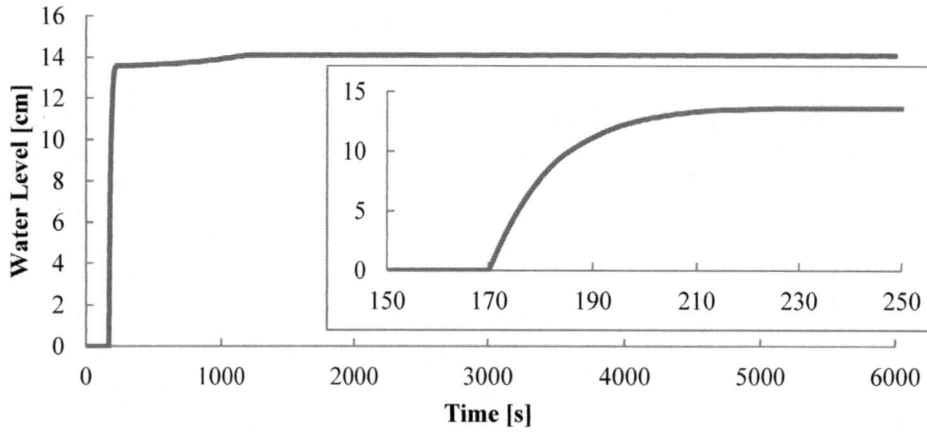

Figure 7. Water level in the final heat sink.

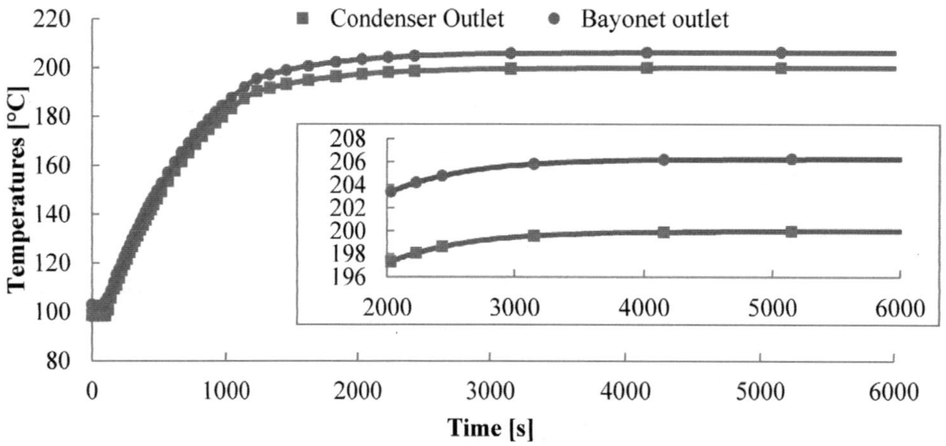

Figure 8. Temperatures in the loop (Bayonet outlet – Condenser outlet).

4.2. Sensitivity study on the initial mass and the water level in the loop

As far as the sensitivity study is concerned, we focused on both the water inventory inside the facility and the level inside the pool of the final heat sink, which defines its active length. The objective is to observe the expected operating pressure at which the system will operate, the temperature of the bayonet and the natural circulation flow rate. These quantities were important to understand both the piping dimension for integrity and the order of magnitude of the physical quantities for measurement.

Simulations were performed as follows. The model of the facility is initialized in saturated conditions with a mixture of liquid and steam. We therefore hypothesize to have completely vented all the noncondensables from the loop that contains a certain amount of water, or, alternatively, that the amount of noncondensables still in the facility do not affect the flow and the heat transfer. Starting from this point, heat is provided on the outermost diameter of the bayonet heat exchanger by means of an imposed heat flux boundary condition simulating the heating cables. By considering the available heating wires on the market we have assumed a power of 1.56 kW uniformly distributed on 97.5 cm of the bayonet. The water level inside the heat sink is kept constant by means of a control logic of the valve model. From this initial condition the system undergoes a transient until a steady state condition is reached. The system then works in a two-phase condition where water boils in the annular riser of the bayonet and condenses in the final heat sink. Steady state pressure, flow rate and temperatures will depend on the volume of the facility, on the mass of water stored and the active length of the final heat sink. The steady state condition has been analyzed to understand the possible operating conditions of the facility.

Cases are grouped according to the amount of water stored inside the loop.

Figure 9 shows the pressure reached in steady state condition by the facility depending on the initial void fraction and the active length of the heat sink. Maximum pressure reached by the system varies between 20 and 50 bar. A small heat sink or higher initial fillings of the facility go in the direction of a higher steady state pressure, because the available volume to expand is lower and a larger temperature difference between the loop and the environment is necessary to evacuate the heat (note that the fluid temperature in the region of the final heat sink is the saturation temperature at the working pressure).

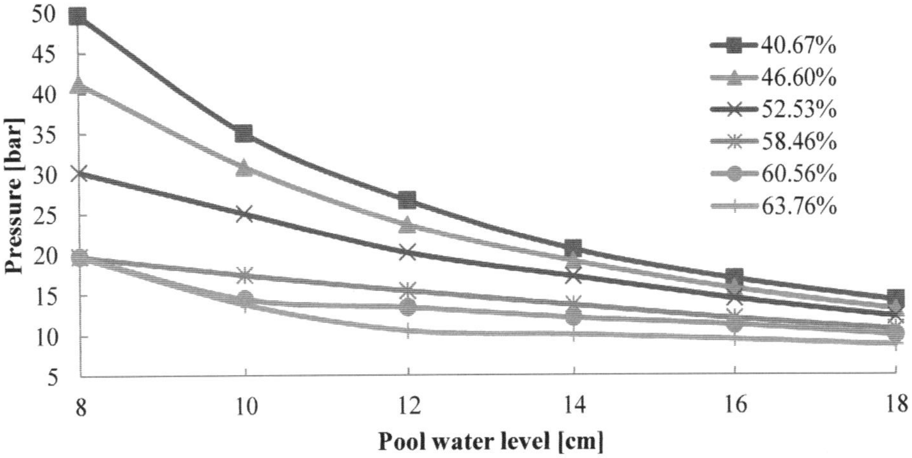

Figure 9. Steady state pressure as a function of the water level in the pool.

Figure 10 shows the flow rate circulating in the system. An increase of the final heat sink active length or the amount of water stored in the system increases the flow rate. The amount of natural circulation depends on the concomitant variation of the density difference between hot and cold legs, and the different height of thermal barycentres of the source and the sink. Also the void fraction of the fluid plays a role as it changes the friction in the loop. By varying the liquid level in the pool between 8 and 18 cm the maximum variation of the barycentres height difference is in the order of 3%, therefore its impact is limited. On the other hand, at a fixed amount of water in the loop, natural circulation increases as the length of the heat sink increases, because at decreasing pressure the density ratio between the liquid and the gas phase increases. By considering a fixed active length for the final heat sink, as the initial mass inventory increases the mass flow rate increases as well. This is due to the fact that despite the higher pressure the average void fraction in the two-phase pipes is larger. An important conclusion to this study is also that considering the low values of flow rates created by natural circulation careful attention shall be devoted to minimize the measurement uncertainty.

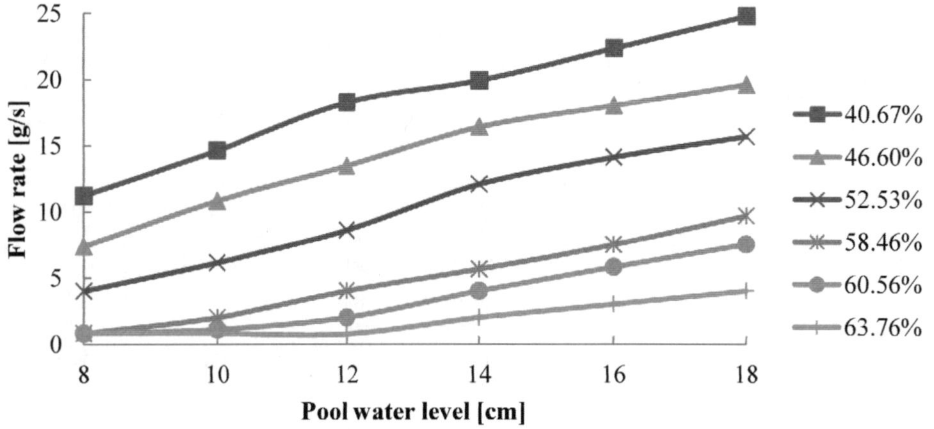

Figure 10. Natural circulation flow rate as a function of the water level in the pool.

Another figure of merit for the loop is represented by the maximum temperature in the heated pipe of the bayonet heat exchanger. The prediction can help to assess whether mechanical problems concerning the pipe can be expected in the region of interest. Figure 11 shows the maximum temperature for the inner wall of the outermost tube. As it is possible to see, the maximum temperatures are below 270 °C, therefore no mechanical problem is foreseen. In addition to that, as the heat transfer coefficient on the water side is considerably higher than the transmittance of the tube the temperature gradient on the pipe is almost constant in all the simulations considered, between 7 and 8 °C (the heat flux used in the simulations is constant).

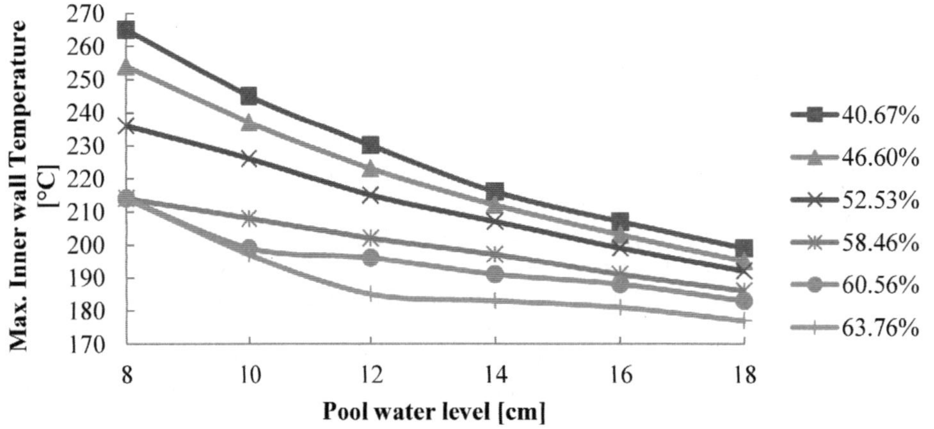

Figure 11. Maximum inner wall temperature as a function of the water level in the pool.

Figure 12 shows the water level in the cold leg. It is the transition front between the two-phase region and the single phase region, and decreases as the initial mass of water in the system decreases. Figure 13 shows the void fraction at the outlet of the bayonet heat exchanger. There is a very low dependence on the active length of the pool, while a large impact is provided by the initial mass of water in the system.

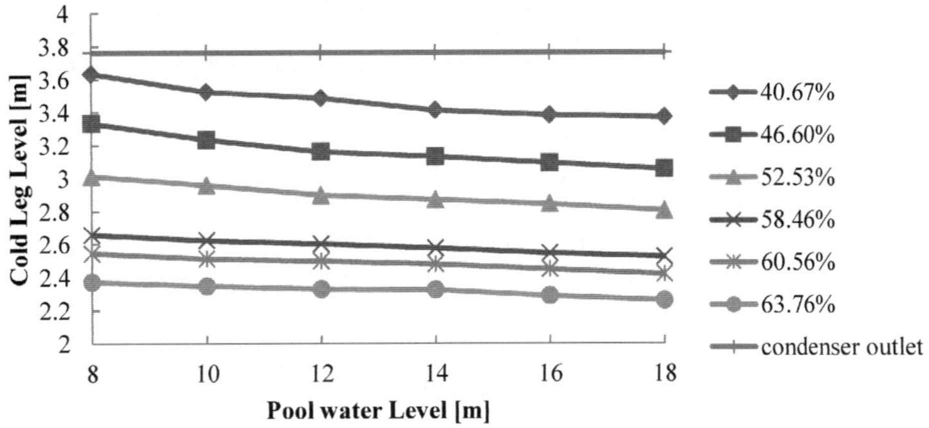

Figure 12. Transition front between two-phase region and single phase region.

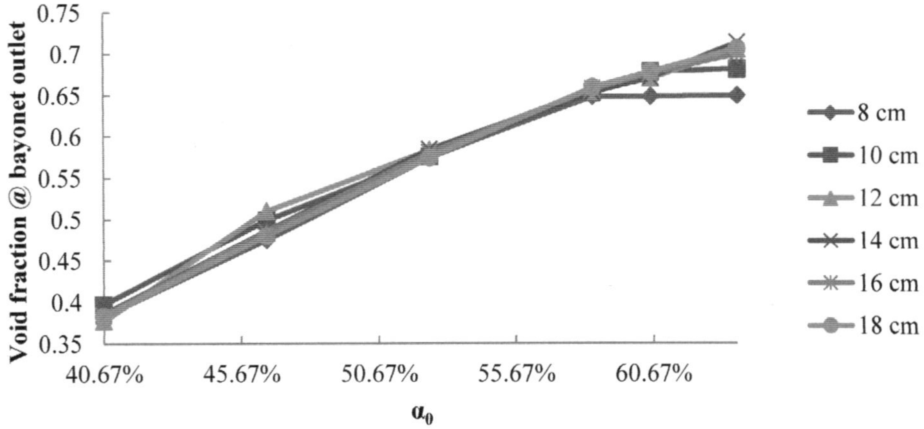

Figure 13. Void fraction.

As a last remark, the values of initial void fraction cover a range between 40% and 63%. We have observed high flow rate oscillations due to intermittent flow patterns outside this region. More detailed studies shall be carried out to understand the origin of instabilities, and a valuable help will be provided by the experimental results.

5. Conclusions and future work

In this paper a new experimental facility under construction at Energy Department of Politecnico di Torino to study the natural circulation in a safety system for innovative nuclear reactors has been presented. A Relap5-3D model developed to represent the operating conditions of the facility has been presented and used to study some steady state conditions for the loop under different hypotheses of water inventory and active length for the final heat sink. The model results have shown the possible application of the facility for a variety of different conditions and have highlighted the impact of the investigated parameters on the behavior of the loop. Operation compliance with maximum pressures and temperatures has been numerically validated. Preliminary simulations suggest to handle the measurement of the flow rate with particular care as it is low, and to foresee operating conditions characterized by unstable flow patterns for certain water inventories and heat sink values. Future work will be devoted to complete the facility, to perform the experimental activity (cold zero power test characterization, test matrix definition, hot maximum power tests) and to compare the results with the code prediction.

References

[1] International Atomic Energy Agency (IAEA) 2015 sti/pub/1710 Vienna
[2] US Department of Energy (DOE) 2008 DOE-STD-1189-2008 Washington
[3] International Atomic Energy Agency (IAEA) 2009 IAEA TECDOC-1624 Vienna
[4] U.S. National Regulatory Commission 2005 NUREG-1826 Washington
[5] Alemberti A, Frogheri M and Mansani L 2013 *International Conference on Fast Reactors and Related Fuel Cycles, Safe Technologies and Sustainable Scenarios* (Paris), France.
[6] Caramello M, Gregorini M, De Salve M, Bertani C, Alemberti A and Panella B 2015 *International Congress on Advances in Nuclear Power Plants (ICAPP)* (Nice) (American Nuclear Society) p 838–45
[7] RELAP5 2015 *RELAP5-3D 4.3.4 Code Manual Volume I: Code Structure, System Models and Solution Methods* Idaho National Laboratory

CFD and Neutron codes coupling on a computational platform

Daniele CERRONI, Roberto DA VIÀ, Sandro MANSERVISI, Filippo MENGHINI and Ruben SCARDOVELLI

University of Bologna, DIN, Lab. of Montecuccolino, Via dei colli 16, 40126 Bologna (BO), Italy

E-mail: daniele.cerroni2@unibo.it

Abstract. In this work we investigate the thermal-hydraulics behavior of a PWR nuclear reactor core, evaluating the power generation distribution taking into account the local temperature field. The temperature field, evaluated using a self-developed CFD module, is exchanged with a neutron code, DONJON-DRAGON, which updates the macroscopic cross sections and evaluates the new neutron flux. From the updated neutron flux the new peak factor is evaluated and the new temperature field is computed. The exchange of data between the two codes is obtained thanks to their inclusion into the computational platform SALOME, an open-source tools developed by the collaborative project NURESAFE. The numerical libraries MEDmem, included into the SALOME platform, are used in this work, for the projection of computational fields from one problem to another. The two problems are driven by a common supervisor that can access to the computational fields of both systems, in every time step, the temperature field, is extracted from the CFD problem and set into the neutron problem. After this iteration the new power peak factor is projected back into the CFD problem and the new time step can be computed. Several computational examples, where both neutron and thermal-hydraulics quantities are parametrized, are finally reported in this work.

1. Introduction

The numerical simulation of a nuclear reactor system is a challenging multiphysics problem. One of most important region in these systems, is the active zone of the core where the neutrons interacts with the fissile materials and the coolant flows removing the heat produced by the nuclear chain reaction [1]. During years lot of codes have been developed in order to study the nuclear interaction and the thermal hydraulics evolution of these systems. The first types, neutron codes, usually investigate the evolution of the neutron population at different energy levels considering all the different materials present in the core region [2]. The material properties that determine the interaction with the neutron flux strongly depend on the local temperature field. In order to produce an accurate evaluation of the energy production, the temperature dependencies of the materials properties must be taken into account by the neutron code. The temperature evolution in the core region is usually investigated with thermal hydraulics codes, that solve the energy and momentum balance equation in the core region under the assumption of a constant energy production rate. The mutual interaction between the neutron flux and the local temperature field is a fundamental aspect in the study of a reactor core but, at the same time, a very challenging phenomena to investigate [3]. In this work we couple, using a

IOP Conf. Series: Journal of Physics: Conf. Series **796** (2017) 012047 doi:10.1088/1742-6596/796/1/012047

numerical platform, the solution of a neutron code with the solution of a CFD code in order to investigate the thermal-hydraulics dynamics of a PWR nuclear reactor evaluating the power generation distribution taking into account the local temperature field. The CFD code that we use is the in house multigrid finite element code FEMLCORE that has been used and validated for the thermal hydraulic simulation of a reactor core in [4–6]. The neutron code used in this work is DONJON-DRAGON, an open source framework developed by the Institute of Nuclear Engineering of the École Polytechnique de Montréal that as been integrated into the numerical platform SALOME. The software is composed by two modules: DRAGON and DONJON. The DRAGON code [7] results from a concerted effort to unify inside a single code various numerical techniques and calculation methods that can be used to solve the neutron transport equation both with the collision probability technique and the method of characteristics. The standard calculation procedure we carry is in two steps. In the first one we perform critical transport calculations on the transport model and generate a consistent set of multigroup properties such as the various cross sections and diffusion coefficients for each different material. In the second step we introduce these nuclear properties in the DONJON full core model and compute the macroscopic flux distribution and the multiplication factor of the core. The DRAGON generated macroscopic properties are stored in COMPO files. The coupling between the neutron and the CFD code is achieved through their inclusion into the numerical platform SALOME, a software developed by the CEA for the design of light water reactor. In this platform all the included codes can exchange data run-time thanks to the MEDmem libraries, a computational toolbox that handles the exchange of data between the codes. During the execution of the codes, the temperature field in the core region, evaluated using the FEMLCORE module, is exchanged, through a MEDmem interface, with the DONJON code. The material macroscopic cross sections are then updated according the new temperature field using the COMPO object previously generated with the code DRAGON and, the consequent neutron flux, is computed. From the updated neutron flux the new peak factor is evaluated and projected, again through the MEDmem interface, into the FEMLCORE problem that evaluate the new temperature field. The work is divided as follows. After a general description of the SALOME platform structure we describe the coupling procedure of a generic code into the platform. In particular we introduce the principal elements that must be developed in order to exchange data with another code that has been previously integrated into this platform. In the following Section a computational example is presented and the results are analyzed. Finally in the last Section we give some conclusion and we draw lines for future works.

2. Numerical Platform

The collaborative project NURESAFE for the development of reliable software usable for safety nuclear reactor analysis, has been funded by the European community based on the open-source SALOME platform. NURESAFE has been created to improve the nuclear safety by developing high level of expertise and collecting the most recent simulation tools in the nuclear field [4, 8]. At the moment the platform is based on computational tools for light water reactors but many of these codes can be adapted to a large range of applications. SALOME integrates a large number of modules each having its own function: KERNEL, GUI, GEOM, SMESH, MED, YACS and Paravis module. Above all the computational tools included into platform, the MED module provides a library for storing and recovering computer data in MED format, associating numerical meshes and fields allowing the exchange between different codes and solvers. Inside the platform these structures are exchanged between solvers at the communication level (CORBA or MPI) offering common read and write functions through MED files. The MED libraries are divided into three groups of functions: MED File, MED Memory and MED CORBA. The first group (MED File) is a C and Fortran api that implements mesh and read/write functions into files with med extension, these files are in HDF5 format: a data model for storing and managing data. The module supports an unlimited variety of data types, it is designed for flexible and efficient input/output and for handle complex data. The MED Memory group, which is a C++ and Python api, creates mesh and field objects in memory, the mesh creation can be done using set functions, or by loading a file. Fields are also created loading files or by initialization with user-defined functions. Finally the group of functions called MED CORBA

2.1. Data exchange.
This platform has been conceived not only to collect a series of codes that have been extensively used in the field of thermal hydraulics of nuclear reactors but also to harmonize them with the aim of solving complex problems by exchanging information among different codes over a common platform and on large multiprocessor architectures. SALOME can also be used as a platform for the integration of external third-party numerical codes to produce a new application with full pre and post processing management of CAD models. The integration of a code on the SALOME platform is obtained by generating an interface with functions available in the MEDMem library that allows the data transfer from the platform to the code and vice versa. Two different codes both with SALOME MEDMem interface can transfer data to the interface and then from the interface to the other code. MED supports different element shapes such as point, line, triangle, quadrangle, tetrahedron, pyramid, hexahedron, polygon and polyhedron. Each element has a different number of nodes, depending on linear or quadratic interpolation. In order to have a working platform, common input and output formats should be harmonized between different codes. This can be achieved by using SALOME as the basic platform taking care of the data exchange between codes and of the distributed computation between different clusters.
In the following part of the Section we describe the integration procedure of a CFD code into the SALOME platform with the aim of using computational modules generated with this library for multiphysics coupling. The integration of an open-source code into the a numerical platform, can be divided into three major steps: the first is the generation of the code-library from the original code, the second is the generation of the MEDMem interface and finally the generation of SALOME-code interface integration. In a non open-source code, the source code is not available but usually the binary version of the library is provided, in this case the first step of generation of the code-library is no longer needed.

2.2. Generation of the code-library
The generation of the dynamic library from a code is pretty straightforward for modern codes since they are already built as libraries. The main code is simply a collection of call functions to libraries where the algorithms are developed. For old codes, especially in Fortran, sometimes a monolithic main program is developed with a few functions in support with experimental data. In this case it should be straightforward, for developers, to rebuild the code using a library structure.

2.3. MEDMem interface
Simulation studies require the manipulation of meshes and fields for data processing or post processing. Corresponding computer codes can be viewed as software components accessing input meshes and fields, with specific constraints, along with parameters and producing output meshes and fields. The MED module aims at pooling operations on those items, facilitating their use by various codes involved in a simulation process. This includes making codes communicate while preserving as much as possible the integrity of their content. In order to fulfill its objective, the MED module includes methods for: handling meshes and fields to satisfy code input requirements, extraction of field information and projections and serialization to exchange meshes and fields between codes. The structure of the MEDMem libraries is shown in the

IOP Conf. Series: Journal of Physics: Conf. Series **796** (2017) 012047 doi:10.1088/1742-6596/796/1/012047

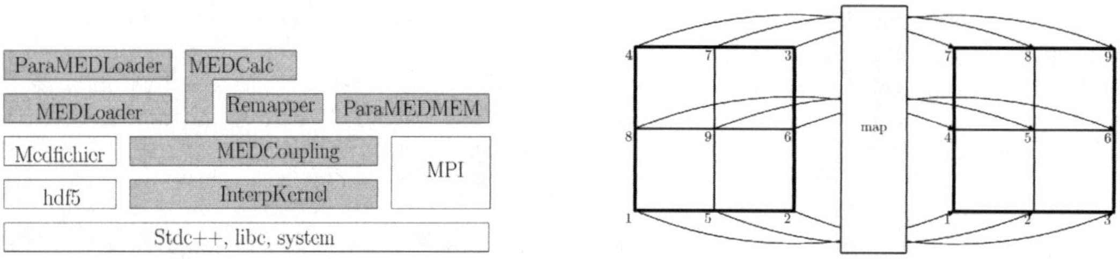

Figure 1. Layer structure of the packages of the MEDMEm library on the left. The operator that map the each node of a generic element into its duplicate in med format on the right.

left part Figure 1, the fundamental set (blue background) consists in three atomic libraries: MEDCoupling that describes data structures used for cross process exchange of meshes and fields, MEDLoader and ParaMEDLoader that provides input output functions to the med file format and interpolation tools that provides mathematical structures and algorithms for interpolation and localization. The main services offered by MEDCoupling are the manipulation of fields and their support mesh and multidimensional interpolation on nodes, cells, Gauss points and nodes by element. Once the code has been build as a library we need to add methods that can export the results into the MED format, in this way, in order to exchange a computational field with another code, we can just extract the solution from the MED file and project it into the different computational domain still in MED format. For this purpose we build a duplicate of the original computational grid in the med format and create a map that associates each computational node from the original mesh to one of the MED duplicate and vice-versa as shown in the right part of Figure 1. The generated map can be used both for projecting a solution coming from the specific code into the MED support but also for transfer a computational field from the MED mesh to the code specific computational domain. The projection into the MED grid consist on a reordering of the solution according to the generated map. When the solution is prepared as just described, it can be attached to the med support and transferred to any other med support using the methods of the med api. Once the field is transferred into a different support it can be extracted and ordered according to the particular map generated for the second program.

2.4. SALOME interface

The final step in the integration of a generic code into the numerical platform, is the construction of different methods that allow the control of the execution and the data input/output of the selected code. There are basically five classes that must be introduced in the selected code: *driver*, *problem*, *equations*, *grids* and *methods*. The *driver* class is the top level class that interface with the SALOME platform as it is shown in the collaboration diagram in Figure 2, where the basic structure of the code that we want to integrate into the platform is illustrated. This diagram show how the *driver* class transfers data to the *problem* class which is specialized for different physics. This class is inherited by all the physical modules used by the selected code. The *equations* class, which uses only MEDMem functions, inherits the system particular class which contains the assembly and solver of the generic code. The data from the *driver* class can be transferred into the assembly routine which is user accessible. The data should also be transferred in the opposite direction from the equations to the MEDMem interface. The data flow from *parent* to *son* class is ensured by the C++ inheritance rules wile the flow in opposite direction is defined by a *dynamic_cast* command that allows to use *son* class functions from the *parent* class. For this reason the *parent* class should be polymorphic with at least one virtual function. The *grids* class is an extension of mesh class of the particular CFD code, it contains

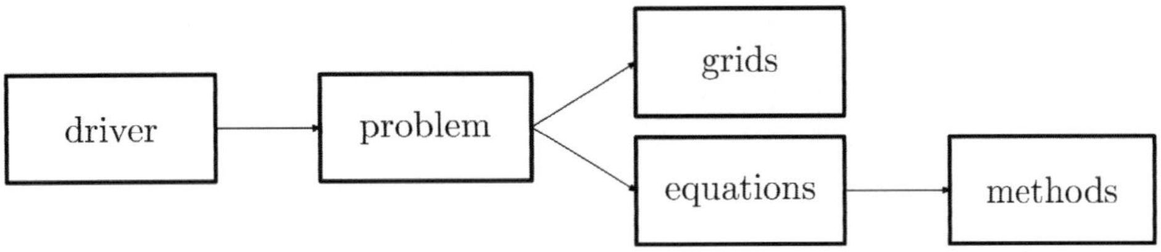

Figure 2. Collaboration diagram of generic CFD interface.

all the routines for mesh handling together with MEDMem methods that are used to extract and manage groups of nodes and cells from the original computational grid. In particular the extension class contains the methods that allow the data transfer from one format to another. As in the previous class, data from the *driver* class can be transferred, by using a *dynamic_cast* statement, into the assembly routine which is user accessible. Data can also be transferred in the opposite directions from the grids to the MEDMem interface through *parent* to *son* class inheritance rules. All the code interfaces must have similar commands to run the program. The mesh boundary names and their flags, used for creating mesh groups, are stored in the *driver* interface class. Over these groups, by using MEDMem functions and the map between the med support and the specific mesh, we are able to extract field from the med support and project it from the MED support, coming from other SALOME platform codes, into the problem specific mesh format. The basics commands that the *problem* class must have are: *setType*, *setMesh* and *solve*. The first command sets the problem type (Navies-Stokes, energy, etc.), the second one sets and prepares the mesh which should be available in MED and code specific formats for data exchange and, finally, the *solve* command controls the solution of the discrete system. The boundary regions of the domain should be controlled for input and output. For this purpose we have to introduce class *methods* that could set and get analytic and numerical solutions in these particular zones. The interface to the selected code is obtained through the *problem* class. For this reason it should access to both *grids* and *equations* classes as shown in the collaboration diagram of the problem class in Figure 2.
While the *grids* class can only handle the data the specific code format, the *equations* class contains two standard maps that associate to any zone of the computational grid a *methods* so that a volumetric field can be extracted and transferred to a *methods* with a MEDMem mesh format by using the *getSource* and the *setSource* function. If the data transfer is between 3D and 1D the average source functions may be used. When codes are organized as just described exchange data is easy. We just have to create a supervisor that access the *driver* class from different problems and, after each *solving* iteration the desired field is extracted from the first *driver* and projected into the common format. This field is then projected by the second *driver* into the computational grid of the second problem and can be taken into account during its execution.

3. Numerical Results

In this Section we show some numerical results obtained coupling the neutron code DONJON with the FEM code FEMLCORE. In the example we investigate the temperature field, the velocity profile and the neutron flux evolution in the active core region of a PWR reactor. The codes are able to exchange data through a MED interface, in particular during a time step iteration firstly we evaluate the temperature and the velocity field according to a given distribution of energy production. Those fields are then accessed by the neutron code that update the material properties according to the new temperature field and a new neutron flux distribution is evaluated. The new energy distribution is the taken into account for the evaluation of the temperature and velocity field in the next time step iteration. In the following part of the Section we give a general description of the reactor core considered and the some results are shown under different working condition.

3.1. General Description

The geometry of the reactor used in this work, is shown in the left part of Figure 3, we consider an annular reactor characterized by a diameter (d) of $3.23m$ and an height (a) of $6.6m$. The active part of the core start form $c = 2.5m$ and ends at $b = 4.5m$. The coolant (water) flows from the lower inlet region to the outlet on the top. The core that is composed of 221 fuel elements, each of which is made up of 428 fuel pins. The discretization of such complex domain with a realistic geometry would require a number of computational cells that is challenging even on huge clusters and supercomputers. It is therefore mandatory to approximate the velocity, pressure and temperature fields inside the core on a coarser grid of computational cells that somehow homogenize the underlying structure of pins and subchannels. We can consider a multiscale approach, where a fine level and a coarse level are considered. The fine level takes into account for the real geometry of the core. The solution on this level is extended to the whole volume of the coarse cell and projected on the degrees of freedom (dofs) of the coarse cell. At this level all geometrical details pertinent to the single subchannels are ignored, while a mean value of the physical fields are computed like in a porous medium, with physical properties taken as mean values of the coolant and structural materials. We denote the coarse grid fields with a ^ sign. We therefore define a porosity coefficient $r = A_C/A_{\mathrm{FA}}$ where A_C is the coolant flow area, while

Figure 3. Geometry of the reactor on the left. DONJON computational grid and active zone of reactor core on the center and right part, respectively.

A_{FA} is the entire section of the fuel assembly. The conservation equations can be written as

$$\begin{aligned}
&\nabla \cdot (r\rho\hat{\boldsymbol{v}}) = 0 \\
&\frac{\partial(r\rho\hat{\boldsymbol{v}})}{\partial t} + \nabla \cdot (r\rho\hat{\boldsymbol{v}}\hat{\boldsymbol{v}}) = -\nabla\hat{p} + \nabla \cdot r\hat{\boldsymbol{\tau}} + r\rho\boldsymbol{g} + \hat{m}_g \\
&\frac{\partial(r\rho c_p\hat{T})}{\partial t} + \nabla \cdot (r\rho c_p\hat{T}\boldsymbol{v}) = -\nabla \cdot (r\kappa\nabla\hat{T}) - \hat{q}_g\,.
\end{aligned} \tag{1}$$

The surface subgrid momentum and heat sources $\hat{m}_g$ and $\hat{q}_g$ are independent of r. The momentum source $\hat{m}_g$ can be modeled by

$$\hat{m}_g = -\boldsymbol{\beta} \cdot \frac{\rho \hat{\boldsymbol{v}} |\hat{\boldsymbol{v}}|}{2\, D_{eq}} \,. \tag{2}$$

This term in the momentum equation represents the loss of pressure in the coolant due to the viscous friction in the subchannels. The vector $\boldsymbol{\beta}$ takes into account the direction of the flow. For pure vertical flow the term is set to the standard pressure losses for a standard channel geometry. For cross flow the term should be evaluated case by case. In (1), as a first approximation, r can be assumed constant. In this case even if system (1) is formally very similar to the standard balance equation system but there are large differences hidden in the definition of the stress tensor $\hat{\boldsymbol{\tau}}$ and of the thermal conductivity κ. The term $\hat{\boldsymbol{\tau}}$ can be split in three components: the standard viscous part $\hat{\boldsymbol{\tau}}^{\text{visc}}$, the turbulent part $\hat{\boldsymbol{\tau}}^{T}$ and an additional term $\hat{\boldsymbol{\tau}}^{\text{eff}}$ that comes from the subgrid (fine) level. The evaluation of $\hat{\boldsymbol{\tau}}^{\text{eff}}$ is a challenging task. It may be computed with a DNS simulation of a subchannel, or can be obtained by experimental data. In the case of lack of data the term could be modeled with simple algebraic standard relations or ignored. Similarly, the thermal conductivity κ can be split in κ^{visc}, κ^{T} and κ^{eff}. The 3D-porous model can simulate naturally an open assembly core but also a closed or mixed open-closed assembly core simply by enforcing no cross flow through the assemblies. For details about this model the interested reader can see [4, 5, 9]. The input parameters, used by DRAGON, for the generation of the macroscopic cross section library needed for solving the multigroup diffusion equation, and the particular fuel pin geometry are reported in the left and right part of Figure 4, respectively. The transport equation is integrated over a lot of tiny energy interval and the results are

r_4	$0.41cm$		
r_1	$\sqrt{0.5}r_4$	$\mathrm{N}^{U^5}_{UOX}$	0.037%
r_2	$\sqrt{0.8}r_4$	$\mathrm{N}^{U^5}_{MOX}$	0.0025%
r_3	$\sqrt{0.95}r_4$	$\%Pu\ r_1$	0.075%
R_{iTG}	$0.56cm$	$\%Pu\ r_2$	0.049%
R_{eTG}	$0.62cm$	$\%Pu\ r_3$	0.028%
R_{iG}	$0.418cm$	T_f	$800.0K$
R_{eG}	$0.48cm$	T_m	$600.0K$
lp	$1.26cm$	ρ_m	$0.659g/cm^3$
ws	$0.04cm$		

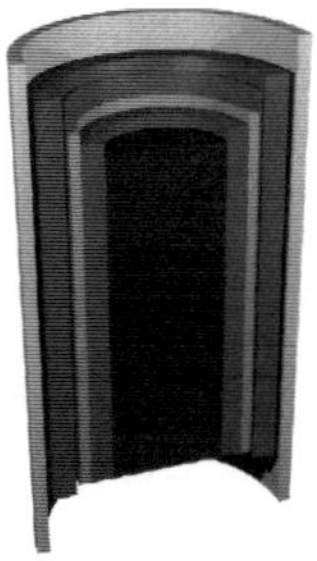

Figure 4. Fuel pin geometry on the right and composition parameters on the left.

then condensed into the energy groups considered into the multigroup diffusion model. The macroscopic cross section are evaluated for different fuel compositions and for different average temperatures according to a range defined by the user. This operation allows the interpolation of the macroscopic cross section between the range values that has been set. We remark that this first step, the generation of macroscopic cross section library, depends only on the fuel pin composition and geometry, the possibility to evaluate the cross section in a range of composition and temperature is meant for avoiding the generation of this library every time iteration during a simulation. Since DONJON, the code that solves the multigorup diffusion problem, can handle only Cartesian structured mesh we have to embed the computational domain shown in Figure 3 into a Cartesian three-dimensional grid specified by a characteristic length u. In our case we choose $u = 0.215m$ and the computational domain is show in central and right part of Figure 3, in particular in the right part of the same Figure we mark the active core region in the domain, in this zone the neutron flux and the peak factor are evaluated considering the particular composition of every fuel element of the computational grid. In particular the domain is composed by 31 sections made of 17 × 17 cells and the composition of each cell is specified in a 17 × 17 matrix. In this example we consider three material types: 0 is a dummy material that mark a zone in which the neutron flux is zero and must not be evaluated, 1 is the fuel material which has the composition shown in the left part of Figure 4 and finally 2 is the reflector material. In the right part of Figure 4 a sketch of the fuel pin element is reported, from that Figure one can appreciate the geometrical complexity that can be considered in the evaluation of the macroscopic cross sections. The firsts and the lasts 10 planes of the neutron computational grid, has the A configuration show in Figure 5, the 11th and 22th planes has the B configuration show in Figure 5 and finally the remaining planes, the active core, has the C configuration shown in Figure 5. Now that we have specified the composition of all the elements of the computational grid we can solve the two groups diffusion equation evaluating the neutron flux inside the domain. In the next Section we describe the interface operator that is used for the projection of the computational fields in the different computational grids.

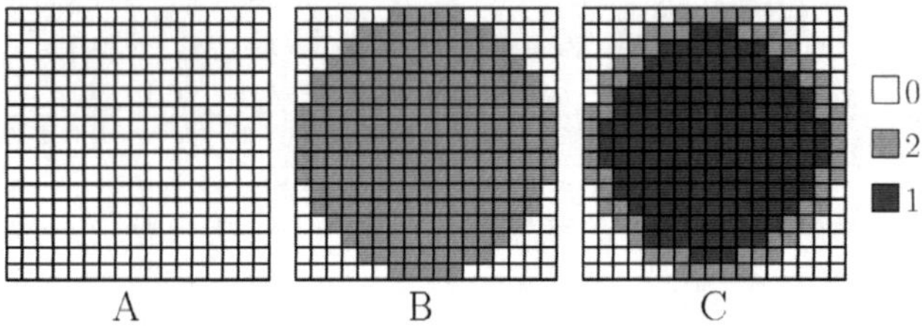

Figure 5. Three possible mix (A, B and C) of materials 1, 2 and 3 in a reactor plane.

3.2. Interface

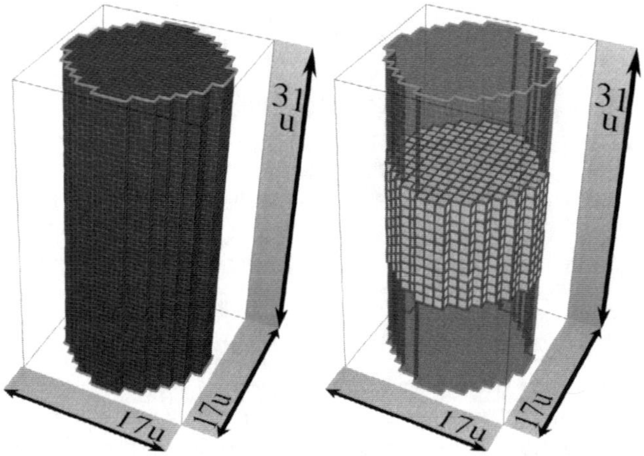

Figure 6. FEMLCORE (solid color) and DONJON (outline box) meshes. Whole meshes on the left and heated zone on the right.

The interface between the neutron and CFD module is created by the construction of a duplicated mesh of the two computational domains in the MEDmem format and creating a map between them. In this case the two meshes have different geometry as we can see in Figure 6. In particular in the left part of that Figure the two computational domains are shown, the bounding box mark the Cartesian structured mesh mesh and the intern grid is the CFD mesh. In the right part of Figure 6 we can see the overlapping between the active core zone in the Cartesian mesh and the CFD domain. The map between the two MEDmem meshes couples all the elements of the CFD computational grid in the nearest cell into the Cartesian grid.

After a *solve* iteration of the CFD problem the temperature field is taken from the problem ad projected, through the MEDmem meshes duplicates and the map operator, to the neutron code. The neutron flux is then calculated by DONJON updating the cross section according according to the new temperature field. From the updated neutron flux the new peak factor is evaluated and projected back to the CFD code which can evaluate the new temperature field.

3.3. Core evolution

Figure 7. Temperature overview at different time steps t_1, t_2 and t_3.

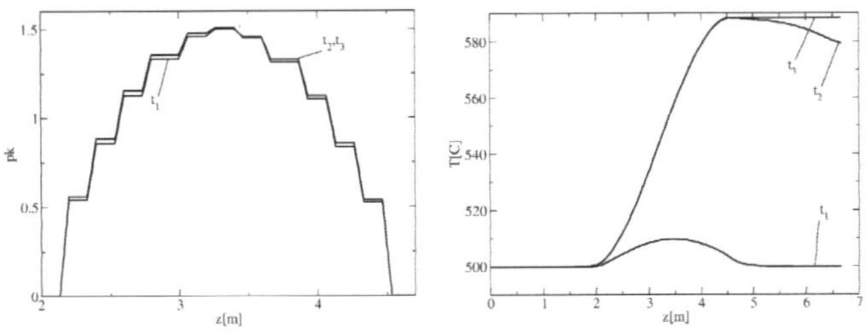

Figure 8. Neutron flux and temperature profile over the center line of the domain at different time steps.

The temperature field in the initial condition is set to a constant value of $500C$ while the axial velocity field to a constant value of $1m/s$. The velocity field remains equal to the initial field during the whole simulation because of the boundary condition and the geometry of the problem. For these reasons is not reported. Due to the energy production in the active zone of the core, the temperature increases from the initial value until the steady state is reached. The temperature field at different time steps t_1, t_2 and t_3 (0.1, 2.5 and $10s$, respectively) over the 3D domain is shown in Figure 7. We can notice that the temperature field increases in the active zone of the reactor in the initial time steps (t_1 and t_2) and then, because of the velocity field, all the upper zone is heated until the steady state is reached (t_3). In Figure 8 the peak factor (on the left) and the temperature field (on the right) over the center line of the domain at different time steps (t_1, t_2 and t_3) is shown. We can notice that as the temperature field increase the peak factor i changes because of the different cross section. This effect cannot be considered without the usage of the neutron code.

4. Conclusion

In this work we have shown a general procedure for the integration of numerical codes into a computational platform and we have show how this computational tool can be used to investigate

the energy production inside the core region of a nuclear reactor by taking into account the local temperature field. The solution of this complex problem has been achieved by coupling the solution of an in-house finite element code that solves the balance system of equations, with a neutron code that evaluate the energy production in the core region. At every time iteration the material properties, used to evaluate the neutron flux distribution, have been evaluated according to the local temperature field. The exchange of data between the different codes has been obtained through the numerical libraries MEDMem included into the SALOME platform where both of the codes have been included. The numerical results have shown the consistency and the robustness of the coupling algorithm. In future works we would like to integrate into the platform a system code such as cathare and couple the solution of the multiphysics problem presented in this work, with the simplified model of the whole primary loop of the reactor.
An attractive characteristic of the coupling technique shown in this work, is that the execution and the interface of every module included into the platform is independent. This modular approach allows to easily increase the number of codes that can exchange data run-time and, moreover, the substitution of one of this module with any other code already included into the platform.

References

[1] James J Duderstadt L J H 1976 *Nuclear Reactor Analysis* ISBN 0471223638,9780471223634
[2] Lamarsh J R 1983 *Introduction to nuclear engineering* 2nd ed Addison-Wesley series in nuclear science and engineering (Addison-Wesley) ISBN 9780201142006,0201142007
[3] Giannetti F, Maio D V D, Naviglio A and Caruso G 2016 *Nuclear Engineering and Design* **305** 168–178 ISSN 0029-5493
[4] Bornia G, Finelli M and Manservisi S 2011 Development and validation of fem-lcore code for the thermal hydraulics of open cores Tech. rep. Report RdS-2011-002
[5] Bornia G, Cerroni D, Manservisi S, Polidori M and Donato F 2012
[6] Olander D R 1976 Fundamental aspects of nuclear reactor fuel elements Tech. rep. California Univ., Berkeley (USA). Dept. of Nuclear Engineering
[7] Marleau G, Roy R and Hébert A 1994 *Report IGE-157, Institut de génie nucléaire, École Polytechnique de Montréal, Montréal, Québec*
[8] Chanaron B, Ahnert C, Crouzet N, Sanchez V, Kolev N, Marchand O, Kliem S and Papukchiev A 2015 *Annals of Nuclear Energy* **84** 166–177
[9] Cervone A and Manservisi M 2008 A three-dimensional cfd program for the simulation of the thermo-hydraulic behaviour of an open core liquid metal reactor Tech. rep. Report RSE/2009/85

IOP Conf. Series: Journal of Physics: Conf. Series **796** (2017) 012048 doi:10.1088/1742-6596/796/1/012048

Experimental thermal characterization of concrete to be used in CP5.2 packaging system

R Lo Frano[1], A Maggini[1], D Aquaro[1]
[1] University of Pisa – DICI (Dipartimento di Ingegneria Civile e Industriale)
Largo L. Lazzarino, 56122 - Pisa
rosa.lofrano@ing.unipi.it

Abstract This work deals with the experimental evaluation of the thermal conductivity of a concrete matrix to be used for embedding LILW bituminised Wastes into the packaging system. Such a type of packaging, identified with the acronym CP5.2, has been also qualified by executing at the Lab. Scalbatraio of Dep. of Civil and Industrial Engineering of the University of Pisa, an experimental test campaign accordingly to the IAEA regulations.
In particular, the knowledge of the thermal conductivity is important because of the fire or furnace test to carry out in oven at 800 °C for 30 minutes. These data allowed to simulate pre-test conditions and to set up safety and operational fire test procedures.
The concrete thermal conductivity was obtained by performing hot wire tests on cylindrical concrete samples for temperatures ranging from 100° to about 800°C.
Thermal conductivity is determined at steady state condition. Results indicate a monotonically reduction of the thermal conductivity as the temperature increases. The comparison with concrete thermal conductivity data available in literature indicates a quite good agreement. Finally, visual and X-ray inspection of sample did not highlight the presence of micro/macro damages that would have affected the thermal performance of the concrete under study.

1. Introduction

Concrete is widely used as a primary structural material both in conventional construction and in nuclear activities due to several advantages, such as strength, durability, fire resistance and non-combustibility properties. With reference to the fire exposure, that represents one of the most severe environmental conditions to which structures may be subjected, providing proper fire safety measures is an important outcome of design to fulfil.

Specifically, the present study deals with the thermal characterization of concrete to be used to immobilise bituminised LILW radioactive waste, stowed in drums, in turn stored in the inert concrete matrix of a CP5.2 packaging.

To this aim, the packaging integrity plays an important role, because a reliable structure must guarantee a safely storage in the repository during and beyond the institutional period. To be reliable, CP5.2 packaging must be qualified by submitting it to the tests foreseen by the IAEA SSR-6 rules [1], like fire test (engulfing fire condition foresees 800 °C for 30 minutes).

This means knowledge of deformations and of thermal and mechanical property changes [2] of the packaging constituting components, among which there is the concrete that, similarly to other materials, changes substantially at elevated temperature.

Although the availability of concrete and steel properties at elevated temperatures can permit a mathematical approach for pre-testing, to be confident that no abnormal situation will arise during the fire test execution, due to the peculiarity of bituminised wastes, the prediction of the thermal conductivity of concrete becomes safety relevant (because of high burning risk) and mandatory. In fact, the bitumen may decompose forming flammable vapour, especially at temperature higher than about 300°C. On this basis, an experimental campaign aimed at the evaluation of the thermal conductivity of concrete was carried out at the Lab. Guerrini of Department of Civil and Industrial Engineering of the University of Pisa. The thermal performances of concrete samples were therefore evaluated for different temperatures up to 800°C.

In the following Section 2, the wire test methodology and the adopted test procedure are described, while the results obtained will be presented and discussed critically in Section 3, also by comparing them with the values of Kodur et al., Vodak et al. or Shin et al. [3, 4, 5, 6, 7, 8], which are available in literature.

2. Description of the experimental device

To thermally characterize the concrete, several cylindrical samples, grouted from the same cement paste, were prepared (Figure 1 a). They are 400 mm long and 100 mm in diameter (Figure 1 b).

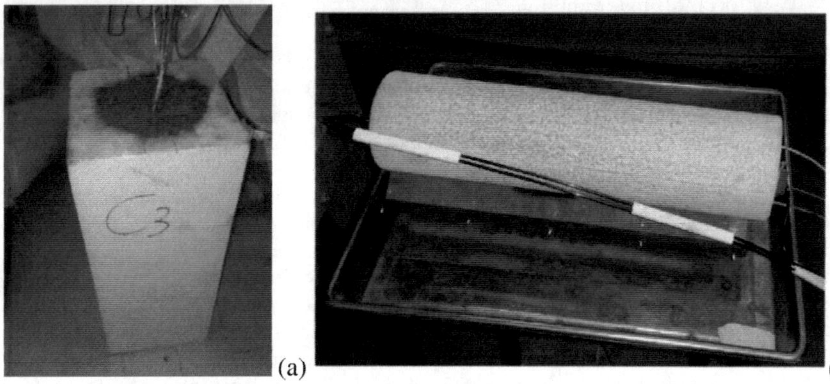

(a) (b)

Fig. 1 – Sample C3 soon after the grouting of the cement (a) and after 28-days curing (b).

Each sample is equipped with two thermocouples (positioned at a radial distance of 0.05 and 0.03 m from the centroid) and a central wire resistor. They were inserted in the mold before the grouting and held in the vertical position by means of a spacer device designed specifically to this purpose. In this way, the verticality/alignment of the sensors is controlled.

The two thermocouples (indicated in the following TC_INT and TC_EXT) are of type K and properly calibrated (calibration error less than 2%). They are immersed in the concrete at two different radial points, as shown in Fig. 2, in order to measure the temperature during the wire test-heating phase. In addition, attention was put to possible concrete de-hydration phenomena that would have degraded concrete properties, if material is exposed for long time at elevated temperature (higher than 500°C).

For this reason, an appropriate inspection of samples in post-test phase was carried out.

After waiting 28 days for the maturation of the cement, thermal tests on samples were performed at the Lab. Scalbatraio according to the following procedure:

a) opening and closure of the oven gate for the positioning of the sample (Fig. 3);
b) setting up of the oven temperature through PID controller;
c) starting of test by heating up uniformly the electric oven until the testing temperature is reached;
d) monitoring and control of temperature ramp by means of an adequate data acquisition system (DAS)
e) acquisition of temperature values by DAS when steady state condition is reached for at least 300 s.

Tests were performed between 100 and up to about 800 °C.

Fig. 2 – Concrete sample with instrumentation: (a) cylindrical sample, (b) the internal and external thermocouples, namely TC_INT, (c) TC_EXT, and (d) through wire resistor.

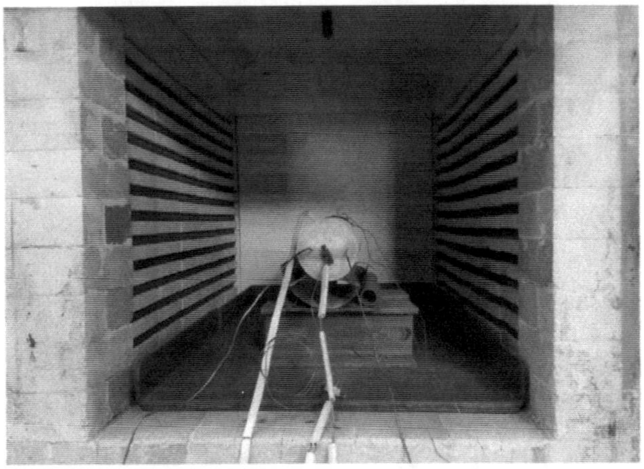

Fig. 3 – Sample positioning in the oven before the closure of tailgate of the oven.

It is noteworthy that there are very few standardized methods available for measuring thermal properties of concrete. Therefore, to the aim of the study, the hot wire method is adopted; a more detailed description of which is given in [9].

Figure 4 exemplifies the acquisition data chain.

Each test is performed by heating up the sample homogeneously (thanks to the thermal resistances, distributed evenly along all walls of the oven, and the central resistor) in the oven until the steady state condition is reached. The temperature increase is adequately controlled through a PID system (GEFRAN© GF Promer) connected to the data acquisition system (DAS). The interface with PC is obtained via Labview©. GEFRAN system is a Microprocessor Controller that supports thermocouple configuration via the complete operator faceplate. It is equipped with a Lexan© membrane faceplate to provide a frontal IP65 rating.

The thermal conductivity is determined measuring the radial temperature gradient in correspondence of the cylinder cross section located at half of its length [10, 11], once steady state condition is reached. Indeed, the superficial temperature of sample, the oven temperature and the current intensity circulating in the wire are measured; this latter is used to calculate the power generated within the cylinder due to the Joule effect.

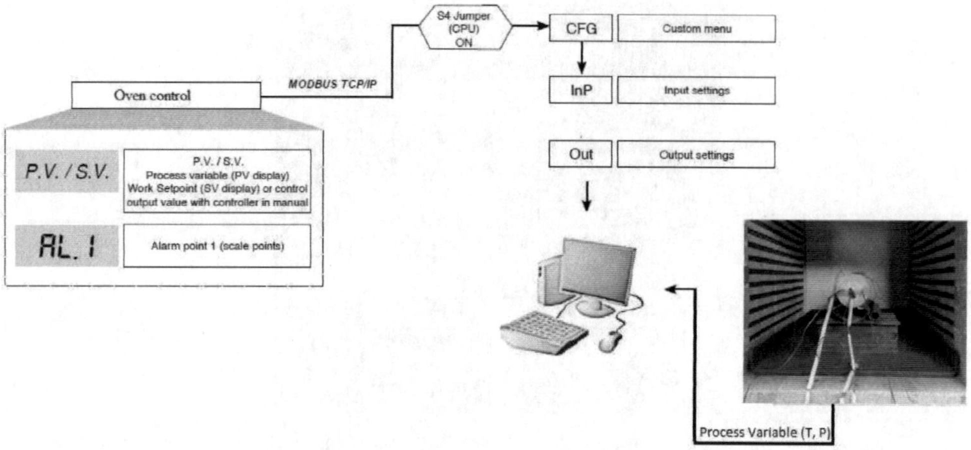

Fig. 4 – Scheme of the data acquisition chain.

The thermal conductivity is so calculated by means of the Fourier law of heat conduction assuming a homogeneous thermal power production within the sample (because of the combined control of samples heating made by means of the imposed ramp of the resistor and oven electrical resistances), in plane geometry:

$$k = \frac{\emptyset}{2\pi R^2 L \Delta T}\left(X_{ext}^2 - X_{int}^2\right) \quad (1)$$

where:
k [W/m °C] is the thermal conductivity;
Ø [W] is the power generated by Joule effect within the wire;
L [m] is the cylinder length;
ΔT [°C] is the difference of temperature between TC_INT and TC_EXT;
R [m] is the radius of cylinder; and
X_{ext} and X_{int} [m] are the radial distances of thermocouples from the cylinder axis.

Moreover, before the execution of wire tests, samples were X-ray inspected in order to exclude, as confirmed, any misalignment of thermocouples from their own generatrix or of the hot wire from its centreline. Figure 5 shows the steady state condition for wire test at 600 °C: for an average temperature of about 615 °C, an average power of 192.4 W, the thermal conductivity of sample is equal to 1.49 W/m °C.

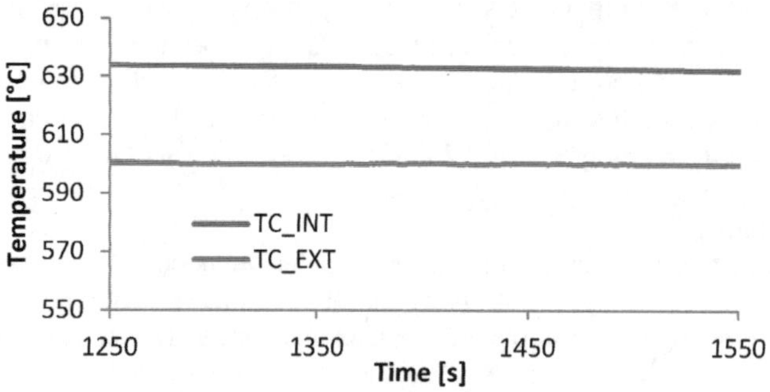

Fig. 5 – Steady state condition of test at 600 °C.

3. Results and discussion

The average values of thermal conductivity at each testing temperature are listed in Table 1.

The (total) maximum error associated to these measurements is lower than 8%. Furthermore, the discrepancy between thermal conductivity values, obtained from different samples and at the same testing temperature, is lesser than 5%.

The thermal conductivity of concrete at room temperature is in the range of 1.6 and 2.2 W/m °C and varies with temperature and depends on concrete composition [8]. It complies with the value of the upper curve of the Eurocode and with that proposed for carbonate and siliceous high-strength concrete (HSC) by Kodur and Sultan [4]. By analysing the behaviour shown in Fig. 6, it is possible to note that the thermal conductivity of the concrete monotonically decreases as the temperature increases. Best-fit curve, based on linear interpolation, see the Eq. (2) below, provides provisions of thermal conductivity mean values.

$$k = 1.9364 - 0.0007\,T \quad (2)$$

This trend is mirrored in similar data available in the literature, e.g. those pointed out by Kodur [3], even if for normal strength concrete.

Table 1 – Concrete thermal conductivity values.

Temperature [°C]	Average value [W/m °C]	Standard deviation
100	1.86	0.04
200	1.80	0.07
300	1.73	0.10
400	1.63	0.12
500	1.56	0.06
600	1.51	0.06

Moreover the behaviour of the thermal conductivity indicates 20% reduction of the thermal conductivity in the temperature range between 100 °C and 600 °C. This reduction can be explained by considering de-hydration process occurring at elevated temperature and that leads to an increase of porosity and simultaneously to a reduction of the thermal conductivity: the same behaviour was found in ASCE and Eurocode data, as reported by Kodur in [3].

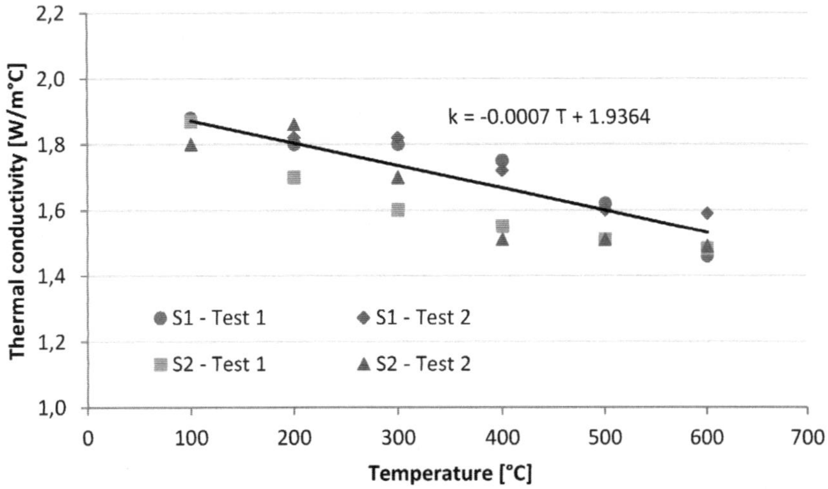

Fig. 6 – Behaviour of the thermal conductivity (the black solid line identifies interpolation curve).

On the basis of the measured data, a qualitative trend of thermal conductivity has been extrapolated up to 1000 °C (Fig. 7) for predictive purposes; it indicates that at this temperature we would have about 36 % reduction in thermal conductivity. This result seems in agreement with the reduction found by Vodak [5] and Kodur [6] in the range between 20 and 800 °C; nevertheless, the greatest deviation respect with the ASCE values appears at elevated temperature.

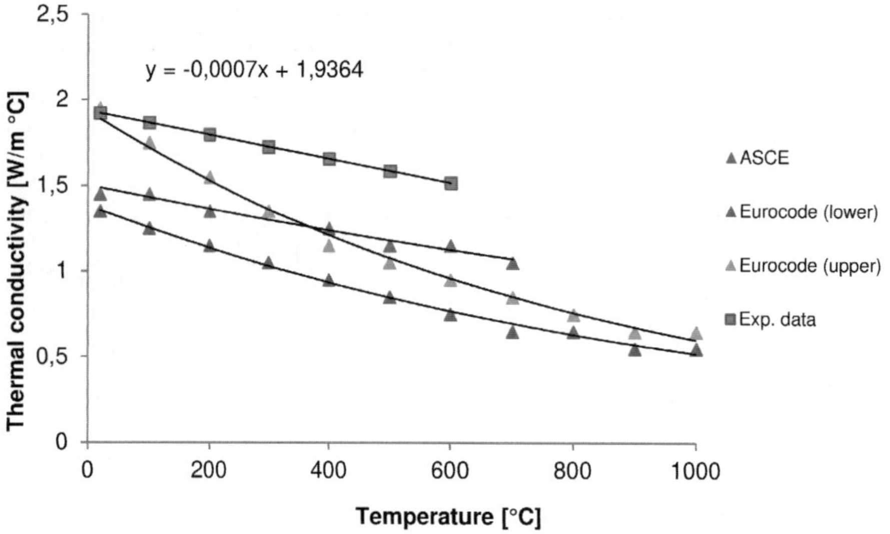

Fig. 7 – Comparison between measured and ASCE and Eurocode thermal conductivity values [3].

Finally, Fig. 8 shows an overview of the C3 sample surfaces before and after testing: they did not indicate significant damage on the cylinder surface, a part from an increased micro porosity, main responsible of the thermal conductivity reduction at high temperature.

a)

b)

Fig. 8 – Overview of C3 cylinder surface before (a) and after (b) the exposition at 800 °C for 30 minutes.

4. Conclusion and future activities

The concrete experimentally investigated by adopting the hot wire method allowed determining the thermal conductivity of this material that will be as matrix for bituminised LILW packaging.

The relevance of the activity relies with the fire test (800 °C for 30 min.) to execute to qualify such type of container. In fact, although concrete and steel properties are known at elevated temperatures and available in literature for conventional application, to be confident that no abnormal situation (high risk of burning of bitumen vapour) will arise during the fire test execution, concrete thermal performances/characteristics must be evaluate. This is also because these latter are dependent on the concrete matrix and constituents.

The hot wire method test performed at the Department of Civil and Industrial Engineering of the University of Pisa allowed to determine the thermal conductivity of concrete in a wide range of temperatures, ranging from 100° to about 800 °C.

The thermal conductivity of concrete at room temperature is between 1.6 and 2.2 W/m °C and monotonically decreases as the temperature increases. Results highlight also about 32.5% reduction in the temperature range between 100 and 600 °C, probably due to de-hydration process. This consideration was supported by visual and X-ray inspection of sample.

Finally, the results have been compared with Eurocode, ASCE and Kodur data, available in literature, highlighting a good agreement (same trend also).

The experimental thermal conductivity of concrete was used in turn for pre-test purposes in planning the fire test of CP5.2 LILW packaging.

References

[1] IAEA SS n° SSR-6 "Regulation for the safe transport of radioactive materials", 2012.
[2] Lo Frano R *et al* 2011 Thermal analysis of a spent fuel cask in different transport conditions *Energy* **36** pp. 2285-2293.
[3] Kodur V 2014 Properties of Concrete at Elevated Temperatures *Hindawi Publishing Corporation* Article ID 468510
[4] Kodur V and Sultan M A 2003 Effect of Temperature on Thermal Properties of High-Strength Concrete *J. Mater. Civ. Eng* **15**(2) pp. 101-107.
[5] Kodur V 2008 High-Temperature Properties of Concrete for Fire Resistance Modelling of Structures *ACI Mater. J.* **105**(5) pp. 517-527.
[6] Vodak F *et al* 1997 Thermophysical Properties of Concrete *Cement and Concrete Research* **27**(3), pp. 415-426.
[7] Shin K Y *et al* 2002 Thermo-Physical Properties and Transient Heat Transfer of Concrete at Elevated Temperatures *Nucl. Eng. Des.* **212**, pp. 233–241.
[8] Lo Frano R and Stefanini L, 2016 Investigation of the behaviour of a LILW superficial repository under aircraft impact *Nucl. Eng. Des.* **300**, pp. 552-562.
[9] Lo Frano R *et al* 2014 Thermo-mechanical test rig for experimental evaluation of thermal conductivity of ceramic pebble beds *Fusion Eng. Des.* **89**, pp 1309-1313.
[19] Lo Frano R *et al* 2014 Numerical-experimental analyses by Hot-Wire method of an alumina cylinder for future studies on thermal conductivity of the fusion breeder materials *IOP J. Phys.: Conf. Series* **547** 012037.
[11] Lo Frano R and Aquaro D 2014 Preliminary experimental evaluation of thermal conductivity of ceramic pebble beds *IOP J. Phys.: Conf. Series* **501** 012031.